Introduction to

Micromechanics and Nanomechanics

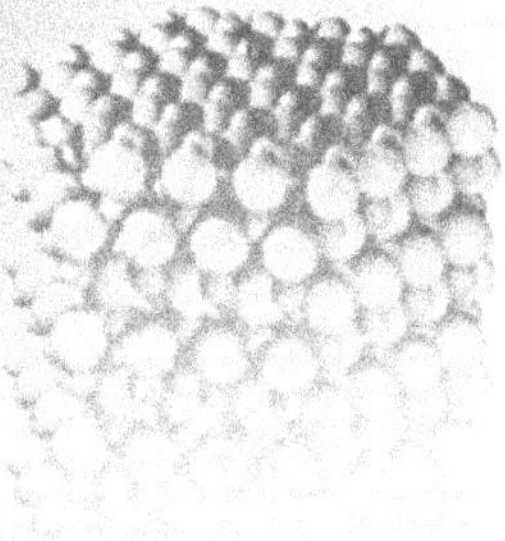

Introduction to Micromechanics and Nanomechanics

Shaofan Li
University of California at Berkeley, USA

Gang Wang
Hong Kong University of Science and Technology, China

World Scientific
NEW JERSEY • LONDON • SINGAPORE • BEIJING • SHANGHAI • HONG KONG • TAIPEI • CHENNAI

Published by

World Scientific Publishing Co. Pte. Ltd.
5 Toh Tuck Link, Singapore 596224
USA office: 27 Warren Street, Suite 401-402, Hackensack, NJ 07601
UK office: 57 Shelton Street, Covent Garden, London WC2H 9HE

Library of Congress Cataloging-in-Publication Data
Li, Shaofan.
Introduction to micromechanics and nanomechanicsl / Shaofan Li, Gang Wang.
p. cm.
Includes bibliographical references and indexes.
ISBN-13 978-981-281-413-5
ISBN-10 981-281-413-2
ISBN-13 978-981-281-414-2 (pbk)
ISBN-10 981-281-414-0 (pbk)
1. Micromechanics -- Textbooks. 2. Nanotechnology -- Textbooks. I. Title. II. Wang, Gang.
QC176.8.M5 L5 2008
2008300032

British Library Cataloguing-in-Publication Data
A catalogue record for this book is available from the British Library.

First published 2008
Reprinted 2011

In memory of my mother,
Yu-zheng Zhang,
for her love and perpetual inspirations.

S. Li

To my mother,
Pei-xue Gong,
for her love and sacrifice.

G. Wang

Preface

My early interests in Micromechanics was largely inspired by Professor Toshio Mura. I had been studied under Professor Mura from 1994-1998, during which I had taken his graduate class *Micromechanics* I and II, and I had worked with him in the same office for almost four years. In specific, Professor Mura taught me equivalent eigenstrain theory (which should be labeled as the Eshelby-Mura theory), dislocation theory, and lattice statics/dynamics. As I can remember, one favorite line of Professor Mura's is: *the eigenstrain method is panacea.* My current interests in Nanomechanics and Computational Nanomechanics researches are mainly motivated by my Ph.D. dissertation advisor, Professor Wing Kam Liu, who is one of the leading experts in Computational Nanomechanics today. Readers may find that this book is greatly influenced by Professor Mura's book, *Micromechanics of Defects in Solids* (Kluwer Academic Publisher, 1987) and Professor Liu's book, *Nano Mechanics and Materials: Theory, Multiscale Methods and Applications* (John Wiley & Sons, Ltd., 2005).

Since spring 2001, I have been regularly teaching a graduate course on *Micromechanics* (CE236) in the University of California at Berkeley. This book is the outcome of the lecture notes as well as research projects of that course. In recent years, more focus of the course has been placed on the presentation of nanomechanics — an emerging field that is still very much under development. Therefore, aside from traditional Micromechanics, a unique feature of this book is its in-depth discussions of the latest topics on Nanomechanics and its applications. This includes: lattice Green's function method (LGFM), embedded atom method (EAM), quasi-continuum method, discrete dislocation dynamics (DDD), the Peierls-Nabarro model, the Gurtin-Murdoch surface elasticity model, and the concept of the virial stress, etc.

Many students who had taken the class have participated in the related class research projects. Most of those researches have been published in peer-reviewed journals, and constitute a significant part of materials presented in this book. My co-author, Dr. Gang Wang, is among the first group of students participating in the class research project. Since then, we have been working together for several years, and he has contributed significantly on many subjects discussed in this book.

I would also like to thank those who have made unique contributions to this book. They are: Dr. Roger Sauer, Dr. Christian Linder, Dr. Chin-long Lee, Dr. Xiaohu Liu, Dr. James Foulk III, Dr. Daniel Simkins, Jr., Dr. Elif Ertekin, Dr. Albert To, Dr. Elisa Morgan, Dr. Anurag Gupta, Dr. Ni Sheng, Ms. Veronique Le Corvec, Mr. Morteza Mahyari, Mr. Noang-Nam Nguyen among others.

During the writing of this book, many colleagues have given us encouragements and suggestions. In particular, Professor Dong Qian of University of Cincinnati, Professor L. Z. Sun of the University of California at Irvine, Professor H. Wang of the Texas A & M University, Professor P. Sharma of University of Houston, Professor X. Markenscoff of the University of California at San Diego, and Dr. L. P. Liu of California Institute of Technology, who have generously provided their own research results or materials helping us writing the book. I would also like to acknowledge the financial support from National Science Foundation through the Career Award (Grant No. CMS-0239130), which makes this book and related researches possible.

The objective of the book is twofold: it can serve as a graduate textbook on Micromechanics and Nanomechanics for the first-year graduate students, and also a research guide book for researchers who want to master the fundamental theories of Micromechanics and Nanomechanics through self-study. One of the main features of this book is to give as many detailed derivations as necessary to assist the readers in understanding the theoretical assumptions, mathematical techniques, and possible limitations. To make the self-learning an enjoyable journey for our readers, our motto is to *spell out all the details even if they may be trivial.* By doing so, we hope to fill the gap between the literature and the actual research notes.

Due to our limitations, the book may contain mistakes, misrepresentations, and errors; Moreover, we are aware the fact that some of the presentations in the book may be biased or limited by our own technical capacities and inadequacies. Readers can send their comments and suggestions to the following email address:

`micro.and.nanomechanics@gmail.com`

which, we hope, can be used to correct and improve the quality of the book in the future.

Finally, we would also like to thank our wives, Yan Zhang (SL) and Furong Wang (GW), and our families. Without their supports and encouragements, this book will never be finished.

Shaofan Li

Spring 2007, Berkeley, California

Contents

Chapter 1

INTRODUCTION

1.1 What are micromechanics and nanomechanics?

Generally speaking, MICROMECHANICS is a scientific discipline that studies mechanical, electrical, and thermodynamical behaviors of materials with microstructure; NANOMECHANICS is a research field that studies material behaviors at nanoscale level.

In recent years, micromechanics has become an indispensable part of theoretical foundation for many engineering fields and emerging technologies such as nanotechnology as well as biomedical and bioenvironmental technologies. Because of its multidisciplinary characteristics, the term *micromechanics* has 'multidisciplinary interpretations', and it has been used with different meanings in different contexts. In the area of applied mechanics, micromechanics is often referred to as a hierarchical mechanics and mathematics paradigm that is mainly used to study the effective material properties of composite materials. A major objective of this kind of study is to find the statistical average material properties of the heterogeneous material through various homogenization methods. In condensed solid state physics and statistical mechanics, this process is called the *Coarse Graining*. One of the fundamental challenge of the contemporary statistical physics is how to construct accurate coarse grain models.

Traditionally, the standard micromechanics methodology in engineering applications treats a composite material as a generic continuum model with a two-level paradigm: microscopic structure and macroscopic structure. The material properties at microscale are usually given as a priori, and the task is to find the material behaviors at macroscale, which are also called as the effective or overall material properties. From this perspective, a material point at the macro-level may be viewed as a microscopic material ensemble. In principle, the constitutive relations at macro level should be able to be derived from the ensemble average of micro-objects that are governed by the microscale physical laws, which can be quantum mechanics, lattice dynamics, microscale plasticity, or elasticity, etc. Two subtle points worth further clarification: (1) the constitutive relations or material behaviors at macroscale may be very much different from their counterparts at microscale, so the task of micromechanics is to find the unknown macroscale constitutive laws. One of such examples is the well-known Gurson's model, in which the microscale constitutive law is the rigid perfectly plasticity whereas the macroscale constitutive law obtained by homogenization is a pressure sensitive damage plasticity; (2) in many other cases, the microscale and macroscale constitutive laws are the same type, e.g. elastic behaviors, however, the detailed elastic stiffness tensors at the different scales are different. The effective material properties at a macroscale point are average material properties of a microscale ensemble or a unit cell.

The conventional two-level paradigm of micromechanics is a special mathematical homogenization model that is usually not associated with any fixed length scale. When studying material properties of a metal, 1 mm may be viewed as macroscale, and the length scale at microlevel may range from nm to μm ; whereas studying the deformation of a dam, the macroscale may be up to $10^3 m$, and the length scale of micro-structure may be around $10^{-2} m$.

In the conventional micromechanics, the classical ergodic assumption is usually adopted: *if a mesoscale is large enough, the underline micro-structure is assumed to statistically homogeneous and stable in both space and in time.* Therefore, one simply uses spatial average to replace the temporal average of a random stochastic process. In this sense, traditional micromechanics is essentially a particular ensemble averaging theory that takes into account the overall effects of microstructure.

In engineering applications, the conventional micromechanics deals with practical engineering problems of a broad spectrum: effective material properties of composite/synthetic materials, such as cementitious materials, geotechnical materials, etc.; constitutive modeling of bio-materials, such as bone, muscle, blood flow, etc.; phase transformations; defects in solids, such as dislocation motion and crack growth; and environmental problems, such as air pollution, ground water flow and chemical transport, etc.

Contemporary condensed matter physics and applied mechanics in general agree that the physics at molecular or atomic level (Å) can be described by the quantum mechanics or related approximation theories, e.g. density functional theory; the physics between the nm scale to sub-μm scale is governed by nanomechanics though presently we are mainly relying on the molecular dynamics simulation; from μm scale to or sub-mm length scale, micromechanics and related mesoscale mechanics are playing more important roles; and the macroscopic phenomenological theory is generally valid at the length scale mm level or up.

In this book, we shall focus on several areas of nanomechanics and micromechanics. Different from traditional micromechanics, a salient feature of nanomechanics is its multiscale and multiphysics characteristics. It has some features presented in quantum mechanics, or quantum statistical mechanics, manifesting the statistical effects at atomic or sub-atomic level; on the other hand, it also shares many features of continuum mechanics, because a nanostructure could contain millions of atoms.

The impetus for contemporary micromechanics and nanomechanics is primarily due to the emergence of nanoscience and biomedical technology. It appears that traditional physics alone is not sufficient to deal with many engineering problems that are emerging from nanotechnologies and nanoengineering. There is a call for nanomechanics and nanocomputational mechanics to serve as the infrastructure of these developing engineering fields. For instance, much attention has been focused recently on material properties of thin films; manufacturing devices and components of a microelectromechanical system (MEMS), such as sub-micro sized sensors, motors; mechanics of nanotubes and nanowires; and micro-biophysics/biochemistry systems, e.g. protein/DNA interaction in biomolecular simulation, etc.

From the perspective of higher learning and intellectual advancement, micromechanics has been developed into a rigorous and beautiful mathematical framework, philosophical methodology, and powerful computational realization. Forty years ago, micro-elasticity started with simple definitions of eigenstrain and inclusion, came along with Eshelby's equivalent homogenization theory [Eshelby (1957, 1959, 1961)]) and Hashin & Shtrikman's variational principle [Hashin and Shtrikman (1962a,b)], it is now the foundation of composite material research. Even though the conventional micromechanics deals with the objects with the length scale around μm, it has been extensively used to estimate or to analyze the behaviors of nanocomposites and nanoscale structures, such as the composite made by nanowires and quantum dots.

Besides homogenization, another main aspect of mciromechanics is the study of defect mechanics at small scale. This includes: crack growth, dislocation motion, and evolution of vacancies and interstitial, etc. In parallel to the development of micro-mechanics, another major paradigm of defect mechanics is the Configurational Force Mechanics. It seems to us that future trend of micro-mechanics is to develop multiscale configurational mechanics that can describe defect motions in a multiscale thermodynamic environment.

The main task of nanomechanics is to establish *coarse-graining models* at small scales or to bridge the gap between the atomic scale and continuum scale. For example, an efficient coarse-graining technique is the so-called Cauchy-Born rule. The Cauchy-Born rule may be viewed as a simple "homogenization approximation" in lattice statics and it serves as a passage or linkage between molecular mechanics and continuum mechanics. The Cauchy-Born rule assumes that under certain kinematic conditions, for instance, uniformity of local deformation gradient, the continuum energy density can be computed directly by using the atomistic potential, which links the continuum elastic potential energy with the atomistic potential. By using the Cauchy-Born rule, one may be able to derive the expressions for stress tensors and elastic stiffness tensors directly from the interatomic potential, which allows the use of the standard nonlinear finite element method in nanoscale computations.

Another useful nanomechanics approach is the Lattice Green's Function (LGF) method. It provides an important limit case for continuum mechanics, which allows us examine the differences between the molecular mechanics and the continuum mechanics.

Presently, nanomechanics is only at its infancy. There are many approaches to be explored and many new phenomena to be studied. In this book, we are attempting to synthesize some recent research results at the forefront of nanomechanics research while presenting traditional micromechanics in a coherent fashion. By doing so, we hope that it may serve as a stepping stone for nanomechanics research in the quest for a multiscale mechanics of our time.

Many research monographs on Micromechanics and Composite Materials have been published over the years, notably the classical treatises by Professor Mura [Mura (1987)], Christensen [Christensen (1979)], Nemat-Nasser and Hori [Nemat-Nasser and Hori (1999)], Teodosiu [Teodosiu (1982)], Hahn and Tsai [Hahn and Tsai (1980)], Kim and Karrila [Kim and Karrila (1991)], and Krajcinovic [Krajcinovic (1996)]. In recent years, quite a number of books have been published focusing on various different aspects of micromechanics and defect mechanics, such as statistical micro-mechanics [Torquato (1997)], translation method and variational bounds for composite materials [Milton (2002)], general introductions to micromechanics and composite materials [Cristescu *et al.* (2004)] and [Qu and Cherkaoui (2006)], micro-poromechanics [Dormieux *et al.* (2006)], and comprehensive treatise and handbook on micromechanics as well [Buryachenko (2007)], among others.

The current literature on Micromechanics and Nanomechanics is either too specialized, too esoteric, to be understood, or too elementary to be applied. The objective of the present book is to fill the gap between the graduate study or self-study and the independent or creative research. To do so, firstly, we would like to provide a self-study guide or a readable graduate textbook on Micromechanics and Nanomechanics that is easy to read without much prerequisites and experiences on applied mathematics, continuum mechanics or elasticity theories; Secondly, we would like to merge the theory of micromechanics into the theory of nanomechanics by find internal links and coherence between the two subjects and making the subject more contemporary and more interesting to readers.

1.2 Vectors and tensors

For self-containedness and easy reference, the presentation starts with an outline of some basic prerequisites: mathematics preliminaries, the element of elasticity theory, and lattice and molecular statics and dynamics.

1.2.1 *Vector algebra*

Consider a Cartesian coordinate in a three dimensional space, $\mathbb{R}^3$ with unit vector basis, $\{\mathbf{e}_i\}, i = 1, 2, 3$. An arbitrary position vector, $\mathbf{x}$, may be expressed as

$$\mathbf{x} = x_1\mathbf{e}_1 + x_2\mathbf{e}_2 + x_3\mathbf{e}_3 = x_i\mathbf{e}_i = (\mathbf{x}\cdot\mathbf{e}_i)\mathbf{e}_i \tag{1.1}$$

where the Einstein convention is used that the repeated indices indicate the summation from 1 to 3.

Consider two vectors, $\mathbf{V} = V_i\mathbf{e}_i$ and $\mathbf{W} = W_j\mathbf{e}_j$. The scalar (dot) product of two vectors, $\mathbf{V}$ and $\mathbf{W}$, is defined as

$$\mathbf{V}\cdot\mathbf{W} := \left(V_i\mathbf{e}_i\right)\cdot\left(W_j\mathbf{e}_j\right) = V_iW_j\left(\mathbf{e}_i\cdot\mathbf{e}_j\right) = V_iW_j\delta_{ij} = V_iW_i \qquad (1.2)$$

where

$$\mathbf{e}_i\cdot\mathbf{e}_j = \begin{Bmatrix} 1, & i=j \\ 0, & i\neq j \end{Bmatrix} =: \delta_{ij} \qquad (1.3)$$

is called the Kronecker delta.

A vector cross product of two vectors, $\mathbf{A} = A_i\mathbf{e}_i, \mathbf{B} = B_j\mathbf{e}_j$, is defined as

$$\mathbf{A}\times\mathbf{B} = (A_i\mathbf{e}_i)\times(B_j\mathbf{e}_j) = A_iB_j\mathbf{e}_i\times\mathbf{e}_j = e_{kij}A_iB_j\mathbf{e}_k \qquad (1.4)$$

where $\mathbf{e}_i\times\mathbf{e}_j = e_{kij}\mathbf{e}_k$, and e_{kij} is called the permutation symbol,

$$e_{ijk} = \begin{cases} 1, & \text{for an even permutation of } 1,2,3 \\ -1, & \text{for an odd permutation of } 1,2,3 \\ 0, & \text{repeated indices} \end{cases} \qquad (1.5)$$

This definition can be explained as a permutation rule that change of any two adjacent indices of the symbol, there is a negative sign, (-1), occurring.

For example, since $e_{123} = 1$, then

$$e_{132} = (-1)e_{123} = (-1)(1) = -1$$

and

$$e_{312} = (-1)e_{132} = (-1)(-1)e_{123} = (-1)(-1)1 = 1$$

The cross product of two vectors can also written as

$$\begin{aligned}\mathbf{A}\times\mathbf{B} &= e_{kij}A_iB_j\mathbf{e}_k = e_{1ij}A_iB_j\mathbf{e}_1 + e_{2ij}A_iB_j\mathbf{e}_2 + e_{3ij}A_iB_j\mathbf{e}_3 \\ &= (A_2B_3 - A_3B_2)\mathbf{e}_1 + (A_3B_1 - A_1B_3)\mathbf{e}_2 + (A_1B_2 - A_2B_1)\mathbf{e}_3 \\ &= \begin{vmatrix} \mathbf{e}_1 & \mathbf{e}_2 & \mathbf{e}_3 \\ A_1 & A_2 & A_3 \\ B_1 & B_2 & B_3 \end{vmatrix}\end{aligned} \qquad (1.6)$$

Therefore

$$\mathbf{e}_i\times\mathbf{e}_j = e_{kij}\mathbf{e}_k, \quad \Rightarrow \quad e_{kij} = (\mathbf{e}_i\times\mathbf{e}_j)\cdot\mathbf{e}_k \qquad (1.7)$$

Since

$$\mathbf{e}_i\times\mathbf{e}_j = \begin{vmatrix} \mathbf{e}_1 & \mathbf{e}_2 & \mathbf{e}_3 \\ \delta_{1i} & \delta_{2i} & \delta_{3i} \\ \delta_{1j} & \delta_{2j} & \delta_{3j} \end{vmatrix} \qquad (1.8)$$

then the permutation symbol may be expressed as

$$e_{kij} = e_{ijk} = (\mathbf{e}_i\times\mathbf{e}_j)\cdot\mathbf{e}_k = \begin{vmatrix} \delta_{1k} & \delta_{2k} & \delta_{3k} \\ \delta_{1i} & \delta_{2i} & \delta_{3i} \\ \delta_{1j} & \delta_{2j} & \delta_{3j} \end{vmatrix} = \begin{vmatrix} \delta_{1i} & \delta_{2i} & \delta_{3i} \\ \delta_{1j} & \delta_{2j} & \delta_{3j} \\ \delta_{1k} & \delta_{2k} & \delta_{3k} \end{vmatrix} \qquad (1.9)$$

This provides a link between permutation symbol and Kronecker delta. Consider the product of two permutation symbols,

$$
\begin{aligned}
e_{ijk}e_{rst} &= \begin{vmatrix} \delta_{1i} & \delta_{2i} & \delta_{3i} \\ \delta_{1j} & \delta_{2j} & \delta_{3j} \\ \delta_{1k} & \delta_{2k} & \delta_{3k} \end{vmatrix} \begin{vmatrix} \delta_{1r} & \delta_{2r} & \delta_{3r} \\ \delta_{1s} & \delta_{2s} & \delta_{3s} \\ \delta_{1t} & \delta_{2t} & \delta_{3t} \end{vmatrix} = \begin{vmatrix} \delta_{1i} & \delta_{2i} & \delta_{3i} \\ \delta_{1j} & \delta_{2j} & \delta_{3j} \\ \delta_{1k} & \delta_{2k} & \delta_{3k} \end{vmatrix} \begin{vmatrix} \delta_{1r} & \delta_{2s} & \delta_{3t} \\ \delta_{1r} & \delta_{2s} & \delta_{3t} \\ \delta_{1r} & \delta_{2s} & \delta_{3t} \end{vmatrix} \\
&= \begin{vmatrix} \delta_{ir} & \delta_{is} & \delta_{it} \\ \delta_{jr} & \delta_{js} & \delta_{jt} \\ \delta_{kr} & \delta_{ks} & \delta_{kt} \end{vmatrix} \qquad (1.10)
\end{aligned}
$$

Using Eq. (1.10), one may show the following e-δ identities:

(1) Let $i = r$, then, $e_{ijk}e_{ist} = \delta_{js}\delta_{kt} - \delta_{jt}\delta_{ks}$;
(2) Let $i = r$, $j = s$, then, $e_{ijk}e_{ij\ell} = 2\delta_{k\ell}$;
(3) Let $i = r$, $j = s$, $k = t$, then, $e_{ijk}e_{ijk} = 3! = 6$.

1.2.2 *Tensor algebra*

Consider two vectors, $\mathbf{A} = A_i\mathbf{e}_i$ and $\mathbf{B} = B_j\mathbf{e}_j$. One can form a second-order tensor, $\mathbb{C}$ by using the tensor product

$$\mathbb{C} = \mathbf{A} \otimes \mathbf{B} = \Big(A_i\mathbf{e}_i\Big) \otimes \Big(B_j\mathbf{e}_j\Big) = A_iB_j\mathbf{e}_i \otimes \mathbf{e}_j \qquad (1.11)$$

The dyad is called the second-order tensor[1], and its basis, $\mathbf{e}_i \otimes \mathbf{e}_j$, is called dyadic basis. In this case, the components of the second-order tensor are $C_{ij} = A_iB_j$. In fact, every second-order tensor can be expressed in a dyadic basis, such as the

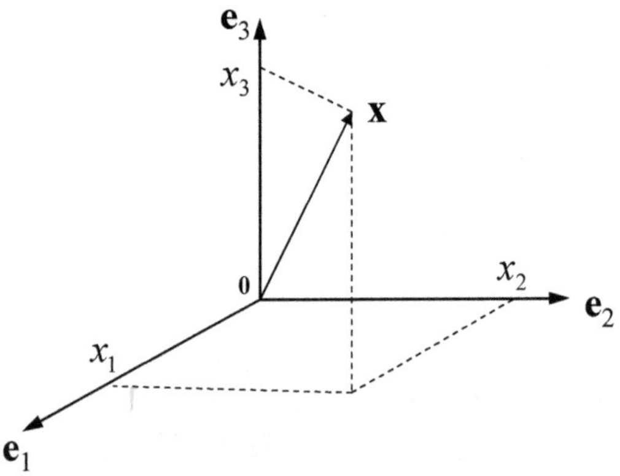

Fig. 1.1 Cartesian Coordinate.

[1]One may call the vector as the first-order tensor.

Cauchy stress tensor and the infinitesimal strain tensor,

$$\boldsymbol{\sigma} = \sigma_{ij}\mathbf{e}_i \otimes \mathbf{e}_j \tag{1.12}$$

$$\boldsymbol{\epsilon} = \epsilon_{ij}\mathbf{e}_i \otimes \mathbf{e}_j \tag{1.13}$$

A conjugate of a dyad (second-order tensor) is defined as

$$(\boldsymbol{\epsilon})^T := \epsilon_{ji}\mathbf{e}_i \otimes \mathbf{e}_j \tag{1.14}$$

We say that a second-order tensor is symmetric, if

$$\mathbb{A} = (\mathbb{A})^T, \quad \text{or in component form } A_{ij} = A_{ji} \tag{1.15}$$

A second-order tensor is skew symmetric, if

$$\mathbb{A} = -\mathbb{A}^T, \quad \text{or in component form } A_{ij} = -A_{ji} \tag{1.16}$$

In general, an arbitrary second-order tensor can be expressed as a linear combination of a second-order symmetric tensor and a second-order skew symmetric tensor by following decomposition:

$$A_{ij} = \frac{1}{2}\Big(A_{ij} + A_{ji}\Big) + \frac{1}{2}\Big(A_{ij} - A_{ji}\Big) = A_{(ij)} + A_{[ij]} \tag{1.17}$$

In linear elasticity we define the infinitesimal strain tensor as the symmetric part of displacement gradient,

$$\boldsymbol{\epsilon} = \frac{1}{2}\Big(\nabla \otimes \mathbf{u} + (\nabla \otimes \mathbf{u})^T\Big) = \frac{1}{2}\Big(u_{j,i} + u_{i,j}\Big)\mathbf{e}_i \otimes \mathbf{e}_j \tag{1.18}$$

or in indicial notation $\epsilon_{ij} = \frac{1}{2}(u_{j,i} + u_{i,j})$.

In general, a n-th order tensor is a polyads, or has a polyadic representation, e.g.

$$\mathbb{C} = C_{ijk\ell}\mathbf{e}_i \otimes \mathbf{e}_j \otimes \mathbf{e}_k \otimes \mathbf{e}_\ell \tag{1.19}$$

is a fourth-order tensor. Analogous to the scalar product of vectors, the *double contraction* of two tensors are defined as two dot products among of Cartesian tensor bases, i.e. if $\mathbb{A} = A_{ij}\mathbf{e}_i \otimes \mathbf{e}_j$ and $\mathbb{B} = B_{k\ell}\mathbf{e}_k \otimes \mathbf{e}_\ell$, then

$$\begin{aligned}\mathbb{A} : \mathbb{B} &= (A_{ij}\mathbf{e}_i \otimes \mathbf{e}_j) : (B_{k\ell}\mathbf{e}_k \otimes \mathbf{e}_\ell) = A_{ij}B_{k\ell}(\mathbf{e}_i \cdot \mathbf{e}_k)(\mathbf{e}_j \cdot \mathbf{e}_\ell) \\ &= A_{ij}B_{k\ell}\delta_{ik}\delta_{j\ell} = A_{ij}B_{ij}\end{aligned} \tag{1.20}$$

The trace of a second-order tensor is defined as

$$tr\mathbb{A} := \mathbb{A} : \mathbb{I}^{(2)} = A_{ii} = A_{11} + A_{22} + A_{33} \tag{1.21}$$

In each contraction, two vector bases are annihilated. Consider a fourth-order tensor $\mathbb{C} = C_{ijk\ell}\mathbf{e}_i \otimes \mathbf{e}_j \otimes \mathbf{e}_k \otimes \mathbf{e}_\ell$ and a second-order tensor $\boldsymbol{\epsilon} = \epsilon_{ij}\mathbf{e}_i \otimes \mathbf{e}_j$. There are six bases in total. A double contraction between these two tensors will annihilate four vector bases and lead to a second-order tensor, i.e.

$$\begin{aligned}\boldsymbol{\sigma} = \mathbb{C} : \boldsymbol{\epsilon} &= \Big(C_{ijk\ell}\mathbf{e}_i \otimes \mathbf{e}_j \otimes \mathbf{e}_k \otimes \mathbf{e}_\ell\Big) : \Big(\epsilon_{st}\mathbf{e}_s \otimes \mathbf{e}_t\Big) \\ &= C_{ijk\ell}\epsilon_{st}\mathbf{e}_i \otimes \mathbf{e}_j\delta_{ks}\delta_{\ell t} = C_{ijk\ell}\epsilon_{k\ell}\mathbf{e}_i \otimes \mathbf{e}_j\end{aligned} \tag{1.22}$$

Using indicial notation, we write $\sigma_{ij} = C_{ijk\ell}\epsilon_{k\ell}$.

It may be convenient to define the second-order Cartesian basis as

$$\mathbf{e}_{ij} := \mathbf{e}_i \otimes \mathbf{e}_j \ . \tag{1.23}$$

The second-order unit tensor and the fourth-order unit tensor are constructed based on the following rules:

$$\mathbb{I}^{(2)} := \Big(\mathbf{e}_i \cdot \mathbf{e}_j\Big)\mathbf{e}_i \otimes \mathbf{e}_j = \delta_{ij}\mathbf{e}_i \otimes \mathbf{e}_j = \delta_{ij}\mathbf{e}_{ij} \tag{1.24}$$

$$\begin{aligned}\mathbb{I}^{(4)} &:= \Big(\mathbf{e}_i \otimes \mathbf{e}_j\Big) : \Big(\mathbf{e}_k \otimes \mathbf{e}_\ell\Big)\mathbf{e}_i \otimes \mathbf{e}_j \otimes \mathbf{e}_k \otimes \mathbf{e}_\ell \\ &= (\mathbf{e}_{ij} : \mathbf{e}_{k\ell})\mathbf{e}_{ij} \otimes \mathbf{e}_{k\ell} = \delta_{ik}\delta_{j\ell}\mathbf{e}_i \otimes \mathbf{e}_j \otimes \mathbf{e}_k \otimes \mathbf{e}_\ell \end{aligned} \tag{1.25}$$

The superscript indicates the order. It is interesting to note that the fourth-order unit tensor defined in (1.25) is not symmetric with all indices.

To represent symmetric identity tensor, it may be expedient to first define symmetric tensor basis. The second-order symmetric basis is defined as

$$\mathbf{e}^S_{ij} = \frac{1}{2}\Big(\mathbf{e}_{ij} + \mathbf{e}^T_{ij}\Big) = \frac{1}{2}\Big(\mathbf{e}_i \otimes \mathbf{e}_j + \mathbf{e}_j \otimes \mathbf{e}_i\Big) \tag{1.26}$$

Any symmetric second-order tensor can then be expressed as $\mathbb{S} = S_{ij}\mathbf{e}^S_{ij}$. One may denote the space of all symmetric second-order tensors as

$$T^{(2s)} = \{\mathbb{S} \ \Big| \ \mathbb{S} = S_{ij}\mathbf{e}^S_{ij}\} \tag{1.27}$$

The corresponding second-order symmetric unit tensor is defined as

$$\begin{aligned}\mathbb{I}^{(2s)} &= \frac{1}{2}\Big(\mathbf{e}_i \cdot \mathbf{e}_j + \mathbf{e}_j \cdot \mathbf{e}_i\Big)\mathbf{e}_i \otimes \mathbf{e}_j \\ &= \delta_{ij}\mathbf{e}_i \otimes \mathbf{e}_j = \mathbb{I}^{(2)} \end{aligned} \tag{1.28}$$

One may also define the second-order anti-symmetric tensor basis as

$$\mathbf{e}^A_{ij} = \frac{1}{2}\Big(\mathbf{e}_i \otimes \mathbf{e}_j - \mathbf{e}_j \otimes \mathbf{e}_i\Big) \tag{1.29}$$

The fourth-order symmetric tensor basis is built upon the second order symmetric tensor basis, i.e.

$$\mathbf{e}^S_{ijk\ell} = \mathbf{e}^S_{ij} \otimes \mathbf{e}^S_{k\ell} \tag{1.30}$$

and the fourth-order symmetric tensor space can be defined as

$$T^{(4s)} = \{\mathbb{S} \ \Big| \ \mathbb{S} = S_{ijk\ell}\mathbf{e}^S_{ijk\ell}\} \tag{1.31}$$

The corresponding fourth-order unit tensor is defined as

$$\mathbb{I}^{(4s)} := (\mathbf{e}^S_{ij} : \mathbf{e}^S_{k\ell})\mathbf{e}^S_{ijk\ell} = \frac{1}{2}\Big(\delta_{ik}\delta_{j\ell} + \delta_{i\ell}\delta_{jk}\Big)\mathbf{e}_i \otimes \mathbf{e}_j \otimes \mathbf{e}_k \otimes \mathbf{e}_\ell \tag{1.32}$$

Note that the fourth-order unit tensor can be decomposed to a symmetric part and an anti-symmetric part in terms of the first and second indices, or the of the third and fourth indices,

$$\begin{aligned}\mathbb{I}_{ijk\ell} &:= \delta_{ik}\delta_{j\ell} = \frac{1}{2}(\delta_{ik}\delta_{j\ell} + \delta_{i\ell}\delta_{jk}) + \frac{1}{2}(\delta_{ik}\delta_{j\ell} - \delta_{i\ell}\delta_{jk}) \\ &= \mathbb{I}^{(4s)}_{ijk\ell} + \mathbb{I}^{(4a)}_{ijk\ell} \end{aligned} \tag{1.33}$$

One can show that for a given second-order tensor, $\mathbb{A}$,

$$\mathbb{I}^{(4)} : \mathbb{A} \to \mathbb{A} \tag{1.34}$$

$$\mathbb{I}^{(4s)} : \mathbb{A} \to \frac{1}{2}\left(\mathbb{A} + \mathbb{A}^T\right) \tag{1.35}$$

$$\mathbb{I}^{(4a)} : \mathbb{A} \to \frac{1}{2}\left(\mathbb{A} - \mathbb{A}^T\right) \tag{1.36}$$

Note that $\mathbb{I}^{(4)} \neq \mathbb{I}^{(2)} \otimes \mathbb{I}^{(2)}$.

When two second-order tensors satisfy the relation,

$$\mathbb{A} \cdot \mathbb{B} = \mathbb{I}^{(2)},$$

we say that tensor $\mathbb{A}$ is the inverse of tensor $\mathbb{B}$ or vice versa. The inversion of a second-order Cartesian tensor is similar to the inversion of a square matrix in linear algebra. In mathematics, the two operations are isomorphic. However, for a given regular fourth-order tensor, $\mathbb{A}$, it is not trivial to find its inverse $\mathbb{B}$ such that

$$\mathbb{A} : \mathbb{B} = \mathbb{I}^{(4)} .$$

In the following, we shall derive a very useful inversion formula for the fourth-order *isotropic* tensor. An isotropic tensor is defined as a tensor that has the same components in all rotated coordinate systems[2]. For example, an fourth-order tensor can be written with respect to two different bases

$$\mathbb{A} = A_{ijk\ell}\mathbf{e}_i \otimes \mathbf{e}_j \otimes \mathbf{e}_k \otimes \mathbf{e}_\ell = A_{i'j'k'\ell'}\mathbf{e}'_i \otimes \mathbf{e}'_j \otimes \mathbf{e}'_k \otimes \mathbf{e}'_\ell$$

Isotropy requires that

$$A_{ijk\ell} = A_{i'j'k'\ell'}$$

1.2.3 *Inversion formula for fourth-order isotropic tensor*

It can be shown in [Malvern (1969)] that the general form of a fourth order isotropic tensor is,

$$\mathbb{Q} = m\mathbb{I}^{(2)} \otimes \mathbb{I}^{(2)} + 2w\mathbb{I}^{(4s)} \tag{1.37}$$

Let $\mathbb{Q}^{-1}$ be its inverse tensor. According to the well-known Sherman-Morrison formula (e.g. [Dahlquist and Björck (1974)]),

$$\mathbb{Q}^{-1} = -\frac{m}{2w(3m+2w)}\mathbb{I}^{(2)} \otimes \mathbb{I}^{(2)} + \frac{1}{2w}\mathbb{I}^{(4s)} . \tag{1.38}$$

In indicial notation,

$$Q_{ijk\ell} = m\delta_{ij}\delta_{k\ell} + w(\delta_{ik}\delta_{j\ell} + \delta_{i\ell}\delta_{jk}) \tag{1.39}$$

$$Q^{-1}_{ijk\ell} = -\frac{m}{2w(3m+2w)}\delta_{ij}\delta_{k\ell} + \frac{1}{4w}(\delta_{ik}\delta_{j\ell} + \delta_{i\ell}\delta_{jk}) \tag{1.40}$$

[2]All zero-order tensors (scalars) are isotropic, but no first-order tensors (vectors) are isotropic. The unique second-order isotropic tensor is the Kronecker delta, and the unique third-order isotropic tensor is the permutation symbol.

An efficient approach to invert an isotropic tensor is to adopt the following E-basis orthogonal decomposition. Let

$$\mathbb{E}^{(1)} := \frac{1}{3}\mathbb{I}^{(2)} \otimes \mathbb{I}^{(2)}\ , \quad E^{(1)}_{ijk\ell} = \frac{1}{3}\delta_{ij}\delta_{k\ell} \tag{1.41}$$

$$\mathbb{E}^{(2)} := -\frac{1}{3}\mathbb{I}^{(2)} \otimes \mathbb{I}^{(2)} + \mathbb{I}^{(4s)}\ , \quad E^{(2)}_{ijk\ell} = -\frac{1}{3}\delta_{ij}\delta_{k\ell} + \frac{1}{2}(\delta_{ik}\delta_{j\ell} + \delta_{i\ell}\delta_{jk}) \tag{1.42}$$

The E-bases have the following special properties,

$$\begin{aligned}
\mathbb{E}^{(1)} + \mathbb{E}^{(2)} &= \mathbb{I}^{(4s)} \\
\mathbb{E}^{(1)} : \mathbb{E}^{(1)} &= \mathbb{E}^{(1)}, \text{ and } \mathbb{E}^{(2)} : \mathbb{E}^{(2)} = \mathbb{E}^{(2)} \\
\mathbb{E}^{(1)} : \mathbb{E}^{(2)} &= \mathbb{E}^{(2)} : \mathbb{E}^{(1)} = \mathbf{0}\ .
\end{aligned}$$

We now use E-basis approach to verify the Sherman-Morrison formula. Let,

$$\mathbb{Q} = (3m + 2w)\mathbb{E}^{(1)} + 2w\mathbb{E}^{(2)} \tag{1.43}$$

and

$$\mathbb{Q}^{-1} = h\mathbb{E}^{(1)} + v\mathbb{E}^{(2)} \tag{1.44}$$

By definition,

$$\begin{aligned}
\mathbb{Q} : \mathbb{Q}^{-1} = \mathbb{I}^{(4s)} &= \mathbb{E}^{(1)} + \mathbb{E}^{(2)} \\
(3m + 2w)h\mathbb{E}^{(1)} + 2wv\mathbb{E}^{(2)} &= \mathbb{E}^{(1)} + \mathbb{E}^{(2)}
\end{aligned}$$

which then leads to

$$h = \frac{1}{3m + 2w}, \quad \text{and} \quad v = \frac{1}{2w} \tag{1.45}$$

Consequently, we can write that

$$\begin{aligned}
\mathbb{Q}^{-1} &= (h - v)\mathbb{E}^{(1)} + v(\mathbb{E}^{(1)} + \mathbb{E}^{(2)}) = -\frac{3m}{2w(3m + 2w)}\mathbb{E}^{(1)} + \frac{1}{2w}\mathbb{I}^{(4s)} \\
&= -\frac{m}{2w(3m + 2w)}\mathbb{I}^{(2)} \otimes \mathbb{I}^{(2)} + \frac{1}{2w}\mathbb{I}^{(4s)}
\end{aligned}$$

Let's practice more examples.

Example 1.1. Consider an isotropic elastic tensor,

$$\mathbb{C} = \lambda\mathbb{I}^{(2)} \otimes \mathbb{I}^{(2)} + 2\mu\mathbb{I}^{(4s)} = 3K\mathbb{E}^{(1)} + 2\mu\mathbb{E}^{(2)}$$

Since by definition, $\mathbb{C} : \mathbb{D} = \mathbb{I}^{(4s)}$, it can be readily shown that

$$\mathbb{D} = \frac{1}{3K}\mathbb{E}^{(1)} + \frac{1}{2\mu}\mathbb{E}^{(2)} = -\frac{\lambda}{2\mu(3\lambda + 2\mu)}\mathbb{I}^{(2)} \otimes \mathbb{I}^{(2)} + \frac{1}{2\mu}\mathbb{I}^{(4s)}$$

Example 1.2. The Eshelby tensor for a spherical inclusion is

$$\begin{aligned}\mathbb{S}^{\Omega} &= \frac{5\nu-1}{15(1-\nu)}\mathbb{I}^{(2)}\otimes\mathbb{I}^{(2)} + \frac{2(4-5\nu)}{15(1-\nu)}\mathbb{I}^{(4s)}\\ &= \frac{(1+\nu)}{3(1-\nu)}\mathbb{E}^{(1)} + \frac{2(4-5\nu)}{15(1-\nu)}\mathbb{E}^{(2)} = s_1\mathbb{E}^{(1)} + s_2\mathbb{E}^{(2)}\end{aligned}$$

where $s_1 = \dfrac{1+\nu}{3(1-\nu)}$ and $s_2 = \dfrac{2(4-5\nu)}{15(1-\nu)}$. Then

$$\begin{aligned}(\mathbb{S}^{\Omega})^{-1} &= \frac{3(1-\nu)}{1+\nu}\mathbb{E}^{(1)} + \frac{15(1-\nu)}{2(4-5\nu)}\mathbb{E}^{(2)}\\ &= \frac{3(1-\nu)(1-5\nu)}{2(1+\nu)(4-5\nu)}\mathbb{I}^{(2)}\otimes\mathbb{I}^{(2)} + \frac{15(1-\nu)}{2(4-5\nu)}\mathbb{I}^{(4s)}\end{aligned}$$

Moreover,

$$\begin{aligned}\mathbb{T}^{\Omega} &= \mathbb{I}^{(4s)} - \mathbb{C} : \mathbb{S}^{\Omega} : \mathbb{D}\\ &= (\mathbb{E}^{(1)} + \mathbb{E}^{(2)}) - (3K\mathbb{E}^{(1)} + 2\mu\mathbb{E}^{(2)}) : (s_1\mathbb{E}^{(1)} + s_2\mathbb{E}^{(2)})\\ &: \left(\frac{1}{3K}\mathbb{E}^{(1)} + \frac{1}{2\mu}\mathbb{E}^{(2)}\right) = (1-s_1)\mathbb{E}^{(1)} + (1-s_2)\mathbb{E}^{(2)}\end{aligned}$$

1.2.4 *Tensor analysis*

Define the gradient operator as

$$\nabla = \frac{\partial}{\partial x_i}\mathbf{e}_i \tag{1.46}$$

This is a vector differential operator, meaning that applying the gradient operator to a scalar function, $f \in C^0(\Omega)$, $\Omega \subset \mathbb{R}^d$, will result a vector. In other words, the gradient of a scalar function (zero-th order tensor) is a first order tensor, i.e.

$$grad\ f := \nabla f = \left(\frac{\partial}{\partial x_i}\mathbf{e}_i\right) f = \frac{\partial f}{\partial x_i}\mathbf{e}_i \tag{1.47}$$

For a vector field, or a vector function, $\mathbf{A}(\mathbf{x}) = A_i(\mathbf{x})\mathbf{e}_i$, its gradient is a tensor product between the gradient operator and the vector field,

$$grad\ \mathbf{A} := \nabla\otimes\mathbf{A} == \left(A_i e_i\right)\otimes\left(\frac{\overleftarrow{\partial}}{\partial x_j}e_j\right) = \frac{\partial A_i}{\partial x_j}\mathbf{e}_i\otimes\mathbf{e}_j \tag{1.48}$$

Note that the gradient operator is 'acting from behind'. The gradient of a vector field, a first order tensor field, is thus a second order tensor. In general, the gradient operation increases the order of a tensorial field one order higher. On the other hand, the scalar product or contraction between a gradient operator and a tensorial field is called *divergence* operation, which will result in a new tensorial field with reduced order. Consider a vector field, $\boldsymbol{A} = A_i\boldsymbol{e}_i$. Its divergence is being defined as

$$div\boldsymbol{A} := \nabla\cdot\boldsymbol{A} = \left(A_i e_i\right)\cdot\left(\frac{\overleftarrow{\partial}}{\partial x_j}e_j\right) = \frac{\partial A_i}{\partial x_j}(e_i\cdot e_j) = \frac{\partial A_i}{\partial x_i} \tag{1.49}$$

Again, the divergence operator is also acting from behind.

The cross product between the gradient operator and a tensorial field. $\boldsymbol{A} = A_i \boldsymbol{e}_i$, is called the *curl* or *rot* of the tensorial field.

$$curl\boldsymbol{A} := \nabla \times \boldsymbol{A} = \frac{\partial A_j}{\partial x_i}\Big(\boldsymbol{e}_i \times \boldsymbol{e}_j\Big) = e_{ijk}\partial_i A_j \mathbf{e}_k = e_{ijk}\partial_j A_k \mathbf{e_i} \tag{1.50}$$

Nevertheless, the cross product is acting from the front. In what follows, a few integral transformations are listed. Suppose that there is a continuous function, $f(x) \in C^1(\Omega)$, defined in a domain $\Omega \in \mathbb{R}^d$ with smooth boundary $\partial\Omega$. A well-known integral theorem is

$$\int_\Omega \nabla f d\Omega = \int_{\partial\Omega} f\boldsymbol{n} dS \tag{1.51}$$

or in indicial notation,

$$\int_\Omega \frac{\partial f}{\partial x_i} d\Omega = \int_{\partial\Omega} f n_i dS \tag{1.52}$$

In general for a smooth tensorial field, $\boldsymbol{A}$, we have the following statement,

$$\int_\Omega \nabla \otimes \boldsymbol{A} d\Omega = \int_{\partial\Omega} \boldsymbol{A} \otimes \boldsymbol{n} dS \tag{1.53}$$

which is often referred to as one form of *the divergence theorem.*

Consider a continuous m-order tensorial field, $\mathbf{A}(x) \in [C^1(\Omega)]^m$, the well known divergence theorem is referred to as the following identity,

$$\int_\Omega \nabla \cdot \boldsymbol{A} d\Omega = \int_{\partial\Omega} \boldsymbol{A} \cdot \boldsymbol{n} dS \tag{1.54}$$

If $\boldsymbol{A}$ is a vector field, i.e. $\boldsymbol{A} = A_i \boldsymbol{e}_i$, the divergence theorem is often stated by using indicial notation,

$$\int_\Omega \frac{\partial A_i}{\partial x_i} d\Omega = \int_{\partial\Omega} A_i n_i dS \tag{1.55}$$

Note that consistent with the gradient operator, the surface normal $\mathbf{n}$ comes "from behind" in both Eqs. (1.53) and (1.54).

If we consider the volume integration of a cross product between gradient operator and the tensorial field, we can have the following integral transformation,

$$\int_\Omega \nabla \times \boldsymbol{A} d\Omega = \int_{\partial\Omega} \boldsymbol{n} \times \boldsymbol{A} dS \tag{1.56}$$

Again, if $\boldsymbol{A}$ is a vector field, we may write it by using indicial notation as,

$$\int_\Omega e_{ijk} \frac{\partial A_k}{\partial x_j} d\Omega = \int_{\partial\Omega} e_{ijk} n_j A_k dS \tag{1.57}$$

1.3 Review of linear elasticity theory

1.3.1 *Governing equations*

The traditional micromechanics is primarily based on theory of micro-elasticity. The applicability of linear elasticity theory is believed to be valid up to the scale of 10 nm. To set the stage, we outline some basic formulations of the infinitesimal, linear elasticity theory.

(1)*Equations of motion*

Denote $\boldsymbol{\sigma} = \sigma_{ij}\boldsymbol{e}_i \otimes \boldsymbol{e}_j$ as Cauchy stress tensor, and $\boldsymbol{u} = u_i\boldsymbol{e}_i$ as the infinitesimal displacement field, ρ as the density of the continuum, and $\boldsymbol{b} = b_i\boldsymbol{e}_i$ as the body force per unity volume. The equation of motion of a material particle can be expressed in a Cartesian coordinate as $\forall \mathbf{x} \in \Omega$,

$$\nabla \cdot \boldsymbol{\sigma} + \boldsymbol{b} = \rho \frac{\partial^2 \boldsymbol{u}}{\partial t^2} \tag{1.58}$$

For convenience, it is often expressed in terms of indicial notation,

$$\sigma_{ij,j} + b_i = \rho \frac{\partial^2 u_i}{\partial t^2} \tag{1.59}$$

where $\sigma_{ij,j} = \dfrac{\partial \sigma_{ij}}{\partial x_j}$ is the short-handed notation for spatial derivative.

(2)*Geometric relations*

The infinitesimal strain field $\boldsymbol{\epsilon} = \epsilon_{ij}\boldsymbol{e}_i \otimes \boldsymbol{e}_j$ is defined as

$$\boldsymbol{\epsilon} = \frac{1}{2}\Big(\nabla \otimes \boldsymbol{u} + (\nabla \otimes \boldsymbol{u})^T\Big) \tag{1.60}$$

Note that $\nabla \otimes \boldsymbol{u} = u_{i,j}\boldsymbol{e}_i \otimes \boldsymbol{e}_j$. Hence $(\nabla \otimes \boldsymbol{u})^T = u_{j,i}\boldsymbol{e}_i \otimes \boldsymbol{e}_j$.
Therefore in component form,

$$\epsilon_{ij} = \frac{1}{2}(u_{i,j} + u_{j,i}) \tag{1.61}$$

(3)*Constitutive equations*

For linear elastic solids, the constitutive equations have the following form,

$$\boldsymbol{\sigma} = \mathbb{C} : \boldsymbol{\epsilon} \quad \Rightarrow \quad \sigma_{ij} = C_{ijkl}\epsilon_{kl} \tag{1.62}$$

where $\mathbb{C} = C_{ijkl}\boldsymbol{e}_i \otimes \boldsymbol{e}_j \otimes \boldsymbol{e}_k \otimes \boldsymbol{e}_l$ is the elasticity tensor.
For isotropic elastic media, it has the form,

$$\mathbb{C} = \lambda \mathbb{I}^{(2)} \otimes \mathbb{I}^{(2)} + 2\mu \mathbb{I}^{(4s)} \tag{1.63}$$

where λ, μ are Lame constants. Via indicial notation, it reads

$$C_{ijkl} = \lambda\delta_{ij}\delta_{kl} + \mu(\delta_{ik}\delta_{jl} + \delta_{il}\delta_{jk}) \tag{1.64}$$

Inversely, one may write that

$$\boldsymbol{\epsilon} = \mathbb{C}^{-1} : \boldsymbol{\sigma} = \mathbb{D} : \boldsymbol{\sigma} \quad \epsilon_{ij} = D_{ijkl}\sigma_{kl} \tag{1.65}$$

where the fourth order tensor, $\mathbb{D}$, is called compliance tensor. For isotropic materials, it has the following form:

$$D_{ijkl} = -\frac{\lambda}{2\mu(3\lambda + 2\mu)}\delta_{ij}\delta_{kl} + \frac{1}{4\mu}(\delta_{ik}\delta_{jl} + \delta_{il}\delta_{jk}) \tag{1.66}$$

(4) *Compatibility condition*

Compatibility conditions for infinitesimal deformation field may be expressed as [Malvern (1969)],

$$\nabla \times \boldsymbol{\epsilon} \times \nabla = \mathbf{0} \tag{1.67}$$

In indicial natation, it reads,

$$e_{pki}e_{qlj}\epsilon_{ij,kl} = 0 \tag{1.68}$$

or alternatively

$$\epsilon_{ij,kl} + \epsilon_{kl,ij} - \epsilon_{ik,jl} - \epsilon_{il,jk} = 0 \tag{1.69}$$

(5) *Elastic potential energy*

The elastic strain energy density is defined as

$$U(\boldsymbol{\epsilon}) = \int_0^{\boldsymbol{\epsilon}} \boldsymbol{\sigma}(\boldsymbol{\epsilon}') : d\boldsymbol{\epsilon}' \tag{1.70}$$

The term of *strain energy* may come from the fact that the strain is the integration (dummy) variable in above integration. Based on the fundamental theorem of calculus, one may find its inverse relationship as

$$\frac{\partial U(\boldsymbol{\epsilon})}{\partial \boldsymbol{\epsilon}} = \boldsymbol{\sigma}, \quad \text{or} \quad \frac{\partial U}{\partial \epsilon_{ij}} = \sigma_{ij} \tag{1.71}$$

The complementary strain energy density can be obtained via the Legendre transform,

$$U^*(\boldsymbol{\sigma}) = \boldsymbol{\sigma} : \boldsymbol{\epsilon} - U(\boldsymbol{\epsilon}) \tag{1.72}$$

Or one may define

$$U^*(\boldsymbol{\sigma}) = \int_0^{\boldsymbol{\sigma}} \boldsymbol{\epsilon}(\boldsymbol{\sigma}')d\boldsymbol{\sigma}' \tag{1.73}$$

One may derive that

$$\boldsymbol{\epsilon} = \frac{\partial U^*}{\partial \boldsymbol{\sigma}}, \quad or \quad \epsilon_{ij} = \frac{\partial U^*}{\partial \sigma_{ij}} \tag{1.74}$$

For linear elastic materials,

$$C_{ijkl}\epsilon_{kl} = \frac{\partial U}{\partial \epsilon_{ij}} \quad \Rightarrow \quad C_{ijkl} = \frac{\partial^2 U}{\partial \epsilon_{ij}\partial \epsilon_{kl}} \tag{1.75}$$

For hyperelastic materials, the elastic stiffness tensor can be calculated based on the following formula

$$C_{ijkl} = \frac{\partial^2 U}{\partial \epsilon_{ij}\partial \epsilon_{kl}} \tag{1.76}$$

Similarly, one may find elastic compliance tensor by using complementary strain energy density,

$$D_{ijkl} = \frac{\partial^2 U^*}{\partial \sigma_{ij}\partial \sigma_{kl}} \tag{1.77}$$

Change the order of differentiation in Eq.(1.76),

$$\frac{\partial^2 U}{\partial \epsilon_{ij}\partial \epsilon_{kl}} = \frac{\partial^2 U}{\partial \epsilon_{kl}\partial \epsilon_{ij}} \tag{1.78}$$

One may deduce that $C_{ijkl} = C_{klij}$, which is often called "*major symmetry*". Furthermore, since $\epsilon_{ij} = \epsilon_{ji}$ and $\epsilon_{kl} = \epsilon_{lk}$, one can deduce the so-called "minor symmetry", such that $C_{ijkl} = C_{jikl} = C_{ijlk} = C_{jlik}$. Similar conclusions can be drawn for elastic compliance tensors.

A fourth-order tensor, C_{ijkl}, is *positive-definite*, if

$$C_{ijkl}\epsilon_{ij}\epsilon_{kl} > 0, \quad \text{for } \forall \epsilon_{ij} \neq 0 \tag{1.79}$$

By definition, both elastic tensor $\mathbb{C}$ and compliance tensor $\mathbb{D}$ are positive definite, because for nonzero stress and strain, both strain energy density and complementary strain energy density must be positive, i.e.

$$U(\boldsymbol{\epsilon}) = \frac{1}{2}\boldsymbol{\epsilon} : \mathbb{C} : \boldsymbol{\epsilon} = \frac{1}{2}C_{ijkl}\epsilon_{ij}\epsilon_{kl} > 0$$

$$U^*(\boldsymbol{\sigma}) = \frac{1}{2}\boldsymbol{\sigma} : \mathbb{D} : \boldsymbol{\sigma} = \frac{1}{2}D_{ijkl}\sigma_{ij}\sigma_{kl} > 0$$

1.3.2 *Betti's reciprocal theorem and the Somigliana identity*

Consider two sets of different self-equilibrating states: $\left\{\boldsymbol{u}^{(\alpha)}, \boldsymbol{\epsilon}^{(\alpha)}, \boldsymbol{\sigma}^{(\alpha)}, \boldsymbol{f}^{(\alpha)}\right\}$, $\alpha = 1, 2$,

$$\nabla \cdot \boldsymbol{\sigma}^{(\alpha)} + \boldsymbol{f}^{(\alpha)} = 0 \tag{1.80}$$

with boundary conditions,

$$\boldsymbol{n} \cdot \boldsymbol{\sigma}^{(\alpha)} = \bar{\boldsymbol{t}}^{(\alpha)}, \quad \forall \mathbf{x} \in \Gamma_t^{(\alpha)} \tag{1.81}$$

$$\boldsymbol{u}^{(\alpha)} = \bar{\boldsymbol{u}}^{(\alpha)}, \quad \forall \mathbf{x} \in \Gamma_u^{(\alpha)}, \quad \alpha = 1, 2 \tag{1.82}$$

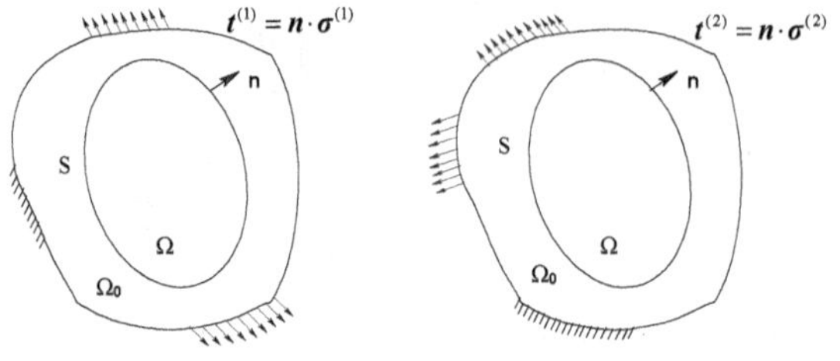

Fig. 1.2 Two sets of different self-equilibrating states.

acting in a same object Ω_0. The Betti's reciprocal theorem[3] states that: the work done by the first set of self-equilibrating surface traction, $\boldsymbol{t}^{(1)}$, and body force $\boldsymbol{f}^{(1)}$ in any interior region $\Omega \subset \Omega_0$, going through the displacement field, $\boldsymbol{u}^{(2)}$, of the second self-equilibrating system, equals the work done by the second set of traction, $\boldsymbol{t}^{(2)}$, and the body force, $\boldsymbol{f}^{(2)}$, in the same interior region going through the displacement field, $\boldsymbol{u}^{(1)}$, of the first self-equilibrating system, i.e.

$$\int_\Omega f_i^{(1)} u_i^{(2)} d\Omega + \int_{\partial\Omega} t_i^{(1)} u_i^{(2)} dS = \int_\Omega f_i^{(2)} u_i^{(1)} d\Omega + \int_{\partial\Omega} t_i^{(2)} u_i^{(1)} dS \tag{1.83}$$

Proof. Consider both states being equilibrium states. It has

$$\begin{aligned}\int_\Omega f_i^{(1)} u_i^{(2)} d\Omega &= -\int_\Omega \sigma_{ji,j}^{(1)} u_i^{(2)} d\Omega = -\int_{\partial\Omega} \sigma_{ji}^{(1)} n_j u_i^{(2)} dS + \int_\Omega \sigma_{ji}^{(1)} u_{i,j}^{(2)} d\Omega \\ &= -\int_{\partial\Omega} t_i^{(1)} u_i^{(2)} dS + \int_\Omega \sigma_{ji}^{(1)} \epsilon_{ji}^{(2)} d\Omega \end{aligned} \tag{1.84}$$

Moving the first term of the right-hand side of the above expression to the left-hand side yields

$$\int_\Omega f_i^{(1)} u_i^{(2)} d\Omega + \int_{\partial\Omega} t_i^{(1)} u_i^{(2)} dS = \int_\Omega \sigma_{ij}^{(1)} \epsilon_{ij}^{(2)} d\Omega \tag{1.85}$$

Similarly, one may show that

$$\int_\Omega f_i^{(2)} u_i^{(1)} d\Omega + \int_{\partial\Omega} t_i^{(2)} u_i^{(1)} dS = \int_\Omega \sigma_{ij}^{(2)} \epsilon_{ij}^{(1)} d\Omega \tag{1.86}$$

Consider the fact that the two systems have the same material property,

$$\int_\Omega \sigma_{ij}^{(1)} \epsilon_{ij}^{(2)} d\Omega = \int_\Omega C_{ijkl} \epsilon_{kl}^{(1)} \epsilon_{ij}^{(2)} d\Omega = \int_\Omega C_{klij} \epsilon_{kl}^{(1)} \epsilon_{ij}^{(2)} d\Omega = \int_\Omega \epsilon_{kl}^{(1)} \sigma_{kl}^{(2)} d\Omega$$

Comparing both sides of (1.85) and (1.86), it leads to Betti's (second) reciprocal theorem (1.83).

In addition, the equality

$$\int_\Omega \sigma_{ij}^{(1)} \epsilon_{ij}^{(2)} d\Omega = \int_\Omega \sigma_{ij}^{(2)} \epsilon_{ij}^{(1)} d\Omega \tag{1.87}$$

is called Betti's first reciprocal theorem. □

[3]Strictly speaking, it is the Betti's second reciprocal theorem.

To derive Somigliana identity, we first consider Dirac's delta function, which is defined as the limit of the following piece-wise function,

$$\delta(x) = \lim_{\epsilon\to 0} \delta_\epsilon(x), \quad \text{where} \quad \delta_\epsilon(x) = \begin{cases} 0; & x < -\epsilon/2 \\ 1/\epsilon; & -\epsilon/2 < x < \epsilon/2 \\ 0; & x > \epsilon/2 \end{cases} \tag{1.88}$$

A graph of Dirac's delta function is shown in Fig. 1.3.

Dirac delta function has following properties

$$(1) \qquad \int_{-\infty}^{\infty} \delta(x)dx = 1 \tag{1.89}$$

$$(2) \qquad \int_{-\infty}^{\infty} \delta(x-y)f(y)dy = f(x) \tag{1.90}$$

The first property can be easily shown by definition that

$$\int_{-\infty}^{\infty} \delta(x)dx = \int_{-\epsilon/2}^{\epsilon/2} \frac{1}{\epsilon}dx = 1 \tag{1.91}$$

To show the second property, we let $x - y = z$ and $dy = -dz$. Thus

$$\begin{aligned} \int_{-\infty}^{\infty} \delta(x-y)f(y)dy &= -\int_{\infty}^{-\infty} \delta(z)f(x-z)dz = \int_{-\infty}^{\infty} \delta(z)f(x-z)dz \\ &= \frac{1}{\epsilon}\int_{-\epsilon/2}^{\epsilon/2} f(x-z)dz = \frac{1}{\epsilon}f(x-\zeta\frac{\epsilon}{2})\int_{-\epsilon/2}^{\epsilon/2} dz \\ &= f(x), \quad as \ \ \epsilon \to 0 \end{aligned} \tag{1.92}$$

where $-1 < \zeta < 1$.

Consider an infinitely space filled with homogeneous elastic medium. The body force is in the form of a concentrated load at a fixed point $\mathbf{y}$,

$$\boldsymbol{f}(\mathbf{x}) = \delta(\mathbf{x}-\mathbf{y})\delta_{mk}\boldsymbol{e}_k \tag{1.93}$$

The subscript index m indicates that the concentrated force is in the direction of m. The differential equilibrium equations then have the form,

$$\nabla \cdot \boldsymbol{\sigma}(\mathbf{x}) + \delta(\mathbf{x}-\mathbf{y})\delta_{mk}\boldsymbol{e}_k = 0, \ \forall \mathbf{x} \in \mathbb{R}^3 \tag{1.94}$$

The displacement solution of this problem is called the fundamental solution of the Navier equation, or the Green's function for an infinitely extended homogeneous elastic domain. Denote the displacement solution at $\mathbf{x}$ due to concentrated force at $\mathbf{y}$ in the direction of m as

$$\boldsymbol{u}(\mathbf{x}) = \mathbf{G}^{\infty}_{m}(\mathbf{x},\mathbf{y}) = G^{\infty}_{mi}(\mathbf{x},\mathbf{y})\boldsymbol{e}_i \tag{1.95}$$

The corresponding strain and stress fields are:

$$\epsilon^{G^{\infty}_m}_{ij} = \frac{1}{2}\Big(G^{\infty}_{mi,j} + G^{\infty}_{mj,i}\Big), \quad \sigma^{G^{\infty}_m}_{ij} = C_{ijkl}\epsilon^{G^{\infty}_m}_{k\ell} \tag{1.96}$$

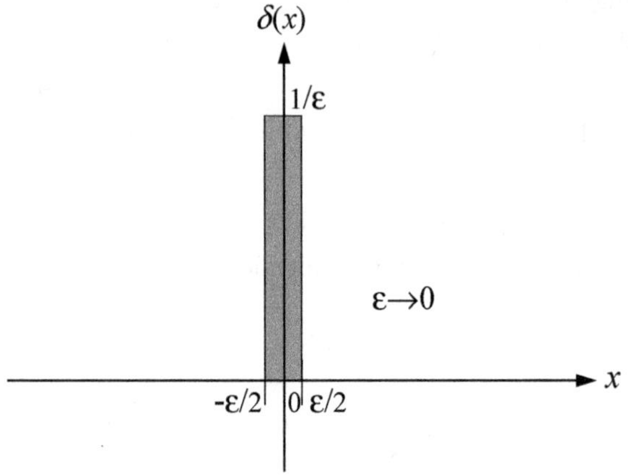

Fig. 1.3 Dirac's delta function.

Next, we consider a singly connected finite region $\Omega \subset \mathbb{R}^3$. The finite region Ω is in a self-equilibrating state, i.e., there is a body force distribution: $\nabla \cdot \boldsymbol{\sigma} + \boldsymbol{f} = \mathbf{0}, \ \forall \mathbf{x} \in \Omega$, and a traction force distribution: $\boldsymbol{t} = \boldsymbol{n} \cdot \boldsymbol{\sigma}, \ \forall \mathbf{x} \in \partial\Omega$.

Consider these two self equilibrium systems,

$$\boldsymbol{f}^{(1)}(\mathbf{x}) = \delta(\mathbf{x} - \mathbf{y})\delta_{mk}\boldsymbol{e}_k, \ \boldsymbol{u}^{(1)}(\mathbf{x}) = G^{\infty}_{mi}(\mathbf{x}, \mathbf{y})\boldsymbol{e}_i, \ \boldsymbol{t}^{(1)}(\mathbf{x}) = \sigma^{G^{\infty}_m}_{ij}(\mathbf{x})n_j\boldsymbol{e}_i \quad (1.97)$$

$$\boldsymbol{f}^{(2)}(\mathbf{x}) = f_i(\mathbf{x})\boldsymbol{e}_i, \qquad \boldsymbol{u}^{(2)}(\mathbf{x}) = u_i(\mathbf{x})\boldsymbol{e}_i, \qquad \boldsymbol{t}^{(2)}(\mathbf{x}) = \sigma_{ij}(\mathbf{x})n_j\boldsymbol{e}_i \quad (1.98)$$

Apply Betti's reciprocal theorem,

$$\int_\Omega \delta(\mathbf{x} - \mathbf{y})\delta_{mi}u_i(x)d\Omega_x + \int_{\partial\Omega} n_j\sigma^{G^{\infty}_m}_{ji}u_i(\mathbf{x})dS_x$$
$$= \int_\Omega f_i(\mathbf{x})G^{\infty}_{mi}(\mathbf{x}, \mathbf{y})d\Omega_x + \int_{\partial\Omega} n_j\sigma_{ji}G^{\infty}_{mi}(\mathbf{x}, \mathbf{y})dS_x \quad (1.99)$$

Considering the property of the Dirac delta function, one can obtain

$$u_m(\mathbf{y}) = \int_\Omega f_i(\mathbf{x})G^{\infty}_{mi}(\mathbf{x}, \mathbf{y})d\Omega_x + \int_{\partial\Omega} t_i(\mathbf{x})G^{\infty}_{mi}(\mathbf{x}, \mathbf{y})dS_x$$
$$- \int_{\partial\Omega} t^{G^{\infty}_m}_i(\mathbf{x}, \mathbf{y})u_i(\mathbf{x})dS_x, \quad m = 1, 2, 3 \quad (1.100)$$

Equation (1.100) is the well-known Somigliana identity [Somigliana (1886)].

1.4 Review of finite elasticity

In this section, we list some basic formulas of finite elasticity. Consider a set of material points that are belong to a domain or body $\Omega_0 \subset \mathbb{R}^3$. The motion of the body is described by a map,

$$\boldsymbol{\phi} : \Omega_0 \to \Omega, \quad \mathbf{x} = \boldsymbol{\phi}(\mathbf{X}, t) \ , \quad \forall \mathbf{X} \in \Omega_0 \quad (1.101)$$

The motion $\boldsymbol{\phi}$ carries points $\mathbf{X}$ located in Ω_0 of reference configuration to places $\mathbf{x} \in \Omega$ of the current configuration. We assume subsequently that $\boldsymbol{\phi}$ possesses continuous derivatives with respect to both space and time.

When the reference configuration is identical to the initial configuration, the position vector $\mathbf{x}$ of any point at $t = 0$ is called as the material coordinates,

$$\mathbf{X} = \mathbf{x}(\mathbf{X}, 0) \quad \text{or in component form} \quad X_i = x_i(\mathbf{X}, 0) = \phi_i(\mathbf{X}, 0) \ . \tag{1.102}$$

The displacement of a material point is defined as the difference between its current position $\mathbf{x}$ and its original position $\mathbf{X}$ in the material configuration,

$$\mathbf{u}(\mathbf{X}, t) = \boldsymbol{\phi}(\mathbf{X}, t) - \boldsymbol{\phi}(\mathbf{X}, t) = \mathbf{x} - \mathbf{X} \tag{1.103}$$

If we express both $\mathbf{x}$ and $\mathbf{X}$ with the same basis vectors in a same coordinate system, we can write the displacement field in the following component form,

$$u_i = x_i - X_i$$

An important variable in characterizing deformation state is the deformation gradient, which is defined as

$$\mathbf{F} = \frac{\partial \boldsymbol{\phi}}{\partial \mathbf{X}} = \frac{\partial \mathbf{x}}{\partial \mathbf{X}} \tag{1.104}$$

Deformation gradient is often referred to as a two-point tensor. If $\mathbf{x} = x_i \mathbf{e}_i$ and $\mathbf{X} = X_I \mathbf{E}_I$, we can then write

$$\mathbf{F} = \frac{\partial x_i}{\partial X_J} \mathbf{e}_i \otimes \mathbf{E_J} \tag{1.105}$$

where $\mathbf{e}_i$ and $\mathbf{E}_J$ are bases vectors in the current configuration and the reference configuration. The determinant of $\mathbf{F}$ is called the Jacobian determinant, which is often denoted as

$$J = det(\mathbf{F}) = det\{\frac{\partial x_i}{\partial X_J}\} \ . \tag{1.106}$$

Many strain measures are expressed in terms of $\mathbf{F}$. We call

$$\mathbf{C} = \mathbf{F}^T \cdot \mathbf{F} \tag{1.107}$$

as the right Cauchy-Green tensor, and we call

$$\mathbf{E} = \frac{1}{2}\Big(\mathbf{F}^T \cdot \mathbf{F} - \mathbf{I}\Big) \tag{1.108}$$

as the Green-Lagrange strain tensor.

In finite deformation theory, stress measures are often related to deformation gradient as well, e.g. the first Piola-Kirchhoff stress tensor $\mathbf{P}$ is related to the Cauchy stress tensor $\boldsymbol{\sigma}$ as

$$\mathbf{P} = J\mathbf{F}^{-1} \cdot \boldsymbol{\sigma} \tag{1.109}$$

and the second Piola-Kirchhoff stress tensor $\mathbf{S}$ is related to the Cauchy stress tensor as

$$\mathbf{S} = J\mathbf{F}^{-1} \cdot \boldsymbol{\sigma} \cdot \mathbf{F}^{-T} \tag{1.110}$$

and to the first Piola-Kirchhoff stress tensor as

$$\mathbf{S} = \mathbf{P} \cdot \mathbf{F}^{-T} \tag{1.111}$$

A hyperelastic material is a sub-class of elastic materials, whose stress response function can be derived from a potential function of elastic strain measures. For instance, there exists a potential function $\Psi(\mathbf{F})$ so that

$$\mathbf{P} = \frac{\partial \Psi(\mathbf{F})}{\partial \mathbf{F}} \tag{1.112}$$

or $\Phi(\mathbf{C}) = w(\mathbf{E})$ such that

$$\mathbf{S} = 2\frac{\partial \Phi(\mathbf{C})}{\partial \mathbf{C}} = \frac{\partial w(\mathbf{E})}{\partial \mathbf{E}} \ . \tag{1.113}$$

For detailed discussions on finite elasticity and continuum mechanics in general, readers may consult books by [Marsden and Hughes (1983)] or [Holzapfel (2000)].

1.5 Review of molecular dynamics

In this section, we briefly outline the fundamental theory of conventional molecular dynamics and the lattice dynamics.

There are two basic assumptions made in standard molecular dynamics (MD) simulations:

(1) Molecules or atoms are described as a system of interacting material points, whose motion is described dynamically with a vector of instantaneous positions and velocities. The atomic interaction has a strong dependence on the spatial orientation and distances between separate atoms. This model is often referred to as the soft sphere model, where the softness is analogous to the electron clouds of atoms.
(2) No mass changes in the system. Equivalently, the number of atoms in the system remains the same. The simulated system is usually treated as an isolated domain system with conserved energy.

1.5.1 *Lagrangian equations of motion*

The equation of motion of a system of interacting material points (particles, atoms), having s degrees of freedom in total, can be most generally written in terms of a Lagrangian function L, e.g. [Landau and Lifshitz (1965)],

$$\frac{d}{dt}\frac{\partial L}{\partial \dot{q}_\alpha} - \frac{\partial L}{\partial q_\alpha} = 0, \qquad \alpha = 1, 2, \cdots, s. \tag{1.114}$$

Here q_α are the generalized coordinate, the arbitrary observer that uniquely defines spatial positions of the atoms. The superposed dot denotes time derivatives. As will be discussed later, Eq. (1.114) can be rewritten in terms of the generalized coordinates and momenta. This scheme has been extensively used in the statistical mechanics formulation. Molecular dynamics simulations are usually run with the reference to a Cartesian coordinate system, where equations (1.114) can be simplified as

$$\frac{d}{dt}\frac{\partial L}{\partial \dot{\mathbf{r}}_i} - \frac{\partial L}{\partial \mathbf{r}_i} = 0, \quad i = 1, 2, \cdots, N. \tag{1.115}$$

where $\mathbf{r}_i = (x_i, y_i, z_i)$ is the position vector of atom i (see Fig. 1.4), N is the total number of simulated atoms and $N = s/3$ in three-dimensional space. The spatial volume occupied by these N atoms is usually referred to as the MD domain. Due to the homogeneity of time and space, and also isotropy of space in inertial coordinate systems, the equations of motion (1.115) must not depend on the choice of initial time of observation, the origin of the coordinate system, and directions of its axes. These basic principles are equivalent to the requirements that the Lagrangian function cannot explicitly depend on time, directions of the radius and velocity vectors $\mathbf{r}_i$ and $\dot{\mathbf{r}}_i$, and it can only depend on the absolute value of the velocity vector $\dot{\mathbf{r}}_i$. In order to provide identical equations of motions in all inertial coordinate systems, the Lagrangian function must also comply with the Galilean relativity principle. One function satisfying all these requirements reads [Landau and Lifshitz (1965)]

$$L = \sum_{i=1}^{N} \frac{m_i}{2}(\dot{x}_i^2 + \dot{y}_i^2 + \dot{z}_i^2) \equiv \sum_{i=1}^{N} \frac{m_i \dot{\mathbf{r}}_i^2}{2} \tag{1.116}$$

for a system of free, non-interacting particles, where m_i is the mass of particle i. Interaction between the particles can be described by adding to (1.116) a particular function of atomic coordinates U, depending on the properties of this interaction. Such a function is defined with a negative sign, so that the system's Lagrangian acquires the form:

$$L = \sum_{i=1}^{N} \frac{m_i \dot{\mathbf{r}}_i^2}{2} - U(\mathbf{r}_1, \mathbf{r}_2, \cdots, \mathbf{r}_N), \tag{1.117}$$

where the two terms represent the system's kinetic and potential energy, respectively. This gives the general structure of Lagrangian for a conservative system of interacting material points in Cartesian coordinates. It is important to note two features of this Lagrangian: the additivity of the kinetic energy term and the absence of explicit dependence on time. The fact that the potential energy term only depends on spatial configuration of the particles implies that any change in this configuration results in an immediate effect on the motion of all particles within the simulated domain.

By substituting the Lagrangian (1.117) to equations (1.115), the equations of motion can finally be written in the Newtonian form,

$$m_i \ddot{\mathbf{r}}_i = -\frac{\partial U(\mathbf{r}_1, \mathbf{r}_2, \cdots, \mathbf{r}_N)}{\partial \mathbf{r}_i} \equiv \mathbf{F}_i, \quad i = 1, 2, \cdots, N \tag{1.118}$$

The force $\mathbf{F}_i$ is usually referred to as the internal force, i.e. the force exerted on atom i due to specifics of the environment it is exposed to. Eqs. (1.118) are further solved for a given set of initial conditions to get trajectories of the atomic motion in the simulated system.

1.5.2 *Hamiltonian equations of motion*

The Lagrangian formulation for the MD equations of motion discussed in previous section assumes description of the mechanical state of simulated system by means of generalized coordinates and velocities. This description, however, is not the only description. An alternative description, in terms of the generalized coordinates and momentum, is called the Hamiltonian formulation, e.g. [Landau and Lifshitz (1965)].

Transition to the new set of independent variables can be accomplished as follows: First, employ the complete differential of the Lagrangian function of Eq. (1.114),

$$dL = \sum_\alpha \frac{\partial L}{\partial q_\alpha} dq_\alpha + \sum_\alpha \frac{\partial L}{\partial \dot{q}_\alpha} d\dot{q}_\alpha, \quad \alpha = 1, 2, \cdots, s, \tag{1.119}$$

and rewrite this as

$$dL = \sum_\alpha \dot{q}_\alpha dq_\alpha + \sum_\alpha p_\alpha d\dot{q}_\alpha, \tag{1.120}$$

where the generalized momenta are defined to be

$$p_\alpha = \frac{\partial L}{\partial \dot{q}_\alpha} \ . \tag{1.121}$$

Based on the Lagrangian equation, one can deduce that

$$\frac{\partial L}{\partial q_\alpha} = \dot{p}_\alpha$$

The right-hand side of Eq. (1.119) can be rearranged as

$$dL = \sum_\alpha \dot{p}_\alpha dq_\alpha + d\Big(\sum_\alpha p_\alpha \dot{q}_\alpha\Big) - \sum_\alpha \dot{q}_\alpha dp_\alpha, \tag{1.122}$$

and

$$d\Big(\sum_\alpha p_\alpha \dot{q}_\alpha - L\Big) = \sum_\alpha \dot{q}_\alpha dp_\alpha - \sum_\alpha \dot{p}_\alpha dq_\alpha, \tag{1.123}$$

where the function

$$H(p, q, t) = \sum_\alpha p_\alpha \dot{q}_\alpha - L \tag{1.124}$$

is referred to as the (classical) Hamiltonian of the system. The value of the Hamiltonian function is an integral of motion for conservative systems, and it is defined to be the total energy of the system in terms of the generalized coordinates and momenta. Thus, we have obtained

$$dH = \sum_{\alpha} \dot{q}_\alpha dp_\alpha - \sum_{\alpha} \dot{p}_\alpha dq_\alpha \tag{1.125}$$

and therefore

$$\dot{q}_\alpha = \frac{\partial H}{\partial p_\alpha}, \quad \dot{p}_\alpha = -\frac{\partial H}{\partial q_\alpha} \, . \tag{1.126}$$

These are the Hamiltonian equations of motion in terms of new variable p_α and q_α. They comprise a system of $2s$ first-order ODEs on $2s$ unknown functions $p_\alpha(t)$ and $q_\alpha(t)$. A complete set of these vectors, observed in the course of temporal evolution of the system, defines a hyper-surface in the phase space, known as the phase space trajectory. The phase space trajectory provides a complete description of the system's dynamics. Although both the kinetic and potential energies do usually vary or fluctuate in time, the phase space trajectory determined from Eqs. (1.126) conserves the total energy of the system. Indeed, the time rate of change of the Hamiltonian is equal to zero,

$$\frac{dH}{dt} = \frac{\partial H}{\partial t} + \sum_{\alpha} \frac{\partial H}{\partial q_\alpha} \dot{q}_\alpha + \sum_{\alpha} \frac{\partial H}{\partial p_\alpha} \dot{p}_\alpha = \frac{\partial H}{\partial t} = 0, \tag{1.127}$$

since it has no explicit dependence on time in the case of a conservative system, as follows from (1.124) and (1.126). For a conservative system of N interacting atoms in a Cartesian coordinate system, the Hamiltonian description acquires the following form:

$$H(\mathbf{r}_1, \mathbf{r}_2, \cdots, \mathbf{r}_N; \mathbf{p}_1, \mathbf{p}_2, \cdots, \mathbf{p}_N) = \sum_i \frac{\mathbf{p}_i^2}{2m_i} + U(\mathbf{r}_1, \mathbf{r}_2, \cdots, \mathbf{r}_N) \tag{1.128}$$

and

$$\dot{\mathbf{r}}_i = \frac{\partial H}{\partial \mathbf{p}_i}, \quad \dot{\mathbf{p}}_i = -\frac{\partial H}{\partial \mathbf{r}_i}, \tag{1.129}$$

where the momenta are related to the radius vectors as $\mathbf{p}_i = m_i \dot{\mathbf{r}}_i$. If the Hamiltonian function and an initial state of the atoms in the system are known, one can compute the instantaneous positions and momentums of the atoms at all successive times, by solving Eqs. (1.129). That gives the phase space trajectory of the atomic motion, which can be of particular importance to study the dynamic evolution of atomic structure and bonds, as well as the thermodynamic states of a system. We note, however, that the Newtonian equations (1.118), following from the Lagrangian formulation (1.114), can be more appropriate in studying particular details of the atomic processes, especially in solids. Newtonian formulation is usually more convenient in terms of imposing external forces and constraints (for instance, periodic boundary conditions), as well as the post-processing and visualization of the results.

1.5.3 *Interatomic potentials*

The general structure of the governing equations for molecular dynamics simulations is given by a straightforward second order ODE. However, the potential function in Eq.(1.118) can be extremely complicated, if one wishes to accurately represent the atomic interaction of the simulated system. The nature of this interaction is due to complicated quantum effects taking place at the subatomic level that are responsible for chemical properties such as valence and bond energy; quantum effects also are responsible for the spatial arrangement (topology) of the interatomic bonds, their formation and breakage. In order to obtain reliable results in molecular dynamic simulations, the classical interatomic potential should accurately account for these quantum mechanical processes, though in an averaged sense. This issue is related

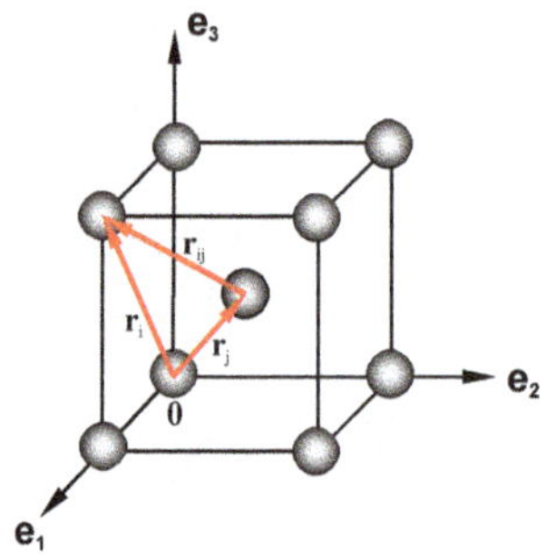

Fig. 1.4 Coordination in atomic systems.

to the form of the potential function for various classes of atomic systems, which have been extensively discussed in literature. General structure of this function is presented by the following equation:

$$U(\mathbf{r}_1, \mathbf{r}_2, \cdots, \mathbf{r}_N) = \sum_i \Phi_1(\mathbf{r}_i) + \sum_{i,j>i} \Phi_2(\mathbf{r}_i, \mathbf{r}_j) + \sum_{i,j>i,k>j} \Phi_3(\mathbf{r}_i, \mathbf{r}_j, \mathbf{r}_k) + \cdots , \tag{1.130}$$

where $\mathbf{r}_n$ is the radius vector of the n-th particle, and functions Φ_m is called the m-body potential. The first term of Eq.(1.130) represents the energy due an external force field, such as gravitational or electrostatic, into which the system is immersed. The second term shows pairwise interaction of the particles, and the third term gives the three-body components, etc.

1.5.4 *Two-body (pair) potentials*

At the subatomic level, the electrostatic field due to the positively charged atomic nucleus is neutralized by the negatively charged electron clouds surrounding the nucleus. The quantum mechanical description of electron motion employs a probabilistic approach to evaluate the probability densities at which the electrons can occupy particular spatial locations. The term "electron cloud" is typically used in relation to spatial distributions of these densities. The negatively charged electron clouds, however, experience cross-atomic attraction, which grows as the distance between the nuclei decreases. On reaching some particular distance, which is referred to as the bond length, this attraction is equilibrated by the repulsive force due to the positively charged nuclei. A further decrease in the inter-nuclei distance results in a quick growth of the resultant repulsive force.

Interatomic forces or intermolecular forces are often called London-Van der Waals force or dispersion forces, which are resulted from instantaneous dipoles or momentary polarization of particles due to complex electron motions and their interaction with nucleus. Fig. 1.5 shows a schematic illustration of the origin of

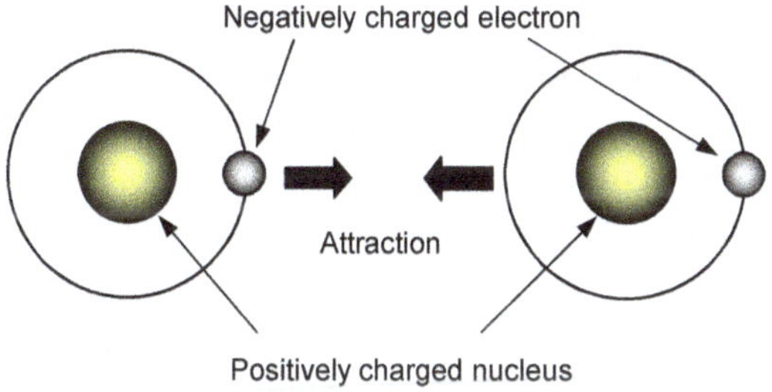

Fig. 1.5 The origin of the London-Van der Waals interatomic forces.

London -Van der Waals force. The main mechanism is that the outer electrons are rotating rapidly around its nucleus, and it can create an instantaneous induced dipole; this is because the electron is negatively charged, the nucleus is positively charged, and at any instance there is a separation between the two. If you line up two such instantaneous dipoles along a line, they will interact each other like a magnet. This instantaneous dipole-dipole interaction is very much different from simple electrostatic interaction because it dropped off more rapidly[4], inversely proportional to the distance between two atoms to the power of 7. Although some people often refer the attractive part of the interatomic force as the Van der Waals force, the Van der Waals force is the totality of the interatomic or intermolecular force including both the attractive part and the repulsive part. The inverse power of the repulsive term is between 9 and 20, and it is typically chosen at 12 or 13. When this power is high e.g. 20, it is called as a "hard-sphere" potential. There exist a variety of mathematical models to describe the interatomic potential or the interatomic force.

[4]Please note that the Coulomb force (electrostatic force) between two point electric charges is inversely proportional to the square of the distance between the charges

In 1924, Jones [Jones (1924a,b)] proposed the following potential function to describe pair-wise atomic interactions:

$$\Phi(\mathbf{r}_i, \mathbf{r}_j) = \Phi(r) = 4\epsilon\left[\left(\frac{\sigma}{r}\right)^{12} - \left(\frac{\sigma}{r}\right)^{6}\right], \quad r = |\mathbf{r}_{ij}| = |\mathbf{r}_i - \mathbf{r}_j|. \tag{1.131}$$

This model is known as the Lennard-Jones (LJ) potential, and it is used in simulations of a great variety of atomic systems and processes. Here, $\mathbf{r}_{ij}$ is the inter-atomic radius-vector, see Fig. 1.4, σ is the collision diameter, the distance at which $\Phi(r) = 0$, and ϵ shows the bond-ing/dislocation energy — the minimum of function (1.131) to occur for an atomic pair in equilibrium. The first term of this potential represents atomic repulsion, dominating at small separation distances while the second term shows attraction (bonding) between two atoms. Since the square bracket quantity is dimensionless, the choice of units for Φ depends on the definition of ϵ. Typically, it is more convenient to use a smaller energy unit, such as electron volt (eV), rather than joule(J). $1\text{eV} = 1.602 \times 10^{-19}\text{J}$, which represents the work done if an elementary charge is accelerated by an electrostatic field of a unit voltage. The energy ϵ represents the amount of work that needs to be done in order to remove one of two coupled atoms from its equilibrium position ρ to infinity. The value ρ is also known as the equilibrium bond length, and it is linked to the collision diameter σ by the relationship $\rho = \sqrt[6]{2}\sigma$. In a typical atomic system, the collision diameter as is equal to several angstrom ($\mathring{A}$), $1\mathring{A} = 10^{-10}m$. The corresponding force between

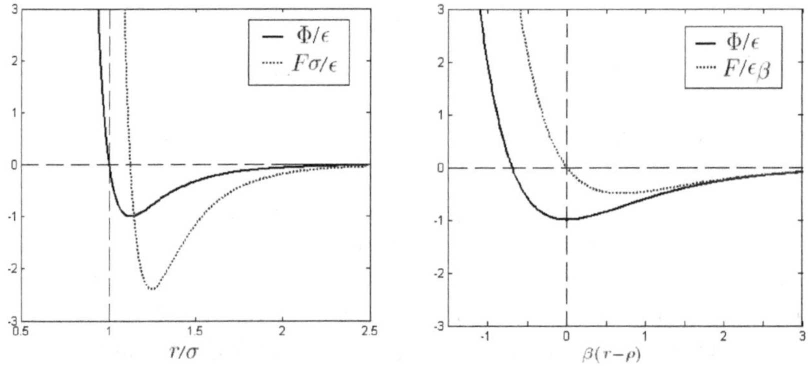

(a) (b)

Fig. 1.6 Pair-wise potentials and the interatomic forces: (a) the Lennard-Jones, (b) the Morse.

the two atoms can be expressed as a function of the inter-atomic distance,

$$F(r) = -\frac{\partial \Phi(r)}{\partial r} = 24\frac{\epsilon}{\sigma}\left[2\left(\frac{\sigma}{r}\right)^{13} - \left(\frac{\sigma}{r}\right)^{7}\right]. \tag{1.132}$$

The potential and force functions (1.131) and (1.132) are plotted versus the inter-atomic distance in Fig. 1.6 (a), using dimensionless quantities. Another popular model for pairwise interaction is known as the Morse potential, also shown in Fig. 1.6 (b):

$$\Phi(r) = \epsilon\Big[e^{2\beta(\rho-r)} - 2e^{\beta(\rho-r)}\Big], \quad F(r) = 2\epsilon\beta\Big[e^{2\beta(\rho-r)} - e^{\beta(\rho-r)}\Big], \tag{1.133}$$

where ρ and ϵ are the equilibrium bond length and dislocation energy respectively; β is an inverse length scaling factor. Similar to the Lennard-Jones model, the first term of this potential is repulsive and the second is attractive, which is interpreted as representing bonding. The Morse potential (1.133) has been adapted for modeling atomic interaction in various types of materials and interfaces; examples can be found elsewhere, e.g. [Wang *et al.* (1991)].

1.5.5 *Embedded-Atom-Method (EAM)*

For an inhomogeneous system embedded with impurities, the total potential energy should be the sum of host potential and impurity potentials, e.g. [Stott and Zarembra (1980)]. One may view that each atom in a solid is an "impurity" embedded in a host lattice of atoms, From this viewpoint, Daw and Baskes [Daw and Baskes (1984); Foiles *et al.* (1986); Daw *et al.* (1993)] find a way to construct the potential energy of atomistic system, in which each atom is embedded in an electron cloud permeated ambient space. Therefore, the total energy of the system is

$$E_{tot} = \sum_{i=1}^{N} F_i(\rho_{h,i}) + \frac{1}{2}\sum_{i=1}^{N}\sum_{j\neq i}^{N} \phi_{ij}(r^{ij}), \quad r^{ij} = |\mathbf{x}_i - \mathbf{x}_j| \tag{1.134}$$

where F_i is the embedding energy functional of atom i, and $\rho_{h,i}$ is the density of the host at the position $\mathbf{x}_i$, $\phi_{ij}(r^{ij})$ is the short-range pair potential and r^{ij} is the distance between the atom i and the atom j. Daw and Baskes further simplified the electron density of the host at the position $\mathbf{x}_i$ as a linear superposition of spherically averaged atomic electron densities,

$$\rho_{h,i} = \sum_{j\neq i}^{N} \rho^a_j(r^{ij}) \tag{1.135}$$

Note that the form of F_i may be obtained through by calculation based on first principles[5] or from curve fitting the experimental data.

If we know the form of F_i and ϕ, we can derive information about lattice constant, elastic constants (tensor), etc. In the following, we show how to relate them to lattice constant or stresses.

Assume that in a homogeneous crystal all atoms are the same, and

$$F_i = F, \ \phi_{ij} = \phi, \ \text{and} \ \ \rho^a_j(r^{ij}) = \rho_j$$

[5] In physics, a calculation is said based on first principle, or *ab initio*, if it is directly based on the established physical laws, for example, Schrödinger's equation

Denote that $\bar{\rho}$ and $\bar{\phi}$ to be the density and the total potential at equilibrium, and r^{ij} is the length for bond between atom i and j, i.e. $r^{ij} = |\mathbf{x}_i - \mathbf{x}_j|$ at the equilibrium. We can then write

$$\bar{\rho} = \rho_{h,i} = \sum_{j \neq i}^{N} \rho(r^{ij}) \quad \text{and} \quad \bar{\phi} = \sum_{j=1}^{N} \phi(r^{ij}) \tag{1.136}$$

and

$$E_{tot} = \sum_{i=1}^{N} F(\bar{\rho}) + \frac{1}{2} \sum_{i=1}^{N} \sum_{m \neq i} \phi(r^{im}) \tag{1.137}$$

Let the volume of the local region be Ω_0, which is the region inside the cut-off radius of the atom interested. We may define the local Cauchy stress as

$$\boldsymbol{\sigma} = \frac{\partial}{\partial \boldsymbol{\epsilon}} \left(\frac{E_{tot}}{\Omega_0} \right) \tag{1.138}$$

Note that this definition is under an assumption that the lattice deformation is uniform, or an assumption called the Cauchy-Born rule. The so-called Cauchy-Born rule states that if a lattice is subjected to uniform deformation or homogeneous constant strain, the positions of the atoms follows the constant strain of the material. Thus, based on Eq.(1.138), we can find the stress in a local region Ω_0 (which is not necessarily a unit cell) as

$$\boldsymbol{\sigma} = \frac{NF'(\bar{\rho})}{\Omega_0} \sum_{m \neq i} \rho'_m \frac{\partial r^{mi}}{\partial \boldsymbol{\epsilon}} + \frac{N}{2\Omega_0} \sum_{m \neq i} \phi'_m \frac{\partial r^{mi}}{\partial \boldsymbol{\epsilon}} \tag{1.139}$$

where

$$\rho'_m = \left[\frac{d\rho(r)}{dr} \right]_{r=r^{im}} \quad \text{and} \quad \phi'_m = \left[\frac{d\phi_m(r)}{dr} \right]_{r=r^{im}}$$

Let

$$\mathbf{r}^{mi} = \mathbf{x}_m - \mathbf{x}_i = \mathbf{R}^{mi} + \mathbf{u}^{mi}, \quad \text{where} \quad \mathbf{u}^{mi} := \mathbf{u}^m - \mathbf{u}^i$$

Since in uniform deformation the strain is constant, and we further assume that $curl(\mathbf{u}) = 0$, this allows us write

$$\mathbf{u}^{mi} = \boldsymbol{\epsilon} \cdot \mathbf{r}^{mi}, \quad \text{or in component form} \quad u_i^{mi} = \epsilon_{ij} r_j^{mi} \tag{1.140}$$

Hence

$$\frac{\partial u_k^{mi}}{\partial \epsilon_{ij}} = \delta_{ki} r_j^{mi} \tag{1.141}$$

and subsequently,

$$\frac{\partial r^{mi}}{\partial \epsilon_{st}} = \frac{\partial r^{mi}}{\partial r_k^{mi}} \frac{\partial r_k^{mi}}{\partial \epsilon_{st}} = \frac{r_s^{mi} r_t^{mi}}{r^{mi}} \quad \Rightarrow \quad \frac{\partial r^{mi}}{\partial \boldsymbol{\epsilon}} = \frac{\mathbf{r}^{mi} \otimes \mathbf{r}^{mi}}{r^{mi}} \tag{1.142}$$

We can finally express the Cauchy stress at an atom site as

$$\sigma_{st} \Big|_{\mathbf{x}_i} = \frac{1}{\Omega_0} \Big(A_{st} + F'(\bar{\rho}) V_{st} \Big) \tag{1.143}$$

where

$$A_{st} = \frac{1}{2}\sum_{m\neq i} \phi'_m \frac{r_s^{mi} r_t^{mi}}{r^{mi}}, \quad \text{and} \quad V_{st} = \sum_{m\neq i} \rho'_m \frac{r_s^{mi} r_t^{mi}}{r^{mi}}, \quad s,t = 1,2,3 \tag{1.144}$$

In finite deformation, we can write the first Piola-Kirchhoff stress in an unit cell as

$$\mathbf{P} = \frac{\partial}{\partial \mathbf{F}}\Big(\frac{E_{tot}}{\Omega_0}\Big) \tag{1.145}$$

Assume that there are N atoms in the local region Ω_0, and each atom links to all other $N-1$ atoms in a local region. Based on the Cauchy-Born rule, the local stress is constant and we only need the evaluate the stress at any point $\mathbf{x}_i, i = 1, 2, \cdots, N$ and make an average

$$\mathbf{P} = \frac{N}{\Omega_0}\sum_{m\neq i}\Big(\phi'_m + F'(\bar{\rho})\rho'_m\Big)\frac{\partial r^{mi}}{\partial \mathbf{F}} \tag{1.146}$$

The Cauchy-Born explicitly states

$$\mathbf{r}^{mi} = \mathbf{F}\cdot\mathbf{R}^{mi}, \quad \text{or} \quad r_k^{mi} = F_{k\ell}^m R_\ell^m \ .$$

Therefore,

$$\frac{\partial r^{mi}}{\partial F_{st}} = \frac{\partial r^{mi}}{\partial r_k^{mi}}\frac{\partial r_k^{mi}}{\partial F_{st}} = \frac{r_k^{mi}}{r^{mi}}\frac{\partial (F_{k\ell}R_\ell^{mi})}{\partial F_{st}} = \frac{r_k^{mi}}{r^{mi}}\delta_{ks}R_t^{mi} = \frac{r_s^{mi} r_p^{mi}}{r^{mi}}F_{tp}^{-1} \tag{1.147}$$

and finally,

$$\mathbf{P} = \frac{N}{\Omega_0}\sum_{m\neq i}\Big(\phi'_m + F'(\bar{\rho})\rho'_m\Big)\frac{\mathbf{r}^{mi}\otimes\mathbf{r}^{mi}}{r^{mi}}\cdot(\mathbf{F}^{-1})^T \tag{1.148}$$

The results similar to the above expression have been derived long before by Milstein and Hill [Milstein and Hill (1977, 1978, 1979b,a)]. In fact, under the assumption of the Cauchy-Born rule, one can also derive the expressions for elastic stiffness tensor, vacancy-formation energy, and sublimation energy, etc. by using EAM, and readers can find detailed calculation examples in [Milstein (1982); Daw and Baskes (1984); Foiles *et al.* (1986)]. We shall come back to this subject again when we discuss the so-called quasi-continuum method in Chapter 10.

1.6 Elements of lattice dynamics

1.6.1 *Crystal lattice structures*

Lattice statics/dynamics was an active research field thirty or forty years ago. The recent emergence of nanotechnology renews the interests in this area in order to seek guidance for molecular dynamics simulations. A good reference book on Lattice dynamics is Maradudin *et al.*'s "*Theory of Lattice Dynamics in the Harmonic Approximation*" [Maradudin *et al.* (1971)].

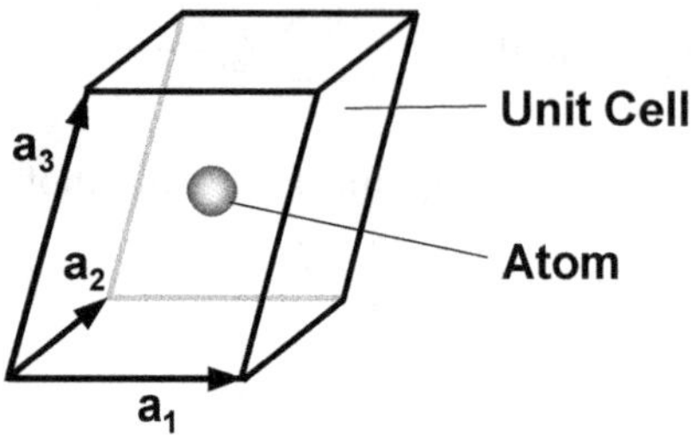

Fig. 1.7 The unit cell of a Bravais crystal.

Consider a crystal composed of infinitely many unit cells and each of them is a parallelepiped region bounded by three non-coplanar vectors, $\mathbf{a}_1$, $\mathbf{a}_2$, and $\mathbf{a}_3$. The equilibrium position of the $\boldsymbol{\ell}$-th unit cell relative to the origin of a coordinate may be expressed as,

$$\mathbf{x}_c(\boldsymbol{\ell}) = \ell_1\mathbf{a}_1 + \ell_2\mathbf{a}_2 + \ell_3\mathbf{a}_3, \quad \text{where} \quad \ell_i = 0, \pm 1, \pm 2, \cdots \tag{1.149}$$

The vectors $\mathbf{a}_1, \mathbf{a}_2$ and $\mathbf{a}_3$ are called the primitive translation vector of the lattice. If there is only one atom per unit cell, we call the lattice as *Bravais lattice*, and we call the unit cell as the *primitive cell*. If there are more than one atom in a unit cell, say r, we call the crystal as a non-primitive crystal. Based on this definition, the position of the α-th atom inside the ℓ-th primitive cell may be written as

$$\mathbf{x}(\boldsymbol{\ell}, \alpha) = \mathbf{x}(\boldsymbol{\ell}) + \mathbf{x}(\alpha), \quad \alpha = 1, 2, \cdots, r \tag{1.150}$$

According to the theory of crystallography, there are 5 Bravais lattices in two-dimensional space (see Fig. 1.9), and there are 14 three-dimensional Bravais lattices (see Fig. 1.11). Note that the unit cells shown in Figs. 1.9 and 1.11 are not

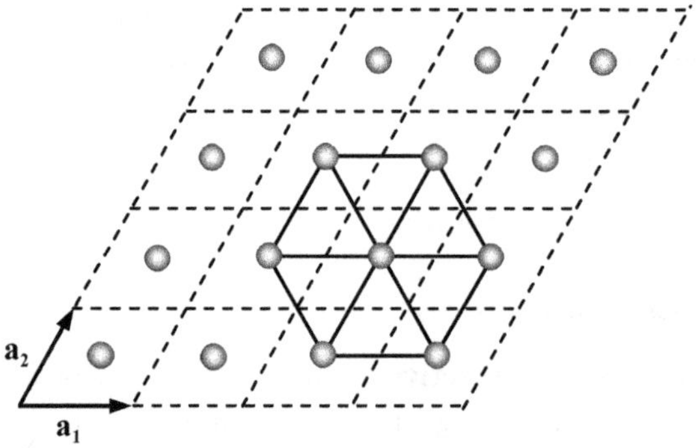

Fig. 1.8 Unit cell for hexagonal lattice.

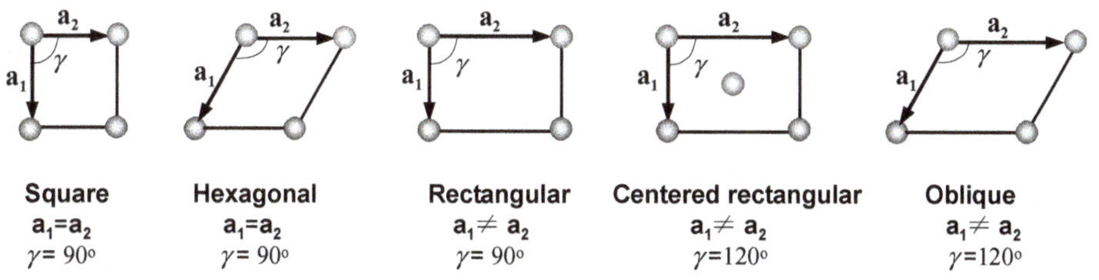

Fig. 1.9 The five Bravais lattices in two-dimensional space.

primitive cells, although they are the equivalent presentation of the Bravais cell. One very efficient way to construct the primitive cell is the so-called Wigner-Seitz cell algorithm, which is named after the Hungarian physicist Eugene Paul Wigner (1902-1995), a Nobel Laureate, and his student, Frederick Seitz, an American physicist. The Wigner-Seitz cell algorithm consists of three steps [Seitz (1987)]:

(1) Draw lines to connect a given lattice point to all nearby lattice points, which refer to as the nearest neighbors at different directions;
(2) At the midpoint and normal to these lines, draw new lines (or planes in three-dimensional space);
(3) The smallest volume enclosed in this procedure is called the Wigner-Seitz primitive cell.

In computational mathematics and geometry, a more general concept embodied in a Wigner-Seitz cell is often called as the Voronoi cell. In crystallography, the

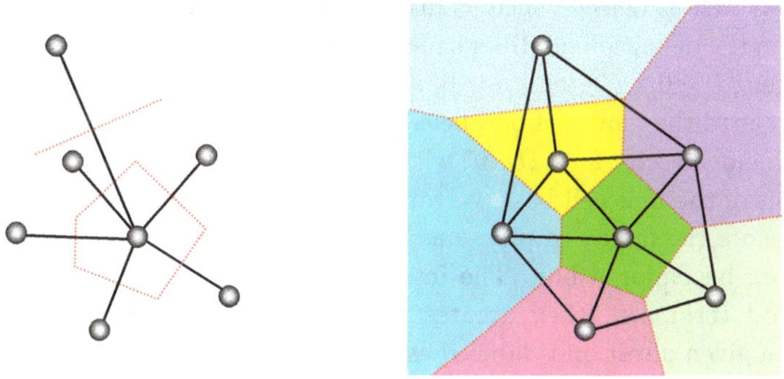

Fig. 1.10 The concept of the Wigner-Seitz primitive cell.

reciprocal lattice space of a Bravais lattice is defined as the wave number space. By wave number, we refer to as the reciprocal of wave length. If the lattice position vector is expressed as $\mathbf{r}$, the reciprocal lattice is the set of all wave number vectors that satisfy the relation,

$$\exp(i\mathbf{k} \cdot \mathbf{r}) = 1 \tag{1.151}$$

If the bases of the unit cell in real physical space are $\mathbf{a}_1, \mathbf{a}_2$ and $\mathbf{a}_3$, the bases for the reciprocal lattice are the dual bases that can be determined by vector cross product as follows,

$$\mathbf{b}_i = 2\pi \frac{\mathbf{a}_j \times \mathbf{a}_k}{\mathbf{a}_i \cdot (\mathbf{a}_j \times \mathbf{a}_k)}, \quad \text{where} \quad i, j, k = 1, 2, 3 \tag{1.152}$$

where both $\{\mathbf{a}_i\}_{i=1,2,3}$ and $\{\mathbf{b}_i\}_{i=1,2,3}$ form a right-handed triad. Obviously, the above reciprocal lattice basis is defined in the wave-number space, and it has clear physical meanings or implications.

In differential geometry, one can always define the dual vector bases for any triad of non-planar vector bases as

$$\mathbf{b}^i = \frac{\mathbf{a}_j \times \mathbf{a}_k}{\mathbf{a}_i \cdot (\mathbf{a}_j \times \mathbf{a}_k)}, \quad \text{where} \quad i, j, k = 1, 2, 3 \tag{1.153}$$

such that $\mathbf{b}^i \cdot \mathbf{a}_j = \delta^i_j$. The reciprocal lattice defined based on the dual vector bases is completely mathematical or geometrical. Many refer to such definition as the crystallographer's definition, whereas the definition (1.152) is referred to as the physicist's definition.

1.6.2 *Crystallographic system*

A systematic approach to describe lattice planes and crystallographic orientation is known as the Bravais-Miller indices, which was developed by W. H. Miller in 1839. The Miller indices notation has proven to be very efficient in describe microstructures as well as defects, such as dislocations, in crystalline.

Based on the definition of Miller's indices, a family of lattice planes is denoted by three integers, $\ell \geq 0$, $m \geq$, and $n \geq 0$, and they represent a normal vector that are perpendicular to the family of paralleled planes. For this purpose, they are written as a short-handed vector form (ℓmn) with bases vectors as the geometrical reciprocal lattice bases of the unit cell basis vectors. For the negative integer component, say $-m$, we denote the family of planes as $(\ell \bar{m} n)$. The negative sign moves on top of the letter m, becoming a bar. The integers, ℓ, m, and n, are expressed in terms of the lowest term, i.e. their greatest common factor should be 1. The Miller indices for a given direction is labeled as $[\ell mn]$ by using a square bracket instead of parenthesis. It should be noted that the vector bases for $[\ell, m, n]$ are lattice triad, $\mathbf{a}_1, \mathbf{a}_2$, and $\mathbf{a}_3$, instead of its reciprocal bases $\mathbf{b}_1$, $\mathbf{b}_2$, and $\mathbf{b}_3$. The notation $\{\ell, m, n\}$ denotes the set of all planes that are equivalent to the plane (ℓmn), for instance, in the cubic lattice, the set of planes $\{100\}$ are: $(100), (010)$ and (001), because all three planes are indistinguishable due to orthogonal rotation symmetry. The family of planes of $\{110\}$ in cubic lattice are: $(110), (011), (101), (\bar{1}10), (\bar{1}01)$, and $(0\bar{1}1)$. Similarly, the notation $< \ell mn >$ denotes the set of all similar directions that relate

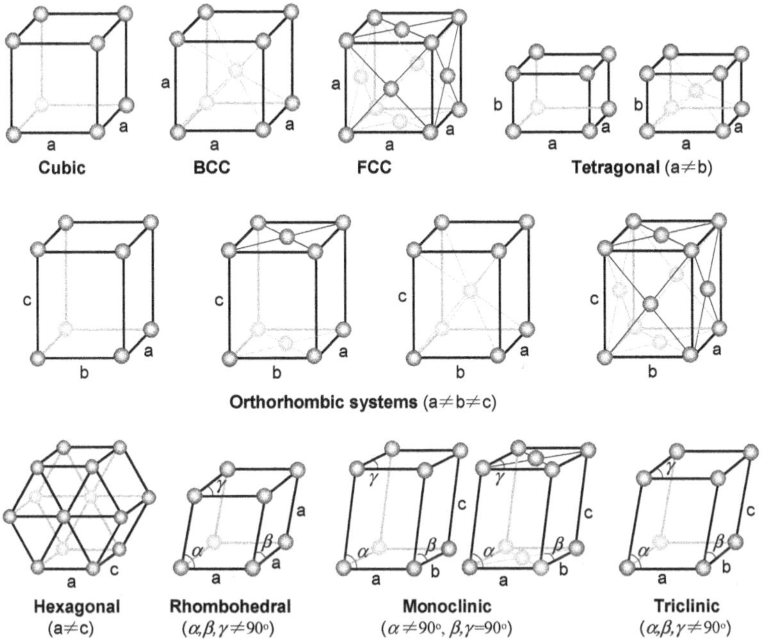

Fig. 1.11 The fourteen Bravais lattices in three-dimensional space.

to it by symmetry. For example in cubic lattices, $< 100 >$ contains six directions: $[100], [\bar{1}00], [010], [0\bar{1}0], [001]$ and $[00\bar{1}]$. However, in orthorhombic lattices, $[100]$ and $[010]$ may belong to the set $< 100 >$, but $[001]$ does not, if the lattice constants, $a_1 = a_2 \neq a_3$.

In applications, Miller's index is mainly used in cubic or hexagonal lattices. Therefore in the following, we restrict our following discussions on how to calculate Miller's indices for simple orthogonal or hexagonal crystal structures. We first consider the principal directions of a general orthorhombic unit cell by three primitive lattice vectors, $\mathbf{a}_i = a_i \mathbf{e}_i, i = 1, 2, 3$, where $\mathbf{e}_i, i = 1, 2, 3$ are basis vectors of a Cartesian coordinate system. Obviously, its geometrical reciprocal lattice basis vectors are $\mathbf{b}_i = \dfrac{1}{a_i}\mathbf{e}_i$. We now consider that the family of planes is denoted by its normal vector (ℓ, m, n), or

$$\mathbf{n} = \ell \mathbf{b}_1 + m\mathbf{b}_2 + n\mathbf{b}_3 = \frac{\ell}{a_1}\mathbf{e}_1 + \frac{m}{a_2}\mathbf{e}_2 + \frac{n}{a_3}\mathbf{e}_3$$

then the crystal plane intercepts the axes of the Cartesian coordinate axes at points $(c_1, 0, 0), (0, c_2, 0)$ and $(0, 0, c_3)$ in the real lattice space with

$$c_1 = \frac{a_1}{\ell}, \quad c_2 = \frac{a_2}{m}, \quad \text{and } c_3 = \frac{a_3}{n} \tag{1.154}$$

and the crystal plane is represented by equation

$$\frac{x_1}{c_1} + \frac{x_2}{c_2} + \frac{x_3}{c_3} - 1 = 0 \, .$$

Often times, we are given the coordinates of the intercept points first, i.e. c_1, c_2 and c_3 first, and we are asked to determine the corresponding normal direction, i.e. ℓ, m, and n. This process is often not a simple reciprocal calculation of Eq. (1.154). For example, suppose that $a_1 = a_2 = a_3 = 1$ if a crystal plane intercepts the x_1-, x_2- and x_3- axes at $(2,0,0), (0,1,0)$ and $(0,0,3)$, the Miller indices are calculated as

(1) Calculate $\dfrac{a_i}{c_i}, i = 1,2,3 \Rightarrow \Big(\dfrac{1}{2}, \dfrac{1}{1}, \dfrac{1}{3}\Big)$;
(2) Multiplying their least common multiplier, which is 6 in this case: $\Rightarrow \quad (3,6,2)$;
(3) Reduce to the lowest terms: in this case, they are already there.

So the Miller indices for this family of planes are (362). When a lattice plane is in

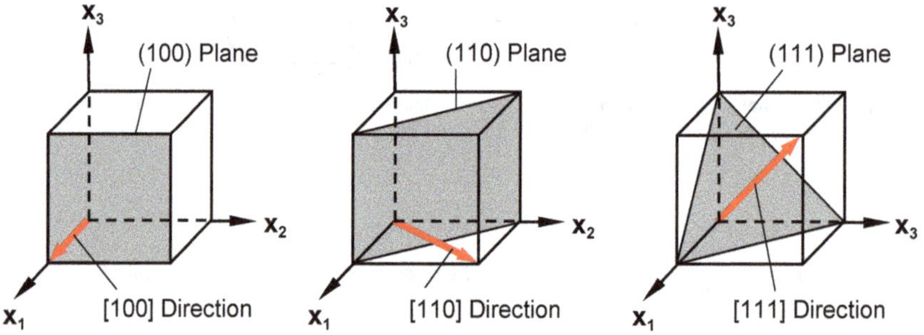

Fig. 1.12 Examples of Miller's indices for cubic lattice.

parallel to an axis, its intercept in the axis will be an infinite integer $int[\infty]$. For instance, a plane that is parallel to X_2OX_3 and intercepts the x_1 axis at the unit cell apex point $(a, 0, 0)$ will have the following Miller indices,

$$(\ell mn) = \Big(\frac{a}{a}, \frac{a}{\infty}, \frac{a}{\infty}\Big) = (100)$$

In Fig. 1.12, three lattice planes and lattice directions are illustrated with the Miller indices.

In hexagonal lattices, cf. Fig. 1.13, the symmetry condition ensures that there exist three equivalent axes on the basal plane. We label them as OX_1, OX_2, and OX_3, whose orientation space 120° apart among each other. The axis perpendicular to the basal plane is ladled as OZ. A four-index system (ℓmnp) contains the lowest integer components of four reciprocal lattice bases, which is used to represent a family of parallel planes. For instance, (0001) is the basal plane, $(1\bar{1}00)$ a prism plane, and $(10\bar{1}1)$ is a pyramidal plane, etc. Since the first three axes are not

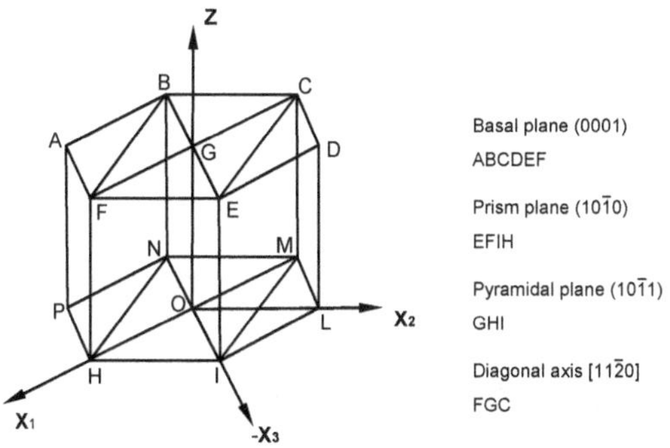

Fig. 1.13 Examples of Miller's indices for hexagonal lattice.

orthogonal, and they are not independent with each other. Therefore the integer components of a plane have to satisfy the following relationship,

$$\ell + m + n = 0 \ . \tag{1.155}$$

1.6.3 *Lattice dynamics*

We now discuss Lattice dynamics, or molecular dynamics in a lattice. For Bravais lattices, the initial position of the atom in the $\boldsymbol{\ell}$-th unit cell can be written as

$$\mathbf{x}(\boldsymbol{\ell}) = \mathbf{x}_0 + \sum_{i=1}^{3} \ell_i \mathbf{a}_i \tag{1.156}$$

Suppose that the atom's current position is

$$\mathbf{r}(\boldsymbol{\ell}) = \mathbf{x}(\boldsymbol{\ell}) + \mathbf{u}(\boldsymbol{\ell}) \tag{1.157}$$

The total kinetic energy of a Bravais lattice can be written as

$$T = \frac{1}{2} \sum_{\boldsymbol{\ell},i} M_\ell \dot{u}_i^2(\boldsymbol{\ell}) \ . \tag{1.158}$$

where M_ℓ is the mass of the atom in the $\boldsymbol{\ell}$-th cell.The total potential energy of the lattice may be written as

$$\Phi = \Phi(\mathbf{r}_{\boldsymbol{\ell}_1}, \mathbf{r}_{\boldsymbol{\ell}_2}, \cdots, \mathbf{r}_{\boldsymbol{\ell}_n}, \cdots) \tag{1.159}$$

Using Taylor expansion, we can write

$$\Phi = \Phi(\mathbf{x}_{\boldsymbol{\ell}_1}, \mathbf{x}_{\boldsymbol{\ell}_2}, \cdots, \mathbf{x}_{\boldsymbol{\ell}_n}, \cdots) + \sum_{\boldsymbol{\ell},i} \frac{\partial \Phi}{\partial u_i(\boldsymbol{\ell})} \Big|_{\mathbf{u}=0,\forall \boldsymbol{\ell}} u_i(\boldsymbol{\ell})$$

$$+ \frac{1}{2!} \sum_{\boldsymbol{\ell},i} \sum_{\boldsymbol{\ell}',j} \frac{\partial^2 \Phi}{\partial u_i(\boldsymbol{\ell}) \partial u_j(\boldsymbol{\ell}')} \Big|_{\mathbf{u}=0,\forall \boldsymbol{\ell}} u_i(\boldsymbol{\ell}) u_j(\boldsymbol{\ell}') + \cdots \tag{1.160}$$

Under the harmonic approximation, one can drop all the higher order terms. In fact the harmonic approximation originates from lattice dynamics not quantum mechanics.

Denote that

$$\Phi_0 := \Phi(\mathbf{x}_{\boldsymbol{\ell}_1}, \mathbf{x}_{\boldsymbol{\ell}_2}, \cdots, \mathbf{x}_{\boldsymbol{\ell}_n}, \cdots) \tag{1.161}$$

$$\Phi_i(\boldsymbol{\ell}) = \frac{\partial \Phi}{\partial u_i(\boldsymbol{\ell})}\Big|_{\mathbf{u}(\boldsymbol{\ell})=0, \forall \boldsymbol{\ell}} \tag{1.162}$$

$$\Phi_{ij}(\boldsymbol{\ell}; \boldsymbol{\ell}') = \frac{\partial^2 \Phi}{\partial u_i(\boldsymbol{\ell}) \partial u_j(\boldsymbol{\ell}')}\Big|_{\mathbf{u}(\boldsymbol{\ell})=0, \forall \boldsymbol{\ell}} \tag{1.163}$$

Consider that at equilibrium position, $\mathbf{r}(\boldsymbol{\ell}) = \mathbf{x}(\boldsymbol{\ell})$,

$$\frac{\partial \Phi}{\partial u_i(\boldsymbol{\ell})}\Big|_{\mathbf{u}(\boldsymbol{\ell})=0, \forall \boldsymbol{\ell}} = f_i(\boldsymbol{\ell}) \equiv 0 \tag{1.164}$$

and define the linear momentum as

$$p_i(\boldsymbol{\ell}) = M_\ell \dot{u}_i(\boldsymbol{\ell}) \ . \tag{1.165}$$

Under the harmonic approximation, the Hamiltonian for a Bravais lattice becomes

$$H = \sum_{\boldsymbol{\ell}, i} \frac{p_i(\boldsymbol{\ell}) p_i(\boldsymbol{\ell})}{2M_\ell} + \Phi_0 + \frac{1}{2} \sum_{\boldsymbol{\ell}, i} \sum_{\boldsymbol{\ell}', j} \Phi_{ij}(\boldsymbol{\ell}; \boldsymbol{\ell}') u_i(\boldsymbol{\ell}) u_j(\boldsymbol{\ell}') \tag{1.166}$$

The Hamiltonian equations,

$$\frac{\partial H}{\partial u_i} = -\dot{p}_i \tag{1.167}$$

yield the equations of motion

$$M_\ell \ddot{u}_i = -\sum_{\boldsymbol{\ell}', j} \Phi_{ij}(\boldsymbol{\ell}; \boldsymbol{\ell}') u_j(\boldsymbol{\ell}') \tag{1.168}$$

Example 1.3. In one-dimensional case, the vector index $\boldsymbol{\ell}$ becomes a scalar, and there is no difference between i and j. We may write $\Phi_{ij}(\boldsymbol{\ell}, \boldsymbol{\ell}')$ as $\Phi_{xx}(\ell, \ell')$. Consider the nearest neighbor interaction. One may denote that

$$\Phi_{xx}(\ell, \ell) = 2k \tag{1.169}$$

$$\Phi_{xx}(\ell - 1, \ell) = -k \tag{1.170}$$

$$\Phi_{xx}(\ell, \ell + 1) = -k \tag{1.171}$$

The equations of motion for 1D harmonic lattice are,

$$M_\ell \ddot{u}_\ell = k(u_{\ell-1} - 2u_\ell + u_{\ell+1}) \tag{1.172}$$

Example 1.4. For two-dimensional anti-plane motion, $u_1(\boldsymbol{\ell}) = u_2(\boldsymbol{\ell}) \equiv 0$ and $u_3(\boldsymbol{\ell}) \not\equiv 0$, $\forall \boldsymbol{\ell}$, we can write $\boldsymbol{\ell} = \ell_{m,n}$. For a square lattice under nearest neighbor interaction, one can choose,

$$\Phi_{33}(\ell_{m,n};\ell_{m,n}) = 2(k_1 + k_2) \tag{1.173}$$

$$\Phi_{33}(\ell_{m,n};\ell_{m-1,n}) = -k_1 \tag{1.174}$$

$$\Phi_{33}(\ell_{m,n};\ell_{m+1,n}) = -k_1 \tag{1.175}$$

$$\Phi_{33}(\ell_{m,n};\ell_{m,n-1}) = -k_2 \tag{1.176}$$

$$\Phi_{33}(\ell_{m,n};\ell_{m,n+1}) = -k_2 \tag{1.177}$$

The equation of motion for an anti-plane harmonic lattice is

$$\begin{aligned} M_\ell \ddot{u}_3 = {} & k_1\Big(u_3(\ell_{m-1,n}) - 2u_3(\ell_{m,n}) + u_3(\ell_{m+1,n})\Big) \\ & + k_2\Big(u_3(\ell_{m,n-1}) - 2u_3(\ell_{m,n}) + u_3(\ell_{m,n+1})\Big) \end{aligned} \tag{1.178}$$

1.7 Exercises

Problem 1.1. *Let $\delta \boldsymbol{u}$ be a virtual displacement field and $\boldsymbol{\sigma}$ be a self-equilibrium stress field. Show*

$$\Big(\nabla \cdot \boldsymbol{\sigma}\Big) \cdot \delta \boldsymbol{u} = \nabla \cdot \Big(\boldsymbol{\sigma} \cdot \delta \boldsymbol{u}\Big) - \boldsymbol{\sigma} : (\nabla \otimes \delta \boldsymbol{u}) \tag{1.179}$$

Problem 1.2. *Assume body force $\boldsymbol{f} = 0$. The elastostatics equilibrium equation takes the form:*

$$\sigma_{ji,j} = 0, \quad \textit{or} \quad \nabla \cdot \boldsymbol{\sigma} = \mathbf{0} \tag{1.180}$$

Show

$$\int_\Omega \boldsymbol{\sigma} : \boldsymbol{\epsilon} d\Omega = \int_{\partial\Omega} \boldsymbol{t} \cdot \boldsymbol{u} dS \tag{1.181}$$

where $\boldsymbol{t} = \boldsymbol{n} \cdot \boldsymbol{\sigma}$ (Hint: use the Gauss theorem, i.e. the divergence theorem).

Problem 1.3. *Suppose that there are two different solutions of equilibrium equation,*

$$\nabla \cdot \boldsymbol{\sigma}_1 = \mathbf{0}, \quad \nabla \cdot \boldsymbol{\sigma}_2 = \mathbf{0} \tag{1.182}$$

which satisfy the same boundary conditions,

$$\begin{cases} \boldsymbol{u}_1 = \boldsymbol{u}^0, \\ \boldsymbol{u}_2 = \boldsymbol{u}^0; \end{cases} \quad \forall \mathbf{x} \in \Gamma_u \tag{1.183}$$

$$\begin{cases} \boldsymbol{n} \cdot \boldsymbol{\sigma}_1 = \boldsymbol{t}^0, \\ \boldsymbol{n} \cdot \boldsymbol{\sigma}_2 = \boldsymbol{t}^0; \end{cases} \quad \forall \mathbf{x} \in \Gamma_t \tag{1.184}$$

where $\Gamma_u \bigcup \Gamma_t = \partial\Omega$. By using the positive-definiteness of elastic tensor and compliance tensor, show:

$$\Delta \boldsymbol{\sigma} = \boldsymbol{\sigma}_1 - \boldsymbol{\sigma}_2 = \mathbf{0} \tag{1.185}$$

$$\Delta \boldsymbol{\epsilon} = \boldsymbol{\epsilon}_1 - \boldsymbol{\epsilon}_2 = \mathbf{0} \tag{1.186}$$

Problem 1.4. *Show that for a given second-order tensor,* $\mathbb{A}$,

$$\mathbb{I}^{(4)} : \mathbb{A} \rightarrow \mathbb{A} \tag{1.187}$$

$$\mathbb{I}^{(4s)} : \mathbb{A} \rightarrow \frac{1}{2}\Big(\mathbb{A} + \mathbb{A}^T\Big) \tag{1.188}$$

$$\mathbb{I}^{(4a)} : \mathbb{A} \rightarrow \frac{1}{2}\Big(\mathbb{A} - \mathbb{A}^T\Big) \tag{1.189}$$

Problem 1.5. *Let* $\mathbb{A} = A_{ij}\mathbf{e}_i \otimes \mathbf{e}_j$ *be an arbitrary second order tensor. Show*

$$\mathbb{I}^{(4)} : \mathbb{A} = \mathbb{A}, \tag{1.190}$$

$$\mathbb{I}^{(4s)} : \mathbb{A} = \frac{1}{2}(\mathbb{A} + \mathbb{A}^T), \tag{1.191}$$

$$\mathbb{I}^{(4a)} : \mathbb{A} = \frac{1}{2}(\mathbb{A} - \mathbb{A}^T)\ . \tag{1.192}$$

Problem 1.6. *Show that*

1. $\mathbb{E}^{(1)} + \mathbb{E}^{(2)} = \mathbb{I}^{(4s)}$ (1.193)
2. $\mathbb{E}^{(1)} : \mathbb{E}^{(1)} = \mathbb{E}^{(1)}$ (1.194)
3. $\mathbb{E}^{(2)} : \mathbb{E}^{(2)} = \mathbb{E}^{(2)}$ (1.195)
4. $\mathbb{E}^{(2)} : \mathbb{E}^{(1)} = \mathbb{E}^{(1)} : \mathbb{E}^{(2)} = \mathbf{0}$. (1.196)

Problem 1.7. *Write the Lagrangian for both simple and double pendulums and derive the equations of motion for both simple and double pendulums.*

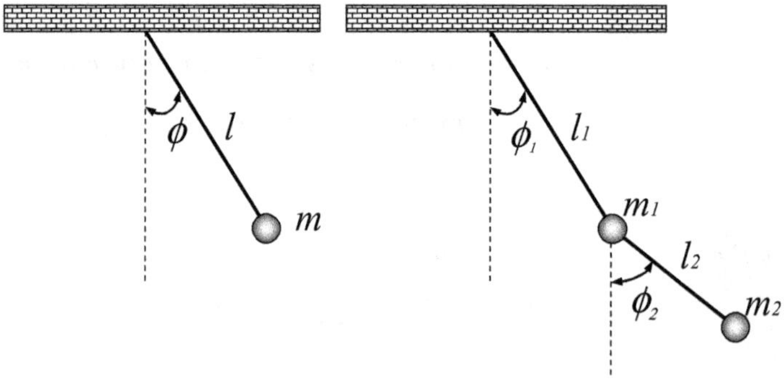

Fig. 1.14 Single and double pendulums.

Chapter 2

GREEN'S FUNCTION AND FOURIER TRANSFORM

An important approach in both micromechanics and nanomechanics is the Green's function method. In conventional micro-mechanics, the main homogenization procedure, the Eshelby inclusion/eigenstrain theory, is based on Green's function method; while in both lattice statics and lattice dynamics, lattice Green's function method has been a very useful method in solid state physics and material science for almost half century.

To have a better understanding of these theories, we first review basic theory of Green's function and Fourier transform.

2.1 Basics of Green's function

We first consider the Green's function for differential equations. Suppose $\boldsymbol{L}$ is an one-dimensional (1D) differential operator, and

$$L[u] = f(\mathbf{x}), \quad \forall \mathbf{x} \in \mathbb{R} \tag{2.1}$$

Choose $f(\mathbf{x}) = \delta(\mathbf{x}-\mathbf{y})$ (Dirac's delta function) and the solution of the following equation is called its Green's function, and it is denoted as $G(\mathbf{x}, \mathbf{y})$, i.e.

$$L[G(\mathbf{x},\mathbf{y})] = \delta(\mathbf{x}-\mathbf{y}), \quad \forall \mathbf{x}, \mathbf{y} \in \mathbb{R} \tag{2.2}$$

Why are we so interested in Green's function ? and what makes it so special ? To answer this question, let us re-examine the differential operator L. Suppose that there exists an inverse operator to L, and it is denoted as L^{-1}, such that,

$$LL^{-1} = L^{-1}L = \mathbf{I} \tag{2.3}$$

where $\mathbf{I}$ is the identity operator.

For example, the simplest 1D differential operator is,

$$L = \frac{d}{dx}\Big(\cdot\Big) \quad \text{and} \quad \Rightarrow \quad L^{-1} = \int (\cdot) dy \tag{2.4}$$

For a general differential operator L, its inverse operator may be written as

$$L^{-1}(f(\mathbf{x})) = \int \mathcal{K}(\mathbf{x}-\mathbf{y})(f(\mathbf{y}))d\mathbf{y}$$

where $\mathcal{K}$ is the so-called kernel function. Once the kernel function is determined, the inverse operator L^{-1} is determined.

Suppose that we have already known the inverse operator of L in Eq.(2.1). We then can solve the differential equation by applying the inverse operation,

$$\begin{aligned} L^{-1}L[u] &= L^{-1}(f(\mathbf{x})) \\ u(\mathbf{x}) &= L^{-1}(f(\mathbf{x})) = \int \mathcal{K}(\mathbf{x}-\mathbf{y})f(\mathbf{y})d\mathbf{y} \end{aligned} \tag{2.5}$$

Next question: what is the kernel function? Or how to find the kernel function for a differential operator L?

Considering

$$\begin{aligned} u(\mathbf{x}) &= \mathbf{I}u(\mathbf{x}) = LL^{-1}(u(\mathbf{x})) = L\int \mathcal{K}(\mathbf{x}-\mathbf{y})u(\mathbf{y})d\mathbf{y} \\ &= \int L\mathcal{K}(\mathbf{x}-\mathbf{y})u(\mathbf{y})d\mathbf{y} \end{aligned} \tag{2.6}$$

comparing (2.2) with

$$u(\mathbf{x}) = \int \delta(\mathbf{x}-\mathbf{y})u(\mathbf{y})d\mathbf{y},$$

one may find that $L\mathcal{K}(\mathbf{x}-\mathbf{y}) = \delta(\mathbf{x}-\mathbf{y})$. Therefore, one can deduce that the kernel function of the differential operator L is its Green's function:

$$\mathcal{K}(\mathbf{x}-\mathbf{y}) = G(\mathbf{x}-\mathbf{y}) \tag{2.7}$$

In principle, if the Green's function of a BVP has been found, the BVP is considered to be solved. This is because one can obtain the general solution of the differential equation $L[u] = f(\mathbf{x})$ via superposition through certain reciprocal formula. Equation (2.5) is usually termed as "the superposition principle".

Example 2.1. We consider the Euler-Bernoulli beam equation with simply supported boundary conditions

$$L[u] = \frac{d^2}{dx^2}\Big(EI\frac{d^2u}{dx^2}\Big) = f(x),\ \forall x \in (0,l) \tag{2.8}$$

$$u(0) = u(l) = 0, \quad \text{and} \quad u''(0) = u''(l) = 0 \tag{2.9}$$

Suppose that we have found the Green's function related to this problem, i.e.

$$L[G] = \frac{d^2}{dx^2}\Big(EI\frac{d^2G(x,y)}{dx^2}\Big) = \delta(x-y),\ \forall x,y \in (0,l) \tag{2.10}$$

$$G(0,y) = G(l,y) = 0, \quad \text{and} \quad G''(0,y) = G''(l,y) = 0 \tag{2.11}$$

Via integration by parts, one can show that

$$\begin{aligned} \int_0^l u\Big(\frac{d^2}{dx^2}EI\frac{d^2v}{dx^2}\Big)dx &= \Big[u\Big(\frac{d}{dx}EI\frac{d^2v}{dx^2}\Big)\Big]_0^l - \Big[\Big(\frac{du}{dx}\Big)\Big(EI\frac{d^2v}{dx^2}\Big)\Big]_0^l \\ &+ \int_0^l \Big(\frac{d^2u}{dx^2}\Big)EI\Big(\frac{d^2v}{dx^2}\Big)dx \end{aligned} \tag{2.12}$$

Let $v = G(x,y)$. We have the following reciprocal formula

$$\int_0^l uL[G]dx - \int_0^l GL[u]dx = \Big[u\Big(\frac{d}{dx}EI\frac{d^2G}{dx^2}\Big)\Big]_0^l - \Big[\Big(\frac{du}{dx}\Big)\Big(EI\frac{d^2G}{dx^2}\Big)\Big]_0^l$$
$$-\Big[G\Big(\frac{d}{dx}EI\frac{d^2u}{dx^2}\Big)\Big]_0^l + \Big[\Big(\frac{dG}{dx}\Big)\Big(EI\frac{d^2u}{dx^2}\Big)\Big]_0^l \tag{2.13}$$

Consider the fact that both $u(x)$ and $G(x,y)$ satisfy the same homogeneous essential boundary conditions. Because L is self-adjoint, a simple reciprocal formula holds

$$\int_0^l uL(G)dx = \int_0^l GL(u)dx \tag{2.14}$$

which leads to

$$\int_0^l u(y)\delta(x-y)dy = \int_0^l G(x-y)f(y)dy \tag{2.15}$$

and consequently,

$$u(x) = \int_0^l G(x-y)f(y)dy \tag{2.16}$$

In structural engineering, the Green's function solution represents the concentrated load solution, and the Green's function is called the influence function. In this case, it can be obtained by straightforward integration,

$$G(x-y) = \frac{<x-y>^3}{6EI} + \frac{(l-y)x^3}{6EIl} - \frac{x((l-y)^3 + (l-y)\ell^2)}{6EIl} \tag{2.17}$$

where the bracket function,

$$<x-y>^3 := \begin{cases} (x-y)^3 & , \quad x > y \\ 0 & , \quad x \le y \; . \end{cases}$$

And the general solution, Eq.(2.16), can be obtained by the argument of superposition, i.e.

$$u(x) = \int_0^l G(x-y)f(y)dy \; .$$

Example 2.2. In the second example, we consider Poisson's equation,

$$\nabla^2 u = f_1(\mathbf{x}), \quad \text{and} \quad \nabla^2 v = f_2(\mathbf{x}), \quad \forall \mathbf{x} \in \Omega \tag{2.18}$$

One can derive the following identity via integration by parts,

$$\int_\Omega u\nabla\cdot(\nabla v)d\Omega = \int_\Omega \{\nabla\cdot(u\nabla v) - (\nabla u)\cdot(\nabla v)\}d\Omega$$
$$= \int_{\partial\Omega}\Big(\frac{\partial v}{\partial n}\Big)udS - \int_\Omega (\nabla u)\cdot(\nabla v)d\Omega \tag{2.19}$$

Similarly, by interchanging the position of u and v,

$$\int_\Omega v\nabla\cdot(\nabla u)d\Omega = \int_{\partial\Omega}\Big(\frac{\partial u}{\partial n}v\Big)dS - \int_\Omega (\nabla v)\cdot(\nabla u)d\Omega \tag{2.20}$$

Subtracting (2.19) from (2.20) yields the so-called Green's reciprocal theorem:

$$\int_\Omega \Big(u\nabla^2 v - v\nabla^2 u \Big) d\Omega = \int_{\partial\Omega} \Big\{ \Big(u\frac{\partial v}{\partial n} \Big) - \Big(v\frac{\partial u}{\partial n} \Big) \Big\} dS \tag{2.21}$$

Let $v(\mathbf{x}) = G(\mathbf{x},\mathbf{y})$, $f_1(\mathbf{x}) = f(\mathbf{x})$, and $f_2(\mathbf{x}) = \delta(\mathbf{x}-\mathbf{y})$. Eq. (2.21) can be rearranged as

$$u(\mathbf{x}) = \int_{\partial\Omega} \Big\{ \Big(\frac{\partial G}{\partial n} \Big) u - G\Big(\frac{\partial u}{\partial n} \Big) \Big\} dS_{\mathbf{y}} + \int_\Omega G(\mathbf{x},\mathbf{y}) f(\mathbf{y}) d\Omega_{\mathbf{y}} \tag{2.22}$$

Note that in 2.22, the Green's function solution does not necessarily satisfy the same boundary condition as the true solution, $u(\mathbf{x})$, as shown in the previous example. Often times, the Green's function in the infinite domain is chosen in a reciprocal representation.

2.2 Fourier transform

Consider a function, $f(x) \in L^1(\mathbb{R})$, or $\displaystyle\int_{-\infty}^{\infty} |f(x)|dx < \infty$. We define the Fourier transform as

$$\bar{f}(\xi) = \mathbf{F}[f] = \frac{1}{2\pi}\int_{-\infty}^{\infty} f(x)\exp(-i\xi x)dx \tag{2.23}$$

$$f(x) = \mathbf{F}^{-1}[\bar{f}] = \int_{-\infty}^{\infty} \bar{f}(\xi)\exp(i\xi x)d\xi \tag{2.24}$$

In generalized Fourier transform, ξ can be a complex number. Assume that function $f(x)$ has the property such that $\exp(C_1 x)|f(x)| \to 0$ as $x \to \infty$ and $\exp(-C_2 x)|f(x)| \to 0$ as $x \to -\infty$. The inversion formula may be expressed as the following contour integral

$$f(x) = \int_{-\infty - i\gamma}^{\infty - i\gamma} \bar{f}(\xi)\exp(i\xi x)d\xi \tag{2.25}$$

where $C_1 > \gamma > C_2$. The integration contour is usually referred as the Bromwich contour named after Thomas John l'Anson Bromwich (1875-1929).

Lemma 2.1 (Jordan Lemma). *Suppose that on the upper semi-circular arc C_R shown in Fig. 2.1 we have $f(\xi) \to 0$ uniformly as $R \to \infty$. Then*

$$\lim_{R\to\infty} \exp(ix\xi)f(\xi)d\xi = 0, \quad (x > 0)$$

We note that if $x < 0$, similar result holds for the integration along a semi-circular arc contour in lower half space.

Theorem 2.1 (Cauchy-Goursat). *If $f(z)$ is an analytical complex function*[1] *at each point within and on a closed contour C, then*

$$\oint_C f(z)dz = 0 \tag{2.26}$$

[1]In complex analysis, the complex numbers are customarily represented by the symbol z.

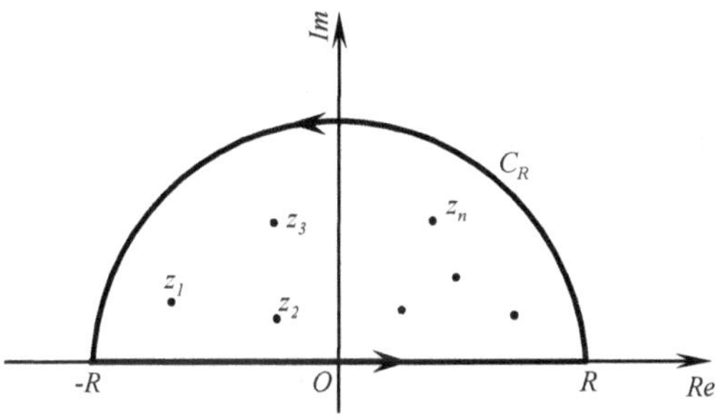

Fig. 2.1 Inversion paths of Fourier transform on complex plane.

Theorem 2.2 (Cauchy's residue theorem). *If complex function $f(z)$ is analytical inside a closed contour C (taken the counter clockwise direction as the positive direction) except at points, $z_1, z_2, \cdots, z_n$, where $f(z)$ has singularities, then*

$$\oint_C f(z)dz = 2\pi i \sum_{j=1}^{n} \text{Residue of } f(z) \text{ at } z_j = 2\pi i \sum_{j=1}^{n} \text{Res}(f(z)) \Big|_{z=z_j} \tag{2.27}$$

Now, the question becomes what is a residue and how to calculate it. The answer involves the singularity of $f(z)$. For a function of complex variable, $f(z)$, one may express $f(z)$ in a local region by its Laurent expansion, which is an extension of Taylor expansion of real variable. For example, around a fixed point z_j, we may write

$$f(z) = \sum_{n=0}^{\infty} a_n (z - z_j)^n + \sum_{n=1}^{\infty} a_{-n} (z - z_j)^{-n}, \quad 0 < |z - z_j| < a \tag{2.28}$$

The residue is defined as the coefficient a_{-1}.

There are three types of singularities: (1) essential singularity, (2) removable singularity, and (3) pole of order n.

- The essential singularity refers to a singularity, or a pole of infinity order. For instance, the pole at $z = 0$ is an essential singularity for the following function,
$$\cos\left(\frac{1}{z}\right) = 1 - \frac{1}{2!z^2} + \frac{1}{4!z^4} - \frac{1}{6!z^6} + \cdots .$$
- The removable singularity is an unsubstantial singularity, i.e., the alleged singularity disappears in the Laurent expansion. For instance, at $z = 0$,
$$f(z) = \frac{\sin z}{z} = 1 - \frac{z^2}{3!} + \frac{z^4}{5!} - \frac{z^6}{7!} + \cdots$$
- Pole of order n: Consider the function,
$$f(z) = \frac{1}{z+1} + \frac{1}{(z-1)^3}$$

This function has two singularities at $z = -1$ and $z = 1$. For singularity at $z = -1$, its order is one, and it is called a pole of order one. For singularity at $z = 1$, its order is three, and it is called a pole of order 3.
The formula to calculate the residue for a pole, z_j, of order n is

$$\text{Res}(f(z))\Big|_{z=z_j} = \frac{1}{(n-1)!}\lim_{z\to z_j}\frac{d^{n-1}}{dz^{n-1}}\Big[(z-z_j)^n f(z)\Big] \tag{2.29}$$

We call the pole of order one as *simple pole*. If $f(z)$ has a simple pole z_j,

$$\text{Res}(f(z))\Big|_{z=z_j} = \lim_{z\to z_j}(z-z_j)f(z) \tag{2.30}$$

If $f(z) = p(z)/q(z)$ and has a simple pole, one may also write

$$\text{Res}(f(z))\Big|_{z=z_j} = \frac{p(z_j)}{q'(z_j)} \tag{2.31}$$

Example 2.3. In this example, we apply Cauchy's residue theorem to evaluate the following line integral,

$$\int_{-\infty}^{\infty}\frac{\exp(ikt)}{(t-x)(t-ia)}dt$$

where $k > 0$ and $a > 0$.

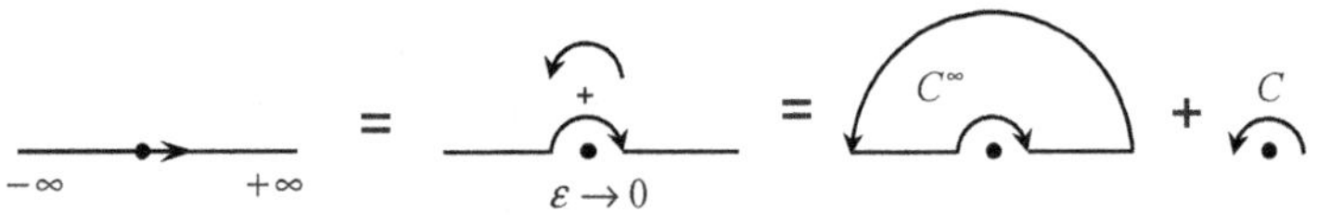

Fig. 2.2 Contour integral and the count of residues.

Since $k > 0$, based on Jordan's lemma, we can use the following contour integral to replace the line integral,

$$\int_{-\infty}^{\infty}\frac{\exp(ikt)}{(t-x)(t-ia)}dt = \int_{C^\infty}\frac{\exp(ikt)}{(t-x)(t-ia)}dt + \int_{C}\frac{\exp(ikt)}{(t-x)(t-ia)}dt$$

where the contour integral C^∞ is the enclosed upper half circle in counter-clockwise direction, and the contour integral C is the upper half circle with radius $\epsilon \to 0$ center at the point x shown in Fig. 2.2. One can evaluate the first contour integral by evaluating its residue, which only has one simple inside the contour, i.e. $z = ia$. To evaluate the second integral, one has to first evaluate a closed contour integral surrounding the center $z = x$. Since the contour integral is only half of it, its value will be the half of its residue,i.e.

$$\begin{aligned}\int_{-\infty}^{\infty}\frac{\exp(ikt)}{(t-x)(t-ia)}dt &= 2\pi i\ \text{Res}(f(ia)) + \pi i\ \text{Res}(f(x))\\ &= -2\pi i\frac{\exp(-ka)(x+ia)}{x^2+a^2} + \pi i\frac{\exp(ikx)(x+ia)}{x^2+a^2}\end{aligned} \tag{2.32}$$

The simple pole at x is only counted for half of the residue is because that it has only half circle.

Theorem 2.3 (Cauchy's Integral Formula). *Let $f(z)$ be analytical interior to and on a simple closed contour C. Then at any interior point z*

$$f(z) = \frac{1}{2\pi i} \oint_C \frac{f(\zeta)}{\zeta - z} d\zeta \tag{2.33}$$

Theorem 2.4 (Convolution). *If $f(x), g(x) \in L^1(\mathbb{R}) \cap L^2(\mathbb{R})$, the following identity holds*

$$\int_{-\infty}^{\infty} \bar{f}(\xi)\bar{g}(\xi)\exp(i\xi x) d\xi = \frac{1}{2\pi} \int_{-\infty}^{\infty} g(x-y) f(y) dy \tag{2.34}$$

Proof:
by definition,

$$\begin{aligned}
\int_{-\infty}^{\infty} \bar{f}(\xi)\bar{g}(\xi)\exp(i\xi x) d\xi &= \int_{-\infty}^{\infty} \Big[\bar{g}(\xi)\Big(\frac{1}{2\pi}\int_{-\infty}^{\infty} f(y)\exp(-i\xi y)dy\Big)\Big]\exp(i\xi x)d\xi \\
&= \int_{-\infty}^{\infty} f(y)\Big[\frac{1}{2\pi}\int_{-\infty}^{\infty} \bar{g}(\xi)\exp(i\xi(x-y))d\xi\Big]dy \\
&= \frac{1}{2\pi}\int_{-\infty}^{\infty} g(x-y)f(y)dy
\end{aligned} \tag{2.35}$$

In 3D, we have

$$\int_{\mathbb{R}^3} \bar{f}(\boldsymbol{\xi})\bar{g}(\boldsymbol{\xi})\exp(i\boldsymbol{\xi}\cdot\mathbf{x}) d\boldsymbol{\xi} = \frac{1}{(2\pi)^3}\int_{\mathbb{R}^3} g(\mathbf{x}-\mathbf{y})f(\mathbf{y})d\mathbf{y} \tag{2.36}$$

Example 2.4. Consider Heaviside function,

$$H(x) = \begin{cases} 1, & x > 0 \\ 0, & x < 0 \end{cases} \tag{2.37}$$

Note that at $x = 0$ the Heaviside function is not defined.

The Fourier transform of the Heaviside function is

$$\begin{aligned}
\bar{H}(\xi) &= \frac{1}{2\pi}\int_{-\infty}^{\infty} H(x)\exp(-i\xi x)dx \\
&= \frac{1}{2\pi}\int_0^{\infty} \exp(-i\xi x)dx = \frac{1}{2\pi}\frac{(-1)}{i\xi}\exp(-i\xi x)\Big|_0^{\infty} = \frac{1}{2\pi i\xi}
\end{aligned} \tag{2.38}$$

The result implies that $\exp(-i\xi\infty) \to 0$, which requires that $\mathrm{Im}(\xi) < 0$. Lighthill [Lighthill (1964)] showed that in the sense of generalized function,

$$\bar{H}(\xi) = \exp\Big(-\frac{\pi i}{2}\mathrm{sgn}(\xi)\Big)\frac{1}{2\pi|\xi|}, \quad \text{where } \mathrm{sgn}(\xi) := \begin{cases} 1, & \xi > 0 \\ -1, & \xi < 0 \end{cases} \tag{2.39}$$

Note that $H(x) \notin L^1(\mathbb{R})$. Therefore, the Fourier transform of Heaviside function does not really exit for $f \in L^1$. The condition $\int_{-\infty}^{\infty} |f(x)| < \infty$ is very stringent. It is why many functions that has Laplace transform do not possess Fourier transform, which is the reason why sometimes we use Laplace transform instead of Fourier transform. By the way, if ξ is taken as a complex number, Fourier transform is equivalent to bilateral Laplace transform.

Example 2.5. To find the Fourier transform of the Dirac's delta function, we first calculate $\bar{\delta}$ by definition,

$$\bar{\delta}(\xi) = \frac{1}{2\pi}\int_{-\infty}^{\infty} \delta(x)\exp(-ix\xi)dx = \frac{1}{2\pi} \tag{2.40}$$

Then inversely,

$$\delta(x) = \int_{-\infty}^{\infty} \bar{\delta}(\xi)\exp(i\xi x)d\xi = \frac{1}{2\pi}\int_{-\infty}^{\infty} \exp(i\xi x)d\xi \tag{2.41}$$

Example 2.6. On the other hand, based on the inversion Fourier transform formula,

$$\int_{-\infty}^{\infty} \delta(\xi)\exp(i\xi x)d\xi = \exp(i0x) = 1\ . \tag{2.42}$$

one can deduce that the Fourier transform of real number 1 is,

$$\bar{1}(\xi) = \delta(\xi) = \frac{1}{2\pi}\int_{-\infty}^{\infty} \exp(-i\xi x)dx \tag{2.43}$$

Combining the Euler formula, $\exp(i(\cdot)) = \cos(\cdot) + i\sin(\cdot)$, and (2.45), one may conclude that

$$\delta(\xi) = \frac{1}{(2\pi)^3}\int_{-\infty}^{\infty} \cos(\xi x)d\mathbf{x} \tag{2.44}$$

because $\sin x$ is an odd function.
For $\mathbf{x} \in \mathbb{R}^3$, we have the identity,

$$\delta(\boldsymbol{\xi}) = \frac{1}{(2\pi)^3}\int_{\mathbb{R}^3} \exp(-i\boldsymbol{\xi}\cdot\mathbf{x})d\mathbf{x} \tag{2.45}$$

Example 2.7. Suppose that the Fourier transform of $f(x)$ is

$$\bar{f}(\xi) = \frac{1}{2\pi}\frac{ia}{\xi(\xi^2 - ia\xi - a)} \tag{2.46}$$

Find $f(x)$?

Solution: $\bar{f}(\xi)$ has three poles in the complex plane:

$$\xi_1 = 0,\ and\ \xi_{2,3} = \frac{ia}{2} \pm \sqrt{a - \frac{a^2}{4}} = \frac{ia}{2} \pm \beta,\ \text{where}\ \beta := \sqrt{a - a^2/4} \tag{2.47}$$

Therefore,

$$
\begin{aligned}
f(x) &= \int_{-\infty}^{\infty} \bar{f}(\xi)\exp(i\xi x)d\xi = \oint_{C^\infty + C} \frac{1}{2\pi}\frac{ia\exp(i\xi x)}{(\xi-\xi_1)(\xi-\xi_2)(\xi-\xi_3)}d\xi \\
&= \pi i \mathrm{Res} f(\xi_1) + 2\pi i \sum_{j=2}^{3} \mathrm{Res} f(\xi_j) \\
&= -a\Big\{\frac{\exp(i\xi_1 x)}{2\xi_2\xi_3} + \frac{\exp(i\xi_2 x)}{\xi_2(\xi_2-\xi_3)} + \frac{\exp(i\xi_3 x)}{\xi_3(\xi_3-\xi_2)}\Big\} \\
&= (-a)\Big\{\frac{1}{-2a} + \frac{\exp[ix(\frac{ia}{2}+\beta)]}{\Big(\frac{ia}{2}+\beta\Big)2\beta} - \frac{\exp[ix(\frac{ia}{2}-\beta)]}{\Big(\frac{ia}{2}-\beta\Big)2\beta}\Big\} \\
&= \frac{1}{2}\left(1 + \frac{\exp\Big(-\frac{xa}{2}\Big)}{\sqrt{a-\frac{a^2}{4}}}\Big\{\Big(\beta-\frac{ia}{2}\Big)\exp(i\beta x) + \Big(\beta+\frac{ia}{2}\Big)\exp(-i\beta x)\Big\}\right) \\
&= \frac{1}{2} + \frac{\exp\Big(-\frac{xa}{2}\Big)}{2\beta}\Big(2\beta\cos x - a\sin\beta x\Big)
\end{aligned}
\tag{2.48}
$$

2.3 Examples of Green's functions

Example 2.8. Find the Green's function of two-dimensional Poisson's equation in infinite domain,

$$
\nabla^2 G(\mathbf{x},\mathbf{y}) + \delta(\mathbf{x}-\mathbf{y}) = 0, \quad \forall \mathbf{x} \in \mathbb{R}^2 \tag{2.49}
$$

Use the polar coordinate $\nabla^2 = \frac{1}{r}\frac{d}{dr}\Big(r\frac{d}{dr}\Big)$ and denote $\mathbf{x}' = \mathbf{x}-\mathbf{y}$. We have

$$
\frac{1}{r}\frac{d}{dr}\Big(r\frac{d}{dr}G\Big) = -\delta(x_1')\delta(x_2') \tag{2.50}
$$

and

$$
\int_0^{2\pi}\int_0^r \frac{1}{\rho}\frac{d}{d\rho}\Big(\rho\frac{d}{d\rho}G\Big)\rho d\rho d\theta = -\int_\Omega \delta(x_1')\delta(x_2')dx_1'dx_2' \tag{2.51}
$$

where ρ is the integration dummy variable, and $r = |\mathbf{x}-\mathbf{y}| = \sqrt{(x_1-y_1)^2+(x_2-y_2)^2}$. The integration domain is a circular region centered at $\mathbf{x} = \mathbf{y}$.

Therefore,

$$
2\pi\Big(r\frac{d}{dr}G\Big) = -1, \quad \Rightarrow \frac{d}{dr}G = -\frac{1}{2\pi}\frac{1}{r} \tag{2.52}
$$

Finally, we find that

$$
G(\mathbf{x}-\mathbf{y}) = -\frac{1}{2\pi}\ln r \tag{2.53}
$$

Example 2.9. Consider one dimensional harmonic equation,

$$\frac{d^2u}{dx^2} - k^2 u = \delta(|x-y|) \tag{2.54}$$

Apply Fourier transform,

$$\bar{u}(\xi) = \frac{1}{2\pi}\int_{-\infty}^{\infty} u(x)\exp(-i\xi x)dx$$

$$\mathcal{F}\Big(\frac{d^2u}{dx^2}\Big) = \frac{1}{2\pi}\int_{-\infty}^{\infty} \frac{d^2u}{dx^2}\exp(-i\xi x)dx = -\xi^2\bar{u}(\xi)$$

$$\bar{\delta}(x-y) = \frac{1}{2\pi}\int_{-\infty}^{\infty} \delta(x-y)\exp(-i\xi x)d\xi = \frac{1}{2\pi}\exp(-i\xi y) \tag{2.55}$$

and

$$\bar{u}(\xi) = -\frac{1}{2\pi}\frac{1}{\xi^2+k^2}\exp(-i\xi y) \tag{2.56}$$

Therefore,

$$\begin{aligned} u(x,y) &= \int_{-\infty}^{\infty} \bar{u}(\xi)\exp(i\xi x)d\xi \\ &= -\frac{1}{2\pi}\int_{-\infty}^{\infty}\frac{1}{(\xi-ik)(\xi+ik)}\exp(i\xi(x-y))d\xi \\ &= -\frac{1}{2\pi}\oint_{C^\infty}\frac{\exp(i\xi|x-y|)}{(\xi-ik)(\xi+ik)}d\xi \end{aligned} \tag{2.57}$$

Inside the upper half semi-circle, C^∞, there is only one simple pole, ik. Thus,

$$\begin{aligned} u(x,y) &= -\frac{1}{2\pi}\oint_{C^\infty}\frac{\exp(i\xi|x-y|)}{(\xi-ik)(\xi+ik)}d\xi = -i\sum_{i=1}\mathrm{Res}(f(\xi))\Big|_{\xi=ik} \\ &= -\Big\{\frac{1}{2k}\exp(-k|x-y|)\Big\} \end{aligned} \tag{2.58}$$

Example 2.10. Find the Green's function for one-dimensional (1D) Helmholtz equation,

$$\Big(\mu\frac{d^2}{dx^2} + \rho\omega^2\Big)u(\mathbf{x},\omega) = -\delta(\mathbf{x}-\mathbf{y}) \tag{2.59}$$

Apply Fourier transform to Eq. (2.59). We then obtain,

$$\bar{u}(\xi,\omega) = \frac{1}{2\pi\mu}\frac{\exp(-i\xi y)}{(\xi^2-k^2)} \tag{2.60}$$

where $k = \dfrac{\omega}{c}$ and $c = \sqrt{\dfrac{\mu}{\rho}}$.

Via inverse Fourier transform,

$$\begin{aligned} u(x,y) &= \int_{-\infty}^{\infty} \bar{u}(\xi)\exp(i\xi x)d\xi \\ &= \frac{1}{2\pi\mu}\int_{-\infty}^{\infty}\frac{1}{(\xi-k)(\xi+k)}\exp(i\xi(x-y))d\xi \\ &= \frac{1}{2\pi\mu}\oint_{C^\infty}\frac{\exp(i\xi|x-y|)}{(\xi-k)(\xi+k)}d\xi \end{aligned} \tag{2.61}$$

Inside the upper half semi-circle, C^∞, there is only one simple pole, k. Thus,

$$u(x,y) = \frac{1}{2\pi\mu}\oint_{C^\infty}\frac{\exp(i\xi|x-y|)}{(\xi-k)(\xi+k)}d\xi = \frac{i}{\mu}\sum_{i=1}\text{Res}f(\xi)\Big|_{\xi=k}$$

$$= \left\{\frac{i}{2k\mu}\exp(ik|x-y|)\right\} = \frac{i}{2\rho c^2 k}\exp(ik|x-y|) \quad (2.62)$$

Example 2.11. Find Green's function for three-dimensional Poisson's equation,

$$\nabla^2 G + \delta(\mathbf{x}-\mathbf{x}') = 0 \;\Rightarrow\; G_{,ii} + \delta(\mathbf{x}-\mathbf{x}') = 0 \quad (2.63)$$

where $\nabla^2 = \dfrac{\partial^2}{\partial x_i \partial x_i}, i = 1,2,3$ and $\delta(\mathbf{x}-\mathbf{x}') = \delta(x_1-x_1')\delta(x_2-x_2')\delta(x_3-x_3')$

Consider the fact that

$$\delta(\mathbf{x}-\mathbf{x}') = \frac{1}{(2\pi)^3}\int_{\mathbb{R}^3}\exp\Big(i\boldsymbol{\xi}\cdot(\mathbf{x}-\mathbf{x}')\Big)d\boldsymbol{\xi}$$

and based on the definition,

$$G(\mathbf{x}-\mathbf{x}') = \int_{\mathbb{R}^3}\bar{G}(\boldsymbol{\xi})\exp\Big(i\boldsymbol{\xi}\cdot(\mathbf{x}-\mathbf{x}')\Big)d\boldsymbol{\xi}$$

one may derive that

$$G_{,ii}(\mathbf{x}-\mathbf{x}') = -\int_{\mathbb{R}^3}\bar{G}(\boldsymbol{\xi})\xi_i\xi_i\exp(i\boldsymbol{\xi}\cdot(\mathbf{x}-\mathbf{x}'))d\boldsymbol{\xi} \quad (2.64)$$

and the Fourier transform of the equation, $G_{,ii} + \delta(\mathbf{x}-\mathbf{x}') = 0$, leads to

$$\bar{G}(\boldsymbol{\xi})\xi_i\xi_i = \frac{1}{(2\pi)^3} \;\Rightarrow\; \bar{G}(\boldsymbol{\xi}) = \frac{1}{(2\pi)^3}\left(\frac{1}{\xi_i\xi_i}\right) \quad (2.65)$$

Let $\xi^2 := \xi_i\xi_i, \bar{\mathbf{x}} = \mathbf{x}-\mathbf{x}', r := |\mathbf{x}-\mathbf{x}'|$, and $\boldsymbol{\xi}\cdot(\mathbf{x}-\mathbf{x}') = \xi r\cos\theta$. Then,

$$G(\mathbf{x}-\mathbf{x}') = \frac{1}{(2\pi)^3}\int_{\mathbb{R}^3}\frac{1}{\xi^2}\exp(i\boldsymbol{\xi}\cdot(\mathbf{x}-\mathbf{x}'))d\boldsymbol{\xi}$$

$$= \frac{1}{(2\pi)^3}\int_0^{2\pi}\int_0^{\pi}\int_0^{\infty}\frac{1}{\xi^2}\exp(i\boldsymbol{\xi}\cdot(\mathbf{x}-\mathbf{x}'))\xi^2 d\xi\sin\theta d\theta d\phi$$

$$= \frac{1}{(2\pi)^3}\int_0^{2\pi}d\phi\int_1^{-1}\int_0^{\infty}\exp(i\xi r\cos\theta)d\xi(-d\cos\theta)$$

$$= \frac{1}{(2\pi)^2}\int_{-1}^{1}\int_0^{\infty}\exp(i\xi rt)d\xi dt$$

$$= \frac{1}{(2\pi)^2}\int_0^{\infty}\int_{-1}^{1}\Big[\cos(\xi rt) + i\sin(\xi rt)\Big]dtd\xi \quad (2.66)$$

Considering the fact that t^2 is an even function and $t\sqrt{1-t^2}$ is an odd function, we have

$$\int_{-1}^{1}\cos(\xi rt)dt = \frac{1}{\xi r}\sin(\xi rt)\Big|_{-1}^{1} = \frac{2\sin\xi r}{\xi r} \quad (2.67)$$

$$\int_{-1}^{1}\sin(\xi rt)dt = 0 \quad (2.68)$$

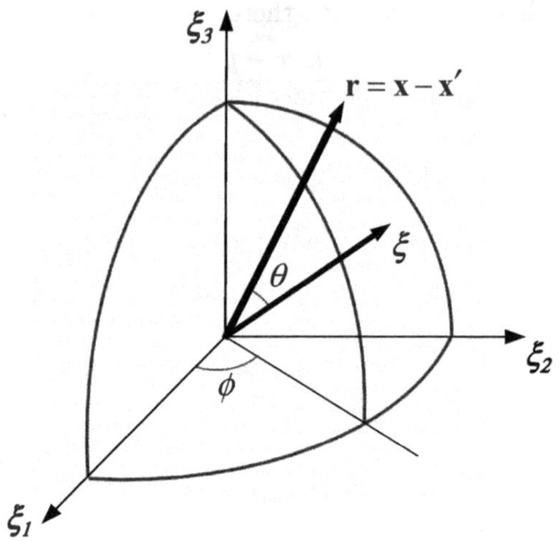

Fig. 2.3 Inversion of three-dimensional Fourier transform.

Hence

$$G(\mathbf{x}-\mathbf{x}') = \frac{1}{2\pi^2}\int_0^\infty \frac{\sin \xi r}{\xi r} d\xi = \frac{1}{2\pi^2 r}\mathrm{Si}(\infty) \tag{2.69}$$

where $\mathrm{Si}(x) := \int_0^x \frac{\sin t}{t} dt$ and $\mathrm{Si}(\infty) = \frac{\pi}{2}$. Finally, we have

$$G(\mathbf{x}-\mathbf{x}') = \frac{1}{4\pi}\frac{1}{|\mathbf{x}-\mathbf{x}'|} \tag{2.70}$$

2.4 Static Green's function for 3D linear elasticity

The Green's function for static, linear, isotropic elasticity was first derived by Lord Kelvin in 1848 [Kelvin (1948)]. The derivation shown below employs the Fourier integral transform technique, which is a systematic and elegant procedure to find Green's function for partial differential equations. Consider the Navier equation,

$$\sigma_{ij,j} + f_i = 0 \tag{2.71}$$

Denote Green's function of the displacement field as

$$u_i^m(\mathbf{x},\mathbf{y}) = G_{mi}^\infty(\mathbf{x},\mathbf{y}) \tag{2.72}$$

where $u_i^m(\mathbf{x},\mathbf{y})$ is the displacement in i-th direction at point $\mathbf{x}$ due to the concentrated load at the point $\mathbf{y}$ in the m-th direction. Subsequently, the induced stress field due to the concentrated load can be written as

$$\sigma_{ij}^{G_m^\infty} = C_{ijkl}\epsilon_{kl}^{G^\infty} = C_{ijkl}G_{mk,l}^\infty \tag{2.73}$$

and the body force becomes

$$f_i^m = \delta(\mathbf{x} - \mathbf{y})\delta_{mi} \tag{2.74}$$

where $\delta(\mathbf{x} - \mathbf{y}) := \delta(x_1 - y_1)\delta(x_2 - y_2)\delta(x_3 - y_3)$. Again, the integer m is a free index indicating the direction of the concentrated load.

Then, the equilibrium equation states that

$$\sigma_{ij,j}^{G_m^\infty} + f_i^m = 0, \quad \text{where } \sigma_{ij,j}^{G_m^\infty} = C_{ijkl}G_{mk,lj}^\infty \tag{2.75}$$

Therefore, Green's function for an infinite linear elastic medium is the solution of the following equilibrium equation,

$$C_{ijkl}G_{mk,lj}^\infty + \delta(\mathbf{x} - \mathbf{y})\delta_{mi} = 0 \tag{2.76}$$

In the following sections, we will find the exact solution for the Green's function. Apply the Fourier integral transform to the above equilibrium equation,

$$\delta(\mathbf{x} - \mathbf{y}) = \frac{1}{(2\pi)^3}\int_{\mathbb{R}^3} \exp(i\boldsymbol{\xi} \cdot (\mathbf{x} - \mathbf{y}))d\boldsymbol{\xi} \tag{2.77}$$

$$G_{mk}^\infty(\mathbf{x} - \mathbf{y}) = \int_{\mathbb{R}^3} \bar{G}_{mk}^\infty(\boldsymbol{\xi}) \exp(i\boldsymbol{\xi} \cdot (\mathbf{x} - \mathbf{y}))d\boldsymbol{\xi} \tag{2.78}$$

where $d\boldsymbol{\xi} = d\xi_1 d\xi_2 d\xi_3$. The integration is over the whole 3D Euclidean space ($\mathbb{R}^3$). Taking derivatives of Green's function with respect to $\mathbf{x}$, we have

$$G_{mk,lj}^\infty(\mathbf{x} - \mathbf{y}) = -\int_{\mathbb{R}^3} \bar{G}_{mk}^\infty(\boldsymbol{\xi})\xi_l\xi_j \exp(i\boldsymbol{\xi} \cdot (\mathbf{x} - \mathbf{y}))d\boldsymbol{\xi} \tag{2.79}$$

So the equilibrium equation Eq. (2.76) can be written as the following algebraic equations in Fourier space,

$$C_{ijkl}\bar{G}_{mk}^\infty(\boldsymbol{\xi})\xi_l\xi_j = \frac{1}{(2\pi)^3}\delta_{im} \tag{2.80}$$

which can be solved as the followings. Denote

$$K_{ik} = C_{ijkl}\xi_j\xi_l \quad \Rightarrow \quad K_{ik}\bar{G}_{mk}^\infty = \frac{1}{(2\pi)^3}\delta_{im} \tag{2.81}$$

Consider Laplace expansion, or the Cramer's rule,

$$N_{ji}(\boldsymbol{\xi})K_{ik}(\boldsymbol{\xi}) = D(\boldsymbol{\xi})\delta_{jk} \tag{2.82}$$

where N_{ji} is the cofactor of K_{ji} and $D(\boldsymbol{\xi}) = det\{K_{ij}(\boldsymbol{\xi})\}$.
Multiplying Eq. (2.82) with N_{ji} yields

$$N_{ji}(\boldsymbol{\xi})K_{ik}(\boldsymbol{\xi})\bar{G}_{mk}^\infty(\boldsymbol{\xi}) = \frac{1}{(2\pi)^3}N_{ji}(\boldsymbol{\xi})\delta_{im} \tag{2.83}$$

$$D(\boldsymbol{\xi})\delta_{jk}\bar{\mathbf{G}}_{mk}^\infty(\boldsymbol{\xi}) = \frac{1}{(2\pi)^3}N_{jm}(\boldsymbol{\xi}) \tag{2.84}$$

which leads to

$$\bar{G}_{mj}^\infty(\boldsymbol{\xi}) = \frac{1}{(2\pi)^3}\frac{N_{jm}(\boldsymbol{\xi})}{D(\boldsymbol{\xi})} \tag{2.85}$$

Change indices $j \leftrightarrow i$ and $m \leftrightarrow j$. Through inverse Fourier transform, one may find that

$$G_{ij}^{\infty}(\mathbf{x}-\mathbf{y}) = \frac{1}{(2\pi)^3}\int_{\mathbb{R}^3} \frac{N_{ji}(\boldsymbol{\xi})}{D(\boldsymbol{\xi})}\exp(i\boldsymbol{\xi}\cdot(\mathbf{x}-\mathbf{y}))d\boldsymbol{\xi} \tag{2.86}$$

For linear isotropic material, one may find that

$$\begin{aligned} N_{ij}(\boldsymbol{\xi}) = N_{ji}(\boldsymbol{\xi}) &= \frac{1}{2}e_{ik\ell}e_{jmn}K_{km}(\boldsymbol{\xi})K_{\ell n}(\boldsymbol{\xi}) \\ &= \mu\xi^2\Big((\lambda+2\mu)\delta_{ij}\xi^2 - (\lambda+\mu)\xi_i\xi_j\Big) \end{aligned} \tag{2.87}$$

$$D(\boldsymbol{\xi}) = e_{mn\ell}K_{m1}(\boldsymbol{\xi})K_{n2}(\boldsymbol{\xi})K_{\ell 3}(\boldsymbol{\xi}) = \mu^2(\lambda+2\mu)\xi^6 \tag{2.88}$$

and

$$K_{ij}^{-1}(\boldsymbol{\xi}) = \frac{N_{ji}(\boldsymbol{\xi})}{D(\boldsymbol{\xi})} = \frac{1}{\mu(\lambda+2\mu)\xi^4}\Big((\lambda+2\mu)\delta_{ij}\xi^2 - (\lambda+\mu)\xi_i\xi_j\Big) \tag{2.89}$$

where $\xi = |\boldsymbol{\xi}|$. Denote $\mathbf{z} = \mathbf{x}-\mathbf{y}$, then we arrive the following expression for Green's function,

$$G_{ij}^{\infty}(\mathbf{z}) = \frac{1}{(2\pi)^3}\int_{\mathbb{R}^3} K_{ji}^{-1}(\boldsymbol{\xi})\exp(i\boldsymbol{\xi}\cdot\mathbf{z})d\boldsymbol{\xi} \tag{2.90}$$

Note that Eq.(2.90) is integrated over the infinite space. The integration may be

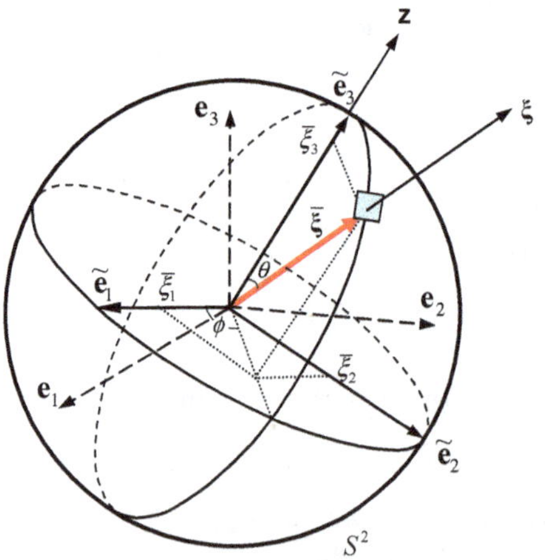

Fig. 2.4 The unit sphere S^2 in the ξ-space.

easier to solve in a spherical coordinate, as shown in Fig. 2.4. Here, we denote S^2 as a unit sphere, and there are two set of orthogonal coordinates: The basis ($\mathbf{e}_1$, $\mathbf{e}_2$, $\mathbf{e}_3$) is associated with the Fourier space, so $\boldsymbol{\xi} = \xi_i \boldsymbol{e}_i$. The other is a rotated basis ($\tilde{\mathbf{e}}_1$, $\tilde{\mathbf{e}}_2$, $\tilde{\mathbf{e}}_3$), so that the unit vector $\tilde{\mathbf{e}}_3$ is in align with the vector $\mathbf{z}$, so $\mathbf{z} = z\tilde{\mathbf{e}}_3$.

By applying the Radon decomposition, the volume integration can be converted to surface integration over the unit sphere S^2,

$$d\boldsymbol{\xi} = d\xi_1 d\xi_2 d\xi_3 = \xi^2 d\xi dS \tag{2.91}$$

where $\xi = |\boldsymbol{\xi}|$ and dS is the surface element on the unit sphere S^2.

Denote $\bar{\boldsymbol{\xi}} = \boldsymbol{\xi}/|\boldsymbol{\xi}|$ as a unit vector pointing from the origin to the S^2 surface in $\boldsymbol{\xi}$ direction, and denote $\bar{\mathbf{z}} = \tilde{e}_3$ as another unit vector point from the origin to the S^2 surface in $\mathbf{z}$ direction. Written in component form, $\bar{\xi}_i = \xi_i/\xi$ and $\bar{z}_i = z_i/z$.

Then Eq.(2.90) can be written as

$$G_{ij}^{\infty}(\mathbf{z}) = \frac{1}{(2\pi)^3}\int_0^{\infty}\oint_{S^2}\frac{1}{\mu(\lambda+2\mu)}\Big((\lambda+2\mu)\delta_{ij} - (\lambda+\mu)\bar{\xi}_i\bar{\xi}_j\Big) \cdot \exp\{i\xi z\bar{\boldsymbol{\xi}}\cdot\bar{\mathbf{z}}\} dS d\xi \tag{2.92}$$

Consider the symmetry property, i.e.

$$K_{ji}^{-1}(\boldsymbol{\xi}) = K_{ji}^{-1}(-\boldsymbol{\xi}) \tag{2.93}$$

Changing $\boldsymbol{\xi} \to -\boldsymbol{\xi}$ in Eq.(2.90), we may also have

$$G_{ij}^{\infty}(\mathbf{z}) = \frac{1}{(2\pi)^3}\int_0^{\infty}\oint_{S^2}\frac{1}{\mu(\lambda+2\mu)}\Big((\lambda+2\mu)\delta_{ij} - (\lambda+\mu)\bar{\xi}_i\bar{\xi}_j\Big) \cdot \exp\{-i\xi z\bar{\boldsymbol{\xi}}\cdot\bar{\mathbf{z}}\} dS(\bar{\boldsymbol{\xi}}) d\xi \tag{2.94}$$

Change the scalar variable $\xi \to -\xi$. Eq.(2.94) yields

$$G_{ij}^{\infty}(\mathbf{z}) = \frac{1}{(2\pi)^3}\int_{-\infty}^{0}\oint_{S^2}\frac{1}{\mu(\lambda+2\mu)}\Big((\lambda+2\mu)\delta_{ij} - (\lambda+\mu)\bar{\xi}_i\bar{\xi}_j\Big) \cdot \exp\{i\xi z\bar{\boldsymbol{\xi}}\cdot\bar{\mathbf{z}}\} dS(\bar{\boldsymbol{\xi}}) d\xi \tag{2.95}$$

Combining (2.94) with (2.95) yields

$$G_{ij}^{\infty}(\mathbf{z}) = \frac{1}{2(2\pi)^3}\int_{-\infty}^{\infty}\oint_{S^2}\frac{1}{\mu(\lambda+2\mu)}\Big((\lambda+2\mu)\delta_{ij} - (\lambda+\mu)\bar{\xi}_i\bar{\xi}_j\Big) \cdot \exp\{i\xi z\bar{\boldsymbol{\xi}}\cdot\bar{\mathbf{z}}\} dS(\bar{\boldsymbol{\xi}}) d\xi \tag{2.96}$$

Considering

$$\int_{-\infty}^{\infty}\exp(i\xi z\bar{\boldsymbol{\xi}}\cdot\bar{\mathbf{z}})d\xi = 2\pi\delta(z\bar{\boldsymbol{\xi}}\cdot\bar{\mathbf{z}}) \tag{2.97}$$

one has

$$G_{ij}^{\infty}(\mathbf{z}) = \frac{1}{2(2\pi)^2}\oint_{S^2}\delta(z\bar{\boldsymbol{\xi}}\cdot\bar{\mathbf{z}})\frac{[(\lambda+2\mu)\delta_{ij} - (\lambda+\mu)\bar{\xi}_i\bar{\xi}_j]}{\mu(\lambda+2\mu)}dS(\bar{\boldsymbol{\xi}}) \tag{2.98}$$

To integrate Eq. (2.98), one has to evaluate the following two integrals:

$$\int_{S^2}\delta(z\bar{\boldsymbol{\xi}}\cdot\bar{\mathbf{z}})dS? \quad \text{and} \quad \int_{S^2}\bar{\xi}_i\bar{\xi}_j\delta(z\bar{\boldsymbol{\xi}}\cdot\bar{\mathbf{z}})dS?$$

The surface integration on the unit sphere can be evaluated conveniently using the spherical coordinate as shown in Fig. 2.4. Considering $\bar{\boldsymbol{\xi}}\cdot\bar{\mathbf{z}} = \cos\theta$, $d(\bar{\boldsymbol{\xi}}\cdot\bar{\mathbf{z}}) = -\sin\theta d\theta$, one may decompose the surface element on S^2 into: $dS(\bar{\boldsymbol{\xi}}) = \sin\theta d\theta d\phi = -d(\bar{\boldsymbol{\xi}}\cdot\bar{\mathbf{z}})d\phi$, where $\theta \in [0,\pi]$ and $\phi \in [0,2\pi]$. Change of variable $t = \bar{\boldsymbol{\xi}}\cdot\bar{\mathbf{z}} = \cos\theta$, $t \in [-1,1]$,

$$\int_{S^2}\delta(z\bar{\boldsymbol{\xi}}\cdot\bar{\mathbf{z}})dS = \int_{-1}^{1}\delta(zt)dt\int_0^{2\pi}d\phi = \frac{2\pi}{z} \tag{2.99}$$

To evaluate the second integral, we use two different approaches in order to give readers a better understanding.

(1) The first approach: To evaluate

$$\int_{S^2} \bar{\xi}_i\bar{\xi}_j\delta(z\bar{\boldsymbol{\xi}}\cdot\bar{\mathbf{z}})dS = \int_{-1}^{1}\int_{0}^{2\pi}\delta(zt)\bar{\xi}_i\bar{\xi}_j d\phi dt \tag{2.100}$$

we consider the ramp function

$$\int_{-1}^{1}\langle zt\rangle dt = \int_{-1}^{1}\langle\bar{\boldsymbol{\xi}}\cdot\mathbf{z}\rangle dt \tag{2.101}$$

where $< x >$ is the MaCaulay's ramp function, i.e.

$$< x >= \begin{cases} x & x > 0 \\ 0 & x \leq 0 \end{cases} \tag{2.102}$$

By definition,

$$\frac{d}{dx} < x >= H(x), \quad \text{and} \quad \frac{d}{dx}H(x) = \delta(x) \tag{2.103}$$

where $H(x)$ and $\delta(x)$ are the Heaviside function and the Dirac's delta function respectively.
Then

$$\int_{-1}^{1}\langle\bar{\boldsymbol{\xi}}\cdot\mathbf{z}\rangle dt = \int_{-1}^{1} < zt > dt = z\int_{0}^{1} t dt = \frac{|\mathbf{z}|}{2} \tag{2.104}$$

Taking the derivative with respect to z_i and z_j, by chain rule we have

$$\frac{|\mathbf{z}|_{,i}}{2} = \frac{\partial}{\partial z_i}\int_{-1}^{1}\langle\bar{\boldsymbol{\xi}}\cdot\mathbf{z}\rangle dt = \int_{-1}^{1} H(\bar{\boldsymbol{\xi}}\cdot\mathbf{z})\bar{\xi}_i dt \tag{2.105}$$

and

$$\frac{|\mathbf{z}|_{,ij}}{2} = \frac{\partial}{\partial z_j}\int_{-1}^{1} H(\bar{\boldsymbol{\xi}}\cdot\mathbf{z})\bar{\xi}_i dt = \int_{-1}^{1}\delta(\bar{\boldsymbol{\xi}}\cdot\mathbf{z})\bar{\xi}_i\bar{\xi}_j dt \tag{2.106}$$

It is straightforward to show that

$$|\mathbf{z}|_{,ij} = \frac{1}{|\mathbf{z}|}\left(\delta_{ij} - \frac{z_i z_j}{|\mathbf{z}|^2}\right) \tag{2.107}$$

Therefore,

$$\begin{aligned}\int_{S^2}\bar{\xi}_i\bar{\xi}_j\delta(z\bar{\boldsymbol{\xi}}\cdot\bar{\mathbf{z}})dS &= \int_{0}^{2\pi}\int_{-1}^{1}\delta(\bar{\boldsymbol{\xi}}\cdot\mathbf{z})\bar{\xi}_i\bar{\xi}_j dt d\phi = \int_{0}^{2\pi}|\mathbf{z}|_{,ij}d\phi \\ &= \frac{\pi}{|\mathbf{z}|}\left(\delta_{ij} - \frac{z_i z_j}{|\mathbf{z}|^2}\right)\end{aligned} \tag{2.108}$$

(2) The second approach: To evaluate

$$\oint_{S^2} \bar{\xi}_i \bar{\xi}_j \delta(z\bar{\boldsymbol{\xi}} \cdot \bar{\mathbf{z}}) dS = \int_{-1}^{1} \int_{0}^{2\pi} \delta(zt) \bar{\xi}_i \bar{\xi}_j d\phi dt \tag{2.109}$$

The unit vector $\bar{\boldsymbol{\xi}}$ can be decomposed in the ($\tilde{\mathbf{e}}_1$,$\tilde{\mathbf{e}}_2$,$\tilde{\mathbf{e}}_3$) coordinate. It is straightforward to show,

$$\begin{aligned} \bar{\boldsymbol{\xi}} &= \sin\theta\cos\phi\ \tilde{\mathbf{e}}_1 + \sin\theta\sin\phi\ \tilde{\mathbf{e}}_2 + \cos\theta\ \tilde{\mathbf{e}}_3 \\ &= \sqrt{1-t^2}\cos\phi\ \tilde{\mathbf{e}}_1 + \sqrt{1-t^2}\sin\phi\ \tilde{\mathbf{e}}_2 + t\ \tilde{\mathbf{e}}_3 = \hat{\xi}_i \tilde{\mathbf{e}}_i \end{aligned} \tag{2.110}$$

where $\hat{\xi_1} = \sqrt{1-t^2}\cos\phi$, $\hat{\xi_2} = \sqrt{1-t^2}\sin\phi$, and $\hat{\xi_3} = t$ are three components in $(\tilde{\mathbf{e}}_1, \tilde{\mathbf{e}}_2, \tilde{\mathbf{e}}_3)$ coordinate. The components can be rotated into $(\mathbf{e}_1, \mathbf{e}_2, \mathbf{e}_3)$ coordinate through the rotation matrix $Q_{k\ell} = \mathbf{e}_k \cdot \tilde{\mathbf{e}}_\ell$. So the dyad $\bar{\xi}_i \bar{\xi}_j$ can be evaluated as,

$$\bar{\xi}_i \bar{\xi}_j = \hat{\xi}_m \hat{\xi}_n (\tilde{\mathbf{e}}_m \cdot \mathbf{e}_i)(\tilde{\mathbf{e}}_n \cdot \mathbf{e}_j) = \hat{\xi}_m \hat{\xi}_n Q_{im} Q_{jn} \tag{2.111}$$

Therefore, the integral can be written as

$$\oint_{S^2} \delta(zt) \bar{\xi}_i \bar{\xi}_j dS = \int_{-1}^{1} \int_{0}^{2\pi} \delta(zt) \hat{\xi}_m \hat{\xi}_n Q_{im} Q_{jn} d\phi dt \tag{2.112}$$

We note that the rotation matrix $Q_i m$ does not depend on variable t and ϕ. Inspecting each component of $\hat{\xi}_m \hat{\xi}_n$ can be written in a matrix form,

$$[\hat{\xi}_i \hat{\xi}_j] = \begin{bmatrix} \underline{\underline{\cos^2\phi}} - t^2\cos^2\phi & (1-t^2)\sin\phi\cos\phi & t\sqrt{1-t^2}\cos\phi \\ (1-t^2)\sin\phi\cos\phi & \underline{\underline{\sin^2\phi}} - t^2\sin^2\phi & t\sqrt{1-t^2}\sin\phi \\ t\sqrt{1-t^2}\cos\phi & t\sqrt{1-t^2}\sin\phi & t^2 \end{bmatrix} \tag{2.113}$$

and considering the fact that

$$\begin{aligned} &\int_{-1}^{1} t^2 \delta(zt) dt = 0 \\ &\int_{0}^{2\pi} \sin\phi\cos\phi = \int_{0}^{2\pi} \sin\phi = \int_{0}^{2\pi} \cos\phi = 0 \end{aligned} \tag{2.114}$$

It is easy to identify that only underlined terms contribute, and all other terms vanish after integration. So, we finally have

$$\begin{aligned} \oint_{S^2} \delta(zt) \bar{\xi}_i \bar{\xi}_j dS &= \int_{-1}^{1} \delta(zt) dt \int_{0}^{2\pi} (\cos^2\phi Q_{i1} Q_{j1} + \sin^2\phi Q_{i2} Q_{j2}) d\phi \\ &= \frac{\pi}{z}(Q_{i1}Q_{j1} + Q_{i2}Q_{j2}) = \frac{\pi}{z}(\delta_{ij} + \bar{z}_i \bar{z}_j) \end{aligned} \tag{2.115}$$

The last expression was derived by considering the orthogonal property of the rotation matrix, $Q_{ik} Q^T_{kj} = Q_{ik} Q_{jk} = \delta_{ij}$. It is trivial to show $Q_{i1}Q_{j1} + Q_{i2}Q_{j2} = \delta_{ij} - Q_{i3}Q_{j3}$, and $Q_{i3} = \mathbf{e}_i \cdot \tilde{\mathbf{e}}_3 = \mathbf{e}_i \cdot \mathbf{z}/|\mathbf{z}| = \bar{z}_i$.

Consequently,

$$\begin{aligned} G^{\infty}_{ij}(\mathbf{z}) &= \frac{1}{2(2\pi)^2 z}\Big[\frac{2\pi(\lambda+2\mu)\delta_{ij} - \pi(\lambda+\mu)(\delta_{ij} - \bar{z}_i\bar{z}_j)}{\mu(\lambda+2\mu)}\Big] \\ &= \frac{(\lambda+\mu)}{8\pi\mu z(\lambda+2\mu)}\Big\{\frac{\lambda+3\mu}{\lambda+\mu}\delta_{ij} + \bar{z}_i\bar{z}_j\Big\} \\ &= \frac{1}{16\pi\mu(1-\nu)|\mathbf{x}-\mathbf{y}|}\Big\{(3-4\nu)\delta_{ij} + \frac{(x_i-y_i)(x_j-y_j)}{|\mathbf{x}-\mathbf{y}|^2}\Big\} \end{aligned} \tag{2.116}$$

2.5 Green's function for Stokes equations

Green's function technique has also been extensively used in fluid mechanics. In fact, the Green's function theory has been important theoretical foundation for *Micro-hydrodynamics*, e.g. in [Kim and Karrila (1991)] and [Phan-Thien and Kim (1994)], and found many applications in bioengineering and nanotechnology.

In the following section, we outline the derivation of the Green's function for the Stokes problem that obeys the following governing equations,

$$\begin{cases} \nabla\cdot\boldsymbol{\sigma} = -\nabla p + \mu\nabla^2\mathbf{v} = -\rho\mathbf{f}, & (a) \\ \nabla\cdot\mathbf{v} = 0, & (b) \end{cases} \tag{2.117}$$

where $\boldsymbol{\sigma}$ is the Cauchy stress tensor, p is the hydrostatic pressure, μ is the dynamic viscosity, ρ is the density, $\mathbf{f}$ is the force per unit mass (body force), and $\mathbf{v}$ is the velocity field. The Stokes flow is incompressible, i.e. $\nabla\cdot\mathbf{v} = 0$, which is reflected in Eq.(2.117) (b).

To find the Green's function, we consider a point load i.e. $-\rho\mathbf{f} = -\mathbf{P}\delta(\mathbf{x})$ where $\mathbf{P}$ is a constant vector. Equations (2.117) become

$$-\nabla p + \mu\nabla^2\mathbf{v} = -\mathbf{P}\delta(\mathbf{x}),\quad (a) \quad \text{and} \quad \nabla\cdot\mathbf{v} = 0 \quad (b). \tag{2.118}$$

Applying Fourier transform to above Stokes equations and considering the fact that

$$\delta(\mathbf{x}) = \frac{1}{(2\pi)^3}\int_{\mathbb{R}^3} \exp(i\boldsymbol{\xi}\cdot\mathbf{x})d\boldsymbol{\xi},$$

we can transform Eqs. (2.118) into a set of algebraic equations in the Fourier domain

$$i\xi_j\hat{p}(\boldsymbol{\xi}) + \mu\xi^2\hat{v}_j(\boldsymbol{\xi}) = \frac{P_j}{(2\pi)^3},\quad (a) \quad \text{and} \quad i\xi_j\hat{v}_j(\boldsymbol{\xi}) = 0, \quad (b) \tag{2.119}$$

where $i = \sqrt{-1}$ and $\xi = \sqrt{\xi_1^2+\xi_2^2+\xi_3^2}$ and

$$\hat{p}(\boldsymbol{\xi}) = \frac{1}{(2\pi)^3}\int_{\mathbb{R}^3} p(\mathbf{x})\exp(-i\boldsymbol{\xi}\cdot\mathbf{x})d\mathbf{x} \tag{2.120}$$

$$\hat{\mathbf{v}}(\boldsymbol{\xi}) = \frac{1}{(2\pi)^3}\int_{\mathbb{R}^3} \mathbf{v}(\mathbf{x})\exp(-i\boldsymbol{\xi}\cdot\mathbf{x})d\mathbf{x} \tag{2.121}$$

Multiplying (2.119)(a) with $i\xi_j$ and utilizing (2.119)(b), we can solve for $\hat{p}(\boldsymbol{\xi})$

$$\hat{p}(\boldsymbol{\xi}) = -\frac{i\xi_j P_j}{(2\pi)^3\xi^2} \tag{2.122}$$

Substituting (2.122) back to (2.119), one can find that

$$i\xi_j\Big(-\frac{i\xi_k P_k}{(2\pi)^3\xi^2}\Big) + \mu\xi^2\hat{v}_j(\boldsymbol{\xi}) = \frac{P_j}{(2\pi)^3} \quad \Rightarrow \quad \hat{v}_j(\boldsymbol{\xi}) = \frac{P_k}{\mu\xi^2}\Big(\frac{\delta_{jk}}{(2\pi)^3} - \frac{\xi_j\xi_k}{(2\pi)^3\xi^2}\Big) \tag{2.123}$$

Utilizing the following identities in Fourier transform, e.g. [Jones (1966)],

$$\mathbf{F}\Big[\frac{1}{4\pi r}\Big] = \frac{1}{(2\pi)^3\xi^2} \tag{2.124}$$

$$\mathbf{F}\Big[\frac{r}{8\pi}\Big] = -\frac{1}{(2\pi)^3\xi^4} \tag{2.125}$$

and

$$\frac{\partial^2 r}{\partial x_i \partial x_j} = \frac{\delta_{ij}}{r} - \frac{x_i x_j}{r^3} \quad \Rightarrow \quad \mathbf{F}\Big(\frac{1}{8\pi}\frac{\partial^2 r}{x_i x_j}\Big) = \frac{-\xi_i\xi_j}{(2\pi)^3\xi^4}$$

Hence,

$$v_j(\mathbf{x}) = \frac{P_k}{8\pi}\Big(\frac{\delta_{kj}}{r} + \frac{x_j x_k}{r^3}\Big) \quad \text{or} \quad \mathbf{v}(\mathbf{x}) = \frac{\mathbf{P}}{8\mu\pi}\cdot\boldsymbol{\mathcal{G}}(\mathbf{x}) \tag{2.126}$$

where

$$\boldsymbol{\mathcal{G}} = \mathcal{G}_{ij}(\mathbf{x})\mathbf{e}_i\otimes\mathbf{e}_j, \quad \text{and} \quad \mathcal{G}_{ij} = \frac{\delta_{ij}}{r} + \frac{x_i x_j}{r^3} \tag{2.127}$$

Similarly,

$$\frac{\partial}{\partial x_i}\Big(\frac{1}{r}\Big) = -\frac{x_i}{r^3} \quad \Rightarrow \quad \mathbf{F}\Big[\frac{x_i}{8\pi r^3}\Big] = -\frac{i\xi_i}{(2\pi)^3\xi^2} \tag{2.128}$$

we then have

$$p(\mathbf{x}) = \frac{P_j x_j}{4\pi r^3} + \frac{P_j P_j^\infty}{8\mu\pi} \tag{2.129}$$

where P_j^∞ is a constant. In general, we can write

$$p(\mathbf{x}) = \frac{\mathbf{P}}{8\mu\pi}\cdot\boldsymbol{\mathcal{P}}(\mathbf{x}), \quad \text{where } \boldsymbol{\mathcal{P}} = \mathcal{P}_j(\mathbf{x})\mathbf{e}_j, \quad \text{and } \mathcal{P}_j = 2\mu\frac{x_j}{r^3} + P_j^\infty \tag{2.130}$$

In literature, we call $\boldsymbol{\mathcal{G}}(\mathbf{x})$ as the *Oseen tensor*; and we call $\boldsymbol{\mathcal{P}}(\mathbf{x})$ as the Oseen tensor of the pressure field.

By direct differentiation, we can show that

$$\mathcal{G}_{ij,k} = -\frac{1}{r^3}\delta_{ij}x_k + \frac{1}{r^3}(\delta_{ik}x_j + \delta_{jk}x_i) - \frac{3}{r^5}x_i x_j x_k \tag{2.131}$$

$$\nabla^2\mathcal{G}_{ij} = \mathcal{G}_{ij,kk} = \frac{2}{r^3} - \frac{6}{r^5}x_i x_j \tag{2.132}$$

One can further verify that

$$8\pi\mu\nabla\cdot\mathbf{v} = P_j\mathcal{G}_{ij,i} = P_j\Big[-\frac{1}{r^3}\delta_{ij}x_i + \frac{1}{r^3}(\delta_{ii}x_j + \delta_{ji}x_i) - \frac{3}{r^5}x_i x_j x_i\Big] = 0\,. \tag{2.133}$$

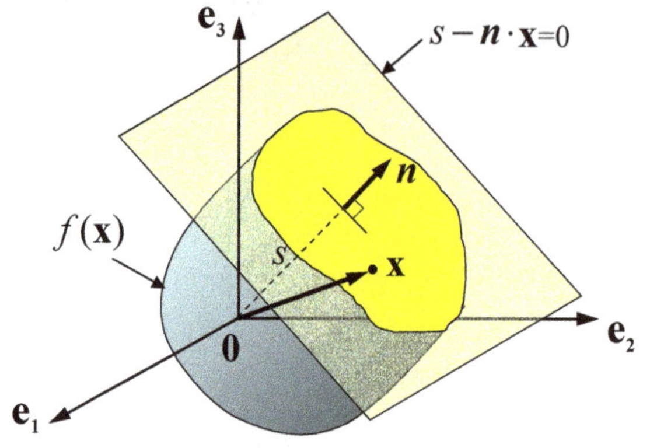

Fig. 2.5 Illustration of three-dimensional Radon transform.

2.6 Radon transform

Radon transform is named after Austrian mathematician Johann Karl August Radon (December 16, 1887 - May 25, 1956). Radon transform has been widely used in image processing, for example, in reconstructing images from computed tomography scans (CTA).

As shown in Fig. 2.5, a regular function $f(\mathbf{x})$ is defined in $\mathbb{R}^3$, where $\mathbf{x} = x_1\mathbf{e}_1 + x_2\mathbf{e}_2 + x_3\mathbf{e}_3$ is the position vector of a spatial point in $\mathbb{R}^3$. We can define a plane, with a unit vector $\boldsymbol{n}$ and a distance s from the plane to the origin of the coordinate. Mathematically, the plane can be written as $s - \boldsymbol{n} \cdot \mathbf{x} = 0$.

The Radon transform of $f(\mathbf{x})$ is defined as the integration of $f(\mathbf{x})$ over the plane,

$$\hat{f}(s, \boldsymbol{n}) = \boldsymbol{R}\{f(\mathbf{x})\} = \int_{\mathbb{R}^3} f(\mathbf{x})\delta(s - \boldsymbol{n} \cdot \mathbf{x})d\mathbf{x} \tag{2.134}$$

Therefore, the Radon transform maps $f(\mathbf{x})$ in the Cartesian coordinate to $\hat{f}(s, \boldsymbol{n})$ in a polar coordinate. The collection of all $\hat{f}(s, \boldsymbol{n})$ for all unit vector $\boldsymbol{n}$ is called the Radon transform.

The inverse Radon transform is carried out by the following two steps:
(1) Take the second order partial derivatives of $\hat{f}(s, \boldsymbol{n})$ with respect to s,

$$\tilde{f}(s, \boldsymbol{n}) = \partial_s^2 \hat{f}(s, \boldsymbol{n}) \tag{2.135}$$

(2) substitute $s = \boldsymbol{n} \cdot \mathbf{x}$ in $\tilde{f}(s, \boldsymbol{n})$, and integrate it over the surface of a unit sphere,

$$f(\mathbf{x}) = \boldsymbol{R}^{-1}(\hat{f}) = -\frac{1}{8\pi^2} \int_{|\boldsymbol{n}|=1} \tilde{f}(\boldsymbol{n} \cdot \mathbf{x}, \boldsymbol{n})dS(\boldsymbol{n}) \tag{2.136}$$

The Radon transform has the following properties:
(1) $\hat{f}(s, \boldsymbol{n})$ is an even and homogeneous function of order -1, i.e.

$$\hat{f}(\alpha s, \alpha \boldsymbol{n}) = |\alpha|^{-1} \hat{f}(s, \boldsymbol{n})$$

(2) linearity:

$$\boldsymbol{R}(c_1 f + c_2 g) = c_1 \hat{f} + c_2 \hat{g}$$

(3) transform of derivatives:

$$\boldsymbol{R}(\partial_i f(\mathbf{x})) = n_i \partial_s \hat{f}(s, \boldsymbol{n})$$
$$\boldsymbol{R}(\partial_i \partial_j f(\mathbf{x})) = n_i n_j \partial_s^2 \hat{f}(s, \boldsymbol{n})$$

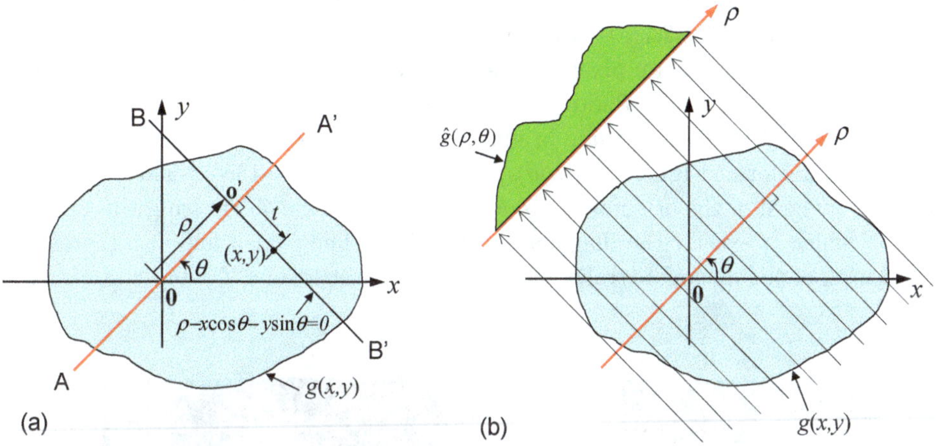

Fig. 2.6 Illustration of two-dimensional Radon transform.

Example 2.12. (Two-dimensional Radon Transform) Fig.2.6 shows the Radon transform of an image density function, $g(x, y)$, defined in two-dimensional Cartesian coordinate. As shown in Fig. 2.6(a), for a line AA' of a given slope angle θ, Radon transform can be viewed as the *projection* of the image function onto AA'. The *projection*here is understood as integrating $g(x, y)$ along a perpendicular line BB', whose function form can be written as $\rho - x\cos\theta - y\sin\theta = 0$.Therefore, the two-dimensional Radon transform is defined as

$$\hat{g}(\rho, \theta) = \int_{-\infty}^{+\infty} \int_{-\infty}^{+\infty} g(x, y)\delta(\rho - x\cos\theta - y\sin\theta)dxdy \tag{2.137}$$

As shown in Fig. 2.6(a), x and y can be further parameterized by θ and another parameter t,

$$x = \rho\cos\theta + t\sin\theta, \quad \text{and} \quad y = \rho\sin\theta - t\cos\theta \tag{2.138}$$

Therefore, Eq.(2.137) is identical to the following line integral

$$\hat{g}(\rho, \theta) = \int_{-\infty}^{\infty} g(\rho\cos\theta + t\sin\theta, \rho\sin\theta - t\cos\theta)dt \tag{2.139}$$

Fig.2.6(b) schematically illustrates the interpolation of Radon transform as a projection of the function into a radial direction. The concept of Radon transform is analogous to X-ray absorption tomography, where X-ray radiation from an emitter is received by a detector. The radiation attenuation is a measurement of the integrated density of the object, which corresponds to the line integral that is calculated in the Radon transform.

Example 2.13. (Radon Transform used in Line Detection) This simple example demonstrates the use of Radon transform in image processing, particularly, for line detection. Fig. 2.7(a) shows an original image $g(x, y)$, which contains a letter 'Z' in white. The white pixels are assigned value 1, while the black background pixels are 0. The image in Radon transform space, $\hat{g}(\rho, \theta)$, is shown in Fig. 2.7(b). It can be seen that each of three bright spots represents one straight line of letter 'Z'. The transformed function $\hat{g}(\rho, \theta)$ peaks at $(\rho = 0, \theta = 135^o)$. This represents the line integral of the original image along a line of angle 45° passing through the origin, which is exactly the diagonal stroke of 'Z'. The peak value also represents the length of the diagonal stroke, which is the longest line in 'Z'.

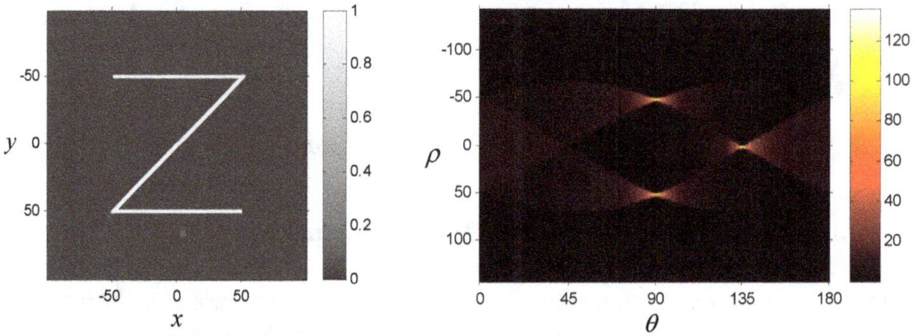

(a) Original image $g(x, y)$

(b) Image in Radon transform space $\hat{g}(\rho, \theta)$

Fig. 2.7 Radon transform of an image.

Example 2.14. (Radon Transform of Dirac's delta function) The Radon transform of Dirac's delta function is

$$\hat{\delta}(s, \boldsymbol{n}) = \boldsymbol{R}(\delta) = \int_{\mathbb{R}^3} \delta(\mathbf{x})\delta(s - \boldsymbol{n} \cdot \mathbf{x}) d\mathbf{x} = \delta(s) \tag{2.140}$$

where $s = n_i x_i$.

Subsequently,

$$\tilde{\delta}(s, \boldsymbol{n}) = \delta''(s) \tag{2.141}$$

and the inverse Radon transform is

$$\delta(\mathbf{x}) = -\frac{1}{8\pi^2}\int_{S^2} \delta''(n_k x_k)dS \tag{2.142}$$

One can verify this by considering the identity (2.99), i.e.

$$\int_{S^2} \delta(n_k x_k)dS = \frac{2\pi}{|\mathbf{x}|} \tag{2.143}$$

Applying the harmonic operator $\nabla^2 = \dfrac{\partial^2}{\partial x_i \partial x_i}$ to the above identity and considering Example 2.11 Eq.((2.70) yields the following expression

$$\int_{S^2} \delta''(n_k x_k)n_i n_i dS = 2\pi\nabla^2\Big(\frac{1}{|\mathbf{x}|}\Big) = -8\pi^2\delta(\mathbf{x}) \tag{2.144}$$

Example 2.15. Now we use the Radon transform to derive 3D static Green's function of a linear elastic medium. Consider the equilibrium equation for concentrated load is acting at the origin of the coordinate ($\mathbf{y} = \mathbf{0}$).

$$C_{ijkl}G^{\infty}_{km,lj} + \delta(\mathbf{x})\delta_{im} = 0 \tag{2.145}$$

Assume that the Green's function can be written as a form of inverse Radon transform,

$$G^{\infty}_{km}(\mathbf{x}) = -\frac{1}{8\pi^2}\int_{S^2} \tilde{G}^{\infty}_{km}(\bar{\xi}_n x_n)dS \tag{2.146}$$

where $\boldsymbol{\xi}$ is a unit vector, S^2 is the surface of a unit sphere. Then

$$G^{\infty}_{km,lj}(\mathbf{x}) = -\frac{1}{8\pi^2}\int_{S^2} \tilde{G}^{\infty}_{km}{}''(\bar{\xi}_n x_n)\bar{\xi}_l\bar{\xi}_j dS \tag{2.147}$$

On the other hand,

$$\delta(\mathbf{x}) = \boldsymbol{R}^{-1}\Big(\tilde{\delta}(s)\Big) = -\frac{1}{8\pi^2}\int_{S^2} \delta''(\bar{\xi}_n x_n)dS \tag{2.148}$$

then, the equilibrium equation reduces to

$$C_{ijkl}\bar{\xi}_j\bar{\xi}_l\tilde{G}^{\infty}_{km}{}''(\bar{\xi}_n x_n) = -\delta_{im}\delta''(\bar{\xi}_n x_n) \tag{2.149}$$

which leads to the solution

$$\tilde{G}^{\infty}_{ij}(\bar{\boldsymbol{\xi}}) = -K^{-1}_{ij}(\bar{\boldsymbol{\xi}})\delta(\bar{\xi}_n x_n) + C_1\bar{\xi}_n x_n + C_0 \tag{2.150}$$

where

$$K^{-1}_{ri}(\bar{\boldsymbol{\xi}})C_{ijkl}\bar{\xi}_l\bar{\xi}_j = \delta_{rk}, \quad \text{or} \quad K^{-1}_{ij}(\bar{\boldsymbol{\xi}}) = \frac{N_{ij}(\bar{\boldsymbol{\xi}})}{D(\bar{\boldsymbol{\xi}})} \tag{2.151}$$

Note that $C_1 = C_0 = 0$ because it is required that $G^{\infty}_{ij}(\mathbf{x}) \to 0$, as $\mathbf{x} \to \infty$.

For isotropic materials,

$$K^{-1}_{ij}(\bar{\boldsymbol{\xi}}) = \frac{1}{\mu}\Big[\delta_{ij} - \frac{(\lambda+\mu)\bar{\xi}_i\bar{\xi}_j}{(\lambda+2\mu)}\Big] \tag{2.152}$$

and accordingly,

$$G^{\infty}_{ij}(\mathbf{x}) - \frac{1}{8\pi^2}\int_{S^2} K^{-1}_{ij}(\bar{\boldsymbol{\xi}})\delta(\bar{\xi}_n x_n)dS \tag{2.153}$$

Subsequently, we can obtain the static Green's function for 3D linearly elasticity as

$$G^{\infty}_{ij}(\mathbf{x}) = \frac{1}{4\pi\mu}\left[\frac{\delta_{ij}}{|\mathbf{x}|} - \frac{(\lambda+\mu)}{2(\lambda+2\mu)}|\mathbf{x}|_{,ij}\right] . \tag{2.154}$$

In general, Eq. (2.153) is valid for anisotropic materials. Let the unit vector $\bar{\boldsymbol{\xi}} = \mathbf{z}$ and replace the spatial argument $\mathbf{x} \to \mathbf{r} - \mathbf{r}'$ and denote $\mathbf{r} - \mathbf{r}' = |\mathbf{r} - \mathbf{r}'|\mathbf{T}$, where $\mathbf{T}$ is the unit vector along the direction $\mathbf{r} - \mathbf{r}'$. Eq. (2.153) can be rewritten as

$$\begin{aligned} G^{\infty}_{ij}(\mathbf{x}) &= \frac{1}{8\pi^2}\int_{S^2} K^{-1}_{ij}(\mathbf{z})\delta(|\mathbf{r}-\mathbf{r}'|z_nT_n)dS \\ &= \frac{1}{8\pi^2|\mathbf{r}-\mathbf{r}'|}\int_0^{2\pi} K^{-1}_{ij}(\mathbf{z}(\psi))d\psi \end{aligned} \tag{2.155}$$

where the integrand must be evaluated on the plane where $\mathbf{z}\cdot\mathbf{T} = 0$. Barnett [Barnett (1972)] provided an explicit evaluation of $\mathbf{z}$ on the plane $\mathbf{z}\cdot\mathbf{T} = 0$. Consider variables θ and ϕ being the angular spherical coordinates of $\mathbf{T}$ related to the Cartesian coordinate $[x_1, x_2, x_3]$ or $[\mathbf{e}_1, \mathbf{e}_2, \mathbf{e}_3]$. Let

$$T_1 = \sin\phi\cos\theta, \quad T_2 = \sin\phi\sin\theta, \quad \text{and} \quad T_3 = \cos\phi \tag{2.156}$$

Then the two fixed orthogonal unit vector bases on the plane $\mathbf{z}\cdot\mathbf{T} = 0$ will be

$$a_1 = \sin\theta, \quad a_2 = -\cos\theta, \quad a_3 = 0, \tag{2.157}$$

$$b_1 = \cos\phi\cos\theta, \quad b_2 = \cos\phi\sin\theta, \quad b_3 = -\sin\phi \tag{2.158}$$

It is straightforward to very that

$$\mathbf{a}\cdot\mathbf{T} = 0, \quad \mathbf{b}\cdot\mathbf{T} = 0, \quad \text{and} \quad \mathbf{a}\cdot\mathbf{b} = 0\,.$$

Therefore, on the plane $\mathbf{z}\cdot\mathbf{T} = 0$, the vector $\mathbf{z}$ can be expressed as

$$\mathbf{z} = \cos\psi\mathbf{a} + \sin\psi\mathbf{b} = \cos\psi a_i\mathbf{e}_i + \sin\psi b_i\mathbf{e}_i, \quad i = 1,2,3 \tag{2.159}$$

Recently, the above techniques have been used to develop Green's function for functionally graded materials, e.g. [Martin *et al.* (2002)] .

2.7 Green's function for elastodynamics

In this section, we shall illustrate how to use Radon transform and Fourier transform to find the Green's function for dynamics problems.

Example 2.16. We first solve the Green's function for 3D Helmholtz equation,

$$\left\{\mu\Big(\frac{\partial^2}{\partial x_1^2} + \frac{\partial^2}{\partial x_2^2} + \frac{\partial^2}{\partial x_3^2}\Big) + \rho\omega^2\right\} G(\mathbf{x},\omega) = -\delta(\mathbf{x}) \tag{2.160}$$

Applying the Radon transform to the both sides of (2.160) yields,

$$\left(\mu\frac{\partial^2}{\partial s^2} + \rho\omega^2\right)\hat{G}(s) = -\delta(s) \tag{2.161}$$

To obtain (2.161), we have used the properties: $n_in_i = 1, i = 1,2,3$ and $\boldsymbol{R}\{\delta(\mathbf{x})\} = \delta(s)$.

One may find that the transformed 3D-Helmholtz equation looks exactly the same as the 1D-Helmholtz equation (2.59), and therefore it has the following solution,

$$\hat{G}(s) = \frac{i}{2\rho c^2 k} \exp\left(ik|s|\right) \tag{2.162}$$

where $c = \sqrt{\dfrac{\mu}{\rho}}$ and $k = \dfrac{\omega}{c}$. By applying the inverse Radon transform, we shall be able to obtain the real-space solution for the Green's function.

The first step of the inverse Radon transform is,

$$\tilde{G}(s) = \frac{\partial^2}{\partial s^2}\hat{G} = \frac{i}{2\rho c^2 k}\frac{\partial^2}{\partial s^2}\exp(ik|s|) = \frac{-1}{2\rho c^2}\Big(2\delta(s) + ik\exp(ik|s|)\Big)$$

Then we can separate $\tilde{G}(s, \mathbf{n})$ into two parts,

$$\tilde{G}(s,\mathbf{n}) = \tilde{G}^S(s,\mathbf{n}) + \tilde{G}^R(s,\mathbf{n}) \tag{2.163}$$

where

$$\tilde{G}^S(s,\mathbf{n}) = \frac{-1}{\rho c^2}\delta(\mathbf{n}\cdot\mathbf{x}) \tag{2.164}$$

$$\tilde{G}^R(s,\mathbf{n}) = \frac{-k}{2\rho c^2}\exp(ik|\mathbf{n}\cdot\mathbf{x}|) \tag{2.165}$$

Let $\mathbf{x} = r\bar{\mathbf{x}}$ where $\bar{\mathbf{x}}$ is a unit vector along $\mathbf{x}$ direction. We can then find the inverse of $\tilde{G}^S$ as

$$\begin{aligned}
G^S(\mathbf{x}) &= \mathbf{R}^{-1}(\tilde{G}^S) = \left(\frac{-1}{8\pi^2}\right)\int_{|\mathbf{n}|=1}\tilde{G}(s,\mathbf{n})dS(\mathbf{n}) \\
&= \left(\frac{-1}{8\pi^2}\right)\left(\frac{-1}{\rho c^2}\right)\int_{S^2}\delta(\mathbf{n}\cdot\mathbf{x})dS = \frac{1}{8\pi^2\rho c^2}\int_{S^2}\delta(r\mathbf{n}\cdot\bar{\mathbf{x}})dS \\
&= \frac{1}{4\pi\rho c^2 r}
\end{aligned} \tag{2.166}$$

in the last line Eq. (2.99) is used. Similarly,

$$G^R(\mathbf{x}) = \mathbf{R}^{-1}(\tilde{G}^S) = \left(\frac{-1}{8\pi^2}\right)\left(\frac{-ik}{2\rho c^2}\right)\int_{S^2}\exp(ikr|\mathbf{n}\cdot\bar{\mathbf{x}}|)dS(\mathbf{n}) \tag{2.167}$$

Let $\mathbf{n}\cdot\bar{\mathbf{x}} = \cos\theta = t$ and $dS(\mathbf{n}) = \sin\theta d\theta d\phi = -dtd\phi$. Hence,

$$\begin{aligned}
\int_{S^2}\exp(ikr|\mathbf{n}\cdot\bar{\mathbf{x}}|)dS &= \int_{-1}^{1}\int_0^{2\pi}\exp(ikr|t|)d\phi dt \\
&= 2\int_0^1\int_0^{2\pi}\exp(ikrt)d\phi dt = \frac{4\pi}{ikr}\Big(\exp(ikr) - 1\Big)
\end{aligned} \tag{2.168}$$

Finally, we have

$$G^R(\mathbf{x}) = \frac{1}{4\pi\rho c^2}\frac{\exp(ikr) - 1}{r} \tag{2.169}$$

and hence

$$G(\mathbf{x},\omega) = G^S(\mathbf{x},\omega) + G^R(\mathbf{x},\omega) = \frac{\exp(ikr)}{4\pi\rho c^2 r} \tag{2.170}$$

Example 2.17. In this example, we solve the Green's function for an elastodynamics field under a time-harmonic point load. The general linear elastodynamics equations of motion can be written as

$$C_{ijpq}u_{p,qj} - \rho\frac{\partial^2 u_i}{\partial t^2} = -f_i \tag{2.171}$$

where C_{ijpq} are the components of linear elastic tensor, ρ is the mass density, u_i and f_i are the components of displacement and body force. Consider a time-harmonic point load applied at the origin of the coordinate in x_k direction starting at the time $t = -\infty$,

$$f_i(\mathbf{x}, t) = \delta_{ik}\delta(\mathbf{x})\exp(-i\omega t) \tag{2.172}$$

One would expect a steady state displacement field solution admits the following form,

$$u_p(\mathbf{x}, t) = G_{pk}(\mathbf{x}, \omega)\exp(-i\omega t) \tag{2.173}$$

Substitution of (2.172) and (2.173) into (2.171) and omission of the common factor $\exp(-i\omega t)$ leads to the following equations,

$$C_{ijpq}G_{pk,qj} + \rho\omega^2 G_{ik} = -\delta_{ik}\delta(\mathbf{x}) \tag{2.174}$$

Applying Fourier transform, we can obtain,

$$K_{ip}(\boldsymbol{\xi})\bar{G}_{pk}(\boldsymbol{\xi}) = \frac{\delta_{ik}}{(2\pi)^3}, \quad \text{where } K_{ip} = (\lambda + \mu)\xi_i\xi_p + \mu\delta_{ip}|\boldsymbol{\xi}|^2 - \rho\omega^2\delta_{ip} \tag{2.175}$$

We can then find the Fourier transform of the Green's function as

$$\bar{G}_{ij}(\boldsymbol{\xi}) = \frac{K^*_{ij}(\boldsymbol{\xi})}{D(\boldsymbol{\xi}, \omega)} \tag{2.176}$$

where $K^*_{ij}(\boldsymbol{\xi})$ is the co-factor of $K_{ij}(\boldsymbol{\xi})$, and

$$D(\boldsymbol{\xi}, \omega) = \det\{K_{ij}\} = (\mu|\boldsymbol{\xi}|^2 - \rho\omega^2)\Big[(\lambda + 2\mu)|\boldsymbol{\xi}|^2 - \rho\omega^2\Big] \tag{2.177}$$

In fact, we can rewrite (2.176) in an explicit expression,

$$\bar{G}_{ip}(\boldsymbol{\xi}, \omega) = \frac{\xi_i\xi_p}{|\boldsymbol{\xi}|^2(\lambda + 2\mu)(|\boldsymbol{\xi}|^2 - k_L^2)} + \frac{\delta_{ip}|\boldsymbol{\xi}|^2 - \xi_i\xi_p}{|\boldsymbol{\xi}|^2\mu(|\boldsymbol{\xi}|^2 - k_T^2)} \tag{2.178}$$

where

$$k_L = \frac{\omega}{c_L}, \quad \text{and } c_L = \sqrt{\frac{\lambda + 2\mu}{\rho}} \tag{2.179}$$

$$k_T = \frac{\omega}{c_T}, \quad \text{and } c_T = \sqrt{\frac{\mu}{\rho}} \tag{2.180}$$

k_L and k_T are wave numbers corresponding to the longitudinal and transverse wave speeds c_L and c_T. Let $r = |\mathbf{x}|$. Via inverse Fourier transform, we can obtain the spatial distribution of the Green's function under harmonic excitation,

$$G_{ij}(\mathbf{x}, \omega) = \frac{1}{4\pi\rho\omega^2 r}\left\{k_T^2\delta_{ij}\exp(ik_T r) - r\frac{\partial^2}{\partial x_i \partial x_j}\left[\frac{\exp(ik_L r) - \exp(ik_T r)}{r}\right]\right\} . \tag{2.181}$$

The above solution can be obtained by applying the Radon transformation as well. In mid-1990s, Wang and Achenbach published series papers [Wang and Achenbach (1995, 1996)], in which they used the Radon transform obtaining the Green's function of time-harmonic elastodynamics problems for general anisotropic solids in three-dimensional space.

2.8 Lattice statics Green's function (LSGF)

A powerful method in nanoscale computation is the so-called Lattice Green's Function Method.

Consider the general lattice dynamics equation under harmonic approximation,

$$M_{\boldsymbol{\ell}}\ddot{u}_i(\boldsymbol{\ell}) = -\sum_{\boldsymbol{\ell}',j} \Phi_{ij}(\boldsymbol{\ell};\boldsymbol{\ell}')u_j(\boldsymbol{\ell}') \tag{2.182}$$

Consider a two-dimensional scalar field, e.g. anti-plane displacement field ($u_1 = u_2 \equiv 0; u_3(\boldsymbol{\ell}) = u(m,n) \not\equiv 0$, $\forall \boldsymbol{\ell} = (m,n) \in \mathbb{Z}^2$. Assume the crystal has a square lattice, with lattice elastic stiffness, $k_x = k_y = k$. We can then calculate the lattice coefficients,

$$\Phi_{33}(\ell_{m,n};\ell_{m,n}) = 2(k_x + k_y) = 4k; \tag{2.183}$$
$$\Phi_{33}(\ell_{m,n};\ell_{m-1,n}) = -k_x = -k; \tag{2.184}$$
$$\Phi_{33}(\ell_{m,n};\ell_{m+1,n}) = -k_x = -k; \tag{2.185}$$
$$\Phi_{33}(\ell_{m,n};\ell_{m,n-1}) = -k_y = -k; \tag{2.186}$$
$$\Phi_{33}(\ell_{m,n};\ell_{m,n+1}) = -k_y = -k; \tag{2.187}$$

For lattice statics, the governing equation becomes,

$$\begin{aligned} &k\Big(u(m+1,n) + u(m-1,n) + u(m,n+1) + u(m,n-1) - 4u(m,n)\Big) \\ &= -f_{ext}(m,n), \quad \forall (m,n) \in \mathbb{Z}^2 \end{aligned} \tag{2.188}$$

The so-called Lattice Green's Function Method (LGFM) is to find the Green's function that satisfies the following equation,

$$\begin{aligned} &k\Big(G(m+1,n) + G(m-1,n) + G(m,n+1) + G(m,n-1) - 4G(m,n)\Big) \\ &= -\delta_{mp}\delta_{nq} \quad \text{for given } p,q \ \ \forall (m,n) \in \mathbb{Z}^2 \end{aligned} \tag{2.189}$$

such that

$$u(m,n) = \sum_{(p,q)\in\mathbb{Z}^2} G(m-p,n-q) f_{ext}(p,q) \tag{2.190}$$

Note that in (2.189) δ_{mp} and δ_{nq} are Kronecker delta (not Dirac's delta).

Consider the discrete Fourier transform,

$$\bar{G}(\boldsymbol{\xi}) = \bar{G}(\xi,\eta) = \sum_{(m,n)\in\mathbb{Z}^2} G(m,n)\exp\Big[-i(m\xi + n\eta)\Big] \tag{2.191}$$

where $\boldsymbol{\xi} = (\xi, \eta)$ and $-\pi < \xi, \eta \leq \pi$.

The inverse Fourier transform is

$$G(m,n) = \frac{1}{(2\pi)^2} \int\int_{-\pi}^{\pi} \bar{G}(\boldsymbol{\xi}) \exp\Big[i(m\xi + n\eta)\Big] d\xi d\eta \tag{2.192}$$

Considering the orthogonality condition,

$$\delta_{mp}\delta_{nq} = \frac{1}{(2\pi)^2} \int\int_{-\pi}^{\pi} \exp\Big[i\Big((m-p)\xi + (n-q)\eta\Big)\Big] d\xi d\eta \tag{2.193}$$

and substituting Eqs. (2.192) and (2.193) into (2.189), we have

$$\begin{aligned} &k\bar{G}(\xi,\eta)\Big(4 - (\exp(i\xi) + \exp(-i\xi) + \exp(i\eta) + \exp(-i\eta))\Big) \\ &= \exp(-ip\xi)\exp(-iq\eta) \end{aligned} \tag{2.194}$$

and hence,

$$\bar{G}(\xi,\eta) = \frac{\exp(-ip\xi)\exp(-iq\eta)}{\sigma(\xi,\eta)} \tag{2.195}$$

where $\sigma(\xi,\eta) = 4k(\sin^2\frac{\xi}{2} + \sin^2\frac{\eta}{2})$.

The lattice Green's function can then be obtained as

$$G(m-p, n-q) = \frac{1}{(2\pi)^2} \int\int_{-\pi}^{\pi} \frac{\exp\Big[i\Big((m-p)\xi + (n-q)\eta\Big)\Big]}{4k(\sin^2\frac{\xi}{2} + \sin^2\frac{\eta}{2})} d\xi d\eta \tag{2.196}$$

One can easily find asymptotic solution of the above integral. Let $\mathbf{n} = (p-m, q-n)$ and consider

$$\frac{1}{\sigma(\boldsymbol{\xi})} = \frac{1}{|\boldsymbol{\xi}|^2} + \frac{\xi^4 + \eta^4}{12|\boldsymbol{\xi}|^4} + \mathcal{O}(|\boldsymbol{\xi}|^2) \tag{2.197}$$

where $|\boldsymbol{\xi}|^2 = \xi^2 + \eta^2$. We have

$$\begin{aligned} G(m-p, n-q) &\approx \frac{1}{(2\pi)^2} \int\int_{I_2} \exp(-i\mathbf{n}\cdot\boldsymbol{\xi}) \left\{ \frac{1}{|\boldsymbol{\xi}|^2} + \frac{\xi^4+\eta^4}{|\boldsymbol{\xi}|^4} \right\} d\xi d\eta \\ &= -\frac{1}{2\pi}\ln z + C + \frac{(x-x')^4 + 6(x-x')^2(y-y')^2 + (y-y')^4}{24\pi z^6} \\ &=: G(\mathbf{x}-\mathbf{x}') \end{aligned} \tag{2.198}$$

where $\mathbf{x} = (x,y)$, $z = \sqrt{(x-x')^2 + (y-y')^2}$, and $C = -0.257343\cdots\cdots$. For more detailed discussions, readers may consult [Martinsson and Rodin (2002)].

2.8.1 *Maradudin's solution for the screw dislocation*

In this section, we present a classical lattice statics solution for a screw dislocation in a cubic lattice — the Maradudin's solution [Maradudin (1958)]. The lattice points (m,n) are equilibrium positions of atoms. We choose the coordinate system

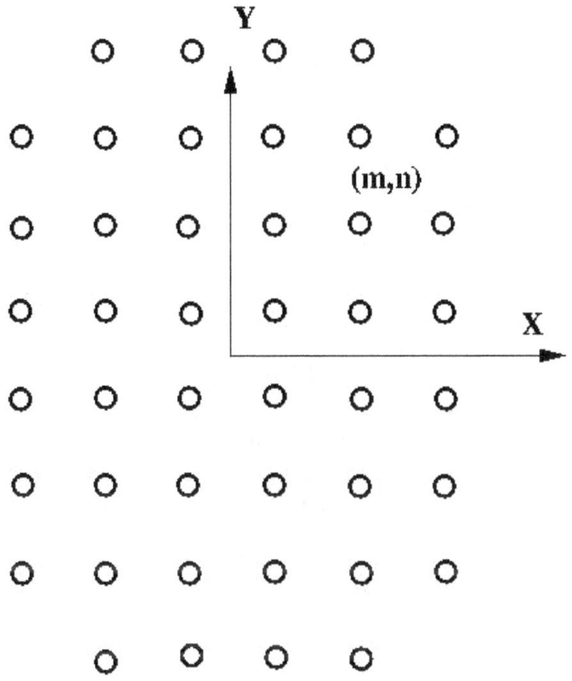

Fig. 2.8 A 2D cubic lattice with half-odd integer coordinates.

in such a way m and n are half-odd integers as shown in Fig 2.8. This configuration corresponds to the symmetric location for the core of a screw dislocation (see Chapter 8) as slip occurs on a $\{110\}$ plane in an alkali-halide crystal. Consider a screw dislocation in a lattice. Let $A = k_1$ and $B = k_2$ and $w(m,n) = u_3(m,n)$. As indicated in Eq. (eq:anti-plane1), the lattice statics equilibrium equations are

$$A(w_{m+1,n} + w_{m-1,n}) + B(w_{m,n+1} + w_{m,n-1}) - 2(A+B)w_{m,n} = 0 \qquad (2.199)$$

Assume that this displacement jump in the Z axis or the x_3 direction occurs along the positive direction of X-axis ($x > 0$) or x_1 axis. The equation (2.199) holds everywhere except at the lattice points $(k + \frac{1}{2}, \pm\frac{1}{2}$ for $k = 0, 1, 2, 3, \cdots$. In these locations, atoms are on the opposite side of the dislocation gliding plane over which there is a discontinuous displacement $\mathbf{b} = b\mathbf{e}_Z$ that we call as the Burgers' vector.

If we write the equilibrium equations involved with the atoms that are on two different sides of the discontinuity or cut, the vertical displacements for the atoms with positive Y-index should be for instance:

$$\xi_{m,\frac{1}{2}} = w_{m,\frac{1}{2}} + \frac{b}{2} \qquad (2.200)$$

where as for the atoms with negative Y-index should be for instance

$$\xi_{m,-\frac{1}{2}} = w_{m,-\frac{1}{2}} - \frac{b}{2} \qquad (2.201)$$

Therefore at the locations of these points Eq. (2.199) has to be modified as

$$A(w_{k+\frac{3}{2},\frac{1}{2}} + w_{k-\frac{1}{2},\frac{1}{2}}) + B(w_{k+\frac{1}{2},\frac{3}{2}} + w_{k+\frac{1}{2},-\frac{1}{2}}) - 2(A+B)w_{k+\frac{1}{2},\frac{1}{2}} = Bb \quad (2.202)$$

and

$$A(w_{k+\frac{3}{2},-\frac{1}{2}} + w_{k+\frac{1}{2},-\frac{1}{2}}) + B(w_{k+\frac{1}{2},\frac{1}{2}} + w_{k+\frac{1}{2},-\frac{3}{2}}) - 2(A+B)w_{k+\frac{1}{2},-\frac{1}{2}} = -Bb. \quad (2.203)$$

Combining Eqs. (2.199), (2.202), and (2.203), we can write the governing equation for a screw dislocation on a cubic lattice as

$$A(w_{m+1,n} + w_{m-1,n}) + B(w_{m,n+1} + w_{m,n-1}) - 2(A+B)w_{m,n} = \sum_{k=0}^{\infty} Bb(\delta_{m,k+\frac{1}{2}}\delta_{n,\frac{1}{2}} - \delta_{m,k+\frac{1}{2}}\delta_{n,-\frac{1}{2}}) \quad (2.204)$$

where $\delta^{'}$s are the usual Kronecker delta symbols.

To solve this problem, we are first looking for the following lattice Green's function, which satisfies the following partial difference equation,

$$A(G_{m+1,n}(k) + G_{m-1,n}(k)) + B(G_{m,n+1}(k) + G_{m,n-1}(k)) - 2(A+B)G_{m,n}(k) = (\delta_{m,k+\frac{1}{2}}\delta_{n,\frac{1}{2}} - \delta_{m,k+\frac{1}{2}}\delta_{n,-\frac{1}{2}}) \, . \quad (2.205)$$

Consider the identity,

$$\delta_{m,k+\frac{1}{2}}\delta_{n,\pm\frac{1}{2}} = \frac{1}{4\pi^2}\int_{-\pi}^{\pi}\int_{-\pi}^{\pi} \exp\{i\Big((m-k-\frac{1}{2})x + (n\mp\frac{1}{2})y\Big)\}dxdy \quad (2.206)$$

where $i = \sqrt{-1}$; and introduce an auxiliary function

$$G_{m,n}(k) = \int_{\pi}^{\pi}\int_{-\pi}^{\pi} f(x,y,k)\exp\{i(mx+ny)\} \quad (2.207)$$

Substituting Eqs. (2.206) and (2.207) into (2.204), we have the following equation

$$\int_{-\pi}^{\pi}\int_{-\pi}^{\pi} f(x,y,k)\exp i(mx+ny)\Big[2A\cos x + 2B\cos y - 2(A+B)\Big]dxdy = \frac{Bb}{4\pi^2}\int^{\pi}\int_{-\pi} \exp i\Big((m-(k+\frac{1}{2}))x + ny\Big)[2i\sin\frac{y}{2}]dxdy \quad (2.208)$$

from which, it can be found that

$$f(x,y,k) = \frac{Bb}{4\pi^2}\frac{2i\sin\frac{y}{2}\exp i(k+\frac{1}{2})x}{2(A+B) - 2(A\cos x + B\cos y)}dxdy \quad (2.209)$$

We immediately have

$$G_{m,n}(k) = \frac{1}{4\pi^2}\int_{-\pi}^{\pi}\int_{-\pi}^{\pi}\frac{2i\sin\frac{y}{2}\exp -i(k+\frac{1}{2})x\exp i(mx+ny)}{2(A+B) - 2(A\cos x + B\cos y)}dxdy \quad (2.210)$$

Based on the formula (page 480 [Knopp (1990)]),

$$2i\sum_{k}\exp\Big[i(k+\frac{1}{2})x\Big] = \frac{1}{\sin\frac{x}{2}}, \quad (2.211)$$

we have

$$w_{m,n} = \sum_{k=0}^{\infty} f_{ext}(k) G_{m,n}(k) = \sum_k BbG_{m,n}(k)$$
$$= \frac{Bb}{4\pi^2} \int_{-\pi}^{\pi} \int_{-\pi}^{\pi} \frac{\sin \frac{y}{2}}{\sin \frac{x}{2}} \frac{\exp i(mx+ny)}{2(A+B)-2(A\cos x + B\cos y)} dxdy \quad (2.212)$$

The integrand must be an even function of x and y, so that

$$w_{m,n} = -\frac{b}{\pi^2} \int_{-\pi/2}^{\pi/2} \int_{-\pi/2}^{\pi/2} \frac{\sin y}{\sin x} \frac{\sin 2mx \sin 2ny}{(A/B)\sin^2 x + \sin^2 y} dxdy \quad (2.213)$$

For our purpose, it will be more convenient to superpose a constant displacement to make the displacement along X-axis to be zero. It is accomplished by adding $b/4$ to the value of $w_{m,n}$,

$$w_{m,n} = \frac{b}{4} - \frac{b}{\pi^2} \int_{-\pi/2}^{\pi/2} \int_{-\pi/2}^{\pi/2} \frac{\sin y}{\sin x} \frac{\sin 2mx \sin 2ny}{(A/B)\sin^2 x + \sin^2 y} dxdy \,. \quad (2.214)$$

Considering asymptotic solution and neglecting the higher order terms $\mathcal{O}([Bm^2 + An^2]^{-1})$, we replace

$$\sin x \sim x, \quad \sin^2 x \sim x^2, \quad \sin y \sim y, \quad \sin^2 y \sim y^2 \quad (2.215)$$

in the integrand, so that

$$w_{m,n} \sim \frac{b}{4} - \frac{b}{\pi^2} \int_0^{\infty} \int_0^{\infty} \frac{y \sin 2ny}{y^2 + (A/B)x^2} \frac{\sin 2mx}{x} dxdy$$
$$= \frac{b}{4} - \frac{b}{2\pi} \int_0^{\infty} \exp(-2n\sqrt{A/B}x) \frac{\sin 2mx}{x} dx$$
$$= \frac{b}{4} - \frac{b}{2\pi} \tan^{-1} \sqrt{\frac{B}{A}} \frac{m}{n} \quad (2.216)$$

Using the trigonometry identity,

$$\frac{\pi}{2} - \tan^{-1} \frac{1}{x} = \tan^{-1} x \quad (2.217)$$

we finally obtain

$$w_{m,n} = \frac{b}{2\pi} \tan^{-1}\Big(\sqrt{\frac{A}{B}} \cdot \frac{n}{m}\Big) + \mathcal{O}([Bm^2 + An^2]^{-1}) \quad (2.218)$$

in which the first term agrees with the continuum limit,

$$w(x,y) = \frac{b}{2\pi} \tan^{-1}\Big(\sqrt{\frac{A}{B}} \cdot \frac{y}{x}\Big) \,. \quad (2.219)$$

2.9 Exercises

Problem 2.1. *Find the Green's function for a both end clamped Euler-Bernoulli beam, i.e.*

$$\frac{d^2}{dx^2}EI\frac{d^2G(x,y)}{dx^2} = \delta(x-y), \quad \forall x, y \in (0,\ell) \tag{2.220}$$

and

$$G(0,y) = G(\ell,y) = 0, \quad G'(0,y) = G'(\ell,y) = 0\ . \tag{2.221}$$

Problem 2.2. *For isotropic materials, elasticity tensor has the form*

$$C_{ijk\ell} = \lambda\delta_{ij}\delta_{k\ell} + \mu(\delta_{i\ell}\delta_{jk} + \delta_{ik}\delta_{j\ell}) \tag{2.222}$$

Show

(1).

$$K_{ik}(\boldsymbol{\xi}) = C_{ijk\ell}\xi_j\xi_\ell = (\lambda+\mu)\xi_i\xi_k + \mu\delta_{ik}\xi_j\xi_j \tag{2.223}$$

(2).

$$N_{ij}(\boldsymbol{\xi}) = \frac{1}{2}e_{ik\ell}e_{jmn}K_{km}K_{\ell n} = \mu\xi^2((\lambda+2\mu)\delta_{ij}\xi^2 - (\lambda+\mu)\xi_i\xi_j) \tag{2.224}$$

Hint : use $e_{ijk}e_{imn} = \delta_{jm}\delta_{kn} - \delta_{jn}\delta_{km}$.

(3).

$$D(\boldsymbol{\xi}) = \mu^2(\lambda+2\mu)\xi^6 \tag{2.225}$$

Problem 2.3. *The Green's function,* $G^\infty(\mathbf{x},\mathbf{x}')$*, satisfies the 2D Laplace equation,*

$$\nabla^2 G^\infty(\mathbf{x},\mathbf{x}') + \delta(\mathbf{x}-\mathbf{x}') = \mathbf{0}, \quad \forall \mathbf{x} \in \mathbb{R}^2 \tag{2.226}$$

where $\nabla^2 = \dfrac{\partial^2}{\partial x_1^2} + \dfrac{\partial^2}{\partial x_2^2} = \dfrac{\partial^2}{\partial x_\alpha \partial x_\alpha}$, $\alpha = 1,2$. *And* $\delta(\mathbf{x}-\mathbf{x}') = \delta(x_1 - x_1')\delta(x_2 - x_2')$. *Use Fourier transform method to derive*

$$G^\infty(\mathbf{x}-\mathbf{x}') = -\frac{1}{2\pi}\ln|\mathbf{x}-\mathbf{x}'|\ . \tag{2.227}$$

Hint: considering

$$\delta(\mathbf{x}-\mathbf{x}') = \frac{1}{(2\pi)^2}\int_{-\infty}^{\infty}\int_{-\infty}^{\infty}\exp\Big(i\boldsymbol{\xi}\cdot(\mathbf{x}-\mathbf{x}')\Big)d\boldsymbol{\xi} \tag{2.228}$$

and

$$\int_{-\infty}^{\infty}\int_{-\infty}^{\infty}\frac{\exp(i(\xi_1 x_1 + \xi_2 x_2))}{\xi_1^2+\xi_2^2}d\xi_1\xi_2 = -2\pi\ln R \tag{2.229}$$

where $R = \sqrt{(x_1-x_1')^2 + (x_2-x_2')^2}$.

Problem 2.4. *In isotropic materials, the static Green's function of linear elasticity is*

$$G_{ij}^{\infty}(\mathbf{x},\mathbf{x}') = \frac{1}{4\pi\mu}\frac{\delta_{ij}}{|\mathbf{x}-\mathbf{x}'|} - \frac{1}{16\pi\mu(1-\nu)}\frac{\partial^2}{\partial x_i \partial x_j}|\mathbf{x}-\mathbf{x}'| \tag{2.230}$$

Let $\bar{\mathbf{x}} = \mathbf{x} - \mathbf{x}'$ *and* $\bar{x} = |\bar{\mathbf{x}}| = |\mathbf{x}-\mathbf{x}'|$. *Show that for isotropic materials,*

$$C_{j\ell mn}G_{ij,\ell} = \frac{-1}{8\pi(1-\nu)}\left\{(1-2\nu)\frac{\delta_{mi}\bar{x}_n + \delta_{ni}\bar{x}_m - \delta_{mn}\bar{x}_i}{\bar{x}^3} + 3\frac{\bar{x}_m\bar{x}_n\bar{x}_i}{\bar{x}^5}\right\} \tag{2.231}$$

where ν *is the Poisson ratio, and* μ, λ *are the Lamé constants with*

$$\lambda = \frac{2\mu\nu}{1-2\nu}, \quad \mu = \frac{\lambda(1-2\nu)}{2\nu}, \quad \nu = \frac{\lambda}{2(\lambda+\mu)} \tag{2.232}$$

Hint: $C_{j\ell mn} = \lambda\delta_{j\ell}\delta_{mn} + \mu(\delta_{jm}\delta_{\ell n} + \delta_{jn}\delta_{\ell m})$.

Chapter 3

MICROMECHANICAL HOMOGENIZATION THEORY

In natural world as well as in engineering world, there is almost no perfectly homogeneous material. Continuum objects ranging from minerals, geotechnical materials, bio-organic materials, and synthetic composite materials all have rich and complex micro-structures, whose statistical behaviors determine overall material responses at macroscale.

In general, we may characterize material micro-structure as either a definite structure or a random structure. For instance, microstructure of various materials have been categorized as: (1) deterministic micro-structure with periodicity and (2) the stochastic micro-structure with random distribution. In engineering applications, some of synthetic or man-made composites have definite micro-structures, or components, such as sandwich or reticulated foam structured beams and plates. Before the micromechanics approach, there have been well-developed composite structural mechanics theories to determine the overall behaviors of composite structures, such as the classical laminated plate theory e.g. [Reddy (2003)]. However, for complex micro-structures, the simple structural mechanics approach ceases to be effective. For instance, a current favorite in composite structures for aerospace and automobile industries is the woven and braided fabric reinforced composite. This type of composites does not have clearly defined interfaces, so its strength is significantly improved. However, one may not be able to use simple structural mechanics approach to analyze their overall material properties as well as structural responses. In fact, there is a well-established mathematical theory for homogenization of periodic structures. A material with periodic micro-structures consists of many identical and repetitive unit cells. To study micro-mechanics of such materials, there has been developed a highly sophisticated and rigorous mathematical homogenization theory, e.g. [Bensoussan *et al.* (1978); Sanchez-Palencia (1980); Bakhvalov and Panasenko (1984)], which is based on mathematical perturbation theory and asymptotic theory of partial differential equations.

On the other hand, the so-called mechanical homogenization theory primarily deals with materials or media whose have random micro-structure distribution. The main focus of this book is on materials with random micro-structures, even though the approach developed here may be suitable for materials with definite or periodic

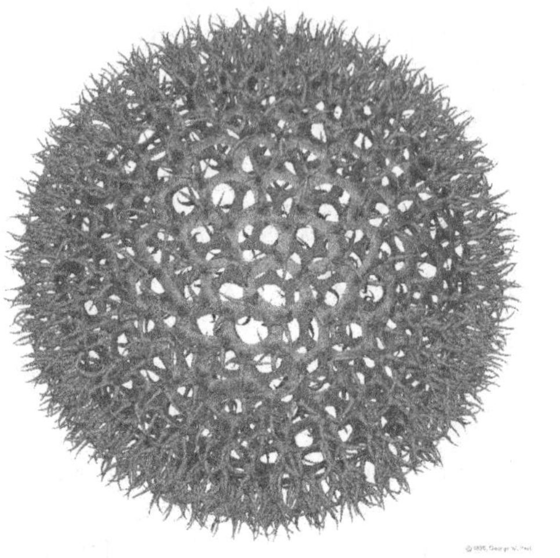

Fig. 3.1 An example of periodic reticulated structure (Photo reproduced with permission of George W. Hart, http://www.georgehart.com).

micro-structures as well.

3.1 Ergodicity principle and representative volume element(RVE)

It is generally assumed that any sample of a random heterogenous material is a realization of a specific random or stochastic process. Mathematically speaking, a realization is an event, α, that belongs to a sample space, $\mathcal{S}$. Second, an *ensemble* is the collection of all the possible realizations of a random medium generalized by a specific stochastic process.

Consider a sample space $\mathcal{S}$ over which a probability density function, $p(\alpha)$, is defined for an event $\alpha \in \mathcal{S}$. Then any particular property, f, of a composite (such as mass density, volume fraction density) is a function of α, and its *ensemble* average can defined as

$$< f >= \int_{\mathcal{S}} f(\alpha)p(\alpha)d\alpha \tag{3.1}$$

In reality, different events, α, in the sample space, $\mathcal{S}$, happen at different times. The ensemble average in (3.1) is essentially in a sense of averaging over a long time period if one keeps repeating, realizing the event, or measuring the event. Therefore,

$$< f >_T= \lim_{T\to\infty} \frac{1}{T} \int_0^T f(t)dt \tag{3.2}$$

When a system is statistically homogeneous, or when a stochastic process is homogeneous both in space and time, one can relate ensemble (time) average to the volume (spatial) average. This is because material properties in every regions of the space are similar, and hence any realization of a statistical ensemble must contain all statistical information or details as the rest of other realizations do, provided that the spatial realization space is large enough to render a stable statistical interpretation.

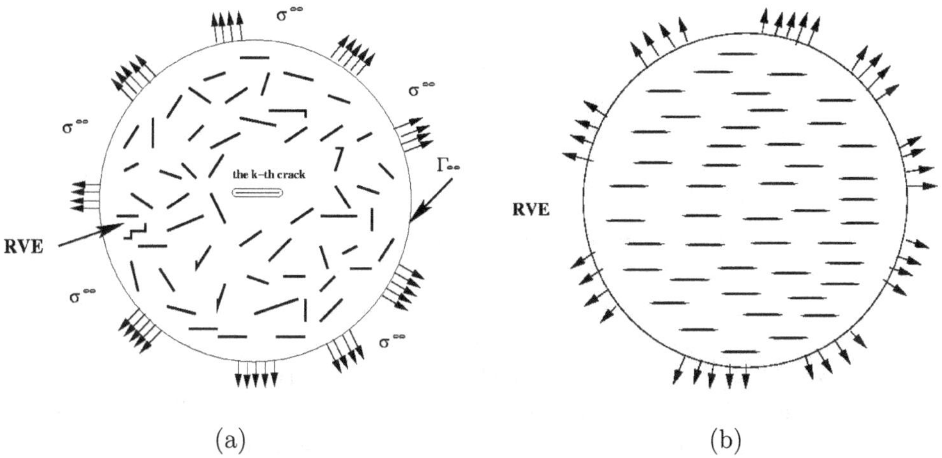

Fig. 3.2 Illustrations of macroscopically homogeneous isotropic micro-structure distribution (a), and macroscopically homogeneous anisotropic microstructure distribution (b).

3.1.1 *Ergodic principle*

The intuitive concept of ergodic principle is that the result of averaging over all realizations of the ensemble is equivalent to averaging over the volume of one realization in an infinite-volume limit. Under the ergodic assumption, the complete probabilistic information can be obtained from a single realization of an infinite domain. By letting

$$\alpha = \mathbf{y}, \;\; p(\alpha) = \frac{1}{V}, \;\; \text{and} \;\; d\alpha = dV_{\mathbf{y}}$$

the ergodic hypothesis enables us to replace ensemble averaging with volume averaging in the limit that the volume approaches to infinity, i.e.,

$$< f >_V = \lim_{V \to \infty} \frac{1}{V} \int_V f(\mathbf{y}) dV_{\mathbf{y}} = < f >_T \tag{3.3}$$

We refer to such systems as ergodic media. In practice, instead of using the infinite spatial space, we often use finite sampling spaces, which is much large than the characteristic length scale that one wishes to capture. This finite spatial sampling space is usually called as *representative volume element* (RVE).

The RVE is usually regarded as a volume of heterogeneous material that is sufficiently large to be statistically representative of the material, i.e. it contains a representative sampling of all microstructural heterogeneities that occur in the medium. This is generally the principle adopted (Kanit *et al.*, 2003). Based on this definition, an RVE must contains a large number of inhomogeneities such as inclusions, grains, voids, defects (cracks and dislocations), etc. On the other hand, it must however remain small enough to be considered as a volume element of a valid continuum mechanics description. In reality, finite-size RVE will display statistical scatter, and it may depend on scale and boundary conditions as well. Readers may find useful discussions on effect of finite size RVE, e.g., in Ostoja-Starzewski (2002).

In literature, there is another concept, the *unit cell*, that is often confused or mixed up with the concept of RVE. Unit cell is a concept that originally comes from material science, and it refers to the primitive cell in a periodic lattice structure. By translating such cell unit, one can populate the lattice structure into the whole space. On the one hand, the unit cell is a representative element of the space; on the other hand, it has a different statistical meaning from the notion of RVE, because a unit cell is not a sampling of many statistical events. A unit cell is a definite micro-structure. The same exact unit cell concept has also been used in asymptotic homogenization or asymptotic multiscale analysis in mathematics[1]. However, this unit cell concept has been extended to quasi-crystals, polycrystallines, and other non-periodic composites as the unit of some sort of "superlattices" of cellular structures. Meanwhile, engineers who study random media find the following simplistic view appears very attractive: *an isotropic random medium may be viewed as a "lattice" packed by many RVEs.*

From this perspective, Drugan and Willis (1996) proposed an alternative definition of the RVE : *It is the smallest material volume element of the composite for which the usual spatially constant overall modulus macroscopic constitutive representation is a sufficiently accurate model to represent mean constitutive response.* This definition is affinitive to the unit cell concept in asymptotic homogenization for periodic media, and it does not consider statistical fluctuations of the effective properties over the finite domains. In contrast to the first interpretation of the RVE, this definition of RVE requires a much smaller size than the previous definition, or interpretation.

In this book, we will use both interpretations of RVE. However, in most cases, when we say RVE we refer to the first interpretation, i.e. statistical interpretation. The theoretical foundation of statistical interpretation of the notion RVE is ergodic theory. Ergodicity is a mathematical term, meaning " space filling". Ergodic theory can find its origin from the work of Boltzmann in statistical physics. Ergodic theory in statistical mechanics refers to where time- and space-distribution averages are equal. The mathematical origins of ergodic theory are due to von Neumann, Birkhoff, and Koopman.

In his best-seller, *Mathematical Snapshots* (Steinhaus, 1983) , Hugo Steinhaus gave an insightful while amusing illustration of the ergodic principle as to keeping one's feet dry ("in most cases, stormy weather" are excluded) when walking along

[1]See: Chapter 11

a shoreline without having to constantly turn one's head to anticipate incoming waves.

Fig. 3.3 Instantaneous wave edges along a sandy beach.

He wrote, "*When strolling along a sandy beach in shores most people choose the wet strip left by retreating waves, which is hard and smooth enough to make the walk more comfortable than the dry part of the beach. On the other hand, to avoid their shoes and socks being soaked they must constantly watch the play the surf licking the strip. This steady twisting of the neck becomes disagreeable after a few minutes. There is, however, a remedy. Instead of looking sidewise one keeps looking straight ahead; in every instant he sees the instantaneous water edge and he directs his steps tangentially; he walks along a line touching the edge in a single point without cutting contact lies far enough away to render the variations small and easily accounted for: neither looking to the left, nor sudden jumping to the right is necessary.*

The background for the behavior I recommend here (after having tried it) is the **'ergodic principle'**: *the distribution of water tongues licking the shore in a fixed point observed during a long time is the same as the distributions shown in a fixed moment by a long portion of the water edge — the principle involved is the identity of time-distribution and space-distribution. To apply it here the walker has to limit his observation to the part of the shore he will cover in the next minute — in most cases such tactics keep him on the safe side without leading him out of the wet strip of the beach.* ······"

Some elaboration may be needed to correctly understand Steinhaus' analogy. What Steinhaus was trying to say is the following: for a set of infinitely many good weather days at a beach, if a person comes to the beach, say, every afternoon at 2:00 pm, he can measure the sea water line at a fixed spatial location, which will

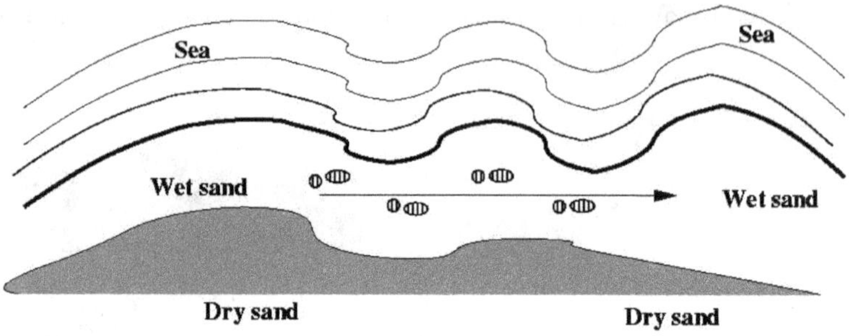

Fig. 3.4 Footprints on a sandy beach.

be a stochastic event, and all the measurements of the sea water line position on different days consists of a statistical ensemble. The ergodic principle suggests that if a system is homogeneous both in space and in time, one can then find that the average value of the sea water front position without measuring the water line on infinite many days. Instead, he can just walk along a path that is tangential to the water (shore) line on the beach, which is also assumed to be "infinite". By doing so, the average position along his path on the beach may be equal to the statistical average of the time ensemble. On the other hand, one can confidently walk along such tangential path without getting one's feet wet, because the egordic principle guarantees that it is the time average of the sea water front, there is no risk of getting one's feet wet. Note that we do not consider the surge or recede of sea water line due to the effect of tide. Hence, the person who is in charge of the measurement has to come to the beach every afternoon at the same time (e.g. 2:00 pm), provided that the weather is always good.

3.1.2 *Representative volume element*

The classical micromechanics paradigm is a two-level hierarchical mechanical structure: macro-level and micro-level, or it consists of two elements: macro-element and micro-element. At macro-level, a continuum is made of many material points, and each material point is related to a micro-space. A macro material point is also called a macro-element, or volume element. Its associated micro-space contains many micro-elements. In fact, it is a microscopic continuum. If a material is statistically homogeneous at macro-level, to study material behaviors, we only need to examine material properties at an arbitrary (typical) macro-point, and the micro-space associated with that macro-point is called the *Representative Volume Element* (RVE).

RVE is one of the fundamental concepts in classical micromechanics. An RVE for a material point of a continuum mass is a statistical ensemble of microscale objects surrounding or constituting the macro material point. This means that an RVE should contain a very large number of micro-elements such that it can be a

"Curiouser and curiouser!" cried Alice, "Now I'm opening out like the largest telescope that ever was !"

— Lewis Carroll, *Alice in Wonderland*

Fig. 3.5 Microscopic View of World: Alice in Wonderland. Illustration is credited to Sir John Tenniel (1820-1914).

statistically representative of the local continuum properties, or it is statistically stable.

In essence, the concept of representative volume element in classical micromechanics is a notion in mathematics rather than a notion in physics. It has no fixed length scale associated with each level. The length scales associated with the macro-level and micro-level are relative. If we study effective material properties of a heterogenous metal, the length scale of the micro-level maybe from a few nm to μm, and the length scale of the macro-level may be from a few millimeters to centimeters. If you study concrete aggregate, the length scale of the micro-level could be several centimeters, whereas the length scale of macro-level could be meters.

In classical mechanics, the material properties at the macro-level are assumed to be homogeneous but unknown, which is often called *overall material properties* or *effective material properties.* Whereas at the micro-level, i.e., inside the RVE, the material properties are heterogeneous but known. One of the major interests in micromechanics study is to derive the effective material properties and the physical laws at the macro-level, from the knowledge of the microstructure properties.

The methodology to find effective material properties is called *homogenization.* Homogenization is a term that has been used in many different contexts. In this book, the term "homogenization" is used as the synonym of statistical averaging. There are two major homogenization methods: mathematical homogenization and mechanical homogenization.

The first subject of continuum micromechanics is micro-elasticity. The basic premises of micro-elasticity is to assume that inside an RVE, the micro-constitutive relation of a material is elastic, and in more simplified cases, it is assumed to be linearly elastic. The concept of the RVE is used to derive the effective material properties due to micro-structures. In most simplified scenarios, the micro-structures are often studied in the absence of gravity or other types of body forces. Therefore, in micro-continuum mechanics, the body force effect is often neglected. The equilibrium equations inside an RVE is often written as

$$\nabla \cdot \boldsymbol{\sigma} = 0 \quad \Rightarrow \quad \sigma_{ij,j} = 0. \tag{3.4}$$

3.2 Average field in an RVE

Suppose that $\mathbf{T}(\mathbf{x}, \boldsymbol{X})$ is a general tensor field defined in an RVE. Note that here $\mathbf{x}$ is the position vector for a material point inside the RVE, or the coordinates for micro variables inside an RVE. Throughout this book, we always assume that at micro-level the medium that we are interested in is always heterogeneous, and as mentioned above, an RVE is an ensemble of many micro material points.

In general, an arbitrary material does not necessarily have long-range homogeneity, i.e. it may not be homogeneous at macro-level. Therefore, we may have to distinguish different RVEs at different macro-level positions. For a fixed macro-level point, we use $\mathbf{X}$ to denote the position of an RVE, or $\mathbf{X}$ is the spatial coordinate of a macro material point with respect to a macroscale coordinate. If at macro-level, material is homogeneous, i.e. material properties at macro-level do no change from place to place, $\mathbf{X}$ is often dropped out. We simply write $\mathbf{T} = \mathbf{T}(\mathbf{x})$, which means that one RVE is sufficient to represent all the macro material points in the object that is under investigation.

To associate a micro-level tensor field with a tensorial quantity at macro-level is called homogenization. To do so, we first define the average operator $< \cdot >$. The average value of a microscale tensor field $\mathbf{T}(\mathbf{x})$ of a material point ensemble at the macro point $\boldsymbol{X}$ is defined as

$$< \mathbf{T} >_{\mathbf{X}} := \frac{1}{V} \int_V \mathbf{T}(\mathbf{x}, \boldsymbol{X}) dV_{\mathbf{x}} \tag{3.5}$$

where the integration is over the volume of the RVE V with respect to $\mathbf{x}$. If the material is homogeneous at macro-level, we write

$$< \mathbf{T} > := \frac{1}{V} \int_V \mathbf{T}(\mathbf{x}) dV_{\mathbf{x}} \tag{3.6}$$

For instance, if $\mathbf{T} = \boldsymbol{\sigma}(\mathbf{x})$ is a micro-stress field, the macro-stress of a material point ensemble is denoted as $< \boldsymbol{\sigma} >$ or $< \boldsymbol{\sigma} >_V$ in emphasizing the averaging domain. Similarly, if $\mathbf{T} = \boldsymbol{\epsilon}(\mathbf{x})$ is a micro-strain field, the macro-strain of a material point ensemble is denoted as $< \boldsymbol{\epsilon} >$.

A very useful theorem that relates the average micro-Cauchy stress field to its remote value is stated as follows:

Theorem 3.1. *Suppose an RVE is subjected to a natural boundary condition and the traction on remote boundary of the RVE (∂V) is prescribed by a constant stress tensor, $\boldsymbol{\Sigma}$. That is*

$$\boldsymbol{\sigma}(\mathbf{x}) \cdot \mathbf{n} = \boldsymbol{\Sigma} \cdot \mathbf{n}, \forall \mathbf{x} \in \partial V \tag{3.7}$$

Then the average stress (or the macro stress) of the RVE equals the prescribed remote stress $\boldsymbol{\Sigma}$,

$$< \boldsymbol{\sigma} >= \boldsymbol{\Sigma} \tag{3.8}$$

We remark that here one only knows the traction distribution on the remote boundary of an RVE, but material properties and the exact stress distribution inside the RVE remain unknown.

Proof. Consider

$$\frac{\partial x_i}{\partial x_j} = \delta_{ij} \text{ and } \sigma_{ji,j} = 0 \tag{3.9}$$

One then can express Cauchy stress inside an RVE as

$$\sigma_{ij} = \sigma_{ik}\delta_{jk} = \left(\sigma_{ik}\frac{\partial x_j}{\partial x_k}\right) = (\sigma_{ik}x_j)_{,k} - \sigma_{ik,k}x_j = (\sigma_{ik}x_j)_{,k} \tag{3.10}$$

Therefore,

$$\begin{aligned} < \sigma_{ij} > &= \frac{1}{V}\int_V \sigma_{ij} dV = \frac{1}{V}\int_V \Big(\sigma_{ik}x_j\Big)_{,k} dV = \frac{1}{V}\oint_{\partial V} \sigma_{ik}x_j n_k dS \\ &= \frac{1}{V}\oint_{\partial V} \Sigma_{ik}x_j n_k dS = \frac{\Sigma_{ik}}{V}\oint_{\partial V} x_j n_k dS = \frac{\Sigma_{ik}}{V}\int_V \frac{\partial x_j}{\partial x_k} dV = \Sigma_{ij} \end{aligned} \tag{3.11}$$

We remark that in the proof the theorem one only knows the boundary data, i.e. the traction distribution on the remote boundary of an RVE, without knowing the actual material properties and the exact stress distribution inside the RVE. □

Consider a displacement field, $\boldsymbol{u}(\mathbf{x}) = u_i(\mathbf{x})\boldsymbol{e}_i$, inside an RVE. Suppose that on the remote boundary of the RVE, the displacement filed is prescribed,

$$u_i(\mathbf{x}) = u_i^0(\mathbf{x}), \quad \forall \mathbf{x} \in \partial V \tag{3.12}$$

One can find the average displacement gradient field in terms of boundary data, i.e.,

$$< u_{i,j} >_V = \frac{1}{V}\int_V u_{i,j} dV = \frac{1}{V}\int_{\partial V} n_j u_i^0 dS \tag{3.13}$$

Note that you don't know exact distribution of the displacement field inside the RVE.

Moreover, one may find the average strain and rotation fields in terms of boundary displacement data,

$$< \epsilon_{ij} >_V = \frac{1}{2}\Big(< u_{i,j} >_V + < u_{j,i} >_V \Big) = \frac{1}{2V}\oint_{\partial V} (n_j u_i^0 + n_i u_j^0) dS \quad (3.14)$$

$$< \omega_{ij} >_V = \frac{1}{2V}\oint_{\partial V} (n_j u_i^0 - n_i u_j^0) dS \quad (3.15)$$

Remark 3.1. The second term of Eq. (3.14) is the original definition of average strain in an RVE by Bishop and Hill (1951). In general, the average displacement fields of an RVE can not be expressed in terms of remote boundary data. To illustrate this, one may evaluate the average displacement field using the following trick,

$$u_i = u_k \delta_{ki} = u_k \frac{\partial x_i}{\partial x_k} = (u_k x_i)_{,k} - u_{k,k} x_i$$

Hence

$$\begin{aligned} < u_i >_V &= \frac{1}{V}\int_V u_i dV = \frac{1}{V}\int \Big((u_k x_i)_{,k} - u_{k,k} x_i \Big) dV \\ &= \frac{1}{V}\Big(\oint_{\partial V} u_k^0 x_i n_k dS - \int_V u_{k,k} x_i dV \Big) \end{aligned} \quad (3.16)$$

It is clear that $< u_i >$ can not be expressed in terms of boundary data, unless $u_{k,k} = 0$. However, for incompressible materials, such as rubber or plastic zone of ductile materials, it is often true that $u_{k,k} = 0$. Therefore,

$$< u_i >_V = \frac{1}{V}\int_V u_i dV = \frac{1}{V}\oint_{\partial V} u_k^0 x_i n_k dS \quad (3.17)$$

The average theorem for infinitesimal strain can be stated as follows:

Theorem 3.2. *Suppose that an RVE is only subjected to prescribed displacement boundary condition (essential boundary condition). On the remote surface of the RVE, its displacement field is prescribed as*

$$\mathbf{u}^0(\mathbf{x}) = \boldsymbol{\mathcal{E}} \cdot \mathbf{x}, \quad \Rightarrow \quad u_i^0 = \mathcal{E}_{ij} x_j, \quad \forall \mathbf{x} \in \partial V \quad (3.18)$$

where $\mathcal{E}_{ij}$ is a constant strain tensor. Then, the average strain field of the RVE equals the constant strain tensor, i.e.

$$< \boldsymbol{\epsilon} >_V = \boldsymbol{\mathcal{E}} \quad \Rightarrow \quad < \epsilon_{ij} >_V = \mathcal{E}_{ij} \quad (3.19)$$

Proof. By definition,

$$\begin{aligned} < \epsilon_{ij} >_V &= \frac{1}{V}\int_V \epsilon_{ij} dV = \frac{1}{2V}\int_V \Big(u_{i,j} + u_{j,i} \Big) dV = \frac{1}{2V}\oint_{\partial V} (u_i^0 n_j + u_j^0 n_i) dS \\ &= \frac{1}{2V}\oint_{\partial V} (\mathcal{E}_{ik} x_k n_j + \mathcal{E}_{jk} x_k n_i) dS = \frac{1}{2V}\int_V (\mathcal{E}_{ik}\delta_{kj} + \mathcal{E}_{jk}\delta_{ki}) dV = \mathcal{E}_{ij} \end{aligned}$$

□

It is worthy pointing out that the prescribed essential boundary condition does not necessarily generate a constant strain field inside the RVE, i.e.

$$\epsilon_{ij}(\mathbf{x}) \neq \epsilon^0_{ij} = \mathcal{E}_{ij}, \quad \forall \mathbf{x} \in V$$

In fact, the strain field can be expressed as supposition of ϵ^0_{ij} and an unknown disturbance field $\epsilon^d_{ij}(\mathbf{x})$

$$\epsilon_{ij}(\mathbf{x}) = \epsilon^0_{ij} + \epsilon^d_{ij}(\mathbf{x}), \quad \forall \mathbf{x} \in V$$

and the disturbance strain field satisfies the boundary condition such that $\epsilon^d_{ij}(\mathbf{x}) = 0, \ \forall \mathbf{x} \in \partial V$.

A very useful averaging theorem in homogenization is under the name of Hill's lemma.

Theorem 3.3 (Hill's Lemma). *For RVEs that are either subjected to the essential boundary condition,*

$$\mathbf{u}^0(\mathbf{x}) = \boldsymbol{\mathcal{E}} \cdot \mathbf{x}, \ \forall \ \mathbf{x} \in \partial V \tag{3.20}$$

or the prescribed traction boundary condition,

$$\boldsymbol{\sigma}(\mathbf{x}) \cdot \mathbf{n} = \mathbf{t}^0(\mathbf{x}) = \boldsymbol{\Sigma} \cdot \mathbf{n}, \quad \forall \mathbf{x} \in \partial V \tag{3.21}$$

the following identities about the average virtual and complementary works and the average strain energy density hold:

$$(\mathbf{1}). \quad <\boldsymbol{\sigma} : \delta\boldsymbol{\epsilon}> = \frac{1}{V}\oint_{\partial V} \boldsymbol{t} \cdot \delta\boldsymbol{u} \, dS \tag{3.22}$$

$$(\mathbf{2}). \quad <\boldsymbol{\epsilon} : \delta\boldsymbol{\sigma}> = \frac{1}{V}\oint_{\partial V} \boldsymbol{u} \cdot \delta\boldsymbol{t} \, dS \tag{3.23}$$

$$(\mathbf{3}). \quad <\boldsymbol{\sigma} : \boldsymbol{\epsilon}> - <\boldsymbol{\sigma}>:<\boldsymbol{\epsilon}>$$

$$= \frac{1}{V}\oint_{\partial V} \Big(\boldsymbol{u} - \mathbf{x}\cdot <\nabla \otimes \boldsymbol{u}>\Big) \cdot \Big(\boldsymbol{n} \cdot (\boldsymbol{\sigma} - <\boldsymbol{\sigma}>)\Big) \, dS \tag{3.24}$$

$$\text{or} \quad <\boldsymbol{\sigma}>:<\boldsymbol{\epsilon}> = \boldsymbol{\mathcal{E}} : \boldsymbol{\Sigma} \, . \tag{3.25}$$

Proof. To show Eq. (3.22), we consider $\sigma_{ij}\delta\epsilon_{ij} = \frac{1}{2}\sigma_{ij}(\delta u_{i,j} + \delta u_{j,i}) = \sigma_{ij}\delta u_{i,j}$, and then

$$\begin{aligned} \frac{1}{V}\int_V \sigma_{ij}\delta\epsilon_{ij} dV &= \frac{1}{V}\int_V \sigma_{ij}\delta u_{i,j} dV = \frac{1}{V}\int_V \Big(\sigma_{ij}\delta u_i\Big)_{,j} dV \\ &= \frac{1}{V}\oint_{\partial V} \sigma_{ij}\delta u_i n_j dS = \frac{1}{V}\oint_{\partial V} t_i \delta u_i \, dS \end{aligned} \tag{3.26}$$

where $t_i := n_j\sigma_{ji}$. Hence, Eq. (3.22) holds.

Similarly for complementary virtual work,

$$\begin{aligned} <\boldsymbol{\epsilon} : \delta\boldsymbol{\sigma}> &= \frac{1}{V}\int_V u_{i,j}\delta\sigma_{ij} dV = \frac{1}{V}\int_V \Big(u_i\delta\sigma_{ij}\Big)_{,j} dV \\ &= \frac{1}{V}\oint_{\partial V} u_i\delta\sigma_{ij} n_j dS = \frac{1}{V}\oint_{\partial V} u_i \delta t_i dS \end{aligned} \tag{3.27}$$

which is Eq. (3.23).

To show Eq. (3.25), one may start from the right hand side,

$$\begin{aligned}
&\frac{1}{V}\int_{\partial V}\Big(u_i - x_j < u_{i,j} >_V\Big)\Big(n_k(\sigma_{ki} - < \sigma_{ki} >_V)\Big)dS \\
&= \frac{1}{V}\int_{\partial V}\Big(u_i n_k \sigma_{ki} - u_i n_k < \sigma_{ki} >_V - x_j < u_{i,j} >_V n_k \sigma_{ki} \\
&+ x_j < u_{i,j} >_V n_k < \sigma_{ki} >_V\Big)dS \\
&= \frac{1}{V}\int_V \sigma_{ki} u_{i,k} dV - \Big(\frac{1}{V}\int_V u_{i,k} dV\Big) < \sigma_{ki} >_V \\
&- \delta_{jk} < u_{i,j} >_V \frac{1}{V}\int_V \sigma_{ki} dV + < \epsilon_{ij} >_V < \sigma_{ij} >_V \\
&= < \sigma_{ij}\epsilon_{ij} >_V - < \sigma_{ij} >_V < \epsilon_{ij} >_V
\end{aligned} \tag{3.28}$$

Since the RVE satisfies either the prescribed displacement boundary condition (3.20) or the prescribed traction boundary (3.21) the integrand of (3.25) will be identically zero, thus the equality $< \sigma_{ij}\epsilon_{ij} > = < \sigma_{ij} >< \epsilon_{ij} > = \Sigma_{ij}\mathcal{E}_{ij}$ holds. □

An important consequence of above Hill's lemma is the following theorem.

Theorem 3.4. *Consider solids that have the following microscale strain and stress potentials, $\phi(\boldsymbol{\epsilon})$ and $\psi(\boldsymbol{\sigma})$, such that*

$$\boldsymbol{\sigma} = \frac{\partial \phi}{\partial \boldsymbol{\epsilon}}, \quad \text{and} \quad \boldsymbol{\epsilon} = \frac{\partial \psi}{\partial \boldsymbol{\sigma}} \tag{3.29}$$

Suppose an RVE is either under the boundary condition (3.20) or under the boundary condition (3.21). The solids under consideration will have the following macroscale potentials,

$$\Phi := < \phi >, \quad \text{and} \quad \Psi = < \psi > \tag{3.30}$$

such that

$$\boldsymbol{\Sigma} = \frac{\partial \Phi}{\partial \boldsymbol{\mathcal{E}}}, \quad \text{and} \quad \boldsymbol{\mathcal{E}} = \frac{\partial \Psi}{\partial \boldsymbol{\Sigma}} \,. \tag{3.31}$$

Proof. Consider the RVE is subjected to the boundary condition (3.20). The virtual displacement field on the boundary of RVE is

$$\delta \mathbf{u} = \mathbf{x} \cdot \delta \boldsymbol{\mathcal{E}}, \quad \forall \mathbf{x} \in \partial V \tag{3.32}$$

Then the average virtual potential of the solid is,

$$\begin{aligned}
\delta \Phi &= \frac{1}{V}\int_V \Big(\frac{\partial \phi}{\partial \boldsymbol{\epsilon}} : \frac{\partial \boldsymbol{\epsilon}}{\partial \boldsymbol{\mathcal{E}}} : \delta \boldsymbol{\mathcal{E}}\Big) dV \\
&= \frac{1}{V}\int_V \boldsymbol{\sigma} : \delta \boldsymbol{\epsilon} dV = \frac{1}{V}\oint_{\partial V} \mathbf{t} \cdot \delta \mathbf{u} dS = \frac{1}{V}\oint_{\partial V}\Big((\boldsymbol{\sigma}\cdot\mathbf{n})\cdot \mathbf{x} dV\Big)\cdot \delta \boldsymbol{\mathcal{E}} \\
&= \Big(\frac{1}{V}\int_V \boldsymbol{\sigma}\cdot \mathbb{I}^{(2)} dV\Big) : \delta \boldsymbol{\mathcal{E}} = \boldsymbol{\Sigma} : \delta \boldsymbol{\mathcal{E}}
\end{aligned} \tag{3.33}$$

Comparing the above result with

$$\delta\Phi = \frac{\partial\Phi}{\partial\boldsymbol{\mathcal{E}}}\delta\boldsymbol{\mathcal{E}} \tag{3.34}$$

we have

$$\boldsymbol{\Sigma} = \frac{\partial\Phi}{\partial\boldsymbol{\mathcal{E}}}\ . \tag{3.35}$$

On the other hand, if we consider the boundary condition (3.21) and $\delta\boldsymbol{\sigma} = \delta\boldsymbol{\Sigma}, \forall\mathbf{x} \in \partial V$,

$$\begin{aligned}\delta\Psi &= \frac{1}{V}\int_V \Big(\frac{\partial\psi}{\partial\boldsymbol{\sigma}} : \frac{\partial\boldsymbol{\sigma}}{\partial\boldsymbol{\Sigma}} : \delta\boldsymbol{\Sigma}\Big)dV \\ &= \frac{1}{V}\int_V \boldsymbol{\epsilon} : \delta\boldsymbol{\sigma} dV = \frac{1}{V}\oint_{\partial V} \mathbf{u}\cdot\mathbf{n}\cdot\delta\boldsymbol{\Sigma} dS = \frac{1}{V}\oint_{\partial V}\Big(\nabla\times\mathbf{u}\Big)dS : \delta\boldsymbol{\Sigma} \\ &= \Big(\frac{1}{V}\int_V \boldsymbol{\epsilon} dV\Big) : \delta\Sigma = \boldsymbol{\mathcal{E}} : \delta\boldsymbol{\Sigma}\end{aligned} \tag{3.36}$$

Comparing the above expression with

$$\delta\boldsymbol{\mathcal{E}} = \frac{\partial\Psi}{\partial\boldsymbol{\Sigma}} : \delta\boldsymbol{\Sigma}$$

we have the desired result

$$\boldsymbol{\mathcal{E}} = \frac{\partial\Psi}{\partial\boldsymbol{\Sigma}}\ . \tag{3.37}$$

The expression (3.35) was first proved in Hutchinson (1987b) ; and the expression (3.37) was first proved in Nemat-Nasser and Hori (1990) . □

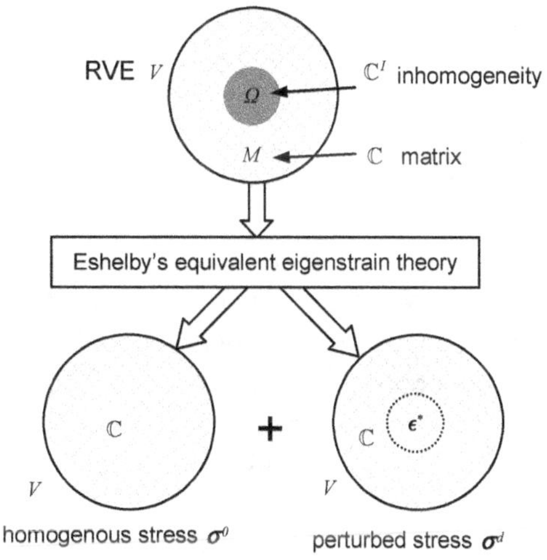

Fig. 3.6 Illustration of Eshelby's equivalent eigenstrain principle.

3.3 Definition of eigenstrain, eigenstress, and inclusion

"Eigenstrain" is a generic name to describe a transformation strain field that can equivalently represent induced strain due to misfit of inhomogeneities, thermal expansion, plastic strain, residual strain, phase transformation, etc., all of which, when homogeneously applied, produce a compatible deformation field without generating stresses. The German word "*eigen*" means characteristic. It is believed that any strain field generated by an inhomogeneity distribution may have a one-to-one correspondence to a fictitious eigenstrain field, which is characteristically equivalent (in the sense of mechanical variables, such as stress, strain, and displacements) to the induced strain field generated by the inhomogeneity distribution. On the other hand, "Eigenstress" is a generic name given to self-equilibrated transformation stress (internal) field that can generate equivalent perturbed stress and strain distributions caused by one or several of eigenstrains in bodies which are free from any other external forces and surface constraints. In a nutshell, the eigenstress field is created by the incompatibility of the eigenstrain field.

The term *inclusion* denotes a subdomain in the matrix subjected to transformation strains (eigenstrains), while the inhomogeneity is a subdomain with properties distinct from those from the matrix.

3.4 Eshelby's equivalent eigenstrain method

Eshelby's equivalent eigenstrain principle is a homogenization method. It establishes the equivalency between an eigenstrain (or eigenstress) field and an inhomogeneity distribution, such that distribution of inhomogeneities may be replaced by the eigenstrain field with the equivalent mechanical effect. This equivalency mapping process translates the heterogeneity of material into an added non-uniform strain distribution, while making the material properties become homogeneous again.

Let's consider an elastic solid, V, with elasticity tensor, $\mathbb{C}$, and compliance tensor, $\mathbb{D}$. Inside the elastic solid, there is an inhomogeneity, a subdomain Ω, with different elastic constants, $\mathbb{C}^I$ and $\mathbb{D}^I$ (Fig. 3.6). The Eshelby's equivalent eigenstrain principle replace the initial heterogeneous body with a homogeneous body, within which an eigenstrain field is prescribed, such that the strain and stress fields in the homogenized field are mechanically equivalent to those of the original inhomogeneous field.

(1)Traction boundary condition Consider that the original inhomogeneous solid is subjected to a traction boundary condition, $\boldsymbol{t} = \boldsymbol{n} \cdot \boldsymbol{\sigma}^0$. The presence of inhomogeneity will produce stress and strain field perturbation,

$$\boldsymbol{\sigma}(\mathbf{x}) = \boldsymbol{\sigma}^0 + \boldsymbol{\sigma}^d(\mathbf{x}), \quad \boldsymbol{\epsilon}(\mathbf{x}) = \boldsymbol{\epsilon}^0 + \boldsymbol{\epsilon}^d(\mathbf{x}) .$$

where the constant stress and strain are related through $\boldsymbol{\sigma}^0 = \mathbb{C} : \boldsymbol{\epsilon}^0$ and $\boldsymbol{\epsilon}^0 = \mathbb{D} : \boldsymbol{\sigma}^0$. The stress and strain distributions inside the inhomogeneous solid are

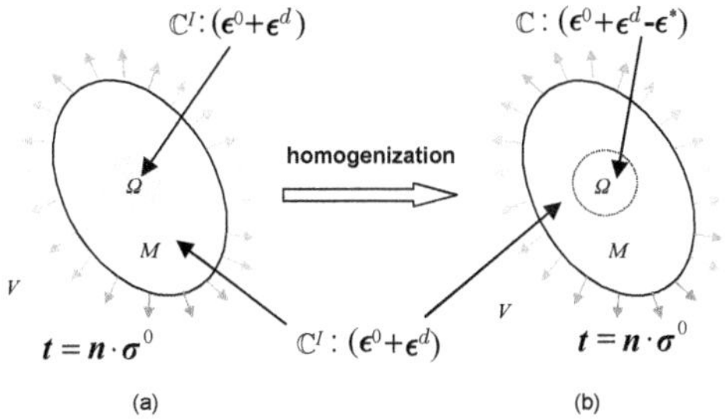

Fig. 3.7 Illustration of Eshelby's equivalent eigenstrain principle. (a) initial heterogeneous body, (b) equivalent homogeneous body ($V = \Omega \cup M$).

$$\boldsymbol{\sigma}(\mathbf{x}) = \begin{cases} \mathbb{C} : (\boldsymbol{\epsilon}^0 + \boldsymbol{\epsilon}^d(\mathbf{x})), & \mathbf{x} \in M \\ \mathbb{C}^I : (\boldsymbol{\epsilon}^0 + \boldsymbol{\epsilon}^d(\mathbf{x})), & \mathbf{x} \in \Omega \end{cases}$$

$$\boldsymbol{\epsilon}(\mathbf{x}) = \begin{cases} \mathbb{D} : (\boldsymbol{\sigma}^0 + \boldsymbol{\sigma}^d(\mathbf{x})), & \mathbf{x} \in M \\ \mathbb{D}^I : (\boldsymbol{\sigma}^0 + \boldsymbol{\sigma}^d(\mathbf{x})), & \mathbf{x} \in \Omega \end{cases} \tag{3.38}$$

The equivalent eigenstrain procedure is depicted in Fig. 3.7. When applying Eshelby's equivalent eigenstrain homogenization method, one has to choose a suitable eigenstrain strain field, usually in the following form to mimic the material misfit, or other defect or inhomogeneity misfit,

$$\boldsymbol{\epsilon}^*(\mathbf{x}) = \begin{cases} 0, & \forall \mathbf{x} \in M \\ \boldsymbol{\epsilon}^*, & \forall \mathbf{x} \in \Omega \end{cases} \tag{3.39}$$

By superposing the eigenstrain with the actual strain field, $\boldsymbol{\epsilon}(\mathbf{x}) = \boldsymbol{\epsilon}^0 + \boldsymbol{\epsilon}^d(\mathbf{x})$, we hope to achieve an equivalency that the total stress field of homogeneous comparison solid is equivalent to the total stress field of the original inhomogeneous solid, i.e.

$$\begin{aligned} \boldsymbol{\sigma}(\mathbf{x}) &= \mathbb{C} : (\boldsymbol{\epsilon}(\mathbf{x}) - \boldsymbol{\epsilon}^*(\mathbf{x})) \\ &= \begin{cases} \mathbb{C} : (\boldsymbol{\epsilon}^0 + \boldsymbol{\epsilon}^d(\mathbf{x})), & \mathbf{x} \in M \\ \mathbb{C} : (\boldsymbol{\epsilon}^0 + \boldsymbol{\epsilon}^d(\mathbf{x}) - \boldsymbol{\epsilon}^*), & \mathbf{x} \in \Omega \end{cases} \\ &= \begin{cases} \mathbb{C} : (\boldsymbol{\epsilon}^0 + \boldsymbol{\epsilon}^d(\mathbf{x})), & \mathbf{x} \in M \\ \mathbb{C}^I : (\boldsymbol{\epsilon}^0 + \boldsymbol{\epsilon}^d(\mathbf{x})), & \mathbf{x} \in \Omega \end{cases} \end{aligned} \tag{3.40}$$

Under the chosen traction boundary condition, the average stress field $< \boldsymbol{\sigma} >= \boldsymbol{\sigma}^0$, but the average strain field $< \boldsymbol{\epsilon} > \neq \boldsymbol{\epsilon}^0$.

From (3.40), one may derive that

$$\boldsymbol{\sigma}^d(\mathbf{x}) = \mathbb{C} : (\boldsymbol{\epsilon}^d(\mathbf{x}) - \boldsymbol{\epsilon}^*(\mathbf{x})), \quad \forall \mathbf{x} \in V \tag{3.41}$$

$$\mathbb{C}^I(\boldsymbol{\epsilon}^0 + \boldsymbol{\epsilon}^d) = \mathbb{C} : (\boldsymbol{\epsilon}^0 + \boldsymbol{\epsilon}^d - \boldsymbol{\epsilon}^*), \quad \forall \mathbf{x} \in \Omega \tag{3.42}$$

where Eq.(3.42) is called "*stress consistency condition*". It is the criterion for choosing suitable eigenstrain field. Note that $\boldsymbol{\epsilon}^0 + \boldsymbol{\epsilon}^d(\mathbf{x}) - \boldsymbol{\epsilon}^*(\mathbf{x})$ is the total *elastic* strain.

Alternatively, Eqs. (3.40) and (3.41) can be recast into following forms,

$$\boldsymbol{\sigma}(\mathbf{x}) = \mathbb{C} : (\boldsymbol{\epsilon}(\mathbf{x}) - \boldsymbol{\epsilon}^*(\mathbf{x})) \;\; \Rightarrow \;\; \boldsymbol{\epsilon}(\mathbf{x}) = \mathbb{D} : \boldsymbol{\sigma}(\mathbf{x}) + \boldsymbol{\epsilon}^*(\mathbf{x}) \tag{3.43}$$

$$\boldsymbol{\sigma}^d(\mathbf{x}) = \mathbb{C} : (\boldsymbol{\epsilon}^d(\mathbf{x}) - \boldsymbol{\epsilon}^*(\mathbf{x})) \;\; \Rightarrow \;\; \boldsymbol{\epsilon}^d(\mathbf{x}) = \mathbb{D} : \boldsymbol{\sigma}^d(\mathbf{x}) + \boldsymbol{\epsilon}^*(\mathbf{x}) \tag{3.44}$$

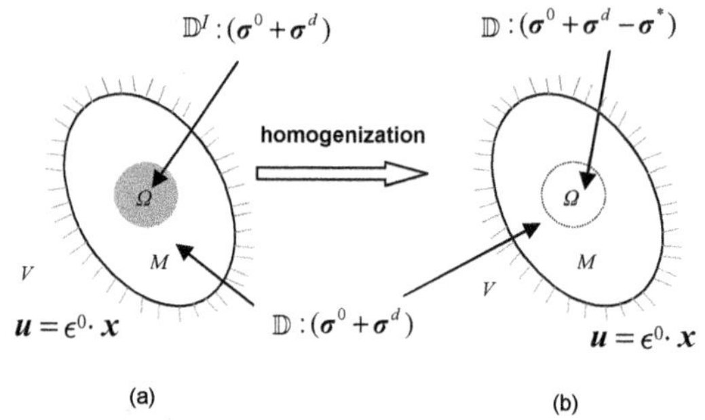

Fig. 3.8 Illustration of Eshelby's equivalent eigenstress principle: (a) Initial heterogeneous body, (b)equivalent homogeneous body ($V = \Omega \cup M$).

(2) Displacement boundary condition Consider the same inhomogeneous solid and following displacement boundary condition

$$\boldsymbol{u}(\mathbf{x}) = \boldsymbol{\epsilon}^0 \cdot \mathbf{x}, \;\; \forall \mathbf{x} \in \partial V \tag{3.45}$$

The inhomogeneity inside the solid will generate a disturbance stress field, $\boldsymbol{\sigma}^d(\mathbf{x})$. The total strain field is

$$\boldsymbol{\epsilon}(\mathbf{x}) = \begin{cases} \mathbb{D} : (\boldsymbol{\sigma}^0 + \boldsymbol{\sigma}^d(\mathbf{x})), & \forall \mathbf{x} \in M \\ \mathbb{D}^I : (\boldsymbol{\sigma}^0 + \boldsymbol{\sigma}^d(\mathbf{x})), & \forall \mathbf{x} \in \Omega \end{cases} \tag{3.46}$$

As proved in previous section, under prescribed boundary condition, the average strain $< \boldsymbol{\epsilon} >= \boldsymbol{\epsilon}^0$. On the other hand, the average stress $< \boldsymbol{\sigma} > \neq \boldsymbol{\sigma}^0$.

To homogenize the heterogeneous medium, we introduce the following eigenstress distribution,

$$\boldsymbol{\sigma}^*(\mathbf{x}) = \begin{cases} 0, & \forall \mathbf{x} \in M \\ \boldsymbol{\sigma}^*, & \forall \mathbf{x} \in \Omega \end{cases} \tag{3.47}$$

such that

$$\boldsymbol{\epsilon}(\mathbf{x}) = \begin{cases} \mathbb{D} : (\boldsymbol{\sigma}^0 + \boldsymbol{\sigma}^d(\mathbf{x})) \\ \mathbb{D} : (\boldsymbol{\sigma}^0 + \boldsymbol{\sigma}^d(\mathbf{x}) - \boldsymbol{\sigma}^*) \end{cases} = \begin{cases} \mathbb{D} : (\boldsymbol{\sigma}^0 + \boldsymbol{\sigma}^d(\mathbf{x})), & \mathbf{x} \in M \\ \mathbb{D}^I : (\boldsymbol{\sigma}^0 + \boldsymbol{\sigma}^d(\mathbf{x})), & \mathbf{x} \in \Omega \end{cases} \tag{3.48}$$

From Eq.(3.48), we can derive that

$$\boldsymbol{\epsilon}^d(\mathbf{x}) = \mathbb{D} : (\boldsymbol{\sigma}^d(\mathbf{x}) - \boldsymbol{\sigma}^*(\mathbf{x})), \;\; \forall \mathbf{x} \in V \tag{3.49}$$

$$\mathbb{D}^I(\boldsymbol{\sigma}^0 + \boldsymbol{\sigma}^d(\mathbf{x})) = \mathbb{D} : (\boldsymbol{\sigma}^0 + \boldsymbol{\sigma}^d - \boldsymbol{\sigma}^*), \;\; \forall \mathbf{x} \in \Omega \tag{3.50}$$

where Eq.(3.50) is called *"strain consistency condition."*

Alternatively,

$$\boldsymbol{\epsilon}^d(\mathbf{x}) = \mathbb{D} : (\boldsymbol{\sigma}^d(\mathbf{x}) - \boldsymbol{\sigma}^*(\mathbf{x})) \quad \Rightarrow \quad \boldsymbol{\sigma}^d(\mathbf{x}) = \mathbb{C} : \boldsymbol{\epsilon}^d(\mathbf{x}) + \boldsymbol{\sigma}^*(\mathbf{x}) \tag{3.51}$$

Comparing Eq.(3.51) with (3.44) yield the following identities,

$$\boldsymbol{\epsilon}^*(\mathbf{x}) + \mathbb{D} : \boldsymbol{\sigma}^*(\mathbf{x}) = 0, \quad \text{or} \quad \boldsymbol{\sigma}^*(\mathbf{x}) + \mathbb{C} : \boldsymbol{\epsilon}^*(\mathbf{x}) = 0 \tag{3.52}$$

which provides connection between eigenstrain and eigenstress.

3.5 Fundamental equations of micro-elasticity

There are mainly two homogenization methods used in engineering applications today. The first one is the Eshelby's eigenstrain theory, which has been mainly used for materials with random micro-structures. The other is the mathematical homogenization method used for materials with periodic micro-structures. The central part of the theory is Eshelby's eigenstrain solution for ellipsoidal inclusion. Eshelby's theory has been further refined, detailed, and applied to many engineering problems by Toshia Mura [Mura (1987)] and his co-workers. It is fair to call it as Eshelby-Mura eigenstrain theory, and it has had widespread applications.

Consider equilibrium equation in an RVE

$$\sigma_{ji,j} = 0 \tag{3.53}$$

After homogenization, inhomogeneities are replaced by an eigenstrain distribution $\epsilon^*_{ij}(\mathbf{x})$. Assuming that material is linear elastic, and the total strain is the sum of elastic strain and eigenstrain,

$$\epsilon_{ij} = e_{ij} + \epsilon^*_{ij} \tag{3.54}$$

The total strain is defined as $\epsilon_{ij} = \dfrac{1}{2}(u_{i,j} + u_{j,i})$. And elastic strain is related with Cauchy stress by Hooke's law

$$\sigma_{ij} = C_{ijk\ell}(\epsilon_{k\ell} - \epsilon^*_{k\ell}) = C_{ijk\ell}(u_{k,\ell} - \epsilon^*_{k\ell}) \tag{3.55}$$

The equilibrium equation then takes the form

$$C_{ijk\ell}u_{k,\ell j} - C_{ijk\ell}\epsilon^*_{k\ell,j} = 0 \tag{3.56}$$

We can interpret the effect of eigenstrain distribution as a type of body force, $f_i = -C_{ijk\ell}\epsilon^*_{k\ell,j}$, and the original equilibrium equation has the familiar form $\sigma_{ji,j} + f_i = 0$.

In the following section, we will show two methods to establish the connection between the (disturbed) displacement/strain/stress field and prescribed eigenstrain field.

3.5.1 *Method of Fourier transform*

Using inverse Fourier transform,

$$u_k(\mathbf{x}) = \int_{\mathbb{R}^3} \bar{u}_k(\boldsymbol{\xi}) \exp(i\boldsymbol{\xi}\cdot\mathbf{x})d\boldsymbol{\xi} = \int_{\mathbb{R}^3} \bar{u}_k(\boldsymbol{\xi}) \exp(i\xi_m x_m)d\boldsymbol{\xi} \tag{3.57}$$

$$\epsilon^*_{k\ell}(\mathbf{x}) = \int_{\mathbb{R}^3} \bar{\epsilon}^*_{k\ell}(\boldsymbol{\xi}) \exp(i\xi_m x_m)d\boldsymbol{\xi} \tag{3.58}$$

one can find that

$$u_{k,\ell j}(\mathbf{x}) = -\int_{\mathbb{R}^3} \bar{u}_k(\boldsymbol{\xi})\xi_\ell\xi_j \exp(i\xi_m x_m)d\boldsymbol{\xi} \tag{3.59}$$

$$\epsilon^*_{k\ell,j}(\mathbf{x}) = i\int_{\mathbb{R}^3} \bar{\epsilon}^*_{k\ell}(\boldsymbol{\xi})\xi_j \exp(i\xi_m x_m)d\boldsymbol{\xi} \tag{3.60}$$

Substituting (3.59) and (3.60) into (3.56) yields

$$\int_{\mathbb{R}^3} (C_{ijk\ell}\bar{u}_k\xi_\ell\xi_j + iC_{ijk\ell}\bar{\epsilon}^*_{k\ell}(\boldsymbol{\xi})\xi_j) \exp(i\xi_m x_m)d\mathbf{x} = 0 \tag{3.61}$$

which leads to

$$C_{ijk\ell}\xi_j\xi_\ell\bar{u}_k = -iC_{ijk\ell}\bar{\epsilon}^*_{k\ell}(\boldsymbol{\xi})\xi_j \tag{3.62}$$

Denote

$$K_{ik}(\boldsymbol{\xi}) = C_{ijk\ell}\xi_j\xi_\ell \tag{3.63}$$

$$\bar{f}_i = -iC_{ijk\ell}\bar{\epsilon}^*_{k\ell}\xi_j \tag{3.64}$$

The system of equations can be written in matrix form as

$$\begin{bmatrix} K_{11} & K_{12} & K_{13} \\ K_{21} & K_{22} & K_{23} \\ K_{31} & K_{32} & K_{33} \end{bmatrix} \begin{pmatrix} \bar{u}_1 \\ \bar{u}_2 \\ \bar{u}_3 \end{pmatrix} = \begin{pmatrix} \bar{f}_1 \\ \bar{f}_2 \\ \bar{f}_3 \end{pmatrix} \tag{3.65}$$

whose solution can be written as,

$$\bar{u}_i(\boldsymbol{\xi}) = K^{-1}_{ij}\bar{f}_j \tag{3.66}$$

Considering

$$\bar{u}_i(\boldsymbol{\xi}) = K^{-1}_{ij}(\boldsymbol{\xi})\bar{f}_j = -iC_{j\ell mn}\bar{\epsilon}^*_{mn}\xi_\ell K^{-1}_{ij}(\boldsymbol{\xi}) \tag{3.67}$$

and applying inverse Fourier transform, we have

$$u_i(\mathbf{x}) = -i\int_{\mathbb{R}^3} C_{j\ell mn}\bar{\epsilon}^*_{mn}(\boldsymbol{\xi})\xi_\ell K^{-1}_{ij}(\boldsymbol{\xi}) \exp(i\boldsymbol{\xi}\cdot\mathbf{x})d\boldsymbol{\xi} \tag{3.68}$$

Substituting

$$\bar{\epsilon}^*_{mn}(\boldsymbol{\xi}) = \frac{1}{(2\pi)^3}\int_{\mathbb{R}^3} \epsilon^*_{mn}(\mathbf{x}') \exp(-i\boldsymbol{\xi}\cdot\mathbf{x}')d\mathbf{x}' \tag{3.69}$$

into (3.68), we obtain *the fundamental formula of 3D micro-elasticity,*

$$u_i(\mathbf{x}) = -\frac{i}{(2\pi)^3}\int_{\mathbb{R}^3}\int_{\mathbb{R}^3} C_{j\ell mn}\epsilon^*_{mn}(\mathbf{x}')\xi_\ell K^{-1}_{ij}(\boldsymbol{\xi}) \exp(i\boldsymbol{\xi}\cdot(\mathbf{x}-\mathbf{x}'))d\mathbf{x}'d\boldsymbol{\xi} \tag{3.70}$$

The above derivation is valid for general anisotropic materials. For isotropic materials, the formulation could be simplified, and it is easy to evaluate

$$\begin{aligned}\mathbb{K}(\boldsymbol{\xi}) &= \boldsymbol{\xi}\cdot\mathbb{C}\cdot\boldsymbol{\xi} = \boldsymbol{\xi}\cdot\left\{\lambda\mathbb{I}^{(2)}\otimes\mathbb{I}^{(2)} + 2\mu\mathbb{I}^{(4s)}\right\}\cdot\boldsymbol{\xi}\\ &= \lambda\boldsymbol{\xi}\otimes\boldsymbol{\xi} + \mu\Big(\boldsymbol{\xi}\otimes\boldsymbol{\xi} + |\boldsymbol{\xi}|^2\mathbb{I}^{(2)}\Big)\\ &= (\lambda+\mu)\boldsymbol{\xi}\otimes\boldsymbol{\xi} + \mu|\boldsymbol{\xi}|^2\mathbb{I}^{(2)}\end{aligned} \tag{3.71}$$

where $\boldsymbol{\xi} = \xi_i\mathbf{e}_i$, and thus $\mathbb{K}^{-1}(\boldsymbol{\xi})$ must also be an isotropic second order tensor in Fourier space as well. Assume that

$$\mathbb{K}^{-1}(\boldsymbol{\xi}) = [\boldsymbol{\xi}\cdot\mathbb{C}\cdot\boldsymbol{\xi}]^{-1} = A\boldsymbol{\xi}\otimes\boldsymbol{\xi} + B\mathbb{I}^{(2)} \tag{3.72}$$

then

$$\Big[(\lambda+\mu)\boldsymbol{\xi}\otimes\boldsymbol{\xi} + \mu|\boldsymbol{\xi}|^2\mathbb{I}^{(2)}\Big]\cdot\Big[A\boldsymbol{\xi}\otimes\boldsymbol{\xi} + B\mathbb{I}^{(2)}\Big] = \mathbb{I}^{(2)} \tag{3.73}$$

subsequently,

$$\Big[A(\lambda+2\mu)|\boldsymbol{\xi}|^2 + B(\lambda+\mu)\Big]\boldsymbol{\xi}\otimes\boldsymbol{\xi} + B\mu|\boldsymbol{\xi}|^2\mathbb{I}^{(2)} = \mathbb{I}^{(2)} \tag{3.74}$$

One can then determine the constants A and B,

$$A = -\frac{(\lambda+\mu)}{\mu(\lambda+2\mu)|\boldsymbol{\xi}|^4}, \quad \text{and} \quad B = \frac{1}{\mu|\boldsymbol{\xi}|^2} \tag{3.75}$$

Hence,

$$\mathbb{K}^{-1}(\boldsymbol{\xi}) = \Big(\boldsymbol{\xi}\cdot\mathbb{C}\cdot\boldsymbol{\xi}\Big)^{-1} = \frac{|\boldsymbol{\xi}|^{-2}}{\mu}\Big\{-\frac{(\lambda+\mu)}{(\lambda+2\mu)|\boldsymbol{\xi}|^2}\boldsymbol{\xi}\otimes\boldsymbol{\xi} + \mathbb{I}^{(2)}\Big\} \tag{3.76}$$

or in component form,

$$K_{ij}^{-1}(\boldsymbol{\xi}) = \frac{1}{\mu|\boldsymbol{\xi}|^2}\Big\{\delta_{ij} - \frac{(\lambda+\mu)}{(\lambda+2\mu)|\boldsymbol{\xi}|^2}\xi_i\xi_j\Big\} \tag{3.77}$$

3.5.2 *Method of Green's function*

Consider Green's function

$$G_{ij}^{\infty}(\mathbf{x}-\mathbf{y}) = \frac{1}{(2\pi)^3}\int_{\mathbb{R}^3} K_{ij}^{-1}(\boldsymbol{\xi})\exp(i\boldsymbol{\xi}\cdot(\mathbf{x}-\mathbf{y}))d\boldsymbol{\xi} \tag{3.78}$$

Taking derivative of Green's function with respect to $\mathbf{x}$

$$G_{ij,\ell}^{\infty}(\mathbf{x}-\mathbf{y}) = \int_{\mathbb{R}^3}\left[\frac{i}{(2\pi)^3}K_{ij}^{-1}(\boldsymbol{\xi})\xi_\ell\right]\exp(i\boldsymbol{\xi}\cdot(\mathbf{x}-\mathbf{y}))d\boldsymbol{\xi} \tag{3.79}$$

$$= \int_{\mathbb{R}^3}\overline{G_{ij,\ell}^{\infty}}(\boldsymbol{\xi})\exp(i\boldsymbol{\xi}\cdot(\mathbf{x}-\mathbf{y}))d\boldsymbol{\xi} \tag{3.80}$$

It is easy to find that Eq. (3.68) could be rewritten as:

$$\begin{aligned}u_i(\mathbf{x}) &= -i\int_{\mathbb{R}^3} C_{j\ell mn}\overline{\epsilon}_{mn}^*(\boldsymbol{\xi})\xi_\ell K_{ij}^{-1}(\boldsymbol{\xi})\exp(i\boldsymbol{\xi}\cdot\mathbf{x})d\boldsymbol{\xi}\\ &= -(2\pi)^3\int_{\mathbb{R}^3} C_{j\ell mn}\overline{\epsilon}_{mn}^*\overline{G_{ij,\ell}^{\infty}}(\boldsymbol{\xi})\exp(i\boldsymbol{\xi}\cdot\mathbf{x})d\boldsymbol{\xi}\end{aligned} \tag{3.81}$$

Recalling Convolution Theorem in Chapter 2, we arrive at the following integral equation to connect displacement field and eigenstrain through Green's function,

$$u_i(\mathbf{x}) = -\int_{\mathbb{R}^3} C_{j\ell mn}\epsilon^*_{mn}(\mathbf{y})G^\infty_{ij,\ell}(\mathbf{x}-\mathbf{y})d\mathbf{y} \tag{3.82}$$

The corresponding expressions for stress and strain are

$$\epsilon_{ij}(\mathbf{x}) = -\frac{1}{2}\int_{\mathbb{R}^3} C_{k\ell mn}\epsilon^*_{mn}(\mathbf{y})\{G^\infty_{ik,\ell j}(\mathbf{x}-\mathbf{y}) + G^\infty_{jk,\ell i}(\mathbf{x}-\mathbf{y})\}dV_{\mathbf{y}} \tag{3.83}$$

$$\sigma_{ij}(\mathbf{x}) = -C_{ijk\ell}\Big\{\int_{\mathbb{R}^3} C_{pqmn}\epsilon^*_{mn}(\mathbf{y})G^\infty_{kp,q\ell}(\mathbf{x}-\mathbf{y})d\mathbf{y} + \epsilon^*_{k\ell}(\mathbf{x})\Big\} \tag{3.84}$$

We emphasize that all the partial derivatives are made with respect to $\mathbf{x}$. If eigenstrain ϵ^*_{mn} is constant inside a domain, say Ω, Eq. (3.83) can be further simplified as an integral in Ω,

$$\epsilon_{ij}(\mathbf{x}) = -\frac{\epsilon^*_{mn}}{2}\int_{\Omega} C_{k\ell mn}\Big(G^\infty_{ik,\ell j}(\mathbf{x}-\mathbf{y}) + G^\infty_{jk,\ell i}(\mathbf{x}-\mathbf{y})\Big)dV_{\mathbf{y}} \tag{3.85}$$

A slight different but related approach is to use the Radon transform. Recall that we can interpret the effect of eigenstrain distribution as a type of body force, such that $f_i = -C_{ijk\ell}\epsilon^*_{k\ell,j}$. Therefore, using Green's function, the displacement field can be determined by the following convolution, or superposition,

$$u_i(\mathbf{x}) = \int_{\mathbb{R}^3} G^\infty_{ij}(\mathbf{x}-\mathbf{y})f_j(\mathbf{y})dV_{\mathbf{y}} \tag{3.86}$$

Recall Eq.(2.153) and write

$$G^\infty_{ij}(\mathbf{x}-\mathbf{y}) = \frac{1}{8\pi^2}\int_{S^2}\delta\Big((\mathbf{x}-\mathbf{y})\cdot\boldsymbol{\xi}\Big)K^{-1}_{ij}(\boldsymbol{\xi})dS(\boldsymbol{\xi}) \tag{3.87}$$

where $\boldsymbol{\xi}$ is the unit vector normal to unit sphere S^2 (we used $\bar{\boldsymbol{\xi}}$ before in (2.153)). Substituting (3.87) into (3.86) yields another form of the fundamental formula of micro-elasticity,

$$\begin{aligned} u_i(\mathbf{x}) &= \frac{1}{8\pi^2}\int_{\mathbb{R}^3}\Big(\int_{S^2}\delta\Big((\mathbf{x}-\mathbf{y})\cdot\boldsymbol{\xi}\Big)K^{-1}_{ij}(\boldsymbol{\xi})dS(\boldsymbol{\xi})\Big)f_j(\mathbf{y})dV_{\mathbf{y}} \\ &= \frac{1}{8\pi^2}\int_{S^2}K^{-1}_{ij}(\boldsymbol{\xi})\Big[\int_{\mathbb{R}^3} f_j(\mathbf{y})\delta(s-y_m\xi_m)dV_{\mathbf{y}}\Big]dS(\boldsymbol{\xi}) \\ &= \frac{1}{8\pi^2}\int_{S^2}K^{-1}_{ij}(\boldsymbol{\xi})\hat{f}_j(s,\boldsymbol{\xi})\ dS(\boldsymbol{\xi}) \end{aligned} \tag{3.88}$$

where $s = x_m\xi_m$ and

$$\hat{f}_j(s,\boldsymbol{\xi}) = \int_{\mathbb{R}^3} f_j(\mathbf{y})\delta(s-y_m\xi_m)dV_{\mathbf{y}}$$

is the Radon transform of $f_j(\mathbf{y})$. Based on Eq.(3.76), for isotropic materials, $K^{-1}_{ij}(\boldsymbol{\xi})$ reduces to

$$K^{-1}_{ij}(\boldsymbol{\xi}) = \frac{1}{\mu}\Big[\delta_{ij} - \frac{(\lambda+\mu)\xi_i\xi_j}{(\lambda+2\mu)}\Big]$$

since $\boldsymbol{\xi}$ is integrated over unit sphere S^2 .

Example 3.1. Assume that a linearly distributed eigenstrain is prescribed in a spherical ball ($|\mathbf{x}| \leq a$).

$$\epsilon^*_{k\ell}(\mathbf{x}) = \begin{cases} \frac{1}{2}(c_k x_\ell + c_\ell x_k), & |\mathbf{x}| \leq a \\ 0, & |\mathbf{x}| > a \end{cases} \tag{3.89}$$

Hence

$$\epsilon^*_{k\ell,j} = \frac{1}{2}\Big(c_k \delta_{\ell j} + c_\ell \delta_{kj}\Big) H(a - |\mathbf{x}|) \tag{3.90}$$

where $H(\cdot)$ is the Heaviside function. For isotropic materials

$$f_i(\mathbf{x}) = -C_{ijk\ell}\epsilon^*_{k\ell,j} = -(\lambda + 4\mu) c_i H(a - |\mathbf{x}|)$$

The area of intersection of the plane $\xi_m x_m = s$ with the sphere of radius a (called S^a) is $\pi(a^2 - s^2)$, if $|s| \leq a$; The area of intersection is zero otherwise. Thus

$$\begin{aligned} \hat{f}_j(s, \boldsymbol{\xi}) &= -\int_{\mathbb{R}^3} (\lambda + 4\mu) c_i H(a - |\mathbf{y}|)\delta(s - y_m \xi_m) dV_{\mathbf{y}} \\ &= -\int_{S^a \cap \{\xi_m x_m = s\}} (\lambda + 4\mu) c_i dS = -(\lambda + 4\mu) c_i \pi (a^2 - s^2) \\ &= -(\lambda + 4\mu) c_i \pi (a^2 - (\xi_m x_m)(\xi_k x_k)) \end{aligned} \tag{3.91}$$

Therefore, based on Eq.(3.88), the induced displacement field inside the sphere is

$$u_i(\mathbf{x}) = -\frac{(\lambda + 4\mu)c_j}{8\pi} \int_{S^2} K^{-1}_{ij}(\boldsymbol{\xi})(a^2 - (\xi_m x_m)(\xi_k x_k)) dS(\boldsymbol{\xi}) \tag{3.92}$$

Considering

$$K^{-1}_{ij}(\boldsymbol{\xi}) = \frac{1}{\mu}\left[\delta_{ij} - \frac{(\lambda + \mu)\xi_i \xi_j}{(\lambda + 2\mu)}\right]$$

and

$$\oint_{S^2} dS = 4\pi, \qquad \oint_{S^2} \xi_i \xi_j dS(\boldsymbol{\xi}) = \frac{4\pi}{3}\delta_{ij},$$

$$\oint_{S^2} \xi_i \xi_j \xi_k \xi_m dS(\boldsymbol{\xi}) = \frac{4\pi}{15}(\delta_{ij}\delta_{km} + \delta_{ik}\delta_{jm} + \delta_{im}\delta_{jk})$$

we have

$$\begin{aligned} u_i(\mathbf{x}) = -\frac{(\lambda + 4\mu)c_j}{8\pi\mu}\Bigg(& 4\pi a^2 \delta_{ij}\frac{(2\lambda + 5)\mu}{3(\lambda + 2\mu)} - \frac{4\pi x_m x_m \delta_{ij}}{15}\frac{(4\lambda + 9\mu)}{\lambda + 2\mu} \\ & + \frac{8\pi x_i x_j}{15}\frac{(\lambda + \mu)}{(\lambda + 2\mu)}\Bigg) \end{aligned} \tag{3.93}$$

and the induced strain field is

$$\epsilon_{ij} = \frac{(\lambda + 4\mu)(3\lambda + 8\mu)}{30(\lambda + 2\mu)\mu}(x_i c_j + x_j c_i) - \frac{(\lambda + 4\mu)(\lambda + \mu)}{15(\lambda + 2\mu)\mu}\delta_{ij} c_m x_m \tag{3.94}$$

Finally, we can express the induced strain field in terms of the prescribed eigenstrain through the following relation

$$\epsilon_{ij} = S_{ijmn}\epsilon^*_{mn} \tag{3.95}$$

where the tensor S_{ijmn} can be evaluated as

$$S_{ijmn} = \frac{(\lambda + 4\mu)(3\lambda + 8\mu)}{15(\lambda + 2\mu)\mu}(\delta_{im}\delta_{jn} + \delta_{in}\delta_{jm}) - \frac{(\lambda + 4\mu)(\lambda + \mu)}{15(\lambda + 2\mu)\mu}\delta_{ij}\delta_{mn} \tag{3.96}$$

In fact, S_{ijmn} is so called interior 'Eshelby tensor'. In this case, the Eshelby tensor is an isotropic constant tensor even though the prescribed eigenstrain is not constant. We shall discuss Eshelby tensor in detail in next section.

3.6 Eshelby's solution to the inclusion problem in an infinite space

From 1957 to 1961, J.D. Eshelby published three seminal papers systematically studying inclusion problem in an elastic medium. Eshelby's solution has been the milestone in the development of micromechanics, since his solution provides complete solution to the elastic field inside and outside of the inclusion due to a prescribed eigenstrain. Eshelby's solution has found extensive applications, since the ellipsoidal shape is very versatile, which can be used to represent various shapes, including spheres, flat crack, and cylindrical wires.

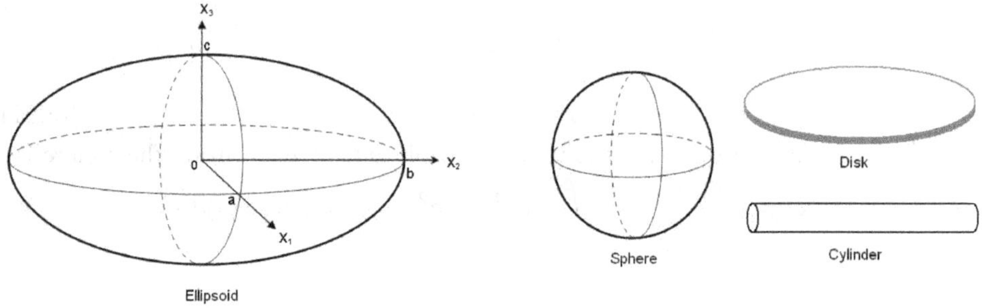

Fig. 3.9 An ellipsoidal inclusion.

Eshelby's ellipsoidal inclusion problem is stated as follows: *Find induced displacement, strain, and stress fields by an ellipsoidal inclusion, Ω, embedded in an isotropic unbounded elastic medium, in which a uniform eigenstrain is prescribed, i.e.*

$$\epsilon^*_{ij}(\mathbf{x}) = \begin{cases} \epsilon^*_{ij}, & \mathbf{x} \in \Omega \\ 0, & \mathbf{x} \in \mathbb{R}^3/\Omega \end{cases} \tag{3.97}$$

Using the fundamental formula of micro-elasticity, the induced displacement field due to an uniformly prescribed eigenstrain field can be written as

$$\begin{aligned} u_i(\mathbf{x}) &= -\int_{\mathbb{R}^3} C_{j\ell mn}\epsilon^*_{mn}(\mathbf{y})G^{\infty}_{ij,\ell}(\mathbf{x}-\mathbf{y})d\Omega_{\mathbf{y}} \\ &= -\epsilon^*_{mn}\int_{\Omega} C_{j\ell mn}G^{\infty}_{ij,\ell}(\mathbf{x}-\mathbf{y})d\Omega_{\mathbf{y}} \end{aligned} \tag{3.98}$$

The above equation applies for $\forall \mathbf{x} \in \mathbb{R}^3$, and the integration is made for the argument $\mathbf{y}$ over the ellipsoidal volume Ω. For isotropic elastic materials, it is easy to evaluate

$$C_{j\ell mn}G^{\infty}_{ij,\ell}(\mathbf{z}) = \frac{-1}{8\pi(1-\nu)}\Big\{(1-2\nu)\frac{\delta_{mi}z_n + \delta_{ni}z_m - \delta_{mn}z_i}{z^3} + 3\frac{z_m z_n z_i}{z^5}\Big\}$$
$$= \frac{g_{imn}(\bar{\mathbf{r}})}{8\pi(1-\nu)|\mathbf{z}|^2} \tag{3.99}$$

where $\mathbf{z} = \mathbf{x} - \mathbf{y}$ and $\bar{\mathbf{r}} = -\mathbf{z}/|\mathbf{z}|$. In literature, the notation $\bar{r} = \boldsymbol{\ell} = \ell_i \mathbf{e}_i$ is often used, and hence we can write

$$g_{imn}(\bar{\mathbf{r}}) = (1-2\nu)(\delta_{mi}\ell_n + \delta_{ni}\ell_m - \delta_{mn}\ell_i) + 3\ell_m\ell_n\ell_i \tag{3.100}$$

3.6.1 *Interior solution of ellipsoidal inclusion*

By seeking the "interior solution", it is meant for the solution *inside* the inclusion. It takes great ingenuity to find an exact solution for the integration Eq.(3.98). The integration scheme is illustrated in Fig.3.10. Given a point $\mathbf{x} \in \Omega$, the volume integration is carried out by placing the center of origin at point $\mathbf{x}$. By turning an infinitesimal volume angle $d\theta d\phi$ around $\mathbf{x}$, the unit surface area on the unit sphere S^2 is $d\omega = \sin\theta d\theta d\phi$. The corresponding volume element at an arbitrary point $\mathbf{y} \in \Omega$ can be found as $d\Omega_{\mathbf{y}} = dz\ dS = z^2\ dz\ d\omega$, where z measures the distance between $\mathbf{x}$ and $\mathbf{y}$, i.e., $z = |\mathbf{z}| = |\mathbf{x} - \mathbf{y}|$. Denote the projection point of vector $\mathbf{y} - \mathbf{x}$ on the ellipsoidal surface is $\mathbf{x}'$. It is easy to identify that the vector $\bar{\mathbf{x}} = \mathbf{x}' - \mathbf{x}$ is aligned with the direction of the unit vector $\bar{\mathbf{r}}$. Given $\mathbf{x}$ inside an inclusion, the distance $r = |\mathbf{x}' - \mathbf{x}|$ is solely dependent on the direction $\bar{\mathbf{r}}$. Finally, we can rewrite the displacement field as

$$u_i(\mathbf{x}) = \frac{-\epsilon^*_{mn}}{8\pi(1-\nu)}\int_{\Omega}\frac{g_{imn}(\bar{\mathbf{r}})}{z^2}d\Omega_{\mathbf{y}} = \frac{-\epsilon^*_{mn}}{8\pi(1-\nu)}\int_{S^2}\int_0^r g_{imn}(\bar{\mathbf{r}})dz d\omega$$
$$= \frac{-\epsilon^*_{mn}}{8\pi(1-\nu)}\int_{S^2} r(\bar{\mathbf{r}})g_{imn}(\bar{\mathbf{r}})d\omega \tag{3.101}$$

The above derivation transforms the volume integration over an ellipsoid into a surface integration on a unit sphere. Since the scalar $r(\bar{\mathbf{r}})$ is the distance between the point $\mathbf{x} \in \Omega$ and the point $\mathbf{x}'$ on the surface of the ellipsoid in the direction of $\bar{\mathbf{r}}$, it can be written as $\mathbf{x}' = \mathbf{x} + \bar{\mathbf{x}} = \mathbf{x} + r\bar{\mathbf{r}}$, i.e.

$$\begin{cases} x'_1 = x_1 + r\ell_1 \\ x'_2 = x_2 + r\ell_2 \\ x'_3 = x_3 + r\ell_3 \end{cases} \tag{3.102}$$

and $\mathbf{x}'$ is on the ellipsoidal surface

$$\frac{{x'_1}^2}{a_1^2} + \frac{{x'_2}^2}{a_2^2} + \frac{{x'_3}^2}{a_3^2} = 1 \tag{3.103}$$

One can substitute (3.102) into (3.103). For a fixed point $\mathbf{x}$ and a fixed direction $\bar{\mathbf{r}}$, it yields a quadratic equation of the unknown variable $r(\bar{\mathbf{r}})$,

$$\frac{(x_1 + r\ell_1)^2}{a_1^2} + \frac{(x_2 + r\ell_2)^2}{a_2^2} + \frac{(x_3 + r\ell_3)^2}{a_3^2} = 1 \tag{3.104}$$

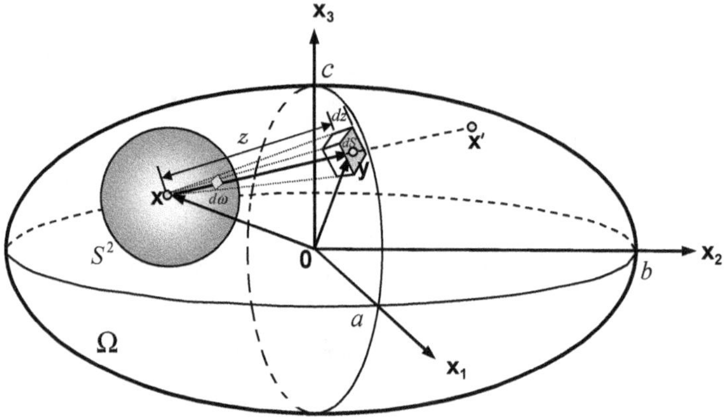

Fig. 3.10 Illustration of integration scheme over an ellipsoidal.

More explicitly,

$$gr^2 + 2fr - e = 0 \tag{3.105}$$

where

$$g := \frac{\ell_1^2}{a_1^2} + \frac{\ell_2^2}{a_2^2} + \frac{\ell_3^2}{a_3^2}$$
$$f := \frac{x_1\ell_1}{a_1^2} + \frac{x_2\ell_2}{a_2^2} + \frac{x_3\ell_3}{a_3^2}$$
$$e := 1 - \left(\frac{x_1^2}{a_1^2} + \frac{x_2^2}{a_2^2} + \frac{x_3^2}{a_3^2}\right)$$

Eq.(3.105) has two roots,

$$r(\bar{\mathbf{r}}) = -\frac{f}{g} \pm \sqrt{\frac{f^2}{g^2} + \frac{e}{g}} \tag{3.106}$$

Since $\sqrt{\dfrac{f^2}{g^2} + \dfrac{e}{g}}$ is even in $\bar{\mathbf{r}}$, while $g_{imn}(\bar{\mathbf{r}})$ is odd in $\bar{\mathbf{r}}$, the following integration vanishes due to antisymmetry

$$\int_{S^2} \sqrt{\frac{f^2}{g^2} + \frac{e}{g}}\; g_{imn}(\bar{\mathbf{r}})d\omega = 0 \tag{3.107}$$

To evaluate the remaining term, let $\lambda_1 = \ell_1/a_1^2$, $\lambda_2 = \ell_2/a_2^2$ and $\lambda_3 = \ell_3/a_3^2$. We have

$$\begin{aligned} u_i(\mathbf{x}) &= \frac{\epsilon^*_{mn}}{8\pi(1-\nu)} \oint_{S^2} \frac{f}{g} g_{imn}(\bar{\mathbf{r}})d\omega = \frac{\epsilon^*_{mn}}{8\pi(1-\nu)} \oint_{S^2} \left(\frac{x_k\lambda_k}{g}\right) g_{imn}(\bar{\mathbf{r}})d\omega \\ &= \frac{\epsilon^*_{mn} x_k}{8\pi(1-\nu)} \oint_{S^2} \left(\frac{\lambda_k}{g}\right) g_{imn}(\bar{\mathbf{r}})d\omega \end{aligned} \tag{3.108}$$

Then

$$u_{i,j}(\mathbf{x}) = \frac{\epsilon^*_{mn}\delta_{kj}}{8\pi(1-\nu)}\oint_{S^2}\left(\frac{\lambda_k}{g}\right)g_{imn}(\bar{\mathbf{r}})d\omega = \frac{\epsilon^*_{mn}}{8\pi(1-\nu)}\oint_{S^2}\left(\frac{\lambda_j}{g}\right)g_{imn}(\bar{\mathbf{r}})d\omega \quad (3.109)$$

One can find induced elastic strain field by symmetrizing the elastic distortion,

$$\epsilon_{ij} = \frac{1}{2}(u_{i,j} + u_{j,i}) = \frac{\epsilon^*_{mn}}{16\pi(1-\nu)}\oint_{S^2}\frac{\lambda_i g_{jmn} + \lambda_j g_{imn}}{g}d\omega \quad (3.110)$$

where $\lambda_i = \dfrac{\ell_i}{a_i^2}$ (no summation over repeated indices).

Finally, we can define a fourth order symmetric tensor, called as "Eshelby tensor", that provided direct connection between disturbance strain and eigenstrain,

$$\epsilon_{ij}(\mathbf{x}) = \left(\text{or } \epsilon^d_{ij}(\mathbf{x})\right) = S^{I,\infty}_{ijmn}\epsilon^*_{mn} \quad (3.111)$$

where

$$S^{I,\infty}_{ijmn} := \frac{1}{16\pi(1-\nu)}\oint_{S^2}\frac{\lambda_i g_{jmn} + \lambda_j g_{imn}}{g}d\omega \quad (3.112)$$

where the superscript I, ∞ indicates that the Eshelby tensor is for induced strain field inside the ellipsoid (*interior* solution), and the inclusion is embedded in an *infinite* medium.

By considering the last two indices of the third order tensor g_{ijk} is symmetric, i.e., $g_{imn} = g_{inm}$, it is easy to verify that Eshelby tensor has the following property of minor symmetry,

$$S^{I,\infty}_{ijmn} = S^{I,\infty}_{ijnm} = S^{I,\infty}_{jimn}$$

We proceed to evaluate the Eshelby tensor. Define the following elliptic integrals,

$$I_I(0) = 2\pi a_1 a_2 a_3 \int_0^\infty \frac{ds}{(a_I^2+s)\Delta(s)} \quad (3.113)$$

$$I_{IJ}(0) = 2\pi a_1 a_2 a_3 \int_0^\infty \frac{ds}{(a_I^2+s)(a_J^2+s)\Delta(s)} \quad (3.114)$$

$$J_{IJ}(0) = a_I^2 I_{IJ} - I_J \quad (3.115)$$

where $\Delta(s) = \sqrt{(a_1^2+s)(a_2^2+s)(a_3^2+s)}$, and the upper case indices $I, J, K = 1, 2, 3$. There is no summation over repeated upper case indices. The argument (0) indicates the lower limit of the elliptic integrals are zero. In particular, one can show that

$$I_1 = \int_{S_2}\frac{\ell_1^2 d\omega}{a_1^2 g} \quad (3.116)$$

$$I_{11} = \int_{S_2}\frac{\ell_1^4 d\omega}{a_1^4 g} \quad (3.117)$$

$$I_{12} = 3\int_{S_2}\frac{\ell_1^2\ell_2^2 d\omega}{a_1^2 a_2^2 g} \quad (3.118)$$

The Eshelby tensor in (3.112) can be explicitly expressed by these integrals through the following identity,

$$
\begin{aligned}
8\pi(1-\nu)S^{I,\infty}_{ijk\ell} &= [2\nu I_I(0) + J_{IK}(0)]\,\delta_{ij}\delta_{k\ell} \\
&+ [(1-\nu)(I_K(0)+I_L(0)) + J_{IJ}(0)]\,(\delta_{ik}\delta_{j\ell}+\delta_{jk}\delta_{i\ell})
\end{aligned}
\tag{3.119}
$$

where the upper case indices are not summed with lower case indices.

Example 3.2. To compute $S^{I,\infty}_{1111}$, we consider

$$
\begin{aligned}
8\pi(1-\nu)S^{I,\infty}_{1111} &= 2\nu I_1(0) + J_{11}(0) + 2(1-\nu)2I_1(0) + 2J_{11}(0) \\
&= (4-2\nu)I_1(0) + 3J_{11}(0) = (1-2\nu)I_1(0) + 3a_1^2 I_{11}(0)
\end{aligned}
\tag{3.120}
$$

which leads to

$$
S^{I,\infty}_{1111} = \frac{3a_1^2}{8\pi(1-\nu)} I_{11}(0) + \frac{(1-2\nu)}{8\pi(1-\nu)} I_1(0)
\tag{3.121}
$$

The integral $I_I(0)$ and $I_{IJ}(0)$ can be expressed in terms of standard elliptic integrals. For example, assuming $a_1 > a_2 > a_3$, we have

$$
\begin{aligned}
I_1(0) &= \frac{4\pi a_1 a_2 a_3}{(a_1^2-a_2^2)(a_1^2-a_2^2)^{1/2}}\{F(\theta,\kappa) - E(\theta,\kappa)\} \\
I_3(0) &= \frac{4\pi a_1 a_2 a_3}{(a_2^2-a_3^2)(a_1^2-a_3^2)^{1/2}}\left\{\frac{a_2(a_1^2-a_3^2)^{1/2}}{a_1 a_3} - E(\theta,\kappa)\right\}
\end{aligned}
$$

where

$$
\begin{aligned}
F(\theta,\kappa) &= \int_0^\theta \frac{dt}{(1-\kappa^2\sin^2 t)^{1/2}} \\
E(\theta,\kappa) &= \int_0^\theta (1-\kappa^2\sin^2 t)^{1/2} dt
\end{aligned}
\tag{3.122}
$$

and $\theta = \sin^{-1}\sqrt{1-a_3^2/a_1^2}$, $\kappa = \sqrt{(a_1^2-a_2^2)/(a_1^2-a_3^2)}$.

In applications, the following invariant formulas are very useful,

$$
\begin{aligned}
& I_1(0) + I_2(0) + I_3(0) = 4\pi \\
& 3I_{11}(0) + I_{12}(0) + I_{13}(0) = 4\pi/a_1^2 \\
& 3a_1^2 I_{11}(0) + a_2^2 I_{12}(0) + a_3^2 I_{13}(0) = 3I_1 \\
& I_{12}(0) = (I_2(0) - I_1(0))/(a_1^2 - a_2^2)
\end{aligned}
$$

When the ellipsoid becomes a sphere, Eshelby tensor become simple numbers. Let $a_1 = a_2 = a_3 = a$ be the radius of the sphere, we have

$$
I_I(0) = \frac{4\pi}{3}, \quad I_{I,J}(0) = \frac{4\pi}{5a^2}, \quad J_{IJ}(0) = -\frac{8\pi}{15}
$$

and hence

$$
S^{I,\infty}_{ijk\ell} = \frac{5\nu-1}{15(1-\nu)}\delta_{ij}\delta_{k\ell} + \frac{(4-5\nu)}{15(1-\nu)}(\delta_{ik}\delta_{j\ell}+\delta_{jk}\delta_{i\ell})
\tag{3.123}
$$

Remark 3.2. The above Eshelby tensor for spherical inclusion is a fourth-order isotropic tensor, which does not depend on the size of the sphere, i.e. it does not depend on its radius a. This remarkable property implies that no matter how large or how small spherical inclusions are, they share the same Eshelby tensor. In other words, there is no embedded length scale or scaling factor for spherical inclusions. This property will lead to some remarkable consequences in ensuing homogenization process.

For other specified shapes of ellipsoidal inclusions, readers may consult Mura's book for detailed information. A systematic documentation on Eshelby's tensor in various cases can be found in [Mura (1987)]

3.6.2 *Eshelby's conjectures*

The most amazing fact of Eshelby's work is that the interior Eshelby tensor for an ellipsoidal inclusion is a *constant* tensor under uniform elastic loading. The solution implies that the induced strain field and stress field inside the inclusion are uniform due to the uniformly prescribed eigenstrain or eigenstress distributions. The general form of the interior Eshelby tensor is a fourth-order isotropic tensor, and the value of its components only depends on the aspect ratio of the ellipsoid. That is: the Eshelby tensor remains the same for ellipsoids with the same aspect ratio scaling. The homogeneous nature of induced stress and strain fields of the ellipsoidal inclusion found by Eshelby is an astonishing discovery. It provides a very convenient and useful tool in engineering homogenization procedure, which provides accurate characterization and prediction for statistical distribution of inclusions or inhomogeneities.

Naturally and curiously, it is compelling to ask the following question of practical significance:

Is the ellipsoid the only shape of inclusions that produce uniform induced stresses or strains inside the inclusion by prescribed uniform eigenstrains or uniform boundary conditions?

In 1961, Eshelby (1961) conjectured "······ *among closed surfaces, the ellipsoid alone has this convenient property* ······". In other words, the ellipsoidal inclusion of uniform eigenstrain in an isotropic elastic material is the only type of inclusions that can incur uniform, constant strain and stress fields inside the inclusion. If we call the inclusion that generates uniform strain/stress fields when it is subjected to uniform eigenstrain and uniform boundary conditions as *the E-inclusion*, will the ellipsoid be the only shape of such inclusions ? The *Yes* answer to this question is the so-called Eshelby's conjecture.

To frame the question in rigorous mathematics terms, we distinguish two Eshelby's conjectures:

1. Strong Eshelby Conjecture:

If the induced elastic fields inside an inclusion are uniform under a single uni-

form loading, the inclusion is of elliptic or ellipsoidal shape.

and

2. Weak Eshelby Conjecture:

If the induced elastic fields inside an inclusion are unform under all (any) uniform loadings, the inclusion is of elliptic or ellipsoidal shape.

Obviously, the strong Eshelby conjecture implies the weak Eshelby conjecture. It was believed that the original Eshelby conjecture is in the sense of the weak Eshelby conjecture [Kang and Milton (2008)]. Since the Eshelby conjecture was proposed in 1961, it has drawn increased attention. In 1970, Sendeckyj (1970) first proved the strong Eshelby conjecture for two-dimensional inclusions. In 1996, Ru and Schiavone (1996) proved the strong Eshelby conjecture for anti-plane problem. Nevertheless, not everyone is convinced with Eshelby's conjecture. Some researchers have tried to find the counterexample to show that there exist inclusions other than ellipsoids that can sustain constant strain fields in their interior regions. For example, a star-shaped inclusion as shown in Fig. 3.11 was thought [Mura (1997, 2000)] to be a possible candidate, but it might not be true in this case [Rodin (1996)]. On the other hand, no rigorous proof has been given to prove or disprove the Eshelby conjectures for many years. The best proof on the Eshelby conjecture before 2000 is that Markenscoff (1998) showed that the only perturbations of any ellipsoid that can preserve the uniformity of the elastic field are those that perturb the one ellipsoid to another ellipsoid.

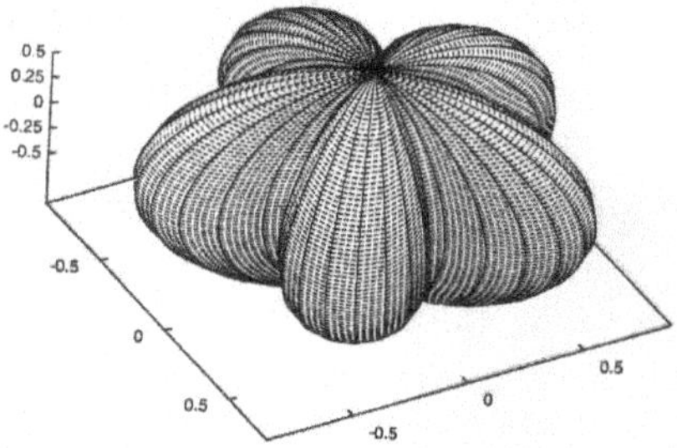

Fig. 3.11 The inclusion thought to be a possible candidate that violates the weak Eshelby conjecture.

Recently, a breakthrough on the Eshelby conjecture was made by Kang and Milton (2008). They proved the weak Eshelby conjecture for three-dimensional inclusions for isotropic materials by using the maximum principle of harmonic potentials. Moreover, under certain conditions Kang and Milton linked the Eshebly conjecture to the Pólya-Szegö conjecture [Pólya and Szegö (1951)]. Within a few months after Kang & Milton's success, by using the technique of variational inequality, Liu (2008) has proved the following general theorem on the Eshelby's conjectures.

Theorem 3.5 (Liu's Theorem). *Let:*

(1) the elastic tensor $\mathbb{L} : \mathbb{R}^{n\times n} \to \mathbb{R}^{n\times n}$ *with the form*

$$(\mathbf{L})_{piqj} = \mu_1 \delta_{ij}\delta_{pq} + \mu_2 \delta_{pj}\delta_{iq} + \lambda \delta_{ip}\delta_{jq} \tag{3.124}$$

in which

$$\mu_1 \geq \mu_2, \quad \mu_1 + \mu_2 > 0, \quad \text{and} \quad \lambda > -\frac{\mu_1 + \mu_2}{n} \tag{3.125}$$

(2) $\Omega \in \mathbb{R}^n$ *be an inclusion, and* χ_Ω *be the characteristic function of* Ω*, and*
(3) $\mathbf{u}(\mathbf{x}) \in W^{1,2}_{loc}(\mathbb{R}^n)$ *be a solution of the following partial differential equation,*

$$\operatorname{div}\left[\mathbb{L} : (\nabla \otimes \mathbf{u}) + \mathbf{p}\chi_\Omega\right] = 0, \quad \forall \mathbf{x} \in \mathbb{R}^n \tag{3.126}$$

in the sense that

$$[\nabla \otimes \mathbf{u}(\mathbf{x})]_{pi} = \frac{-1}{(2\pi)^n} \int_{\mathbb{R}^n} K^{-1}_{pq}(\boldsymbol{\xi})(\mathbf{p})_{qj}\xi_j\xi_i \int_\Omega \exp(i\boldsymbol{\xi}\cdot(\mathbf{x}-\mathbf{x}'))d\mathbf{x}'d\boldsymbol{\xi} \tag{3.127}$$

where $K^{-1}_{pq}(\boldsymbol{\xi})$ *is the inverse of the matrix* $(\mathbf{L})_{piqj}\xi_i\xi_j$*, and* $\mathbf{p}$ *is the prescribed eigenstrain stress.*

The following results hold:

(1) For single inclusion, if $n \geq 2$*,* $\mathbf{u}(\mathbf{x})$ *satisfying (3.126), and*

$$\nabla \otimes \mathbf{u}(\mathbf{x},\mathbf{p}) = const. \quad \text{on } \Omega, \quad \forall\ \mathbf{p} \in \mathbb{R}^{n\times n} \tag{3.128}$$

then Ω *must be an ellipsoidal inclusion.*
(2) For single inclusion, if $n = 2$*, and*

$$\nabla \otimes \mathbf{u}(\mathbf{x},\mathbf{p}) = const. \quad \text{on } \Omega \quad \text{for a single nonzero constant } \mathbf{p} \in \mathbb{R}^{2\times 2} \tag{3.129}$$

then Ω *must be an ellipsoidal inclusion.*
(3) There exist multiply-connected inclusions $\Omega \in \mathbb{R}^n (n \geq 2)$ *such that*

- *The induced displacement gradient field (equivalent strain field) is uniform on* Ω *for* $\forall \mathbf{p} \in \mathbb{R}^{n\times n}$*, if* $\mu_2 + \lambda = 0$*; and*
- *The induced displacement gradient field is uniform on* Ω *for the constant eigenstress* $p\mathbb{I} \in \mathbb{R}^{n\times n}$*, if* $\mu_2 + \lambda \neq 0$*.*

Furthermore, Liu and his co-workers have found several multiple connected E(shelby)-inclusions in both two-dimensional and three-dimensional spaces, which are displayed in Fig. 3.12. In fact, Liu *et al.* (2007) have constructed periodic E-inclusions. The most significant contribution of Liu's work is that he actually constructed a counterexample for 3D strong Eshelby conjecture. Liu showed that the strong Eshelby conjecture is false in three and higher dimensions. Fig. 3.13 displays Liu's counterexample on the strong Eshelby conjecture. Finally, we would like to mention that the Eshelby conjectures in anisotropic linear elastic materials have not been rigorously proved yet at the time when this book is going to print, except that a sketch of proof was made in [Kang and Milton (2008)].

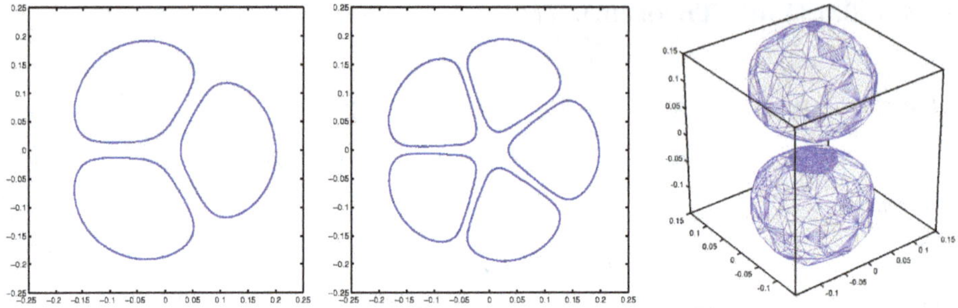

(a) (b) (c)

Fig. 3.12 Examples of E-inclusions that are not ellipsoids (These figures are kindly provided by Dr. L. P. Liu of California Institute of Technology).

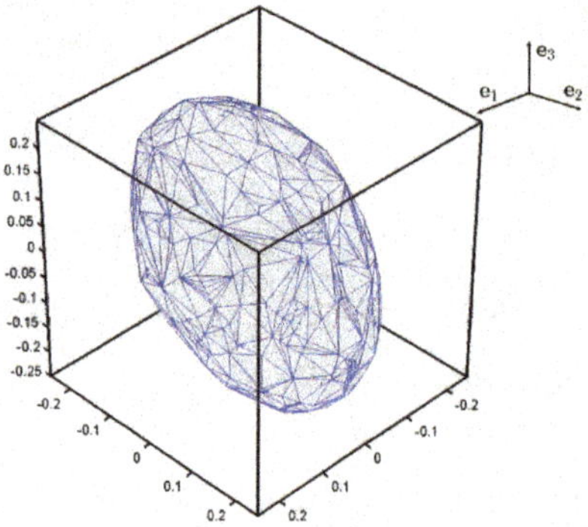

Fig. 3.13 An counterexample of the strong Eshelby conjecture (The figure is kindly provided by Dr. L.P. Liu of California Institute of Technology).

3.6.3 *Exterior solution of ellipsoidal inclusion*

For a point outside the inclusion, $\mathbf{x} \notin \Omega$, the exterior elastic fields due to eigenstrain distribution had also been found by Eshelby, although evaluation of the induced exterior displacement fields and strain fields are often difficult.

Assume that eigenstrain distribution inside the ellipsoid is constant. For any point $\mathbf{x} \in \mathbb{R}^3$, we have

$$u_i(\mathbf{x}) = -\epsilon^*_{mn} \int_\Omega C_{jkmn} G^\infty_{ij,k}(\mathbf{x} - \mathbf{y}) d\Omega_{\mathbf{y}} \tag{3.130}$$

where

$$C_{j\ell mn}G^{\infty}_{ij,\ell}(\mathbf{x}-\mathbf{y}) = \frac{-1}{8\pi(1-\nu)}\Big\{(1-2\nu)\frac{\delta_{mi}z_n + \delta_{ni}z_m - \delta_{mn}z_i}{z^3} + 3\frac{z_m z_n z_i}{z^5}\Big\}$$

$$= \frac{-1}{8\pi(1-\nu)}\left\{\frac{\partial^3}{\partial x_i \partial x_m \partial x_n}|\mathbf{x}-\mathbf{y}| - 2(1-\nu)\left[\delta_{mi}\frac{\partial}{\partial x_n}\frac{1}{|\mathbf{x}-\mathbf{y}|}\right.\right.$$

$$\left.\left. +\delta_{ni}\frac{\partial}{\partial x_m}\frac{1}{|\mathbf{x}-\mathbf{y}|}\right] - 2\nu\delta_{mn}\frac{\partial}{\partial x_i}\frac{1}{|\mathbf{x}-\mathbf{y}|}\right\} \tag{3.131}$$

where $\mathbf{z} = \mathbf{x} - \mathbf{y}$.

Introduce the following potential functions,

$$\psi(\mathbf{x}) = \int_\Omega |\mathbf{x}-\mathbf{y}| d\Omega_{\mathbf{y}} \tag{3.132}$$

$$\phi(\mathbf{x}) = \int_\Omega \frac{1}{|\mathbf{x}-\mathbf{y}|} d\Omega_{\mathbf{y}} \tag{3.133}$$

where $\psi(\mathbf{x})$ is the biharmonic potential, whereas $\phi(\mathbf{x})$ is the Newtonian potential. This is because of the fact

$$\nabla^4 \psi = 2\nabla^2\phi = \begin{cases} -8\pi, & \mathbf{x} \in \Omega \\ 0, & \mathbf{x} \in \mathbb{R}^3/\Omega \end{cases} \tag{3.134}$$

To verify Eq. (3.134), one can first show that

$$\nabla^2\psi = \frac{\partial^2}{\partial x_\ell \partial x_\ell}\int_\Omega |\mathbf{x}-\mathbf{y}| d\Omega_{\mathbf{y}} = \int_\Omega \frac{2}{|\mathbf{x}-\mathbf{y}|} d\Omega_{\mathbf{y}} = 2\phi(\mathbf{x}) \tag{3.135}$$

Subsequently,

$$\begin{aligned}\nabla^4\psi &= \nabla^2\nabla^2\psi = 2\nabla^2\phi \\ &= 2\int_\Omega \nabla^2 \frac{1}{|\mathbf{x}-\mathbf{y}|} d\Omega = 8\pi \int_\Omega \nabla^2 \frac{1}{4\pi|\mathbf{x}-\mathbf{y}|} d\Omega \\ &= 8\pi \int_\Omega \nabla^2 G^P(\mathbf{x}-\mathbf{y}) d\Omega_{\mathbf{y}}\end{aligned} \tag{3.136}$$

where $G^P(\mathbf{x}-\mathbf{y})$ is the Green's function for three-dimensional Poisson's equation, i.e.

$$\nabla^2 G^P(\mathbf{x}-\mathbf{y}) + \delta(\mathbf{x}-\mathbf{y}) = 0 \tag{3.137}$$

Consequently,

$$\begin{aligned}\nabla^4\psi &= 2\nabla^2\phi = -8\pi\int_\Omega \delta(\mathbf{x}-\mathbf{y}) d\Omega_{\mathbf{y}} \\ &= \begin{cases} -8\pi, & \mathbf{x} \in \Omega \\ 0, & \mathbf{x} \in \mathbb{R}^3/\Omega \end{cases}\end{aligned} \tag{3.138}$$

We can then express induced displacement as

$$u_i(\mathbf{x}) = -\int_\Omega \epsilon^*_{mn} C_{j\ell mn} G^\infty_{ij,\ell}(\mathbf{x}-\mathbf{y}) d\Omega_{\mathbf{y}}$$
$$= \frac{\epsilon^*_{mn}}{8\pi(1-\nu)} \Big\{ \frac{\partial^3 \psi}{\partial x_i \partial x_m \partial x_n} - 2(1-\nu)\Big(\delta_{mi}\frac{\partial \phi}{\partial x_n} + \delta_{ni}\frac{\partial \phi}{\partial x_m}\Big)$$
$$-2\nu\delta_{mn}\frac{\partial \phi}{\partial x_i}\Big\} \tag{3.139}$$

Similarly for elastic distortion field and strain field,

$$u_{i,j}(\mathbf{x}) = \frac{\epsilon^*_{mn}}{8\pi(1-\nu)}\Big(\psi_{,mnij} - 2(1-\nu)(\delta_{mi}\phi_{,nj} + \delta_{ni}\phi_{,mj})$$
$$-2\nu\delta_{mn}\phi_{,ij}\Big) \tag{3.140}$$

$$\epsilon_{ij}(\mathbf{x}) = (\frac{1}{2}(u_{i,j}+u_{j,i}) = \frac{\epsilon^*_{mn}}{8\pi(1-\nu)}\Big(\psi_{,mnij} - 2\nu\delta_{mn}\phi_{,ij}$$
$$-(1-\nu)(\delta_{mi}\phi_{,nj} + \delta_{ni}\phi_{,mj} + \delta_{mj}\phi_{,ni} + \delta_{nj}\phi_{,mi})\Big) \tag{3.141}$$

One can rewrite the above expression in a succinct manner,

$$\epsilon^d_{ij}(\mathbf{x}) = S^{E,\infty}_{ijk\ell}(\mathbf{x})\epsilon^*_{k\ell}, \quad \forall \mathbf{x} \in \mathbb{R}^3/\Omega \tag{3.142}$$

which defines the exterior Eshelby tensor, $S^{E,\infty}_{ijk\ell}(\mathbf{x})$.

$$S^{E,\infty}_{ijk\ell}(\mathbf{x}) = \frac{1}{8\pi(1-\nu)}\Big(\psi_{,ijk\ell}(\mathbf{x}) - 2\nu\delta_{k\ell}\phi_{,ij}(\mathbf{x})$$
$$-(1-\nu)(\delta_{ki}\phi_{,\ell j}(\mathbf{x}) + \delta_{\ell i}\phi_{,kj}(\mathbf{x}) + \delta_{kj}\phi_{,\ell i}(\mathbf{x}) + \delta_{\ell j}\phi_{,ki}(\mathbf{x})\Big) \tag{3.143}$$

It depends on where the tensor is being evaluated.

The derivatives of Newtonian potential and biharmonic potential can be also expressed by elliptic integrals. For instance, for $\mathbf{x} \in \mathbb{R}^3$,

$$\phi_{,ij}(\mathbf{x}) = -\delta_{ij} I_I(\lambda) - x_i I_{IJ}(\lambda) \tag{3.144}$$
$$\psi_{,ijk\ell}(\mathbf{x}) = -\delta_{ij}(x_k J_{IK}(\lambda))_{,\ell} + (x_i x_j J_{IJ}(\lambda))_{,k\ell} \tag{3.145}$$

where

$$I_I(\lambda) = 2\pi a_1 a_2 a_3 \int_\lambda^\infty \frac{ds}{(a_I^2+s)\Delta(s)} \tag{3.146}$$
$$I_{IJ}(\lambda) = 2\pi a_1 a_2 a_3 \int_\lambda^\infty \frac{ds}{(a_I^2+s)(a_J^2+s)\Delta(s)} \tag{3.147}$$
$$J_{IJ}(\lambda) = a_I^2 I_{IJ}(\lambda) - I_J(\lambda) \tag{3.148}$$

where λ is zero when $\mathbf{x} \in \Omega$ and λ is the largest positive root of the following equation when $\mathbf{x} \in \mathbb{R}^3/\Omega$,

$$\frac{x_1^2}{(a_1^2+\lambda)} + \frac{x_2^2}{(a_2^2+\lambda)} + \frac{x_3^2}{(a_3^2+\lambda)} = 1 \tag{3.149}$$

A very useful identity that relates $S^{E,\infty}_{ijk\ell}(\mathbf{x})$ with elliptic integrals is

$$
\begin{aligned}
&8\pi(1-\nu)S^{E,\infty}_{ijk\ell}(\mathbf{x}) = 8\pi(1-\nu)S^{I,\infty}_{ijk\ell} + 2\nu\delta_{k\ell}x_i I_{I,j}(\lambda)\\
&+(1-\nu)\Big[\delta_{i\ell}x_k I_{K,j}(\lambda) + \delta_{j\ell}x_k I_{K,i}(\lambda) + \delta_{ik}x_\ell I_{L,j}(\lambda) + \delta_{jk}x_\ell I_{L,i}(\lambda)\Big]\\
&+\delta_{ij}x_k J_{IK,\ell}(\lambda) + (\delta_{ik}x_j + \delta_{jk}x_i)J_{IJ,\ell}(\lambda)\\
&+(\delta_{i\ell}x_j + \delta_{j\ell}x_i)J_{IJ,k}(\lambda) + x_i x_j J_{IJ,k\ell}(\lambda)
\end{aligned}
\tag{3.150}
$$

where

$$
\begin{aligned}
8\pi(1-\nu)S^{I,\infty}_{ijk\ell} &= \left[2\nu I_I(0) + J_{IK}(0)\right]\delta_{ij}\delta_{k\ell}\\
&+ \left[(1-\nu)(I_K(0)+I_L(0)) + J_{IJ}(0)\right](\delta_{ik}\delta_{j\ell} + \delta_{jk}\delta_{i\ell})
\end{aligned}
\tag{3.151}
$$

when $\mathbf{x}\in\Omega$, Eq. (3.150) becomes (3.151).

Ju and Sun [Ju and Sun (1999)] developed a simple and explicit way to evaluate exterior Eshelby tensor. From

$$
u_i(\mathbf{x}) = -\int_\Omega C_{j\ell mn}G^\infty_{ij,\ell}(\mathbf{x}-\mathbf{y})\epsilon^*_{mn}(\mathbf{y})d\Omega_{\mathbf{y}} \tag{3.152}
$$

one can derive that

$$
\begin{aligned}
\epsilon_{ij}(\mathbf{x}) &= -\frac{1}{2}\int_\Omega C_{k\ell mn}\Big(G^\infty_{ik,\ell j}(\mathbf{x}-\mathbf{y}) + G^\infty_{jk,\ell i}(\mathbf{x}-\mathbf{y})\Big)\epsilon^*_{mn}(\mathbf{y})d\Omega_{\mathbf{y}}\\
&= \int_\Omega \mathcal{G}^\infty_{ijmn}(\mathbf{x}-\mathbf{y})\epsilon^*_{mn}(\mathbf{y})d\Omega_{\mathbf{y}}
\end{aligned}
\tag{3.153}
$$

where

$$
\begin{aligned}
\mathcal{G}^\infty_{ijmn}(\mathbf{x}-\mathbf{y}) &= -\frac{1}{2}C_{k\ell mn}\Big(G^\infty_{ik,\ell j}(\mathbf{x}-\mathbf{y}) + G^\infty_{jk,\ell i}(\mathbf{x}-\mathbf{y})\Big)\\
&= \frac{1}{8\pi(1-\nu)r^3}\Big[(1-2\nu)(\delta_{im}\delta_{jn} + \delta_{in}\delta_{jm} - \delta_{ij}\delta_{mn})\\
&+3\nu(\delta_{im}\ell_j\ell_n + \delta_{in}\ell_j\ell_m + \delta_{jm}\ell_i\ell_n + \delta_{jn}\ell_j\ell_m)\\
&+3\delta_{ij}\ell_m\ell_n + 3(1-2\nu)\delta_{mn}\ell_i\ell_j - 15\ell_i\ell_j\ell_m\ell_n\Big]
\end{aligned}
\tag{3.154}
$$

with $r=|\mathbf{x}|$ measuring the distance of $\mathbf{x}$ from the center of the inclusion, and $\mathcal{G}^\infty_{ijmn}$ is called the fourth order Green's function (the second derivative of the Green's function). Note that expression (3.154) is valid only for Eshelby's exterior problem, because in this case the integral is always well behaved.

If $\epsilon^*_{mn}(\mathbf{x})$ is constant inside the inclusion, the exterior Eshelby tensor can be defined as

$$
S^{E,\infty}_{ijmn} = \int_\Omega \mathcal{G}^\infty_{ijmn}(\mathbf{x}-\mathbf{y})d\Omega_{\mathbf{y}} \tag{3.155}
$$

where the superscript, E, indicates that this is an exterior tensor, whereas the superscript, ∞, indicates that the surrounding ambient space (matrix) for the inclusion is unbounded.

For a spherical inclusion ($a_1 = a_2 = a_3 = a$), one may find that the potential functions assume simple forms

$$\phi(\mathbf{x}) = \begin{cases} -\dfrac{2\pi}{3}(|\mathbf{x}|^2 - 3a^2), & \forall \mathbf{x} \in \Omega \\ \dfrac{4\pi a^3}{3|\mathbf{x}|}, & \forall \mathbf{x} \in \mathbb{R}^3/\Omega \end{cases} \tag{3.156}$$

and

$$\psi(\mathbf{x}) = \begin{cases} -\dfrac{\pi}{15}(|\mathbf{x}|^4 - 10a^2|\mathbf{x}|^2 - 15a^4), & \forall \mathbf{x} \in \Omega \\ \\ \dfrac{4\pi a^3}{3}\Big(|\mathbf{x}| + \dfrac{a^2}{5|\mathbf{x}|}\Big), & \forall \mathbf{x} \in \mathbb{R}^3/\Omega \end{cases} \tag{3.157}$$

The exterior Eshelby tensor can then be obtained by straightforward differentiation,

$$\begin{aligned} S^{E,\infty}_{ijmn}(\mathbf{x}) = & \frac{\rho^3}{30(1-\nu)}\Big[(3\rho^2 + 10\nu - 5)\delta_{ij}\delta_{mn} \\ & +(3\rho^2 - 10\nu + 5)(\delta_{im}\delta_{jn} + \delta_{im}\delta_{jn}) \\ & +15(1-\rho^2)\delta_{ij}r_m r_n + 15(1 - 2\nu - \rho^2)\delta_{mn}r_i r_j \\ & +15(\nu - \rho^2)(\delta_{im}r_j r_n + \delta_{in}r_j r_m + \delta_{jm}r_i r_n + \delta_{jn}r_i r_m) \\ & +15(7\rho^2 - 5)r_i r_j r_m r_n\Big]. \end{aligned} \tag{3.158}$$

where $\rho = a/|\mathbf{x}|$, and $r_i = x_i/|\mathbf{x}|$. Note that when $r = |\mathbf{x}| \to a$, the exterior solution $S^{E,\infty}_{ijmn}$ does not approach the interior solution $S^{I,\infty}_{ijmn}$, which indicates that the disturbance strain field is not continuous across the interface of the matrix and inclusion.

In fact, when $\rho \to 1$, the jump in strain across the interface can be evaluated as,

$$\begin{aligned} [\epsilon_{ij}] := \epsilon_{ij}(r \to a^+) - \epsilon_{ij}(r \to a^-) = & \Big(S^{E,\infty}_{ijmn} - S^{I,\infty}_{ijmn}\Big)\epsilon^*_{mn} \\ = & \frac{-1}{(1-\nu)}\Big[\nu\delta_{mn}n_i n_j + \frac{1}{2}(1-\nu)\Big(\delta_{im}n_j n_n + \delta_{in}n_j n_m \\ & +\delta_{jm}n_i n_n + \delta_{jn}n_i n_m\Big) - n_i n_j n_m n_n\Big]\epsilon^*_{mn} \end{aligned} \tag{3.159}$$

which is the weak discontinuity at the interface between the matrix and the inclusion. However, we must emphasize here, that the induced displacement and stress are continuous across the matrix/inclusion interface.

3.6.4 *The second derivatives of Green's function and the Eshelby tensors*

From (3.153), one may observe that the Eshelby tensors are basically the integrations of the second order derivative of Green's function, $\mathcal{G}^{\infty}_{ijmn}$,

$$\mathcal{G}^{\infty}_{ijmn}(\mathbf{x} - \mathbf{y}) = -\frac{1}{2}C_{k\ell mn}\Big(G^{\infty}_{ik,\ell j}(\mathbf{x} - \mathbf{y}) + G^{\infty}_{jk,\ell i}(\mathbf{x} - \mathbf{y})\Big) \tag{3.160}$$

in different domains. However, for the interior Eshelby tensor, the integration (3.153) is hypersingular and it is not well defined, and it is only valid in the sense of the distribution, because $\mathbf{x}$ can reach to $\mathbf{y}$. Therefore, to evaluate the following integral

$$S^{I,\infty}_{ijmn} = \int_\Omega \mathcal{G}^\infty_{ijmn}(\mathbf{x}-\mathbf{y})d\Omega_{\mathbf{y}} \tag{3.161}$$

one has to separate Ω two domains: a small spherical ball B that contains $\mathbf{x}$ and the rest of domain Ω/B.

E. Kröner [Kröner (1986, 1990)] and S. Torquato [Torquato (1997)] showed that if the ball is spherical in shape, which eliminates the singularity, the second order derivatives of the Green's function tensor can be decomposed into two parts,

$$\begin{aligned}\Gamma^\infty_{ijk\ell}(\mathbf{x}-\mathbf{y}) &:= \frac{\partial^2 G^\infty_{ik}}{\partial x_j \partial y_\ell}\Big|_{(ij)(k\ell)}(\mathbf{x}-\mathbf{y}) \\ &= -A_{ijk\ell}\delta(\mathbf{x}-\mathbf{y}) + H_{ijk\ell}(\mathbf{x}-\mathbf{y})\end{aligned} \tag{3.162}$$

where the symbol $(i,j)(k,\ell)$ means that the indices are symmetric between i and j and between k and ℓ, and

$$A_{ijk\ell} = \frac{1}{3K+4\mu}E^{(1)}_{ijk\ell} + \frac{3}{5\mu}\frac{K+2\mu}{3K+4\mu}E^{(2)}_{ijk\ell} \tag{3.163}$$

$$\begin{aligned}H_{ijk\ell} = \frac{1}{2\Omega(3K+4\mu)}\cdot\frac{1}{|\mathbf{x}-\mathbf{y}|^3}\Big[&\alpha\delta_{ij}\delta_{k\ell} - 3(\delta_{ik}\delta_{j\ell}+\delta_{i\ell}\delta_{jk}) \\ &-3\alpha(\delta_{ij}n_kn_\ell + \delta_{k\ell}n_in_j) + \frac{3(2-\alpha)}{2}(\delta_{ik}n_jn_i + \delta_{i\ell}n_jn_k \\ &+\delta_{jk}n_in_\ell + \delta_{j\ell}n_in_k + 15\alpha n_in_jn_kn_\ell\Big]\end{aligned} \tag{3.164}$$

where K and μ are material bulk and shear modulus, $\alpha = 3\frac{K}{\mu}+1$, and $\Omega = \frac{2\pi^{3/2}}{\Gamma(3/2)}$. Subsequently, we can derive that

$$\mathcal{G}^\infty_{ijk\ell}(\mathbf{x}-\mathbf{y}) = \Gamma^\infty_{ijmn}C_{mnk\ell} = \mathcal{E}_{ijk\ell}\delta(\mathbf{x}-\mathbf{y}) + \mathcal{F}_{ijk\ell}(\mathbf{x}-\mathbf{y}) \tag{3.165}$$

For isotropic materials $C_{mnk\ell} = 2\lambda\delta_{mn}\delta_{k\ell}+\mu(\delta_{mk}\delta_{n\ell}+\delta_{m\ell}\delta_{nk})$ and $K = \frac{2\mu(1+\nu)}{3(1-2\nu)}$. It is straightforward to verify that

$$\mathcal{E}_{ijk\ell} = S^{I,\infty}_{ijk\ell} = \frac{5\nu-1}{15(1-\nu)}\delta_{ij}\delta_{k\ell} + \frac{(4-5\nu)}{15(1-\nu)}(\delta_{ik}\delta_{j\ell}+\delta_{jk}\delta_{i\ell}) \tag{3.166}$$

and

$$\begin{aligned}\mathcal{F}_{ijmn}(\mathbf{x}-\mathbf{y}) = \frac{1}{8\pi(1-\nu)r^3}\Big[&(1-2\nu)(\delta_{im}\delta_{jn}+\delta_{in}\delta_{jm}-\delta_{ij}\delta_{mn}) \\ &+3\nu(\delta_{im}\ell_j\ell_n + \delta_{in}\ell_j\ell_m + \delta_{jm}\ell_i\ell_n + \delta_{jn}\ell_j\ell_m) \\ &+3\delta_{ij}\ell_m\ell_n + 3(1-2\nu)\delta_{mn}\ell_i\ell_j - 15\ell_i\ell_j\ell_m\ell_n\Big]\end{aligned} \tag{3.167}$$

where $\mathbf{y}\in\Omega/B$, $\mathbf{x}$ is in the center of B, and $r = |\mathbf{x}-\mathbf{y}|$ measuring the length of the segment $\mathbf{x}-\mathbf{y}$.

3.7 Applications of eigenstrain theory

The eigenstrain method has some useful practical applications, including calculating stress/strain fields induced by inhomogeneities, inclusions, dislocations, and cracks. In what follows, we shall discuss a few examples.

3.7.1 *Strain field in embedded quantum dots*

In recent years, study of quantum dots (QDs) have attracted tremendous attention. Quantum dots, typically semiconductor materials, is spatially confined electron clusters with typical dimensions between nanometers to a few microns. Since quantum dots have a sharper density of states than its ambient material, they usually have superior transport and optical properties. The unique properties of quantum dots has been considered to hold the great promise for developing new nanodevices, and are being researched to use in the next generation imaging probes, quantum-dot based sensors, new light sources, and even solid-state quantum computation.

Quantum dots can occur or be synthesized into different shapes, including sphere, plate, and pyramid. They are often embedded in a host matrix with distinct elastic properties and lattice structures (cf. Fig. 3.14). The lattice mismatch between QDs and the matrix would induce misfit strain, which subsequently has significant impact on the nanostructure and the optoelectronic properties of QDs. A review on strain field calculations of embedded quantum dots can be found in Maranganti and Sharma (2005). In this section, we illustrate a simple example that uses eigenstrain theory to compute the strain field of an embedded quantum dots, which was mainly referred from Maranganti and Sharma (2005). Consider a spherical quantum dot Ω of radius a, embedded in an infinite host matrix M. The induced *misfit strain* due to lattice mismatch can be evaluated as

$$\epsilon_{ij}^{m} = \epsilon^{m}\delta_{ij} \tag{3.168}$$

Considering the spherical symmetry of QDs, the above equation assumes the misfit strain to be volumetric, and it can be related to the lattice parameters of the quantum dot and the matrix,

$$\epsilon^{m} = \frac{L_{\Omega} - L_{M}}{L_{M}} \tag{3.169}$$

where L_{Ω} and L_{M} are lattice parameters of the QD (Ω) and the matrix M, respectively.

Recall the definition of eigenstrain, the misfit strain induced by lattice mismatch is such an example that we assume it is prescribed over the QD domain (Ω). We should point out here, that the setting is different from our previous lecture. The misfit strain $\boldsymbol{\epsilon}^{m}$ is now prescribed over the inclusion which has different material properties from the matrix. We further introduce a fictitious eigenstrain $\boldsymbol{\epsilon}^{f}$ and invoke the equivalent eigenstrain theory to account for the difference in QD and matrix properties. Therefore, after the total eigenstrain $\boldsymbol{\epsilon}^{*} = \boldsymbol{\epsilon}^{m} + \boldsymbol{\epsilon}^{f}$ is prescribed

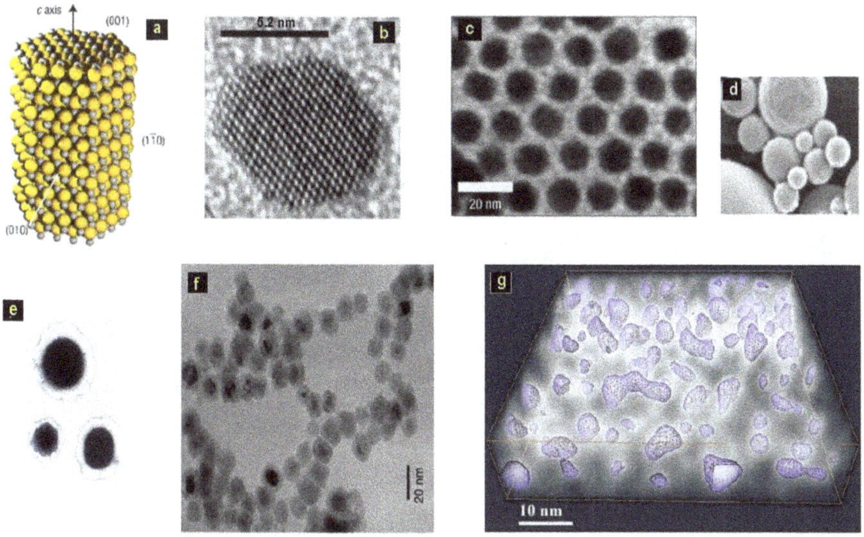

(a) A structural model of idealized wurtzite nanorod. (b) A high-resolution transmission electron micrograph of an individual CdSe wurtzite nanorod, looking down the nanocrystal c-axis. From: Fritz, K.P. *et al.*, Structural characterization of CdSe nanorods. J. Cryst. Growth, 293, 203-208, 2006. Reprinted with permission from Elsevier. (c)TEM image of 10 nm gold QDs (dark areas) bound within the pores of the 9-nm-pore chaperonin protein template. From: R.A. McMillan *et al.*, Ordered nanoparticle arrays formed on engineered chaperonin protein templates. Nature Materials 1, 247-252, 2002. Reproduced with permission from Nature Publishing Group. (d) Scanning electron micrograph of a nanocore synthesized from doxorubicin-conjugated PLGA polymer. (e) TEM image of a cross-section of a nanocell. The dark center is the nanocore entrapped inside the lipid layer. (f) Fe nanoparticles coated with polymer, which stabilizes the nanoparticles. (d) (e) and (f) from: S. Sengupta and R. Sasisekharan, Exploiting nanotechnology to target cancer. British Journal of Cancer, 96, 1315-1319, 2007. Reproduced with permission from Nature Publishing Group. (g) Silicon nanoparticles embedded in silicon oxide. The tomographic reconstruction shows 3D distribution of non-spherical nanoparticles. From: A. Yurtsever*et al.*, 3D imaging of non-spherical silicon nanoparticles embedded in silicon oxide by plasmon tomography. Microsc Microanal 12 (supp 2) 2006. Reproduced with permission from Microscopy Society of America.

Fig. 3.14 Some examples of quantum dots.

over the inclusion domain, the inclusion and the matrix can be assumed to have the same property. The stress consistency condition states that,

$$\left.\begin{array}{l}\mathbb{C}^M : (\boldsymbol{\epsilon} - \boldsymbol{\epsilon}^m - \boldsymbol{\epsilon}^f) = \mathbb{C}^\Omega : (\boldsymbol{\epsilon} - \boldsymbol{\epsilon}^m) \\ \boldsymbol{\epsilon} = \mathbb{S}^{I,\infty} : (\boldsymbol{\epsilon}^m + \boldsymbol{\epsilon}^f)\end{array}\right\} \quad \mathbf{x} \in \Omega. \tag{3.170}$$

We can first link the total eigenstrain with the misfit strain by,

$$\boldsymbol{\epsilon}^m + \boldsymbol{\epsilon}^f = \left[\mathbb{C}^M : (\mathbb{I}^{(4s)} - \mathbb{S}^{I,\infty}) + \mathbb{C}^\Omega : \mathbb{S}^{I,\infty}\right]^{-1} : \mathbb{C}^\Omega : \boldsymbol{\epsilon}^m$$

We can then establish a direct connection between the total strain ($\mathbf{x} \in \Omega$) and the misfit strain,

$$\boldsymbol{\epsilon} = \mathbb{S}^{I,\infty} : \left[\mathbb{C}^M : (\mathbb{I}^{(4s)} - \mathbb{S}^{I,\infty}) + \mathbb{C}^\Omega : \mathbb{S}^{I,\infty}\right]^{-1} : \mathbb{C}^\Omega : \boldsymbol{\epsilon}^m \tag{3.171}$$

It should be noted that the Eshelby tensor is only a function of the matrix properties. Since misfit strain $\boldsymbol{\epsilon}^m = \epsilon^m \mathbb{I}^{(2)}$ is only volumetric, the total strain inside the quantum dot is also volumetric, and also uniform due to the property of Eshelby tensor for spherical inclusion. Straightforward evaluation of Eq. (3.171) yields the following result: $\boldsymbol{\epsilon}^m + \boldsymbol{\epsilon}^f = (\epsilon^m + \epsilon^f)\mathbb{I}^{(2)}$, the tensorial expression (3.7.1) is really a scalar relation

$$\epsilon^m + \epsilon^f = \frac{3K^M + 4\mu^M}{3K^\Omega + 4\mu^M}\left(\frac{K^\Omega}{K^M}\right)\epsilon^m \ .$$

Therefore

$$\begin{aligned}
\boldsymbol{\epsilon} &= \mathbf{S} : (\boldsymbol{\epsilon}^m + \boldsymbol{\epsilon}^f) = (\epsilon^m + \epsilon^f)\mathbf{S} : \mathbb{I}^{(2)} \\
&= (\epsilon^m + \epsilon^f)\begin{cases} s_1 \delta_{ij} \mathbf{e}_i \otimes \mathbf{e}_j, & \text{interior solution} \\ s_1 \rho^3 (\delta_{ij} - 3n_i n_j)\mathbf{e}_i \otimes \mathbf{e}_j, & \text{exterior solution} \end{cases}
\end{aligned}$$

where $\rho = a/r$ and $s_1 = 3K^M/(3K^M + 4\mu^M)$. For the interior solution,

$$\boldsymbol{\epsilon} = \frac{3K^\Omega \epsilon^m}{3K^\Omega + 4\mu^M}\mathbf{e}_i \otimes \mathbf{e}_j, \quad \mathbf{x} \in \Omega \tag{3.172}$$

which depends on the properties of both QD and the host matrix.

The strain field outside the quantum dot could be evaluated directly using exterior Eshelby tensor,

$$\begin{aligned}
\boldsymbol{\epsilon} &= \mathbb{S}^{E,\infty} : (\boldsymbol{\epsilon}^m + \boldsymbol{\epsilon}^f) \\
&= (\epsilon^m + \epsilon^f)\frac{3K^M \rho^3}{3K^M + 4\mu^M}(\delta_{ij} - 3n_i n_j)\mathbf{e}_i \otimes \mathbf{e}_j, \quad \mathbf{x} \in M
\end{aligned} \tag{3.173}$$

The strain fields, Eqs. (3.172) and (3.173), were expressed in Cartesian coordinate, and they could be easily converted to spherical coordinates ($\boldsymbol{e}_r$, $\boldsymbol{e}_\theta$, $\boldsymbol{e}_\phi$), considering $\delta_{ij}\boldsymbol{e}_i \otimes \boldsymbol{e}_j = \boldsymbol{e}_r \otimes \boldsymbol{e}_r + \boldsymbol{e}_\theta \otimes \boldsymbol{e}_\theta + \boldsymbol{e}_\phi \otimes \boldsymbol{e}_\phi$, and $n_i \boldsymbol{e}_i = \boldsymbol{e}_r$. The radial and tangential components are obtained as follows

$$\epsilon_{rr}(r) = \epsilon_{\theta\theta}(r) = \epsilon_{\phi\phi}(r) = \frac{3K^\Omega \epsilon^m}{3K^\Omega + 4\mu^M}; \quad r < a \tag{3.174}$$

$$\left.\begin{aligned}
\epsilon_{rr}(r) &= \frac{-6K^\Omega \epsilon^m}{3K^\Omega + 4\mu^M}\left(\frac{a^3}{r^3}\right) \\
\epsilon_{\theta\theta}(r) = \epsilon_{\phi\phi}(r) &= \frac{3K^\Omega \epsilon^m}{3K^\Omega + 4\mu^M}\left(\frac{a^3}{r^3}\right)
\end{aligned}\right\}; \quad r > a \tag{3.175}$$

Due to the spherical symmetry of this problem, the displacement field only admits radial component, i.e., $\mathbf{u} = u(r)\boldsymbol{e}_r$. The displacement and strain are related through the following equation

$$\boldsymbol{\epsilon} = \frac{\partial u(r)}{\partial r}\boldsymbol{e}_r \otimes \boldsymbol{e}_r + \frac{u(r)}{r}\boldsymbol{e}_\theta \otimes \boldsymbol{e}_\theta + \frac{u(r)}{r}\boldsymbol{e}_\phi \otimes \boldsymbol{e}_\phi \tag{3.176}$$

It is easy to derive the displacement field inside and outside the quantum dot

$$\mathbf{u}(r) = \begin{cases} \dfrac{3K^\Omega \epsilon^m}{3K^\Omega + 4\mu^M} r, & r < a \\ \dfrac{3K^\Omega \epsilon^m}{3K^\Omega + 4\mu^M} \dfrac{a^3}{r^2}, & r > a \end{cases} \tag{3.177}$$

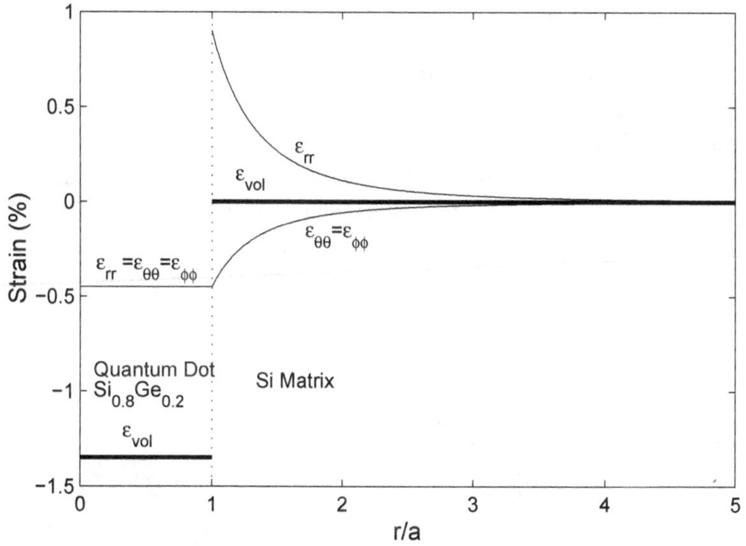

Fig. 3.15 Strain components for a $Si_{0.8}Ge_{0.2}$ quantum dot embedded in an infinite Si matrix.

Remark 3.3. In this example, we have evaluated the strain field inside and outside the quantum dot by using equivalent eigenstrain theory. The simple idealized example underscores several interesting features:

(1) The displacement field is continuous across inclusion and matrix interface, which verified our assumption that the inclusion and matrix are perfectly bonded.
(2) The induced strain inside the quantum dot is uniform. This property is a direct consequence that the interior Eshelby tensor is uniform for a spherical inclusion. One may expect the same observation also holds true for a general ellipsoidal shaped quantum dot.

(3) The tangential strain ($\epsilon_{\theta\theta}$, $\epsilon_{\phi\phi}$) is continuous along the QD and matrix interface, while the radial strain component (ϵ_{rr}) experienced a jump at the material interface. The strain outside the dot is rapidly decreasing towards zero (in the order of r^{-3}). From Fig. 3.15, it can be concluded that the strain is negligible at a distance as low as 3.5 times of the QD radius.

(4) Fig. 3.15 also shows the volumetric strain (dilation) inside the QD is uniform, and more importantly, the dilation outside the QD is always zero everywhere. This finding is manifested in the fact that the volumetric part of the exterior Eshelby tensor $S_{iijj}^{E,\infty} = 0$. In [Maranganti and Sharma (2005)], it has also been pointed out that this salient attribute actually hold true for *all quantum dot shapes*, provided that the QD and the matrix is perfectly bounded.

3.7.2 *Dislocation problems*

A dislocation is a distorted region among substantially perfect crystal lattice environment. In other words, a dislocation is a linear defect around which some of the atoms are misaligned or crystal lattice being distorted. There are two types of dislocations: (1) edge dislocation, and (2) screw dislocation (see Fig. 3.16). Use of eigenstrain theory to describe the effect of dislocations and their induced disturbance mechanical fields is a success. Eigenstrain theory has been an important approach in the development of dislocation theory. Here we only introduce some simple examples.

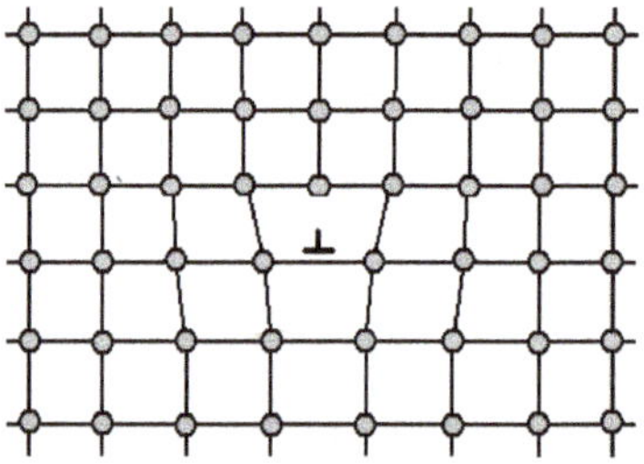

(a)

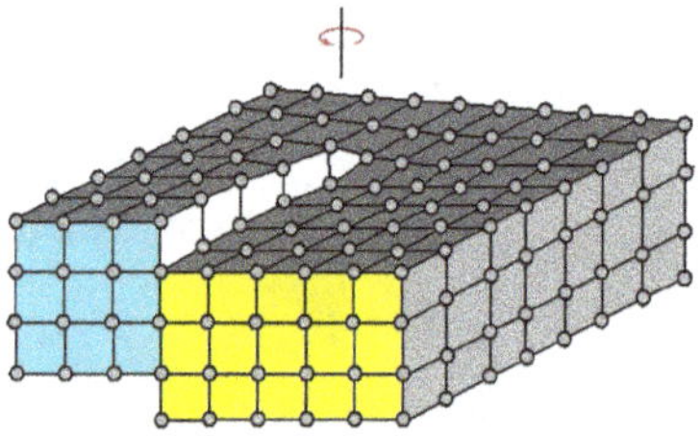

(b)

Fig. 3.16 Illustrations of dislocations: (a)edge dislocation, and (b) screw dislocation.

Consider a straight screw dislocation on a half space (Fig. 3.17). There is a jump or discontinuity in displacement at $x_2 = 0$ and $-\infty < x_1 < 0$, with the magnitude of b, the Burger's vector. A fictitious eigenstrain field is prescribed on the slip plane to mimic the mechanical effect of dislocation,

$$\epsilon_{23}^* = \begin{cases} \frac{1}{2} b\delta(x_2)H(-x_1), & \mathbf{x} \in \Omega \\ 0. & \mathbf{x} \in \mathbb{R}^3/\Omega \end{cases} \tag{3.178}$$

where the slip surface may be described as

$$\Omega = \Big\{(x_1, 0, x_3)\Big| x_1 < 0, -\infty < x_3 < \infty\Big\}$$

and $H(\cdot)$ is the heaviside function.

The eigenstrain field may be considered as the consequence of the displacement field,

$$u_3^*(\mathbf{x}) = bH(x_2)H(-x_1) \tag{3.179}$$

since

$$\epsilon_{23}^* = \frac{1}{2}\Big(\frac{\partial u_3^*}{\partial x_2} + \frac{\partial u_2^*}{\partial x_3}\Big) = \frac{b}{2}\delta(x_2)H(-x_1)$$

(Question: what about ϵ_{31}^* ? and what is the definition of shear strain ?)

Apply Fourier transform

$$\begin{aligned}\bar{\epsilon}_{23}^*(\boldsymbol{\xi}) &= \frac{1}{(2\pi)^3}\int_{\mathbb{R}^3} \epsilon_{23}^*(\mathbf{x}) \exp(-i\boldsymbol{\xi}\cdot\mathbf{x}))d\mathbf{x} \\ &= \frac{1}{(2\pi)^3}\int_{\mathbb{R}^3} \frac{b}{2}\delta(x_2)H(-x_1)\exp(-i\boldsymbol{\xi}\cdot\mathbf{x})d\mathbf{x}\end{aligned} \tag{3.180}$$

Consider

$$\int_{\mathbb{R}} \delta(x_2)\exp(-i\xi_2 x_2)dx_2 = 1 \tag{3.181}$$

$$\begin{aligned}&\int_{\mathbb{R}} H(-x_1)\exp(-i\xi_1 x_1)dx_1 = \int_{-\infty}^{0} \exp(-i\xi_1 x_1)dx_1 \\ &= \frac{i}{\xi_1}, \quad \forall Im(\xi_1) > 0, \quad \text{and} \\ &\frac{1}{2\pi}\int_{\mathbb{R}} \exp(-i\xi_3 x_3)dx_3 = \delta(\xi_3)\end{aligned} \tag{3.182}$$

Therefore,

$$\bar{\epsilon}_{23}^* = \frac{1}{(2\pi)^2}\frac{b}{2}\Big(\frac{i}{\xi_1}\Big)\delta(\xi_3) \tag{3.183}$$

Substituting (3.183) into the general formula of micro-elasticity,

$$\begin{aligned}u_i(\mathbf{x}) &= -i\int_{\mathbb{R}^3} C_{j\ell mn}\bar{\epsilon}_{mn}^*\xi_\ell K_{ij}^{-1}(\boldsymbol{\xi})\exp(i\boldsymbol{\xi}\cdot\mathbf{x})d\boldsymbol{\xi} \\ &= -2i\int_{\mathbb{R}^3} C_{j\ell 23}\bar{\epsilon}_{23}^*\xi_\ell K_{ij}^{-1}(\boldsymbol{\xi})\exp(i\boldsymbol{\xi}\cdot\mathbf{x})d\boldsymbol{\xi} \\ &= \Big(\frac{b}{(2\pi)^2}\Big)\int_{\mathbb{R}^3}\Big(\frac{\delta(\xi_3)}{\xi_1}\Big)C_{j\ell 23}\xi_\ell K_{ij}^{-1}(\boldsymbol{\xi})\exp(i\boldsymbol{\xi}\cdot\mathbf{x})d\boldsymbol{\xi}\end{aligned} \tag{3.184}$$

where the factor 2 is due to the presence of ϵ_{32}^*, if the minor symmetry is being considered. For isotropic materials,

$$\begin{aligned}C_{j\ell 23} &= \lambda\delta_{j\ell}\delta_{23} + \mu(\delta_{j2}\delta_{\ell 3} + \delta_{j3}\delta_{\ell 2}) \\ &= \mu(\delta_{j2}\delta_{\ell 3} + \delta_{j3}\delta_{\ell 2})\end{aligned}$$

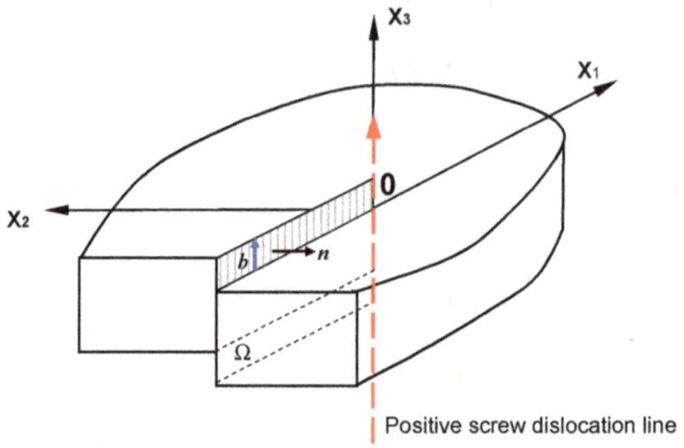

Fig. 3.17 A screw dislocation.

The only non-zero components are $C_{2323} = \mu$ and $C_{3223} = \mu$. Therefore,

$$u_1(\mathbf{x}) = \left(\frac{b}{(2\pi)^2}\right) \int_{\mathbb{R}^3} \left(\frac{\delta(\xi_3)}{\xi_1}\right) \left(C_{2323} K_{12}^{-1}(\boldsymbol{\xi})\xi_3 + C_{3223} K_{13}^{-1}(\boldsymbol{\xi})\xi_2\right) \exp(i\xi \cdot \mathbf{x}) d\boldsymbol{\xi}$$

$$u_2(\mathbf{x}) = \left(\frac{b}{(2\pi)^2}\right) \int_{\mathbb{R}^3} \left(\frac{\delta(\xi_3)}{\xi_1}\right) \left(C_{2323} K_{22}^{-1}(\boldsymbol{\xi})\xi_3 + C_{3223} K_{23}^{-1}(\boldsymbol{\xi})\xi_2\right) \exp(i\xi \cdot \mathbf{x}) d\boldsymbol{\xi}$$

$$u_3(\mathbf{x}) = \left(\frac{b}{(2\pi)^2}\right) \int_{\mathbb{R}^3} \left(\frac{\delta(\xi_3)}{\xi_1}\right) \left(C_{2323} K_{32}^{-1}(\boldsymbol{\xi})\xi_3 + C_{3223} K_{33}^{-1}(\boldsymbol{\xi})\xi_2\right) \exp(i\xi \cdot \mathbf{x}) d\boldsymbol{\xi}$$

in which,

$$K_{12}^{-1}(\boldsymbol{\xi}) = -\frac{(\lambda+\mu)}{\mu(\lambda+2\mu)} \frac{\xi_1\xi_2}{\xi^4}$$

$$K_{22}^{-1}(\boldsymbol{\xi}) = \frac{[(\lambda+2\mu)\xi^2 - (\lambda+\mu)\xi_2^2]}{\mu(\lambda+2\mu)\xi^4}$$

$$K_{13}^{-1}(\boldsymbol{\xi}) = -\frac{(\lambda+\mu)}{\mu(\lambda+2\mu)} \frac{\xi_1\xi_3}{\xi^4}$$

$$K_{23}^{-1}(\boldsymbol{\xi}) = -\frac{(\lambda+\mu)}{\mu(\lambda+2\mu)} \frac{\xi_2\xi_3}{\xi^4}$$

$$K_{32}^{-1}(\boldsymbol{\xi}) = K_{23}^{-1}(\boldsymbol{\xi})$$

$$K_{33}^{-1}(\boldsymbol{\xi}) = \frac{[(\lambda+2\mu)\xi^2 - (\lambda+\mu)\xi_3^2]}{\mu(\lambda+2\mu)\xi^4} \tag{3.185}$$

Obviously,

$$\int_{\mathbb{R}} \delta(\xi_3) K_{12}^{-1}(\boldsymbol{\xi}) \xi_3 d\xi_3 = 0$$
$$\int_{\mathbb{R}} \delta(\xi_3) K_{13}^{-1}(\boldsymbol{\xi}) \xi_2 d\xi_3 = 0$$
$$\int_{\mathbb{R}} \delta(\xi_3) K_{22}^{-1}(\boldsymbol{\xi}) \xi_3 d\xi_3 = 0$$
$$\int_{\mathbb{R}} \delta(\xi_3) K_{23}^{-1}(\boldsymbol{\xi}) \xi_2 d\xi_3 = 0$$
$$\int_{\mathbb{R}} \delta(\xi_3) K_{32}^{-1}(\boldsymbol{\xi}) \xi_3 d\xi_3 = 0$$
$$\int_{\mathbb{R}} \delta(\xi_3) K_{33}^{-1}(\boldsymbol{\xi}) \xi_2 d\xi_3 = \frac{1}{\mu} \frac{\xi_2}{(\xi_1^2 + \xi_2^2)}$$

Thereby, $u_1(\mathbf{x}) = u_2(\mathbf{x}) = 0$, and

$$u_3(\mathbf{x}) = \frac{b}{(2\pi)^2} \int_{\mathbb{R}^2} \frac{\xi_2}{\xi_1(\xi_1^2 + \xi_2^2)} \exp\Big(i(\xi_1 x_1 + \xi_2 x_2)\Big) d\xi_1 d\xi_2$$
$$= \frac{b}{2\pi} \tan^{-1}\Big(\frac{x_2}{x_1}\Big) \tag{3.186}$$

according to the inverse Fourier transform e.g. [Mura (1987)],

$$\int_{\mathbb{R}^2} \frac{\xi_2}{\xi_1(\xi_1^2 + \xi_2^2)} \exp\Big(i(\xi_1 x_1 + \xi_2 x_2)\Big) d\xi_1 d\xi_2 = 2\pi \tan^{-1}\Big(\frac{x_2}{x_1}\Big)$$

3.7.3 *Stress intensity factor for a flat ellipsoidal crack*

In late 1960s, J. R. Willis (1967a, 1968) used the eigenstrain method solving a class of crack and contact problems in anisotropic space. In the following, we illustrate Willis' solution procedure in the case of a 3D ellipsoidal crack in an isotropic space.

Consider an ellipsoidal crack embedded in an infinite space. Suppose that the crack region Ω is:

$$\Omega = \Big\{ \mathbf{x} \ \Big| \ \frac{x_1^2}{a_1^2} + \frac{x_2^2}{a_2^2} \leq 1, \quad \text{and} \ \ x_3 = 0 \ \Big\}. \tag{3.187}$$

For simplicity, we assume that the crack opening has the following form:

$$[u_3] = b\sqrt{1 - \frac{x_1^2}{a_1^2} - \frac{x_2^2}{a_2^2}}\ \chi(\Omega) \tag{3.188}$$

where parameter b is the Burger's vector, and $\chi(\Omega)$ is the characteristic function of crack region, which can be defined as interpreted as

$$\chi(\Omega) = H(\mathbf{x} - \Omega) = \begin{cases} 1, \ \forall \mathbf{x} \in \Omega \\ 0, \ \forall \mathbf{x} \in \mathbb{R}^3/\Omega \end{cases} \tag{3.189}$$

where $H(\cdot)$ is the Heaviside function.

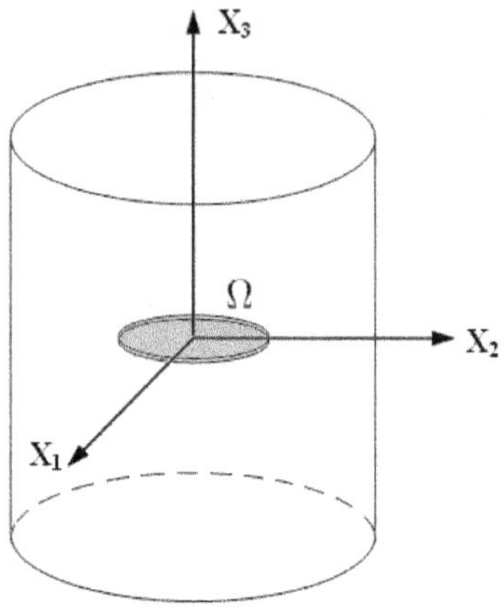

Fig. 3.18 A three-dimensional ellipsoidal crack.

Because $\dfrac{d}{dx_3}H(\mathbf{x}-\Omega) = \delta(\mathbf{x}-\Omega)$, we can prescribe the following eigenstrain on the crack surface to simulating the crack opening,

$$\epsilon^*_{33} = b\sqrt{1 - \frac{x_1^2}{a_1^2} - \frac{x_2^2}{a_2^2}}\ \delta(\mathbf{x}-\Omega)\ . \tag{3.190}$$

Therefore,

$$\int_{\mathbb{R}^3} \epsilon^*_{33}(\mathbf{x}') \exp\Big(i\boldsymbol{\xi}\cdot(\mathbf{x}-\mathbf{x}')\Big) d\mathbf{x}' =$$

$$\int_\Omega b\sqrt{1 - \frac{x_1'^2}{a_1^2} - \frac{x_2'^2}{a_2^2}} \exp(i\boldsymbol{\xi}\cdot(\mathbf{x}-\mathbf{x}')\Big) dx_1' dx_2' \tag{3.191}$$

where in the right hand side of the above equation, all vectors become 2D vectors, i.e. $\boldsymbol{\xi} = \xi_1\mathbf{e}_1 + \xi_2\mathbf{e}_2$, $\mathbf{x} = x_1\mathbf{e}_1 + x_2\mathbf{e}_2$, and $\mathbf{x}' = x_1'\mathbf{e}_1 + x_2'\mathbf{e}_2$.

Employing the fundamental formula of micro-elasticity,

$$u_i(\mathbf{x}) = -\frac{i}{(2\pi)^3}\int_{\mathbb{R}^3}\left\{\int_{\mathbb{R}^3} C_{j\ell mn}\epsilon^*_{mn}(\mathbf{x}')\xi_\ell K^{-1}_{ij}(\boldsymbol{\xi}) \exp\Big(i\boldsymbol{\xi}\cdot(\mathbf{x}-\mathbf{x}')\Big) d\mathbf{x}'\right\} d\boldsymbol{\xi} \tag{3.192}$$

and changing the dummy indices $i \to k, j \to p, m \to 3, n \to 3, \ell \to q$, we have

$$u_k(x_1, x_2, 0) = \frac{-ib}{(2\pi)^3}\int_{\mathbb{R}^3}\int_\Omega C_{pq33}\sqrt{1 - \frac{x_1'^2}{a_1^2} - \frac{x_2'^2}{a_2^2}}\,\xi_q K^{-1}_{kp}(\boldsymbol{\xi}) \cdot \exp\Big(i\boldsymbol{\xi}\cdot(\mathbf{x}-\mathbf{x}')\Big) dx_1' dx_2' d\boldsymbol{\xi} \tag{3.193}$$

and

$$u_{k,\ell}(x_1, x_2, 0) = \frac{b}{(2\pi)^3}\int_{\mathbb{R}^3}\int_\Omega C_{pq33}\sqrt{1 - \frac{x_1'^2}{a_1^2} - \frac{x_2'^2}{a_2^2}}\,\xi_q \xi_\ell K^{-1}_{kp}(\boldsymbol{\xi}) \cdot \exp\Big(i\boldsymbol{\xi}\cdot(\mathbf{x}-\mathbf{x}')\Big) dx_1' dx_2' d\boldsymbol{\xi} \tag{3.194}$$

Subsequently,

$$\sigma_{ij} = C_{ijk\ell}u_{k,\ell} = \frac{b}{(2\pi)^3}\int_{\mathbb{R}^3} C_{ijk\ell}C_{pq33}\xi_q\xi_\ell K^{-1}_{kp}(\boldsymbol{\xi})$$
$$\cdot\left\{\int_\Omega \sqrt{1-\frac{(x_1')^2}{a_1^2}+\frac{(x_2')^2}{a_2^2}}\exp\Big(i\boldsymbol{\xi}\cdot(\mathbf{x}-\mathbf{x}')\Big)dx_1'dx_2'\right\}d\boldsymbol{\xi} \tag{3.195}$$

We first calculate the inverse Fourier transform along ξ_3 axis, i.e. evaluating the following integral,

$$\int_{-\infty}^{\infty} C_{ijk\ell}C_{pq33}\xi_q\xi_\ell K^{-1}_{kp}(\boldsymbol{\xi})\exp\Big(i\boldsymbol{\xi}\cdot(\mathbf{x}-\mathbf{x}')\Big)d\xi_3\,. \tag{3.196}$$

For isotropic materials,

$$K^{-1}_{kp}(\boldsymbol{\xi}) = \frac{(\lambda+2\mu)\delta_{kp}\xi^2-(\lambda+\mu)\xi_k\xi_p}{\mu(\lambda+2\mu)\xi^4} \tag{3.197}$$

where the denominator may be decomposed into

$$\xi^4 = \left(\xi_1^2+\xi_2^2+\xi_3^2\right)^2 = \left(\xi_3 - i\sqrt{\xi_1^2+\xi_2^2}\right)^2\left(\xi_3 + i\sqrt{\xi_1^2+\xi_2^2}\right)^2 \tag{3.198}$$

Since the problem is symmetric with respect to X_1OX_2 plane, we only consider the upper half space ($x_3 > 0$). Because of the convergence requirement of Fourier transform, we are only interested in the root with a negative imaginary part, i.e.

$$\xi_3^N = -i\sqrt{\xi_1^2+\xi_2^2} \tag{3.199}$$

which is a double root as shown in Eq. (3.198).

Suppose z_j is a n-th pole of $f(z)$, its residue is then

$$\mathrm{Res}(f(z))\Big|_{z=z_j} = \frac{1}{(n-1)!}\lim_{z\to z_j}\frac{d^{n-1}}{dz^{n-1}}\Big[(z-z_j)^n f(z)\Big] \tag{3.200}$$

By the Residue Theorem, we have

$$\int_{-\infty}^{\infty}\Big(\text{Integrand of (3.196)}\Big)d\xi_3 = 2\pi i\ \mathrm{Res}\Big(\text{Integrand of (3.196)}\Big)\Big|_{\xi_3=\xi_3^N} \tag{3.201}$$

which can be expressed as

$$F_{ij3} = 2\pi i C_{ijk\ell}C_{pq33}\frac{\partial}{\partial\xi_3}\left\{(\xi_3-\xi_3^N)^2\xi_q\xi_\ell K^{-1}_{kp}(\boldsymbol{\xi})\exp(i\boldsymbol{\xi}\cdot(\mathbf{x}-\mathbf{x}'))\right\} \tag{3.202}$$

After some tedious calculation, we find that on the plane $x_3 = 0$,

$$F_{311} = \frac{2\pi\mu(\lambda+\mu)}{(\lambda+2\mu)}\Big(\xi_1^2+\frac{1}{2}\frac{\lambda+2\mu}{\lambda+\mu}\xi_2^2\Big)(\xi_1^2+\xi_2^2)^{-1/2}\exp(i\boldsymbol{\xi}\cdot(\mathbf{x}-\mathbf{x}')) \tag{3.203}$$

$$F_{322} = \frac{2\pi\mu(\lambda+\mu)}{(\lambda+2\mu)}\Big(\xi_2^2+\frac{1}{2}\frac{\lambda+2\mu}{\lambda+\mu}\xi_1^2\Big)(\xi_1^2+\xi_2^2)^{-1/2}\exp(i\boldsymbol{\xi}\cdot(\mathbf{x}-\mathbf{x}')) \tag{3.204}$$

$$F_{333} = \frac{2\pi\mu(\lambda+\mu)}{(\lambda+2\mu)}\sqrt{\xi_1^2+\xi_2^2}\exp(i\boldsymbol{\xi}\cdot(\mathbf{x}-\mathbf{x}')) \tag{3.205}$$

Note that the vectors in the exponential are planar vectors.

Hence,

$$\sigma_{33}(x_1,x_2,0) = \frac{b\mu(\lambda+\mu)}{4\pi^2(\lambda+2\mu)} \int_\Omega \sqrt{1-\frac{x_1'^2}{a_1^2}-\frac{x_2'^2}{a_2^2}}$$

$$\left\{ \int_{-\infty}^{\infty}\int_{-\infty}^{\infty} \left(\xi_1^2+\xi_2^2\right)^{1/2} \exp\big(i\boldsymbol{\xi}\cdot(\mathbf{x}-\mathbf{x}')\big) d\xi_1 d\xi_2 \right\} dx_1' dx_2' \qquad (3.206)$$

Let $y_1 = x_1/a_1, y_2 = x_2/a_2$, $y = \sqrt{y_1^2+y_2^2}$; $y_1' = x_1'/a_1, y_2' = x_2'/a_2$, $y' = \sqrt{(y_1')^2+(y_2')^2}$; $\zeta_1 = a_1\xi_1, \zeta_2 = a_2\xi_2$; and $\eta_1 = \zeta_1/\zeta, \eta_2 = \zeta_2/\zeta$, where $\zeta = \sqrt{\zeta_1^2+\zeta_2^2}$. Then

$$\boldsymbol{\xi}\cdot(\mathbf{x}-\mathbf{x}') = \boldsymbol{\zeta}\cdot(\mathbf{y}-\mathbf{y}') \qquad (3.207)$$

$$dx_1'dx_2'd\xi_1 d\xi_2 = dy_1'dy_2'd\zeta_1 d\zeta_2 \qquad (3.208)$$

$$\sqrt{1-\frac{x_1'^2}{a_1^2}-\frac{x_2'^2}{a_2^2}} = \sqrt{1-y_1'^2-y_2'^2} = \sqrt{1-y'^2} \qquad (3.209)$$

$$\sqrt{\xi_1^2+\xi_2^2} = \zeta\sqrt{\frac{\eta_1^2}{a_1^2}+\frac{\eta_2^2}{a_2^2}} \qquad (3.210)$$

Thus in Eq. (3.206)

$$\int_{-\infty}^{\infty}\int_{-\infty}^{\infty} \xi \exp(i\boldsymbol{\xi}\cdot(\mathbf{x}-\mathbf{x}')d\xi_1 d\xi_2$$

$$= \int_{-\infty}^{\infty}\int_{-\infty}^{\infty} \zeta\sqrt{\frac{\eta_1^2}{a_1^2}+\frac{\eta_2^2}{a_2^2}} \exp\Big(i\zeta\boldsymbol{\eta}\cdot(\mathbf{y}-\mathbf{y}')\Big)d\zeta_1 d\zeta_2$$

$$= \int_0^{2\pi}\int_0^{\infty} \zeta^2\sqrt{\left(\frac{\eta_1}{a_1}\right)^2+\left(\frac{\eta_2}{a_2}\right)^2} \exp\Big(i\zeta\boldsymbol{\eta}\cdot(\mathbf{y}-\mathbf{y}')\Big)d\zeta d\phi \qquad (3.211)$$

Denote $g = -\boldsymbol{\eta}\cdot\mathbf{y}$. The above integral becomes

$$\int_0^{2\pi}\int_0^{\infty} \zeta^2\sqrt{\left(\frac{\eta_1}{a_1}\right)^2+\left(\frac{\eta_2}{a_2}\right)^2} \exp\Big(i\zeta\boldsymbol{\eta}\cdot(\mathbf{y}-\mathbf{y}')\Big)d\zeta d\phi$$

$$= \int_0^{2\pi} \sqrt{\left(\frac{\eta_1}{a_1}\right)^2+\left(\frac{\eta_2}{a_2}\right)^2} \left\{ \Big(-\frac{\partial^2}{\partial g^2}\Big) \int_0^{\infty} \exp(-i\zeta(g+\boldsymbol{\eta}\cdot\mathbf{y}')d\zeta \right\} d\phi$$

$$= \int_0^{2\pi} \sqrt{\left(\frac{\eta_1}{a_1}\right)^2+\left(\frac{\eta_2}{a_2}\right)^2} \Big(\frac{\partial^2}{\partial g^2}\Big(\frac{i}{g+\boldsymbol{\eta}\cdot\mathbf{y}'}\Big)\Big) d\phi \qquad (3.212)$$

Denoting $\boldsymbol{\eta}\cdot\mathbf{y}' = y'\cos(\theta-\phi)$ and considering (3.206), we have

$$\sigma_{33}\Big|_{x_3=0} = \frac{ib\mu(\lambda+\mu)}{(2\pi)^2(\lambda+2\mu)} \int_0^{2\pi} \sqrt{\left(\frac{\eta_1}{a_1}\right)^2+\left(\frac{\eta_2}{a_2}\right)^2}$$

$$\cdot\left\{ \frac{\partial^2}{\partial g^2} \int_\Omega \frac{\sqrt{1-y'^2}}{g+y'\cos(\theta-\phi)} dy_1' dy_2' \right\} d\phi \qquad (3.213)$$

Using the polar coordinate,

$$\int_\Omega \frac{\sqrt{1-y'^2}}{g+y'\cos(\theta-\phi)} dy_1'dy_2' = \int_0^1\int_0^{2\pi} \frac{y'\sqrt{1-y'^2}}{g+y'\cos(\theta-\phi)} d\theta dy' \qquad (3.214)$$

and considering following integral identity,

$$\int_0^{2\pi} \frac{d(\theta - \phi)}{g + y' \cos(\theta - \phi)} = \frac{2\pi}{\sqrt{g^2 - y'^2}} \ . \tag{3.215}$$

we further derive that

$$\sigma_{33} \Big|_{x_3=0} = \frac{ib\mu(\lambda + \mu)}{2\pi(\lambda + \mu)} \int_0^{2\pi} \sqrt{\left(\frac{\eta_1}{a_1}\right)^2 + \left(\frac{\eta_2}{a_2}\right)^2} \cdot \left\{ \frac{\partial^2}{\partial g^2} \int_0^1 \frac{y' \sqrt{1 - y'^2} dy'}{\sqrt{g^2 - y'^2}} \right\} d\phi \tag{3.216}$$

Let

$$I = \int_0^1 \frac{y' \sqrt{1 - y'^2} dy'}{\sqrt{g^2 - y'^2}}$$

Changing of variables, we define w in terms of g and y',

$$(y')^2 = 1 - \frac{g^2 - 1}{4} \left(w - \frac{1}{w} \right)^2 \tag{3.217}$$

and

$$2y' dy' = -\frac{(g^2 - 1)}{2} (w - \frac{1}{w})(w + \frac{1}{w}) \frac{dw}{w} \tag{3.218}$$

Hence,

$$I = \left(\frac{g^2 - 1}{4} \right) \int_0^{\sqrt{\frac{g+1}{g-1}}} \left(w - \frac{2}{w} + \frac{1}{w^3} \right) dw \tag{3.219}$$

subsequently

$$\frac{\partial^2 I}{\partial g^2} = -\frac{1}{2} \ln \left(\frac{g+1}{g-1} \right) + \frac{g}{g^2 - 1} \ . \tag{3.220}$$

(1). Interior solution:

When $y < 1$, $x_3 = 0$, it is (interior) crack region. Obviously $|g| = |\boldsymbol{\eta} \cdot \mathbf{y}| < 1$. Since

$$\frac{g+1}{g-1} = -\left(\frac{1+g}{1-g} \right) = \exp(-i\pi) \left(\frac{1+g}{1-g} \right)$$

then,

$$\frac{\partial^2 I}{\partial g^2} = -\frac{1}{2} \ln \left| \frac{1+g}{1-g} \right| - i\frac{\pi}{2} + \frac{g}{g^2 - 1} \ . \tag{3.221}$$

Both $\ln|(1+g)/(1-g)|$ and $g/(g^2-1)$ are odd function of ϕ. We let $\eta_1 = \sin\phi, \eta_2 = \cos\phi$, and hence

$$\sqrt{\left(\frac{\eta_1}{a_1}\right)^2 + \left(\frac{\eta_2}{a_2}\right)^2} = \left(\frac{\sin^2\phi}{a_1^2} + \frac{\cos^2\phi}{a_2^2} \right)^{1/2}$$

is an even function of ϕ.

Assume that $a_1 > a_2$ and when $y < 1$

$$\sigma_{33}(x_1, x_2, 0) = \frac{b\mu(\lambda+\mu)}{4(\lambda+2\mu)} \int_0^{2\pi} \left(\frac{\sin^2\phi}{a_1^2} + \frac{\cos^2\phi}{a_2^2}\right)^{1/2} d\phi$$
$$= \frac{b\mu E(k)}{2a_2(1-\nu)} \tag{3.222}$$

where identities $\dfrac{\mu(\lambda+\mu)}{(\lambda+2\mu)} = \dfrac{\mu}{2(1-\nu)}$ and

$$E(k) = \int_0^{\pi/2} \sqrt{1-k^2\sin^2\phi d\phi}, \quad k^2 := \frac{a_1^2 - a_2^2}{a_1^2} > 0\,. \tag{3.223}$$

are used. If

$$\sigma_{33}(\Omega) = \sigma_{33}^0 = \frac{b\mu E(k)}{2a_2(1-\nu)} \tag{3.224}$$

it then links the Burgers' vector with the prescribed stress on the crack surfaces,

$$b = \frac{2(1-\nu)a_2\sigma_{33}^0}{\mu E(k)}\,. \tag{3.225}$$

This suggests that the type of prescribed eigenstrain is equivalent to prescribed constant stress on crack surfaces.

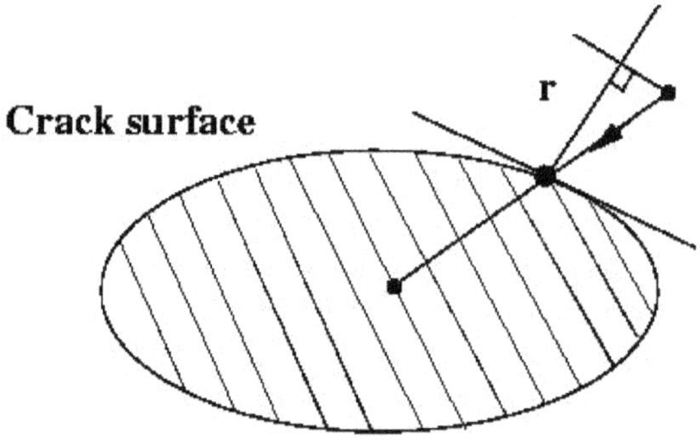

Fig. 3.19 The shortest distance between the crack surface and a point.

(2). Exterior solution:

We are only interested in the asymptotic solution at the crack tip, i.e. $y \to 1$, and $|g| \to 1$. In other words when $y \to 1$, the term $|g/(g^2-1)| > ln|(g+1)/(g-1)| \to \infty$ is the leading term of asymptotic expansion. Therefore

$$\sigma_{33}(x_1, x_2, 0) = \frac{ib\mu(\lambda+\mu)}{2\pi(\lambda+2\mu)} \int_0^{2\pi} \sqrt{\left(\frac{\eta_1}{a_1}\right)^2 + \left(\frac{\eta_2}{a_2}\right)^2} \frac{gd\phi}{g^2-1} + \mathcal{O}(1) \tag{3.226}$$

Let

$$f(\boldsymbol{\eta}) = g\left(\frac{\eta_1^2}{a_1^2} + \frac{\eta_2^2}{a_2^2}\right)^{1/2}, \text{ and } \hat{\mathbf{y}} = \frac{\mathbf{y}}{y}\,. \tag{3.227}$$

$$\begin{aligned}\sigma_{33}(x_1,x_2,0) &= \frac{ib\mu(\lambda+\mu)}{2\pi(\lambda+2\mu)}\int_0^{2\pi}\frac{f(\boldsymbol{\eta})-f(\hat{\mathbf{y}})}{g^2-1}d\phi+\frac{ib\mu(\lambda+\mu)}{2\pi(\lambda+2\mu)}\int_0^{2\pi}\frac{f(\hat{\mathbf{y}})}{g^2-1}d\phi \\ &= \frac{ib\mu(\lambda+\mu)}{2\pi(\lambda+2\mu)}\int_0^{2\pi}\frac{f(\hat{\mathbf{y}})}{g^2-1}d\phi+\mathcal{O}(1)\end{aligned} \tag{3.228}$$

Assume that $g=-\boldsymbol{\eta}\cdot\mathbf{y}=y\cos\psi$. Then

$$\int_0^{2\pi}\frac{d\phi}{g^2-1}=\int_0^{2\pi}\frac{d(\phi-\psi)}{y^2\cos^2(\phi-\psi)-1}=\frac{-2\pi}{\sqrt{1-y^2}}=\frac{2i\pi}{\sqrt{y^2-1}}\,. \tag{3.229}$$

Considering

$$f(\hat{\mathbf{y}})=-\hat{\mathbf{y}}\cdot\mathbf{y}\Big(\frac{\hat{y}_1}{a_1^2}+\frac{\hat{y}_2}{a_2^2}\Big)^{1/2} \tag{3.230}$$

we have

$$\sigma_{33}(x_1,x_2,0)=\frac{b\mu(\lambda+\mu)}{(\lambda+2\mu)}\frac{\hat{\mathbf{y}}\cdot\mathbf{y}}{\sqrt{y^2-1}}\Big(\frac{\hat{y}_1}{a_1^2}+\frac{\hat{y}_2}{a_2^2}\Big)^{1/2}\Big|_{y\to\hat{y}} \tag{3.231}$$

The stress intensity factor is defined as

$$k_1:=\lim_{r\to 0}\frac{\sigma_{33}}{2\pi r} \tag{3.232}$$

For an ellipsoidal crack (see Fig. 3.19),

$$r=\frac{(y-1)y^2}{\Big(\frac{x_1^2}{a_1^4}+\frac{x_2^2}{a_2^4}\Big)^{1/2}} \tag{3.233}$$

and

$$k_1=\frac{\sqrt{2\pi}b\mu(\lambda+\mu)}{(\lambda+2\mu)}\frac{\sqrt{y-1}}{\sqrt{y^2-1}}\Big(\frac{x_1^2}{a_1^4}+\frac{x_2^2}{a_2^4}\Big)^{1/4},\quad y\to 1 \tag{3.234}$$

Substituting $b=\dfrac{2(1-\nu)a_2\sigma_{33}^0}{\mu E(k)}$ into the above expression, we finally obtain the stress intensity factor

$$k_1=\frac{\sqrt{\pi}a_2\sigma_{33}^0}{E(k)}\Big(\frac{x_1^2}{a_1^4}+\frac{x_2^2}{a_2^4}\Big)^{1/4}\,. \tag{3.235}$$

This example shows that in principle the eigenstrain approach can solve fracture problems as well.

3.8 John Douglas Eshelby (1916-1981)

John Douglas Eshelby, or 'Jock' as he came to be known to a wide circles of scientists and engineers throughout the world, was born at Puddington in Cheshire on 21 December 1916 and died suddenly in Sheffield, South Yorkshire, on 10 December 1981.

Early schooling began for Jock at St Cyprian's, Eastbourne, but when he was about 13 he was taken very seriously ill with rheumatic fever. He bore the legacy of this illness for the rest of his life and in later years was also under medication for enhanced blood pressure. About this time the family moved from London to Somerset, first taking up residence in the Manor House at Farrington Gurney in 1930. After his rheumatic fever, Jock did not go to school again and his education was conducted privately at home through tutors in science and classics. These tutors were the village schoolmaster and the parson from a village nearby.

Jock often used to say that, as a consequence of this private education, he had to work out many things for himself. It is an interesting speculation that perhaps this lack of formal conventional group instruction may have helped to make him such an original and creative thinker. He told a story of this period which shows that he already alter to the intriguing phenomena of nature that appear in the course of our day to day activities. He noticed while watching the diesel generator which his father ran to supply electricity to the house, that an indentation made in the moving belt by striking it was remarkably persistent. In later years he would recall how he evolved an understanding of this phenomenon.

Fig. 3.20 John Douglas Eshelby (1916-1981) (Photo reproduced by permission of Sheffield University library).

Eshelby went as an undergraduate to the Physics Department of the University of Bristol; an article in the Bristol Evening World seems to have played some part in this decision. He obtained First Class Honors in Physics there in 1937. As a student, he had made up his mind that he would be a theoretical physicist. But against his own inclination, he began some research work in experimental physics on the soft X-ray spectra of solids and on magnetism, working in the laboratory under H.W.B. Skinner and W. Sucksmith.

From the anecdotes that he later used to tell of this period, his main impressions seem to have been of the patience needed to master the vacuum systems and of a requirement that the floor be swept. This work was interrupted in 1939 by World Wall II.

Eshelby spent an initial period in the Admiralty (with H.M.S. Vernon), dealing with the degaussing of ships, but on 4 May 1940, he joined the Technical Branch of the Royal Air Force.

Eshelby returned to the Physics department at Bristol after the War to resume research for the degree of Ph.D., which he obtained in 1950. By now, however, he had decided that he would use the additional freedom granted to an older student returning after war service to 'try to convert himself to a theoretician': a process which he seems to have assisted when off duty during the War by studying applied mathematics and theoretical physics. Characteristically, for he was also much interested in languages, he did this partly by learning Russian and reading texts in that language.

At Professor Mott's suggestion he began some calculations on the theory of dielectric loss, a topic which, incidentally, had also played a part in arousing the interest of F.C. Frank in crystal defects. This period was an exciting time for the development of the particular corner of solid stat physics in which Eshelby was finally to become one of the ultimate authorities: the theory of dislocations and internal stresses in solids.

From a later perspective, however, the culminating work of Eshelby in the Bristol period was his definitive account of forces on elastic singularities. Notable in this paper of 1951 is his derivation of results by the use of imaginary cutting, straining and welding operations, a technique that he used frequently with great effect. The method was not regarded as quite respectable by some with a more formal mathematical training. Jock, who will long be remembered for his dry wit, would poke fun where the cap fitted: 'These results have been derived elsewhere but are obtained more elaborately here', but always with a shrewd sense of values. He had a number of stories about applied mathematicians and theoretical physicists. In one of his more publishable comparisons he likened the applied mathematician to a pilot who habitually in all weathers flies blind with the cockpit covered; the theoretical physicist looks about him whenever he can but is glad of help when landing on a foggy night. In another, the physicist finds his work taken away and returned beautifully laundered and starched, but almost unwearable.

Already, then Eshelby was making an effective attack on difficult problems at the frontier of dislocation theories; this selection of useful problems which are just tractable is characteristic, as is also his ability to use ideas from one discipline to solve problems in another.

Eshelby left Bristol to become a Research Associate in the Physics Department of the University of Illinois at Urbana. On arrival, he described himself to two colleague who met him at the airport as 'a theoretical physicist - someone who performs no useful work'.

On his return from the U.S.A. in 1953, Eshelby spent some time at the family home in West Luccombe... but he was soon to be offered an appointment as a Lecturer in the department of Physical Metallurgy at the University of Birmingham, where he remained for ten years. This stay at Birmingham was interrupted for a period in 1963, when he went as Visiting Professor to the Technische Hochschule and Max Planck Institute at Stuttgart.

In his work on the forces on elastic singularities and on point defects, Eshelby had already made use of the 'sphere in hole' model. He was now to generalize this in three very significant papers concerned with elastic inclusions and inhomogeneities. These have had very wide application[2]*. The papers solve the problem of finding the elastic field in a medium containing a region that has changed its form (an inclusion) or that has elastic constants differing from those of the remainder (an inhomogeneity). Eshelby obtained his results, as he often did, by a sequence of imaginary cutting, straining and welding operations.*

Eshelby went to Cambridge in October 1964 to take a post as Senior Visiting Fellow in the Metal Physics Group at the Cavendish Laboratory. He held a Fellowship and College Lectureship at Churchill College from October 1965 to September 1966. In 1966 Eshelby left Cambridge to join the Department of the Theory of Materials at the University of Sheffield. The title of Reader was conferred immediately and he was elected to a Personal Chair in 1971.

During this period at Sheffield, much of his work was on the theory of fracture where he made important contributions. In 1968 he published papers with C. Atkinson and B.A. Bilby applying the energy-momentum tensor to obtain expressions for the crack extension force in an elastic material. Quite independently, however, at this time, two other workers gave path impendent integrals of the energy-momentum tensor form and noted their connection with the crack extension force. Rice's integral rapidly become used very widely by engineers in the analysis of the fracture of tough ductile materials and it is now known throughout the world as the 'J' integral.

Eshelby was a very private man. His needs were simple and he was anxious to be of no trouble; a true gentleman, always ready to serve. Much steeped in the works of the classical physicists of the past century, he was a great admirer of Lord Rayleigh. Above all a scholar, his lack of concern for personal advancement left him from time to time in situations where his ability was not matched to his position in the scientific community; and he was sometimes a little put out by remarks which revealed that his true capacity was sadly misjudge.

He was a constant focus for discussion and at his best with colleagues and stu-

[2]Remark by the authors: Eshelby's paper of 1957 was cited 3981 times by the Science Citation Index, during 1955-2007

dents in a small group. At one time his room at Sheffield became an unofficial coffee house where many useful arguments took place and much real work was done. At all times he mad us laugh; 'It is obvious!' he would say, 'I forget exactly why'. And he would 'go to any lengths to avoid a bit of trouble'.

As a lecturer and teacher he was clear and amusing. He prepared carefully for lectures or colloquia, sometimes approaching them with diffidence surprising in one so knowledgeable. Although formal teaching was not his first interest he was conscientious in trying to distil the essence of a subject and always helpful to the genuine questioner.

Eshelby was elected a Fellow of the Royal Society in 1974. He was always being consulted by colleagues from all over the world and particularly by those in the United States. The American Society of Mechanical Engineers awarded him the 1977 Timoshenko Medal.

Never truly in full health, he was spared the agony of long illness and was active to the last. In the period just before his death [in 1981] he was preparing to spend Christmas in Canada and to lecture in California in the New Year. He was arguing with colleagues about his work on forces on defects in liquid crystals; and he had just made an interesting proposal about the origin of cracks in fatigue.

Eshelby liked to regard some of his important works as 'Amusing applications of the theorem of Gauss'. Yet his scholarship was always devoted to useful ends and has had a major influence on many areas of science and technology.

It is alarming to realize how difficult it would be for someone now to follow in his way.

— This article is excerpted from *B.A. Bilby, John Douglas Eshelby. Biographical Memoirs of Fellows of the Royal Society, Vol. 36, 127-150, 1990.* Reproduced with kind permission from Royal Society.

3.9 Exercises

Problem 3.1. *Show that the integral*

$$\int_{V_0} \exp\{i\boldsymbol{\xi}\cdot\mathbf{x}\}dV_x = 4\pi\sqrt{\frac{\pi}{2}}a^3\frac{J_{3/2}(\eta)}{\eta^{3/2}} \tag{3.236}$$

where V_0 is a sphere with radius a; $\eta = a|\xi|$, and $|\xi| = \sqrt{\xi_1^2+\xi_2^2+\xi_3^2}$.

Hint:

(1) Consider the identity

$$\nabla_{\mathbf{x}}\exp\Big(i\boldsymbol{\xi}\cdot\mathbf{x}\Big) = i\boldsymbol{\xi}\exp\Big(i\boldsymbol{\xi}\cdot\mathbf{x}\Big)$$

(2)

$$\int_0^1 t\sin(a|\xi|t)dt = \Gamma(1)(a|\xi|)^{-1/2}J_{3/2}(a|\xi|)$$

where $\Gamma(1) = \sqrt{\dfrac{\pi}{2}}$, $J_{3/2}(\eta)$ *is the Bessel function of the first kind.*

Problem 3.2. *Derive the displacement field inside an inclusion in which prescribed eigenstrain is a linear function of coordinates, i.e. Example 5.1.*

Problem 3.3. *Derive Green's function for plane strain problem by solving the following Navier equations,*

$$\sigma_{\beta\alpha,\beta} + \delta(\mathbf{x} - \mathbf{y})\delta_{\alpha\gamma} = 0 \tag{3.237}$$

where γ *is the direction that the concentrated force points to.*

Assume that the 2D elastic tensor is

$$C_{\alpha\beta\zeta\eta} = \lambda\delta_{\alpha\beta}\delta_{\zeta\eta} + \mu(\delta_{\alpha\zeta}\delta_{\beta\eta} + \delta_{\alpha\eta}\delta_{\beta\zeta}), \quad \alpha, \beta, \zeta, \eta = 1, 2 \tag{3.238}$$

define 2D permutation symbol

$$e_{\alpha\beta}: \quad e_{11} = 0, \; e_{12} = 1, \; e_{21} = -1, \; e_{22} = 0 \tag{3.239}$$

The corresponding e-δ identities are:

$$(1) \qquad e_{\alpha\beta} = \begin{vmatrix} \delta_{\alpha 1} & \delta_{\alpha 2} \\ \delta_{\beta 1} & \delta_{\beta 2} \end{vmatrix}$$

$$(2) \qquad e_{\alpha\zeta}e_{\beta\eta} = \delta_{\alpha\beta}\delta_{\zeta\eta} - \delta_{\alpha\eta}\delta_{\beta\zeta}$$

$$(3) \qquad e_{\alpha\eta}e_{\beta\eta} = \delta_{\alpha\beta} \tag{3.240}$$

$$(4) \qquad e_{\alpha\eta}e_{\alpha\eta} = \delta_{\alpha\alpha} = 2! \tag{3.241}$$

Hints:

$$\int_{\mathbb{R}^2} \frac{\exp(i(\xi_1 x_1 + \xi_2 x_2))}{\xi_1^2 + \xi_2^2} d\xi_1\xi_2 = -2\pi \ln R \tag{3.242}$$

$$\int_{\mathbb{R}^2} \frac{\xi_\alpha\xi_\beta}{\xi^4} \exp(i\boldsymbol{\xi} \cdot \mathbf{x}) d\boldsymbol{\xi} = -\pi\delta_{\alpha\beta} \ln R - \pi\frac{x_\alpha x_\beta}{R^2} \tag{3.243}$$

where $R = \sqrt{x_1^2 + x_2^2}$.

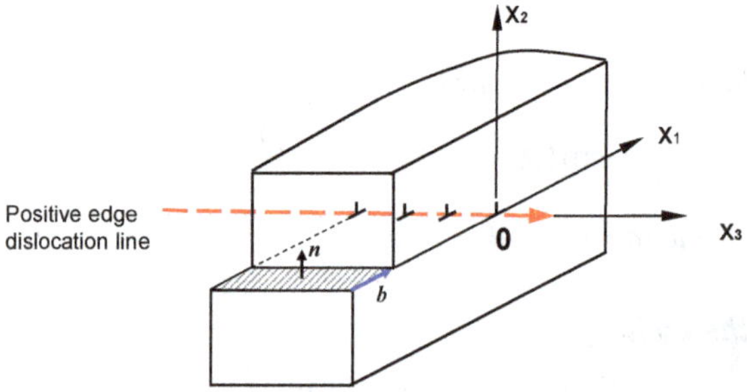

Fig. 3.21 A straight edge dislocation.

Problem 3.4. *Let Ω be the half plane $(x_2 = 0, \quad x_1 < 0)$, and ϵ^*_{21} be prescribed as*

$$\epsilon^*_{21}(\mathbf{x}) = \frac{b}{2}\delta(x_2)H(-x_1) \tag{3.244}$$

Show

$$u_1(\mathbf{x}) = \frac{b}{2\pi}\tan^{-1}\Big(\frac{x_2}{x_1}\Big) + \frac{b}{4\pi}\Big(\frac{1}{1-\nu}\Big)\frac{x_1 x_2}{x_1^2 + x_2^2} \tag{3.245}$$

where ν is the Poisson's ratio.

Hint: ([Mura (1987)] page 17)

$$\int_{-\infty}^{\infty}\int_{-\infty}^{\infty}\frac{\xi_2}{\xi_1(\xi_1^2+\xi_2^2)}\exp\{i(\xi_1 x_1 + \xi_2 x_2)\}d\xi_1 d\xi_2 = 2\pi\tan^{-1}\Big(\frac{x_2}{x_1}\Big)$$

$$\int_{-\infty}^{\infty}\int_{-\infty}^{\infty}\frac{\xi_1\xi_2}{(\xi_1^2+\xi_2^2)^2}\exp\{i(\xi_1 x_1 + \xi_2 x_2)\}d\xi_1 d\xi_2 = -\frac{\pi x_1 x_2}{x_1^2 + x_2^2}$$

Problem 3.5. *The 2D Green's function for plane strain problem is*

$$G^{\infty}_{\alpha\beta}(\mathbf{x}-\mathbf{x}') = \frac{1}{8\pi}\frac{1}{\mu(1-\nu)}\left\{\frac{(x_\alpha - x'_\alpha)(x_\beta - x'_\beta)}{R^2} - (3-4\nu)\delta_{\alpha\beta}\ln R\right\}$$
$$\alpha, \beta = 1, 2 \tag{3.246}$$

where $R = \sqrt{(x_1 - x'_1)^2 + (x_2 - x'_2)^2}$.

Consider the following elliptical inclusion problem,

$$\epsilon^*_{\alpha\beta}(\mathbf{x}) = \begin{cases} \epsilon^*_{\alpha\beta}; & \forall\ \mathbf{x} \in \Omega \\ 0; & \forall\ \mathbf{x} \in \mathbb{R}^2/\Omega \end{cases} \tag{3.247}$$

where $\epsilon^*_{\alpha\beta}$ is a constant tensor, and $\Omega := \left\{\mathbf{x} \;\middle|\; \frac{x_1^2}{a_1^2} + \frac{x_2^2}{a_2^2} \le 1\right\}$. Find the Eshelby

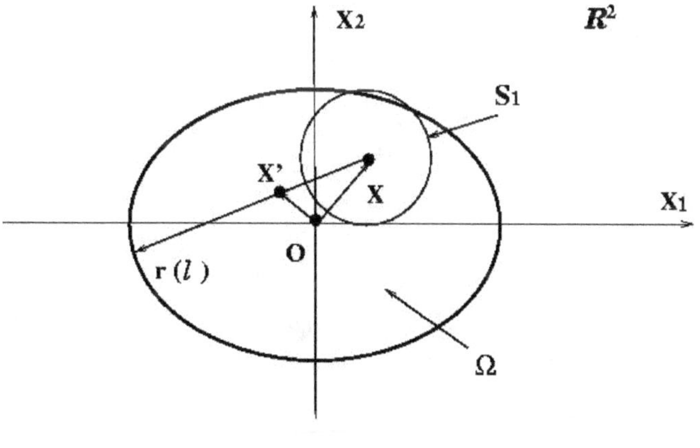

Fig. 3.22 2D elliptical inclusion.

tensor for interior problem ($\mathbf{x} \in \Omega$). Hint: see [Li (2000)] pages 5606-5607.

Problem 3.6. *Consider a spherical inclusion with radius a. Use identities*

$$\oint_{S^2} \ell_i \ell_j dS = \frac{4\pi a^2}{3}\delta_{ij} \tag{3.248}$$

$$\oint_{S^2} \ell_i \ell_j \ell_m \ell_n dS = \frac{4\pi a^2}{15}(\delta_{ij}\delta_{mn} + \delta_{im}\delta_{jn} + \delta_{in}\delta_{jm}) \tag{3.249}$$

to show that

$$\begin{aligned} \mathbb{S}^{I,\infty}_{ijmn} &= \frac{1}{16\pi(1-\nu)} \oint_{S^2} \frac{\lambda_i g_{jmn} + \lambda_j g_{imn}}{g} dS \\ &= \frac{5\nu - 1}{15(1-\nu)}\delta_{ij}\delta_{mn} + \frac{4 - 5\nu}{15(1-\nu)}(\delta_{im}\delta_{jn} + \delta_{jm}\delta_{in}) \end{aligned} \tag{3.250}$$

where $g_{ijk} = (1-2\nu)(\delta_{ij}\ell_k + \delta_{ik}\ell_j - \delta_{jk}\ell_i) + 3\ell_i\ell_j\ell_k$, $g = \ell_i\ell_i/a^2 = 1/a^2$, *and* $\lambda_i = \ell_i/a^2$.

Problem 3.7. *Show that*

$$\begin{aligned} \mathcal{G}^{\infty}_{ijmn}(\mathbf{x}-\mathbf{y}) &= -\frac{1}{2}C_{klmn}\Big(G^{\infty}_{ik,lj}(\mathbf{x}-\mathbf{y}) + G^{\infty}_{jk,li}(\mathbf{x}-\mathbf{y})\Big) \\ &= \frac{1}{8\pi(1-\nu)r^3}\Big[(1-2\nu)(\delta_{im}\delta_{jn} + \delta_{in}\delta_{jm} - \delta_{ij}\delta_{mn}) \\ &\quad +3\nu\Big(\delta_{im}\ell_j\ell_n + \delta_{in}\ell_j\ell_m + \delta_{jm}\ell_i\ell_n + \delta_{jn}\ell_j\ell_m\Big) \\ &\quad +3\delta_{ij}\ell_m\ell_n + 3(1-2\nu)\delta_{mn}\ell_i\ell_j - 15\ell_i\ell_j\ell_m\ell_n\Big] \end{aligned}$$

where $\mathcal{G}^{\infty}_{ijmn}$ *is called the fourth-order Green's function (the second derivative of the Green's function).*

Problem 3.8. *Let*

$$w(\mathbf{x}) = \frac{1}{\pi R^3}\exp(-\frac{\mathbf{x}\cdot\mathbf{x}}{R^2}), \tag{3.251}$$

representing a Gaussian distribution .

For any smooth vector field, $\mathbf{A} \in \mathbb{R}^3$, *define weighted average operation,*

$$< \mathbf{A} > (\mathbf{x}) := \int_{\mathbb{R}^3} w(\mathbf{x}-\mathbf{x}')\mathbf{A}(\mathbf{x}')d\Omega_{\mathbf{x}'} \tag{3.252}$$

where $d\Omega_{\mathbf{x}'} := dx'_1 dx'_2 dx'_3$.

Show that

$$\nabla\cdot < \mathbf{A} > = < \nabla\cdot\mathbf{A} > \tag{3.253}$$

(Hint: Use Gauss theorem (divergence theorem), and the fact that $w(\mathbf{x}) \to 0$ *as* $|\mathbf{x}| \to \infty$.*)*

Problem 3.9. *Use identity*

$$e_{ijk}e_{rst} = \begin{vmatrix} \delta_{ir} & \delta_{is} & \delta_{it} \\ \delta_{jr} & \delta_{js} & \delta_{jt} \\ \delta_{kr} & \delta_{ks} & \delta_{kt} \end{vmatrix} \tag{3.254}$$

show:

$$e_{ijk}e_{ijk} = 3! = 6 \tag{3.255}$$

$$e_{ijk}e_{ij\ell} = 2\delta_{k\ell} \tag{3.256}$$

$$e_{ijk}e_{i\ell m} = \delta_{j\ell}\delta_{km} - \delta_{jm}\delta_{k\ell} \tag{3.257}$$

Problem 3.10. *Prove*

$$\mathbb{S}^{I,\infty} + \mathbb{D} : \mathbb{T}^{I,\infty} : \mathbb{C} = \mathbb{I}^{(4s)} \tag{3.258}$$

$$\mathbb{T}^{I,\infty} + \mathbb{C} : \mathbb{S}^{I,\infty} : \mathbb{D} = \mathbb{I}^{(4s)} \tag{3.259}$$

where $\mathbb{S}^I$ and $\mathbb{T}^I$ are the interior Eshelby tensor and its conjugate Eshelby tensor respectively.

Hint: First show that

$$\boldsymbol{\sigma}^d = \mathbb{C} : \boldsymbol{\epsilon}^d - \boldsymbol{\sigma}^* \ , \quad \text{and} \quad \boldsymbol{\sigma}^* + \mathbb{C} : \boldsymbol{\epsilon}^* = 0 \ . \tag{3.260}$$

Problem 3.11. *Consider eigenstress homogenization problem illustrated in Fig. 3.2. Suppose that the disturbance stress field, $\boldsymbol{\sigma}^d$, can be related to the eigenstress field, $\boldsymbol{\sigma}^*$, i.e.*

$$\boldsymbol{\sigma}^d = \mathbb{T}^{I,\infty} : \boldsymbol{\sigma}^*, \quad \forall \mathbf{x} \in \Omega \tag{3.261}$$

where $\mathbb{T}^{I,\infty}$ is the so-called conjugate Eshelby tensor. Show that the effective elastic tensor is equal to

$$\bar{\mathbb{C}} = \left[\mathbb{I}^{(4s)} + f(\mathbb{B}^\Omega - \mathbb{T}^{I,\infty})^{-1}\right] : \mathbb{C} \tag{3.262}$$

where the tensor, $\mathbb{B}^\Omega := (\mathbb{D} - \mathbb{D}^\Omega)^{-1} : \mathbb{D}$, and $\mathbb{D}^\Omega$, $\mathbb{D}$ are elastic compliance tensors of the inclusion and matrix, respectively.

Problem 3.12. *Suppose that an RVE (V) is subjected the following pure traction boundary condition,*

$$\mathbf{n} \cdot \boldsymbol{\sigma} = \bar{\mathbf{t}} = \mathbf{n} \cdot \boldsymbol{\Sigma}, \quad \forall \mathbf{x} \in \partial V \tag{3.263}$$

Show that

$$< \boldsymbol{\sigma} : \delta\boldsymbol{\epsilon} > = \boldsymbol{\Sigma} : < \delta\boldsymbol{\epsilon} > \ . \tag{3.264}$$

Chapter 4

EFFECTIVE ELASTIC MODULUS

In this Chapter, we apply the inclusion theory to derive some of the most commonly used homogenization procedures in estimating the effective material properties of composite materials. For illustration purpose, all the discussions are focused on the two-phase composite materials, and sometimes we do list the general expressions for n-phase ($n > 2$) composites. We start by presenting the homogenization procedure based on Eshelby's equivalent eigenstrain theory.

4.1 Effective modulus for composites with dilute suspension phases

First, we apply the engineering homogenization theory to composites whose second phase concentration or other phase concentrations are small in comparison with the concentration of the matrix. In literature, they are usually referred to as the composites with inhomogeneities of dilute suspension.

4.1.1 *Basic equations for average stress and strain*

Considering a composite with two phases of constituents: matrix phase (M) with elastic stiffness $\mathbb{C}$ and compliance $\mathbb{D}$, and inclusion phase (Ω) with elastic stiffness $\mathbb{C}^I$ and compliance $\mathbb{D}^I$. Be more specific, we define the average operator in the matrix, in the inclusion, and in the whole RVE as

$$<\cdot>_M := \frac{1}{M}\int_M \Big(\cdot\Big) dV, \quad <\cdot>_\Omega := \frac{1}{\Omega}\int_\Omega \Big(\cdot\Big) dV, \tag{4.1}$$

and

$$<\cdot>_V := \frac{1}{V}\int_V \Big(\cdot\Big) dV \tag{4.2}$$

where $V = M \bigcup \Omega$. Further define the volume fraction of inclusion phase as

$$f = \frac{\Omega}{V} \tag{4.3}$$

It is straightforward that the average stress over the RVE can be written as

$$< \boldsymbol{\sigma} >_V = \frac{1}{V}\int_V \boldsymbol{\sigma} dV = \frac{1}{V}\int_{M\cup\Omega} \boldsymbol{\sigma} dV = \frac{1}{V}\Big[\frac{M}{M}\int_M \boldsymbol{\sigma} dV + \frac{\Omega}{\Omega}\int_\Omega \boldsymbol{\sigma} dV\Big]$$
$$= (1-f) < \boldsymbol{\sigma} >_M + f < \boldsymbol{\sigma} >_\Omega \tag{4.4}$$

Therefore,

$$(1-f) < \boldsymbol{\sigma} >_M = < \boldsymbol{\sigma} >_V - f < \boldsymbol{\sigma} >_\Omega = \bar{\mathbb{C}} :< \boldsymbol{\epsilon} >_V - f\mathbb{C}^I :< \boldsymbol{\epsilon} >_\Omega \tag{4.5}$$

Following similar steps, one can show that the average strain over the RVE is

$$< \boldsymbol{\epsilon} >_V = (1-f) < \boldsymbol{\epsilon} >_M + f < \boldsymbol{\epsilon} >_\Omega \tag{4.6}$$

similarly,

$$(1-f) < \boldsymbol{\epsilon} >_M = < \boldsymbol{\epsilon} >_V - f < \boldsymbol{\epsilon} >_\Omega = \bar{\mathbb{D}} :< \boldsymbol{\sigma} >_V - f\mathbb{D}^I :< \boldsymbol{\sigma} >_\Omega \tag{4.7}$$

Therefore, we can write

$$(1-f) < \boldsymbol{\sigma} >_M = (1-f)\mathbb{C} :< \boldsymbol{\epsilon} >_M = \mathbb{C} : \Big(< \boldsymbol{\epsilon} >_V - f < \boldsymbol{\epsilon} >_\Omega\Big) \tag{4.8}$$

Combining Eqs. (4.5) and (4.8) yields

$$\boxed{\left(\bar{\mathbb{C}} - \mathbb{C}\right) :< \boldsymbol{\epsilon} >_V = f\left(\mathbb{C}^I - \mathbb{C}\right) :< \boldsymbol{\epsilon} >_\Omega} \tag{4.9}$$

Similarly,

$$(1-f) < \boldsymbol{\epsilon} >_M = (1-f)\mathbb{D} :< \boldsymbol{\sigma} >_M = \mathbb{D} : (< \boldsymbol{\sigma} >_V - f < \boldsymbol{\sigma} >_\Omega) \tag{4.10}$$

Combining Eqs. (4.7) and (4.10) yields

$$\boxed{\left(\bar{\mathbb{D}} - \mathbb{D}\right) :< \boldsymbol{\sigma} >_V = f\left(\mathbb{D}^I - \mathbb{D}\right) :< \boldsymbol{\sigma} >_\Omega} \tag{4.11}$$

Eqs. (4.9) and (4.11) are often referred to as the basic equations of average stress/strain fields. The equations are derived solely based on geometric properties of the average operators, and are valid regardless of boundary conditions and material properties.

Considering a special case that the macro stress boundary condition is prescribed on the RVE boundary,

$$\boldsymbol{t} = \boldsymbol{\sigma}^0 \cdot \mathbf{n}, \quad \forall\ \mathbf{x} \in \partial V$$

Based on the averaging theorem, $< \boldsymbol{\sigma} >_V = \boldsymbol{\sigma}^0$, we can rewrite Eq. (4.11) as

$$\left(\bar{\mathbb{D}} - \mathbb{D}\right) : \boldsymbol{\sigma}^0 = f\left(\mathbb{D}^I - \mathbb{D}\right) :< \boldsymbol{\sigma} >_\Omega \tag{4.12}$$

One may note that the remote background strain in this case is

$$\boldsymbol{\epsilon}^0 = \mathbb{D} : \boldsymbol{\sigma}^0 = \mathbb{D} :< \boldsymbol{\sigma} >_V \neq < \boldsymbol{\epsilon} >_V \tag{4.13}$$

Similarly, for prescribed macro-strain boundary condition,

$$\boldsymbol{u}(\mathbf{x}) = \mathbf{x} \cdot \boldsymbol{\epsilon}^0, \quad \mathbf{x} \in \partial V$$

The averaging theorem asserts that in this case $< \boldsymbol{\epsilon} >_V = \boldsymbol{\epsilon}^0$, therefore, Eq. (4.9) may be written as

$$\left(\bar{\mathbb{C}} - \mathbb{C}\right) : \boldsymbol{\epsilon}^0 = f\left(\mathbb{C}^I - \mathbb{C}\right) :< \boldsymbol{\epsilon} >_\Omega \tag{4.14}$$

and the corresponding remote background stress is

$$\boldsymbol{\sigma}^0 = \mathbb{C} :< \boldsymbol{\epsilon} >_V \neq < \boldsymbol{\sigma} >_V \ . \tag{4.15}$$

4.1.2 *Homogenization: equivalent stress/strain conditions*

In the following, we shall formulate equivalent stress and strain homogenization schemes in parallel, and each scheme is derived based on a different type of boundary condition. Considering a two-phase composite material, for $\forall \mathbf{x} \in \Omega$, the stress and strain equivalent conditions are

$$\mathbb{C}^I : (\boldsymbol{\epsilon}^0 + \boldsymbol{\epsilon}^d) = \mathbb{C} : (\boldsymbol{\epsilon}^0 + \boldsymbol{\epsilon}^d - \boldsymbol{\epsilon}^*) \tag{4.16}$$

or

$$\mathbb{D}^I : (\boldsymbol{\sigma}^0 + \boldsymbol{\sigma}^d) = \mathbb{D} : (\boldsymbol{\sigma}^0 + \boldsymbol{\sigma}^d - \boldsymbol{\sigma}^*) \tag{4.17}$$

Then one can find the average stress and strain fields inside the inclusion,

$$< \boldsymbol{\epsilon} >_\Omega = \mathbb{A}^I : \boldsymbol{\epsilon}^* \tag{4.18}$$

$$< \boldsymbol{\sigma} >_\Omega = \mathbb{B}^I : \boldsymbol{\sigma}^* \tag{4.19}$$

where

$$\mathbb{A}^I = (\mathbb{C} - \mathbb{C}^I)^{-1} : \mathbb{C} \quad \text{and} \quad \mathbb{B}^I = (\mathbb{D} - \mathbb{D}^I)^{-1} : \mathbb{D} \tag{4.20}$$

are called phase misfit tensors.

Since inclusion population is small, one can neglect the interaction among inclusions. The disturbance field inside each inclusion can then be related to eigenstrain fields through interior Eshelby tensor $\mathbb{S}^{I,\infty}$ and its conjugate Eshelby tensor $\mathbb{T}^{I,\infty}$,

$$\boldsymbol{\epsilon}^d = \mathbb{S}^{I,\infty} : \boldsymbol{\epsilon}^*, \quad \forall \mathbf{x} \in \Omega \tag{4.21}$$

$$\boldsymbol{\sigma}^d = \mathbb{T}^{I,\infty} : \boldsymbol{\sigma}^*, \quad \forall \mathbf{x} \in \Omega \tag{4.22}$$

Note that the interior Eshelby tensor and the conjugate Eshelby tensor are related as $\mathbb{S}^{I,\infty} + \mathbb{D} : \mathbb{T}^{I,\infty} : \mathbb{C} = \mathbb{I}^{(4s)}$. Subsequently, one can decide how much the eigenstrain or eigenstress have to be prescribed by the following conditions,

$$\boldsymbol{\epsilon}^* = (\mathbb{A}^I - \mathbb{S}^{I,\infty})^{-1} : \boldsymbol{\epsilon}^0 \tag{4.23}$$

$$\boldsymbol{\sigma}^* = (\mathbb{B}^I - \mathbb{T}^{I,\infty})^{-1} : \boldsymbol{\sigma}^0 \tag{4.24}$$

Therefore the average strain/stress inside the inclusion can be expressed by eigenstrain/eigenstress, i.e.

$$< \boldsymbol{\epsilon} >_\Omega = \mathbb{A}^I : \boldsymbol{\epsilon}^* = \mathbb{A}^I : (\mathbb{A}^I - \mathbb{S}^{I,\infty})^{-1} : \boldsymbol{\epsilon}^0 \tag{4.25}$$

$$< \boldsymbol{\sigma} >_\Omega = \mathbb{B}^I : \boldsymbol{\sigma}^* = \mathbb{B}^I : (\mathbb{B}^I - \mathbb{T}^{I,\infty})^{-1} : \boldsymbol{\sigma}^0 \tag{4.26}$$

Subsequently, one can relate the average strain and average stress in the α-th inclusion (inhomogeneity) with the background strain and background stress through the so-called *concentration tensors*,

$$< \boldsymbol{\epsilon} >_\Omega = \boldsymbol{\mathcal{A}}^I : \boldsymbol{\epsilon}^0 \tag{4.27}$$

$$< \boldsymbol{\sigma} >_\Omega = \boldsymbol{\mathcal{B}}^I : \boldsymbol{\sigma}^0 \tag{4.28}$$

where the concentration tensors are defined as

$$\boldsymbol{\mathcal{A}}^I = \mathbb{A}^I : (\mathbb{A}^I - \mathbb{S}^{I,\infty})^{-1} \tag{4.29}$$

$$\boldsymbol{\mathcal{B}}^I = \mathbb{B}^I : (\mathbb{B}^I - \mathbb{T}^{I,\infty})^{-1} \tag{4.30}$$

Since by definition $< \boldsymbol{\sigma} >_\Omega = \mathbb{C}^I : < \boldsymbol{\epsilon} >_\Omega$ and $< \boldsymbol{\epsilon} >_\Omega = \mathbb{D}^I :< \boldsymbol{\sigma} >_\Omega$, one can rewrite Eqs. (4.27) and (4.28) as

$$< \boldsymbol{\sigma} >_\Omega = \begin{cases} \mathbb{C}^I : \boldsymbol{\mathcal{A}}^I : \mathbb{D} : \boldsymbol{\sigma}^0 \\ \boldsymbol{\mathcal{B}}^I : \boldsymbol{\sigma}^0 \end{cases} \tag{4.31}$$

or

$$< \boldsymbol{\epsilon} >_\Omega = \begin{cases} \mathbb{D}^I : \boldsymbol{\mathcal{B}}^I : \mathbb{C} : \boldsymbol{\epsilon}^0 \\ \boldsymbol{\mathcal{A}}^I : \boldsymbol{\epsilon}^0 \end{cases} \tag{4.32}$$

By virtue of (4.31) and (4.32), we have the following useful identities,

$$\boxed{\mathbb{C}^I : \boldsymbol{\mathcal{A}}^I = \boldsymbol{\mathcal{B}}^I : \mathbb{C} \;\; \text{and} \;\; \mathbb{D}^I : \boldsymbol{\mathcal{B}}^I = \boldsymbol{\mathcal{A}}^I : \mathbb{D}\ .} \tag{4.33}$$

Suppose that prescribed macro-stress boundary condition is applied. Substituting both expressions in Eq. (4.31) into the basic average equation (4.11) yields,

$$(\bar{\mathbb{D}} - \mathbb{D}) : \boldsymbol{\sigma}^0 = f(\mathbb{D}^I - \mathbb{D}) : \begin{cases} \mathbb{C}^I : \boldsymbol{\mathcal{A}}^I : \mathbb{D} : \boldsymbol{\sigma}^0 \\ \\ \boldsymbol{\mathcal{B}}^I : \boldsymbol{\sigma}^0 \end{cases} \tag{4.34}$$

Therefore, for prescribed traction boundary condition, we have the following estimate on effective compliance tensor,

$$\bar{\mathbb{D}} = \begin{cases} \mathbb{D} + f(\mathbb{D}^I - \mathbb{D}) : \mathbb{C}^I : \boldsymbol{\mathcal{A}}^I : \mathbb{D} \\ \\ \mathbb{D} + f(\mathbb{D}^I - \mathbb{D}) : \boldsymbol{\mathcal{B}}^I \end{cases} \tag{4.35}$$

By considering the identities,

$$(\mathbb{A}^I)^{-1} = (\mathbb{D}^I - \mathbb{D}) : \mathbb{C}^I, \;\; \text{and} \;\; \mathbb{B}^I = (\mathbb{D} - \mathbb{D}^I)^{-1} : \mathbb{D} \tag{4.36}$$

we finally obtain the effective modulus for dilute inclusion distributions,

$$\bar{\mathbb{D}} = \begin{cases} \mathbb{D} + f(\mathbb{A}^I - \mathbb{S}^{I,\infty})^{-1} : \mathbb{D} \\ \\ \mathbb{D} - f\mathbb{D} : (\mathbb{B}^I - \mathbb{T}^{I,\infty})^{-1} \end{cases} \tag{4.37}$$

For $(n+1)$-phase composites, the above expression can be generalized to

$$\bar{\mathbb{D}} = \begin{cases} \mathbb{D} + \sum_{\alpha=1}^{n} f_\alpha (\mathbb{A}^\alpha - \mathbb{S}^{I,\infty})^{-1} : \mathbb{D} \\ \\ \mathbb{D} - \sum_{\alpha=1}^{n} f_\alpha \mathbb{D} : (\mathbb{B}^\alpha - \mathbb{T}^{I,\infty})^{-1} \end{cases} \tag{4.38}$$

Note that the index α ranges from 1 to n, and each α is an inhomogeneous phase.

If prescribed macro-strain boundary condition is applied, one may substitute both expressions of (4.32) into the basic average equation (4.9). It then leads to

$$(\bar{\mathbb{C}} - \mathbb{C}) : \boldsymbol{\epsilon}^0 = f(\mathbb{C}^I - \mathbb{C}) : \begin{cases} \mathbb{D}^I : \boldsymbol{\mathcal{B}}^I : \mathbb{C} : \boldsymbol{\epsilon}^0 \\ \\ \boldsymbol{\mathcal{A}}^I : \boldsymbol{\epsilon}^0 \end{cases} \tag{4.39}$$

The following estimate on effective elastic tensor may be obtained,

$$\bar{\mathbb{C}} = \begin{cases} \mathbb{C} + f(\mathbb{C}^I - \mathbb{C}) : \mathbb{D}^I : \boldsymbol{\mathcal{B}}^I : \mathbb{C} \\ \\ \mathbb{C} + f(\mathbb{C}^I - \mathbb{C}) : \boldsymbol{\mathcal{A}}^I \end{cases} \tag{4.40}$$

Using the identities,

$$(\mathbb{B}^I)^{-1} = (\mathbb{C}^I - \mathbb{C}) : \mathbb{D}^I, \quad \text{and} \quad \mathbb{A}^I = -(\mathbb{C}^I - \mathbb{C})^{-1} : \mathbb{C} \tag{4.41}$$

we then have the following estimate on effective elastic stiffness tensor

$$\bar{\mathbb{C}} = \begin{cases} \mathbb{C} + f(\mathbb{B}^I - \mathbb{T}^{I,\infty})^{-1} : \mathbb{C} \\ \\ \mathbb{C} - f\mathbb{C} : (\mathbb{A}^I - \mathbb{S}^{I,\infty})^{-1} \end{cases} \tag{4.42}$$

For $(n+1)$-phase composites, the above estimate becomes,

$$\bar{\mathbb{C}} = \begin{cases} \mathbb{C} + \sum_{\alpha=1}^{n} f_\alpha (\mathbb{B}^\alpha - \mathbb{T}^{I,\infty})^{-1} : \mathbb{C} \\ \\ \mathbb{C} - \sum_{\alpha=1}^{n} f_\alpha : \mathbb{C} : (\mathbb{A}^\alpha - \mathbb{S}^{I,\infty})^{-1} \end{cases} \tag{4.43}$$

Remark 4.1. The effective stiffness tensor $\bar{\mathbb{C}}$ and effective compliance tensor $\bar{\mathbb{D}}$ are obtained from two different kinds of prescribed boundary conditions. It is assumed in the homogenization process that the concentration of inclusions is dilute, so the interaction between them can be neglected. One of the drawbacks of dilute distribution homogenization is the lack of consistency in the obtained stiffness and compliance tensors, i.e.,

$$\bar{\mathbb{D}} : \bar{\mathbb{C}} \neq \mathbb{I}^{(4s)} \quad \text{or} \quad \bar{\mathbb{D}} \neq \bar{\mathbb{C}}^{-1}.$$

This can be shown as follows

$$\begin{aligned} \bar{\mathbb{D}} : \bar{\mathbb{C}} &= \left(\mathbb{I}^{(4s)} + f(\mathbb{A}^I - \mathbb{S}^{I,\infty})^{-1}\right) : \mathbb{D} : \mathbb{C} : \left(\mathbb{I}^{(4s)} - f(\mathbb{A}^I - \mathbb{S}^{I,\infty})^{-1}\right) \\ &= \mathbb{I}^{(4s)} - f^2(\mathbb{A}^I - \mathbb{S}^{I,\infty})^{-1} : (\mathbb{A}^I - \mathbb{S}^{I,\infty})^{-1} \\ &= \mathbb{I}^{(4s)} + \mathcal{O}(f^2) \neq \mathbb{I}^{(4s)} . \end{aligned} \tag{4.44}$$

Obviously, the effective elastic stiffness tensor is not consistent with the effective compliance tensor.

4.1.3 *Example: elastic modulus of isotropic composites*

In the following, a working example is illustrated on how to derive elastic moduli for macroscopically isotropic composite materials. Suppose that there are n different phases of inhomogeneities in a solid. For prescribed traction boundary condition, the effective elastic compliance tensor obtained based the Eshelby's equivalent eigenstrain method reads as

$$\bar{\mathbb{D}} = \left\{ \mathbb{I} + \sum_{\alpha=1}^{n} f_\alpha (\mathbb{A}^\alpha - \mathbb{S}^{I,\infty})^{-1} \right\} : \mathbb{D}$$

where $\mathbb{A}^\alpha$ is defined as

$$\mathbb{A}^\alpha = (\mathbb{C} - \mathbb{C}^\alpha)^{-1} : \mathbb{C}$$

Here $\mathbb{C}$ is the elastic tensor of the matrix, which is assumed to be isotropic, i.e. $\mathbb{C} = 3K\mathbb{E}^{(1)} + 2\mu\mathbb{E}^{(2)}$. We can then calculate

$$\mathbb{C} - \mathbb{C}^\alpha = 3(K - K^\alpha)\mathbb{E}^{(1)} + 2(\mu - \mu^\alpha)\mathbb{E}^{(2)}$$

and

$$\begin{aligned} \mathbb{A}^\alpha &= (\mathbb{C} - \mathbb{C}^\alpha)^{-1} : \mathbb{C} \\ &= \left(\frac{1}{3(K - K^\alpha)} \mathbb{E}^{(1)} + \frac{1}{2(\mu - \mu^\alpha)} \mathbb{E}^{(2)} \right) : \left(3K\mathbb{E}^{(1)} + 2\mu\mathbb{E}^{(2)} \right) \\ &= \frac{K}{K - K^\alpha} \mathbb{E}^{(1)} + \frac{\mu}{\mu - \mu^\alpha} \mathbb{E}^{(2)} \end{aligned}$$

Since the composite is isotropic, we can use the Eshelby tensor of spherical inclusions. For the spherical inclusion, the interior Eshelby tensor is

$$\begin{aligned} \mathbb{S}^{I,\infty} &= \frac{5\nu - 1}{15(1-\nu)} \mathbb{I}^{(2)} \otimes \mathbb{I}^{(2)} + \frac{2(4 - 5\nu)}{15(1-\nu)} \mathbb{I}^{(4s)} \\ &= \frac{(1+\nu)}{3(1-\nu)} \mathbb{E}^{(1)} + \frac{2(4-5\nu)}{15(1-\nu)} \mathbb{E}^{(2)} = s_1 \mathbb{E}^{(1)} + s_2 \mathbb{E}^{(2)} \end{aligned}$$

where $s_1 = \dfrac{1+\nu}{3(1-\nu)}$ and $s_2 = \dfrac{2(4-5\nu)}{15(1-\nu)}$ are evaluated using the material properties of the matrix. Then

$$\mathbb{A}^\alpha - \mathbb{S}^{I,\infty} = \left(\frac{K}{(K - K^\alpha)} - s_1 \right) \mathbb{E}^{(1)} + \left(\frac{\mu}{(\mu - \mu^\alpha)} - s_2 \right) \mathbb{E}^{(2)}$$

and

$$\left(\mathbb{A}^\alpha - \mathbb{S}^{I,\infty} \right)^{-1} = \left(\frac{1}{\dfrac{K}{(K - K^\alpha)} - s_1} \right) \mathbb{E}^{(1)} + \left(\frac{1}{\dfrac{\mu}{(\mu - \mu^\alpha)} - s_2} \right) \mathbb{E}^{(2)}$$

Hence

$$\bar{\mathbb{D}} = \Big(\mathbb{I} + \sum_{\alpha=1}^{n} f_\alpha(\mathbb{A}^\alpha - \mathbb{S}^{I,\infty})^{-1}\Big) : \mathbb{D}$$

$$= \left\{\mathbb{E}^{(1)} + \mathbb{E}^{(2)} + \sum_{\alpha=1}^{n}\left(\frac{f_\alpha}{\dfrac{K}{(K-K^\alpha)} - s_1}\mathbb{E}^{(1)} + \frac{f_\alpha}{\dfrac{\mu}{(\mu-\mu^\alpha)} - s_2}\mathbb{E}^{(2)}\right)\right\} : \Big(\frac{1}{3K}\mathbb{E}^{(1)} + \frac{1}{2\mu}\mathbb{E}^{(2)}\Big)$$

Finally,

$$\bar{\mathbb{D}} = \frac{1}{3K}\left(1 + \sum_{\alpha=1}^{n}\frac{f_\alpha}{\dfrac{K}{K-K^\alpha} - s_1}\right)\mathbb{E}^{(1)} + \frac{1}{2\mu}\left(1 + \sum_{\alpha=1}^{n}\frac{f_\alpha}{\dfrac{\mu}{\mu-\mu^\alpha} - s_2}\right)\mathbb{E}^{(2)} \quad (4.45)$$

Note that $\bar{\mathbb{D}} = \dfrac{1}{3\bar{K}}\mathbb{E}^{(1)} + \dfrac{1}{2\bar{\mu}}\mathbb{E}^{(2)}$. Assume $f_\alpha << 1$,

$$\frac{\bar{K}}{K} = \left(1 + \sum_{\alpha=1}^{n}\frac{f_\alpha}{\dfrac{K}{K-K^\alpha} - s_1}\right)^{-1} = 1 - \sum_{\alpha=1}^{n}\frac{f_\alpha}{\dfrac{K}{K-K^\alpha} - s_1} + \mathcal{O}(f_\alpha^2)$$

and

$$\frac{\bar{\mu}}{\mu} = \left(1 + \sum_{\alpha=1}^{n}\frac{f_\alpha}{\dfrac{\mu}{\mu-\mu^\alpha} - s_2}\right)^{-1} = 1 - \sum_{\alpha=1}^{n}\frac{f_\alpha}{\dfrac{\mu}{\mu-\mu^\alpha} - s_2} + \mathcal{O}(f_\alpha^2)$$

Similarly, by considering remote traction boundary condition, we have the estimate of effective elastic modulus for solids with dilute suspension of inhomogeneities,

$$\bar{\mathbb{C}} = \left\{\mathbb{I} + \sum_{\alpha=1}^{n} f_\alpha(\mathbb{B}^\alpha - \mathbb{T}^{I,\infty})^{-1}\right\} : \mathbb{C}$$

where $\mathbb{B}^\alpha$ is defined as $\mathbb{B}^\alpha = (\mathbb{D} - \mathbb{D}^\alpha)^{-1} : \mathbb{D}$. Here $\mathbb{D}$ is the elastic compliance tensor of the matrix, i.e. $\mathbb{D} = \dfrac{1}{3K}\mathbb{E}^{(1)} + \dfrac{1}{2\mu}\mathbb{E}^{(2)}$. We can then calculate

$$\begin{aligned}
\mathbb{B}^\alpha &= (\mathbb{D} - \mathbb{D}^\alpha)^{-1} : \mathbb{D} \\
&= \left(\frac{1}{3}\Big(\frac{1}{K} - \frac{1}{K^\alpha}\Big)\mathbb{E}^{(1)} + \frac{1}{2}\Big(\frac{1}{\mu} - \frac{1}{\mu^\alpha}\Big)\mathbb{E}^{(2)}\right)^{-1} : \mathbb{D} \\
&= \left(\frac{K^\alpha - K}{3KK^\alpha}\mathbb{E}^{(1)} + \frac{\mu^\alpha - \mu}{2\mu\mu^\alpha}\mathbb{E}^{(2)}\right)^{-1} : \Big(\frac{1}{3K}\mathbb{E}^{(1)} + \frac{1}{2\mu}\mathbb{E}^{(2)}\Big) \\
&= -\frac{K^\alpha}{K - K^\alpha}\mathbb{E}^{(1)} - \frac{\mu^\alpha}{\mu - \mu^\alpha}\mathbb{E}^{(2)}
\end{aligned}$$

Subsequently,

$$\begin{aligned}
\mathbb{B}^\alpha - \mathbb{T}^I &= \Big(-\frac{K^\alpha}{(K-K^\alpha)} - (1 - s_1)\Big)\mathbb{E}^{(1)} + \Big(-\frac{\mu^\alpha}{(\mu-\mu^\alpha)} - (1 - s_2)\Big)\mathbb{E}^{(2)} \\
&= -\Big(\frac{K}{K-K^\alpha} - s_1\Big)\mathbb{E}^{(1)} - \Big(\frac{\mu}{\mu-\mu^\alpha} - s_2\Big)\mathbb{E}^{(2)}
\end{aligned}$$

and

$$\left(\mathbb{B}^{\alpha}-\mathbb{T}^{I,\infty}\right)^{-1}=-\frac{1}{\dfrac{K}{(K-K^{\alpha})}-s_1}\mathbb{E}^{(1)}-\frac{1}{\dfrac{\mu}{(\mu-\mu^{\alpha})}-s_2}\mathbb{E}^{(2)}$$

Finally,

$$\begin{aligned}\bar{\mathbb{C}}&=\Big(\mathbb{I}+\sum_{\alpha=1}^{n}f_\alpha(\mathbb{B}^{\alpha}-\mathbb{T}^{I,\infty})^{-1}\Big):\mathbb{C}\\&=\left\{\Big(1-\sum_{\alpha=1}^{n}\frac{f_\alpha}{\dfrac{K}{(K-K^{\alpha})}-s_1}\Big)\mathbb{E}^{(1)}+\Big(1-\sum_{\alpha=1}^{n}\frac{f_\alpha}{\dfrac{\mu}{(\mu-\mu_{\alpha})}-s_2}\Big)\mathbb{E}^{(2)}\right\}\\&\quad:(3K\mathbb{E}^{(1)}+2\mu\mathbb{E}^{(2)})\\&=3K\Big(1-\sum_{\alpha=1}^{n}\frac{f_\alpha}{\dfrac{K}{(K-K^{\alpha})}-s_1}\Big)\mathbb{E}^{(1)}+2\mu\Big(1-\sum_{\alpha=1}^{n}\frac{f_\alpha}{\dfrac{\mu}{(\mu-\mu_{\alpha})}-s_2}\Big)\mathbb{E}^{(2)}\end{aligned}$$

Therefore

$$\frac{\bar{K}}{K}=1-\sum_{\alpha=1}^{n}\frac{f_\alpha}{\dfrac{K}{K-K^{\alpha}}-s_1}$$

and

$$\frac{\bar{\mu}}{\mu}=1-\sum_{\alpha=1}^{n}\frac{f_\alpha}{\dfrac{\mu}{\mu-\mu^{\alpha}}-s_2}$$

It is obviously that these results are different from the results obtained from prescribed traction boundary condition. They are only agreeable to the first order of the volume fraction. In other words, these two results (the results obtained from prescribed macro-stress B.C. and the results obtained from prescribed macro-strain B.C.) are not consistent in the homogenization scheme for dilute inhomogeneity distribution.

4.2 Self-consistent method

As shown above, the effective elastic stiffness tensor and compliance tensor obtained via the dilute distribution model are not reciprocal to each other as supposed to be. Such inconsistency is due to the fact that as the volume fraction of inhomogeneity increases the accuracy of dilute suspension homogenization schemes deteriorates, because the interaction among inhomogeneities become strong that cannot be neglected.

To take into account the interaction among inhomogeneities, a so-called self-consistent method was proposed, which is largely attributed to a series of papers by B. Budiansky [Budiansky (1965)] and R. Hill ([Hill (1965a,c,b)]). The following presentation is mainly adapted from a discussion by S. Nemat-Nasser and M. Hori [Nemat-Nasser and Hori (1999)], which is in the spirit of Budiansky and Hill, nevertheless it is more engineering-oriented.

There are two main differences between self-consistent homogenization and dilute suspension homogenization. The first difference is in the treatment of remote (background) strain and stress fields. Consider the prescribed macro stress boundary condition,

$$\boldsymbol{t} = \boldsymbol{\Sigma} \cdot \mathbf{n}, \quad \forall\, \mathbf{x} \in \partial V$$

Based on the averaging theorem, $< \boldsymbol{\sigma} >_V = \boldsymbol{\Sigma}$. In self-consistent homogenization, we first make the remote boundary conditions consistent, such that we define the remote background strain as

$$\boldsymbol{\mathcal{E}} = \bar{\mathbb{D}} : \boldsymbol{\Sigma} \tag{4.46}$$

Therefore in this case, the remote background strain will be the average strain,

$$\boldsymbol{\mathcal{E}} = \bar{\mathbb{D}} :< \boldsymbol{\sigma} >_V = < \boldsymbol{\epsilon} >_V \; .$$

Similarly, for prescribed macro-strain boundary condition,

$$\boldsymbol{u}(\mathbf{x}) = \boldsymbol{\mathcal{E}} \cdot \mathbf{x}, \quad \mathbf{x} \in \partial V$$

the averaging theorem asserts that in this case $\boldsymbol{\mathcal{E}} = < \boldsymbol{\epsilon} >_V$. If $\boldsymbol{\Sigma} = \bar{\mathbb{C}} : \boldsymbol{\mathcal{E}}$, the background remote stress will be the average stress,

$$\boldsymbol{\Sigma} = \bar{\mathbb{C}} :< \boldsymbol{\epsilon} >_V = < \boldsymbol{\sigma} >_V \tag{4.47}$$

The second main difference between the self-consistent method and the dilute suspension method is that Eshelby's equivalent inclusion principle is applied with respect to the homogenized solid, instead of matrix. Suppose that there are $\alpha = 1, 2, \cdots, n$ distinct phases. In the α-th phase, the following stress homogenization condition is adopted,

$$\mathbb{C}^\alpha : (\boldsymbol{\mathcal{E}} + \boldsymbol{\epsilon}^d) = \bar{\mathbb{C}} : (\boldsymbol{\mathcal{E}} + \boldsymbol{\epsilon}^d - \boldsymbol{\epsilon}^*) \tag{4.48}$$

or

$$\mathbb{D}^\alpha : (\boldsymbol{\Sigma} + \boldsymbol{\sigma}^d) = \bar{\mathbb{D}} : (\boldsymbol{\Sigma} + \boldsymbol{\sigma}^d - \boldsymbol{\sigma}^*) \tag{4.49}$$

Moreover, the disturbance field generated by eigenstrain is also calculated with respect to the homogenized solid, i.e.

$$\boldsymbol{\epsilon}^d = \bar{\mathbb{S}}^{I,\infty} : \boldsymbol{\epsilon}^*, \quad \forall \mathbf{x} \in \Omega^\alpha \tag{4.50}$$

$$\boldsymbol{\sigma}^d = \bar{\mathbb{T}}^{I,\infty} : \boldsymbol{\sigma}^*, \quad \forall \mathbf{x} \in \Omega^\alpha \tag{4.51}$$

The bars over the interior Eshelby and the conjugate interior Eshelby tensor emphasize that they shall be evaluated using the effective material properties of the homogenized solid. Therefore the average strain/stress inside the α-th phase inclusion may be expressed by eigenstrain/eigenstress, i.e.

$$< \boldsymbol{\epsilon} >_\alpha = \bar{\mathbb{A}}^\alpha : \boldsymbol{\epsilon}^* \tag{4.52}$$

$$< \boldsymbol{\sigma} >_\alpha = \bar{\mathbb{B}}^\alpha : \boldsymbol{\sigma}^* \tag{4.53}$$

where

$$\bar{\mathbb{A}}^\alpha = (\bar{\mathbb{C}} - \mathbb{C}^\alpha)^{-1} : \bar{\mathbb{C}} \tag{4.54}$$

$$\bar{\mathbb{B}}^\alpha = (\bar{\mathbb{D}} - \mathbb{D}^\alpha)^{-1} : \bar{\mathbb{D}} \tag{4.55}$$

Subsequently, one can relate the average strain and average stress in the inclusion (inhomogeneity) with the macro-strain and macro-stress field by concentration tensors,

$$< \boldsymbol{\epsilon} >_\alpha = \bar{\boldsymbol{\mathcal{A}}}^\alpha : \boldsymbol{\mathcal{E}} \tag{4.56}$$

$$< \boldsymbol{\sigma} >_\alpha = \bar{\boldsymbol{\mathcal{B}}}^\alpha : \boldsymbol{\Sigma} \tag{4.57}$$

where the concentration tensors are defined as

$$\bar{\boldsymbol{\mathcal{A}}}^\alpha = \bar{\mathbb{A}}^\alpha : (\bar{\mathbb{A}}^\alpha - \bar{\mathbb{S}}^{I,\infty})^{-1} \tag{4.58}$$

$$\bar{\boldsymbol{\mathcal{B}}}^\alpha = \bar{\mathbb{B}}^\alpha : (\bar{\mathbb{B}}^\alpha - \bar{\mathbb{T}}^{I,\infty})^{-1} \tag{4.59}$$

Since by definition $< \boldsymbol{\sigma} >_\alpha = \mathbb{C}^\alpha :< \boldsymbol{\epsilon} >_\alpha$ and $< \boldsymbol{\epsilon} >_\alpha = \mathbb{D}^\alpha :< \boldsymbol{\sigma} >_\alpha$, one can rewrite Eqs. (4.56) and (4.57) as

$$< \boldsymbol{\sigma} >_\alpha = \begin{cases} \mathbb{C}^\alpha : \bar{\boldsymbol{\mathcal{A}}}^\alpha : \bar{\mathbb{D}} : \boldsymbol{\Sigma} \\ \bar{\boldsymbol{\mathcal{B}}}^\alpha : \boldsymbol{\Sigma} \end{cases} \tag{4.60}$$

or

$$< \boldsymbol{\epsilon} >_\alpha = \begin{cases} \mathbb{D}^\alpha : \bar{\boldsymbol{\mathcal{B}}}^\alpha : \bar{\mathbb{C}} : \boldsymbol{\mathcal{E}} \\ \bar{\boldsymbol{\mathcal{A}}}^\alpha : \boldsymbol{\mathcal{E}} \end{cases} \tag{4.61}$$

Note that the relationships $\boldsymbol{\mathcal{E}} =< \boldsymbol{\epsilon} >_V$ and $\boldsymbol{\Sigma} =< \boldsymbol{\sigma} >_V$ are used.

Suppose that prescribed macro-stress boundary condition is applied. Substituting Eqs. (4.60) and (4.61) into the basic average equation (4.11) yields,

$$(\bar{\mathbb{D}} - \mathbb{D}) : \boldsymbol{\Sigma} = \sum_{\alpha=1}^{n} f_\alpha (\mathbb{D}^\alpha - \mathbb{D}) : \begin{cases} \mathbb{C}^\alpha : \bar{\boldsymbol{\mathcal{A}}}^\alpha : \bar{\mathbb{D}} : \boldsymbol{\Sigma} \\ \\ \bar{\boldsymbol{\mathcal{B}}}^\alpha : \boldsymbol{\Sigma} \end{cases} \tag{4.62}$$

The self-consistent method then gives the following estimate on effective compliance tensor,

$$\bar{\mathbb{D}} = \begin{cases} \mathbb{D} + \displaystyle\sum_{\alpha=1}^{n} f_\alpha (\mathbb{D}^\alpha - \mathbb{D}) : \mathbb{C}^\alpha : \bar{\boldsymbol{\mathcal{A}}}^\alpha : \bar{\mathbb{D}} \\ \\ \mathbb{D} + \displaystyle\sum_{\alpha=1}^{n} f_\alpha (\mathbb{D}^\alpha - \mathbb{D}) : \bar{\boldsymbol{\mathcal{B}}}^\alpha \end{cases} \tag{4.63}$$

If prescribed macro-strain boundary condition is applied, one may substitute Eqs. (4.31) and (4.32) into the basic average equation (4.9). It leads to

$$(\bar{\mathbb{C}} - \mathbb{C}) : \boldsymbol{\mathcal{E}} = \sum_{\alpha=1}^{n} f_\alpha (\mathbb{C}^\alpha - \mathbb{C}) : \begin{cases} \mathbb{D}^\alpha : \bar{\boldsymbol{\mathcal{B}}}^\alpha : \bar{\mathbb{C}} : \boldsymbol{\mathcal{E}} \\ \bar{\boldsymbol{\mathcal{A}}}^\alpha : \boldsymbol{\mathcal{E}} \end{cases} \tag{4.64}$$

Hence self-consistent method gives the following estimate on effective elastic tensor,

$$\bar{\mathbb{C}} = \begin{cases} \mathbb{C} + \sum_{\alpha=1}^{n} f_\alpha (\mathbb{C}^\alpha - \mathbb{C}) : \mathbb{D}^\alpha : \bar{\boldsymbol{\mathcal{B}}}^\alpha : \bar{\mathbb{C}} \\ \mathbb{C} + \sum_{\alpha=1}^{n} f_\alpha (\mathbb{C}^\alpha - \mathbb{C}) : \bar{\boldsymbol{\mathcal{A}}}^\alpha \end{cases} \tag{4.65}$$

Remark 4.2. **(1)** Note that the index α starts from 1, and each phase α is an inhomogeneous phase including the matrix; **(2)** The above expressions for effective elastic moduli are implicit, and it often requires solving non-linear algebraic equations to obtain effective material constants.

We now show the effective moduli obtained from the prescribed macro-stress B.C. and the prescribed macro-strain B.C. satisfy the self-consistent condition:

$$\bar{\mathbb{D}} : \bar{\mathbb{C}} = \mathbb{I} \quad \text{or} \quad \bar{\mathbb{D}} = \bar{\mathbb{C}}^{-1}.$$

Consider

$$\begin{aligned} \mathbb{D} &= \mathbb{D} : \mathbb{I} = \mathbb{D} : \bar{\mathbb{C}} : \bar{\mathbb{C}}^{-1} \\ &= \mathbb{D} : \Big[\mathbb{C} + \sum_{\alpha=1}^{n} f_\alpha (\mathbb{C}^\alpha - \mathbb{C}) : \bar{\boldsymbol{\mathcal{A}}}^\alpha \Big] : \bar{\mathbb{C}}^{-1} \\ &= \bar{\mathbb{C}}^{-1} + \sum_{\alpha=1}^{n} f_\alpha \mathbb{D} : (\mathbb{C}^\alpha - \mathbb{C}) : \bar{\boldsymbol{\mathcal{A}}}^\alpha : \bar{\mathbb{C}}^{-1} \end{aligned} \tag{4.66}$$

Since

$$\mathbb{D} : (\mathbb{C}^\alpha - \mathbb{C}) = \mathbb{D} : \mathbb{C}^\alpha - \mathbb{I} = -\mathbb{I} + \mathbb{D} : \mathbb{C}^\alpha = -(\mathbb{D}^\alpha - \mathbb{D}) : \mathbb{C}^\alpha$$

the last line of (4.66) may be rewritten as

$$\mathbb{D} = \bar{\mathbb{C}}^{-1} - \sum_{\alpha=1}^{n} f_\alpha (\mathbb{D}^\alpha - \mathbb{D}) : \mathbb{C}^\alpha : \bar{\boldsymbol{\mathcal{A}}}^\alpha : \bar{\mathbb{C}}^{-1} \tag{4.67}$$

which leads to

$$\bar{\mathbb{C}}^{-1} = \mathbb{C}^{-1} + \sum_{\alpha=1}^{n} f_\alpha (\mathbb{D}^\alpha - \mathbb{D}) : \mathbb{C}^\alpha : \bar{\boldsymbol{\mathcal{A}}}^\alpha : \bar{\mathbb{C}}^{-1} \tag{4.68}$$

Compare (4.68) with the first line of Eq. (4.63), one can conclude that

$$\bar{\mathbb{C}}^{-1} = \bar{\mathbb{D}} \tag{4.69}$$

Similar arguments can be made to show that $\bar{\mathbb{D}}^{-1} = \bar{\mathbb{C}}$.

Example 4.1. For isotropic composites, the effective moduli obtained from self-consistent scheme can be further simplified.

Consider

$$\bar{\mathbb{C}} = \mathbb{C} + \sum_{\alpha=1}^{n} f_\alpha (\mathbb{C}^\alpha - \mathbb{C}) : \bar{\boldsymbol{\mathcal{A}}}^\alpha \tag{4.70}$$

Step 1.

$$\mathbb{C} = 3K\mathbb{E}^{(1)} + 2\mu\mathbb{E}^{(2)}, \quad \text{and} \quad (\mathbb{C}^\alpha - \mathbb{C}) = 3(K^\alpha - K)\mathbb{E}^{(1)} + 2(\mu^\alpha - \mu)\mathbb{E}^{(2)}$$

Step 2:

$$\begin{aligned}
\bar{\mathbb{A}}^\alpha &= (\bar{\mathbb{C}} - \mathbb{C}^\alpha)^{-1} : \bar{\mathbb{C}} \\
&= \Big(\frac{1}{3(\bar{K} - K^\alpha)}\mathbb{E}^{(1)} + \frac{1}{2(\bar{\mu} - \mu^\alpha)}\mathbb{E}^{(2)}\Big) : (3\bar{K}\mathbb{E}^{(1)} + 2\bar{\mu}\mathbb{E}^{(2)}) \\
&= \frac{\bar{K}}{\bar{K} - K^\alpha}\mathbb{E}^{(1)} + \frac{\bar{\mu}}{\bar{\mu} - \mu^\alpha}\mathbb{E}^{(2)}
\end{aligned}$$

Then,

$$\begin{aligned}
\bar{\boldsymbol{\mathcal{A}}}^\alpha &= \bar{\mathbb{A}}^\alpha (\bar{\mathbb{A}}^\alpha - \bar{\mathbb{S}}^{I,\infty})^{-1} \\
&= \Big[\frac{\bar{K}}{\bar{K} - K^\alpha}\mathbb{E}^{(1)} + \frac{\bar{\mu}}{\bar{\mu} - \mu^\alpha}\mathbb{E}^{(2)}\Big]\Big[\Big(\frac{\bar{K}}{\bar{K} - K^\alpha} - \bar{s}_1\Big)^{-1}\mathbb{E}^{(1)} \\
&\quad + \Big(\frac{\bar{\mu}}{\bar{\mu} - \mu^\alpha} - \bar{s}_2\Big)^{-1}\mathbb{E}^{(2)}\Big] \\
&= \frac{\bar{K}}{\bar{K} - (\bar{K} - K^\alpha)\bar{s}_1}\mathbb{E}^{(1)} + \frac{\bar{\mu}}{\bar{\mu} - (\bar{\mu} - \mu^\alpha)\bar{s}_2}\mathbb{E}^{(2)}
\end{aligned}$$

Therefore,

$$\begin{aligned}
\bar{\mathbb{C}} = 3\bar{K}\mathbb{E}^{(1)} + 2\bar{\mu}\mathbb{E}^{(2)}\mathbb{C} &+ \sum_{\alpha=1}^{n} f_\alpha (\mathbb{C}^\alpha - \mathbb{C}) : \bar{\boldsymbol{\mathcal{A}}}^\alpha \\
&= 3\Big(K + \sum_{\alpha=1}^{n} f_\alpha \frac{(K^\alpha - K)\bar{K}}{\bar{K} - (\bar{K} - K^\alpha)\bar{s}_1}\Big)\mathbb{E}^{(1)} \\
&\quad + 2\Big(\mu + \sum_{\alpha=1}^{n} f_\alpha \frac{(\mu^\alpha - \mu)\bar{\mu}}{\bar{\mu} - (\bar{\mu} - \mu^\alpha)\bar{s}_2}\Big)\mathbb{E}^{(2)}
\end{aligned}$$

which leads to

$$\frac{\bar{K}}{K} = 1 + \sum_{\alpha=1}^{n} f_\alpha \Big(\frac{K^\alpha}{K} - 1\Big)\Big(1 + (\frac{K^\alpha}{\bar{K}} - 1)\bar{s}_1\Big)^{-1} \tag{4.71}$$

$$\frac{\bar{\mu}}{\mu} = 1 + \sum_{\alpha=1}^{n} f_\alpha \Big(\frac{\mu^\alpha}{\mu} - 1\Big)\Big(1 + (\frac{\mu^\alpha}{\bar{\mu}} - 1)\bar{s}_2\Big)^{-1} \tag{4.72}$$

Note that if the Eshelby tensor of the spherical inclusion is used in above equations, it should be evaluated with respect to the effective properties of the homogenized composite, i.e., $\bar{s}_1 = \dfrac{1+\bar{\nu}}{3(1-\bar{\nu})}$, $\bar{s}_2 = \dfrac{2(4-5\bar{\nu})}{15(1-\bar{\nu})}$, and $\bar{\nu} = \dfrac{3\bar{K}-2\bar{\mu}}{2(3\bar{K}+\bar{\mu})}$. The derived solutions are non-linear implicit functions of $\bar{K}$ and $\bar{\nu}$, so proper numerical scheme needs to be carried out for their evaluation.

4.3 Mori-Tanaka method

4.3.1 *Tanaka-Mori lemma*

In 1972, a technical note less than two-page was published in *Journal of Elasticity* by K. Tanaka and T. Mori [Tanaka and Mori (1972)], which revealed an important consequence of the scalability of the interior Eshelby tensor. That result later became the well-known Tanaka-Mori lemma , and it then led to a very efficient homogenization procedure called *Mori-Tanaka method.* Today, the Mori-Tanaka method has become one the most popular homogenization methods among engineers. Its applications include abraded composite [Tang and Postle (2000, 2001)], nano-composites [Fisher *et al.* (2002); Shi *et al.* (2004)], and reinforced fiber composites. We start the discussion with the proof of the Tanaka-Mori lemma.

Lemma 4.1 (Tanaka and Mori). *Consider two coaxial, similar ellipsoidal domains Ω_0 and Ω, where Ω_0 is contained in Ω, i.e., $\Omega_0 \subset \Omega$, as shown in Fig.4.1(a),*

$$\Omega_0 = \left\{ \mathbf{x} \,\middle|\, \frac{x_1^2}{a_1^2} + \frac{x_2^2}{a_2^2} + \frac{x_3^2}{a_3^2} \le 1 \right\}$$
$$\Omega = \left\{ \mathbf{x} \,\middle|\, \frac{x_1^2}{b_1^2} + \frac{x_2^2}{b_2^2} + \frac{x_3^2}{b_3^2} \le 1 \right\} \tag{4.73}$$

where

$$\frac{a_1}{b_1} + \frac{a_2}{b_2} + \frac{a_3}{b_3} = k$$

Assume that a uniform eigenstrain state, $\epsilon^*_{ij}(\mathbf{x})$, is prescribed in the smaller ellipsoidal region Ω_0, i.e.

$$\epsilon^*_{ij}(\mathbf{x}) = \begin{cases} \epsilon^*_{ij} & \mathbf{x} \in \Omega_0 \\ 0 & \mathbf{x} \in \mathbb{R}^3/\Omega_0 \end{cases}$$

The the average disturbance strain field vanishes in the exterior region $\Omega - \Omega_0$, i.e

$$< \boldsymbol{\epsilon} >_{\Omega-\Omega_0} = \frac{1}{\Omega - \Omega_0} \int_{\Omega-\Omega_0} \epsilon_{ij}(\mathbf{x}) d\Omega = 0 \ . \tag{4.74}$$

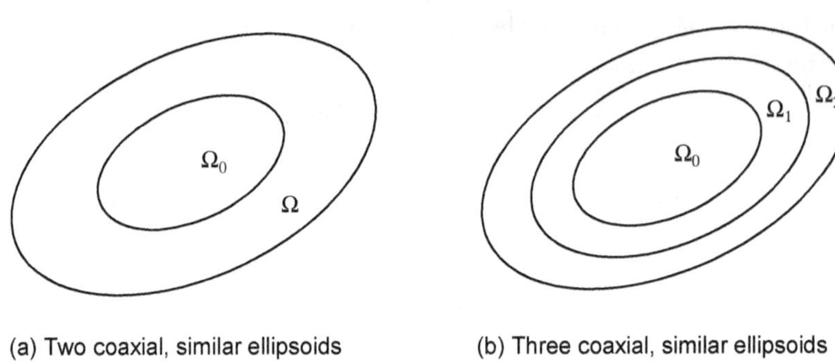

Fig. 4.1 Schematic illustration and proof of the Mori-Tanaka lemma.

Proof. Suppose that there are three coaxial, similar ellipsoids, $\Omega_0 \subset \Omega_1 \subset \Omega_2$ in an infinite homogeneous medium as shown in Fig.4.1(b). A uniform eigenstrain is prescribed in Ω_0, i.e.

$$\epsilon^*_{ij}(\mathbf{x}) = \begin{cases} \epsilon^*_{ij} & \mathbf{x} \in \Omega_0 \\ 0 & \mathbf{x} \in \mathbb{R}^3/\Omega_0 \end{cases}$$

The disturbance displacement field can be then written as

$$u_i(\mathbf{x}) = -\int_{\Omega_0} \epsilon^*_{mn} C_{k\ell mn} G_{ik,\ell}(\mathbf{x}-\mathbf{x}')d\mathbf{x}' \tag{4.75}$$

and the disturbance strain field is

$$\epsilon_{ij}(\mathbf{x}) = -\int_{\Omega_0} \epsilon^*_{mn} \frac{C_{k\ell mn}}{2}\Big(G_{ik,\ell j}(\mathbf{x}-\mathbf{x}') + G_{jk,\ell i}(\mathbf{x}-\mathbf{x}')\Big)d\mathbf{x}' \tag{4.76}$$

where $C_{k\ell mn}$ is the elastic tensor, $G_{ik}(\mathbf{x}-\mathbf{x}')$ is the Green's function in the infinite domain, and

$$\begin{aligned} S_{k\ell mn} &= -\int_{\Omega_0} \frac{C_{k\ell mn}}{2}\Big(G_{ik,\ell j} + G_{jk,\ell i}\Big)d\mathbf{x}' \\ &= \begin{cases} S^{I,\infty}_{k\ell mn}(\Omega_0), & \mathbf{x} \in \Omega_0 \\ S^{E,\infty}_{k\ell mn}(\Omega_0), & \mathbf{x} \in \mathbb{R}^3/\Omega_0 \end{cases} \end{aligned} \tag{4.77}$$

is the Eshelby tensor.

Now consider the average strain in the region $\Omega_1 - \Omega_2$.

$$\int_{\Omega_2-\Omega_1} \epsilon_{ij}(\mathbf{x})d\mathbf{x} = \int_{\Omega_2-\Omega_1}\left[\epsilon^*_{mn}\int_{\Omega_0} -\frac{C_{k\ell mn}}{2}\Big(G_{ik,\ell j}(\mathbf{x}-\mathbf{x}') + G_{jk,\ell i}(\mathbf{x}-\mathbf{x}')\Big)d\mathbf{x}'\right]d\mathbf{x}$$

Since $\mathbf{x} \in \Omega_2 - \Omega_1$, the integrand does contain singularity in either integration domains, Ω_0 and $\Omega_2 - \Omega_1$. Since ϵ^*_{mn} is a constant tensor prescribed on Ω_0, We can then change the order of the integration,

$$\begin{aligned}
&\int_{\Omega_2-\Omega_1}\left[\epsilon^*_{mn}\int_{\Omega_0}-\frac{C_{k\ell mn}}{2}\Big(G_{ik,\ell j}(\mathbf{x}-\mathbf{x}')+G_{jk,\ell i}(\mathbf{x}-\mathbf{x}')\Big)d\mathbf{x}'\right]d\mathbf{x}\\
&=\int_{\Omega_0}\left[\epsilon^*_{mn}\int_{\Omega_2-\Omega_1}-\frac{C_{k\ell mn}}{2}\Big(G_{ik,\ell j}(\mathbf{x}-\mathbf{x}')+G_{jk,\ell i}(\mathbf{x}-\mathbf{x}')\Big)d\mathbf{x}\right]d\mathbf{x}'\\
&=\int_{\Omega_0}\left[\epsilon^*_{mn}\int_{\Omega_2}-\frac{C_{k\ell mn}}{2}\Big(G_{ik,\ell j}(\mathbf{x}-\mathbf{x}')+G_{jk,\ell i}(\mathbf{x}-\mathbf{x}')\Big)d\mathbf{x}\right]d\mathbf{x}'\\
&\quad-\int_{\Omega_0}\left[\epsilon^*_{mn}\int_{\Omega_1}-\frac{C_{k\ell mn}}{2}\Big(G_{ik,\ell j}(\mathbf{x}-\mathbf{x}')+G_{jk,\ell i}(\mathbf{x}-\mathbf{x}')\Big)d\mathbf{x}\right]d\mathbf{x}'\\
&=\epsilon^*_{mn}\int_{\Omega_0}\left[S^{I,\infty}_{k\ell mn}(\Omega_2)-S^{I,\infty}_{k\ell mn}(\Omega_1)\right]d\mathbf{x}'\\
&=\epsilon^*_{mn}\Omega_0\left[S^{I,\infty}_{k\ell mn}(\Omega_2)-S^{I,\infty}_{k\ell mn}(\Omega_1)\right]
\end{aligned}\tag{4.78}$$

where $S^{I,\infty}_{k\ell mn}(\Omega_1)$ and $S^{I,\infty}_{k\ell mn}(\Omega_2)$ are interior Eshelby tensors with constant eigenstrain prescribed on domain Ω_1 and Ω_2, respectively. Since the Eshelby tensors only depend on the material property and the aspect ratio of the ellipsoids, it is straightforward to show that

$$\epsilon^*_{mn}\Omega_0\left[S^{I,\infty}_{k\ell mn}(\Omega_2)-S^{I,\infty}_{k\ell mn}(\Omega_1)\right]=0 \tag{4.79}$$

if Ω_2, Ω_1 are similar. Hence,

$$\int_{\Omega_2-\Omega_1}\epsilon_{ij}(\mathbf{x})d\Omega_{\mathbf{x}}=0\ . \tag{4.80}$$

Let $\Omega_1\to\Omega_0$ and $\Omega_2\to\Omega$, we have the desired result,

$$\int_{\Omega_2-\Omega_1}\epsilon_{ij}(\mathbf{x})d\Omega_{\mathbf{x}}=\int_{\Omega-\Omega_0}\epsilon_{ij}(\mathbf{x})d\Omega_{\mathbf{x}}=0\ . \tag{4.81}$$

□

Remark 4.3.

(1) It is also true that the average disturbance stress field vanished over the domain $\Omega-\Omega_0$, i.e.,

$$\int_{\Omega-\Omega_0}\sigma_{ij}d\Omega_{\mathbf{x}}=0\ . \tag{4.82}$$

(2) Eq. (4.78) is valid as long as $\Omega_1\subset\Omega_2$. They don't need to be confocal, but they definitely need to be similar. They may need to be coaxial, but some researchers has questioned the necessity of this requirement. The fundamental issue underlying the debate is : does Eshelby tensor depend on coordinates ?

(3) This result can be generalized into the cases that the inclusion Ω_0 is not ellipsoidal and the eigenstrain distribution in Ω_0 is not uniform.

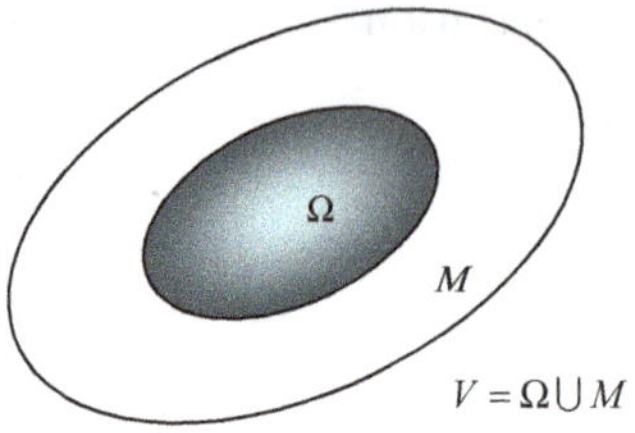

Fig. 4.2 Schematic illustration of two-phase model.

4.3.2 *Mori-Tanaka's mean field theory*

In previous homogenization procedures, the disturbance strain and stress fields due to an inhomogeneity are approximated by Eshelby's single inclusion solution in an infinite space.

In real applications, an RVE is finite, and it is subjected to remote boundary conditions, e.g. prescribed traction condition or prescribed displacement condition, i.e.

$$\boldsymbol{u} = \boldsymbol{\epsilon}^0 \cdot \mathbf{x}, \quad \mathbf{x} \in \partial V \tag{4.83}$$

or

$$\boldsymbol{t} = \boldsymbol{\sigma}^0 \cdot \boldsymbol{n}, \quad \mathbf{x} \in \partial V \tag{4.84}$$

Let $\boldsymbol{\epsilon}^{pt}$ and $\boldsymbol{\sigma}^{pt}$ representing perturbed strain and stress fields due to Eshelby's single inclusion solution in an infinite medium. If we assume

$$\boldsymbol{\epsilon}(\mathbf{x}) = \boldsymbol{\epsilon}^0 + \boldsymbol{\epsilon}^{pt} \tag{4.85}$$

$$\boldsymbol{\sigma}(\mathbf{x}) = \boldsymbol{\sigma}^0 + \boldsymbol{\sigma}^{pt} \tag{4.86}$$

Obviously,

$$\boldsymbol{\sigma}^0 + \boldsymbol{\sigma}^{pt} \neq \boldsymbol{\sigma}^0, \quad \text{or} \quad \boldsymbol{\epsilon}^0 + \boldsymbol{\epsilon}^{pt} \neq \boldsymbol{\epsilon}^0, \quad \forall\, \mathbf{x} \in \partial V \tag{4.87}$$

Therefore, neither boundary condition (4.84) and (4.83) will be satisfied if the RVE is finite. This is because a finite-sized RVE will cause additional interaction between matrix and inclusions, interaction between the boundary and inclusions, and interaction among inclusions themselves. Note that $\boldsymbol{\epsilon}^{pt} \to 0, \boldsymbol{\sigma}^{pt} \to 0$ only if $|\mathbf{x}| \to \infty$.

To take into account the effects of a finite RVE, additional stress and strain fields are needed to faithfully represent the total stress and strain distribution in the RVE, i.e.

$$\boldsymbol{\sigma} = \boldsymbol{\sigma}^0 + \boldsymbol{\sigma}^{im} + \boldsymbol{\sigma}^{pt} \quad \text{and} \quad \boldsymbol{\epsilon} = \boldsymbol{\epsilon}^0 + \boldsymbol{\epsilon}^{im} + \boldsymbol{\epsilon}^{pt}$$

where $\boldsymbol{\sigma}^{im}$ and $\boldsymbol{\epsilon}^{im}$ are the so-called image stress and image strain.

In literature, especially literatures on dislocations, additional stress and strain fields that accommodate the stress solution of an infinite space to satisfy the boundary conditions of a finite domain are called image stress and image strain fields. The terminology comes from the practice that some of these stress and strain fields can be obtained by introducing certain image external sources. Nevertheless, by doing so, the homogenization problem in a finite RVE could become complicated, because it is in general very difficult to know the precise distribution of image stresses and image strains. To circumvent this difficulty, T. Mori and K. Tanaka [Mori and Tanaka (1973)] proposed the following mean field theory, which is proved to be an ingenious and very successful approach. In fact, Mori & Tanaka's theory was later found, in a remarkable paper by G. J. Weng [Weng (1990)] , that it coincides with the Hashin-Shtrikman lower bound.

The following presentation is adapted from Weng's formulation. Suppose that in an RVE there are many inhomogeneities, or the density of inhomogeneities are statistically stable, then the strain or stress field in the matrix may be written as the superposition of a mean field and the disturbance field

$$\boldsymbol{\epsilon}(\mathbf{x}) = \boldsymbol{\epsilon}^m + \boldsymbol{\epsilon}^d, \quad \text{and} \quad \boldsymbol{\sigma}(\mathbf{x}) = \boldsymbol{\sigma}^m + \boldsymbol{\sigma}^d \ .$$

In general we don't know the precise disturbance of the mean field, which may not be directly or explicitly related to the prescribed boundary conditions.

Consider the matrix is the dominant phase in the composite. We denote the average field in the matrix as the mean field, i.e. $< \boldsymbol{\epsilon} >_M = \boldsymbol{\epsilon}^m$ or $< \boldsymbol{\sigma} >_M = \boldsymbol{\sigma}^m$, which include boundary effects and effects of interactions among many other inclusions.

Now we add an inclusion into the spatially averaged ensemble—the RVE. After the inclusion is added, we call the field as the new field in contrast with the old field before the inclusion is added. Therefore, in the matrix,

$$< \boldsymbol{\epsilon}^{new} >_M = < \boldsymbol{\epsilon}^{old} >_M + < \boldsymbol{\epsilon}^{pt} >_M + < \boldsymbol{\epsilon}^{im} >_M, \ \ \forall \mathbf{x} \in M \tag{4.88}$$

where $\boldsymbol{\epsilon}^{pt}$ and $\boldsymbol{\epsilon}^{im}$ are the inclusion solution for infinite space and the corresponding image strain solution due to the finite RVE.

By the Tanaka-Mori lemma, $< \boldsymbol{\epsilon}^{pt} >_M = 0$. Mori and Tanaka then further argued that since there have been so many inclusions inside the RVE, the average effects of the image strain or image stress field due to adding a single inclusion may be negligible as well without altering the mean field of the RVE, i.e. $< \boldsymbol{\epsilon}^{im} >_M = 0$. In short, the mean field, or the average field in the matrix remains the same, which is the essence of Mori-Tanaka mean field theory. Note that $< \boldsymbol{\epsilon} >_M$ does take into account the average effects of the image stress/strain fields of all other inclusions. Consequently, we can write

$$< \boldsymbol{\epsilon}^{new} >_M = < \boldsymbol{\epsilon}^{old} >_M = < \boldsymbol{\epsilon} >_M \ , \forall \mathbf{x} \in M \ . \tag{4.89}$$

Inside the inclusion, we still neglect the effects of image strain or image stress field of the newly added inclusion, we then have

$$\begin{aligned} < \boldsymbol{\epsilon} >_\Omega &= < \boldsymbol{\epsilon} >_M + < \boldsymbol{\epsilon}^{pt} >_\Omega + < \boldsymbol{\epsilon}^{im} >_\Omega \\ &= < \boldsymbol{\epsilon} >_M + < \boldsymbol{\epsilon}^{pt} >_\Omega = < \boldsymbol{\epsilon} >_M + \mathbb{S}^{I,\infty} : \boldsymbol{\epsilon}^*, \ \ \forall \mathbf{x} \in \Omega \end{aligned} \tag{4.90}$$

Similarly, for the stress field,

$$\begin{aligned} <\boldsymbol{\sigma}^{new}>_M &= <\boldsymbol{\sigma}^{old}>_M = <\boldsymbol{\sigma}>_M, \quad \mathbf{x} \in M \\ <\boldsymbol{\sigma}>_\Omega &= <\boldsymbol{\sigma}>_M + \mathbb{T}^{I,\infty} : \boldsymbol{\sigma}^*, \quad \mathbf{x} \in \Omega \end{aligned} \tag{4.91}$$

where the relations $\boldsymbol{\epsilon}^{pt} = \mathbb{S}^{I,\infty} : \boldsymbol{\epsilon}^*$ and $\boldsymbol{\sigma}^{pt} = \mathbb{T}^{I,\infty} : \sigma^*$ are used.

Based on Eshelby's equivalence homogenization conditions,

$$\mathbb{C}^I :< \boldsymbol{\epsilon} >_\Omega = \mathbb{C} : \Big(< \boldsymbol{\epsilon} >_\Omega - \boldsymbol{\epsilon}^* \Big) \tag{4.92}$$

or

$$\mathbb{D}^I :< \boldsymbol{\sigma} >_\Omega = \mathbb{D} : \Big(< \boldsymbol{\sigma} >_\Omega - \boldsymbol{\sigma}^* \Big) \tag{4.93}$$

One may obtain

$$\begin{aligned} & <\boldsymbol{\epsilon}>_\Omega = \mathbb{A}^I : \boldsymbol{\epsilon}^* \quad \Rightarrow \quad <\boldsymbol{\epsilon}>_M + <\boldsymbol{\epsilon}^{pt}>_\Omega = \mathbb{A}^I : \boldsymbol{\epsilon}^* \\ \text{or} \quad & <\boldsymbol{\sigma}>_\Omega = \mathbb{B}^I : \boldsymbol{\sigma}^* \quad \Rightarrow \quad <\boldsymbol{\sigma}>_M + <\boldsymbol{\sigma}^{pt}>_\Omega = \mathbb{A}^I : \boldsymbol{\sigma}^* \end{aligned}$$

where $\mathbb{A}^I := (\mathbb{C} - \mathbb{C}^I)^{-1} : \mathbb{C}$ and $\mathbb{B}^I := (\mathbb{D} - \mathbb{D}^I)^{-1} : \mathbb{D}$. Subsequently, one can obtain that

$$\begin{aligned} <\boldsymbol{\epsilon}>_\Omega &= \boldsymbol{\mathcal{A}}^I :< \boldsymbol{\epsilon} >_M \\ \text{or} \quad <\boldsymbol{\sigma}>_\Omega &= \boldsymbol{\mathcal{B}}^I :< \boldsymbol{\sigma} >_M \end{aligned} \tag{4.94}$$

based on different hominization schemes. In passing, we note that the concentration tensors may be written in different forms,

$$\begin{aligned} \boldsymbol{\mathcal{A}}^I &= \mathbb{A}^I : (\mathbb{A}^I - \mathbb{S}^{I,\infty})^{-1} = \Big[(\mathbb{A}^I - \mathbb{S}^{I,\infty}) : (\mathbb{A}^I)^{-1} \Big]^{-1} \\ &= \Big[\mathbb{I}^{(4s)} - \mathbb{S}^{I,\infty} : \mathbb{A}^{I^{-1}} \Big]^{-1} = \Big[\mathbb{I}^{(4s)} - \mathbb{S}^{I,\infty} : \mathbb{C}^{-1} : (\mathbb{C} - \mathbb{C}^I) \Big]^{-1} \\ &= \Big[\mathbb{I}^{(4s)} + \mathbb{P}^I : (\mathbb{C}^I - \mathbb{C}) \Big]^{-1} \end{aligned} \tag{4.95}$$

and

$$\begin{aligned} \boldsymbol{\mathcal{B}}^I &= \mathbb{B}^I : (\mathbb{B}^I - \mathbb{T}^{I,\infty})^{-1} = \Big[(\mathbb{B}^I - \mathbb{T}^{I,\infty}) : (\mathbb{B}^I)^{-1} \Big]^{-1} \\ &= \Big[\mathbb{I}^{(4s)} - \mathbb{T}^{I,\infty} : \mathbb{B}^{I^{-1}} \Big]^{-1} = \Big[\mathbb{I}^{(4s)} - \mathbb{T}^{I,\infty} : \mathbb{D}^{-1} : (\mathbb{D} - \mathbb{D}^I) \Big]^{-1} \\ &= \Big[\mathbb{I}^{(4s)} + \mathbb{Q}^I : (\mathbb{D}^I - \mathbb{D}) \Big]^{-1} \end{aligned} \tag{4.96}$$

where

$$\mathbb{P}^I = \mathbb{S}^{I,\infty} : \mathbb{C}^{-1} \quad \text{and} \quad \mathbb{Q}^I = \mathbb{T}^{I,\infty} : \mathbb{D}^{-1} \tag{4.97}$$

are called polarization tensors.

By definition,

$$<\boldsymbol{\epsilon}>_V = (1-f) <\boldsymbol{\epsilon}>_M + f <\boldsymbol{\epsilon}>_\Omega \tag{4.98}$$

$$<\boldsymbol{\sigma}>_V = (1-f) <\boldsymbol{\sigma}>_M + f <\boldsymbol{\sigma}>_\Omega \tag{4.99}$$

From (4.98) and (4.99), we may find that

$$\begin{aligned}<\boldsymbol{\sigma}>_V &= (1-f)<\boldsymbol{\sigma}>_M + f<\boldsymbol{\sigma}>_\Omega \\ &= (1-f)\mathbb{C}:<\boldsymbol{\epsilon}>_M + f\mathbb{C}^I:<\boldsymbol{\epsilon}>_\Omega \\ &= \Big((1-f)\mathbb{C} + f\mathbb{C}^I : \boldsymbol{\mathcal{A}}^I\Big):<\boldsymbol{\epsilon}>_M \end{aligned} \tag{4.100}$$

On the other hand,

$$\begin{aligned}<\boldsymbol{\sigma}>_V &= \bar{\mathbb{C}}:<\boldsymbol{\epsilon}>_V = \bar{\mathbb{C}}:\Big((1-f)<\boldsymbol{\epsilon}>_M + f<\boldsymbol{\epsilon}>_\Omega\Big) \\ &= \bar{\mathbb{C}}:\Big((1-f)\mathbb{I}^{(4s)} + f\boldsymbol{\mathcal{A}}^I\Big):<\boldsymbol{\epsilon}>_M \end{aligned} \tag{4.101}$$

Combining (4.100) and (4.101), we have the Mori-Tanaka estimate,

$$\boxed{\bar{\mathbb{C}} = \Big((1-f)\mathbb{C} + f\mathbb{C}^I : \boldsymbol{\mathcal{A}}^I\Big):\Big((1-f)\mathbb{I}^{(4s)} + f\boldsymbol{\mathcal{A}}^I\Big)^{-1}} \tag{4.102}$$

Similarly, we can write,

$$<\boldsymbol{\epsilon}>_V = \begin{cases} \Big((1-f)\mathbb{D} + f\mathbb{D}^I : \boldsymbol{\mathcal{B}}^I\Big):<\boldsymbol{\sigma}>_M \\ \bar{\mathbb{D}}:\Big((1-f)\mathbb{I}^{(4s)} + f\boldsymbol{\mathcal{B}}^I\Big):<\boldsymbol{\sigma}>_M \end{cases} \tag{4.103}$$

which leads to

$$\boxed{\bar{\mathbb{D}} = \Big((1-f)\mathbb{D} + f\mathbb{D}^I : \boldsymbol{\mathcal{B}}^I\Big):\Big((1-f)\mathbb{I}^{(4s)} + f\boldsymbol{\mathcal{B}}^I\Big)^{-1}} \tag{4.104}$$

Define the concentration tensors in the matrix as the fourth-order symmetric identity tensor, i.e.

$$\boldsymbol{\mathcal{A}}^M = \boldsymbol{\mathcal{B}}^M = \mathbb{I}^{(4s)} \tag{4.105}$$

We can rewrite the Mori-Tanaka scheme (4.102) and (4.104) as

$$\bar{\mathbb{C}} = \Big(f_M\mathbb{C}:\boldsymbol{\mathcal{A}}^M + f_I\mathbb{C}^I:\boldsymbol{\mathcal{A}}^I\Big):\Big(f_M\boldsymbol{\mathcal{A}}^M + f_I\boldsymbol{\mathcal{A}}^I\Big)^{-1} \tag{4.106}$$

$$\bar{\mathbb{D}} = \Big(f_M\mathbb{D}:\boldsymbol{\mathcal{B}}^M + f_I\mathbb{D}^I:\boldsymbol{\mathcal{B}}^I\Big):\Big(f_M\boldsymbol{\mathcal{B}}^M + f_I\boldsymbol{\mathcal{B}}^I\Big)^{-1} \tag{4.107}$$

Now we want to show that the Mori-Tanaka estimate is self-consistent, i.e. $\bar{\mathbb{C}}^{-1} = \bar{\mathbb{D}}$. Considering the identity,

$$\mathbb{C}^I:\boldsymbol{\mathcal{A}}^I = \boldsymbol{\mathcal{B}}^I:\mathbb{C} \tag{4.108}$$

we have

$$\bar{\mathbb{C}} = \Big((1-f)\mathbb{C} + f\mathbb{C}^I : \boldsymbol{\mathcal{A}}^I\Big):\Big((1-f)\mathbb{I}^{(4s)} + f\boldsymbol{\mathcal{A}}^I\Big)^{-1} \tag{4.109}$$

$$= \Big((1-f)\mathbb{I}^{(4s)} + f\boldsymbol{\mathcal{B}}^I\Big):\mathbb{C}:\Big((1-f)\mathbb{I}^{(4s)} + f\boldsymbol{\mathcal{A}}^I\Big)^{-1} \tag{4.110}$$

Therefore,

$$\bar{\mathbb{C}}^{-1} = \Big((1-f)\mathbb{I}^{(4s)} + f\boldsymbol{\mathcal{A}}^I\Big) : \mathbb{D} : \Big((1-f)\mathbb{I}^{(4s)} + f\boldsymbol{\mathcal{B}}^I\Big)^{-1} \tag{4.111}$$

$$= \Big((1-f)\mathbb{D} + f\mathbb{D}^I\boldsymbol{\mathcal{B}}^I\Big) : \Big((1-f)\mathbb{I}^{(4s)} + f\boldsymbol{\mathcal{B}}^I\Big)^{-1} = \bar{\mathbb{D}} \tag{4.112}$$

In the last line, the identity

$$\boldsymbol{\mathcal{A}}^I : \mathbb{D} = \mathbb{D}^I : \boldsymbol{\mathcal{B}}^I \tag{4.113}$$

is used. Proof of identities (4.108) and (4.113) are left to the readers.

In general, for a composite with n+1 phases (where the phase $\alpha = 0$ represents the matrix, and non-zero $\alpha = 1 \sim n$ represents the inhomogeneous phases, the Mori-Tanaka mean field theory gives the following estimates,

$$\bar{\mathbb{C}} = \Big(\sum_{\alpha=0}^{n} f_\alpha \mathbb{C}^\alpha : \boldsymbol{\mathcal{A}}^\alpha\Big) : \Big(\sum_{\alpha=0}^{n} f_\alpha \boldsymbol{\mathcal{A}}^\alpha\Big)^{-1}$$

$$\bar{\mathbb{D}} = \Big(\sum_{\alpha=0}^{n} f_\alpha \mathbb{D}^\alpha : \boldsymbol{\mathcal{B}}^\alpha\Big) : \Big(\sum_{\alpha=0}^{n} f_\alpha \boldsymbol{\mathcal{B}}^\alpha\Big)^{-1} \tag{4.114}$$

The above estimate in fact presents the effective modulus as an weighted average of the modulus in each phase. The results derived in (4.114) was first presented by G. J. Weng [Weng (1984, 1990)].

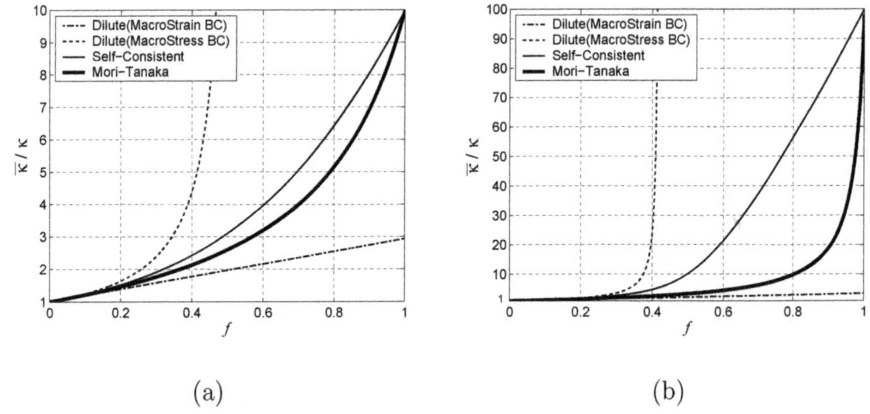

Fig. 4.3 Comparison of effective bulk moduli among various homogenization methods: dilute distribution, self-consistent, and the Mori-Tanaka: (a) composite A, and (b) composite B.

Remark 4.4. Among the above homogenization methods, homogenization under the assumption of dilute suspension of the second phase is simple and explicit, which is also accurate when the second phase distribution is indeed dilute. The self-consistent method considers interactions among the second phase population, but the homogenization algorithm becomes a nonlinear equation, and it often requires implicit numerical methods to solve it. The Mori-Tanaka method is a major

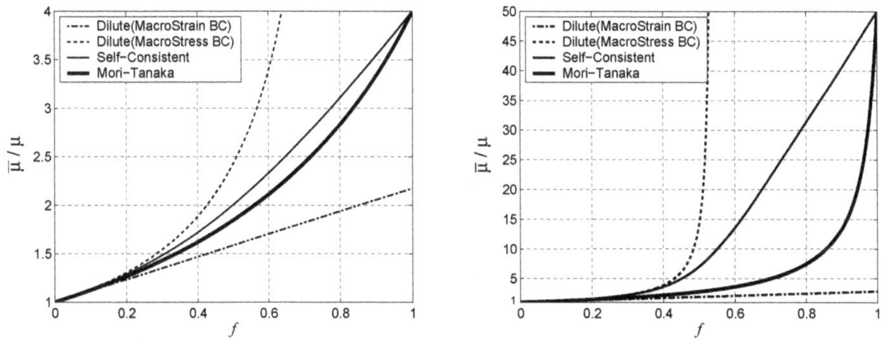

(a) (b)

Fig. 4.4 Comparison of effective shear moduli among various homogenization methods: dilute distribution, self-consistent, and the Mori-Tanaka: (a) composite A, and (b) composite B.

advance in composite material mechanics. It not only takes into account interactions among inclusions and possibly boundary image forces, but also is an explicit homogenization algorithm that is fairly easy to evaluate. It is this very reason that the Mori-Tanaka method has become one of the most popular homogenization methods in the field of composite materials.

In Fig. 4.3 and 4.4, we compare the homogenization results of bulk and shear moduli for two different composites obtained via (1) dilute suspension scheme (two different boundary conditions), the self-consistent scheme, and the Mori-Tanaka method. The bulk modulus ratio of the inclusion and matrix in composite A is 10 : 1, and shear modulus ratio of 4 : 1; while in composite B, modulus contrast is much higher, where the bulk modulus ratio of inclusion and matrix is 100 : 1, and shear modulus ratio is 50 : 1. Poisson's ratio of the matrix is chosen to be 0.1. The plots show the ratios of the effective modulus against the modulus of the matrix as a function of volume fraction of the inclusion. The general trend confirms that the self-consistent method and the Mori-Tanaka method are far more superior to the dilute suspension scheme, which is not convergent at the other end.

To fully take into account the reaction among different inclusions and their interactions with boundaries, one should solve inclusion problem in a finite domain, such that the information of volume fractions of different phases is embedded inside the Eshelby's tensor. We shall discuss this topic in Chapter 6 of this book.

4.4 Rodney Hill

The following biography is excerpted from an article written by Professor Geoffrey Hopkins and Professor Michael Sewell in 1982 for a special volume celebrating Rodney Hill's sixtieth birthday. The special volume was also edited by Professors

G. Hopkins and M. J. Sewell , and it is titled *Mechanics of Solids: The Rodney Hill 60th Anniversary Volume*, which was published by Pergamon Press Ltd. The present authors would like to thank Elsevier for granting us the permission to use part of this article in this book.

Rodney Hill was born on the 11th June 1921 at Stourton, near Leeds, in Yorkshirt. He comes from a family with deep roots in the practical and culture traditions of the West Riding, although with no known mathematical ability in an earlier generation. Rodney's father, Harold Harrison Hill, had been an only child and he was educated at the University of Leeds, gaining an M. A. for postgraduate work in history. He also took an external London degree in economics. After wartime service in the Royal Navy he became a schoolmaster, and was eventually senior History Master at Leeds Boy's Modern School. Rodney's mother had been a student at Leeds School of Art. Rodney himself was also an only child, in an immediate home background which encouraged scholarship and self-sufficiency.

Fig. 4.5 Rodney Hill

Rodney entered Leeds Grammar School with a scholarship in 1932, and there gave regular prize-wining evidence of all-round intellectual ability not only in mathematics, but equally in art, English literature, and other Arts subjects. During this period he taught himself to play the piano, and became proficient at chess in which he was later to represent Cambridge University and town. Thus were developing the powers of accurate observation and analysis to be brought to bear on the mathematics and physics which became his formal specialism from the age of 15. The customary large-team games did not attract him as school, but Rodney enjoyed the one-to-one sports of squash, fencing, and golf. He left school as Head of House, and in December 1938 he was awarded an Open Major Scholarship at Pembroke College, Cambridge. However, it needed the State and County Scholarships gained in the preceding summer to make a financially independent undergraduate.

Hill went up to Cambridge to read Mathematics in October 1939, against a background of external events which must have seemed the least auspicious since the very founding of the University. Major Scholars were expected to take Part II of the Tripos in two years instead of three by omitting all first-year courses. This imposed a heavy workload, to be carried under spartan conditions created by wartime restrictions such as blackout and rationing combined with antique College plumbing. For example, there was no running hot water, the nearest bath was courts away, and the winter allocation of one sack of coal per week fueled a fire in one's room only in the evenings. Hill was not deflected by the adverse general situation from his aim of a first-class honors degree, and he became a Wrangler in June 1941. This entitled him to take Part III of the mathematical Tripos, in the applied mathematical part of which quantum mechanics figured prominently at the time. However, he felt obliged to war-work, and so lost the opportunity for advanced training which those lecture courses would have provided. ······

Problems brought to the Theoretical Research Branch were distributed initially according to specialism of the more senior members, some of whom had acquired relevant experience at Woolwich Arsenal. Those problems which were quite new in context tended to go to the young inexperienced graduates newly arrived from university. This was indeed a baptism of fire for them, but it was a test which was to reveal Hill's true metier. One of his initial assignments was the deep penetration of very thick armor by Munoroe jets and high-velocity shells with tungsten-carbide cores. This required a mechanics of plastic deformation with unlimited magnitude, and thus was aroused Hill's interest in the field in which he later became perhaps the foremost world authority. At this stage, however, he had no prior knowledge of the physics and metallurgy of plasticity, and little of stress, strain or the tensors which the mathematics would eventually require. There was no useful textbook, but G. I. Taylor had written one or two helpful reports on shaped charges and Munros jets. Nevertheless, working at first with Mott and Pack, Hill was soon able to show, for example, that penetration by a tungsten–carbide core with pure ogival head would be seriously degraded if too much of the tip were ground conical (the British practice for manufacturing convenience). The demonstration was achieved not only theoretically, but also in field trials planned by Hill in collaboration with an experimental group under Dr. Charles Sykes, F.R.S.

The problems at Fort Halstead called for simple but effective mathematics guided by physical intuitions and a willingness to communicate with others, including non-mathematicians and experimentalists. There was not time for complicated mathematics, there were no electronic computers to assist it, and the experimental data were usually too crude to warrant it anyway. He acquired a lasting taste for a pragmatic blend of rigor, elegance, and simple realism in the application of mathematics.

The sense of purpose discovered at this time was noticed by colleagues as a cheerful and sparking earnestness. Popular relaxations among the group at Cambridge had included music, books, and lightning chess. At Fort Halstead ballroom dancing was added as a consuming passion for some, and Hill was not slow to find that he had medal-winning ability in this new enthusiasm. He met his future wife, Jeanne Wickens, early in 1945. She had been transferred to work at Fort Halstead from the bombing range at Shoeburyness. Previously she had trained as a dancer and teacher of ballet, but war cut short a promising career. They were married in Cambridge in 1946, and they have one daughter, born in 1955. The strength of his wife's support could already be detected in the Preface to Hill's first book, and the passage of years has happily reinforced this bond.

By this time the applied mechanics of both solid and fluids was being forced to push the boat out onto a sea of nonlinear problems, and away from the haven linearity in which much pre-war work had lingered. The trend was evident not only in England, of course, but in other countries too. Hill found himself in demand as the sole adviser on continuum plasticity in England, not only concerning problems arising from the interests at Fort Halstead, but also for new theories of metal-working processes needed by engineers in the steel industry. He obtained a Cambridge Ph.D. in 1948 for a Thesis entitled "Theoretical studies of the plastic deformation of metals". From the Ph.D. Thesis grew a much more extensive monograph on "The Mathematical Theory of Plasticity", published at the Clarendon Press, Oxford, in 1950. This very rapidly established Hill as an international authority. The final draft was written in his spare time, i.e. in the evenings and weekends. He was then still only in his 28th year, and it is timely to recall a remark from the review of the book in Engineering: "The author has done his work so well that it is difficult to see how it could be bettered. The book should rank for many years as an authoritative source of reference." This prognostication was fully borne out. The book was in print at Oxford for 21 years, Japanese and Russian translations have been made, and total sales currently approach 13,000.

The Journal of Mechanics and Physics of Solids was launched with the encouragement of the infant Pergamon Press in 1952. Hill suggested the title and the general aim of a forum for effective applied mathematics, linked with experimentation, in engineering science. From the onwards the Journal has been regarded as among the foremost in its general field, and unique in flavor. Hill served as Editor-in-Chief until handing over in 1968 to H.G. Hopkins.

The University of Nottinggham had received its Charter, and independence from London, in 1948, and was shortly to embark on two decades substantial expansion. Professor H. R. Pitt was appointed in 1950 to head the existing Mathematics Department, and he was soon instrumental in securing the creation of a new Chair of Applied Mathematics. Rodney Hill applied, and was offered the post in 1953 while still on 31. It was his responsibility to modernize the teaching of applied mathematics. Hill took over some existing course himself, and instigated new ones with the aim of encouraging research students. His undergraduate lectures were characterized by conciseness and tendency to brevity. He would never exceed the time limit. But those students who took the trouble to write down what he said, in addition to what was written on the blackboard, found after reflection that they had a first-class and substantial set of notes.

It may only have been a coincidence that emergence of interest in the so-called rational continuum mechanics was taking place in some American and British universities at this time. Hill's writings demonstrate an independent view of these development, and no taste at all for axiomatics. He was beginning to lay down the basis of general studies of non-uniqueness and instability in continua which were to prove highly influential over the next two decades, and which in due course brought further students and able collaborators.

The University of Cambridge conferred the degree of Sc. D. upon Rodney Hill in 1959. The highest honor to which any British scientist aspires followed in 1961, when he was elected a Fellow of the Royal Society. This gave much pleasure to his colleagues at Nottingham and to his friends elsewhere.

In 1963 Hill was elected to a Berkeley Bye-Fellowship at Gonville and Caius College, Cambridge. This he held for 6 years until the University conferred a personal Readership in Mechanics of Solids. Thus he became a member of the teaching staff of the Department of Applied Mathematics and Theoretical Physics, and in 1972 a personal Professorship was conferred.

During this Cambridge period, properties of heterogeneous media (including fibre composites), single crystals, continuum plasticity, and an independent reformulation of rubber elasticity were explored, ...

His standards of scholarship and intellectual honesty are the highest. He is ready in his appreciation of the good work of others; and he has been sharp in candid criticism of misguided thinking or slack presentation (especially by those mature enough to know better) if he thought the subject-matter would be best served thereby—as some celebrated footnotes and book reviews testify.

The outward character of the man is not unlike his papers: physically tall and slim, with the long fingers of a pianist, and having a quiet but compelling presence. His unusually deep reserve has meant that casual social gatherings and conferences have held less interest and been less rewarding for him than for others.

— Excerpted from Geoffery Hopkins and Michael Sewell [1982] in *Mechanics of Solids* Pergamon Press, reproduced with kind permission from Elsevier.

4.5 Exercises

Problem 4.1. *Consider a n-phase composite material, and each phase has its own elastic tensor $\mathbb{C}^\alpha$, compliance tensor $\mathbb{D}^\alpha$; and matrix has elastic tensor, $\mathbb{C}$, and compliance tensor, $\mathbb{D}$. Assume that in the representative volume element (RVE), each phase only appears as one ellipsoidal inclusion. Under dilute distribution assumption, the corresponding Eshelby tensor and conjugate Eshelby tensor for each phase are $\mathbb{S}^\alpha$ and $\mathbb{T}^\alpha$ respectively. Denote*

$$\mathbb{A}^\alpha = (\mathbb{C} - \mathbb{C}^\alpha)^{-1} : \mathbb{C} \tag{4.115}$$

$$\mathbb{B}^\alpha = (\mathbb{D} - \mathbb{D}^\alpha)^{-1} : \mathbb{D} \tag{4.116}$$

Show

$$\mathbb{C}^\alpha : \mathbb{A}^\alpha : (\mathbb{A}^\alpha - \mathbb{S}^\alpha)^{-1} : \mathbb{D} = \mathbb{B}^\alpha : (\mathbb{B}^\alpha - \mathbb{T}^\alpha)^{-1} \tag{4.117}$$

$$\mathbb{D}^\alpha : \mathbb{B}^\alpha : (\mathbb{B}^\alpha - \mathbb{T}^\alpha)^{-1} : \mathbb{C} = \mathbb{A}^\alpha : (\mathbb{A}^\alpha - \mathbb{S}^\alpha)^{-1} \tag{4.118}$$

Problem 4.2. *For an isotropic two phase material. Assume the inhomogeneity phase is random distributed spherical cavities ($\mu_I = 0; K_I = 0$), and the matrix is an incompressible material ($K \to \infty$). Use the self-consistent scheme,*

$$\frac{\bar{K}}{K} = 1 + \sum_{\alpha=1}^{n} f_\alpha \Big(\frac{K^\alpha}{K} - 1\Big)\Big(1 + (\frac{K^\alpha}{\bar{K}} - 1)\bar{s}_1\Big)^{-1} \tag{4.119}$$

$$\frac{\bar{\mu}}{\mu} = 1 + \sum_{\alpha=1}^{n} f_\alpha \Big(\frac{\mu^\alpha}{\mu} - 1\Big)\Big(1 + (\frac{\mu^\alpha}{\bar{\mu}} - 1)\bar{s}_2\Big)^{-1} \tag{4.120}$$

where

$$\bar{s}_1 = \frac{1+\bar{\nu}}{3(1-\bar{\nu})} \tag{4.121}$$

$$\bar{s}_2 = \frac{2(4-5\bar{\nu})}{15(1-\bar{\nu})} \tag{4.122}$$

to find the effective bulk modulus, $\bar{K}$, and the effective shear modulus, $\bar{\mu}$.
Hint: See references [Willis (1981); Budiansky (1965)]

Problem 4.3. *Assume that in an RVE there are n+1 phases, $\alpha = 0, 1, \cdots, n$. The Mori-Tanaka mean theory states that*

$$\bar{\mathbb{D}} = \sum_{\alpha=0}^{n} f_\alpha \mathbb{D}_\alpha : \boldsymbol{\mathcal{B}}^\alpha : \Big(\sum_{\alpha=0}^{n} f_\alpha \boldsymbol{\mathcal{B}}^\alpha\Big)^{-1} \tag{4.123}$$

$$\bar{\mathbb{C}} = \sum_{\alpha=0}^{n} f_\alpha \mathbb{C}_\alpha : \boldsymbol{\mathcal{A}}^\alpha : \Big(\sum_{\alpha=0}^{n} f_\alpha \boldsymbol{\mathcal{A}}^\alpha\Big)^{-1} \tag{4.124}$$

Show that the Mori-Tanaka scheme is self-consistent, i.e.

$$\bar{\mathbb{C}} = \bar{\mathbb{D}}^{-1} \tag{4.125}$$

Hint: Use the identities

$$\mathbb{C}^\alpha : \boldsymbol{\mathcal{A}}^\alpha = \boldsymbol{\mathcal{B}}^\alpha : \mathbb{C}^0 \tag{4.126}$$

$$\mathbb{D}^\alpha : \boldsymbol{\mathcal{B}}^\alpha = \boldsymbol{\mathcal{A}}^\alpha : \mathbb{D}^0 \tag{4.127}$$

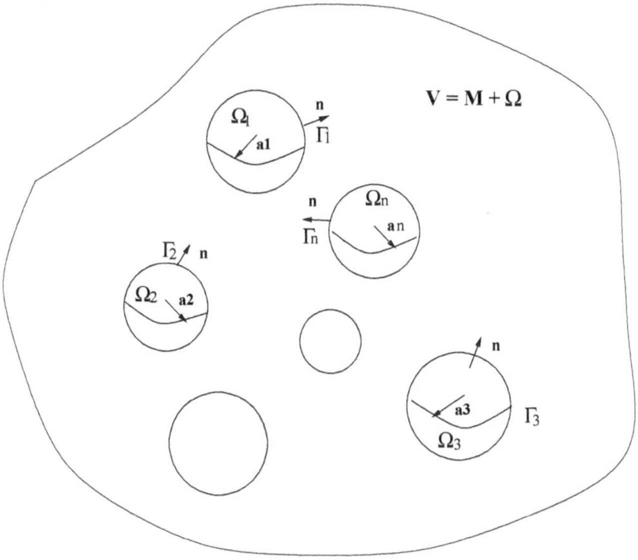

Fig. 4.6 An RVE containing spherical micro-cavities.

Problem 4.4. *Show that for isotropic materials the interior conjugate Eshelby tensor of a spherical inclusion has the form,*

$$\mathbb{T}^{I,\infty} = (1 - s_1)\mathbb{E}^{(1)} + (1 - s_2)\mathbb{E}^{(2)} \tag{4.128}$$

Problem 4.5. *Consider a two-phase composite with randomly distributed spherical inclusions. The ratios of material constants between inhomogeneity and matrix are*

$$\frac{K^{\Omega}}{K} = 25, \quad \text{and} \quad K^{\Omega} = 750MP_a \tag{4.129}$$

$$\frac{\nu^{\Omega}}{\nu} = 4, \quad \text{and} \quad \nu^{\Omega} = 0.4 \tag{4.130}$$

Plot the ratio of $\dfrac{\bar{K}}{K}$, $\dfrac{\bar{\mu}}{\mu}$, *and* $\dfrac{\bar{\nu}}{\nu}$ *verses the volume fraction of inhomogeneity,* f, *by using homogenization methods under the assumption of dilute suspension (both prescribed traction and prescribed displacement), self-consistent method, and the Mori-Tanaka method .*

Problem 4.6. *Scientists found a rare crystal with unknown material properties. The individual size of a crystal is too small to measure its material properties see Fig. 4.6.*

In order to measure the bulk modulus of the crystal, scientists have measured the bulk modulus, $\bar{K}$, *as well as the poisson ratio,* $\bar{\nu}$, *of an 3D aggregated crystal* *specimen with porosity coefficient f ranging from 0.1 to 0.5. Assume the void, or cavity inside the crystal compound is spherical. Find the bulk modulus of the original crystal.*

Chapter 5

COMPARISON VARIATIONAL PRINCIPLES

One of the powerful and yet elegant approaches in mechanical homogenization is the variational approach similar to the case in computational mechanics. In particular, there exists a unique class of variational principles called "*the comparison variational principle*", which has been instrumental in providing accurate estimates on effective material properties. The landmark contributions in comparison variational principles include those by Z. Hashin & S. Shtrikman [Hashin and Shtrikman (1961, 1962a,b)], R. Hill [Hill (1963)], D. Talbot and J. R. Willis [Talbot and Willis (1985, 1987)], and more recently P. Ponte Castañeda [Ponte Castañeda (1991, 1992a,b)]. They will be covered in this Chapter.

5.1 Review of variational calculus

Before discussing comparison variational principles, we first review the basic theories and principles of variational calculus. To put it simple, *a functional is a map from functions to real numbers.* Considering the following functional:

$$I(y) : H^1([x_0, x_1]) \to \mathbb{R} \ , \tag{5.1}$$

it takes function y in the function space $H^1([x_0, x_1])$ as its argument and returns a real number. The function space $H^1([x_0, x_1])$ is defined as

$$H^1([x_0, x_1]) = \left\{ y(x) \ \middle| \ \int_{x_0}^{x_1} \Big(y^2(x) + (y'(x))^2 \Big) dx < \infty \right\}$$

Since $H^1([x_0, x_1])$ is a function space, one can measure the closeness between two functions in the space by defining the following H^1 norm,

$$\|y_1 - y_2\|_{H^1} := \sqrt{\int_{x_0}^{x_1} \Big((y_1 - y_2)^2 + (y_1' - y_2')^2 \Big) dx}$$

There are various forms of functionals. For example, $I(y)$ can be defined as the following integration map,

$$I(y) = \int_{x_0}^{x_1} \Big[p(x)(y')^2 + q(x)y^2 + 2f(x)y \Big] dx \tag{5.2}$$

where $p(x), q(x)$, and $f(x)$ are known continuous functions, i.e. $p(x), q(x)$, and $f(x) \in C^0([x_0, x_1])$, and $p(x) > 0,\ q(x) > 0,\ \forall x \in [x_1, x_2]$. The unknown function $y \in H^1([x_0, x_1])$, and satisfies the following prescribed boundary conditions,

$$y(x_0) = y_0, \quad y(x_1) = y_1 \tag{5.3}$$

Formally, we state that $y(x)$ belongs to the function space $\mathcal{V}$, which is defined as

$$\mathcal{V} := \Big\{ y(x) \ \Big|\ y \in H^1([x_0, x_1]), y(x_0) = y_0 \ \text{ and } \ y(x_1) = y_1 \Big\} \tag{5.4}$$

Consider a perturbation of $y(x)$ in the function space $\mathcal{V}$

$$\tilde{y}(x) = y(x) + \alpha\eta(x), \tag{5.5}$$

if $|\alpha| << 1$ or $\|\eta(x)\|_{H^1} << 1$, this perturbation function will be very close to function $y(x)$ in the sense that the distance between the two functions in a function space, say $H^1([x_0, x_1])$ in this case, will be very small, i.e. $\|\tilde{y} - y\|_{H^1} << 1$.

We usually require that $\eta(x) \in \overset{\circ}{\mathcal{V}}$, i.e.,

$$\overset{\circ}{\mathcal{V}} := \Big\{ \eta(x) \ \Big|\ \eta \in H^1([x_0, x_1]), \eta(x_0) = 0 \ \text{ and } \ \eta(x_1) = 0 \Big\} \tag{5.6}$$

In finite element analysis, $y(x)$ is called the trial function and $\alpha\eta(x)$ or $\eta(x)$ is called the test function, or window function. Note that the requirement that the test function satisfies homogeneous boundary conditions, i.e. $\eta(x) \in \overset{\circ}{\mathcal{V}}$, is due to the fact that $y(x) and \tilde{y}(x) \in \mathcal{V}$.

A fundamental task of variational calculus is to find the extreme values of a given functional, or to find the conditions to achieve the extreme value. It is somewhat similar to finding the extreme values of a function in a given domain. In calculus, the condition that a continuous function has stationary value is

$$\frac{dy}{dx}\Big|_{x=x_0} = 0\ , \quad \text{or} \quad y'\Big|_{x=x_0} = 0\ ,$$

Similarly, to find the function $y(x)$ that yields the extreme value of $I(y)$ in Eq.(5.2), we consider the perturbation of $I(y)$, i.e.,

$$\begin{aligned}
I(y(x) + \alpha\eta(x)) &= \int_{x_0}^{x_1} \{p(x)[y'(x) + \alpha\eta'(x)]^2 + q(x)[y(x) + \alpha\eta(x)]^2 \\
&\quad + 2f(x)[y(x) + \alpha\eta(x)]\}\, dx \\
&= \int_{x_0}^{x_1} \Big[p(x)(y'(x))^2 + q(x)y^2(x) + 2f(x)y(x)\Big] dx \\
&\quad + 2\alpha \int_{x_0}^{x_1} \Big[p(x)y'(x)\eta'(x) + q(x)y(x)\eta(x) + f(x)\eta(x)\Big] dx \\
&\quad + \alpha^2 \int_{x_0}^{x_1} \Big(p(x)(\eta'(x))^2 + q(x)\eta^2(x)\Big) dx
\end{aligned} \tag{5.7}$$

Thereby,

$$\Delta I = I(y(x) + \alpha\eta(x)) - I(y(x)) = \alpha\delta I + \frac{\alpha^2}{2!}\delta^2 I \tag{5.8}$$

where

$$\delta I = 2\int_{x_0}^{x_1} [p(x)y'(x)\eta'(x) + q(x)y(x)\eta(x) + f(x)\eta(x)]dx \tag{5.9}$$

$$\delta^2 I = 2\int_{x_0}^{x_1} \Big[p(x)(\eta'(x))^2 + q(x)\eta^2(x)\Big]dx \tag{5.10}$$

Remark 5.1. The first order variation illustrated above is in the sense of Gâteaux variation . The Gâteaux variation is defined in terms of the so-called Gâteaux derivative

$$\delta_G I = D_G I(y)\eta = \lim_{\alpha\to 0}\frac{I(y+\alpha\eta)-I(y)}{\alpha} = \frac{d}{d\alpha}I(y+\alpha\eta)\Big|_{\alpha=0} \tag{5.11}$$

If the above limit exists, we call it the Gâteaux derivative of functional I at $y \in \mathcal{V}$ in the direction of $\eta \in \overset{\circ}{\mathcal{V}}$.

One may compare this with the so-called Fréchet derivative , $D_F I(y)\eta$, which is defined as a linear functional such that

$$\lim_{\|\eta\|_V\to 0}\frac{I(y+\eta)-I(y)-D_F I(y)\cdot\eta}{\|\eta\|_V} = 0 \tag{5.12}$$

If I is Fréchet differentiable at y, it is also Gâteaux differentiable there. However, not every Gâteaux differentiable functional is Fréchet differentiable. Gâteaux derivative coincides with Fréchet derivative, if $\delta_G I$ is linear in η and uniformly continuous in η.

In general, the n-th order Gâteaux variation is defined as

$$\delta_G^n I = \frac{d^n}{d\alpha^n}I(y+\alpha\eta)\Big|_{\alpha=0}, \quad \forall n \geq 1 \tag{5.13}$$

such that

$$\Delta I = I(y+\alpha\eta) - I(y) = \alpha\delta_G I + \frac{\alpha^2}{2!}\delta_G^2 I + \frac{\alpha^3}{3!}\delta_G^3 I + \frac{\alpha^4}{4!}\delta_G^4 I + \cdots \tag{5.14}$$

In the rest of the book, we omit the subscript G in the variation operator. In fact, if we define a scalar function,

$$\mathcal{I}(\alpha) := I(y+\alpha\eta)$$

then Gâteaux variation can be treated as the "ordinary" derivative with respect to the scalar variable

$$\delta^n I = \frac{d^n\mathcal{I}(\alpha)}{d\alpha^n} = \mathcal{I}^{(n)}\ .$$

Therefore, one nice feature of about the Gâteaux variation is that it can be defined as a scaler differentiation operation. In other words, the variation operation follows the same rule as the differentiation operation in elementary calculus. This can be verified by examining the first order variation of $I(y)$ in the above example,

$$\delta I = 2\int_{x_0}^{x_1} \Big[p(x)y'(x)\eta'(x) + q(x)y(x)\eta(x) + f(x)\eta(x)\Big]dx \tag{5.15}$$

Let $\eta(x) = \delta y$, the Gâteaux variation becomes,

$$\delta I = 2\int_{x_0}^{x_1} \Big[p(x)y'(x)\delta y' + q(x)y(x)\delta y + f(x)\delta y\Big]dx$$
$$= \delta\left\{\int_{x_0}^{x_1} \Big[p(x)(y'(x))^2 + q(x)(y(x))^2 + f(x)y(x)\Big]dx\right\} = \delta I \ .$$

This is to say that one can find the first variation of a functional, $I(y)$, by simply differentiating (taking G-derivative) the unknown function according to the same rule of differentiation in calculus. The only difference is: dy is replaced by δy, which is the variation of the unknown function, or in general, a test function satisfying homogeneous boundary conditions, i.e. $\delta y \in \overset{\circ}{\mathcal{V}}$.

Now we come back to the problem to find the extreme value of functional $I(y)$. We state that : *$I(y)$ is stationary at $y(x)$, if* $\delta I\Big|_{y(x)} = 0$. This is because

$$\frac{d\mathcal{I}(\alpha)}{d\alpha}\Big|_{\alpha=0} = 0 \ \Rightarrow \ \delta I\Big|_{y(x)} = 0 \ .$$

i.e., the stationary condition for $\mathcal{I}(\alpha)$ implies the stationary condition for $I(y)$.

In the above example, since $p(x), q(x) > 0$ and $\delta^2 I > 0$, $I(y)$ will reach a minimum at $\tilde{y} = y(x)$. However, at the present stage, $y(x)$ is still an abstract notation. What is the exact condition that $y(x)$ has to satisfy such that $\delta I = 0$?

Consider the first term in (5.15). Integration by parts yields,

$$\int_{x_0}^{x_1} p(x)y'(x)\eta'(x)dx = p(x)y'(x)\eta(x)\Big|_{x=x_0}^{x_1} - \int_{x_0}^{x_1}(p(x)y'(x))'\eta(x)dx$$
$$= -\int_{x_0}^{x_1}(p(x)y'(x))'\eta(x)dx \tag{5.16}$$

Therefore,

$$\delta I = 2\int_{x_0}^{x_1}\Big[-[p(x)y'(x)]' + q(x)y(x) + f(x)\Big]\eta(x)dx = 0 \tag{5.17}$$

Since this equation must holds for any $\eta(x) \in \overset{\circ}{\mathcal{V}}$, the integrand must vanish, i.e. the solution of the following differential equation

$$-[p(x)y'(x)]' + q(x)y(x) + f(x) = 0, \quad y(x_0) = y_0 \ \text{and} \ y(x_1) = y_1 \ . \tag{5.18}$$

is a minimizer of the functional $I(y)$. Eq. (5.18) is called the Euler-Lagrange equation.

Note that the solution of Eq. (5.18) may not be the only minimizer of the functional $I(y)$. In fact, Eq. (5.18) requires the solution $y(x)$ to be continuously differentiable, i.e., $y(x) \in C^1([x_0, x_1])$. On the other hand, the functional form only requires that $y(x) \in H^1([x_0, x_1])$. Since $C^1([x_0, x_1]) \subset H^1([x_0, x_1])$, Eq. (5.18) is called strong form of the Euler-Lagrange equation. For this purpose, a function is called a week solution if it makes $I(y)$ stationary, but not necessarily satisfy the Euler-Lagrange equation , i.e., the weak solution $y(x)$ only need to satisfy

$$\delta I = 2\int_{x_0}^{x_1}\Big[p(x)y'(x)\delta y' + q(x)y(x)\delta y + f(x)\delta y\Big]dx = 0 \tag{5.19}$$

In general, consider the following functional form,

$$I(y) = \int_{x_0}^{x_1} F(x,y,y')dx, \quad y(x_0) = y_0 \text{ and } y(x_1) = y_1 \ . \tag{5.20}$$

Its first variation is

$$\delta I = \int_{x_0}^{x_1}\left\{\frac{\partial F}{\partial y}\delta y + \frac{\partial F}{\partial y'}\delta y'\right\}dx$$

Integration by parts yields

$$\begin{aligned}\delta I &= \int_{x_0}^{x_1}\Big[\frac{\partial F}{\partial y}\delta y\Big]dx + \frac{\partial F}{\partial y'}\delta y\,\Big|_{x_0}^{x_1} - \int_{x_0}^{x_1}\frac{\partial}{\partial x}\Big[\frac{\partial F}{\partial y'}\Big]\delta y dx \\ &= \int_{x_0}^{x_1}\Big[\frac{\partial F}{\partial y} - \frac{\partial}{\partial x}\Big(\frac{\partial F}{\partial y'}\Big)\Big]\delta y dx \end{aligned} \tag{5.21}$$

One obtains the Euler-Lagrange equation,

$$\frac{\partial F}{\partial y} - \frac{\partial}{\partial x}\Big(\frac{\partial F}{\partial y'}\Big) = 0 \ . \tag{5.22}$$

To this end, we can say that the solution of the Euler-Lagrange equation will ensure the functional (5.20) reaches to a stationary value or stationary point, which may be a minimum point, or maximum point, or a saddle point.

5.2 Extremum variational principles in linear elasticity

In this section, we discuss the variational principles of linear elasticity theory. The variational principles are hen used to derive upper and lower bound estimate of the effective modulus of the composite.

5.2.1 *Minimum potential energy principle*

Consider a linear elastic solid, V. Assume that the solid is subjected to the following boundary conditions,

$$\mathbf{u} = \mathbf{u}^0, \qquad \forall \mathbf{x} \in \Gamma_{\boldsymbol{u}} \tag{5.23}$$

$$\mathbf{t} = \mathbf{n}\cdot\boldsymbol{\sigma} = \mathbf{t}^0, \quad \forall \mathbf{x} \in \Gamma_{\boldsymbol{t}} \tag{5.24}$$

The total potential energy of the elastic solid is

$$\begin{aligned}\Pi(\mathbf{u}) &= \frac{1}{2}\int_V \sigma_{ij}\epsilon_{ij}dV - \int_V f_i u_i dV - \int_{\Gamma_t} t_i^0 u_i dS \\ &= \frac{1}{2}\int_V C_{ijk\ell}u_{i,j}u_{k,\ell}dV - \int_V f_i u_i dV - \int_{\Gamma_t} t_i^0 u_i dS\end{aligned}$$

It is a functional map,

$$\Pi(\mathbf{u}) : \mathcal{V} \to \mathbb{R}$$

where

$$\mathcal{V} := \left\{ u_i(\mathbf{x}) \;\middle|\; u_i(\mathbf{x}) \in H^1(V), \;\; \text{and} \;\; u_i = u_i^0, \;\; \forall \mathbf{x} \in \Gamma_{\boldsymbol{u}} \right\}, \quad i = 1, 2, 3 \qquad (5.25)$$

The statement of a variational principle reads as the following: *Find a minimizer or maximizer, $\mathbf{u}^*$, such that*

$$\Pi(\mathbf{u}^*) = \inf_{\mathbf{u} \in \mathcal{V}} \{\Pi(\mathbf{u})\} \quad \text{or} \quad \mathbf{u}^* = \sup_{\mathbf{u} \in \mathcal{V}} \{\Pi(\mathbf{u})\}$$

or a stationary point, $\mathbf{u}^*$, such that

$$\delta\Pi(\mathbf{u}^*) = 0, \quad \forall \delta\mathbf{u} \in \overset{\circ}{\mathcal{V}}$$

where

$$\overset{\circ}{\mathcal{V}} := \left\{ \eta_i(\mathbf{x}) \;\middle|\; \eta_i(\mathbf{x}) \in H^1(V), \;\; \text{and} \;\; \eta_i(\mathbf{x}) = 0, \;\; \forall \mathbf{x} \in \Gamma_{\boldsymbol{u}} \right\}, \quad i = 1, 2, 3 \qquad (5.26)$$

Since the displacement boundary conditions define the solution space and become a part of the statement of the variational principle, they are therefore called essential boundary conditions. They place constraints on primary variable $\mathbf{u}$ and on the trial function space. For any displacement field that belongs to the trial function space, i.e., $\mathbf{u}(\mathbf{x}) \in \mathcal{V}$, $\mathbf{u}(\mathbf{x})$ is called kinematically admissible. $\overset{\circ}{\mathcal{V}}$ is often referred to as the test function space, i.e. $\delta\mathbf{u} \in \overset{\circ}{\mathcal{V}}$.

In this case, the H^1 norm is defined as

$$\|\mathbf{u}\|_{H^1} = \sqrt{\int_V (u_i u_i + u_{i,j} u_{i,j}) dV}$$

in which the symmetric part of the second term in the integrand can be linked to the so-called energy norm, i.e.

$$\int_V \epsilon_{ij}\epsilon_{ij} dV \;\; \Rightarrow \;\; \|\mathbf{u}\|_E = \int_V \epsilon_{ij} C_{ijk\ell} \epsilon_{k\ell} dV \; .$$

A necessary condition for $\Pi(\mathbf{u})$ to reach to an extreme value is the following stationary condition,

$$\delta\Pi[\mathbf{u}^*] = \int_V C_{ijk\ell} u^*_{i,j} \delta u_{k,\ell} dV - \int_V f_i \delta u_i dV - \int_{\Gamma_t} t_i^0 \delta u_i dS = 0, \quad \forall \delta\mathbf{u} \in \overset{\circ}{\mathcal{V}} \quad (5.27)$$

In computational mechanics, Eq.(5.27) is often called the weak form of Navier equations, which is the starting point of finite element formulation. Eq. (5.27) is also called *the virtual displacement principle* in structural mechanics. The above stationary condition has direct physical meaning: it implies the equilibrium condition in mechanics, i.e. any kinematically admissible $\mathbf{u}$ that satisfies the virtual displacement principle is an equilibrium solution. This fact can be illustrated as follows using integration by parts,

$$\begin{aligned}\delta\Pi &= \int_V \sigma^*_{ij}\delta u_{i,j}dV - \int_V f_i\delta u_i dV - \int_{\Gamma_t} t^0_i\delta u_i dS \\ &= \int_V \Big((\sigma^*_{ij}\delta u_i)_{,j} - \sigma^*_{ij,j}\delta u_i\Big)dV - \int_V f_i\delta u_i dV - \int_{\Gamma_t} t^0_i\delta u_i dS \\ &= \int_{\partial V} \sigma^*_{ij}n_j\delta u_i dS - \int_V (\sigma^*_{ij,j} + f_i)\delta u_i dV - \int_{\Gamma_t} t^0_i\delta u_i dS \\ &= \int_{\Gamma_t} (\sigma^*_{ij}n_j - t^0_i)\delta u_i dS - \int_V (\sigma^*_{ij,j} + f_i)\delta u_i dV + \int_{\Gamma_u} \sigma^*_{ij}n_j\delta u_i dS = 0\end{aligned}$$

Since $\delta\Pi = 0$ holds for $\forall \delta u_i \in \overset{\circ}{\mathcal{V}}$, it yields the Navier equation

$$C_{ijk\ell}u^*_{k,\ell j} + f_i = 0, \quad \forall\, \mathbf{x} \in V \tag{5.28}$$

and the traction boundary conditions,

$$\sigma^*_{ij}n_j = t^0_i = \sigma^0_{ij}n_j, \quad \forall\, \mathbf{x} \in \Gamma_t\,. \tag{5.29}$$

Remark 5.2. (1) Since the traction boundary condition is a natural outcome of the variational principle, it is often called the natural boundary condition; (2) Eqs. (5.28) is a second-order differential equation, which requires the solution to be at least continuously differentiable, i.e., $\mathbf{u} \in C^1(V)$. It is a condition much stringent than the requirement in the original variational statement, where $\mathbf{u} \in H^1(V)$. Therefore, we often call (5.28) *the strong form*, and Eq. (5.27) *the weak form.*

Examine the perturbation of the potential energy $\Delta\Pi(\mathbf{u})$ around an equilibrium configuration,

$$\begin{aligned}\Delta\Pi &= \Pi(\mathbf{u}^* + \delta\mathbf{u}^*) - \Pi(\mathbf{u}) \\ &= \frac{1}{2}\int_V C_{ijk\ell}(u^*_{i,j} + \delta u_{i,j})(u^*_{k,\ell} + \delta u_{k,\ell})dV - \int_V f_i(u^*_i + \delta u_i)dV \\ &\quad - \int_{\Gamma_t} t^0_i(u^*_i + \delta u_i)dV - \frac{1}{2}\int_V C_{ijk\ell}u^*_{i,j}u^*_{k,\ell}dV + \int_V f_i u^*_i dV + \int_{\Gamma_t} t^0_i u^*_i dV \\ &= \underbrace{\int_V C_{ijk\ell}u^*_{i,j}\delta u_{k,\ell}dV - \int_V f_i\delta u_i dV - \int_{\Gamma_t} t^0_i\delta u_i dV}_{} + \underbrace{\frac{1}{2}\int_V C_{ijk\ell}\delta u_{i,j}\delta u_{k,\ell}dV}_{} \\ &= \delta\Pi(\mathbf{u}^*) + \frac{1}{2!}\delta^2\Pi(\mathbf{u}^*)\end{aligned} \tag{5.30}$$

For the equilibrium solution $\delta\Pi(\mathbf{u}^*) = 0$. Because $C_{ijk\ell}$ is assumed to be positive definite, $\Delta\Pi = \frac{1}{2!}\delta^2\Pi(\mathbf{u}^*) > 0$. This means that for all kinematically admissible displacement fields, $\boldsymbol{u} \in \mathcal{V}$, the equilibrium solution is a unique solution, and it is the real solution. For linear elastic problems, it can also be shown that the weak solution equals the strong solution (Lax-Milgram theorem), and it is the minimizer of total potential energy $\Pi(\mathbf{u})$. It leads to the following statement:

Theorem 5.1 (Minimum potential energy principle). *Among all (infinitesimal) kinematically admissible displacement fields, the one that is also statically admissible (equilibrium solution) renders the potential energy* Π *an absolute minimum, that is*

$$\Pi(\boldsymbol{u}^*) \leq \Pi(\boldsymbol{u}), \quad \forall \boldsymbol{u} \in \mathcal{V}$$

or

$$\Pi(\mathbf{u}^*) = \inf_{\mathbf{u} \in \mathcal{V}} \Pi(\mathbf{u}) \ ,$$

5.2.2 *Minimum complementary potential energy principle*

Consider the following complementary potential energy,

$$\Pi^c(\boldsymbol{\sigma}) = \frac{1}{2} \int_V D_{ijk\ell} \sigma_{ij} \sigma_{k\ell} dV - \int_{\Gamma_u} u_i^0 \sigma_{ij} n_j dS \tag{5.31}$$

in which the first term is the complementary strain (stress) energy, and the second term is the external complementary work. We note that in Eq.(5.31), the Cauchy stress ($\boldsymbol{\sigma}$) belongs to the following function space,

$$\mathcal{S} = \Big\{ \sigma_{ij} \ \Big| \ \sigma_{ij} \in H^1(V), \ \ \sigma_{ij,j} = 0, \ \ \forall \mathbf{x} \in V \ \text{ and } \ \sigma_{ij} n_j = t_i^0, \ \ \forall \mathbf{x} \in \Gamma_t \Big\} \tag{5.32}$$

$\mathcal{S}$ is often called the trial function space, and the complementary potential energy may be viewed as a functional map

$$\Pi^c : \mathcal{S} \to \mathbb{R} \tag{5.33}$$

The variation of the Cauchy stress belongs to the test function space

$$\overset{\circ}{\mathcal{S}} = \Big\{ \sigma_{ij} \ \Big| \ \sigma_{ij} \in H^1(V), \ \ \sigma_{ij,j} = 0, \ \ \forall \mathbf{x} \in V \ \text{ and } \ \sigma_{ij} n_j = 0, \ \ \forall \mathbf{x} \in \Gamma_t \Big\} \tag{5.34}$$

We examine complementary potential energy perturbation around a stress state $\boldsymbol{\sigma}^*$,

$$\begin{aligned}
\Delta \Pi^c &= \Pi^c(\sigma_{ij}^* + \delta\sigma_{ij}) - \Pi^c(\sigma_{ij}^*) \\
&= \frac{1}{2} \int_V D_{ijk\ell} (\sigma_{ij}^* + \delta\sigma_{ij})(\sigma_{k\ell}^* + \delta\sigma_{k\ell}) dV - \int_{\Gamma_u} u_i^0 (\sigma_{ij}^* + \delta\sigma_{ij}) n_j dS \\
&\quad - \frac{1}{2} \int_V D_{ijk\ell} \sigma_{ij}^* \sigma_{k\ell}^* dV + \int_{\Gamma_u} u_i^0 \sigma_{ij}^* dS \\
&= \underbrace{\int_V D_{ijk\ell} \sigma_{ij}^* \delta\sigma_{k\ell} dV - \int_{\Gamma_u} u_i^0 \delta\sigma_{ij} n_j dS}_{} + \underbrace{\frac{1}{2} \int_V D_{ijk\ell} \delta\sigma_{ij} \delta\sigma_{k\ell} dV}_{} \\
&= \delta \Pi^c(\boldsymbol{\sigma}^*) + \delta^2 \Pi^c(\boldsymbol{\sigma}^*)
\end{aligned}$$

The necessary condition for $\Pi^c(\boldsymbol{\sigma})$ attaining extreme value is the stationary condition,

$$\delta \Pi^c(\boldsymbol{\sigma}^\star) = 0 \ , \quad \forall \delta\boldsymbol{\sigma} \in \overset{\circ}{\mathcal{S}} \tag{5.35}$$

Eq. (5.35) is often called as *the virtual force principle* in structural mechanics. Since $D_{ijk\ell}$ is positive definite,

$$\Delta\Pi^c = \frac{1}{2!}\delta^2\Pi^c > 0 \tag{5.36}$$

thus, $\Pi^c(\boldsymbol{\sigma})$ has a minimum value at $\boldsymbol{\sigma} = \boldsymbol{\sigma}^*$, where $\boldsymbol{\sigma}^*$ renders the stationary condition $\delta\Pi^c(\boldsymbol{\sigma}^*) = 0$. This fact is the so-called minimum complementary potential energy principle, and its variational statement is

Find $\boldsymbol{\sigma}^*$ such that

$$\Pi^c(\boldsymbol{\sigma}^*) = \inf_{\boldsymbol{\sigma}\in\mathcal{S}} \Pi^c(\boldsymbol{\sigma}), \quad \forall\boldsymbol{\sigma}\in\mathcal{S}\ .$$

Note that in this variational statement, the traction boundary condition

$$\mathbf{n}\cdot\boldsymbol{\sigma} = \mathbf{t}^0, \quad \forall\mathbf{x}\in\Gamma_t \tag{5.37}$$

becomes the essential boundary condition; whereas the displacement boundary condition

$$\mathbf{u} = \mathbf{u}^0, \quad \forall\mathbf{x}\in\Gamma_u\ . \tag{5.38}$$

becomes the natural boundary condition.

Theorem 5.2 (Minimum Complementary Energy Principle). *Among all statically admissible stress fields, the actual stress field (whose corresponding strain field satisfies compatibility condition) renders Π^c an absolute minimum, i.e.*

$$\Pi^c(\boldsymbol{\sigma}^*) \le \Pi^c(\boldsymbol{\sigma}), \quad \forall\boldsymbol{\sigma}\in\mathcal{S} \tag{5.39}$$

or

$$\Pi^c(\boldsymbol{\sigma}^*) = \inf_{\boldsymbol{\sigma}\in\mathcal{S}} \Pi^c(\boldsymbol{\sigma}) \tag{5.40}$$

The stationary condition of complementary energy is also called *virtual force principle* in continuum mechanics, and *the weak form of compatibility condition* in computational mechanics,

$$\delta\Pi^c(\sigma^*_{ij}) = \int_V D_{ijk\ell}\sigma^*_{ij}\delta\sigma_{k\ell}dV - \int_{\Gamma_u} u^0_i\delta\sigma_{ij}n_j dS = 0, \quad \forall\delta\boldsymbol{\sigma}\in\overset{\circ}{\mathcal{S}} \tag{5.41}$$

The above equation can be rewritten as

$$\int_V \epsilon^*_{ij}\delta\sigma_{ij}dV - \frac{1}{2}\int_V\Big(u^*_{i,j}\delta\sigma_{ij} + u^*_{j,i}\delta\sigma_{ij}\Big)dV + \int_V u^*_{i,j}\delta\sigma_{ij}dV - \int_{\Gamma_u} u^0_i\delta\sigma_{ij}n_j dS = 0$$

Integration by parts yields

$$\begin{aligned}&\int_V\Big(\epsilon^*_{ij} - \frac{1}{2}(u^*_{i,j} + u^*_{j,i})\Big)\delta\sigma_{ij}dV + \int_{\partial V} u^*_i\delta\sigma_{ij}n_j dS\\ &\underbrace{\int_V u^*_i\delta\sigma_{ij,j}dV}_{=0} - \int_{\Gamma_u} u^0_i\delta\sigma_{ij}n_j dS = 0\end{aligned}$$

$$\Rightarrow \int_V\Big(\epsilon^*_{ij} - \frac{1}{2}(u^*_{i,j} + u^*_{j,i})\Big)\delta\sigma_{ij}dV + \int_{\Gamma_u}(u^*_i - u^0_i)\delta\sigma_{ij}n_j dS = 0\ ,$$

which holds for $\forall \delta\sigma_{ij} \in \overset{\circ}{\mathcal{S}}$, leading to the following Euler-Lagrange equation (compatibility condition),

$$\epsilon^*_{ij} = D_{ijk\ell}\sigma^*_{k\ell} = \frac{1}{2}(u^*_{i,j} + u^*_{j,i}), \quad \text{or} \quad \epsilon^*_{ij,k\ell} + \epsilon^*_{k\ell,ij} - \epsilon^*_{ik,j\ell} - \epsilon^*_{j\ell,ik} = 0 \ , \tag{5.42}$$

and the natural boundary condition (displacement boundary condition)

$$\mathbf{u} = \mathbf{u}^0, \quad \forall \mathbf{x} \in \Gamma_u \tag{5.43}$$

5.2.3 *Voigt bound and Ruess bound*

Minimum potential energy principle can be applied to estimate the effective properties of a composite elastic solid V. If a particular displacement boundary condition is imposed on entire boundary ∂V, i.e.

$$\boldsymbol{u} = \mathbf{x} \cdot \boldsymbol{\epsilon}^0, \quad \mathbf{x} \in \partial V \tag{5.44}$$

Then $\Gamma_t = \emptyset$ and $\Pi(\boldsymbol{u}) = VW(\boldsymbol{\epsilon})$, where the potential energy density W($\boldsymbol{\epsilon}$) is defined as

$$W(\boldsymbol{\epsilon}) := \frac{1}{2V}\int_V C_{ijk\ell}\epsilon_{ij}\epsilon_{k\ell} dV \tag{5.45}$$

The minimum potential energy principle reads as

$$W(\boldsymbol{\epsilon}^*) = \inf_{\boldsymbol{u}\in\mathcal{V}} W(\boldsymbol{\epsilon}) \tag{5.46}$$

For the actual displacement field, $\boldsymbol{u}^*$,

$$\begin{aligned} W(\boldsymbol{\epsilon}^*) &= \frac{1}{2V}\int_V \boldsymbol{\epsilon}^* : \boldsymbol{\sigma}^* dV = \frac{1}{2} < \boldsymbol{\epsilon}^* >:< \boldsymbol{\sigma}^* >= \frac{1}{2} < \boldsymbol{\epsilon}^* >: \bar{\mathbb{C}} :< \boldsymbol{\epsilon}^* > \\ &= \frac{1}{2}\boldsymbol{\epsilon}^0 : \bar{\mathbb{C}} : \boldsymbol{\epsilon}^0 \end{aligned}$$

On the other hand for any kinematically admissible solution,

$$\begin{aligned} W(\boldsymbol{\epsilon}) &= \frac{1}{2V}\int_V \boldsymbol{\epsilon} : \boldsymbol{\sigma} dV = \frac{1}{2} < \boldsymbol{\epsilon} >_V:< \boldsymbol{\sigma} >_V= \frac{1}{2}\boldsymbol{\epsilon}^0 :< \boldsymbol{\sigma} >_V \\ &= \frac{1}{2}\boldsymbol{\epsilon}^0 : \Big(\sum_{\alpha=0}^{n} f_\alpha \mathbb{C}^\alpha :< \boldsymbol{\epsilon} >_{\Omega_\alpha}\Big) \end{aligned}$$

where the following relation is used,

$$< \boldsymbol{\sigma} >_V = \sum_{\alpha=0}^{n} f_\alpha < \boldsymbol{\sigma} >_{\Omega_\alpha} = \sum_{\alpha=0}^{n} f_\alpha \mathbb{C}^\alpha :< \boldsymbol{\epsilon} >_{\Omega_\alpha}$$

Since $\boldsymbol{\epsilon}^0 \cdot \mathbf{x} \in \mathcal{V}$, we can choose the uniform strain field $\boldsymbol{\epsilon}^0$ prescribed on V as a kinematically admissible solution. Due to the variational principle statement, $W(\boldsymbol{\epsilon}^*) \leq W(\boldsymbol{\epsilon}^0)$, we have the following inequality

$$\frac{1}{2}\boldsymbol{\epsilon}^0 : \bar{\mathbb{C}} : \boldsymbol{\epsilon}^0 \leq \frac{1}{2}\boldsymbol{\epsilon}^0 : \Big(\sum_{\alpha=0}^{n} f_\alpha \mathbb{C}^\alpha : \boldsymbol{\epsilon}^0\Big)$$

which then leads to the following upper bound estimate of the effective moduli

$$\boxed{\bar{\mathbb{C}} \leq \sum_{\alpha=0}^{n} f_\alpha \mathbb{C}^\alpha} \, . \tag{5.47}$$

The above upper bound is also called Voigt upper bound.

Similarly, we can also apply complementary energy principle to the elastic n-phase composite. Consider the prescribed macro-stress boundary condition,

$$\mathbf{n} \cdot \boldsymbol{\sigma} = \mathbf{t}^0, \quad \forall \mathbf{x} \in \partial V. \tag{5.48}$$

In this case, $\Gamma_u = \emptyset$. Therefore,

$$\Pi^c = \frac{1}{2} \int_V D_{ijk\ell} \sigma_{ij} \sigma_{k\ell} dV = W_c(\boldsymbol{\sigma}) V \tag{5.49}$$

where

$$W_c := \frac{1}{2V} \int_V D_{ijk\ell} \sigma_{ij} \sigma_{k\ell} dV \tag{5.50}$$

is the complementary energy density.

The minimum complementary potential energy principle then gives

$$W_c(\boldsymbol{\sigma}^*) = \inf_{\boldsymbol{\sigma} \in \mathcal{S}} W_c(\boldsymbol{\sigma}) \tag{5.51}$$

Recall for the prescribed macro-stress boundary condition (5.48),

$$< \boldsymbol{\sigma} : \boldsymbol{\epsilon} >_V - < \boldsymbol{\sigma} >_V : < \boldsymbol{\epsilon} >_V = \frac{1}{V} \int_{\partial V} \Big(\boldsymbol{u} - \mathbf{x} \cdot < \nabla \otimes \boldsymbol{u} >_V \Big) \Big(\mathbf{n} \cdot (\boldsymbol{\sigma} - < \boldsymbol{\sigma} >_V) \Big) dS = 0$$

The complementary energy density of the actual stress solution becomes

$$\begin{aligned} W_c(\boldsymbol{\sigma}^*) &= \frac{1}{2V} \int_V \boldsymbol{\sigma}^* : \boldsymbol{\epsilon}^* dV = \frac{1}{2} < \boldsymbol{\sigma}^* : \boldsymbol{\epsilon}^* >_V = \frac{1}{2} < \boldsymbol{\sigma}^* >_V : < \boldsymbol{\epsilon}^* >_V \\ &= \frac{1}{2} < \boldsymbol{\sigma}^* >_V : \bar{\mathbb{D}} : < \boldsymbol{\sigma}^* >_V = \frac{1}{2} \boldsymbol{\sigma}^0 : \bar{\mathbb{D}} : \boldsymbol{\sigma}^0 \end{aligned} \tag{5.52}$$

Note that under prescribed remote stress boundary condition,

$$< \boldsymbol{\sigma} >_V = \boldsymbol{\sigma}^0, \quad \forall \, \boldsymbol{\sigma} \in \mathcal{S} \, .$$

For arbitrary statically admissible stress fields,

$$\begin{aligned} W_c(\boldsymbol{\sigma}) &= \frac{1}{2V} \int_V \boldsymbol{\sigma} : \boldsymbol{\epsilon} dV = \frac{1}{2V} < \boldsymbol{\sigma} : \boldsymbol{\epsilon} >_V = \frac{1}{2} < \boldsymbol{\sigma} >_V : < \boldsymbol{\epsilon} >_V \\ &= \frac{1}{2} \boldsymbol{\sigma}^0 : < \boldsymbol{\epsilon} >_V = \frac{1}{2} \boldsymbol{\sigma}^0 : \sum_{\alpha=0}^{n} f_\alpha \mathbb{D}^\alpha < \boldsymbol{\sigma} >_{\Omega_\alpha} \end{aligned}$$

Choose a uniform stress field $\boldsymbol{\sigma}^0$ over V as a statistically admissible solution,

$$W_c(\boldsymbol{\sigma}^0) = \frac{1}{2} \boldsymbol{\sigma}^0 : \sum_{\alpha=0}^{n} f_\alpha \mathbb{D}^\alpha : \boldsymbol{\sigma}^0 \tag{5.53}$$

Complementary energy principle states that $W_c(\boldsymbol{\sigma}^*) \leq W_c(\boldsymbol{\sigma}^0)$, i.e.,

$$\bar{\mathbb{D}} \leq \sum_{\alpha=0}^{n} f_\alpha \mathbb{D}^\alpha \tag{5.54}$$

Since $\bar{\mathbb{D}} : \bar{\mathbb{C}} = \mathbb{I}^{(4s)}$ and both $\bar{\mathbb{D}}$ and $\bar{\mathbb{C}}$ are positive definite, we then have

$$\boxed{\left(\sum_{\alpha=0}^{n} f_\alpha (\mathbb{C}^\alpha)^{-1}\right)^{-1} \leq \bar{\mathbb{C}}} \tag{5.55}$$

which is called Reuss bound. It is a lower bound estimate for the effective elastic moduli.

Assuming that $\mathbb{C}^\alpha = 3K^\alpha \mathbb{E}^{(1)} + 2\mu^\alpha \mathbb{E}^{(2)}$ and hence

$$\mathbb{C}^{\alpha-1} = \frac{1}{3K^\alpha}\mathbb{E}^{(1)} + \frac{1}{2\mu^\alpha}\mathbb{E}^{(2)}$$

one can then derive that

$$\left(\sum_{\alpha=0}^{n} f_\alpha \mathbb{C}^\alpha\right) = 3\sum_{\alpha=0}^{n} f_\alpha K^\alpha \mathbb{E}^{(1)} + 2\sum_{\alpha=0}^{n} f_\alpha \mu^\alpha \mathbb{E}^{(2)}$$

$$\left(\sum_{\alpha=0}^{n} f_\alpha \mathbb{C}^{\alpha-1}\right)^{-1} = \frac{3}{\displaystyle\sum_{\alpha=0}^{n} \frac{f_\alpha}{K^\alpha}}\mathbb{E}^{(1)} + \frac{2}{\displaystyle\sum_{\alpha=0}^{n} \frac{f_\alpha}{\mu^\alpha}}\mathbb{E}^{(2)}$$

Combining Reuss bound with the Voigt bound, we have

$$\left(\sum_{\alpha=0}^{n} f_\alpha \mathbb{C}^{\alpha-1}\right) < \bar{\mathbb{C}} < \left(\sum_{\alpha=0}^{n} f_\alpha \mathbb{C}^\alpha\right)$$

and subsequently,

$$\frac{1}{\displaystyle\sum_{\alpha=0}^{n} \frac{f_\alpha}{K^\alpha}} < \bar{K} < \sum_{\alpha=0}^{n} f_\alpha K^\alpha \quad \text{and} \quad \frac{1}{\displaystyle\sum_{\alpha=0}^{n} \frac{f_\alpha}{\mu^\alpha}} < \bar{\mu} < \sum_{\alpha=0}^{n} f_\alpha \mu^\alpha$$

One can see that the Voigt bound is in fact an arithmetic average and the Reuss bound can be viewed as a geometric average or the harmonic average.

5.3 Hashin-Shtrikman variational principles

To make more accurate prediction on effective material properties and to narrow the gap between the Voigt bound and the Reuss bound, new variational principles for composite materials are needed. One of such principles is the celebrated Hashin-Shtrikman (HS) variational principle. The essence of the HS variational principles is that they are the variational principles specifically designed for composites, or

inhomogeneous solids. To measure the differences between homogeneous solids and inhomogeneous solids, a **homogenous** comparison solid is introduced to profile the inhomogeneous fields.

Let's first consider a boundary value problem of the original *heterogenous* composite (RVE),

$$\begin{aligned} \sigma_{ij,j} &= 0, \\ \sigma_{ij} &= C_{ijk\ell}(\mathbf{x})\epsilon_{k\ell}, \\ U(\boldsymbol{\epsilon}) &= \frac{1}{2}C_{ijk\ell}\epsilon_{ij}\epsilon_{k\ell}, \quad \text{and} \quad W(\boldsymbol{\epsilon}) = < U(\boldsymbol{\epsilon}) >_V \\ u_i &= \bar{u}_i, \ \forall \mathbf{x} \in \Gamma_u, \quad (\Gamma_t = \emptyset, \ \ \Gamma_u = \partial V). \end{aligned}$$

Consider a second BVP in a *homogenous* comparison solid,

$$\begin{aligned} \sigma^{(0)}_{ij,j} &= 0, \\ \sigma^{(0)}_{ij} &= C^{(0)}_{ijk\ell}\epsilon^{(0)}_{k\ell}, \\ U^{(0)}(\boldsymbol{\epsilon}^{(0)}) &= \frac{1}{2}C^{(0)}_{ijk\ell}\epsilon^{(0)}_{ij}\epsilon^{(0)}_{k\ell}, \quad \text{and} \quad W^{(0)}(\boldsymbol{\epsilon}^{(0)}) = < U^{(0)}(\boldsymbol{\epsilon}^{(0)}) >_V \\ u^{(0)}_i &= \bar{u}_i, \ \forall \mathbf{x} \in \Gamma_u, \quad (\Gamma_t = \emptyset, \ \ \Gamma_u = \partial V). \end{aligned}$$

To relate the two BVPs, we introduce the following decomposition in strain field and stress field,

$$u_i = u^{(0)}_i + u^d_i \tag{5.56}$$

$$\epsilon_{ij} = \epsilon^{(0)}_{ij} + \epsilon^d_{ij} \tag{5.57}$$

and

$$\sigma_{ij} = p_{ij} + C^{(0)}_{ijk\ell}\epsilon_{k\ell} = p_{ij} + C^{(0)}_{ijk\ell}(\epsilon^{(0)}_{k\ell} + \epsilon^d_{k\ell}) \tag{5.58}$$

where u^d_i is the disturbance displacement field and p_{ij} is called polarization stress.

On the other hand, the stress polarization can be written as

$$p_{ij} = \sigma_{ij} - C^{(0)}_{ijk\ell}\epsilon_{k\ell} = (C_{ijk\ell} - C^{(0)}_{ijk\ell})\epsilon_{k\ell} \tag{5.59}$$

which indicates that stress polarization is due to heterogeneousness of the composite.

Furthermore, since

$$u_i = \bar{u}_i, \quad \forall \mathbf{x} \in \partial V \ \text{ and } \ u^{(0)}_i = \bar{u}_i, \quad \forall \mathbf{x} \in \partial V$$

it leads to homogeneous boundary conditions for displacement disturbance field

$$u^d_i = 0, \quad \forall \mathbf{x} \in \partial V \tag{5.60}$$

In passing, we note that because $u^d_i = 0, \forall \mathbf{x} \in \partial V$ it can be readily to show that the average work done by the disturbance strain field over any self-equilibrium stress field will be zero, that is

$$\int_V \sigma_{ij}\epsilon^d_{ij} dV = \int_V \sigma_{ij} u^d_{i,j} dV = \int_{\partial V} u^d_i n_j \sigma_{ij} dS + \int_V u^d_i \sigma_{ij,j} dV = 0\,. \tag{5.61}$$

On the other hand, since

$$\sigma_{ij,j} = 0, \quad \sigma^{(0)}_{ij,j} = 0 \ ,$$

one has

$$\sigma_{ij,j} = \sigma^{(0)}_{ij,j} + p_{ij,j} + \left(C^{(0)}_{ijk\ell}\epsilon^d_{k\ell}\right)_{,j} = 0$$

We can see that the stress field can be divided into the homogeneous (or comparison) stress field, $\sigma^{(0)}_{ij}$, and the inhomogeneous stress field,

$$\sigma_{ij} = \sigma^{(0)}_{ij} + t_{ij}, \quad \text{where} \quad t_{ij} = p_{ij} + C^{(0)}_{ijk\ell}\epsilon^d_{k\ell} \tag{5.62}$$

Both homogeneous stress field, $\sigma^{(0)}_{ij}$, and inhomogeneous stress field, t_{ij} satisfy equilibrium equations, i.e.

$$\sigma^{(0)}_{ij,j} = 0, \quad t_{ij,j} = 0 \ . \tag{5.63}$$

In literature, the inhomogeneous equilibrium equation

$$t_{ij,j} = \left(C^{(0)}_{ijk\ell}\epsilon^d_{k\ell}\right)_{,j} + p_{ij,j} = 0 \tag{5.64}$$

is often called "the subsidiary condition."

Theorem 5.3 (Hashin-Shtrikman). *Let $u^d_i \in \mathcal{U}$ and $p_{ij} \in \mathcal{P}$ where*

$$\mathcal{U} = \left\{ u_i \;\middle|\; u_i \in H^1(V), u_i = 0, \forall \mathbf{x} \in \partial V \right\} \tag{5.65}$$

$$\mathcal{P} = \left\{ p_{ij} \;\middle|\; p_{ij} \in L^2(V) \right\} \tag{5.66}$$

Consider the following functional,

$$\Pi : \mathcal{P} \times \mathcal{U} \rightarrow \mathbb{R},$$

where

$$\Pi(p_{ij}, \epsilon^d_{ij}) = \frac{1}{2}\int_V \left(C^{(0)}_{ijk\ell}\epsilon^{(0)}_{ij}\epsilon^{(0)}_{k\ell} - \Delta C^{-1}_{ijk\ell}p_{ij}p_{k\ell} + p_{ij}\epsilon^d_{ij} + 2p_{ij}\epsilon^{(0)}_{ij}\right)dV \tag{5.67}$$

$$\text{where} \quad \begin{cases} \Delta C_{ijk\ell} = C_{ijk\ell} - C^{(0)}_{ijk\ell} \\ p_{ij} \quad\quad = \Delta C_{ijk\ell}\epsilon_{k\ell} \\ \epsilon^d_{ij} \quad\quad = \epsilon_{ij} - \epsilon^{(0)}_{ij} \end{cases} \tag{5.68}$$

We have the following variational statements:
(1) *The functional Π is stationary, i.e. $\delta\Pi = 0$, if the inhomogeneous equilibrium equation (subsidiary condition) is satisfied,*

$$\left(C^{(0)}_{ijk\ell}\epsilon^d_{k\ell}\right)_{,j} + p_{ij,j} = 0 \ ; \tag{5.69}$$

(2) *If $\Delta\mathbb{C}$ is negative definite*[1]*, then $\delta^2\Pi > 0$, functional Π has a global minimum; On the other hand, if $\Delta\mathbb{C}$ is positive definite, then $\delta^2\Pi < 0$, functional Π has a global maximum:*

$$\delta^2\Pi > 0, \quad \text{if} \quad \Delta\mathbb{C} < 0, \quad \Pi \rightarrow Minimum \tag{5.70}$$

$$\delta^2\Pi < 0, \quad \text{if} \quad \Delta\mathbb{C} > 0, \quad \Pi \rightarrow Maximum \tag{5.71}$$

[1]with slightly abuse of notation, negative definite is denoted as $\Delta\mathbb{C} < 0$, positive definite is denoted as $\Delta\mathbb{C} > 0$

Proof.

$$\begin{aligned}\Delta\Pi &= \Pi(p_{ij}+\delta p_{ij},\epsilon^d_{ij}+\delta\epsilon^d_{ij}) - \Pi(p_{ij},\epsilon^d_{ij})\\ &= \frac{1}{2}\int_V\Big(-2\Delta C^{-1}_{ijk\ell}p_{ij}\delta p_{k\ell} + p_{ij}\delta\epsilon^d_{ij} + \delta p_{ij}\epsilon^d_{ij} + 2\delta p_{ij}\epsilon^{(0)}_{ij}\Big)dV\\ &+\frac{1}{2}\int_V\Big(-\Delta C^{-1}_{ijk\ell}\delta p_{ij}\delta p_{k\ell} + \delta p_{ij}\delta\epsilon^d_{k\ell}\Big)dV = \delta\Pi + \frac{1}{2!}\delta^2\Pi\end{aligned}$$

We first show that the first statement is true.

$$\begin{aligned}\delta\Pi &= \Big(-\frac{1}{2}\Big)\int_V\Big(2\Delta C^{-1}_{ijk\ell}p_{k\ell}\delta p_{ij} - 2\epsilon^{(0)}_{ij}\delta p_{ij} - \epsilon_{ij}\delta p_{ij} - p_{ij}\delta\epsilon^d_{ij}\Big)dV\\ &= \Big(-\frac{1}{2}\Big)\int_V\Big(2\Delta C^{-1}_{ijk\ell}p_{k\ell}\delta p_{ij} - 2\underbrace{(\epsilon_{ij}-\epsilon^{(d)}_{ij})}_{=\epsilon^{(0)}_{ij}}\delta p_{ij} - \epsilon^d_{ij}\delta p_{ij} - p_{ij}\delta\epsilon^d_{ij}\Big)dV\\ &= \Big(-\frac{1}{2}\Big)\int_V 2\underbrace{(\Delta C^{-1}_{ijk\ell}p_{k\ell} - \epsilon_{ij})}_{=0}\delta p_{ij} + \epsilon^d_{ij}\delta p_{ij} - p_{ij}\delta\epsilon^d_{ij}\Big)dV\\ &= \Big(-\frac{1}{2}\Big)\int_V\Big(\epsilon^d_{ij}\delta p_{ij} - p_{ij}\delta\epsilon^d_{ij}\Big)dV \qquad (5.72)\end{aligned}$$

If the subsidiary condition is satisfied, i.e.

$$\Big(C^{(0)}_{ijk\ell}\epsilon^d_{k\ell}\Big)_{,j} + p_{ij,j} = 0, \quad \text{or} \quad t_{ij,j} = 0\,. \qquad (5.73)$$

which leads to

$$\delta t_{ij} = \delta p_{ij} + C^{(0)}_{ijk\ell}\delta\epsilon^d_{k\ell}, \quad \text{and} \quad \delta t_{ij,j} = 0\,. \qquad (5.74)$$

Substituting (5.73) and (5.74) into (5.72) yields

$$\begin{aligned}\delta\Pi &= \Big(-\frac{1}{2}\Big)\int_V\Big(\epsilon^d_{ij}(\delta t_{ij} - C^{(0)}_{ijk\ell}\delta\epsilon^d_{k\ell})\Big) - \delta\epsilon^d_{ij}(t_{ij} - C^{(0)}_{ijk\ell}\epsilon^d_{k\ell})\Big)dV\\ &= \Big(-\frac{1}{2}\Big)\int_V\left(\Big(\epsilon^d_{ij}\delta t_{ij} - t_{ij}\delta\epsilon^d_{ij}\Big) - \underbrace{\epsilon^d_{ij}C^{(0)}_{ijk\ell}\delta\epsilon^d_{k\ell} + \delta\epsilon^d_{ij}C^{(0)}_{ijk\ell}\epsilon^d_{k\ell}}_{=0,\ \textit{because}\ \mathbb{C}^{(0)}\ \textit{has major symmetry}}\right)dV\\ &= \Big(-\frac{1}{2}\Big)\int_V\Big(u^d_{i,j}\delta t_{ij} - t_{ij}\delta u^d_{i,j}\Big)dV\end{aligned}$$

Considering the facts

$$\begin{aligned}\int_V \delta t_{ij}u^d_{i,j}dV &= \int_{\partial V}\delta t_{ij}n_j u^d_i dS - \int_V \delta t_{ij,j}u^d_i dV = 0\\ \int_V t_{ij}\delta u^d_{i,j}dV &= \int_{\partial V} t_{ij}n_j\delta u^d_i dS - \int_V t_{ij,j}\delta u^d_i dV \equiv 0,\end{aligned}$$

we just proved that $\delta\Pi = 0$, if the subsidiary condition $t_{ij,j} = 0$ holds.

Now we examine the extreme conditions. Substituting $\delta p_{ij} = \delta t_{ij} - C^{(0)}_{ijk\ell}\delta\epsilon^d_{k\ell}$ into the second order variation,

$$\begin{aligned}\delta^2\Pi &= \Big(\frac{1}{2}\Big)\int_V \Big(-\Delta C^{-1}_{ijk\ell}\delta p_{ij}\delta p_{k\ell} + \delta p_{ij}\delta\epsilon^d_{ij}\Big)dV \\ &= \Big(-\frac{1}{2}\Big)\int_V \Big(\Delta C^{-1}_{ijk\ell}\delta p_{ij}\delta p_{k\ell} + C^{(0)}_{ijk\ell}\delta\epsilon^d_{ij}\delta\epsilon^d_{k\ell} - \delta t_{ij}\delta\epsilon^d_{ij}\Big)dV\end{aligned}$$

Again, the last term

$$\int_V \delta t_{ij}\delta\epsilon^d_{ij}dV = 0.$$

Therefore, we have

$$\delta^2\Pi = \Big(-\frac{1}{2}\Big)\int_V \Big(\Delta C^{-1}_{ijk\ell}\delta p_{ij}\delta p_{k\ell} + C^{(0)}_{ijk\ell}\delta\epsilon^d_{ij}\delta\epsilon^d_{k\ell}\Big)dV \tag{5.75}$$

Obviously if $\Delta\mathbb{C} > 0$ (positive definite), $\delta^2\Pi < 0$, therefore, Π has a global maximum value.

On the other hand, if $\Delta\mathbb{C} < 0$ (negative definite), the judgement is not straightforward. Consider a positive integral,

$$I = \int_V {C^{(0)}_{ijk\ell}}^{-1}\delta p_{ij}\delta p_{k\ell}dV > 0 \tag{5.76}$$

Substitute $\delta p_{ij} = \delta t_{ij} - C^{(0)}_{ijk\ell}\delta\epsilon^d_{k\ell}$ into (5.76). It can be readily shown that

$$\begin{aligned}I &= \int_V \Big({C^{(0)}_{ijk\ell}}^{-1}\delta t_{ij}\delta t_{k\ell} - \underbrace{2\delta t_{ij}\delta\epsilon^d_{k\ell}}_{=0} + C^{(0)}_{ijk\ell}\delta\epsilon^d_{ij}\delta\epsilon^d_{k\ell}\Big)dV \\ &= \int_V \Big({C^{(0)}_{ijk\ell}}^{-1}\delta t_{ij}\delta t_{k\ell} + C^{(0)}_{ijk\ell}\delta\epsilon^d_{ij}\delta\epsilon^d_{k\ell}\Big)dV \end{aligned}\tag{5.77}$$

A direct consequence of (5.76) and (5.77) is

$$\int_V {C^{(0)}_{ijk\ell}}^{-1}\delta p_{ij}\delta p_{k\ell}dV > \int_V C^{(0)}_{ijk\ell}\delta\epsilon^d_{ij}\epsilon^d_{k\ell}dV \tag{5.78}$$

which leads to the following inequality,

$$\begin{aligned}\delta\Pi &= \Big(-\frac{1}{2}\Big)\int_V \Big(\Delta C^{-1}_{ijk\ell}\delta p_{ij}\delta p_{k\ell} + C^{(0)}_{ijk\ell}\delta\epsilon^d_{ij}\epsilon^d_{k\ell}\Big)\,dV \\ &> \Big(-\frac{1}{2}\Big)\int_V \Big(\Delta C^{-1}_{ijk\ell} + {C^{(0)}_{ijk\ell}}^{-1}\Big)\delta p_{ij}\delta p_{k\ell}dV\end{aligned}$$

Consider

$$\begin{aligned}\Delta\mathbb{C}^{-1} + {\mathbb{C}^{(0)}}^{-1} &= \Delta\mathbb{C}^{-1} + {\mathbb{C}^{(0)}}^{-1} : (\mathbb{C} - \mathbb{C}^{(0)}) : (\mathbb{C} - \mathbb{C}^{(0)})^{-1} \\ &= \Delta\mathbb{C}^{-1} + {\mathbb{C}^{(0)}}^{-1} : \mathbb{C} : \Delta\mathbb{C}^{-1} - \Delta\mathbb{C}^{-1} \\ &= {\mathbb{C}^{(0)}}^{-1} : \mathbb{C} : \Delta\mathbb{C}^{-1}\,.\end{aligned}$$

One can evaluate that

$$\delta^2\Pi > \Big(-\frac{1}{2}\Big)\int_V \delta\mathbf{p} : {\mathbb{C}^{(0)}}^{-1} : \mathbb{C} : \Delta\mathbb{C}^{-1} : \delta\mathbf{p}\; dV \tag{5.79}$$

Since both ${\mathbb{C}^{(0)}}^{-1}$ and $\mathbb{C}$ are positive definite, it is clear now that if $\Delta\mathbb{C}^{-1} < 0$, $\delta^2\Pi > 0$ and hence Π has a global minimum. To sum up, we have the following extreme conditions,

$$\begin{aligned}
&\delta^2\Pi < 0, \text{ if } \Delta\mathbb{C} > 0, \ \Pi \to \text{maximum} ;\\
&\delta^2\Pi > 0, \text{ if } \Delta\mathbb{C} < 0, \ \Pi \to \text{minimum} .
\end{aligned}$$

This is the proof of the Hashin-Shtrikman principles. □

A natural question to ask is: *What is the physical meaning of the potential* $\Pi(p_{ij}, \epsilon^d_{ij})$ *in (5.67) ?* or *How did Hashin and Shtrikman come up with such variational principles ?*

Consider that σ_{ij} and $\sigma^{(0)}_{ij}$ are self-equilibrium stress field. We can show that

$$\begin{aligned}
\int_V \sigma_{ij}\epsilon^d_{ij} dV &= \int_V \sigma_{ij} u^d_{i,j} dV = 0\\
\int_V \sigma^{(0)}_{ij}\epsilon^d_{ij} dV &= \int_V \sigma^{(0)}_{ij} u^d_{i,j} dV = 0
\end{aligned}$$

because $u^d_i = 0, \quad \forall \mathbf{x} \in \partial V$. Therefore the total potential energy of a composite can be written as

$$\Pi(\boldsymbol{\epsilon}) = \frac{1}{2}\int_V \sigma_{ij}\epsilon_{ij} dV = \frac{1}{2}\int_V \Big(\sigma_{ij}\epsilon^{(0)}_{ij} - \underbrace{\sigma_{ij}\epsilon^d_{ij}}_{=0}\Big) dV = \frac{1}{2}\int_V \sigma_{ij}\epsilon^{(0)}_{ij} dV$$

Consider the fact that

$$\begin{aligned}
\frac{1}{2}\int_V \sigma_{ij}\epsilon^{(0)}_{ij} dV &= \frac{1}{2}\int_V \Big(\sigma^{(0)}_{ij} + p_{ij} + C^{(0)}_{ijk\ell}\epsilon^d_{k\ell}\Big)\epsilon^{(0)}_{ij} dV\\
&= \frac{1}{2}\int_V \sigma^{(0)}_{ij}\epsilon^{(0)}_{ij} + p_{ij}\epsilon^{(0)}_{ij} + C^{(0)}_{ijk\ell}\epsilon^d_{k\ell}\epsilon^{(0)}_{ij} + \underbrace{p_{ij}\epsilon^{(0)}_{ij} - p_{ij}\epsilon^{(0)}_{ij}}_{=0} dV\\
&= \frac{1}{2}\int_V C^{(0)}_{ijk\ell}\epsilon^{(0)}_{k\ell}\epsilon^{(0)}_{ij} + C^{(0)}_{ijk\ell}\epsilon^d_{k\ell}\epsilon^{(0)}_{ij} + 2p_{ij}\epsilon^{(0)}_{ij} - p_{ij}(\epsilon_{ij} - \epsilon^d_{ij}) dV\\
&= \frac{1}{2}\int_V C^{(0)}_{ijk\ell}\epsilon^{(0)}_{k\ell}\epsilon^{(0)}_{ij} + 2p_{ij}\epsilon^{(0)}_{ij} - p_{ij}\epsilon_{ij} + p_{ij}\epsilon^d_{ij} dV
\end{aligned}$$

Under prescribed remote strain boundary condition, we have

$$\begin{aligned}
\Pi(\boldsymbol{\epsilon}) &= \frac{1}{2}\int_V \sigma_{ij}\epsilon_{ij} dV = W(\boldsymbol{\epsilon})V\\
&= \frac{1}{2}\int_V \Big(C^{(0)}_{ijk\ell}\epsilon^{(0)}_{ij}\epsilon^{(0)}_{k\ell} - \Delta C^{-1}_{ijk\ell} p_{ij} p_{k\ell} + p_{ij}\epsilon^d_{ij} + 2p_{ij}\epsilon^{(0)}_{ij}\Big) dV\\
&= W^{(0)}(\boldsymbol{\epsilon}^{(0)})V + \frac{1}{2}\int_V \Big(-\Delta C^{-1}_{ijk\ell} p_{ij} p_{k\ell} + p_{ij}\epsilon^d_{ij} + 2p_{ij}\epsilon^{(0)}_{ij}\Big) dV\\
&= W^{(0)}(\boldsymbol{\epsilon}^{(0)})V + R_\pi V
\end{aligned} \tag{5.80}$$

where $R_\pi := \dfrac{1}{2V}\displaystyle\int_V \left(-\Delta C^{-1}_{ijk\ell}p_{ij}p_{k\ell} + p_{ij}\epsilon^d_{ij} + 2p_{ij}\epsilon^{(0)}_{ij}\right)dV$. This indicates that for the real solution $(\boldsymbol{\epsilon}^*, \mathbf{p}, \boldsymbol{\epsilon}^d)$ Eq. (5.80) holds, i.e. $\Pi(\mathbf{p}, \boldsymbol{\epsilon}^d) = \Pi(\boldsymbol{\epsilon}^*)$ for the real solution. In other words, the physical meaning of potential of $\Pi(\mathbf{p}, \boldsymbol{\epsilon}^d)$ in (5.67) is the strain energy stored inside the composite when the strain field and polarization field are exact solutions.

Hence based on Hashin-Shtrikman principles , it is possible to choose the comparison solid such that both upper and lower bounds for the potential energy of the composite can be evaluated. If $\Delta\mathbb{C} > 0$, Π has a global maximum; whereas if $\Delta\mathbb{C} < 0$, Π has a global minimum. Therefore, the Hashin-Shtrikman principle provides the following bounds,

$$\underline{R_\pi}(\tilde{\mathbf{p}}, \tilde{\boldsymbol{\epsilon}}^d) \leq W(\boldsymbol{\epsilon}^*) - W^{(0)}(\boldsymbol{\epsilon}^{(0)}) \leq \bar{R}_\pi(\tilde{\mathbf{p}}, \tilde{\boldsymbol{\epsilon}}^d) \tag{5.81}$$

5.4 Hashin-Shtrikman bounds

A direct application of the Hashin-Shtrikman principles is to apply them to find the effective material properties of the composite material through some rigorous analysis that provide estimations on the range of the effective material properties, which are called Hashin-Shtrikman variational bounds.

Consider prescribed macro strain boundary condition for both the composite and the comparison solid,

$$\begin{aligned} \mathbf{u} &= \bar{\mathbf{u}} = \mathbf{x}\cdot\bar{\boldsymbol{\epsilon}}, \quad \forall \mathbf{x} \in \partial V \quad (\Gamma_t = \emptyset) \\ \mathbf{u}^{(0)} &= \bar{\mathbf{u}} = \mathbf{x}\cdot\bar{\boldsymbol{\epsilon}}, \quad \forall \mathbf{x} \in \partial V \quad (\Gamma_t = \emptyset) \end{aligned}$$

by the averaging theorem $\bar{\boldsymbol{\epsilon}} = <\boldsymbol{\epsilon}>$.

Under such a condition, the Hashin-Shtrikman variational principles read as

$$\underbrace{\underline{I}}_{\Delta\mathbb{C}>0} \leq \inf_{\boldsymbol{\epsilon}^d \in E} W(\boldsymbol{\epsilon}^d) \leq \underbrace{\bar{I}}_{\Delta\mathbb{C}<0} \tag{5.82}$$

where $\Delta\mathbb{C} = \mathbb{C} - \mathbb{C}^{(0)}$, and

$$\underline{I}\ \left(\text{or } \bar{I}\right) = W^{(0)}(\boldsymbol{\epsilon}^{(0)}) - \frac{1}{2V}\int_V \left[\Delta C^{-1}_{ijk\ell}p_{ij}p_{k\ell} - p_{ij}\epsilon^d_{ij} - 2p_{ij}\epsilon^{(0)}_{ij}\right]dV \tag{5.83}$$

Assume that there are n-phase in the composite (including the matrix). In each phase (inclusion), the elastic tensor as well as stress polarization tensor is constant, i.e.

$$\mathbb{C}(\mathbf{x}) = \sum_{r=1}^{n} \mathbb{C}^r \chi_r(\mathbf{x}) \tag{5.84}$$

$$\mathbf{p}(\mathbf{x}) = \sum_{r=1}^{n} \mathbf{p}^r \chi_r(\mathbf{x}) \tag{5.85}$$

where $\chi_r(\mathbf{x})$ is the characteristic function of the domain Ω_r for the r-th phase,

$$\chi_r(\mathbf{x}) = H(\Omega_r) = \begin{cases} 1, \forall \mathbf{x} \in \Omega_r \\ \\ 0, \forall \mathbf{x} \notin \Omega_r \end{cases}$$

and $H(\cdot)$ is the Heaviside function.

We now calculate each term in (5.82).

(1)

$$\begin{aligned} \inf_{\boldsymbol{\epsilon}^d \in \mathcal{E}} W(\boldsymbol{\epsilon}^d) &= \frac{1}{2V} \int_V \boldsymbol{\sigma} : \boldsymbol{\epsilon} dV = \frac{1}{2} < \boldsymbol{\sigma} >:< \boldsymbol{\epsilon} > \\ &= \frac{1}{2} < \boldsymbol{\epsilon} >: \bar{\mathbb{C}} :< \boldsymbol{\epsilon} >= \frac{1}{2} \bar{\boldsymbol{\epsilon}} : \bar{\mathbb{C}} : \bar{\boldsymbol{\epsilon}} \end{aligned} \tag{5.86}$$

(2)

$$\begin{aligned} W^{(0)}(\boldsymbol{\epsilon}^{(0)}) &= \frac{1}{2V} \int_V \boldsymbol{\sigma}^{(0)} : \boldsymbol{\epsilon}^{(0)} dV = \frac{1}{2} < \boldsymbol{\sigma}^{(0)} >:< \boldsymbol{\epsilon}^{(0)} > \\ &= \frac{1}{2} < \boldsymbol{\epsilon}^{(0)} >: \mathbb{C}^{(0)} :< \boldsymbol{\epsilon}^{(0)} >= \frac{1}{2} \bar{\boldsymbol{\epsilon}} : \mathbb{C}^{(0)} : \bar{\boldsymbol{\epsilon}} \end{aligned} \tag{5.87}$$

(3)

$$\begin{aligned} \frac{1}{2V} \int_V \boldsymbol{p} : \Delta \mathbb{C}^{-1} : \boldsymbol{p} dV &= \frac{1}{2} \sum_{r=1}^{n} \frac{1}{V} \int_{\Omega_r} \boldsymbol{p}^r : (\mathbb{C}^r)^{-1} : \boldsymbol{p}^r dV \\ &= \frac{1}{2} \sum_{r=1}^{n} f_r \boldsymbol{p}^r : (\Delta \mathbb{C}^r)^{-1} : \boldsymbol{p}^r \end{aligned} \tag{5.88}$$

(4)

$$\frac{1}{V} \int_V \boldsymbol{p} : \boldsymbol{\epsilon}^{(0)} dV = \Big(\frac{1}{V} \int_V \boldsymbol{p} dV \Big) : \bar{\boldsymbol{\epsilon}} =< \boldsymbol{p} >: \bar{\boldsymbol{\epsilon}} = \sum_{r=1}^{n} f_r \boldsymbol{p}^r : \bar{\boldsymbol{\epsilon}} \tag{5.89}$$

(5)

$$\frac{1}{2V} \int_V \boldsymbol{p} : \boldsymbol{\epsilon}^d dV = -\frac{1}{2} \sum_{r=1}^{n} f_r \boldsymbol{p}^r : \mathbb{P}^r : \Big(\boldsymbol{p}^r - < \boldsymbol{p} > \Big) \tag{5.90}$$

where

$$\mathbb{P}^r := \int_{\Omega_r} \boldsymbol{\Gamma}^\infty (\mathbf{x}' - \mathbf{x}) dV_{\mathbf{x}'}$$

and

$$\Gamma^\infty_{ijk\ell} := -\frac{1}{4} \Big(G^\infty_{ki,j\ell}(\mathbf{x}' - \mathbf{x}) + G^\infty_{kj,i\ell}(\mathbf{x}' - \mathbf{x}) + G^\infty_{\ell i,jk}(\mathbf{x}' - \mathbf{x}) + G^\infty_{\ell j,ik}(\mathbf{x}' - \mathbf{x}) \Big)$$

It is straightforward to demonstrate (1)-(4). To this point, the only equation that needs further elaboration is (5). To evaluate

$$\frac{1}{2V}\int_V \mathbf{p} : \boldsymbol{\epsilon}^d dV \tag{5.91}$$

we employ the subsidiary condition,

$$C^{(0)}_{ijk\ell} u^d_{k,\ell j} + p_{ij,j} = 0 \tag{5.92}$$

We first solve u^d_k in terms of p_{ij} by using Green's function method. Consider the Green's function of the comparison solid in an infinite medium, i.e.

$$C^{(0)}_{ijk\ell} G^{\infty}_{km,\ell j} + \delta_{im}\delta(\mathbf{x}-\mathbf{x}') = 0, \quad \forall\ \mathbf{x}, \mathbf{x}' \in \mathbb{R}^3$$

Multiplying $G^{\infty}_{im}(\mathbf{x}'-\mathbf{x})$ with (5.92) and integrating it over V, one has

$$\int_V \left[C^{(0)}_{ijk\ell} u^d_{k,\ell} + p_{ij}\right]_{,j} G^{\infty}_{im}(\mathbf{x}'-\mathbf{x}) dV_{\mathbf{x}'} = 0$$

Let $t_{ij} = C^{(0)}_{ijk\ell} u^d_{k,\ell}$. Integration by parts yields,

$$\int_{\partial V} G^{\infty}_{im}(\mathbf{x}'-\mathbf{x}) \Big[\underbrace{C^{(0)}_{ijk\ell} u^d_{k,\ell}}_{t_{ij}} + p_{ij}\Big] n_j dS - \int_V \frac{\partial G^{\infty}_{im}(\mathbf{x}'-\mathbf{x})}{\partial x'_j}\Big[C^{(0)}_{ijk\ell} u^d_{k,\ell} + p_{ij}\Big] dV_{\mathbf{x}'}$$

$$= \int_{\partial V} G^{\infty}_{im}(\mathbf{x}'-\mathbf{x})[t_{ij} + p_{ij}] n_j dS - \int_V \frac{\partial G^{\infty}_{im}(\mathbf{x}'-\mathbf{x})}{\partial x'_j} p_{ij}(\mathbf{x}') dV_{\mathbf{x}'}$$

$$- \int_{\partial V} \frac{\partial G^{\infty}_{im}(\mathbf{x}'-\mathbf{x})}{\partial x'_j} \underbrace{\Big[C^{(0)}_{ijk\ell} u^d_k n_\ell\Big]}_{=0} dS + \underbrace{\int_V \frac{\partial^2 G^{\infty}_{im}(\mathbf{x}'-\mathbf{x})}{\partial x'_j \partial x'_\ell} C^{(0)}_{ijk\ell} u^d_k\, dV_{\mathbf{x}'}}_{\text{interchange indices}}$$

$$= \int_{\partial V} G^{\infty}_{im}(\mathbf{x}'-\mathbf{x})[t_{ij} + p_{ij}] n_j dS - \int_V \frac{\partial G^{\infty}_{im}(\mathbf{x}'-\mathbf{x})}{\partial x'_j} p_{ij}(\mathbf{x}') dV_{\mathbf{x}'}$$

$$+ \int_V \underbrace{C^{(0)}_{ijk\ell} G^{\infty}_{km,j\ell}(\mathbf{x}'-\mathbf{x})}_{-\delta_{im}\delta(\mathbf{x}'-\mathbf{x})} u^d_i(\mathbf{x}') dV_{\mathbf{x}'}$$

To derive the last term in above expression, one interchanged indices ($k \leftrightarrow i$ and $j \leftrightarrow \ell$), and used the major symmetry of $\mathbb{C}^{(0)}$. Therefore,

$$u^d_m(\mathbf{x}) = \int_{\partial V} G^{\infty}_{im}(\mathbf{x}'-\mathbf{x})[t_{ij}(\mathbf{x}') + p_{ij}(\mathbf{x}')] n_j dS - \int_V G^{\infty}_{im,j}(\mathbf{x}'-\mathbf{x}) p_{ij}(\mathbf{x}') dV \tag{5.93}$$

In order to drop out the surface integral in (5.93), additional manipulation is needed to modify the volume integral in (5.93). To accomplish this goal, we consider identity,

$$< p_{ij} >_{,j} = 0 \ \Rightarrow\ \int_V G^{\infty}_{im}(\mathbf{x}'-\mathbf{x}) < p_{ij} >_{,j} dV_{\mathbf{x}'} = 0$$

Integration by parts yields,

$$\int_{\partial V} < p_{ij} > n_j G^{\infty}_{im}(\mathbf{x}' - \mathbf{x}) dS = \int_V G^{\infty}_{im,j}(\mathbf{x}' - \mathbf{x}) < p_{ij} > dV_{\mathbf{x}'} \tag{5.94}$$

Thus subtracting (5.94) from (5.93) will lead to the following expression,

$$\begin{aligned} u^d_m(\mathbf{x}) = & \int_{\partial V} G^{\infty}_{im}(\mathbf{x}' - \mathbf{x})[t_{ij}(\mathbf{x}') + (p_{ij}(\mathbf{x}') - < p_{ij} >)] n_j dS \\ & - \int_V G^{\infty}_{im,j}(\mathbf{x}' - \mathbf{x})(p_{ij}(\mathbf{x}') - < p_{ij} >) dV_{\mathbf{x}'} \end{aligned} \tag{5.95}$$

The surface integral in above equation can be neglected, since $\boldsymbol{u}^d(\mathbf{x}') = 0, \ \forall \ \mathbf{x}' \in \partial V$, therefore t_{ij} oscillates around zero on the boundary. Then its average $< C^{(0)}_{ijk\ell} u^d_{k,\ell} >_{\partial V}$ along the boundary should be very small. We assume that

$$< C^{(0)}_{ijk\ell} u^d_{k,\ell} >_{\partial V} \approx 0$$

In addition, the stress polarization $p_{ij} - < p_{ij} >$ also oscillates around zero, since its volume average is zero, i.e. $\langle p_{ij} - < p_{ij} > \rangle = 0$. We can then neglect the boundary term, and finally we have

$$u^d_m(\mathbf{x}) \approx - \int_V G^{\infty}_{im,j}(\mathbf{x}' - \mathbf{x})(p_{ij}(\mathbf{x}') - < p_{ij} >) dV_{\mathbf{x}'} \tag{5.96}$$

The gradient of the disturbance displacement field is

$$u^d_{m,\ell}(\mathbf{x}) = \int_V G^{\infty}_{im,j\ell}(\mathbf{x}' - \mathbf{x})(p_{ij}(\mathbf{x}') - < p_{ij} >) dV_{\mathbf{x}'}$$

Hence

$$\epsilon^d_{m\ell}(\mathbf{x}) = \frac{1}{2} \int_V \Big[G^{\infty}_{im,j\ell} + G^{\infty}_{i\ell,jm} \Big](\mathbf{x}' - \mathbf{x}) \Big(p_{ij}(\mathbf{x}') - < p_{ij} > \Big) dV_{\mathbf{x}'} \tag{5.97}$$

Since p_{ij} is symmetric, we can also write that

$$\begin{aligned} \epsilon^d_{m\ell}(\mathbf{x}) &= \frac{1}{4} \int_V \Big[G^{\infty}_{im,j\ell} + G^{\infty}_{i\ell,jm} + G^{\infty}_{jm,i\ell} + G^{\infty}_{j\ell,im} \Big](\mathbf{x}' - \mathbf{x}) \Big(p_{ij}(\mathbf{x}') - < p_{ij} > \Big) dV_{\mathbf{x}'} \\ &= - \int_V \Gamma^{\infty}_{m\ell ij}(\mathbf{x}' - \mathbf{x}) \Big(p_{ij}(\mathbf{x}') - < p_{ij} > \Big) dV_{\mathbf{x}'} \end{aligned} \tag{5.98}$$

where

$$\Gamma^{\infty}_{m\ell ij}(\mathbf{y} - \mathbf{x}) := - \frac{1}{4} \Big[G^{\infty}_{im,j\ell} + G^{\infty}_{i\ell,jm} + G^{\infty}_{jm,i\ell} + G^{\infty}_{j\ell,im} \Big](\mathbf{y} - \mathbf{x}) \tag{5.99}$$

Consider a bounded and simply-connected region, $\Omega_r \in V$. We define a new tensor, $\mathbb{P}$,

$$\mathbb{P}^r(\mathbf{x}) := \int_{\Omega_r} \boldsymbol{\Gamma}^{\infty}(\mathbf{x}' - \mathbf{x}) dV_{\mathbf{x}'}, \quad \forall \mathbf{x} \in \Omega_r \tag{5.100}$$

and in components form,

$$\begin{aligned} P^r_{ijk\ell}(\mathbf{x}) &= \int_{\Omega_r} \Gamma^{\infty}_{ijk\ell}(\mathbf{x}' - \mathbf{x}) dV_{\mathbf{x}'} \\ &= - \frac{1}{4} \int_{\Omega_r} \Big[G^{\infty}_{im,j\ell} + G^{\infty}_{i\ell,jm} + G^{\infty}_{jm,i\ell} + G^{\infty}_{j\ell,im} \Big](\mathbf{x}' - \mathbf{x}) dV_{\mathbf{x}'} \end{aligned} \tag{5.101}$$

One may verify that if Ω_r is an ellipsoid, $\mathbb{P}^r$ is constant. In fact, if one recalls the general definition of Eshelby tensor, for $\mathbf{x} \in \Omega_r$,

$$S^{\infty}_{ijk\ell} = \int_{\Omega_r} \mathcal{G}^{\infty}_{ijk\ell}(\mathbf{x}' - \mathbf{x})dV_{\mathbf{x}'} \tag{5.102}$$

$$= -\frac{1}{2}\int_{\Omega_r} C_{mnk\ell}\Big[G^{\infty}_{im,nj} + G^{\infty}_{jm,ni}\Big](\mathbf{x}' - \mathbf{x})dV_{\mathbf{x}'}$$

$$= -\frac{1}{4}\int_{\Omega_r} \Big[G^{\infty}_{im,nj} + G^{\infty}_{jm,ni} + G^{\infty}_{in,mj} + G^{\infty}_{jn,mj}\Big](\mathbf{x}' - \mathbf{x})C_{mnk\ell}dV_{\mathbf{x}'}$$

$$= \int_{\Omega_r} \Gamma^{\infty}_{ijmn}(\mathbf{x}' - \mathbf{x})C_{mnk\ell}dV_{\mathbf{x}'} = P^{r}_{ijmn}C_{mnk\ell} \tag{5.103}$$

We now come back to evaluate (5.91). Consider

$$\frac{1}{2V}\int_V \boldsymbol{p} : \boldsymbol{\epsilon}^d dV = -\frac{1}{2V}\int_V \boldsymbol{p}(\mathbf{x}) : \Big(\int_{V'} \Gamma^{\infty}(\mathbf{x}' - \mathbf{x}) : (\boldsymbol{p}(\mathbf{x}') - <\boldsymbol{p}>)\, dV_{\mathbf{x}'}\Big)dV_{\mathbf{x}}\ . \tag{5.104}$$

Let stress polarization $\boldsymbol{p}(\mathbf{x})$ be a piecewise constant function, i.e.

$$\boldsymbol{p}(\mathbf{x}) = \sum_{r=1}^{n} \boldsymbol{p}^r \chi_r, \quad \text{and} \quad <\boldsymbol{p}> = \sum_{r=1}^{n} f_r \boldsymbol{p}^r$$

Therefore, $\boldsymbol{p} - <\boldsymbol{p}>$ is piecewise constant over the integration domain. Thus,

$$\int_{V'} \boldsymbol{\Gamma}^{\infty}(\mathbf{x}' - \mathbf{x}) : \Big(\boldsymbol{p} - <\boldsymbol{p}>\Big)dV_{\mathbf{x}'} = \int_{V'} \boldsymbol{\Gamma}^{\infty}(\mathbf{x}' - \mathbf{x}) : \Big(\sum_{r=1}^{n}(\boldsymbol{p}^r - <\boldsymbol{p}>)\chi_r(\mathbf{x}')\Big)dV_{\mathbf{x}'}$$

$$= \sum_{r=1}^{n}\Big(\int_{\Omega_r} \boldsymbol{\Gamma}^{\infty}(\mathbf{x}' - \mathbf{x})dV_{\mathbf{x}'}\Big) : \Big(\boldsymbol{p}^r - <\boldsymbol{p}>\Big)$$

Now (5.104) becomes to

$$-\frac{1}{2V}\int_V \mathbf{p}(\mathbf{x}) : \Big[\sum_{r=1}^{n}\Big(\int_{\Omega_r} \boldsymbol{\Gamma}^{\infty}(\mathbf{x}' - \mathbf{x})dV_{\mathbf{x}'}\Big)dV_{\mathbf{x}}\Big] : \Big(\boldsymbol{p}^r - <\boldsymbol{p}>\Big), \tag{5.105}$$

Neglecting the factor $-\dfrac{1}{2V}$, it can be split into two integrals:

$$\Big(\int_{\Omega_r} + \int_{V-\Omega_r}\Big)\Big(\mathbf{p}(\mathbf{x}) : \Big[\sum_{r=1}^{n}\Big(\int_{\Omega_r} \boldsymbol{\Gamma}^{\infty}(\mathbf{x}' - \mathbf{x})dV_{\mathbf{x}'}\Big)dV_{\mathbf{x}}\Big]\Big) : \Big(\boldsymbol{p}^r - <\boldsymbol{p}>\Big), \tag{5.106}$$

The second integral in (5.106) is the controversy part, which needs special justification. Changing the order of integration, we can rewrite this term as

$$\sum_{r=1}^{n}\int_{\Omega_r}\Big(\int_{V-\Omega_r}\Big(\mathbf{p}(\mathbf{x}) : \boldsymbol{\Gamma}^{\infty}(\mathbf{x}' - \mathbf{x})\Big)dV_{\mathbf{x}}\Big)dV_{\mathbf{x}'} : \Big(\boldsymbol{p}^r - <\boldsymbol{p}>\Big), \tag{5.107}$$

If $\mathbf{p}(\mathbf{x}) = const.\ \forall \mathbf{x} \in V - \Omega_r$, which is certainly true for two phase composites, by assuming the RVE is a gigantic ellipsoidal ball, the Tanaka-Mori lemma states that

$$\int_{V-\Omega_r} \boldsymbol{\Gamma}^{\infty}(\mathbf{x}' - \mathbf{x})dV_{\mathbf{x}} = 0\ . \tag{5.108}$$

In fact, for $\mathbf{x} \in \Omega_r$

$$\int_V \mathbf{\Gamma}^\infty(\mathbf{x}' - \mathbf{x})dV_{\mathbf{x}} = \int_{\Omega_r} \mathbf{\Gamma}^\infty(\mathbf{x}' - \mathbf{x})dV_{\mathbf{x}}$$

because the integral over an ellipsoidal ball does not dependent on the size of inclusion (recall $\mathbb{P} = \mathbb{S} : \mathbb{D}^{(0)}$). Then the second term of (5.106) is then eliminated. For multiphase composites ($n > 2$), we have to either assume that the $\mathbf{p}(\mathbf{x})$ is either slowly changing or assume that the mean polarization outside an inclusion is close to zero, i.e.

$$\mathbf{p}(\mathbf{x}) \approx const. \quad \text{or} \quad \int_{V-\Omega_r} \mathbf{p}(\mathbf{x})dV_{\mathbf{x}} \approx 0$$

such that a generalized Tanaka-Mori statement holds true,

$$\int_{\Omega_r}\int_{V-\Omega_r} \mathbf{p}(\mathbf{x})\mathbf{\Gamma}^\infty(\mathbf{x}' - \mathbf{x})dV_{\mathbf{x}}dV_{\mathbf{x}'} = 0 \ \ \forall\, r = 1, 2, \cdots, n \tag{5.109}$$

To circumvent such difficulties, [Willis (1981)] employed a statistical approach by arguing that the correlation function of two distinct phases of a composite vanishes.

Here, we provide a purely mechanical justification: The second term in (5.106) is basically the interaction term among all the inclusions. Since the exterior Eshelby tensor is proportional to r^{-3} where r is the distance between two inclusions (see (3.154), we can therefore neglect the long range interactions. Within the short range, we assume that the interaction between a considering phase and the rest of phases of the composite is locally self-similar, which can be represented by a series of confocal ellipsoidal shells (see Fig. 5.1), and within each shell the stress polarization will be constant, so that a generalized Tanaka-Mori lemma holds (5.109). This is because that

$$\int_{\Omega_r}\int_{V-\Omega_r} \mathbf{p}(\mathbf{x}) : \mathbf{\Gamma}^\infty(\mathbf{x}' - \mathbf{x})dV_{\mathbf{x}}dV_{\mathbf{x}'} \approx$$
$$\int_{\Omega_r}\sum_{s=1}^{n} \mathbf{p}^s : \Big(\int_{\Omega_s-\Omega_r} \mathbf{\Gamma}^\infty(\mathbf{x}' - \mathbf{x})dV_{\mathbf{x}}\Big)dV_{\mathbf{x}'} \equiv 0 \ \ \forall\, r, s = 1, 2, \cdots, n \tag{5.110}$$

Let

$$\mathbb{P}^r := \int_{\Omega_r} \mathbf{\Gamma}^\infty(\mathbf{x}' - \mathbf{x})dV_{\mathbf{x}'} \ . \tag{5.111}$$

Finally, we can write the term (5.104) as

$$\frac{1}{2V}\int_V \boldsymbol{p} : \boldsymbol{\epsilon}^d dV = -\frac{1}{2V}\sum_{r=1}^{n}\int_{\Omega_r} \boldsymbol{p}^r : \mathbb{P}^r : (\boldsymbol{p}^r - <\boldsymbol{p}>)dV_{\mathbf{x}}$$
$$= -\frac{1}{2}\sum_{r=1}^{n} f_r \boldsymbol{p}^r : \mathbb{P}^r : (\boldsymbol{p}_r - <\boldsymbol{p}>) - -\frac{1}{2}\sum_{r=1}^{n} f_r p^r_{ij} P^r_{ijk\ell}\Big(p^r_{k\ell} - <p_{k\ell}>\Big)$$

where $<p_{k\ell}> = \sum_{r=1}^{n} f_r p^r_{k\ell}$.

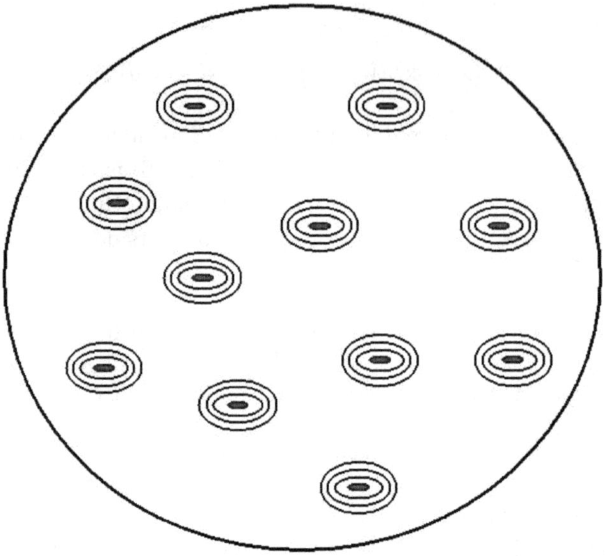

Fig. 5.1 The composite model with locally self-similar interactions.

Now we are in a position to establish Hashin-Shtrikman bounds . To derive Hashin-Shtrikman bound, we first evaluate $\mathbb{P}$ tensor, which can be written as

$$\mathbb{P} = \mathbb{S} : \mathbb{D}^{(0)}$$

For spherical inclusion,

$$\mathbb{S} = s_1^{(0)} \mathbb{E}^{(1)} + s_2^{(0)} \mathbb{E}^2$$

where

$$s_1^{(0)} = \frac{1 + \nu^{(0)}}{3(1 - \nu^{(0)})}, \quad s_2^{(0)} = \frac{2(4 - 5\nu^{(0)})}{15(1 - \nu^{(0)})}$$

and for isotropic comparison solid,

$$\mathbb{D}^{(0)} = \frac{1}{3K^{(0)}} \mathbb{E}^{(1)} + \frac{1}{2\mu^{(0)}} \mathbb{E}^{(2)}$$

Therefore,

$$\begin{aligned} \mathbb{P} &= \frac{s_1^{(0)}}{3K^{(0)}} \mathbb{E}^{(1)} + \frac{s_2^{(0)}}{2\mu^{(0)}} \mathbb{E}^{(2)} \\ &= \frac{1 + \nu^{(0)}}{9K^{(0)}(1 - \nu^{(0)})} \mathbb{E}^{(1)} + \frac{(4 - 5\nu^{(0)})}{15\mu^{(0)}(1 - \nu^{(0)})} \mathbb{E}^{(2)} \\ &= \frac{1}{2\mu^{(0)}(1 - \nu^{(0)})} \left\{ -\frac{1}{15} \mathbb{I}^{(2)} \otimes \mathbb{I}^{(2)} + \frac{2(4 - 5\nu^{(0)})}{15} \mathbb{I}^{(4s)} \right\} \end{aligned} \tag{5.112}$$

Considering $\nu^{(0)} = \dfrac{3K^{(0)} - 2\mu^{(0)}}{2(3K^{(0)} + \mu^{(0)})}$, one can also show that

$$\mathbb{P} = \frac{1}{3K^{(0)} + 4\mu^{(0)}}\mathbb{E}^{(1)} + \frac{3(K^{(0)} + 2\mu^{(0)})}{5\mu^{(0)}(3K^{(0)} + 4\mu^{(0)})}\mathbb{E}^{(2)}$$

For simplicity, we only illustrate the Hashin-Shtrikman bound for a two-phase composite material. Consider a two-phase well ordered composite. Without lose of generality, it is assumed that $K_2 > K_1$ and $\mu_2 > \mu_1$.

Step 1. Let the comparison solid assume the modulus of phase 1

$$K^{(0)} = K_1, \quad K = K_2, \quad \text{and} \quad \mu^{(0)} = \mu_1, \quad \mu = \mu_2 \ .$$

It is obvious that

$$\Delta\mathbb{C} = \mathbb{C} - \mathbb{C}^{(0)} = 3(K_2 - K_1)\mathbb{E}^{(1)} + 2(\mu_2 - \mu_1)\mathbb{E}^{(2)} > 0$$

Choose piecewise constant stress polarization distribution as,

$$p_{ij}^{(1)} = 0 \ \text{ in phase 1}, \qquad \text{and} \qquad p_{ij}^{(2)} = p\delta_{ij} \quad \text{in phase 2.}$$

and remote macro strain distribution

$$\bar{\epsilon}_{ij} = \bar{\epsilon}\delta_{ij}$$

We now calculate each terms in $\underline{I}$.

(1)

$$\begin{aligned}\inf_{\boldsymbol{\epsilon}^d \in \mathcal{E}} W(\boldsymbol{\epsilon}^d) &= \frac{1}{2}\bar{C}_{ijk\ell}\Big(\bar{\epsilon}\delta_{ij}\Big)\Big(\bar{\epsilon}\delta_{k\ell}\Big) \\ &= \frac{1}{2}\Big[3\bar{K}E_{ijk\ell}^{(1)} + 2\bar{\mu}E_{ijk\ell}^{(2)}\Big](\bar{\epsilon})^2\delta_{ij}\delta_{k\ell} = \frac{9}{2}\bar{K}\bar{\epsilon}^2\end{aligned}$$

Note that $E_{ijk\ell}^{(1)}\delta_{ij}\delta_{k\ell} = 3$ and $E_{ijk\ell}^{(2)}\delta_{ij}\delta_{k\ell} = 0$.

(2)

$$\begin{aligned}W^{(0)}(\boldsymbol{\epsilon}^0) &= \frac{1}{2}C_{ijk\ell}^{(0)}(\bar{\epsilon}\delta_{ij})(\bar{\epsilon}\delta_{k\ell}) \\ &= \frac{1}{2}\Big[3K_1E_{ijk\ell}^{(1)} + 2\mu_2E_{ijk\ell}^{(2)}\Big](\bar{\epsilon})^2\delta_{ij}\delta_{k\ell} = \frac{9}{2}K_1\bar{\epsilon}^2\end{aligned}$$

(3)

$$\frac{1}{V}\int_V \mathbf{p} : \boldsymbol{\epsilon}^{(0)} dV = f_1\boldsymbol{p}^{(1)} : \bar{\boldsymbol{\epsilon}} + f_2\boldsymbol{p}^{(2)} : \bar{\boldsymbol{\epsilon}} = 3f_2p\bar{\epsilon}$$

(4) Because $p_{ij}^{(1)} = 0$ and $p_{ij}^{(2)} = p\delta_{ij}$,

$$\frac{1}{2V}\int_V \boldsymbol{p} : \Delta\mathbb{C}^{-1} : \boldsymbol{p} dV = \frac{1}{2}\sum_{r=1}^{2} f_r\boldsymbol{p}^r : \Delta\mathbb{C}_r^{-1} : \boldsymbol{p}^r$$

$$= \frac{1}{2}\left(\frac{f_2}{3(K_2 - K_1)}\mathbb{E}^{(1)} + \frac{f_2}{2(\mu_2 - \mu_1)}\mathbb{E}^{(2)}\right)p^2\delta_{ij}\delta_{k\ell} = \frac{f_2p^2}{2(K_2 - K_1)}$$

(5) Because $< p_{k\ell} >= f_2 p \delta_{k\ell}$,

$$\frac{1}{2V}\int_V p_{ij}\epsilon^d_{ij}dV = -\frac{1}{2}\sum_{r=1}^{2} f_r P^r_{ijk\ell}p^r_{ij}p^r_{k\ell} + \frac{1}{2}\sum_{r=1}^{2} f_r P^r_{ijk\ell}p^r_{ij} < p^r_{k\ell} >$$

$$= -\frac{f_2}{2}\Big(\frac{3p^2}{3K_1+4\mu_1}\Big) + \frac{1}{2}\frac{3f_2^2p^2}{3K_1+4\mu_1}$$

$$= -\frac{1}{2}\frac{3f_1f_2p^2}{3K_1+4\mu_1} = -\frac{1}{2}\frac{f_1f_2p^2}{K_1+\frac{4}{3}\mu_1}$$

Therefore, when $\Delta\mathbb{C} > 0$,

$$\underline{I}(p) = \frac{9}{2}K_1\bar{\epsilon}^2 - \frac{f_2p^2}{2(K_2-K_1)} - \frac{f_1f_2p^2}{2(K_1+\frac{4}{3}\mu_1)} + 3f_2p\bar{\epsilon} \le \frac{9}{2}\bar{K}\bar{\epsilon}^2 \tag{5.113}$$

One can check that $\frac{\partial^2(I)}{\partial p^2 < 0}$, therefore, the maximum value of $\underline{I}$ gives the tightest lower bound. We check the stationary condition,

$$\frac{\partial \underline{I}}{\partial p} = 0 \;\Rightarrow\; -\frac{f_2p}{(K_2-K_1)} - \frac{f_1f_2p}{K_1+\frac{4}{3}\mu_1} + 3\bar{\epsilon}f_2 = 0$$

$$\Rightarrow\; p_{sta} = \frac{3\bar{\epsilon}}{\dfrac{1}{K_2-K_1} + \dfrac{f_1}{K_1+\frac{4}{3}\mu_1}} \tag{5.114}$$

Substituting (5.114) into (5.113) yields a lower bound for effective bulk modulus

$$\bar{K} \ge K_1 + \frac{f_2}{\dfrac{1}{K_2-K_1} + \dfrac{f_1}{K_1+\frac{4}{3}\mu_1}} \tag{5.115}$$

Step 2: Let the comparison solid assume the modulus of phase 2

$$K^{(0)} = K_2, K = K_1, \;\text{ and }\;]\mu^{(0)} = \mu_2, \mu = \mu_1$$

and choose

$$p^{(1)}_{ij} = p\delta_{ij},\;\; p^{(2)}_{ij} = 0\;.$$

One can find an upper bound,

$$\bar{I}(p) = \frac{9}{2}K_2\bar{\epsilon}^2 - \frac{f_1p^2}{2(K_1-K_2)} - \frac{1}{2}\frac{f_1f_2p^2}{K_2+\frac{4}{3}\mu_2} + 3f_1p\bar{\epsilon} \ge \frac{9}{2}\bar{K}\bar{\epsilon}^2 \tag{5.116}$$

To find the minimum value of $\bar{I}(p)$, we examine the stationary condition,

$$\frac{\partial \bar{I}}{\partial p} = 0, \;\;\Rightarrow\; p_{sta} = \frac{3\bar{\epsilon}}{\dfrac{1}{K_1-K_2} + \dfrac{f_2}{K_2+\frac{4}{3}\mu_2}} \tag{5.117}$$

Substituting (5.117) into (5.116), one will find an upper bound for effective bulk modulus

$$\bar{K} \le K_2 + \frac{f_2}{\dfrac{1}{K_1-K_2} + \dfrac{f_2}{K_2+\frac{4}{3}\mu_2}} \tag{5.118}$$

By combining (5.115) and (5.118), we will have the Hashin-Shtrikman bound on bulk modulus,

$$K_1 + \frac{f_2}{\dfrac{1}{K_2 - K_1} + \dfrac{f_1}{K_1 + \frac{4}{3}\mu_1}} \le \bar{K} \le K_2 + \frac{f_1}{\dfrac{1}{K_1 - K_2} + \dfrac{f_2}{K_2 + \frac{4}{3}\mu_2}} \tag{5.119}$$

Following similar procedure, it is readily to show that the following Hashin-Shtrikman bounds are held for shear modulus,

$$\mu_1 + \frac{f_2}{\dfrac{1}{\mu_2 - \mu_1} + \dfrac{6(K_1 + 2\mu_1)f_1}{5(3K_1 + 4\mu_1)\mu_1}} \le \bar{\mu} \le \mu_2 + \frac{f_1}{\dfrac{1}{\mu_1 - \mu_2} + \dfrac{6(K_2 + 2\mu_2)f_2}{5(3K_2 + 4\mu_2)\mu_2}} \tag{5.120}$$

Figs. (5.2) and (5.3) illustrate the variational bounds of effective modulus for

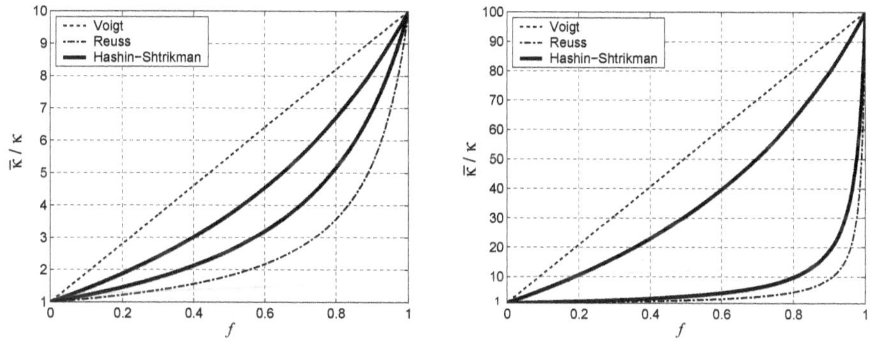

(a) (b)

Fig. 5.2 Variational bounds for bulk modulus: (a) composite A, and (b) composite B.

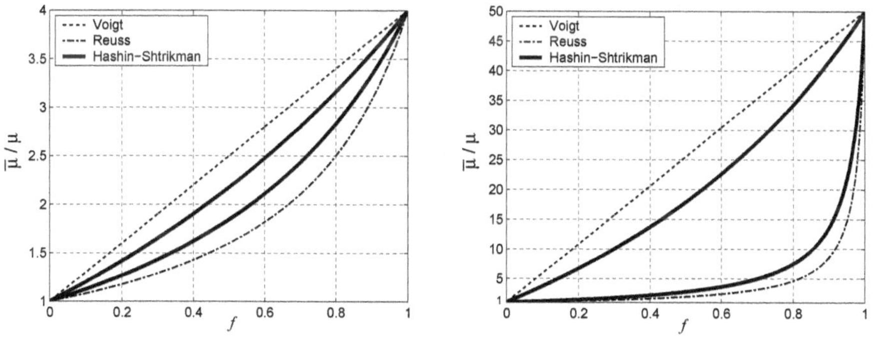

(a) (b)

Fig. 5.3 Variational bounds for shear modulus: (a) composite A, and (b) composite B.

two-phase composite with different modulus contrast. The bulk modulus ratio of the inclusion and matrix in composite A is 10 : 1, and shear modulus ratio of 4 : 1; while in composite B, modulus contrast is much higher, where the bulk modulus ratio of inclusion and matrix is 100 : 1, and shear modulus ratio is 50 : 1. Poisson's ratio of the matrix is chosen to be 0.1. The plots show the ratios of the effective modulus against the modulus of the matrix as a function of volume fraction of the inclusion. The figures shows Voigt bound, Reuss bound, and Hashin-Shtrikman upper/lower bounds. For both cases, Hashin-Shtrikman bounds provide better estimate compared with Voigt bound and Reuss bound. The same problem has been studied in previous Chapter using Mori-Tanaka method. It worths pointing out that the Hashin-Shtrikman lower bound coincides with bound derived from Mori-Tanaka method [Weng (1990)] .

5.5 Review of functional analysis and convex analysis

The above sections provide some important variational principles and variational bounds for linear composite materials. More comprehensive treatment of variational principles require additional mathematical background on functional analysis and convex analysis. Because this is an engineering treatise, this section will briefly review some basics of functional analysis and convex analysis that would be sufficient for the scope of the book. For advanced study of these subjects, we would recommend two excellent monographs that are well written and easy to read: [Gao (2000); Ekeland and Temam (1976)]. In fact, the mathematical presentation in this section are strongly influenced by Professor David Gao's work [Gao (2000)].

Definition 5.1 (Vector Space). *Let F be a field, whose elements are referred to as scalars. A vector space over F is a nonempty set V, whose elements are referred to as vectors, together with two operations. The first operation, called addition and denoted by $+$, assigns to each pair $(\boldsymbol{u}, \mathbf{v}) \in V \times V$ of vectors in V a vector $\boldsymbol{u} + \mathbf{v}$ in V. The second operation, called multiplication and denoted by juxtaposition, assigns to each pair $(r, \boldsymbol{u}) \in F \times V$ a vector $r\mathbf{v} \in V$. Furthermore, the following properties must be satisfied,*

(1) Associativity of addition

$$\boldsymbol{u} + (\mathbf{v} + \boldsymbol{w}) = (\boldsymbol{u} + \mathbf{v}) + \boldsymbol{w}, \quad \forall \boldsymbol{u}, \mathbf{v}, \boldsymbol{w} \in V$$

(2) Commutative of addition

$$\boldsymbol{u} + \mathbf{v} = \mathbf{v} + \boldsymbol{u}, \quad \forall \boldsymbol{u},\ \mathbf{v} \in V$$

(3) Existence of a zero vector, $\mathbf{0} \in V$ such that

$$\mathbf{0} + \boldsymbol{u} = \boldsymbol{u} + \mathbf{0} = \boldsymbol{u}, \quad \forall \boldsymbol{u} \in V$$

(4) Existence of additive inverse: i.e. $\forall \boldsymbol{u} \in V$, $\exists - \boldsymbol{u} \in V$, such that

$$\boldsymbol{u} + (-\boldsymbol{u}) = (-\boldsymbol{u}) + \boldsymbol{u} = \mathbf{0}$$

(5) Properties of scalar multiplication. $\forall r, s \in F$ *and* $\boldsymbol{u}, \mathbf{v} \in V$,

$$\begin{aligned} r(\boldsymbol{u}+\mathbf{v}) &= r\boldsymbol{u}+r\mathbf{v} \\ (r+s)\boldsymbol{u} &= r\boldsymbol{u}+r\mathbf{v} \\ rs\boldsymbol{u} &= r(s\boldsymbol{u}) \\ 1\boldsymbol{u} &= \boldsymbol{u} \end{aligned}$$

Remark 5.3.

(1) The first four properties in the definitions of vector space can be summarized that V is an abelian group under addition;
(2) Any expression of the form

$$r_1\mathbf{v}_1 + r_2\mathbf{v}_2 + \cdots + r_n\mathbf{v}_n$$

where $r_i \in F$ and $\mathbf{v}_i \in V\ \forall i = 1, 2, \cdots, n$ is called a linear combination of the vectors $\mathbf{v}_1, \mathbf{v}_2, \cdots, \mathbf{v}_n$, and

$$r_1\mathbf{v}_1 + r_2\mathbf{v}_2 + \cdots + r_n\mathbf{v}_n \in V$$

(3) The addition operation

$$V \times V \to V : (\boldsymbol{u}, \mathbf{v}) \to \boldsymbol{u} + \mathbf{v} \in V$$

(4) and the scalar multiplication operation,

$$F \times V \to V : (\alpha, \boldsymbol{u}) \to \alpha\boldsymbol{u} \in V$$

are closed.
(5) When the operations

$$f : (\boldsymbol{u}, \mathbf{v}) \to \boldsymbol{u} + \mathbf{v} \in V$$

$$g : (\alpha, \boldsymbol{u}) \to \alpha\boldsymbol{u} \in V$$

are continuous, the vector space is called topological vector space.

Example 5.1 ($\mathbb{R}^n$ space). *Let* $F = \mathbb{R}$. *The set of all ordered n-tuples, i.e.*

$$\boldsymbol{u} = (u_1, u_2, \cdots, u_n), \quad u_i \in \mathbb{R}$$

with addition and scalar multiplication defined component-wise,

$$(a_1, \cdots, a_n) + (b_1, \cdots, b_n) = (a_1 + b_1, \cdots, a_n + b_n)$$

and

$$\alpha(a_1, \cdots, a_n) = (\alpha a_1, \cdots, \alpha a_n)$$

is a vector space, and it is denoted as $\mathbb{R}^n$. *Note that in general vector space (a mathematical concept) is still a primitive set. It may have some algebraic structures, but it does not have topological structures, or geometric structures, such as distance between two elements.*

Example 5.2 (C^0 space). *The set of all continuous function, $C^0(\mathbb{R})$, is a vector space under the operations of addition and scalar multiplication, i.e.*

$$(f+g)(x) = f(x) + g(x), \quad \forall f, g \in C^0(\mathbb{R})$$

and

$$(\alpha f)(x) = \alpha f(x), \quad \forall \alpha \in \mathbb{R}, \quad \forall f \in C^0(\mathbb{R})$$

Definition 5.2 (Bilinear form). *Let X be a vector space and X^* is its dual space. A mapping g of $X \times X^*$ into $\mathbb{R}$ is called a bilinear functional or a bilinear form if*

(1) For fixed $\mathbf{y}$, $g(\mathbf{x}, \mathbf{y})$ is a linear functional in $\mathbf{x}$, i.e.

$$g(\alpha\mathbf{x} + \beta\mathbf{y}, \mathbf{z}) = \alpha g(\mathbf{x}, \mathbf{z}) + \beta g(\mathbf{x}, \mathbf{z}), \quad \forall \mathbf{x}, \mathbf{y} \in X, \quad \mathbf{z} \in X^*$$

(2) For fixed $\mathbf{x}$, $g(\mathbf{x}, \mathbf{y})$ is a linear functional in $\mathbf{y}$, i.e.

$$g(\mathbf{x}, \alpha\mathbf{y} + \beta\mathbf{z}) = \alpha g(\mathbf{x}, \mathbf{y}) + \beta g(\mathbf{x}, \mathbf{z}), \quad \forall \mathbf{x} \in X, \quad \mathbf{y}, \mathbf{z} \in X^*$$

A bilinear form is denoted as

$$g(\mathbf{x}, \mathbf{y}) :=< \mathbf{x}, \mathbf{y} >$$

Definition 5.3 (Inner product). *Choose $X^* = X$. The bilinear form of $X \times X$ is called inner product, denoting $< \cdot, \cdot >$ as $(\cdot, \cdot)$, such that*

$$(\cdot, \cdot) : X \times X \to \mathbb{R}$$

with properties:

(1) $(\mathbf{x}, \mathbf{x}) \geq 0, \forall \mathbf{x} \in X$ and $(\mathbf{x}, \mathbf{x}) = 0$ iff $\mathbf{x} = \mathbf{0}$;
(2) Symmetry $(\mathbf{x}, \mathbf{y}) = (\mathbf{y}, \mathbf{x})$;
(3) Linearity

$$(\alpha\mathbf{x} + \beta\mathbf{y}, \mathbf{z}) = \alpha(\mathbf{x}, \mathbf{z}) + \beta(\mathbf{y}, \mathbf{z}),$$

and

$$(\mathbf{x}, \alpha\mathbf{y} + \beta\mathbf{z}) = \alpha(\mathbf{x}, \mathbf{y}) + \beta(\mathbf{x}, \mathbf{z}) \quad \forall \mathbf{x}, \mathbf{y}, \mathbf{z} \in X \quad \text{and} \quad \alpha\beta \in \mathbb{R}.$$

Example 5.3 (E^n space). *For $\forall \mathbf{x} = (x_1, x_2, \cdots, x_n) \in \mathbb{R}^n$ and $\forall \mathbf{y} = (y_1, y_2, \cdots, y_n) \in \mathbb{R}^n$, we define an inner product*

$$(\mathbf{x}, \mathbf{y}) = \sum_{i=1}^{n} x_i y_i$$

This particular inner product space is Euclidean space, denoted as $E^n = \{\mathbb{R}^n, (\cdot, \cdot)\}$. It generates a norm,

$$\|\mathbf{x}\|_{\ell_2} := \Big(\sum_{i=1}^{n} x_i x_i\Big)^{1/2} = \sqrt{(\mathbf{x}, \mathbf{x})}$$

This norm is called Euclidean norm on $\mathbb{R}^n$. *The space is therefore a normed space as well — called n-dimensional Euclidean space,* $E_n = \{\mathbb{R}^n, \|\cdot\|_{\ell_2}\}$. *One can show that*

$$
\begin{aligned}
&(1)\ \ \|\mathbf{x}\|_{\ell_2} \geq 0, \quad \forall \mathbf{x} \in E^n; \quad \text{and} \quad \|\mathbf{x}\|_{\ell_2} = 0, \quad iff \quad \mathbf{x} = \mathbf{0}\ ; \\
&(2)\ \ \|\alpha \mathbf{x}\|_{\ell_2} = |\alpha| \|\mathbf{x}\|_{\ell_2}, \quad \forall \mathbf{x} \in E^n, \quad \alpha \in \mathbb{R} \\
&(3)\ \ \|\mathbf{x} + \mathbf{y}\|_{\ell_2} \leq \|\mathbf{x}\|_{\ell_2} + \|\mathbf{y}\|_{\ell_2} \qquad (\textit{triangle inequality}); \\
&(4)\ \ \|(\mathbf{x}, \mathbf{y})\|_{\ell_2} \leq \|\mathbf{x}\|_{\ell_2} \|\mathbf{y}\|_{\ell_2} \qquad (\textit{Cauchy-Schwartz inequality}).
\end{aligned}
$$

Based on the ℓ_2*-norm, one can measure the distance between two vectors in* E^n,

$$\rho(\mathbf{x}, \mathbf{y}) := \|\mathbf{x} - \mathbf{y}\|_{\ell_2};$$

One can also show that

$$
\begin{aligned}
&(1)\ \ \rho(\mathbf{x}, \mathbf{y}) = \rho(\mathbf{y}, \mathbf{x}); \\
&(2)\ \ \rho(\mathbf{x}, \mathbf{y}) \geq 0, \quad \text{and} \quad \rho(\mathbf{x}, \mathbf{y}) = 0, \quad \text{iff} \quad \mathbf{x} = \mathbf{y}; \\
&(3)\ \ \rho(\mathbf{x}, \mathbf{y}) \leq \rho(\mathbf{x}, \mathbf{z}) + \rho(\mathbf{z}, \mathbf{y}), \quad \forall \mathbf{x}, \mathbf{y}, \mathbf{z} \in E_n
\end{aligned}
$$

The distance function $\rho(\mathbf{x}, \mathbf{y})$ *is called a metric, and the associated vector space is called metric space.*

Remark 5.4.

(1) A normed space or a metric space is not necessarily an inner product space, but an inner product vector space is a normed space, because inner product can generate a norm, not vice versa.
(2) A complete normed vector space is called Banach space and a complete inner product space is called Hilbert space .

Note that a metric space V is said to be *complete* (or Cauchy) if every Cauchy sequence of points in V has a limit that is also in V. For a metric space, a Cauchy sequence is one such that $\rho(\mathbf{v}_j, \mathbf{v}_k) \to 0$, as $j, k \to \infty$.

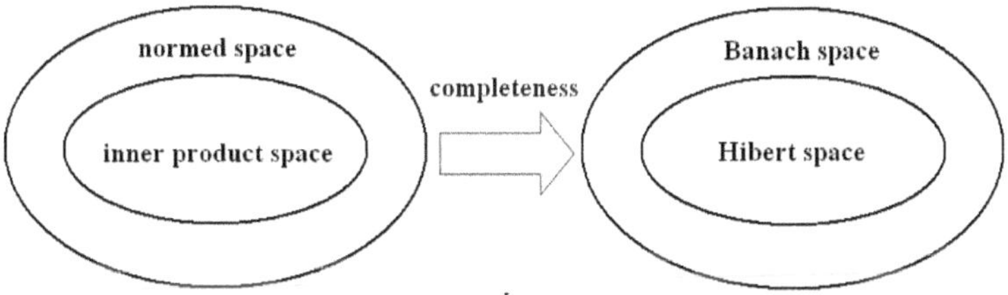

Fig. 5.4 Banach space and Hilbert space.

Example 5.4 (L^2 Space). *Consider a real value function $f(x)$, $x \in [a, b]$. Define an inner product,*

$$(f, g) = \int_a^b f(x)g(x)dx$$

Subsequently, one can define L^2 norm as

$$\|f\|_{L^2([a,b])} = \sqrt{(f, f)} = \sqrt{\int_a^b f^2(x)dx}$$

$L^2([a, b])$space is the set that contains all $f(x)$ such that

$$L^2([a, b]) = \left\{ f \;\middle|\; \|f\|_{L^2([a,b])} < +\infty \right\}$$

Therefore, $L^2([a, b])$ is an inner product vector space, and of course, normed space (metric space).

Example 5.5 (Lebesgue Space $L^p(\Omega)$). *Let Ω be an open set in $\mathbb{R}^n$. For $1 < p < \infty$, one can define a L^p-norm for a measurable function f,*

$$\|f\|_{L^p(\Omega)} := \left(\int_\Omega |f(x)|^p dx \right)^{1/p}$$

and a Lebesgue space is defined as

$$L^p(\Omega) := \left\{ f \;\middle|\; \|f\|_{L^p(\Omega)} < \infty \right\}$$

It has the following properties,

(1) $\|f\|_{L^p(\Omega)} \geq 0$, $\|f\|_{L^p(\Omega)} = 0$, *if and only if* $f = 0$ *almost everywhere;*

(2) $\|cf\|_{L^p(\Omega)} \leq |c| \|f\|_{L^p(\Omega)}$, $\forall f \in L^p(\Omega)$, $c \in \mathbb{R}$

(3) $\|f + g\|_{L^p(\Omega)} \leq \|f\|_{L^p(\Omega)} + \|g\|_{L^p(\Omega)}$ *(Minkowski's inequality)*

(4) For $1 \leq p, q \leq \infty$, such that $\dfrac{1}{p} + \dfrac{1}{q} = 1$,
if $f \in L^p(\Omega)$ and $g \in L^q(\Omega)$, then for finite Ω, $fg \in L^1(\Omega)$, and
$\|fg\|_{L^1(\Omega)} \leq \|f\|_{L^p(\Omega)} \|g\|_{L^q(\Omega)}$ *(Hölder's inequality)*

In particular, if $p = q = 2$, then $f \cdot g \in L^1(\Omega)$ because

$$\int_\Omega |f(x)g(x)|dx \leq \|f\|_{L^2(\Omega)} \|g\|_{L^2(\Omega)} < \infty$$

Note that in general $L^p(\Omega)$ is not an inner product space, except $p = 2$. $L^p(\Omega)$ is, nevertheless, a complete normed space, therefore, a Banach space. The space $L^2(\Omega)$ is the only Hilbert space of this class.

Example 5.6 (Sobolev Space). *Define Sobolev norm*

$$\|f\|_{W_p^k(\Omega)} = \Big(\sum_{\alpha=0}^{k} \|D^\alpha f\|^p_{L^p(\Omega)}\Big)^{1/p}$$

Note that the Sobolev norm is not generated by an inner product in general. A Sobolev space is defined as

$$W_p^k(\Omega) = \{f \;\Big|\; \|f\|_{W_p^k(\Omega)} < \infty\}$$

For $p = 2$*, Sobolev spaces become inner product spaces. In particular,*

(1) For $p = 2, k = 0$, $W_2^0(\Omega) = L^2(\Omega)$,

$$(f, g)_{L^2(\Omega)} = \int_\Omega f(\mathbf{x})g(\mathbf{x})dV$$

(2) For $p = 2, k = 1$, $W_2^1(\Omega) = H^1(\Omega)$,

$$(f, g)_{H^1(\Omega)} = \int_\Omega \Big[f(\mathbf{x})g(\mathbf{x}) + \nabla f(\mathbf{x}) \cdot \nabla g(\mathbf{x})\Big]dV$$

and

$$\|f\|_{H^1(\Omega)} = \sqrt{\int_\Omega \Big[f(\mathbf{x})^2 + \nabla f(\mathbf{x}) \cdot \nabla f(\mathbf{x})\Big]dV}$$

(3) For $p = 2, k = 2$, $W_2^2(\Omega) = H^2(\Omega)$,

$$(f, g)_{H^2(\Omega)} = \int_\Omega \Big[f(\mathbf{x})g(\mathbf{x}) + \nabla f(\mathbf{x}) \cdot \nabla g(\mathbf{x}) + \nabla \otimes \nabla f(\mathbf{x}) : \nabla \otimes \nabla g(\mathbf{x})\Big]dV$$

and

$$\|f\|_{H^2(\Omega)} = \sqrt{\int_\Omega \Big[f(\mathbf{x})^2 + \nabla f(\mathbf{x}) \cdot \nabla f(\mathbf{x}) + \nabla \otimes \nabla f(\mathbf{x}) : \nabla \otimes \nabla f(\mathbf{x})\Big]dV}$$

5.5.1 *Concept of convexity*

Definition 5.4. Let $\mathcal{U}$ be a linear vector space over $\mathbb{R}$. A subset (subspace) $\mathcal{K} \subset \mathcal{U}$ is said to be convex, if it contains the line segment between any two of its elements, i.e.

$$\theta \boldsymbol{u} + (1 - \theta)\mathbf{v} \in \mathcal{K}, \quad \forall \boldsymbol{u}, \mathbf{v} \in \mathcal{K}, \quad \forall \theta \in [0, 1] \tag{5.121}$$

Example 5.7. Let $\mathcal{U} = \mathbb{R} \times \mathbb{R}$, and $\mathcal{K} \subset \mathcal{U}$. We say $\mathcal{K}$ is convex, if for any $\boldsymbol{u}_1 = (x_1, x_2) \in \mathcal{K}$, $\boldsymbol{u}_2 = (y_1, y_2) \in \mathcal{K}$, such that $\theta \boldsymbol{u}_1 + (1-\theta)\boldsymbol{u}_2 \in \mathcal{K}$, $\forall \theta \in [0, 1]$. We say $\mathcal{K}$ is non-convex, if $\exists \boldsymbol{u}_1$, $\boldsymbol{u}_2 \in \mathcal{K}$, and $\exists \theta \in [0, 1]$, such that $\theta \boldsymbol{u}_1 + (1-\theta)\boldsymbol{u}_2 \notin \mathcal{K}$. A graphic illustration is demonstrated in Fig. (5.5).

h

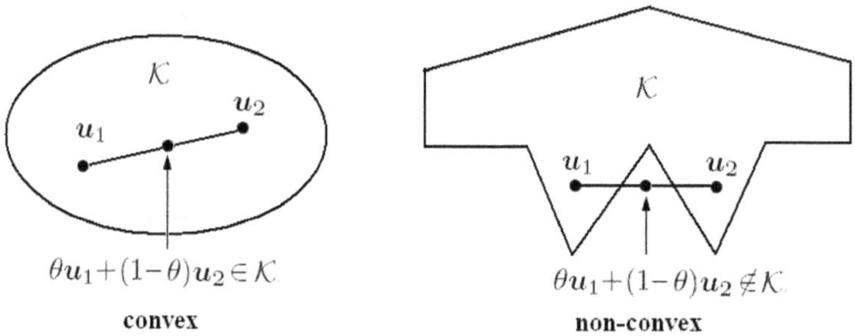

Fig. 5.5 Convex set and non-convex set in $\mathbb{R}^2$.

Definition 5.5 (Convex and concave functionals). *(1) A functional $P : \mathcal{U} \to \mathbb{R}$ is said to be convex on $\mathcal{U}$ if*

$$P(\theta \boldsymbol{u}_1 + (1-\theta)\boldsymbol{u}_2) \leq \theta P(\boldsymbol{u}_1) + (1-\theta)P(\boldsymbol{u}_2), \quad \forall \boldsymbol{u}_1, \boldsymbol{u}_2 \in \mathcal{U}, \quad \forall \theta \in [0,1]$$

whenever the right-hand side is defined.

(2) P is said to be strictly convex if the strict form of the inequality holds for any $\boldsymbol{u}_1 \neq \boldsymbol{u}_2$;

(3) P is said to be concave if $-P$ is convex.

Example 5.8. Let $\mathcal{U} = \mathbb{R}$, the quadratic function $P(x) = (x-a)^2 + b$ is a convex function. See Fig.(5.6).

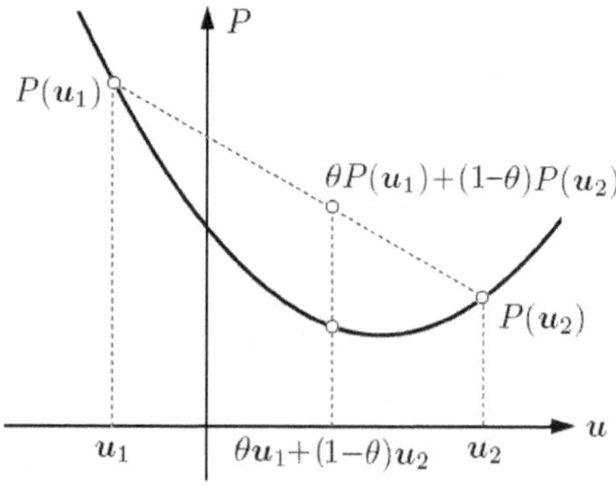

Fig. 5.6 An example of convex function.

Example 5.9. Consider a 1D elastic string bounded by interval $I = [0, \ell]$. Let $\mathcal{U} = \mathcal{E}$ and $\mathcal{U}^* = \mathcal{E}^* = \mathcal{S}$ where

$$\mathcal{E} = \{\epsilon \,\Big|\, \epsilon \in L^\alpha(I), \epsilon = \frac{du}{dx}\}$$
$$\mathcal{S} = \{\sigma \,\Big|\, \sigma \in L^\beta(I), \frac{d\sigma}{dx} = 0\}$$
$$1 < \alpha, \beta < \infty, \quad \text{and} \quad \frac{1}{\alpha} + \frac{1}{\beta} = 1.$$

Define the stain energy density and complementary strain energy density as

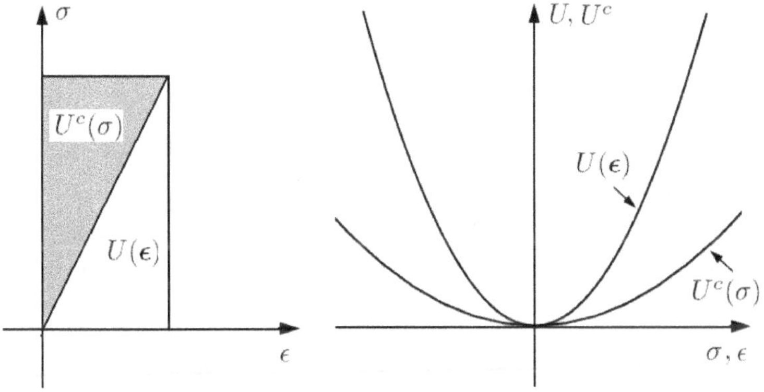

Fig. 5.7 Strain energy density and complementary strain energy.

$$U : \mathcal{E} \to \mathrm{I\!R}, \qquad U(\epsilon) = \int_0^\epsilon \sigma(\tilde{\epsilon}) d\tilde{\epsilon}$$
$$U^c : \mathcal{S} \to \mathrm{I\!R}, \qquad U^c(\sigma) = \int_0^\sigma \epsilon(\tilde{\sigma}) d\tilde{\sigma}$$

Both strain energy density and complementary strain energy density are convex, and they are plotted in Fig. (5.7).

5.5.2 *Gâteaux variation and convex functional*

The Gâteaux variation of a functional in a linear space is the generalized directional derivative of a real-value function in vector calculus.

Definition 5.6 (Gâteaux variation). *(1) Let $P : \mathcal{U} \to \mathrm{I\!R}$ be a real-valued functional and $\mathcal{U}_a \subset \mathcal{U}$ a subspace. For a given $\bar{\mathbf{u}} \in \mathcal{U}_a$, if the limit,*

$$\delta P(\bar{\mathbf{u}}, \boldsymbol{u}) := \lim_{\lambda \to 0^+} \frac{P(\bar{\mathbf{u}} + \lambda \boldsymbol{u}) - P(\bar{\mathbf{u}})}{\lambda}, \quad \forall \boldsymbol{u} \in \mathcal{U}_a$$

exists as $\lambda \to 0^+$ (i.e. $\lambda \to 0, \lambda > 0$), then $\delta P(\bar{\mathbf{u}}; \boldsymbol{u}) \in \mathrm{I\!R}$ is called the Gâteaux variation of P at $\bar{\mathbf{u}}$ in the direction of $\boldsymbol{u}$.

(2) If the Gâteaux variation is a linear operator in $\boldsymbol{u}$ *such that*

$$\delta P(\bar{u}, \boldsymbol{u}) =< \boldsymbol{u}, DP(\bar{\mathbf{u}}) >, \quad \forall \boldsymbol{u} \in \mathcal{U}_a$$

we say that P *is Gâteaux differentiable at* $\bar{\mathbf{u}}$*. The linear operator* $DP(\bar{\mathbf{u}}) : \mathcal{U}_a \to \mathcal{U}^*$*, which generally depends on* $\bar{\mathbf{u}}$*, is called the Gâteaux derivative of* P *at* $\bar{\mathbf{u}}$*.*

(3) The functional $P : \mathcal{U} \to \mathbb{R}$ *is said to be Gâteaux differentiable on* $\mathcal{U}_a$ *if it is Gâteaux differentiable at each* $\boldsymbol{u} \in \mathcal{U}_a$*.*

Note that

$$\begin{aligned} \delta P(\bar{\mathbf{u}}, \boldsymbol{u}) &= \frac{d}{d\lambda} P(\bar{\mathbf{u}} + \lambda \boldsymbol{u}) \Big|_{\lambda=0} \\ \frac{\delta P}{\delta u} &:= DP(\bar{\mathbf{u}}) \end{aligned}$$

Why are convex functionals so special ? The following theorem answers this question:

Theorem 5.4. *If* $P : \mathcal{U}_k \subset \mathcal{U} \to \mathbb{R}$ *is Gâteaux differentiable, then, the following statements are equivalent to each other*

$$\begin{aligned} &(1) \quad P : \mathcal{U}_k \subset \mathcal{U} \to \mathbb{R} \text{ is convex} \\ &(2) \quad P(\mathbf{v}) - P(\boldsymbol{u}) \geq \big\langle \mathbf{v} - \boldsymbol{u}, DP(\boldsymbol{u}) \big\rangle, \ \forall \mathbf{v}, \boldsymbol{u} \in \mathcal{U}_k \\ &(3) \ \big\langle \mathbf{v} - \boldsymbol{u}, DP(\mathbf{v}) - DP(\boldsymbol{u}) \big\rangle \geq 0 \ , \quad \forall \mathbf{v}, \boldsymbol{u} \in \mathcal{U}_k \end{aligned}$$

Remark 5.5. The statement (3) shows that Gâteaux derivative of a convex function is a monotone operator of $\mathcal{U}$ into $\mathcal{U}^*$. By the mean value theorem,

$$\big\langle \mathbf{v} - \boldsymbol{u}, DP(\mathbf{v}) - DP(\boldsymbol{u}) \big\rangle = \big\langle \mathbf{v} - \boldsymbol{u}, D^2 P(\bar{\boldsymbol{u}}) \cdot (\mathbf{v} - \boldsymbol{u}) \big\rangle \geq 0$$

where $\bar{\boldsymbol{u}} = \mathbf{v} + \theta(\boldsymbol{u} - \mathbf{v}), \ \theta \in [0, 1]$.
Hence, a sufficient condition for P being convex on $\mathcal{U}$ is that

$$D^2 P(\boldsymbol{u}) \geq 0, \quad \forall \boldsymbol{u} \in \mathcal{U}_k$$

Recall the total potential energy for a linear elastic solid is

$$\begin{aligned} \Pi(\boldsymbol{u}, \nabla \boldsymbol{u}) &= \int_V U(\epsilon) dV - \int_{\Gamma_t} t_i^0 u_i dS \\ \delta \Pi(\boldsymbol{u}, \nabla \boldsymbol{u}) &= \int_V \frac{\partial U}{\partial \epsilon_{ij}} \delta \epsilon_{ij} dV - \int_{\Gamma_t} t_i^0 \delta u_i dS \\ \delta^2 \Pi(\boldsymbol{u}, \nabla \boldsymbol{u}) &= \int_V \frac{\partial^2 U}{\partial \epsilon_{ij} \partial \epsilon_{k\ell}} \delta \epsilon_{ij} \delta \epsilon_{k\ell} dV = \int_V C_{ijk\ell} \delta \epsilon_{ij} \delta \epsilon_{k\ell} dV \geq 0 \ . \end{aligned}$$

This is to say that if elastic tensor is positive definite, the elastic potential energy is convex. Similar statement can be made for complementary potential energy, if the compliance tensor is positive definite.

5.5.3 *Primal variational problems*

We consider the following primal variational problems:

Let $P : \mathcal{U}_\kappa \subset U \to \mathbb{R}$ be a given functional.

(1) The infimum (or inf) primal variational problems is to find a global minimizer $\tilde{\boldsymbol{u}} \in \mathcal{U}_\kappa$ such that

$$\left(\mathcal{P}_{inf}\right): \quad P(\tilde{\boldsymbol{u}}) = \inf P(\boldsymbol{u}), \quad \forall \boldsymbol{u} \in \mathcal{U}_\kappa$$

(2) The supremum (or *sup*) primal problem is to find a global maximizer $\tilde{\boldsymbol{u}} \in \mathcal{U}_\kappa$ such that

$$\left(\mathcal{P}_{sup}\right): \quad P(\tilde{\boldsymbol{u}}) = \sup P(\boldsymbol{u}), \quad \forall \boldsymbol{u} \in \mathcal{U}_\kappa$$

(3) The stationary (or *sta*) primal variational problem is to find a stationary point $\boldsymbol{u} \in \mathcal{U}_\kappa$ such that

$$\left(\mathcal{P}_{sta}\right): \quad P(\tilde{\boldsymbol{u}}) = \text{sta } P(\boldsymbol{u}), \quad \forall \boldsymbol{u} \in \mathcal{U}_\kappa$$

Remark 5.6.

(1) A stationary point is also called critical point. The critical point condition,

$$\delta P(\tilde{\boldsymbol{u}}, \boldsymbol{u}) = 0, \quad \forall \boldsymbol{u} \in \mathcal{U}_\kappa$$

leads to the Euler-Lagrange equation.

(2) The problem $(\mathcal{P}_{\text{inf}})$ is called realizable if there exists a vector $\tilde{\boldsymbol{u}} \in \mathcal{U}_\kappa$ such that the infimum of P is achieved at $\tilde{\boldsymbol{u}}$ and is not $-\infty$. Then $\tilde{\boldsymbol{u}}$ is called the minimizer of $(\mathcal{P}_{inf})$ and we write $P(\tilde{\boldsymbol{u}}) = \min_{\boldsymbol{u} \in \mathcal{U}_\kappa} P(\boldsymbol{u})$.

Similarly, a vector $\tilde{\boldsymbol{u}} \in \mathcal{U}_\kappa$ is called the maximizer of $(\mathcal{P}_{sup})$ if the supermum is achieved at $\tilde{\boldsymbol{u}}$ and is not $+\infty$. We write $\mathcal{P}(\tilde{\nu}) = \max_{\boldsymbol{u} \in \mathcal{U}_\kappa}(\boldsymbol{u})$.

Example 5.10. The real-value function, $P(x) = \exp(x)$ is convex on $\mathcal{U} = \mathbb{R}$ and

$$\inf_{x \in \mathcal{U}} P(x) = 0, \quad \sup_{x \in \mathcal{U}} P(x) = +\infty$$

therefore, $(\mathcal{P}_{\text{inf}})$ and $(\mathcal{P}_{\text{sup}})$ problems are not realizable since the infimum and supremum are archived only if $x \to \pm\infty$. However on the closed interval, $\mathcal{U}_\kappa = [a, b]$ with $-\infty < a < b < +\infty$, $(\mathcal{P}_{\text{inf}})$ and $(\mathcal{P}_{\text{sup}})$ problems are realizable and

$$\inf_{x \in \mathcal{U}_\kappa} P(x) = \min_{x \in \mathcal{U}_\kappa} P(x) = P(a) = e^a,$$

$$\sup_{x \in \mathcal{U}_\kappa} P(x) = \min_{x \in \mathcal{U}_\kappa} P(x) = P(b) = e^b .$$

5.6 Legendre transformation and duality

In continuum mechanics, for a given strain energy density $U(\boldsymbol{\epsilon})$ such that the strain-stress relation $\boldsymbol{\sigma} = \dfrac{\partial U}{\partial \boldsymbol{\epsilon}}$ is invertible, then one can define the complementary energy density of $U^c(\boldsymbol{\sigma})$ by

$$U^c(\boldsymbol{\sigma}) = \boldsymbol{\sigma} : \boldsymbol{\epsilon}(\boldsymbol{\sigma}) - U(\boldsymbol{\epsilon}(\boldsymbol{\sigma})) \tag{5.122}$$

Note that here

$$U = U(\boldsymbol{\epsilon}) : \ \mathcal{E} \to \mathbb{R} \tag{5.123}$$

$$U^c = U^c(\boldsymbol{\sigma}) : \ \mathcal{S} \to \mathbb{R} \tag{5.124}$$

$$< \boldsymbol{\sigma}, \boldsymbol{\epsilon} > = \boldsymbol{\sigma} : \boldsymbol{\epsilon} : \ \mathcal{E} \times \mathcal{E}^* \to \mathbb{R} \tag{5.125}$$

where the space $\mathcal{S}$ may be viewed as $\mathcal{E}^*$ (dual space of $\mathcal{E}$).

In mathematics, this is the well-known Legendre transformation. Generally speaking, the classical Legendre transformation can be viewed as a conversion from one continuous real-valued function to another one. If the transformation is reversible, then we say that each function is the dual of the other. The reversible Legendre transformation is also called the Legendre conjugate transformation, or simply the Legendre transformation.

For infinitesimal continuum deformations, both $\boldsymbol{\sigma}$ and $\boldsymbol{\epsilon}$ are symmetric. For simplicity, we may view both $\boldsymbol{\sigma}$ and $\boldsymbol{\epsilon}$ as vectors in a six-dimensional space. Let $\mathcal{E} = \mathcal{E}^* = \mathbb{R}^n$. The element $\boldsymbol{\epsilon} = \{\epsilon_i\} \in \mathcal{E}$ and $\boldsymbol{\sigma} = \{\sigma_i\} \in \mathcal{E}^*, (i = 1, 2, \cdots, n)$ are vectors in $\mathbb{R}^n$. The bilinear form

$$< \boldsymbol{\epsilon}, \boldsymbol{\sigma} > = \boldsymbol{\epsilon} \cdot \boldsymbol{\sigma} = \sum_{i=1}^{n} \epsilon_i \sigma_i \tag{5.126}$$

is then the inner product on $\mathbb{R}^n$.

Let $X : \mathcal{E} \to \mathbb{R}$ be a real-valued function. Its graph $\{(\boldsymbol{\epsilon}, X) \in \mathbb{R}^{n+1}\}$ is a manifold (or hypersurface) in $\mathbb{R}^{n+1}$. Similarly, one can define $Y : \mathcal{S} \to \mathbb{R}$.

Let any particular point $(\bar{\boldsymbol{\sigma}}, Y(\bar{\boldsymbol{\sigma}})) \in \mathbb{R}^{n+1}$ be called the pole. Then the linear function

$$X(\boldsymbol{\epsilon}) = \boldsymbol{\epsilon} \cdot \bar{\boldsymbol{\sigma}} - Y(\bar{\boldsymbol{\sigma}}) \tag{5.127}$$

is called the polar, which is a hyperplane in $\mathbb{R}^{n+1}$.

Thus, given a pole at a finite point, the polar is well-defined by (5.127), Conversely, given a polar of finite slope, a finite pole can be read off from Eq. (5.127). This correspondence is called the duality between points and planes.

The duality comes to live when the graph of a paraboloid is blended into the picture. The paraboloid is defined as

$$X(\boldsymbol{\epsilon}) = \frac{1}{2}\boldsymbol{\epsilon}^2 \tag{5.128}$$

Theorem 5.5 (Duality between the pole and polar). *(T1) If the pole is outside the paraboloid, the points of contact of tangents drawn from the pole to the paraboloid lie on the polar.*
(T2) If the pole is inside of the paraboloid, the polar lies outside it.

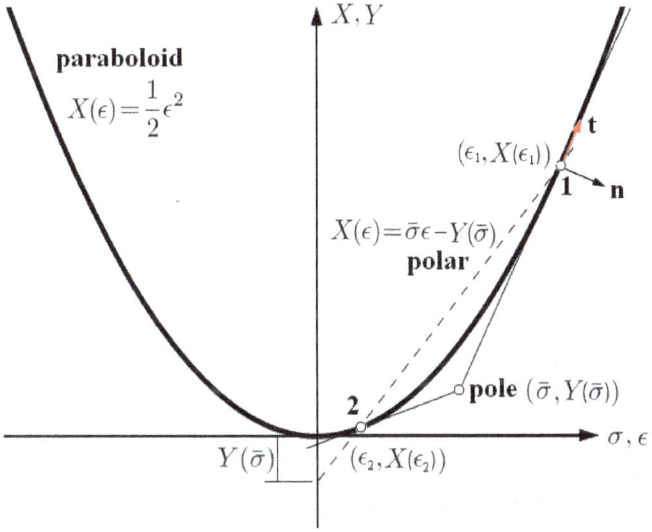

Fig. 5.8 Duality between the pole and polar.

Proof. We only prove the theorem in $\mathbb{R}^2$, which has the full flavor of a rigorous proof.

We first show statement (T1). If the pole is outside the paraboloid, there exist two points of contact of tangents drawn from the pole to the paraboloid. See Fig.5.8. For contact point 1, $(\epsilon_1, X(\epsilon_1))$, the tangential vector from the pole to the paraboloid is

$$\boldsymbol{t} = (\epsilon_1 - \bar{\sigma}, X(\epsilon_1) - Y(\bar{\sigma}))$$

the normal vector of graph $G = X - \frac{1}{2}\epsilon^2 = 0$ at point 1 is

$$\mathbf{n} = \left(-\frac{\partial G}{\partial \epsilon}, -\frac{\partial G}{\partial X}\right)\Big|_{(\epsilon_1, X(\epsilon_1))} = (\epsilon_1, -1)$$

Consider the orthogonal condition $\mathbf{t} \cdot \mathbf{n} = 0$.

$$\begin{aligned}\mathbf{t} \cdot \mathbf{n} &= (\epsilon_1 - \bar{\sigma}, X(\epsilon_1) - Y(\bar{\sigma}))(\epsilon_1, -1) \\ &= -\epsilon_1\bar{\sigma} + \epsilon_1^2 + Y(\bar{\sigma}) - X(\epsilon_1) \\ &= -\epsilon_1\bar{\sigma} + 2X(\epsilon_1) + Y(\bar{\sigma}) - X(\epsilon_1) = -\epsilon_1\bar{\sigma} + X(\epsilon_1) + Y(\bar{\sigma}) = 0\end{aligned}$$

We just showed that $X(\epsilon_1) = \epsilon_1\bar{\sigma} - Y(\bar{\sigma})$. Similarly, we can prove $X(\epsilon_2) = \epsilon_2\bar{\sigma} - Y(\bar{\sigma})$. Hence we have proved the points of contact are on the polar : $X(\epsilon) = \bar{\sigma}\epsilon - Y(\bar{\sigma})$.

We now show (T2). Suppose the pole is inside the paraboloid. We want to show that the polar is outside the paraboloid region.

Assume that part of the polar is inside or on the paraboloid, i.e.

$$X \geq \frac{1}{2}\epsilon^2$$

Since the pole is also inside the paraboloid, i.e.

$$Y(\bar{\sigma}) > \frac{1}{2}\bar{\sigma}^2$$

Therefore,

$$\begin{aligned} X + Y(\bar{\sigma}) &> \frac{1}{2}\left(\bar{\sigma}^2 + \epsilon^2\right) \\ \bar{\sigma}\epsilon &> \frac{1}{2}\left(\bar{\sigma}^2 + \epsilon^2\right) \\ 0 &> \frac{1}{2}\left(\bar{\sigma}^2 - 2\bar{\sigma}\epsilon + \epsilon^2\right) = \frac{1}{2}(\bar{\sigma} - \epsilon)^2 > 0 \end{aligned}$$

which leads to contradiction. Hence, polar must be outside the paraboloid, if the pole is inside the paraboloid. □

Definition 5.7 (Regular point and regular domain). *Let $U : \mathcal{E} \to \mathbb{R}$ be a piecewise C^2 function.*
(D1) A regular point of the function $U(\boldsymbol{\epsilon})$ is a point $\boldsymbol{\epsilon} \in E$ where the determinant of the Hessian matrix $D^2U = \{\frac{\partial^2 U}{\partial\epsilon_i\partial\epsilon_j}\}$ satisfies,

$$det\left\{\frac{\partial^2 U}{\partial\epsilon_i\partial\epsilon_j}\right\} \neq 0, \quad \text{or} \quad \pm\infty$$

(D2) A regular domain, denoted by $\mathcal{E}_r$ is a continuous subset of regular points.

Now we let $U^c : \mathbb{R}^n \to \mathbb{R}$ be a given continuous function such that the graph,

$$G_{U^c} = \{(\boldsymbol{\sigma}, Y) \in \mathbb{R}^n \;\Big|\; Y = U^c(\boldsymbol{\sigma}), \boldsymbol{\sigma} \in \mathbb{R}^n\}$$

of U^c is a continuous surface in $\mathbb{R}^{n+1}$.

When the pole, $(\boldsymbol{\sigma}, Y)$, moves on the graph of U^c, each point on G_{U^c} is corresponding to a polar hyperplane. The collective of these polars hyperplanes will envelop another continuous surface, the graph of $X = U(\boldsymbol{\epsilon})$, described as $U : \mathbb{R}^n \to \mathbb{R}$, which is the conjugate Legendre pair of $U^c(\sigma)$. This is the geometric interpretation of Legendre transformation, see Fig. 5.9. In other words, the correspondence between the functions $U(\boldsymbol{\epsilon})$ and $U^c(\boldsymbol{\sigma})$ is called Legendre transformation.

Now we state the important Legendre duality theorem.

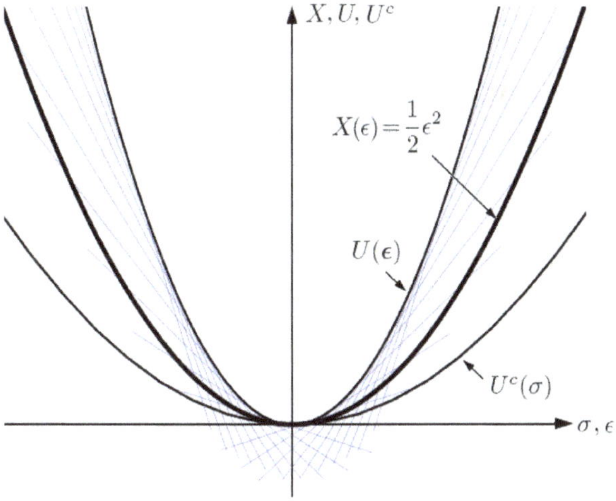

Fig. 5.9 Geometric interpretation of Legendre transformation.

Theorem 5.6 (Legendre Duality Theorem). *Let $U(\epsilon) \in C^2(\mathcal{E})$. If $\mathcal{E}_r \subset \mathcal{E}$ is an open, finite subset of the regular domain of U and $\mathcal{E}_r^* \subset \mathbb{R}^n$ is the range of the mapping $DU : \mathcal{E}_r \to \mathcal{E}^*$. Then there exists a unique C^2 function U^c on $\mathcal{E}^*$, which is dual to U on $\mathcal{E}_r$ in the sense that the Legendre duality relates*

$$U(\boldsymbol{\epsilon}) + U^c(\boldsymbol{\sigma}) = \boldsymbol{\sigma} \cdot \boldsymbol{\epsilon} \Leftrightarrow \boldsymbol{\sigma} = \partial U(\boldsymbol{\epsilon}), \Leftrightarrow \boldsymbol{\epsilon} = \partial U^c(\boldsymbol{\sigma}) \tag{5.129}$$

hold. Moreover, for $(\boldsymbol{\epsilon}, \boldsymbol{\sigma}) \in \mathcal{E}_r \times \mathcal{E}_r^*$ satisfying above relationship,

$$\frac{\partial^2 U}{\partial \epsilon_i \partial \epsilon_k} \frac{\partial^2 U^c}{\partial \sigma_k \partial \sigma_j} = \delta_{ij} \, .$$

The proof of this theorem is basically application of implicit function theorem. It is omitted here. The readers who are interested in the proof may consult an excellent monograph [Gao (2000)].

Now we move to the essential technical ingredient of convex analysis.

Theorem 5.7 (Duality between the regular manifolds). *Let U and U^c be Legendre dual functions over the duality domain $\mathcal{E}$ and $\mathcal{E}^*$ respectively.*

(S1) If U is convex on $\mathcal{E}$, U^c is convex on E^* and

$$U^c(\boldsymbol{\sigma}) = \max_{\boldsymbol{\epsilon} \in \mathcal{E}} \{\boldsymbol{\sigma} \cdot \boldsymbol{\epsilon} - U(\boldsymbol{\epsilon})\}$$

(S2) If U is concave on $\mathcal{E}$, U^c is concave on $\mathcal{E}^*$ and

$$U^c(\boldsymbol{\sigma}) = \min_{\boldsymbol{\epsilon} \in \mathcal{E}} \{\boldsymbol{\sigma} \cdot \boldsymbol{\epsilon} - U(\boldsymbol{\epsilon})\}$$

Proof. For simplicity, we only prove it for case $\mathcal{E} \subset \mathbb{R}$, which contains the essential substance of a general, rigorous proof.

Since $\sigma = \dfrac{\partial U}{\partial \epsilon}$, by Taylor expansion,

$$\sigma = \frac{\partial U}{\partial \epsilon}\Big|_{\epsilon=\bar{\epsilon}} + \frac{\partial^2 U}{\partial \epsilon^2}\Big(*\Big)(\epsilon - \bar{\epsilon}) \tag{5.130}$$

where

$$\frac{\partial^2 U}{\partial \epsilon^2}\Big(*\Big) = \frac{\partial^2 U}{\partial \epsilon^2}\Big|_{\epsilon=\bar{\epsilon}+\theta\Delta\epsilon}$$

and $0 \leq \theta \leq 1$.

Eq. (5.130) can be rewritten as

$$(\sigma - \bar{\sigma}) = \frac{\partial^2 U}{\partial \epsilon^2}\Big(*\Big)(\epsilon - \bar{\epsilon}) \tag{5.131}$$

Similarly, due to $\epsilon = \dfrac{\partial U^c}{\partial \sigma}$, one can write

$$(\epsilon - \bar{\epsilon}) = \frac{\partial^2 U^c}{\partial \sigma^2}\Big(*\Big)(\sigma - \bar{\sigma}) \tag{5.132}$$

where

$$\frac{\partial^2 U^c}{\partial \sigma^2}\Big(*\Big) = \frac{\partial^2 U^c}{\partial \sigma^2}\Big|_{\epsilon=\bar{\sigma}+\theta\Delta\sigma}$$

and $0 \leq \theta \leq 1$, $\Delta\sigma = \sigma - \bar{\sigma}$. Therefore,

$$\begin{aligned}(\sigma - \bar{\sigma})(\epsilon - \bar{\epsilon}) &= \frac{\partial^2 U}{\partial \epsilon^2}\Big(*\Big)(\epsilon - \bar{\epsilon})^2 \\ &= \frac{\partial^2 U^c}{\partial \sigma^2}\Big(*\Big)(\sigma - \bar{\sigma})^2\end{aligned} \tag{5.133}$$

Eq. (5.133) indicates that if $\dfrac{\partial^2 U}{\partial \epsilon^2}\Big(*\Big)$ is positive definite, $\dfrac{\partial^2 U^c}{\partial \sigma^2}\Big(*\Big)$ is also positive; whereas if $\dfrac{\partial^2 U}{\partial \epsilon^2}\Big(*\Big)$ is negative definite, $\dfrac{\partial^2 U^c}{\partial \sigma^2}\Big(*\Big)$ is also negative definite, or both being indefinite.

To prove the Legendre inequality, we consider a special 1D example, $U(\epsilon) = \dfrac{1}{2}k_0\epsilon^2$, with $k_0 > 0$. For a given point $\bar{\epsilon}$ on horizontal axis, the associated stress $\bar{\sigma} = k_0\bar{\epsilon}$ is the slope of the polar, the straight line $X = \bar{\sigma}\epsilon - Y(\bar{\sigma})$, which is tangent to the graph of U at $\bar{\epsilon}$ (see Fig. (5.10).

Therefore, the point $(\bar{\epsilon}, U(\bar{\epsilon})$ is in both polar $X = \bar{\sigma}\epsilon - Y$ and on $U = 1/2k_0\epsilon^2$, which is to say that $X(\bar{\epsilon}) = U(\bar{\epsilon})$ and

$$Y = \bar{\sigma}\bar{\epsilon} - U(\bar{\epsilon}) =: U^c(\bar{\sigma})$$

For any given $\epsilon \in \mathcal{E}_r$, we define a continuous function,

$$y(\epsilon) = \bar{\sigma}\epsilon - U(\epsilon)$$

we want to show that $Y = U^c(\bar{\sigma}) \geq y(\epsilon)$.

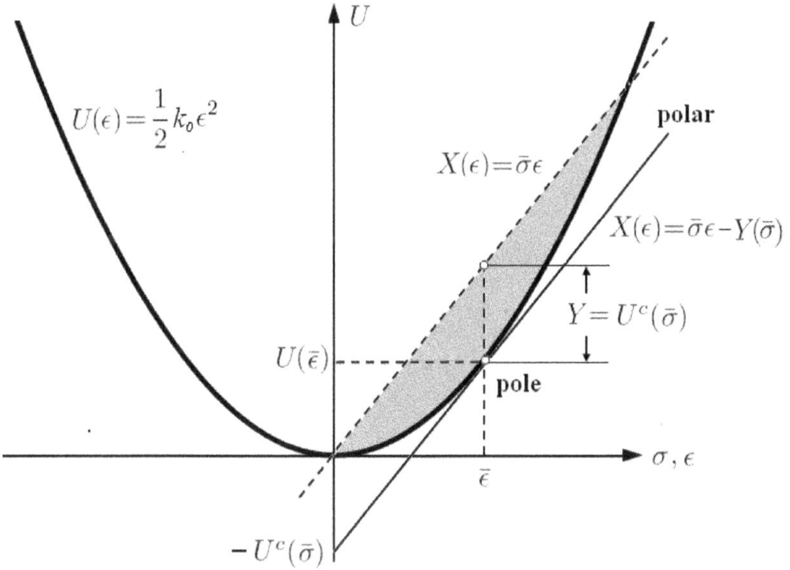

Fig. 5.10 Graphic illustration of Legendre transformation.

Since the polar $X(\epsilon)$ is always below the parabola, e.g. $U(\epsilon) \geq X(\epsilon)$,

$$U(\epsilon) - X \geq 0 \Rightarrow U(\epsilon) - (\bar{\sigma}\epsilon - Y) \geq 0$$
$$\Rightarrow Y \geq \bar{\sigma}\epsilon - U(\epsilon)$$

Since $U(\epsilon)$ is convex, $y(\epsilon)$ is then concave because $\dfrac{\partial^2 y}{\partial \epsilon^2} < 0$. It then takes its maximum value at $\bar{\epsilon}$ because $y'(\bar{\epsilon}) = 0$. That is

$$Y = U^c(\bar{\sigma}) = \max_{\epsilon \in \mathcal{E}_r} \{\bar{\sigma}\epsilon - U(\epsilon)\} \tag{5.134}$$

One drop the bar on σ, because domain of $\bar{\sigma}$ is the same as σ.

Similarly, for concave functions, one can show that

$$U^c(\sigma) = \min_{\epsilon \in \mathcal{E}_r} \{\sigma\epsilon - U(\epsilon)\}$$

□

In an infinite-dimension functional space E, Eq. (5.134) is called the Legendre-Fenchel transformation , and it reads as

$$U^*(\sigma) = \sup_{\epsilon \in \mathcal{E}} \{\sigma\epsilon - U(\epsilon)\}$$

where the superscript $*$ replaces the superscript c to emphasize its meaning as the dual function.

Accordingly, if U is concave, its Legendre-Fenchel conjugate is defined as

$$U_*(\sigma) = \inf_{\epsilon \in \mathcal{E}} \{\sigma\epsilon - U(\epsilon)\}$$

The reason that we add the name Fenchel is because when U is defined as

$$U : \mathcal{E} \to \mathbb{R} \cup \{+\infty\} \cup \{-\infty\} \tag{5.135}$$

and the Fenchel transformations

$$U^*(\sigma) = \sup_{\epsilon \in \mathcal{E}} \{\sigma\epsilon - U(\epsilon) \quad \text{and} \quad U_*(\sigma) = \inf_{\epsilon \in \mathcal{E}} \{\sigma \cdot \epsilon - U(\epsilon)\}$$

are defined as

$$U^* : \mathcal{S} \to \mathbb{R} \cup \{+\infty\} \quad \text{and} \quad U_* : \mathcal{S} \to \mathbb{R} \cup \{-\infty\}$$

The Fenchel duality plays an important role in homogenization of non-linear composites. Let $\{(\mathcal{E}, \mathcal{S}), \langle \cdot, \cdot \rangle\}$ be a duality paring, where $\mathcal{E}$ and $\mathcal{S}$ are two real vector spaces placed in duality with the bi-linear form $\langle \cdot, \cdot \rangle : \mathcal{E} \times \mathcal{S} \to \mathbb{R}$. On $\mathcal{E}$, any continuous affine functional can be written as

$$\ell(\sigma) = < \epsilon, \sigma > -\alpha, \quad \sigma \in S, \quad \alpha \in \mathbb{R} \tag{5.136}$$

Given

$$P : \mathcal{U} \to \mathbb{R} \cup \{+\infty\} \cup \{-\infty\}$$

For a fixed $q \in S$, ℓ belongs to a family of continuous affine functionals such that

$$\ell(\sigma) \leq P(\epsilon) \;\; \& \;\; \alpha \geq \langle \epsilon, \sigma \rangle - P(\epsilon), \quad \forall \epsilon \in \mathcal{U} \;\; \text{and} \;\; \alpha \in \mathbb{R} \tag{5.137}$$

For fixed σ, the $\ell(\sigma)$ corresponds to the lowest value of α can be written as

$$P^*(\sigma) = \sup_{\epsilon \in \mathcal{E}} \{< \epsilon, \sigma > -P(\epsilon)\} \tag{5.138}$$

and by definition,

$$P(\epsilon) \geq \sup_{\sigma \in S} \{< \epsilon, \sigma > -P^*(\sigma)\} \tag{5.139}$$

As has been shown in the previous Section, the equality holds if $P(\epsilon)$ is convex. In this case, $P^* : \to \mathbb{R} \cup \{+\infty\}$ is also convex, and it is called *the super-conjugate functional* of P.

Similarly, if

$$\ell(\sigma) \geq P(\epsilon), \;\; \& \;\; \alpha \leq \langle \epsilon, \sigma \rangle - P(\epsilon), \;\; \forall \epsilon \in \mathcal{E} \tag{5.140}$$

For fixed σ the $\ell(\sigma)$ of the highest value of α can be written as

$$P_*(\sigma) = \inf_{\epsilon \in \mathcal{E}} \{< \epsilon, \sigma > -P(\epsilon)\} \tag{5.141}$$

and

$$P(\epsilon) \leq \langle \epsilon, \sigma \rangle - P_*(\sigma) \tag{5.142}$$

The equality holds only if $P(\epsilon)$ is concave. In that case, $P_* : S \to \mathbb{R} \cup \{-\infty\}$ is also concave, and we call it *the sub-conjugate functional* of P.

5.7 Legendre-Fenchel transformation in linear elasticity

In a classical paper [Hill (1963)] , R. Hill illustrated how to use the Legendre-Fenchel transformation in variational principles of linear elastic system, and he applied them to study micromechanics problems.

Consider the prescribed displacement boundary condition (prescribed macro strain condition),

$$\boldsymbol{u}^0 = \mathbf{x} \cdot \boldsymbol{\epsilon}^0, \quad \forall \mathbf{x} \in \partial V$$

Under such a condition, we have shown previously that

$$\boldsymbol{\epsilon}^0 = < \boldsymbol{\epsilon} >_V = < \boldsymbol{\epsilon}^* >_V, \quad \forall \boldsymbol{\epsilon} \in \mathcal{E}$$

where $\mathcal{E}$ is the space of compatible strain, $\sigma^*, \boldsymbol{\epsilon}^*$ are the real stress/strain solution.

Therefore, the potential energy and complementary energy take the form

$$\begin{aligned}\Pi^c(\boldsymbol{\sigma}) &= \frac{1}{2}\int_V D_{ijk\ell}\sigma_{ij}\sigma_{k\ell}dV - \int_{\partial V} x_k \epsilon^0_{ki}\sigma_{ij}n_j dS \\ &= \frac{1}{2}\int_V D_{ijk\ell}\sigma_{ij}\sigma_{k\ell}dV - \int_V \Big[\delta_{kj}\epsilon^0_{ki}\sigma_{ij} + \underbrace{x_k\epsilon^0_{ki}\sigma_{ij,j}}_{=0}\Big]dV \\ &= \frac{1}{2}\int_V D_{ijk\ell}\sigma_{ij}\sigma_{k\ell}dV - \int_V \epsilon^0_{ij}\sigma_{ij}dV\end{aligned}$$

Based on the minimum complementary energy principle, for any statically admissible stress field, $\forall \boldsymbol{\sigma} \in \mathcal{S}$, $\Pi^c(\boldsymbol{\sigma}) \geq \Pi^c(\boldsymbol{\sigma}^*)$, where

$$\begin{aligned}\Pi^c(\boldsymbol{\sigma}^*) &= \frac{1}{2}\int_V D_{ijk\ell}\sigma^*_{ij}\sigma^*_{k\ell}dV - \int_{\partial V} u^0_i\sigma^*_{ij}n_j dS \\ &= \frac{1}{2}\int_V D_{ijk\ell}\sigma^*_{ij}\sigma^*_{k\ell}dV - \int_V \epsilon^0_{ij}\sigma^*_{ij}dV = -\frac{1}{2}\int_V C_{ijk\ell}\epsilon^*_{ij}\epsilon^*_{k\ell}dV\end{aligned}$$

In the last line, the following equality under prescribed macro-strain B.C. is used,

$$\int_V \epsilon^0_{ij}\sigma^*_{ij}dV = \int_V \epsilon^*_{ij}\sigma^*_{ij}dV$$

Therefore,

$$\frac{1}{2}\int_V C_{ijk\ell}\epsilon^*_{ij}\epsilon^*_{k\ell}dV \geq \int_V \sigma_{ij}\epsilon^0_{ij}dV - \frac{1}{2}\int_V D_{ijk\ell}\sigma_{ij}\sigma_{k\ell}dV$$

which is essentially

$$W(\boldsymbol{\epsilon}^*) = \sup_{\boldsymbol{\sigma}\in\mathcal{S}} \left\{\boldsymbol{\epsilon}^0 :< \boldsymbol{\sigma} >_V - W^*(\boldsymbol{\sigma})\right\} \tag{5.143}$$

where

$$\begin{aligned}W(\boldsymbol{\epsilon}) &= \frac{1}{2V}\int_V C_{ijk\ell}\epsilon_{ij}\epsilon_{k\ell}dV \\ W^*(\boldsymbol{\sigma}) &= \frac{1}{2V}\int_V D_{ijk\ell}\sigma_{ij}\sigma_{k\ell}dV\end{aligned}$$

One may further tighten the bound

$$W(\epsilon^*) = \sup_{\boldsymbol{\sigma}\in\mathcal{S}} \{\boldsymbol{\epsilon}^0 :< \boldsymbol{\sigma} >_V - W^*(\boldsymbol{\sigma}^*)\} \tag{5.144}$$

Remark 5.7. **(1)** Note that Eq. (5.143) looks like Legendre-Fenchel transformation . However, there is a subtle difference. If W is a convex functional of $\boldsymbol{\epsilon} \in \mathcal{E}$, the Legendre-Fenchel transformation assures that

$$W^*(\boldsymbol{\sigma}) = \sup_{\boldsymbol{\epsilon}\in\mathcal{E}} \{\boldsymbol{\sigma} : \boldsymbol{\epsilon} - W(\boldsymbol{\epsilon})\}$$

If the space $\mathcal{E} = \mathcal{E}^{**}$ is reflexive (all the $L^p(V)$ spaces are reflexive, see e.g. [Rudin (1991)], the inverse Legendre-Fenchel transformation exists,

$$W(\boldsymbol{\epsilon}) = (W^*)^*(\boldsymbol{\epsilon}) = W^{**}(\boldsymbol{\epsilon}) = \sup_{\boldsymbol{\sigma}\in\mathcal{S}} \{\boldsymbol{\epsilon} : \boldsymbol{\sigma} - W^*(\boldsymbol{\sigma})\}$$

(2) Choose

$$\boldsymbol{\sigma} =< \boldsymbol{\sigma} >_V = \sum_{\alpha=0}^{n} \frac{1}{V}\int_{\Omega_\alpha} \Big(\mathbb{C}^\alpha : \boldsymbol{\epsilon}^0\Big) dV = \sum_{\alpha=0}^{n} f_\alpha \mathbb{C} : \boldsymbol{\epsilon}^0 \ .$$

One can show that

$$\frac{1}{2}\boldsymbol{\epsilon}^0 : \bar{\mathbb{C}} : \boldsymbol{\epsilon}^0 \geq \boldsymbol{\epsilon}^0 : \left\{ \Big(\sum_{\alpha=0}^{n} f_\alpha \mathbb{C}^\alpha\Big) - \frac{1}{2}\Big(\sum_{\alpha=0}^{n} f_\alpha \mathbb{C}^\alpha\Big) : \Big(\sum_{\alpha=0}^{n} f_\alpha \mathbb{D}^\alpha\Big) : \Big(\sum_{\alpha=0}^{n} f_\alpha \mathbb{C}^\alpha\Big) \right\} : \boldsymbol{\epsilon}^0$$

Hence

$$\bar{\mathbb{C}} \geq 2\Big(\sum_{\alpha=0}^{n} f_\alpha \mathbb{C}^\alpha\Big) : \left\{ \mathbb{I}^{(4s)} - \frac{1}{2}\Big(\sum_{\alpha=0}^{n} f_\alpha \mathbb{D}^\alpha\Big) : \Big(\sum_{\alpha=0}^{n} f_\alpha \mathbb{C}^\alpha\Big) \right\} \tag{5.145}$$

where $\mathbb{D}^\alpha = (\mathbb{C}^\alpha)^{-1}$, and (5.145) is referred to as the Sachs bound.

5.8 Talbot-Willis variational principles

Hashin-Shtrikman variational principles are constructed for *linear* composite materials. In a series papers [Talbot and Willis (1985, 1987, 1998)] , Talbot and Willis generalized to a class of *nonlinear* composite materials.

Consider a composite with nonlinear strain potential energy density, $U(\boldsymbol{\epsilon})$,

$$\begin{aligned}
\nabla \cdot \boldsymbol{\sigma} &= \mathbf{0}, \\
\boldsymbol{\sigma} &= \partial_{\boldsymbol{\epsilon}} U, \\
\boldsymbol{\epsilon} &= \frac{1}{2}(\nabla \otimes \boldsymbol{u} + (\nabla \otimes \boldsymbol{u})^T) \\
\boldsymbol{u} &= \mathbf{x} \cdot \bar{\boldsymbol{\epsilon}}, \quad \forall \mathbf{x} \in \partial V \quad (\Gamma_t = \emptyset)
\end{aligned}$$

Consider a *linear homogeneous* comparison solid,

$$\nabla \cdot \boldsymbol{\sigma}^{(0)} = \mathbf{0}, \tag{5.146}$$

$$\boldsymbol{\sigma}^{(0)} = \partial_{\boldsymbol{\epsilon}^{(0)}} U^{(0)}, \tag{5.147}$$

$$\boldsymbol{\epsilon}^{(0)} = \frac{1}{2}(\nabla \otimes \boldsymbol{u}^{(0)} + (\nabla \otimes \boldsymbol{u}^{(0)})^T) \tag{5.148}$$

$$\boldsymbol{u}^{(0)} = \mathbf{x} \cdot \bar{\boldsymbol{\epsilon}}, \quad \forall \mathbf{x} \in \partial V \quad (\Gamma_t = \emptyset) \tag{5.149}$$

Compare the difference in potential energy density between the composite and the comparison solid,

$$\mathcal{U}(\boldsymbol{\epsilon}) = U(\boldsymbol{\epsilon}) - U^{(0)}(\boldsymbol{\epsilon}) \tag{5.150}$$

Depending on whether the difference potential is a convex function or a concave function, we define

$$\mathcal{U}_p(\boldsymbol{\epsilon}) := U(\boldsymbol{\epsilon}) - U^{(0)}(\boldsymbol{\epsilon}), \quad \text{if} \quad \partial^2_{\boldsymbol{\epsilon}}\mathcal{U} > 0 \tag{5.151}$$

$$\mathcal{U}^p(\boldsymbol{\epsilon}) := U(\boldsymbol{\epsilon}) - U^{(0)}(\boldsymbol{\epsilon}), \quad \text{if} \quad \partial^2_{\boldsymbol{\epsilon}}\mathcal{U} < 0 \tag{5.152}$$

Consider the following kinematic decomposition,

$$\boldsymbol{u} = \boldsymbol{u}^{(0)} + \boldsymbol{u}^d \tag{5.153}$$

$$\boldsymbol{\epsilon} = \boldsymbol{\epsilon}^{(0)} + \boldsymbol{\epsilon}^d \tag{5.154}$$

Assume that the stress and strain fields in the comparison solid are uniquely determined by the boundary condition. The total potential energy difference is a functional of ϵ^d, i.e.

$$\Pi_p(\boldsymbol{\epsilon}^d) = V\Big(W(\boldsymbol{\epsilon}^d) - W^{(0)}(\boldsymbol{\epsilon}^d)\Big) = \int_V \mathcal{U}_p(\boldsymbol{\epsilon}^d)dV \tag{5.155}$$

$$\Pi^p(\boldsymbol{\epsilon}^d) = V\Big(W(\boldsymbol{\epsilon}^d) - W^{(0)}(\boldsymbol{\epsilon}^d)\Big) = \int_V \mathcal{U}^p(\boldsymbol{\epsilon}^d)dV \tag{5.156}$$

where

$$W(\boldsymbol{\epsilon}^d) = \frac{1}{2V}\int_V U(\boldsymbol{\epsilon}^d)dV$$

$$W^{(0)}(\boldsymbol{\epsilon}^d) = \frac{1}{2V}\int_V U^{(0)}(\boldsymbol{\epsilon}^d)dV$$

Obviously, $\Pi_p(\boldsymbol{\epsilon})$ is convex and $\Pi^p(\boldsymbol{\epsilon})$ is concave.

Define stress polarization

$$p_{ij} = \frac{\partial \mathcal{U}}{\partial \epsilon^d_{ij}} \tag{5.157}$$

Subsequently, we can form the following Legendre-Fenchel transformation,

$$\Pi^*_p = \sup_{\epsilon^d \in \mathcal{E}} \left\{ < \boldsymbol{p} : \boldsymbol{\epsilon}^d > - \Pi_p(\boldsymbol{\epsilon}^d) \right\} V \tag{5.158}$$

$$\Pi^{p*} = \inf_{\epsilon^d \in \mathcal{E}} \left\{ < \boldsymbol{p} : \boldsymbol{\epsilon}^d > - \Pi^p(\boldsymbol{\epsilon}^d) \right\} V \tag{5.159}$$

where

$$< \boldsymbol{p} : \boldsymbol{\epsilon}^d > = \frac{1}{V}\int_V \boldsymbol{p} : \boldsymbol{\epsilon}^d dV$$

and

$$\mathcal{E} := \left\{ \epsilon_{ij} \;\middle|\; \epsilon_{ij} \in L^2(V), \epsilon_{ij} = \frac{1}{2}(u_{i,j} + u_{j,i}), \text{ and } u_i \in \overset{\circ}{\mathcal{V}} \right\}$$

$$\mathcal{V} := \left\{ u_i \;\middle|\; u_i \in L^2(V), W(u_{i,j}), W^{(0)}(u_{i,j}) < \infty, \; u_i = x_j \bar{\epsilon}_{ij}, \; \forall \mathbf{x} \in \partial V \right\}$$

$$\overset{\circ}{\mathcal{V}} := \left\{ u_i \;\middle|\; u_i \in L^2(V), W(u_{i,j}), W^{(0)}(u_{i,j}) < \infty, \; u_i = 0, \; \forall \mathbf{x} \in \partial V \right\}$$

In plain terms, Eqs. (5.158) and (5.159) are just

$$\Pi^*_p(\boldsymbol{p})/V = (W - W^{(0)})^*(\boldsymbol{p}) = \sup_{\boldsymbol{\epsilon}^d \in \mathcal{E}} \Big\{ < \boldsymbol{p} : \boldsymbol{\epsilon}^d > -(W - W^{(0)})(\boldsymbol{\epsilon}^d) \Big\}, \tag{5.160}$$

when $\partial^2_{\boldsymbol{\epsilon}} \mathcal{U} > 0$, and $(W - W^{(0)})$ is convex;

$$\Pi^{p*}(\boldsymbol{p})/V = (W - W^{(0)})^*(\boldsymbol{p}) = \inf_{\boldsymbol{\epsilon}^d \in \mathcal{E}} \Big\{ < \boldsymbol{p} : \boldsymbol{\epsilon}^d > -(W - W^{(0)})(\boldsymbol{\epsilon}^d) \Big\}, \tag{5.161}$$

when $partial^2_{\boldsymbol{\epsilon}} \mathcal{U} < 0$, and $(W - W^{(0)})$ is concave.

Now we consider

(**1**) Assume $\partial^2_{\boldsymbol{\epsilon}} \mathcal{U} > 0$. From Eq. (5.160)

$$\Pi^*_p(\boldsymbol{p})/V \geq \Big(< \boldsymbol{p} : \boldsymbol{\epsilon}^d > -W(\boldsymbol{\epsilon}^d) + W^{(0)}(\boldsymbol{\epsilon}^d) \Big)$$

$$\Rightarrow W(\boldsymbol{\epsilon}^d) \geq \{ < \boldsymbol{p} : \boldsymbol{\epsilon}^d > +W^{(0)}(\boldsymbol{\epsilon}^d) \} - \Pi^*_p(\boldsymbol{p})/V$$

Take an infimum through the both sides of the inequality,

$$\inf_{\boldsymbol{\epsilon}^d \in \mathcal{E}} W(\boldsymbol{\epsilon}^d) \geq \inf_{\boldsymbol{\epsilon}^d \in E} \{ < \boldsymbol{p} : \boldsymbol{\epsilon}^d > +W^{(0)}(\boldsymbol{\epsilon}^d) \} - \Pi^*_p(\boldsymbol{p})/V \tag{5.162}$$

(**2**) Assume $\partial^2_{\boldsymbol{\epsilon}} \mathcal{U} < 0$. From Eq. (5.161)

$$\Pi^{p*}(\boldsymbol{p})/V \leq \Big(< \boldsymbol{p} : \boldsymbol{\epsilon}^d > -W(\boldsymbol{\epsilon}^d) + W^{(0)}(\boldsymbol{\epsilon}^d) \Big)$$

$$\Rightarrow W(\boldsymbol{\epsilon}^d) \leq \{ < \boldsymbol{p} : \boldsymbol{\epsilon}^d > +W^{(0)}(\boldsymbol{\epsilon}^d) \} - \Pi^*_p(\boldsymbol{p})/V$$

Take an infimum through the both sides of the above inequality

$$\inf_{\boldsymbol{\epsilon}^d \in \mathcal{E}} W(\boldsymbol{\epsilon}^d) \leq \inf_{\boldsymbol{\epsilon}^d \in \mathcal{E}} \{ < \boldsymbol{p} : \boldsymbol{\epsilon}^d > +W^{(0)}(\boldsymbol{\epsilon}^d) \} - \Pi^{p*}(\boldsymbol{p})/V \tag{5.163}$$

The statement of the prime variational principle is

$$\text{(The primal problem)} \;\; \mathcal{P} : \inf_{\boldsymbol{\epsilon}^d \in \mathcal{E}} W(\boldsymbol{\epsilon}^d)$$

Combining Eqs. (5.163) and (5.163), we have the original form of Talbot-Willis variational principle

$$\begin{aligned} &\inf_{\boldsymbol{\epsilon}^d \in \mathcal{E}} \{ < \boldsymbol{p} : \boldsymbol{\epsilon}^d > +W^{(0)}(\boldsymbol{\epsilon}^d) \} - \Pi^*_p(\boldsymbol{p})/V \\ &\qquad \leq \inf_{\boldsymbol{\epsilon}^d \in \mathcal{E}} W(\boldsymbol{\epsilon}^d) \leq \\ &\inf_{\boldsymbol{\epsilon}^d \in \mathcal{E}} \{ < \boldsymbol{p} : \boldsymbol{\epsilon}^d > +W^{(0)}(\boldsymbol{\epsilon}^d) \} - \Pi^{p*}(\boldsymbol{p})/V \end{aligned} \tag{5.164}$$

which is the generalization of Hashin-Shtrikman principle.

If both the original composite and the comparison solid are *linear* elastic materials, we easily calculate,

$$\begin{aligned} \Pi^*_p(\boldsymbol{p})/V \;\; \Big(\text{or } \Pi^{p*}(\boldsymbol{p})/V\Big) &= \frac{1}{V}\int_V \Big(\epsilon^d_{ij} p_{ij} - \frac{1}{2}\Delta C_{ijk\ell}\epsilon_{ij}\epsilon_{k\ell} \Big) dV \\ &= \frac{1}{V}\int_V \Big((\epsilon_{ij} - \epsilon^{(0)}_{ij}) p_{ij} - \frac{1}{2} p_{ij}\epsilon_{ij} \Big) dV \\ &= \frac{1}{2V}\int_V \Big(\epsilon_{ij} p_{ij} - 2\epsilon^{(0)}_{ij} p_{ij} \Big) dV \\ &= \frac{1}{2V}\int_V \Big(\Delta C^{-1}_{ijk\ell} p_{ij} p_{k\ell} - 2\epsilon^{(0)}_{ij} p_{ij} \Big) dV \end{aligned}$$

Denote

$$\underline{I}(\boldsymbol{\epsilon}^d, \boldsymbol{p}) = \inf_{\boldsymbol{\epsilon}^d \in \mathcal{E}} \{< \boldsymbol{p} : \boldsymbol{\epsilon}^d > + W^{(0)}(\boldsymbol{\epsilon}^d)\} - \Pi_p^*(\boldsymbol{p})/V \tag{5.165}$$

$$\bar{I}(\boldsymbol{\epsilon}^d, \boldsymbol{p}) = \inf_{\boldsymbol{\epsilon}^d \in \mathcal{E}} \{< \boldsymbol{p} : \boldsymbol{\epsilon}^d > + W^{(0)}(\boldsymbol{\epsilon}^d)\} - \Pi^{p*}(\boldsymbol{p})/V \tag{5.166}$$

We can find that

$$\begin{aligned}
\underline{I} \text{ (or } \bar{I}) &= \frac{1}{V}\int_V \Big(p_{ij}\epsilon^d_{ij} + \frac{1}{2}C^{(0)}_{ijk\ell}\epsilon_{k\ell}(\epsilon^{(0)}_{ij} + \epsilon^d_{ij}) - \frac{1}{2}\Delta C^{-1}_{ijk\ell}p_{ij}p_{k\ell} + \epsilon^{(0)}_{ij}p_{ij}\Big)dV \\
&= \underbrace{\frac{1}{2V}\int_V \Big(p_{ij} + C^{(0)}_{ijkl}\epsilon_{k\ell}\Big)\epsilon^d_{ij}dV}_{=0} + \frac{1}{2V}\int_V C^{(0)}_{ijk\ell}(\epsilon^{(0)}_{k\ell} + \epsilon^d_{k\ell})\epsilon^{(0)}_{ij}dV \\
&+ \frac{1}{V}\int_V \Big(\frac{1}{2}\epsilon^d_{ij}p_{ij} - \frac{1}{2}\Delta C^{-1}_{ijk\ell}p_{ij}p_{k\ell} + \epsilon^{(0)}_{ij}p_{ij}\Big)dV = \underbrace{\frac{1}{2V}\int_V C^{(0)}_{ijk\ell}\epsilon^{(0)}_{k\ell}\epsilon^d_{ij}dV}_{=0} \\
&+ \frac{1}{V}\int_V \Big(\frac{1}{2}C^{(0)}_{ijk\ell}\epsilon^{(0)}_{ij}\epsilon^{(0)}_{k\ell} + \frac{1}{2}\epsilon^d_{ij}p_{ij} - \frac{1}{2}\Delta C^{-1}_{ijk\ell}p_{ij}p_{k\ell} + \epsilon^{(0)}_{ij}p_{ij}\Big)dV
\end{aligned}$$

Hence

$$\begin{aligned}
\underline{I}, \text{ (or } \bar{I}) &= \frac{1}{V}\int_V \Big(\frac{1}{2}C^{(0)}_{ijk\ell}\epsilon^{(0)}_{ij}\epsilon^{(0)}_{k\ell} + \frac{1}{2}\epsilon^d_{ij}p_{ij} - \frac{1}{2}\Delta C^{-1}_{ijk\ell}p_{ij}p_{k\ell} + \epsilon^{(0)}_{ij}p_{ij}\Big)dV \\
&= W^{(0)}(\boldsymbol{\epsilon}^{(0)}) + \underline{R_\pi} \text{ (or } \bar{R}_\pi)
\end{aligned}$$

where

$$\underline{R_\pi}, \text{ (or} \bar{R}_\pi) = \frac{1}{2V}\int_V \Big(-\Delta C^{-1}_{ijk\ell}p_{ij}p_{k\ell} + p_{ij}\epsilon^d_{ij} + 2p_{ij}\epsilon^{(0)}_{ij}\Big)dV$$

We then recover the Hashin-Shtrikman variational principle

$$\underline{R_\pi}(\boldsymbol{p}, \boldsymbol{\epsilon}^d) \le \inf_{\boldsymbol{\epsilon}^d \in \mathcal{E}} \Big(W(\boldsymbol{\epsilon}^d) - W^{(0)}(\boldsymbol{\epsilon}^{(0)})\Big) \le \bar{R}_\pi(\boldsymbol{p}, \boldsymbol{\epsilon}^d)$$

5.9 Ponte Castañeda variational principle

We now discuss the Ponte Castañeda variational principle, which is a variational principle for a class of nonlinear composites. The main difference between the Talbot-Willis principles and the Ponte Castañeda principle is the former using a non-linear comparison solid whereas the latter using a linear comparison solid.

5.9.1 *Effective property and nonlinear potential*

Around 1991, P. Ponte Castañeda [Ponte Castañeda (1991, 1992a,b)] proposed new variational principle using a *linear heterogenous* comparison solid to gauge the overall effective material properties of a class of *nonlinear* composite. Distinct from previous variational principles, the proposed linear comparison solid has the same phase structure as the composite, and therefore is called the linear *comparison composite*. The resulted Ponte Castañeda variational principle is one of the few approaches that provide rigorous bound estimate on effective properties of nonlinear materials. In the following, we briefly introduce Ponte Castañeda's variational principle for a specific class of nonlinear composites that follow power-law constitutive relationships. Readers are encouraged to consult the literature on its latest developments and applications to general nonlinear systems.

Consider a nonlinear composite with N isotropic phases that each occupies a domain Ω_r, $r = 1, 2, \cdots, N$. The composite occupies the domain $V = \bigcup_{r=1}^{N} \Omega_r$. Assume the response of the phases is linear for pure hydrostatic loadings and nonlinear in shear, therefore the constitutive relationship for each phase may be characterized by the following potentials,

$$w_r(\mathbf{x}, \boldsymbol{\epsilon}) = \chi_r(\mathbf{x})\Big(\frac{9}{2}k_r(\mathbf{x})\epsilon_m^2 + \varphi_r(\mathbf{x}, \epsilon_{eq})\Big) \tag{5.167}$$

where χ_r is the characteristic function of sub-domain Ω_r; $k_r(\mathbf{x})$ is the bulk modulus of the phase r, which is linear but may not be homogeneous; and φ_r is the deviatoric potential for phase r. The volumetric strain ϵ_m and the equivalent strain $\boldsymbol{\epsilon}_{dev}$ are defined as

$$\epsilon_m := \frac{1}{3}tr(\boldsymbol{\epsilon}), \quad \text{and} \quad \epsilon_{eq} := \Big(\frac{2}{3}\boldsymbol{\epsilon}_{dev} : \boldsymbol{\epsilon}_{dev}\Big)^{1/2}, \tag{5.168}$$

with $\boldsymbol{\epsilon}_{dev} = \boldsymbol{\epsilon} - \epsilon_m \mathbb{I}^{(2)}$. Because the phases are locally isotropic, the nonlinear potential can be expressed in terms of stress invariants. Here we mainly consider the following power-law type of deviatoric potential

$$\varphi(\epsilon_{eq}) = \frac{\sigma_0 \epsilon_0}{m+1}\Big(\frac{\epsilon_{eq}}{\epsilon_0}\Big)^{m+1}, \quad 0 \le m \le 1 \tag{5.169}$$

in which ϵ_0 and σ_0 are reference strain and stress.

Alternatively, the constitutive information of the r-th phase material may be drawn from the following stress potential,

$$\varpi_r^c(\boldsymbol{\sigma}) = \chi_r(\mathbf{x})\big(\frac{1}{2k_r}\sigma_m^2 + \psi(\sigma_{eq})\big) \tag{5.170}$$

where the hydrostatic stress and the equivalent stress are defined as

$$\sigma_m = \frac{1}{3}tr(\boldsymbol{\sigma}), \quad \text{and} \quad \sigma_{eq} = \Big(\frac{3}{2}\boldsymbol{\sigma}_{dev} : \boldsymbol{\sigma}_{dev}\Big)^{1/2} \tag{5.171}$$

with $\boldsymbol{\sigma}_{dev} = \boldsymbol{\sigma} - \sigma_m \mathbb{I}^{(2)}$. The deviatoric stress potential takes the form,

$$\psi(\sigma_{eq}) = \frac{\sigma_0 \epsilon_0}{n+1}\Big(\frac{\sigma_{eq}}{\sigma_0}\Big)^{1+n}, \quad n = \frac{1}{m} \tag{5.172}$$

As a reminder, we use the following notations,

$$\bar{\boldsymbol{\epsilon}}_r := \frac{1}{\Omega_r}\int_\Omega \chi_r(\mathbf{x})\boldsymbol{\epsilon}(\mathbf{x})dV, \quad \text{and} \quad \bar{\boldsymbol{\sigma}}_r := \frac{1}{\Omega_r}\int_\Omega \chi_r(\mathbf{x})\sigma(\mathbf{x})dV, \tag{5.173}$$

and

$$f_r := \frac{\Omega_r}{V} = \frac{1}{V}\int_V \chi_r(\mathbf{x})dV = \langle \chi_r \rangle \tag{5.174}$$

$$\boldsymbol{\mathcal{E}} = \langle \boldsymbol{\epsilon} \rangle = \frac{1}{V}\int_V \boldsymbol{\epsilon} dV = \sum_{r=1}^{N} f_r \bar{\boldsymbol{\epsilon}}_r \tag{5.175}$$

$$\boldsymbol{\Sigma} = \langle \boldsymbol{\sigma} \rangle = \frac{1}{V}\int_V \boldsymbol{\sigma} dV = \sum_{r=1}^{N} f_r \bar{\boldsymbol{\sigma}}_r \tag{5.176}$$

The total potential function densities of the composite can be written

$$w(\mathbf{x}, \boldsymbol{\epsilon}) = \sum_{r=1}^{N} \chi_r(\mathbf{x}) w_r(\boldsymbol{\epsilon}), \quad \text{and} \quad w^c(\mathbf{x}, \boldsymbol{\epsilon}) = \sum_{r=1}^{N} \chi_r(\mathbf{x}) w_r^c(\boldsymbol{\epsilon}) \tag{5.177}$$

The average potential and complementary potential energies are respectively

$$W(\boldsymbol{\epsilon}) := \langle w(\mathbf{x}, \boldsymbol{\epsilon}) \rangle_V, \quad \text{and} \quad W^c(\boldsymbol{\sigma}) := \langle w^c(\mathbf{x}, \boldsymbol{\sigma}) \rangle_V, \tag{5.178}$$

Consider the prescribed displacement boundary condition and the prescribed traction boundary condition for an RVE. The minimum potential energy principle can be stated as

$$W(\boldsymbol{\epsilon}^*) = \inf_{\boldsymbol{u} \in \mathcal{V}(\boldsymbol{\mathcal{E}})} W(\boldsymbol{\epsilon}) =: W^{\text{eff}}(\boldsymbol{\mathcal{E}}) \tag{5.179}$$

where $\mathcal{V}(\boldsymbol{\mathcal{E}}) = \{\boldsymbol{u} \mid \boldsymbol{u} = \boldsymbol{\mathcal{E}} \cdot \mathbf{x}, \text{ on } \partial V\}$, $\boldsymbol{\epsilon}^* = \frac{1}{2}\Big(\nabla \otimes \boldsymbol{u}^* + (\nabla \otimes \boldsymbol{u}^*)^T\Big)$, and $\boldsymbol{u}^*$ is the real solution of the problem.

Similarly, the minimum complementary energy principle can be re-stated as

$$W^c(\boldsymbol{\sigma}^*) = \inf_{\boldsymbol{\sigma} \in \mathcal{S}(\boldsymbol{\Sigma})} \langle w^c(\mathbf{x}, \boldsymbol{\sigma}) \rangle =: (W^c)^{\text{eff}}(\boldsymbol{\Sigma}) \tag{5.180}$$

where $\mathcal{S}(\boldsymbol{\Sigma}) = \{\boldsymbol{\sigma} \mid \nabla \cdot \boldsymbol{\sigma} = 0, \boldsymbol{\sigma} \cdot \mathbf{n} = \boldsymbol{\Sigma} \cdot \mathbf{n}, \text{ on } \partial V\}$, $\boldsymbol{\sigma}^*$ is the real stress solution of the problem.

Eqs.(5.179) (5.180) define the effective strain-energy potential W^{eff} and effective stress-energy potential $(W^c)^{\text{eff}}$. Using the effective potentials, the effective stress-strain relation for the composite can be determined as follows:

$$\frac{\partial W^{\text{eff}}(\boldsymbol{\mathcal{E}})}{\partial \boldsymbol{\mathcal{E}}} = \boldsymbol{\Sigma} \tag{5.181}$$

$$\frac{\partial (W^c)^{\text{eff}}(\boldsymbol{\Sigma})}{\partial \boldsymbol{\Sigma}} = \boldsymbol{\mathcal{E}} \tag{5.182}$$

5.9.2 Variational method based on a linear comparison solid

The nonlinear composite potential is a *convex* function of deviatoric stress. By changing variable $p = \epsilon_{eq}^2$, Ponte Castañeda reformulated the potential into a *concave* function in p:

$$w(\mathbf{x}, \boldsymbol{\epsilon}) = \frac{9}{2}k(\mathbf{x})\epsilon_m^2 + \mathcal{F}(\mathbf{x}, p) \tag{5.183}$$

where

$$k(\mathbf{x}) = \sum_{r=1}^{N} \chi_r(\mathbf{x})k_r, \quad \mathcal{F}(\mathbf{x}, p) = \sum_{r=1}^{N} \chi(\mathbf{x})\mathcal{F}^r(p) \tag{5.184}$$

As shown earlier, φ_r is convex. However, $\mathcal{F}_r(p)$ is not convex. This can be seen from the definition

$$\mathcal{F}^r(p) = \frac{\sigma_0\epsilon_0}{m+1}\left(\frac{p}{\epsilon_0^2}\right)^{\frac{m+1}{2}}, \quad 0 \leq \frac{m+1}{2} \leq 1 \tag{5.185}$$

that $\mathcal{F}^r(p)$ is concave in $[0, \infty]$. To utilize the Fenchel transformation, we extend the definition domain of $\mathcal{F}^r(p)$ to $\mathbb{R}$,

$$\mathcal{F}^r(p) = \begin{cases} \varphi_r(\epsilon_{eq}), & p = \epsilon_{eq}^2 > 0 \\ 0, & p = 0 \\ -\infty, & p < 0 \end{cases} \tag{5.186}$$

and now $\mathcal{F}^r(p)$ is concave in $\mathbb{R}$.

Then we can use the Fenchel transformation to construct its sub-conjugate functional,

$$\mathcal{F}_*^r(q) = \inf_{p \in \mathbb{R}} \{pq - \mathcal{F}^r(p)\} = \inf_{p>0} \{pq - \mathcal{F}^r(p)\} \tag{5.187}$$

where the last equality follows from the fact that $-\mathcal{F}^r(p) = \infty$, $\forall p < 0$. Then $\mathcal{F}_*^r(q)$ is a concave, nonpositive function, and from (5.187) one can deduce that $\mathcal{F}_*^r(q) = -\infty$, $\forall q < 0$. Therefore,

$$\mathcal{F}^r(p) = \mathcal{F}_{**}^r(p) = \inf_{q \in \mathbb{R}} \{pq - \mathcal{F}_*^r(q)\} = \inf_{q>0} \{pq - \mathcal{F}_*^r(q)\} \tag{5.188}$$

Although this is a valid procedure, the question is: What is q ? Ponte Castañeda introduced a *linear comparison composite solid* with potential w_0, such that

$$w_0(\mathbf{x}, \boldsymbol{\epsilon}) = \frac{9}{2}k(\mathbf{x})\epsilon_m^2 + \frac{3}{2}\mu_0(\mathbf{x})\epsilon_{eq}^2 \tag{5.189}$$

Note that the bulk modulus of the linear comparison composite is constant within each phase, the same as for the nonlinear material. The shear modulus is linear but heterogeneous, and the precise variation of which is determined as the solution of the variational principle. To this end, we choose $q = \frac{3}{2}\mu_0$, such that

$$\mathcal{F}_*^r(\frac{3}{2}\mu_0) = \inf_{p>0} \left\{ \left(\frac{3}{2}\mu_0\right)p - \frac{\sigma_0^r\epsilon_0^r}{m_r+1}\left(\frac{\sqrt{p}}{\epsilon_0^r}\right)^{1+m_r} \right\}, \quad p = \epsilon_{eq}^2 \tag{5.190}$$

Multiplying Eq. (5.188) with $\chi_r(\mathbf{x})$ and summing it among index $r = 1, 2, \cdots, N$, we have

$$\sum_{r=1}^{N} \chi_r(\mathbf{x}) \mathcal{F}^r(\epsilon_{eq}^2) = \inf_{\mu_0 > 0} \left\{ \left(\frac{3}{2}\mu_0\right) \epsilon_{eq}^2 - \sum_{r=1}^{N} \chi_r(\mathbf{x}) \mathcal{F}_*^r(\frac{3}{2}\mu_0) \right\} \tag{5.191}$$

Finally, we have

$$w(\mathbf{x}, \boldsymbol{\epsilon}) = \inf_{\mu_0 > 0} \left\{ w_0(\mathbf{x}, \boldsymbol{\epsilon}) - \sum_{r=1}^{N} \chi_r(\mathbf{x}) \mathcal{F}_*^r(\frac{3}{2}\mu_0) \right\} \tag{5.192}$$

where $w(\mathbf{x}, \boldsymbol{\epsilon})$ (see (5.183)-(5.184)) is the total nonlinear potential function for the composite

Following [Ponte Castañeda and Suquet (1998)], we denote the nonlinear part of the potential as

$$v(\mathbf{x}, \mu_0) = -\sum_{r=1}^{N} \chi_r(\mathbf{x}) f_*^r(\frac{3}{2}\mu_0) \tag{5.193}$$

We then have

$$w(\mathbf{x}, \boldsymbol{\epsilon}) = \inf_{\mu_0(\mathbf{x}) > 0} \{ w_0(\mathbf{x}, \boldsymbol{\epsilon}) + v(\mathbf{x}, \mu_0) \} \text{ and} \tag{5.194}$$

$$v(\mathbf{x}, \mu_0) = \sup_{\boldsymbol{\epsilon}} \{ w(\mathbf{x}, \boldsymbol{\epsilon}) - w_0(\mathbf{x}, \boldsymbol{\epsilon}) \} \tag{5.195}$$

Computing the average potential function,

$$W(\boldsymbol{\epsilon}) = \frac{1}{V} \int_V w(\mathbf{x}, \boldsymbol{\epsilon}) dV,$$

and applying the minimum potential energy principle, we have

$$\boxed{W^{\text{eff}}(\boldsymbol{\mathcal{E}}) = \inf_{\boldsymbol{u} \in \mathcal{V}(\boldsymbol{\mathcal{E}})} \inf_{\mu_0(\mathbf{x}) > 0} \{ \langle w_0(\mathbf{x}, \boldsymbol{\epsilon}(\boldsymbol{u})) + < v(\mathbf{x}, \mu_0) > \}} \tag{5.196}$$

Denote the effective potential for the linear comparison composite and the effective difference potential as

$$W_0^{\text{eff}}(\boldsymbol{\mathcal{E}}) = \inf_{\boldsymbol{u} \in \mathcal{V}(\boldsymbol{\mathcal{E}})} \langle w_0(\mathbf{x}, \boldsymbol{\epsilon}(\boldsymbol{u})) \rangle \tag{5.197}$$

$$V(\mu_0) = \langle v(\mathbf{x}, \mu_0) \rangle \tag{5.198}$$

The variational principle (5.196) can be re-written as

$$W^{eff}(\boldsymbol{\mathcal{E}}) = \inf_{\mu_0(\mathbf{x}) > 0} \left\{ W_0^{eff}(\boldsymbol{\mathcal{E}}) + V(\mu_0) \right\} \tag{5.199}$$

Consider the complementary potentials for both the nonlinear composite and the linear comparison composite in terms of the Fenchel transformation,

$$w^c(\mathbf{x}, \boldsymbol{\sigma}) = \sup_{\boldsymbol{\epsilon}} \{ \boldsymbol{\epsilon} : \boldsymbol{\sigma} - w(\mathbf{x}, \boldsymbol{\epsilon}) \} \text{ and } w_0^c(\mathbf{x}, \boldsymbol{\sigma}) = \sup_{\boldsymbol{\epsilon}} \{ \boldsymbol{\epsilon} : \boldsymbol{\sigma} - w_0(\mathbf{x}, \boldsymbol{\epsilon}) \} \tag{5.200}$$

where

$$w^c(\mathbf{x},\boldsymbol{\sigma}) = \frac{1}{2k(\mathbf{x})}\sigma_m^2 + \sum_{r=1}^{N} \chi_r(\mathbf{x})\psi_r(\sigma_{eq}) \tag{5.201}$$

$$w_0^c(\mathbf{x},\boldsymbol{\sigma}) = \frac{1}{2k(\mathbf{x})}\sigma_m^2 + \frac{1}{6\mu_0(\mathbf{x})}\sigma_{eq}^2 \tag{5.202}$$

One can then derive the dual principle of (5.199) from the minimum complementary potential principle (5.180). Consider (5.200) and (5.194)

$$\begin{aligned} w^c(\mathbf{x},\boldsymbol{\sigma}) &= \sup_{\boldsymbol{\epsilon}} \{\boldsymbol{\epsilon} : \boldsymbol{\sigma} - w(\mathbf{x},\boldsymbol{\epsilon})\} \\ &= \sup_{\boldsymbol{\epsilon}} \left\{ \boldsymbol{\epsilon} : \boldsymbol{\sigma} - \inf_{\mu_0 > 0} \{ w_0(\mathbf{x},\boldsymbol{\epsilon}) + v(\mathbf{x},\mu_0 \} \right\} \\ &= \sup_{\mu_0>0} \left\{ \sup_{\boldsymbol{\epsilon}} \{\boldsymbol{\epsilon} : \boldsymbol{\sigma} - -w_0(\mathbf{x},\boldsymbol{\epsilon})\} - v(\mathbf{x},\mu_0 \right\} \\ &= \sup_{\mu_0>0} \{ w_0^c(\mathbf{x},\boldsymbol{\sigma}) - v(\mathbf{x},\mu_0)\} \end{aligned} \tag{5.203}$$

Substituting (5.203) into the minimum complementary potential energy principle (5.180), we have

$$\begin{aligned} (W^c)^{\mathit{eff}}(\boldsymbol{\Sigma}) &= \inf_{\boldsymbol{\sigma}\in\mathcal{S}(\boldsymbol{\Sigma})} \langle w^c(\mathbf{x},\boldsymbol{\sigma})\rangle = \inf_{\boldsymbol{\sigma}\in\mathcal{S}} \langle \sup_{\mu_0>0} \{ w_0^c(\mathbf{x},\boldsymbol{\sigma}) - v(\mathbf{x},\mu_0)\}\rangle \\ &= \sup_{\mu_0>0} \left\{ \inf_{\boldsymbol{\sigma}\in\mathcal{S}} \langle w_0^c(\mathbf{x},\boldsymbol{\sigma})\rangle - < v(\mathbf{x},\mu_0) > \right\} \\ &= \sup_{\mu_0>0} \{ (W_0^c)^{\mathit{eff}}(\boldsymbol{\Sigma}) - V(\mu_0)\} \ . \end{aligned} \tag{5.204}$$

To find applications of variational principles (5.199) and (5.204) to homogenization of nonlinear composites, one can consult the original papers by Ponte Castañeda [Ponte Castañeda (1991, 1992a,b); Ponte Castañeda and Suquet (1998)] .

5.10 Zvi Hashin

Professor Zvi Hashin is the Nathan Cummings Chair of mechanics of solids at Tel-Aviv University, Israel. He was elected as a foreign associate member of National Academy of Engineering of USA in 1998, for his contributions to the theory and technology of advanced composite materials.

5.11 Exercises

Problem 5.1.

Consider a functional

$$P : H^1([a,b]) \rightarrow \mathbb{R}$$

Fig. 5.11 Zvi Hashin

where

$$P(u) = \int_a^b \sqrt{1 + [u'(x)]^2} dx \ .$$

with essential boundary condition $u(a) = \bar{u}_a$ *and* $u(b) = \bar{u}_b$.

Find the first variation, second variation, and Gâteaux derivative. Derive associated the Euler-Lagrange equation.

Problem 5.2. *Let* $\Gamma_u = \emptyset$, $\partial V = \Gamma_t$, *and* $f_i = 0$. *Assume that the RVE has the prescribed traction boundary condition,*

$$\mathbf{n} \cdot \bar{\boldsymbol{\sigma}} = \mathbf{t}^0(\mathbf{x}), \qquad \forall \mathbf{x} \in \partial V \tag{5.205}$$

where $\bar{\boldsymbol{\sigma}} > 0$ *is a constant tensor.*

Show that

$$W^*(\boldsymbol{\sigma}^*) = \sup_{\{\boldsymbol{\epsilon} \in \mathcal{E}\}} \left\{ \bar{\boldsymbol{\sigma}} : \langle \boldsymbol{\epsilon} \rangle - \tilde{W}(< \boldsymbol{\epsilon} >) \right\} \tag{5.206}$$

where $\mathcal{E} := \{\epsilon_{ij} \,\Big|\, \epsilon_{ij,kl} + \epsilon_{kl,ij} - \epsilon_{ik,jl} - \epsilon_{jl,ik} = 0, \;\; \textit{and} \;\; \epsilon_{ij} \in L^2(V)\}$,

$$W^*(\boldsymbol{\sigma}^*) := \frac{1}{2V}\int_V D_{ijkl}\sigma^*_{ij}\sigma^*_{kl}dV = \frac{1}{2}\int_V \epsilon^*_{ij}\bar{\sigma}_{kl}dV \tag{5.207}$$

$$\tilde{W}(<\boldsymbol{\epsilon}>) := \frac{1}{2V}\inf_{\boldsymbol{\epsilon}\in\mathcal{E}\}}\int_V C_{ijkl}\epsilon_{ij}\epsilon_{kl}dV \tag{5.208}$$

Note that σ^*_{ij} *and* ϵ^*_{ij} *are the real solutions.*

Problem 5.3. *Let* $\Gamma_u = \emptyset$ *and* $\partial V = \Gamma_t$. *Consider the following the boundary-value problem,*

$$\sigma_{ij,j} = 0, \qquad \forall \mathbf{x} \in V \tag{5.209}$$

$$n_j\sigma_{ij} = t^0_i, \quad \forall \mathbf{x} \in \Gamma_t, \quad \text{and} \quad \Gamma_u = \emptyset \tag{5.210}$$

$$\epsilon_{ij} = \frac{1}{2}\Big(u_{i,j} + u_{j,i}\Big) \tag{5.211}$$

$$\epsilon_{ij} = \frac{\partial U_c}{\partial \sigma_{ij}}, \quad U_c(\boldsymbol{\sigma}) := \frac{1}{2}D_{ijk\ell}\sigma_{ij}\sigma_{k\ell}. \tag{5.212}$$

Consider a comparison elastic solid with compliance tensor, $D^0_{ijk\ell}$ *and*

$$\sigma^{(0)}_{ij,j} = 0, \qquad \forall \mathbf{x} \in V \tag{5.213}$$

$$n_j\sigma^{(0)}_{ij} = t^0_i, \quad \forall \mathbf{x} \in \Gamma_t, \quad \text{and} \quad \Gamma_u = \emptyset \tag{5.214}$$

$$\epsilon^{(0)}_{ij} = \frac{1}{2}\Big(u^{(0)}_{i,j} + u^{(0)}_{j,i}\Big) \tag{5.215}$$

$$\epsilon^{(0)}_{ij} = \frac{\partial U^{(0)}_0}{\partial \sigma^{(0)}_{ij}}, \quad U^{(0)}_0(\boldsymbol{\sigma}) := \frac{1}{2}D^{(0)}_{ijk\ell}\sigma^{(0)}_{ij}\sigma^{(0)}_{k\ell}\,. \tag{5.216}$$

Let

$$\sigma_{ij} = \sigma^{(0)}_{ij} + \sigma^d_{ij} \tag{5.217}$$

$$\epsilon_{ij} = D^{(0)}_{ijk\ell}\sigma_{k\ell} + q_{ij} \tag{5.218}$$

where σ^d_{ij} *is called disturbance stress, and* q_{ij} *is called the polarization strain (eigenstrain).*

They are connected by the following subsidiary conditions: **1.** *the weak form of subsidiary condition (complementary virtual work principle),*

$$\int_V \epsilon_{ij}\sigma^d_{ij}dV = 0 \tag{5.219}$$

or **2.** *the strong form of subsidiary condition*

$$\epsilon' := D^{(0)}_{ijk\ell}\sigma^d_{k\ell} + q_{ij}\,, \quad \mathcal{C}(\epsilon'_{ij}) = \epsilon'_{ij,k\ell} + \epsilon'_{k\ell,ij} - \epsilon'_{ik,j\ell} - \epsilon'_{i\ell,jk} = 0, \quad \forall \mathbf{x} \in V \tag{5.220}$$

Consider the following variational problem

$$\text{(The primal problem :)} \quad \mathcal{P}: \qquad \inf_{\boldsymbol{\sigma}^d\in\mathcal{S}(V)} \Pi_c(\boldsymbol{\sigma}^d) \tag{5.221}$$

or

$$\text{(The primal problem :)} \quad \mathcal{P}: \quad \inf_{\boldsymbol{\sigma}^d \in \mathcal{S}(V)} W_c(\boldsymbol{\sigma}^d) \tag{5.222}$$

where

$$W_c(\boldsymbol{\sigma}^d) := \frac{1}{2|V|} \int_V D_{ijk\ell} \sigma_{ij} \sigma_{k\ell} dV = \int_V D_{ijk\ell} (\sigma_{ij}^{(0)} + \sigma_{ij}^d)(\sigma_{k\ell}^{(0)} + \sigma_{k\ell}^d) dV, \tag{5.223}$$

$\Pi_c(\boldsymbol{\sigma}^d) = V W_c(\boldsymbol{\sigma}^d)$ *and*

$$\mathcal{S} := \left\{ \boldsymbol{\sigma} \;\middle|\; n_j \sigma_{ij} = 0, \quad \forall \mathbf{x} \in \Gamma_t, \quad \text{and} \quad \sigma_{ij} \in C^0(V) \right\} \tag{5.224}$$

Derive the Hashin-Shtrikman variational principle.

Hints: read references [Hashin and Shtrikman (1962a); Talbot and Willis (1985)].

Chapter 6

ESHELBY TENSORS IN A FINITE VOLUME AND THEIR APPLICATIONS

6.1 Introduction

One of the corner stones of contemporary micro-mechanics and nano-mechanics is Eshelby's inclusion theory [Eshelby (1957, 1959, 1961)]. Eshelby's ellipsoidal inclusion solution was obtained based on the assumption that an inclusion is embedded in unbounded ambient space. This is a good approximation if the size effect of the inclusion is negligible, i.e. the size of the inclusion is small compared to the size of the representative volume element. In many engineering applications, the size of the representative volume element (RVE) is finite. Therefore, certain approximations have to be made in order to utilize Eshelby's classical solution in homogenization. This limitation becomes obvious, when size effects and interfacial boundary effects of a second phase in a composite, or the size effect and boundary effects of an inhomogeneity, become prominent issues, which is one of main focuses of the nano-composite mechanics and materials (e.g. [Thostenson *et al.* (2001); Shi *et al.* (2004)]).

Inclusion problems in a finite domain have been considered before, e.g. see [Kinoshita and Mura (1984); Kröner (1986, 1990)] and [Mazilu (1972)]. A common approach adopted is to first find the Green's function of Navier's equations for a finite domain, and then to find the solution of the corresponding inclusion problem. However, attempts based on this approach have been futile, we believe, because of the mathematical difficulties involved in obtaining a closed form solution of the finite Green's function. This is true even for a highly symmetrical spherical domain. In fact, the Green's function of the Navier equation for a finite spherical domain has not been found yet. To the best of the authors' knowledge, there has never been any exact, closed form solution of the inclusion problem in a finite domain published in the literature. A solution has been obtained by [Luo and Weng (1987)], which coincides with our solution in a special case. Their solution, however, is not in closed form and is without expressions for the Eshelby tensors.

Based on our work [Li *et al.* (2007b,c); Sauer *et al.* (2007)] , in this Chapter we shall present a precise characterization of the elastic fields due to a spherical inclusion embedded within a spherical representative volume element. The RVE is considered having finite size, with either a prescribed uniform displacement or a prescribed uniform traction boundary condition. Based on symmetry and group theoretic arguments, we identify that the Eshelby tensor for a spherical inclusion admits a unique decomposition, which we coin the "*radial transversely isotropic tensor*". Based on this notion, a novel solution procedure is found to solve the resulting Fredholm type integral equations. By using this technique, exact and closed form solutions are obtained for the elastic disturbance fields.

In the solution two new tensors appear, which are termed the Dirichlet-Eshelby tensor and the Neumann-Eshelby tensor. In contrast to the classical Eshelby tensor, they both are position dependent and contain information about the boundary condition of the RVE as well as the volume fraction of the inclusion. The Eshelby tensors derived in this Chapter have some profound consequences for both homogenization and the study of inhomogeneities in finite elastic solids. In the remaining part of this Chapter, applications to homogenization of composites are discussed.

6.2 The inclusion problem of a finite RVE

We consider Eshelby's homogeneous inclusion problem in a finite domain. Figure 6.1 shows a spherical inclusion Ω_I with radius a embedded at the center of a spherical representative volume element V with radius A. Suppose that a constant eigenstrain field is prescribed inside the inclusion,

$$\epsilon^*_{ij}(\mathbf{x}) = \begin{cases} \epsilon^*_{ij} , & \mathbf{x} \in \Omega_I , \\ 0 , & \mathbf{x} \in \Omega_E = V/\Omega_I . \end{cases} \tag{6.1}$$

On the boundary of the RVE, two types of boundary conditions are considered: a prescribed displacement (Dirichlet) boundary condition, or a prescribed traction (Neumann) boundary condition.

$$\text{Dirichlet}: \ u_i = \epsilon^0_{ij} x_j , \quad \forall \mathbf{x} \in \partial V , \tag{6.2}$$

$$\text{Neumann}: \ t_i = \sigma^0_{ij} n_j , \quad \forall \mathbf{x} \in \partial V , \tag{6.3}$$

where ε^0_{ij} and σ^0_{ij} are the background strain and stress fields. The elastic fields inside the RVE can be decomposed into the background field from the remote boundary loads and a disturbance field, which arises due to the presence of the inclusion. The solution for the background field depends on the macro problem. Here we are concerned with the solution of the disturbance fields, and we obtain either the Dirichlet-Eshelby boundary value problem (BVP),

$$\begin{aligned} \mathbb{C}_{ijk\ell} u^d_{k,\ell j}(\mathbf{x}) - \mathbb{C}_{ijk\ell} \epsilon^*_{k\ell,j}(\mathbf{x}) = 0 , \quad \forall \mathbf{x} \in V , \\ u^d_i(\mathbf{x}) = 0 , \quad \forall \mathbf{x} \in \partial V . \end{aligned} \tag{6.4}$$

or the Neumann-Eshelby BVP,

$$\begin{aligned} \mathbb{C}_{ijk\ell} u^d_{k,\ell j}(\mathbf{x}) - \mathbb{C}_{ijk\ell} \epsilon^*_{k\ell,j}(\mathbf{x}) = 0 , \quad \forall \mathbf{x} \in V , \\ t^d_i(\mathbf{x}) = n_j \mathbb{C}_{ijk\ell} u^d_{k,\ell}(\mathbf{x}) = 0 , \quad \forall \mathbf{x} \in \partial V . \end{aligned} \tag{6.5}$$

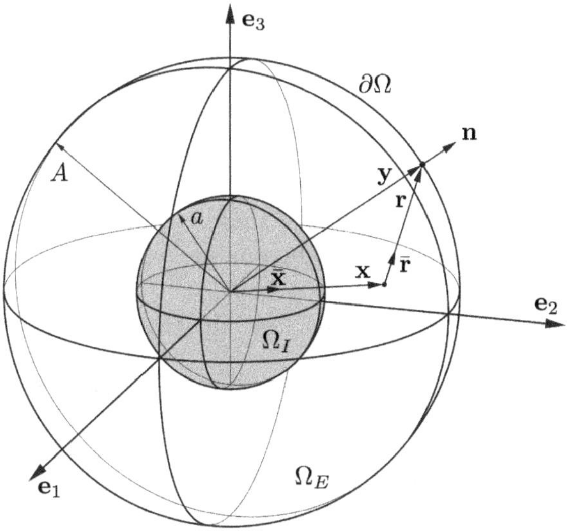

The following notations are used in this Chapter: The RVE is composed of an interior and an exterior region, $V = \Omega_I \cup \Omega_E$; $\mathbf{x}, \mathbf{y} \in V$ are two arbitrary points, and vector $\boldsymbol{R} = \mathbf{y} - \mathbf{x}$; Each vector $\mathbf{x}$, $\mathbf{y}$, $\boldsymbol{R}$ can be expressed as its length multiplied by a unit direction vector, we shall denote them as $\mathbf{x} = |\mathbf{x}|\bar{\mathbf{x}}$, $\mathbf{y} = |\mathbf{y}|\bar{\mathbf{y}}$ and $\boldsymbol{R} = r\bar{\mathbf{r}}$, with $r = |\boldsymbol{R}|$. Note that if $\mathbf{y} \in \partial V$ we have $|\mathbf{y}| = A$ and $\bar{\mathbf{y}} = \mathbf{n}$, i.e. the direction of $\mathbf{y}$ is equal to the outward surface normal $\boldsymbol{n}$. Furthermore we define the ratios $\rho = a/|\mathbf{x}|$, $\rho_0 = a/A$ and $t = |\mathbf{x}|/A = \rho_0/\rho$ to allow for a non-dimensional description.

Fig. 6.1 A spherical representative element containing a spherical inclusion.

By using Somigliana's identity, the displacement field solution of BVPs Eqs.(6.4) and (6.5) may be expressed as

$$u_m^d(\mathbf{x}) = -\int_{\Omega_I} \mathbb{C}_{ijk\ell} G^{\infty}_{im,j}(\mathbf{x}-\mathbf{y})\epsilon^*_{k\ell}(\mathbf{y}) d\Omega_y + \int_{\partial V} \mathbb{C}_{ijk\ell} u^d_{k,\ell}(\mathbf{y}) G^{\infty}_{im}(\mathbf{x}-\mathbf{y}) n_j(\mathbf{y}) dS_y$$
$$+\int_{\partial V} \mathbb{C}_{ijk\ell} u^d_k(\mathbf{y}) G^{\infty}_{im,j}(\mathbf{x}-\mathbf{y}) n_\ell(\mathbf{y}) dS_y \ . \tag{6.6}$$

where we have denoted the derivative of Green's function as $G^{\infty}_{im,j} := \partial G^{\infty}_{im}/\partial x_j = -\partial G^{\infty}_{im}/\partial y_j$. The two surface integrals represent the effects of Dirichlet boundary or Neumann boundary, respectively.

For the Dirichlet Eshelby problem , the Somigliana's identity becomes

$$u_m^d(\mathbf{x}) = -\int_{\Omega_I} \mathbb{C}_{ijk\ell} G^{\infty}_{im,j}(\mathbf{x}-\mathbf{y})\epsilon^*_{k\ell}(\mathbf{y}) d\Omega_y + \int_{\partial V} \mathbb{C}_{ijk\ell} u^d_{k,\ell}(\mathbf{y}) G^{\infty}_{im}(\mathbf{x}-\mathbf{y}) n_j(\mathbf{y}) dS_y \ . \tag{6.7}$$

The disturbance strain field can be derived from the above equation as

$$\epsilon^d_{ij}(\mathbf{x}) = -\frac{1}{2}\epsilon^*_{mn}\int_{\Omega_I}\mathbb{C}_{k\ell mn}\Big(G^\infty_{ki,\ell j}(\mathbf{x}-\mathbf{y}) + G^\infty_{kj,\ell i}(\mathbf{x}-\mathbf{y})\Big)d\Omega_y$$
$$+\frac{1}{2}\int_{\partial V}\mathbb{C}_{k\ell pq}\epsilon^d_{pq}(\mathbf{y})\Big(G^\infty_{ki,j}(\mathbf{x}-\mathbf{y}) + G^\infty_{kj,i}(\mathbf{x}-\mathbf{y})\Big)n_\ell(\mathbf{y})dS_y \,. \quad (6.8)$$

We need to solve Eq. (6.8) which is an *implicit* integral equation for the unknown strain field ϵ^d_{ij}. We note that Eq. (6.8) becomes a hyper-singular integral equation if $\mathbf{x} \in \partial V$.

For the Neumann-Eshelby problem, , the Somigliana's identity reduces to

$$u^d_m(\mathbf{x}) = -\int_{\Omega_I}\mathbb{C}_{ijk\ell}G^\infty_{im,j}(\mathbf{x}-\mathbf{y})\epsilon^*_{k\ell}(\mathbf{y})d\Omega_y + \int_{\partial V}\mathbb{C}_{ijk\ell}u^d_k(\mathbf{y})G^\infty_{im,j}(\mathbf{x}-\mathbf{y})n_\ell(\mathbf{y})dS_y \,. \quad (6.9)$$

Eq. (6.9) is also an *implicit* integral equation for the unknown displacement field u^d_i that needs to be solved.

To illustrate our solution procedure, we re-examine the classical Eshelby tensors. For inclusion problems in unbounded space, the boundary term in (6.7), (6.8), and (6.9) drops out. One can then find the disturbance strain fields in terms of the Eshelby tensors,

$$\epsilon^d_{ij}(\mathbf{x}) = \mathcal{S}^{\bullet,\infty}_{ijk\ell}(\mathbf{x})\epsilon^*_{k\ell} \,, \quad \forall \mathbf{x} \in \mathbb{R}^3 \,, \quad (6.10)$$

where the superscript $\bullet$ represents the interior solution ($\bullet = I$) or the exterior solution ($\bullet = E$), depending on the location of $\mathbf{x}$; ∞ is the notation to signify the solution is derived for RVE in an *infinite* space.

$$\mathcal{S}^{\bullet,\infty}_{ijk\ell}(\mathbf{x}) = \begin{cases} \mathcal{S}^{I,\infty}_{ijk\ell}(\mathbf{x}) \,, & \forall \mathbf{x} \in \Omega_I \,, \\ \mathcal{S}^{E,\infty}_{ijk\ell}(\mathbf{x}) \,, & \forall \mathbf{x} \in \mathbb{R}^3/\Omega_I \,. \end{cases} \quad (6.11)$$

For spherical inclusions in an infinite elastic medium, the Eshelby tensors have the elementary form (e.g. [Mura (1987)], and [Ju and Sun (1999)]),

(1) Interior solution:

$$\mathcal{S}^{I,\infty}_{ijmn}(\mathbf{x}) = \frac{(5\nu-1)}{15(1-\nu)}\delta_{ij}\delta_{mn} + \frac{(4-5\nu)}{15(1-\nu)}\Big(\delta_{im}\delta_{jn} + \delta_{in}\delta_{jm}\Big) \,, \quad \mathbf{x} \in \Omega_I \,, \quad (6.12)$$

(2) Exterior solution:

$$\mathcal{S}^{E,\infty}_{ijmn}(\mathbf{x}) = \frac{\rho^3}{30(1-\nu)}\Big[(3\rho^2 + 10\nu - 5)\delta_{ij}\delta_{mn} + (3\rho^2 - 10\nu + 5)$$
$$(\delta_{im}\delta_{jn} + \delta_{in}\delta_{jm}) + 15(1-\rho^2)\delta_{ij}\bar{x}_m\bar{x}_n + 15(1-2\nu-\rho^2)\delta_{mn}\bar{x}_i\bar{x}_j$$
$$+15(\nu-\rho^2)\Big(\delta_{im}\bar{x}_j\bar{x}_n + \delta_{jm}\bar{x}_i\bar{x}_n + \delta_{in}\bar{x}_j\bar{x}_m + \delta_{jn}\bar{x}_i\bar{x}_m\Big)$$
$$+15(7\rho^2-5)\bar{x}_i\bar{x}_j\bar{x}_m\bar{x}_n\Big] \,, \quad \mathbf{x} \in \mathbb{R}^3/\Omega_I \,, \quad (6.13)$$

where $\rho := a/|\mathbf{x}|$ and $|\mathbf{x}| = \sqrt{x_i x_i},\ i = 1, 2, 3$.

Inspired by Eq. (6.10), we postulate the following form of the two considered BVP's,

$$\epsilon^d_{ij}(\mathbf{x}) = \mathcal{S}^{\bullet;\star}_{ijk\ell}(\mathbf{x})\epsilon^*_{k\ell}\ , \quad \forall \mathbf{x} \in V\ , \tag{6.14}$$

where $\mathcal{S}^{\bullet;\star}_{ijk\ell}(\mathbf{x})$ is an unknown fourth-order tensor (the finite Eshelby tensor), we are seeking to obtain. As before the superscript $\bullet$ represents the interior solution or the exterior solution. The superscript $\star$ stands for the Dirichlet-Eshelby Tensor ($\star = D$) or the Neumann-Eshelby Tensor ($\star = N$). As a special case we expect to obtain the original infinite Eshelby tensor ($\star = \infty$).

In principle, one may be able to use spherical harmonics to represent the general solution of Eqs. (6.7) and (6.9) based on symmetry, but the solution procedure is very much involved. In fact, no explicit solution has been found as shown in a particular case worked out by Luo and Weng [Luo and Weng (1987)] .

6.3 Properties of the radially isotropic tensor

To solve the Dirichlet and Neumann BVP problems (Eqs. (6.8) and (6.9)), we first introduce a novel concept of the *radial transversely isotropic tensor*, or in short, the *radially isotropic tensor*.

It may be observed from the expressions of $\mathcal{S}^{I,\infty}_{ijmn}(\mathbf{x})$ and $\mathcal{S}^{E,\infty}_{ijmn}(\mathbf{x})$ above that there appear six independent tensorial bases, which may be arranged in an array as follows,

$$\mathbf{\Theta}_{ijmn}(\bar{\mathbf{x}}) := \begin{bmatrix} \delta_{ij}\delta_{mn} \\ \delta_{im}\delta_{jn} + \delta_{in}\delta_{jm} \\ \delta_{ij}\bar{x}_m\bar{x}_n \\ \delta_{mn}\bar{x}_i\bar{x}_j \\ \delta_{im}\bar{x}_j\bar{x}_n + \delta_{in}\bar{x}_j\bar{x}_m + \delta_{jm}\bar{x}_i\bar{x}_n + \delta_{jn}\bar{x}_i\bar{x}_m \\ \bar{x}_i\bar{x}_j\bar{x}_m\bar{x}_n \end{bmatrix} . \tag{6.15}$$

We term this array as the circumference basis of the Eshelby tensor. By using $\mathbf{\Theta}_{ijmn}(\bar{\mathbf{x}})$, both the original interior and exterior Eshelby tensor can be recast into a canonical form, the dot product of two arrays, i.e.

$$\begin{aligned} \mathcal{S}^{\bullet,\infty}_{ijmn}(\mathbf{x}) = & S^{\bullet,\infty}_1(t)\delta_{ij}\delta_{mn} + S^{\bullet,\infty}_2(t)\big(\delta_{im}\delta_{jn} + \delta_{in}\delta_{jm}\big) + S^{\bullet,\infty}_3(t)\delta_{ij}\bar{x}_m\bar{x}_n \\ & + S^{\bullet,\infty}_4(t)\delta_{mn}\bar{x}_i\bar{x}_j + S^{\bullet,\infty}_5(t)\big(\delta_{im}\bar{x}_j\bar{x}_n + \delta_{in}\bar{x}_j\bar{x}_m + \delta_{jm}\bar{x}_i\bar{x}_n + \delta_{jn}\bar{x}_i\bar{x}_m\big) \\ & + S^{\bullet,\infty}_6(t)\bar{x}_i\bar{x}_j\bar{x}_m\bar{x}_n \quad = \mathbf{\Theta}^T_{ijmn}(\bar{\mathbf{x}})\mathbf{S}^{\bullet,\infty}(t)\ . \end{aligned} \tag{6.16}$$

The arrays, $\mathbf{S}^{I,\infty}(t)$ and $\mathbf{S}^{E,\infty}(t)$, are termed the radial basis of the infinite Eshelby tensor. In accordance to Eqs. (6.12) and (6.13) they are given as

$$\mathbf{S}^{I,\infty}(t) = \frac{1}{15(1-\nu)}\begin{bmatrix} 5\nu-1 \\ 4-5\nu \\ 0 \\ 0 \\ 0 \\ 0 \end{bmatrix}, \quad \mathbf{S}^{E,\infty}(t) = \frac{\rho_0^3/t^3}{30(1-\nu)}\begin{bmatrix} 3\rho_0^2/t^2+10\nu-5 \\ 3\rho_0^2/t^2-10\nu+5 \\ 15(1-\rho_0^2/t^2) \\ 15(1-2\nu-\rho_0^2/t^2) \\ 15(\nu-\rho_0^2/t^2) \\ 15(7\rho_0^2/t^2-5) \end{bmatrix}, \tag{6.17}$$

where $t = |\mathbf{x}|/A = \rho_0/\rho$, with $\rho = a/|\mathbf{x}|$, and $\rho_0 = a/A$.

The above heuristic discussion reveals an important fact, that the Eshelby tensor is a so-called "*radially isotropic tensor*", which is a generalization of an isotropic tensor. Here, we define the *radially isotropic tensor* as a transversely isotropic tensor along a given radial direction, i.e. a tensor whose properties in all directions perpendicular to the radial direction, $\bar{\mathbf{x}}$, are the same. In general, the radially isotropic tensor, depending on $\mathbf{x} = tA\bar{\mathbf{x}}$, can be expressed in the following canonical form,

$$\begin{aligned} \mathcal{S}_{ijmn}(\mathbf{x}) = & S_1(t)\delta_{ij}\delta_{mn} + S_2(t)\big(\delta_{im}\delta_{jn}+\delta_{in}\delta_{jm}\big) + S_3(t)\delta_{ij}\bar{x}_m\bar{x}_n \\ & + S_4(t)\delta_{mn}\bar{x}_i\bar{x}_j + S_5(t)\big(\delta_{im}\bar{x}_j\bar{x}_n + \delta_{in}\bar{x}_j\bar{x}_m + \delta_{jm}\bar{x}_i\bar{x}_n + \delta_{jn}\bar{x}_i\bar{x}_m\big) \\ & + S_6(t)\bar{x}_i\bar{x}_j\bar{x}_m\bar{x}_n = \mathbf{\Theta}^T_{ijmn}(\bar{\mathbf{x}})\mathbf{S}(t) . \end{aligned} \tag{6.18}$$

This canonical form decomposes $\mathcal{S}_{ijmn}$ into the circumference basis $\mathbf{\Theta}_{ijmn}$, which is only a function of the direction vector $\bar{\mathbf{x}}$, and into the radial basis $\mathbf{S}$, which is only a function of the dimensionless radial distance t.

A transversely isotropic tensor has some interesting symmetric properties. Consider the unit vector $\bar{\mathbf{r}} = \bar{r}_i\mathbf{e}_i$ Using the definition $a_{ij} := \delta_{ij} - \bar{r}_i\bar{r}_j$, $b_{ij} := \bar{r}_i\bar{r}_j$, Walpole [Walpole (1981)] has constructed the following six radially isotropic bases:

$$\mathcal{E}^1_{ijmn} = \frac{1}{2}a_{ij}a_{mn} , \tag{6.19}$$

$$\mathcal{E}^2_{ijmn} = b_{ij}b_{mn} = \bar{r}_i\bar{r}_j\bar{r}_m\bar{r}_n , \tag{6.20}$$

$$\mathcal{E}^3_{ijmn} = \frac{1}{2}\Big(a_{im}a_{jn} + a_{jm}a_{in} - a_{ij}a_{mn}\Big) , \tag{6.21}$$

$$\mathcal{E}^4_{ijmn} = \frac{1}{2}\Big(a_{im}b_{jn} + a_{jn}b_{jm} + a_{jn}b_{im} + a_{jm}b_{in}\Big) , \tag{6.22}$$

$$\mathcal{E}^5_{ijmn} = a_{ij}b_{mn} , \tag{6.23}$$

$$\mathcal{E}^6_{ijmn} = b_{ij}a_{mn} . \tag{6.24}$$

It can be shown that the above bases form a finite non-Abelian group. Furthermore,

$$\mathbb{E}^p : \mathbb{E}^q = \mathbb{E}^p, \text{ if } p = q, \quad \mathbb{E}^p : \mathbb{E}^q = 0, \text{ if } p \neq q, \quad p, q = 1, 2, 3, 4, \tag{6.25}$$

where $\mathbb{E}^p = \mathcal{E}^p_{ijmn}\mathbf{e}_i \otimes \mathbf{e}_j \otimes \mathbf{e}_m \otimes \mathbf{e}_n$. Subsequently, for $p = 1, 2, 3, 4, 5, 6$, one can find the "*less congenial multiplication table*" shown in Table 6.1 (see [Walpole (1981)]).

	$\mathbb{E}^1$	$\mathbb{E}^2$	$\mathbb{E}^3$	$\mathbb{E}^4$	$\mathbb{E}^5$	$\mathbb{E}^6$
$\mathbb{E}^1$	$\mathbb{E}^1$	0	0	0	$\mathbb{E}^5$	0
$\mathbb{E}^2$	0	$\mathbb{E}^2$	0	0	0	$\mathbb{E}^6$
$\mathbb{E}^3$	0	0	$\mathbb{E}^3$	0	0	0
$\mathbb{E}^4$	0	0	0	$\mathbb{E}^4$	0	0
$\mathbb{E}^5$	0	$\mathbb{E}^5$	0	0	0	$2\mathbb{E}^1$
$\mathbb{E}^6$	$\mathbb{E}^6$	0	0	0	$2\mathbb{E}^2$	0

.

Nevertheless, to the best of the authors' knowledge, we are the first to show that the circumference basis $\mathbf{\Theta}_{ijmn}$ of a spherical inclusion in a finite domain is a transversely isotropic tensor[1]. Instead of using the above partially idempotent canonical form to represent the circumference basis $\mathbf{\Theta}_{ijmn}$ of a radial transversely isotropic tensor, we use an equivalent but different description introduced in Eq. (6.15).

Based on the symmetry of the problem, we now postulate that *the Eshelby tensor for a finite RVE, $S^{\bullet,\star}_{ijmn}$, should also be a radially isotropic tensor.* It therefore admits the following multiplicative decomposition

$$S^{\bullet,\star}_{ijmn}(\mathbf{x}) = \mathbf{\Theta}^T_{ijmn}(\bar{\mathbf{x}})\mathbf{S}^{\bullet,\star}(t) \ , \tag{6.26}$$

with the superscripts, $\bullet = I$ or E and $\star = D$ or N. Here $\mathbf{\Theta}_{ijmn}(\bar{\mathbf{x}})$ is the circumference basis according to Eqs.. (6.15) and $\mathbf{S}^{\bullet,\star}(t)$ is the radial basis given as

$$\mathbf{S}^{\bullet,\star}(t) = [S^{\bullet,\star}_1(t), S^{\bullet,\star}_2(t), S^{\bullet,\star}_3(t), S^{\bullet,\star}_4(t), S^{\bullet,\star}_5(t), S^{\bullet,\star}_6(t)]^T \ . \tag{6.27}$$

The scalar entries $S^{\bullet,\star}_J(t)$, $J = 1,2,3,4,5,6$, are unknown functions of the non-dimensional radial variable $t = |\mathbf{x}|/A$, which are to be determined.

The postulate above is motivated by the following two considerations. Due to the concentric and spherical symmetry of inclusion and RVE the tensorial basis of the finite Eshelby tensor can only depend on the radial direction vector $\bar{x}_i$ (and the second order identity δ_{ij}). Therefore its tensorial basis, can only consist of combinations of zero-th, second and fourth order homogeneous functions of $\bar{\mathbf{x}}$. Furthermore due to the symmetry of the strain tensor the finite Eshelby tensor must have minor symmetries. Its tensorial basis can therefore only admit the six tensorial bases listed in $\mathbf{\Theta}_{ijmn}(\bar{\mathbf{x}})$. We note that one should expect more than six bases for problems described by more that one vector, such as ellipsoidal inclusions or non-concentrically placed inclusions within the RVE. Such problems may also be solvable with a similar procedure to ours. Due to the postulate the search for the finite Eshelby tensors reduces to the search for their radial basis $\mathbf{S}^{\bullet,\star}(t)$. We will see in the subsequent section, that the two solutions we obtain satisfy the governing equations exactly, thereby justifying postulate (6.26).

[1]We first derived the result in a 2004 manuscript that was submitted to *Proceedings of Royal Society of London.*

In analogy to Eq. (6.14), we can express the disturbance displacement field as

$$u_i^d(\mathbf{x}) = \begin{cases} \mathbb{U}^{I,\star}_{imn}(\mathbf{x})\ \epsilon^*_{mn}\ , & \forall \mathbf{x} \in \Omega_I\ , \\ \mathbb{U}^{E,\star}_{imn}(\mathbf{x})\ \epsilon^*_{mn}\ , & \forall \mathbf{x} \in \Omega_E\ , \end{cases} \tag{6.28}$$

where $\mathbb{U}^{I,\star}_{imn}(\mathbf{x})$ is a third-order radially isotropic tensor, whose relation to $\mathbb{S}^{I,\star}_{ijmn}(\mathbf{x})$ is discussed next. The disturbance strain is linked to the displacement field by the relation,

$$\begin{aligned} \epsilon^d_{ij}(\mathbf{x}) &= \frac{1}{2}\Big(u^d_{i,j}(\mathbf{x}) + u^d_{j,i}(\mathbf{x})\Big) \\ &= \frac{1}{2}\Big(\mathbb{U}^{\bullet,\star}_{imn,j}(\mathbf{x}) + \mathbb{U}^{\bullet,\star}_{jmn,i}(\mathbf{x})\Big)\epsilon^*_{mn} = \mathbb{S}^{\bullet,\star}_{ijmn}(\mathbf{x})\ \epsilon^*_{mn}\ . \end{aligned} \tag{6.29}$$

It can be shown that $\mathbb{U}^{\bullet,\star}_{imn}(\mathbf{x})$ can only admit the following multiplicative decomposition, so that the related Eshelby tensors $\mathbb{S}^{I,\star}_{ijmn}(\mathbf{x})$ are radially isotropic tensors,

$$\mathbb{U}^{I,\star}_{imn}(\mathbf{x}) = \boldsymbol{\Xi}^T_{imn}(\bar{\mathbf{x}})\mathbf{U}^{I,\star}(t)\ , \quad \forall \mathbf{x} \in \Omega_I\ , \tag{6.30}$$

$$\mathbb{U}^{E,\star}_{imn}(\mathbf{x}) = \boldsymbol{\Xi}^T_{imn}(\bar{\mathbf{x}})\mathbf{U}^{E,\star}(t)\ , \quad \forall \mathbf{x} \in \Omega_E\ , \tag{6.31}$$

with the appearing arrays defined as

$$\mathbf{U}^{I,\star}(t) = \begin{bmatrix} U_1^{I,\star}(t) \\ U_2^{I,\star}(t) \\ U_3^{I,\star}(t) \end{bmatrix}, \quad \mathbf{U}^{E,\star}(t) = \begin{bmatrix} U_1^{E,\star}(t) \\ U_2^{E,\star}(t) \\ U_3^{E,\star}(t) \end{bmatrix}, \quad \text{and } \boldsymbol{\Xi}_{imn}(\bar{\mathbf{x}}) = \begin{bmatrix} \bar{x}_i\delta_{mn} \\ \bar{x}_m\delta_{in} + \bar{x}_n\delta_{im} \\ \bar{x}_i\bar{x}_m\bar{x}_n \end{bmatrix}. \tag{6.32}$$

Here $\mathbf{U}^{I,\star}(t)$ and $\mathbf{U}^{E,\star}(t)$ are the radial basis arrays of the displacement field. $\boldsymbol{\Xi}_{imn}(\bar{\mathbf{x}})$ is the circumference basis array of the displacement field, whose third order tensorial entries can only be first or third order homogeneous function of $\bar{\mathbf{x}}$. Hence the disturbance displacement field has the following canonical form,

$$u_i^d(\mathbf{x}) = u_i^d(\bar{\mathbf{x}}, t) = \begin{cases} \epsilon^*_{mn}\ \boldsymbol{\Xi}^T_{imn}(\bar{\mathbf{x}})\mathbf{U}^{I,\star}(t)\ , & \forall \mathbf{x} \in \Omega_I\ , \\ \epsilon^*_{mn}\ \boldsymbol{\Xi}^T_{imn}(\bar{\mathbf{x}})\mathbf{U}^{E,\star}(t)\ , & \forall \mathbf{x} \in \Omega_E\ . \end{cases} \tag{6.33}$$

Furthermore, the kinematic relation (6.29) yields a differential mapping, which uniquely determines the relationship between the radial basis array of the strain field and the radial basis array of the displacement field. We also remark that the displacements are only uniquely determinable from the strain up to a rigid body displacement, which is set to zero here.

6.4 Eshelby tensors for finite domains

For simplicity, in the rest of the paper, we term the Eshelby tensor for a finite domain as the finite Eshelby tensor.

6.4.1 *Dirichlet-Eshelby tensor*

We first consider the Dirichlet BVP (6.4) in which case $\star = D$. Substituting (6.14) into (6.8), one obtains a tensorial integral equation for the unknown finite Eshelby tensor,

$$\mathcal{S}^{\bullet,D}_{ijmn}(\mathbf{x}) = \mathcal{S}^{\bullet,\infty}_{ijmn}(\mathbf{x}) + \frac{1}{2}\int_{\partial V}\Big(G^{\infty}_{ik,j}(\mathbf{x}-\mathbf{y}) + G^{\infty}_{jk,i}(\mathbf{x}-\mathbf{y})\Big)n_\ell(\mathbf{y})\mathbb{C}_{k\ell pq}\mathcal{S}^{E,D}_{pqmn}(\mathbf{y})dS_y \,. \tag{6.34}$$

This integral equation has two different forms, depending on whether $\mathbf{x}$ is inside or outside the inclusion,

$$\mathcal{S}^{I,D}_{ijmn}(\mathbf{x}) = \mathcal{S}^{I,\infty}_{ijmn}(\mathbf{x}) + \frac{1}{2}\int_{\partial V}\Big(G^{\infty}_{ik,j}(\mathbf{x}-\mathbf{y}) + G^{\infty}_{jk,i}(\mathbf{x}-\mathbf{y})\Big)n_\ell(\mathbf{y})\mathbb{C}_{k\ell pq}\mathcal{S}^{E,D}_{pqmn}(\mathbf{y})dS_y \,, \quad \forall \mathbf{x}\in\Omega_I \,, \tag{6.35}$$

$$\mathcal{S}^{E,D}_{ijmn}(\mathbf{x}) = \mathcal{S}^{E,\infty}_{ijmn}(\mathbf{x}) + \frac{1}{2}\int_{\partial V}\Big(G^{\infty}_{ik,j}(\mathbf{x}-\mathbf{y}) + G^{\infty}_{jk,i}(\mathbf{x}-\mathbf{y})\Big)n_\ell(\mathbf{y})\mathbb{C}_{k\ell pq}\mathcal{S}^{E,D}_{pqmn}(\mathbf{y})dS_y \,, \quad \forall \mathbf{x}\in\Omega_E \,. \tag{6.36}$$

By postulating that the finite Eshelby Tensor is a radially isotropic tensor that admits a set of known circumference basis, the system of tensorial equations can be simplified to a system of algebraic equation. The radial basis arrays of the Dirichlet-Eshelby tensors is given by

$$\mathbf{S}^{I,D}(t) = \mathbf{S}^{I,\infty}(t) + \mathbf{S}^{B,D}(t)\,, \quad 0 \le t < \rho_0\,, \tag{6.37}$$
$$\mathbf{S}^{E,D}(t) = \mathbf{S}^{E,\infty}(t) + \mathbf{S}^{B,D}(t)\,, \quad \rho_0 \le t \le 1\,, \tag{6.38}$$

The finite Eshelby tensors are composed of solution in infinite space and an additional term that represents boundary contribution. With the aid of some fundamental integrals we have derived (see Appendix A.), we can solve the above system of equations analytically. The boundary contribution to the finite RVE to the can be solved as

$$\mathbf{S}^{B,D}(t) = -\frac{\rho_0^3}{15(1-\nu)}\begin{bmatrix}5\nu-1\\4-5\nu\\0\\0\\0\\0\end{bmatrix} + \frac{\rho_0^3(1-\rho_0^2)}{20(1-\nu)(7-10\nu)}\begin{bmatrix}2(7-10\nu t^2)\\7(5t^2-3)-20\nu t^2\\-10t^2(7-10\nu)\\-40\nu t^2\\30\nu t^2\\0\end{bmatrix}. \tag{6.39}$$

With this the considered Dirichlet-Eshelby Problem of a spherical RVE is fully solved. We finally obtain the Dirichlet-Eshelby tensors considering multiplicative decomposition Eq. (6.26). The interior solution ($\mathbf{x} \in \Omega_I$) is

$$\begin{aligned}
\mathcal{S}^{I,D}_{ijmn}(\mathbf{x}) = \frac{1}{1-\nu}\Bigg[&\Big(\frac{5\nu-1}{15}(1-\rho_0^3) + \frac{7-10\nu t^2}{10(7-10\nu)}\rho_0^3(1-\rho_0^2)\Big)\delta_{ij}\delta_{mn} \\
&+ \Big(\frac{4-5\nu}{15}(1-\rho_0^3) + \frac{7(5t^2-3)-20\nu t^2}{20(7-10\nu)}\rho_0^3(1-\rho_0^2)\Big)(\delta_{im}\delta_{jn}+\delta_{in}\delta_{jm}) \\
&- \frac{t^2}{2}\rho_0^3(1-\rho_0^2)\,\delta_{ij}\bar{x}_m\bar{x}_n - \frac{2\nu t^2}{7-10\nu}\rho^3(1-\rho_0^2)\delta_{mn}\bar{x}_i\bar{x}_j + \frac{3\nu t^2}{2(7-10\nu)} \\
&\cdot\rho_0^3(1-\rho_0^2)(\delta_{im}\bar{x}_j\bar{x}_n + \delta_{in}\bar{x}_j\bar{x}_m + \delta_{jm}\bar{x}_i\bar{x}_n + \delta_{jn}\bar{x}_i\bar{x}_m)\Bigg], \qquad (6.40)
\end{aligned}$$

and the exterior solution ($\mathbf{x} \in \Omega_E$) is

$$\begin{aligned}
\mathcal{S}^{E,D}_{ijmn}(\mathbf{x}) = \frac{\rho_0^3}{1-\nu}\Bigg[&\Big(\frac{3\rho_0^2/t^2+10\nu-5}{30t^3} - \frac{5\nu-1}{15} + \frac{7-10\nu t^2}{10(7-10\nu)}(1-\rho_0^2)\Big)\delta_{ij}\delta_{mn} \\
&+ \Big(\frac{3\rho_0^2/t^2-10\nu+5}{30t^3} - \frac{4-5\nu}{15} + \frac{7(5t^2-3)-20\nu t^2}{20(7-10\nu)}(1-\rho_0^2)\Big) \\
&\cdot(\delta_{im}\delta_{jn}+\delta_{in}\delta_{jm}) - \Big(\frac{\rho_0^2/t^2-1}{2t^3} + \frac{t^2}{2}(1-\rho_0^2)\Big)\delta_{ij}\bar{x}_m\bar{x}_n \\
&- \Big(\frac{\rho_0^2/t^2+2\nu-1}{2t^3} + \frac{2\nu t^2}{7-10\nu}(1-\rho_0^2)\Big)\delta_{mn}\bar{x}_i\bar{x}_j \\
&- \Big(\frac{\rho_0^2/t^2-\nu}{2t^3} - \frac{3\nu t^2}{2(7-10\nu)}(1-\rho_0^2)\Big)(\delta_{im}\bar{x}_j\bar{x}_n + \delta_{in}\bar{x}_j\bar{x}_m \\
&+ \delta_{jm}\bar{x}_i\bar{x}_n + \delta_{jn}\bar{x}_i\bar{x}_m) + \frac{7\rho_0^2/t^2-5}{2t^3}\,\bar{x}_i\bar{x}_j\bar{x}_m\bar{x}_n\Bigg]. \qquad (6.41)
\end{aligned}$$

We can see that both the interior and the exterior Dirichlet Eshelby tensor are neither constant nor isotropic. The dependency on the position $\mathbf{x}$ is captured by the dependency on $\bar{\mathbf{x}}$ and t. Furthermore, both tensors depend explicitly on the ratio ρ_0 between inclusion and RVE. If we let $\rho_0 \to 0$ we recover the original infinite Eshelby tensors exactly since the boundary contribution then vanishes. To visualize the Dirichlet-Eshelby tensors the profiles of the components of the radial basis arrays $\mathbf{S}^{\bullet,\infty}(t), \mathbf{S}^{\bullet,D}(t)$ and $\mathbf{S}^{B,D}(t)$ are shown in Fig. 6.2. Here the relative size of the inclusion is chosen as $\rho_0 = 0.4$, so that the volume fraction becomes $\rho_0^3 = 0.064$. Poisson's ratio of the matrix phase is picked as $\nu = 0.3$. One can clearly observe that the boundary term $\mathbf{S}^{B,D}$, which can be understood as a correction of Eshelby's original result, is substantial. It can also be noted that there is a discontinuity across the interface between the inclusion and the matrix.

Remark 6.1. The kinematic relation between displacement and strain Eq. (6.29) yields the following differential mapping, which uniquely determines the relationship between the radial basis array of the strain field and the radial basis array of the displacement field,

$$\mathbf{S}^{\bullet,\star}(t) = \mathfrak{D}(t)\mathbf{U}^{\bullet,\star}(t)\,, \qquad (6.42)$$

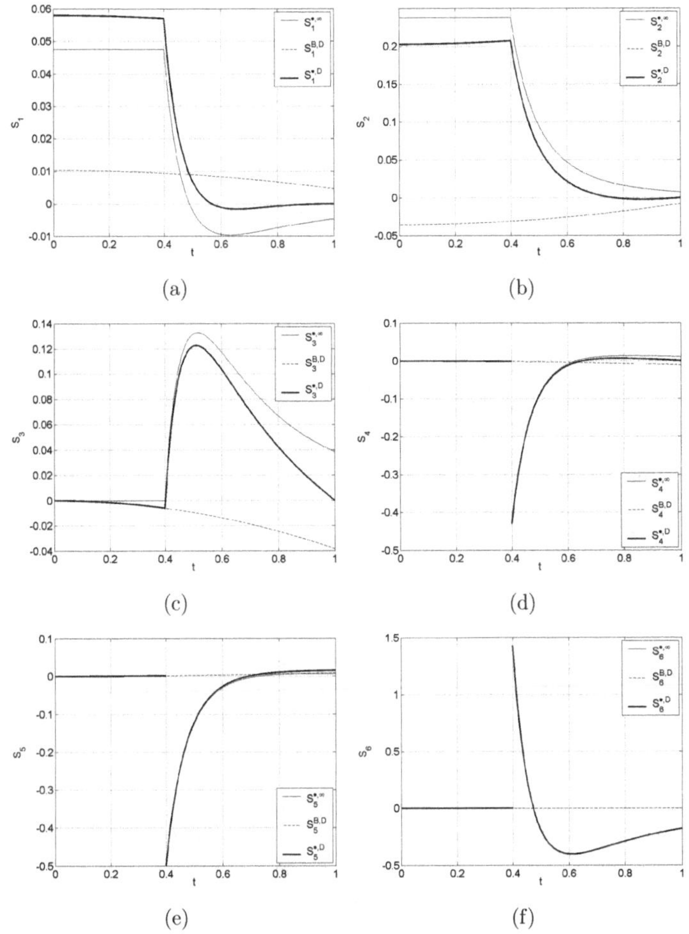

Fig. 6.2 The components of the radial basis arrays $\mathbf{S}^{\bullet,\infty}$, $\mathbf{S}^{B,D}$ and $\mathbf{S}^{\bullet,D}$.

where $\mathfrak{D}(t)$ is a differential operator that is defined in matrix form,

$$\mathfrak{D}(t) = \frac{1}{H_0}\begin{bmatrix} \dfrac{1}{t} & 0 & 0 \\ 0 & \dfrac{1}{t} & 0 \\ 0 & 0 & \dfrac{1}{t} \\ -\dfrac{1}{t}+\dfrac{d}{dt} & 0 & 0 \\ 0 & -\dfrac{1}{2}\dfrac{1}{t}+\dfrac{1}{2}\dfrac{d}{dt} & \dfrac{1}{2}\dfrac{1}{t} \\ 0 & 0 & -\dfrac{3}{t}+\dfrac{d}{dt} \end{bmatrix}^{6\times 3} . \tag{6.43}$$

Likewise, if $\mathbf{S}^{\bullet,\star}$ is given $\mathbf{U}^{\bullet,\star}$ can be determined from

$$\mathbf{U}^{\bullet,\star}(t) = \mathfrak{I}(t)\mathbf{S}^{\bullet,\star}(t) , \tag{6.44}$$

where $\mathfrak{I}(t)$ is the integration operator

$$\mathfrak{I}(t) = H_0 t \begin{bmatrix} 1\,0\,0\,0\,0\,0 \\ 0\,1\,0\,0\,0\,0 \\ 0\,0\,1\,0\,0\,0 \end{bmatrix}^{3\times 6} . \tag{6.45}$$

We remark that the displacements are only uniquely determinable from the strain up to a rigid body displacement, which is set to zero here.

The disturbance displacement field $u_i^d(\mathbf{x})$ is now given by Eqs. (6.28), (6.30) and (6.31), i.e.

$$u_i^d(\mathbf{x}) = \Xi^T_{imn}(\bar{\mathbf{x}})\mathbf{U}^{\bullet,D}(t)\,\epsilon^*_{mn} , \quad \forall \mathbf{x} \in V , \tag{6.46}$$

where the arrays $\Xi_{imn}(\bar{\mathbf{x}})$, $\mathbf{U}^{I,D}(t)$ and $\mathbf{U}^{E,D}(t)$ follow from Eqs. (6.32)$_3$ and (6.44). Applying operator (6.45) to the expressions (6.37), (6.38), (6.17) and (6.39) we easily obtain

$$\mathbf{U}^{I,D}(t) = \mathbf{U}^{I,\infty}(t) + \mathbf{U}^{B,D}(t) , \quad 0 \le t < \rho_0 , \tag{6.47}$$

$$\mathbf{U}^{E,D}(t) = \mathbf{U}^{E,\infty}(t) + \mathbf{U}^{B,D}(t) , \quad \rho_0 \le t \le 1 , \tag{6.48}$$

with

$$\mathbf{U}^{I,\infty}(t) = \frac{tA}{15(1-\nu)} \begin{bmatrix} 5\nu - 1 \\ 4 - 5\nu \\ 0 \end{bmatrix} , \tag{6.49}$$

$$\mathbf{U}^{E,\infty}(t) = \frac{\rho_0^3 A}{30t^2(1-\nu)} \begin{bmatrix} 3\rho_0^2/t^2 + 10\nu - 5 \\ 3\rho_0^2/t^2 - 10\nu + 5 \\ 15 - 15\rho_0^2/t^2 \end{bmatrix} , \tag{6.50}$$

and

$$\mathbf{U}^{B,D}(t) = -\frac{\rho_0^3 tA}{15(1-\nu)} \begin{bmatrix} 5\nu - 1 \\ 4 - 5\nu \\ 0 \end{bmatrix} + \frac{\rho_0^3(1-\rho_0^2)tA}{20(1-\nu)(7-10\nu)} \begin{bmatrix} 2(7 - 10\nu t^2) \\ 7(5t^2 - 3) - 20\nu t^2 \\ -10t^2(7 - 10\nu) \end{bmatrix} . \tag{6.51}$$

Here $\mathbf{U}^{I,\infty}(t)$ and $\mathbf{U}^{E,\infty}(t)$ are the radial basis array of Eshelby's classical solution in unbounded space and $\mathbf{U}^{B,D}(t)$ is the radial basis contribution from the Dirichlet boundary of the RVE.

We remark that u_i^d given by the equations above satisfies the Fredholm-type integral equation of the Dirichlet BVP (6.7) exactly. Furthermore it is readily verified that when $t = 1$,

$$\mathbf{U}^{E,D}(1) = \begin{bmatrix} 0 \\ 0 \\ 0 \end{bmatrix} \quad \rightarrow \quad u_i^d(\mathbf{y}) = \epsilon^*_{mn}\Xi^T_{imn}(\mathbf{n})\mathbf{U}^{E,D}(1) = 0 , \ \forall \mathbf{y} \in \partial V . \tag{6.52}$$

This confirms that the obtained displacement solution does indeed satisfy the Dirichlet boundary condition.

6.4.2 *Neumann-Eshelby tensor*

The solution of the Neumann-Eshelby Problem (now $\star = N$) is different from the Dirichlet-Eshelby Problem tensor; Here the solution is based on the displacement field. For the Neumann-Eshelby problem, we need to solve the following integral equation

$$u^d_m(\mathbf{x}) = -\int_V \mathbb{C}_{ijk\ell} G^{\infty}_{im,j}(\mathbf{x}-\mathbf{y})\epsilon^*_{k\ell}(\mathbf{y})d\Omega_y + \int_{\partial V} \mathbb{C}_{ijk\ell} u^d_k(\mathbf{y}) G^{\infty}_{im,j}(\mathbf{x}-\mathbf{y}) n_\ell(\mathbf{y}) dS_y \ . \tag{6.53}$$

where the displacements on the boundary of the RVE are non-zero. According to (6.33) we have

$$u^d_k(\mathbf{y}) = \epsilon^*_{mn} \boldsymbol{\Xi}^T_{kmn}(\mathbf{n})\, \mathbf{U}^{E,N}(1)\ , \quad \forall \mathbf{y} \in \partial V \ . \tag{6.54}$$

By substituting (6.54) into the integral equation corresponding to the Neumann BVP (6.9), we obtain an equation for the unknown radial basis, $\mathbf{U}^{\bullet,N}(t)$,

$$\begin{aligned} \epsilon^*_{mn} \boldsymbol{\Xi}^T_{imn}(\bar{\mathbf{x}}) \mathbf{U}^{\bullet,N}(t) = & -\epsilon^*_{mn} \int_{\Omega_I} \mathbb{C}_{pqmn} G^{\infty}_{pi,q}(\mathbf{x}-\mathbf{y}) d\Omega_y \\ & + \epsilon^*_{mn} \int_{\partial V} \mathbb{C}_{pqk\ell} G^{\infty}_{pi,q}(\mathbf{x}-\mathbf{y}) \boldsymbol{\Xi}^T_{kmn}(\mathbf{n}) \mathbf{U}^{E,N}(1) n_\ell(\mathbf{y}) dS_y \ , \end{aligned} \tag{6.55}$$

where $\bullet = I$, or E. Depending on whether $\mathbf{x}$ is inside or outside the inclusion, the domain integral in (6.55) has two different forms, which can be expressed in the canonical form,

$$-\int_{\Omega_I} \mathbb{C}_{pqmn} G^{\infty}_{pi,q}(\mathbf{x}-\mathbf{y}) d\Omega_y = \begin{cases} \boldsymbol{\Xi}^T_{imn}(\bar{\mathbf{x}}) \mathbf{U}^{I,\infty}(t)\ , & \forall \mathbf{x} \in \Omega_I\ , \\ \boldsymbol{\Xi}^T_{imn}(\bar{\mathbf{x}}) \mathbf{U}^{E,\infty}(t)\ , & \forall \mathbf{x} \in \Omega_E\ , \end{cases} \tag{6.56}$$

Here $\mathbf{U}^{I,\infty}(t)$ and $\mathbf{U}^{E,\infty}(t)$ are the radial basis arrays of Eshelby's classical solution for unbounded space (see Eqs. (6.49)-(6.50)). In analogy to the Dirichlet case, we stipulate that a similar canonical form holds for the Neumann boundary contribution in (6.55),

$$\int_{\partial V} \mathbb{C}_{pqk\ell} G^{\infty}_{pi,q}(\mathbf{x}-\mathbf{y}) \boldsymbol{\Xi}^T_{kmn}(\mathbf{n}) \mathbf{U}^{E,N}(1) n_\ell(\mathbf{y}) dS_y = \boldsymbol{\Xi}^T_{imn}(\bar{\mathbf{x}}) \mathbf{U}^{B,N}(t), \quad \forall \mathbf{x} \in V\ , \tag{6.57}$$

where $\mathbf{U}^{B,N}(t)$ denotes the radial basis array arising from the Neumann boundary. Substituting Eqs. (6.56)-(6.57) into (6.55) and eliminating ϵ^*_{mn} and the circumference basis $\boldsymbol{\Xi}^T_{imn}(\bar{\mathbf{x}})$, one may reduce Eq. (6.55) into a pair of parametric, algebraic equations for the radial basis arrays, $\mathbf{U}^{\bullet,N}(t)$, i.e.

$$\mathbf{U}^{I,N}(t) = \mathbf{U}^{I,\infty}(t) + \mathbf{U}^{B,N}(t)\ , \quad 0 \le t \le \rho_0\ , \tag{6.58}$$

$$\mathbf{U}^{E,N}(t) = \mathbf{U}^{E,\infty}(t) + \mathbf{U}^{B,N}(t)\ , \quad \rho_0 \le t \le 1\ . \tag{6.59}$$

Here $\mathbf{U}^{I,\infty}(t)$ and $\mathbf{U}^{E,\infty}(t)$ are the radial basis vectors of Eshelby's classical solution of unbounded space (see Eqs. (6.49)-(6.50)). After lengthy derivation, the boundary contribution, $\mathbf{U}^{B,N}(t)$ can be evaluated,

$$\mathbf{U}^{B,N}(t) = \frac{\rho_0^3 t A}{30(1-\nu)} \begin{bmatrix} 2-10\nu \\ 7-5\nu \\ 0 \end{bmatrix} - \frac{\rho_0^3(1-\rho_0^2) t A}{5(1-\nu)(7+5\nu)} \begin{bmatrix} 2(7-10\nu t^2) \\ 7(5t^2-3) - 20\nu t^2 \\ -10t^2(7-10\nu) \end{bmatrix} \tag{6.60}$$

Note the similarity between the two boundary contributions $\mathbf{U}^{B,N}(t)$ and $\mathbf{U}^{B,D}(t)$ (see Eq. (6.51)). With the above result we can now find $\mathbf{U}^{I,N}(t)$ and $\mathbf{U}^{E,N}(t)$ from Eqs. (6.58) and (6.59).

With the radial basis arrays of the displacement field given, one can differentiate it to obtain the radial basis array of the strain field,

$$\mathbf{S}^{B,N}(t) = \frac{\rho_0^3}{30(1-\nu)} \begin{bmatrix} 2-10\nu \\ 7-5\nu \\ 0 \\ 0 \\ 0 \\ 0 \end{bmatrix} - \frac{\rho_0^3(1-\rho_0^2)}{5(1-\nu)(7+5\nu)} \begin{bmatrix} 2(7-10\nu t^2) \\ 7(5t^2-3)-20\nu t^2 \\ -10t^2(7-10\nu) \\ -40\nu t^2 \\ 30\nu t^2 \\ 0 \end{bmatrix} . \tag{6.61}$$

In analogy to Eqs. (6.58) and (6.59) the radial basis arrays $\mathbf{S}^{\bullet,N}$ of the Neumann-Eshelby tensors now follows from the following decomposition,

$$\mathbf{S}^{I,N}(t) = \mathbf{S}^{I,\infty}(t) + \mathbf{S}^{B,N}(t) , \quad 0 \le t \le \rho_0 , \tag{6.62}$$

$$\mathbf{S}^{E,N}(t) = \mathbf{S}^{E,\infty}(t) + \mathbf{S}^{B,N}(t) , \quad \rho_0 \le t \le 1 . \tag{6.63}$$

The Neumann-Eshelby tensors for a spherical inclusion embedded in a spherical RVE under the prescribed traction boundary condition can now be obtained from Eq. (6.15) which yields the following exact and elementary expressions. The interior solution ($\mathbf{x} \in \Omega_I$) is

$$\begin{aligned} \mathcal{S}^{I,N}_{ijmn}(\mathbf{x}) = \frac{1}{1-\nu}\Bigg[& \left(\frac{5\nu-1}{15}(1-\rho_0^3) - \frac{2(7-10\nu t^2)}{5(7+5\nu)}\rho_0^3(1-\rho_0^2)\right)\delta_{ij}\delta_{mn} \\ & + \left(\frac{1-\nu}{2} + \frac{5\nu-7}{30}(1-\rho_0^3) - \frac{7(5t^2-3)-20\nu t^2}{5(7+5\nu)}\rho_0^3(1-\rho_0^2)\right)(\delta_{im}\delta_{jn}+\delta_{in}\delta_{jm}) \\ & + \frac{2t^2(7-10\nu)}{7+5\nu}\rho_0^3(1-\rho^2)\,\delta_{ij}\bar{x}_m\bar{x}_n + \frac{8\nu t^2}{7+5\nu}\rho^3(1-\rho_0^2)\delta_{mn}\bar{x}_i\bar{x}_j \\ & - \frac{6\nu t^2}{7+5\nu}\rho_0^3(1-\rho_0^2)(\delta_{im}\bar{x}_j\bar{x}_n + \delta_{in}\bar{x}_j\bar{x}_m + \delta_{jm}\bar{x}_i\bar{x}_n + \delta_{jn}\bar{x}_i\bar{x}_m)\Bigg] , \end{aligned} \tag{6.64}$$

and the exterior solution ($\mathbf{x} \in \Omega_E$) is

$$\begin{aligned} \mathcal{S}^{E,N}_{ijmn}(\mathbf{x}) = \frac{\rho_0^3}{1-\nu}\Bigg[& \left(\frac{3\rho^2/t^2+10\nu-5}{30t^3} + \frac{5\nu-1}{15} - \frac{2(7-10\nu t^2)}{5(7+5\nu)}(1-\rho_0^2)\right)\delta_{ij}\delta_{mn} \\ & + \left(\frac{3\rho_0^2/t^2-10\nu+5}{30t^3} + \frac{7-5\nu}{30} - \frac{7(5t^2-3)-20\nu t^2}{5(7+5\nu)}(1-\rho_0^2)\right) \\ & \cdot(\delta_{im}\delta_{jn}+\delta_{in}\delta_{jm}) - \left(\frac{\rho_0^2/t^2-1}{2t^3} - \frac{2t^2(7-10\nu)}{7+5\nu}(1-\rho_0^2)\right)\delta_{ij}\bar{x}_m\bar{x}_n \\ & - \left(\frac{\rho_0^2/t^2+2\nu-1}{2t^3} - \frac{8\nu t^2}{7+5\nu}(1-\rho_0^2)\right)\delta_{mn}\bar{x}_i\bar{x}_j \\ & - \left(\frac{\rho_0^2/t^2-\nu}{2t^3} + \frac{6\nu t^2}{7+5\nu}(1-\rho_0^2)\right)(\delta_{im}\bar{x}_j\bar{x}_n + \delta_{in}\bar{x}_j\bar{x}_m \\ & + \delta_{jm}\bar{x}_i\bar{x}_n + \delta_{jn}\bar{x}_i\bar{x}_m) + \frac{7\rho_0^2/t^2-5}{2t^3}\,\bar{x}_i\bar{x}_j\bar{x}_m\bar{x}_n\Bigg] . \end{aligned} \tag{6.65}$$

Figure 6.3 shows a comparison of the Neumann-Eshelby tensor with the original Eshelby tensor of unbounded domain in the cases of $\rho_0 = 0.4$ and $\nu = 0.3$. Here we display the six coefficients of the radial basis arrays of the finite Eshelby tensors, $\mathbf{S}^{\bullet,N}$ and the original Eshelby tensors, $\mathbf{S}^{\bullet,\infty}$. One can see that there are significant differences in the first three coefficients.

3

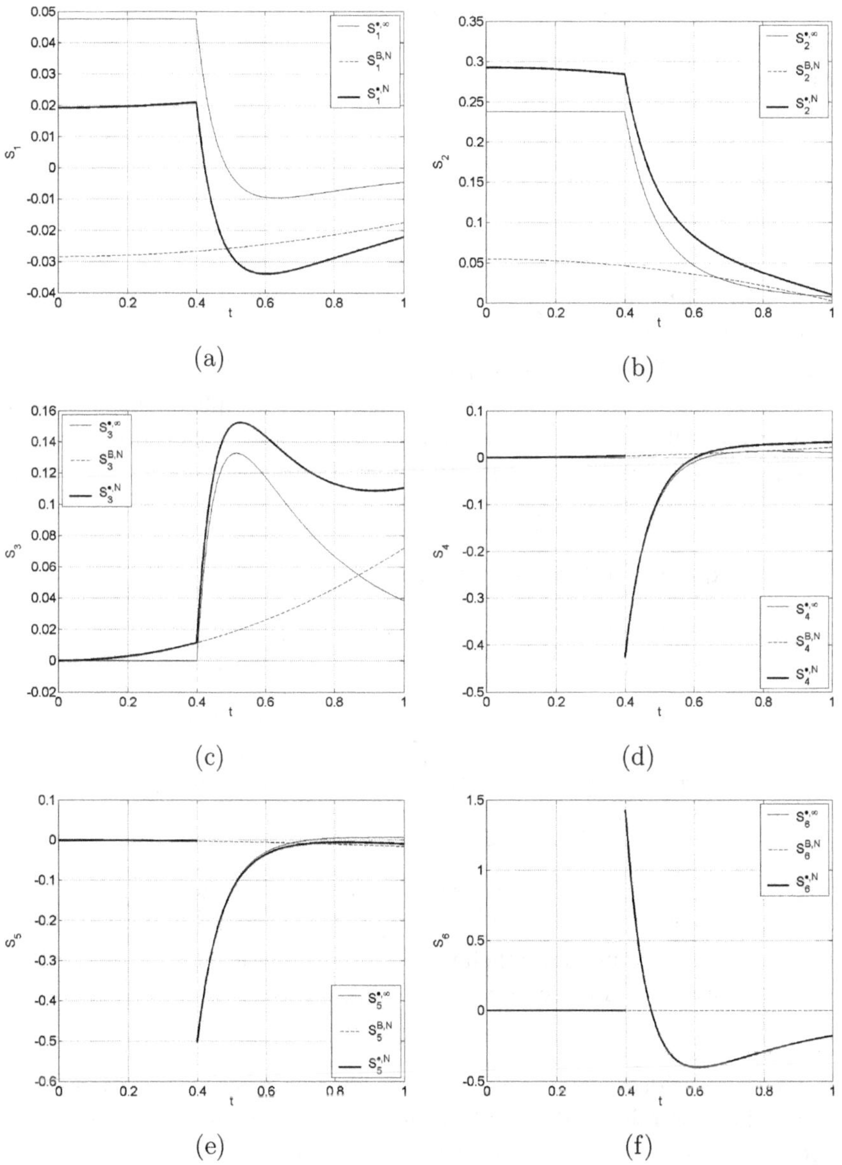

Fig. 6.3 The components of the radial basis arrays $\mathbf{S}^{\bullet,\infty}$, $\mathbf{S}^{B,N}$ and $\mathbf{S}^{\bullet,N}$.

Remark 6.2. Although the finite Eshelby tensors derived here are functions of the position vector $\mathbf{x}$, the following dilatational contractions remain constant

$$\mathbb{S}^{I,D}_{iijj}(\mathbf{x}) = \frac{(1-f)(1+\nu)}{1-\nu}\ , \qquad \mathbb{S}^{E,D}_{iijj}(\mathbf{x}) = -\frac{f(1+\nu)}{1-\nu}\ , \tag{6.66}$$

$$\mathbb{S}^{I,N}_{iijj}(\mathbf{x}) = \frac{(1+\nu)+2f(1-2\nu)}{1-\nu}\ , \quad \mathbb{S}^{E,N}_{iijj} = \frac{2f(1-2\nu)}{1-\nu}\ , \tag{6.67}$$

where $f = \rho_0^3$ is the volume fraction of the inclusion phase. Dilatational eigenstrains are associated with for example thermal expansion. It is also interesting to note that the Dirichlet and the Neumann Eshelby tensors have the ordering

$$\mathbb{S}^{I,N}_{iijj} \geq \mathbb{S}^{I,D}_{iijj}\ , \quad \mathbb{S}^{E,N}_{iijj} \geq \mathbb{S}^{E,D}_{iijj}\ , \qquad \text{“}=\text{” holds iff } f = 0\ , \tag{6.68}$$

and that the difference between interior and exterior solution is

$$\mathbb{S}^{I,D}_{iijj} - \mathbb{S}^{E,D}_{iijj} = \mathbb{S}^{I,N}_{iijj} - \mathbb{S}^{E,N}_{iijj} = \frac{1+\nu}{1-\nu} = \mathbb{S}^{I,\infty}_{iijj}\ . \tag{6.69}$$

Remark 6.3. With Finite Eshelby Tensor obtained, the radial projection of the disturbance stress field can be readily evaluated. The physical meaning of such a stress projection field is a set of parametric traction fields on the surfaces of successive concentric spheres. Any point, $\mathbf{x}$, inside the spherical RVE lies on a spherical surface whose normal $\bar{\mathbf{x}}$ is along the direction of the position vector $\mathbf{x}$. Thus the parametric traction field is defined as,

$$t_i^d(\mathbf{x}) = \sigma_{ji}^d(\mathbf{x})\ \bar{x}_j(\mathbf{x})\ , \tag{6.70}$$

which can be expressed in terms of the eigenstrain,

$$t_i^d(\mathbf{x}) = \begin{cases} \bar{x}_j(\mathbf{x})\ \mathbb{C}_{ijk\ell}\left(\mathbb{S}^{I,\star}_{k\ell mn}(\mathbf{x}) - \mathbb{I}^s_{k\ell mn}\right)\epsilon^*_{mn}\ , & \forall \mathbf{x} \in \Omega_I\ , \\ \bar{x}_j(\mathbf{x})\ \mathbb{C}_{ijk\ell}\ \mathbb{S}^{E,\star}_{k\ell mn}(\mathbf{x})\epsilon^*_{mn}\ , & \forall \mathbf{x} \in \Omega_E\ . \end{cases} \tag{6.71}$$

Here $\mathbb{I}^s_{k\ell mn}$ is the fourth-order symmetric identity tensor, which also falls into our definition of a *fourth-order radial isotropic tensor.* In analogy to the displacement field (see Eq. (6.33)) the disturbance traction can also be written as

$$t_i^d(\mathbf{x}) = \Xi^T_{imn}(\bar{\mathbf{x}})\mathbf{T}^{\bullet,\star}(t)\epsilon^*_{mn}\ , \tag{6.72}$$

where $\mathbf{T}^{\bullet,\star}$ is the radial basis array of the traction field and Ξ_{imn} is given by Eq. (6.32). It is readily verified that on the RVE boundary, the traction basis corresponding to the Neumann-Eshelby Problem is identically zero. Therefore the prescribed Neumann boundary condition is indeed satisfied by the solution presented.

Remark 6.4. In this section, the elastic fields due to a spherical inclusion subjected to prescribed eigenstrains and embedded in a finite spherical RVE are studied. On the outer surface of the RVE, uniform boundary conditions are prescribed, which are either a displacement (Dirichlet) boundary condition or a prescribed traction (Neumann) boundary condition.

The notion of a *radial isotropic tensor* is a generalization of the isotropic tensor. It has been argued that if a spherical inclusion is placed concentrically within a spherical RVE, the finite Eshelby tensors which map the prescribed eigenstrain to the disturbance strain field are radial isotropic tensors. In other words, the tensorial circumference basis for the finite Eshelby tensors is the same as the basis for the infinite Eshelby tensors.

By utilizing this property, we have solved a pair of Fredholm type integral equations, and we have obtained, for the first time, the exact, closed form solutions for both the interior and exterior Eshelby tensors for an inclusion in a finite, three-dimensional RVE. It has been revealed that the finite Eshelby tensors depend on both the location and the volume fraction of the inclusion, which accurately captures both the size effect of the inclusion and the boundary image contribution to the original Eshelby tensors.

We remark that our solution procedure circumvents the use of a finite Green's Function. This work may reveal some further insight into the search of a finite Green's Function, however, we believe, not without some added difficulties. We further note that, by using our solution technique, one may be able to extend the present spherical solution to elliptical geometry. The difficulty then will be to find the symmetry group of the circumference basis of the elliptical geometry, which is still invariant under the boundary integrals.

Remark 6.5. The combined Eshelby tensor The two fundamental solutions corresponding to the Dirichlet and the Neumann boundary conditions form a basis for the finite Eshelby tensors under general boundary value problem. Readers may consult [Sauer *et al.* (2007)] for detailed discussion on this matter. It should also be pointed out that even though the two basic finite Eshelby tensors obtained here are the solutions of the homogeneous inclusion problems, they are two fundamental elements for the finite Eshelby tensors of a general RVE with more complex micro-structures. By using superposition, they can be readily used to construct the solutions for n-inclusion ($n \geq 2$) problem, and they can be used to solve various homogenization problems as well as the problem of inhomogeneity induced elastic fields in a finite spherical domain.

6.5 Average Eshelby tensors and average disturbance fields

In previous sections, the exact solutions of the elastic fields of a spherical inclusion embedded in a finite spherical representative volume (RVE) are obtained under both the prescribed displacement (Dirichlet) boundary condition and the prescribed traction (Neumann) boundary condition.

For simplicity, we refer to the Dirichlet- and Neumann-Eshelby tensors of a finite domain as the finite Eshelby tensors. A salient feature of the finite Eshelby tensors is their ability to capture both the boundary effect, or image force effect, of an RVE and the size effect, i.e. the dependency on the volume fraction of the different phases of a composite. This offers great advantages and flexibilities in homogenization procedures, which is the focus of this second part of our work. Using the new finite Eshelby tensors we can modify the classical homogenization schemes and obtain some remarkable results. Furthermore several new homogenization schemes can be constructed by the application of the finite Eshelby tensors.

In recent years, nanocomposites have emerged as promising materials for future technologies e.g. [Calvert (1999); Thostenson *et al.* (2001)], because of their high strength, excellent conductivity in both heat transfer and electricity. Considerable attention has been devoted to study the interfacial strength, size effects, and agglomeration effects of nanocomosites e.g. [Fisher *et al.* (2002); Odegard *et al.* (2003); Shi *et al.* (2004); Sharma and Ganti (2004)]. The classical homogenization techniques have shown limitations to deal with the above issues. There is a call for a refined micromechanics theory for nano-composites e.g [Ovid'ko and Scheinerman (2005)]. One of the objectives of this research is towards establishing a refined micromechanics homogenization theory for nano-composites.

In previous sections, we derived the finite Eshelby tensors, $\mathbb{S}^{\bullet,D}$ and $\mathbb{S}^{\bullet,N}$, which are valid for a spherical inclusion Ω_I embedded at the center of a finite, spherical RVE V (see Fig. 6.1). In accordance with previous sections, we adopt the following nomenclature to describe the problem: The radii of inclusion and RVE are denoted by a and A, their ratio by $\rho_0 = a/A$. Any point $\mathbf{x}$ inside the RVE can be written as $\mathbf{x} = tA\bar{\mathbf{x}}$, where $t = |\mathbf{x}|/A$ and $\bar{\mathbf{x}} = \mathbf{x}/|\mathbf{x}|$ denote the normalized radial distance and direction of $\mathbf{x}$. The elasticity tensors of the two domains Ω_I and Ω_E are denoted by $\mathbb{C}^I$ and $\mathbb{C}^E = \mathbb{C}$.

For clarity, we first derive the expression of the average finite Eshelby tensors and discuss their relation with the average disturbance strain field.

6.5.1 *Average Eshelby tensors*

The spatial averaging operator is defined as

$$\langle ... \rangle_V = \frac{1}{|V|} \int_V ... \, dV \,, \tag{6.73}$$

where $|V|$ denotes the volume of the spatial domain V. Due to the radial isotropic structure of the finite Eshelby tensors, $\mathbb{S}^{\bullet,\star}_{ijmn}(\mathbf{x}) = \mathbf{\Theta}^T_{ijmn}(\bar{\mathbf{x}})\mathbf{S}^{\bullet,\star}(t)$ ($\bullet = I, E$; $\star = D, N$), their average over the RVE domain V can be written as

$$\langle \mathbb{S}^{\bullet,\star}_{ijmn}(\mathbf{x}) \rangle_\Omega = \frac{1}{|V|} \int_V \mathbf{S}^{\bullet,\star}(t) \cdot \mathbf{\Theta}_{ijmn}(\bar{\mathbf{x}}) \, dV \tag{6.74}$$

For each boundary condition (Dirichlet or Neumann), we have two Eshelby tensors, interior $\mathbb{S}^{I,\star}(\mathbf{x})$ for $\mathbf{x} \in \Omega_I$, or exterior $\mathbb{S}^{E,\star}(\mathbf{x})$ for $\mathbf{x} \in V/\Omega_I := \Omega_E$. Their average over the respective domains follows as

$$\langle \mathbb{S}^{I,\star}_{ijmn} \rangle_{\Omega_I} = s^{I,\star}_1 \, \mathbb{E}^{(1)}_{ijmn} + s^{I,\star}_2 \, \mathbb{E}^{(2)}_{ijmn} \,, \tag{6.75}$$

$$\langle \mathbb{S}^{E,\star}_{ijmn} \rangle_{\Omega_E} = s^{E,\star}_1 \, \mathbb{E}^{(1)}_{ijmn} + s^{E,\star}_2 \, \mathbb{E}^{(2)}_{ijmn} \,, \tag{6.76}$$

where $\mathbb{E}^{(1)}_{ijmn}$ and $\mathbb{E}^{(2)}_{ijmn}$ are the following isotropic basis tensors,

$$\mathbb{E}^{(1)}_{ijmn} = \tfrac{1}{3}\delta_{ij}\delta_{mn}\ , \quad \mathbb{E}^{(2)}_{ijmn} = \tfrac{1}{2}\left(\delta_{im}\delta_{jn} + \delta_{in}\delta_{jm}\right) - \tfrac{1}{3}\delta_{ij}\delta_{mn}\ . \tag{6.77}$$

The coefficients $s_1^{I,\star}$, $s_2^{I,\star}$ and $s_1^{E,\star}$, $s_2^{E,\star}$ depend on the volume fraction $f := \rho_0^3$ and are given as

$$s_1^{I,D} = \frac{(1+\nu)(1-f)}{3(1-\nu)}, \quad s_2^{I,D} = \frac{2(4-5\nu)(1-f)}{15(1-\nu)} - 21\gamma_u[f](1-f^{2/3})\ , \tag{6.78}$$

$$s_1^{E,D} = -\frac{(1+\nu)f}{3(1-\nu)}, \quad s_2^{E,D} = -\frac{2(4-5\nu)f}{15(1-\nu)} + 21\gamma_u[f]f\frac{1-f^{2/3}}{1-f}, \tag{6.79}$$

for the Dirichlet B.C. and

$$s_1^{I,N} = \frac{1+\nu+2(1-2\nu)f}{3(1-\nu)}, \quad s_2^{I,N} = \frac{2(4-5\nu)+(7-5\nu)f}{15(1-\nu)} + 21\gamma_t[f](1-f^{2/3}), \tag{6.80}$$

$$s_1^{E,N} = \frac{2(1-2\nu)f}{3(1-\nu)}, \quad s_2^{E,N} = \frac{(7-5\nu)f}{15(1-\nu)} - 21\gamma_t[f]f\frac{1-f^{2/3}}{1-f}, \tag{6.81}$$

for the Neumann B.C. Here we have denoted

$$\gamma_u[f] := \frac{f(1-f^{2/3})}{10(1-\nu)(7-10\nu)}\ , \quad \gamma_t[f] := \frac{4f(1-f^{2/3})}{10(1-\nu)(7+5\nu)}\ . \tag{6.82}$$

In fact, Eqs. (6.78)-(6.81) are the precise formulas of the size-effect characterization of the inclusion problem. One can find that this effect is linear for the bulk modulus, whereas it is nonlinear in the shear modulus. In contrast to the average finite Eshelby tensors we recall the average Eshelby tensor for a spherical inclusion in an unbounded medium,

$$\langle \mathbb{S}^{\bullet,\infty}_{ijmn}\rangle_{\Omega_I} = s_1^{\bullet,\infty}\,\mathbb{E}^{(1)}_{ijmn} + s_2^{\bullet,\infty}\,\mathbb{E}^{(2)}_{ijmn}\ , \quad \bullet = I \text{ or } E \tag{6.83}$$

$$s_1^{I,\infty} = \frac{1+\nu}{3(1-\nu)}\ , \quad s_2^{I,\infty} = \frac{2(4-5\nu)}{15(1-\nu)}\ , \tag{6.84}$$

$$s_1^{E,\infty} = 0, \quad s_2^{E,\infty} = 0\ . \tag{6.85}$$

Figure 6.4 displays the behaviors of all the coefficients $s_i^{\bullet,\star}$ in dependance of f. The Poisson's ratio is chosen as $\nu = 0.2$. We observe that for the Dirichlet case the coefficients decrease, while for the Neumann case they increase with growing f. The classical Eshelby tensors do not depend on f. Note that when $f \to 0$ in Eqs. (6.78) - (6.81) we recover the expressions for the average of the classical Eshelby tensors. The fact that $s_i^{E,\infty} = 0$ implies the well-known Tanaka-Mori Lemma (see below).

Let us define the difference $\Delta s_i^\star = s_i^{I,\star} - s_i^{E,\star}$; we have

$$\Delta s_1^D = \Delta s_1^N = \Delta s_1^\infty = \frac{1+\nu}{3(1-\nu)}\ , \tag{6.86}$$

$$\Delta s_2^D = \Delta s_2^\infty - 21\gamma_u[f]\frac{1-f^{2/3}}{1-f}\ , \quad \Delta s_2^\infty = \frac{2(4-5\nu)}{15(1-\nu)}\ , \tag{6.87}$$

$$\Delta s_2^N = \Delta s_2^\infty + 21\gamma_t[f]\frac{1-f^{2/3}}{1-f}\ . \tag{6.88}$$

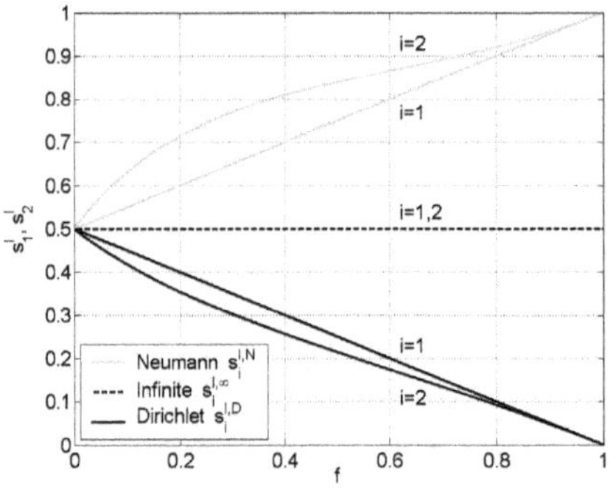

(a)

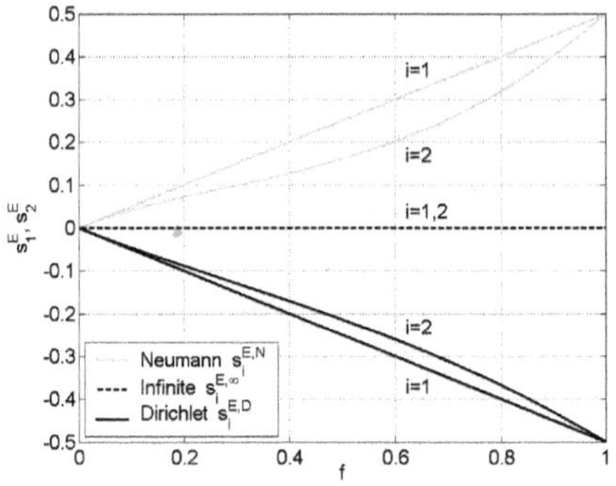

(b)

Fig. 6.4 Average Eshelby tensor coefficients s_1^I, s_1^E $(i = 1)$ and s_2^I, s_2^E $(i = 2)$.

6.5.2 *Average disturbance fields*

The finite Eshelby tensors can be conveniently used to represent the average disturbance fields. Recall the classical Tanaka-Mori Lemma [Tanaka and Mori (1972)]: the exterior average disturbance strain in the exterior domain is zero (see Eq. (6.85))

$$\langle \epsilon^d \rangle_{\Omega_E} = \langle \mathbb{S}^{E,\infty} \rangle_{\Omega_E} : \epsilon^* = (s_1^{E,\infty} \mathbb{E}^{(1)} + s_2^{E,\infty} \mathbb{E}^{(2)}) : \epsilon^* = 0 \,, \tag{6.89}$$

A similar result holds for the disturbance stress field for a linear elastic medium. Using the new finite Eshelby tensors $\mathbb{S}^D$ and $\mathbb{S}^N$ the original Tanaka-Mori Lemma result is modified. The exterior average disturbance strain field is neither zero for the Dirichlet problem

$$\langle \epsilon^d \rangle_{\Omega_E} = \langle \mathbb{S}^{E,D} \rangle_{\Omega_E} : \epsilon^* = (s_1^{E,D} \mathbb{E}^{(1)} + s_2^{E,D} \mathbb{E}^{(2)}) : \epsilon^* \neq 0 \,, \tag{6.90}$$

nor is it zero for the Neumann problem

$$\langle \epsilon^d \rangle_{\Omega_E} = \langle \mathbb{S}^{E,N} \rangle_{\Omega_E} : \epsilon^* = (s_1^{E,N} \mathbb{E}^{(1)} + s_2^{E,N} \mathbb{E}^{(2)}) : \epsilon^* \neq 0 \,, \tag{6.91}$$

(unless $f = 0$). However, in view of (6.79) and (6.81), we find that for both problems,

$$\langle \epsilon^d \rangle_{\Omega_E} = \mathcal{O}(f) \,, \tag{6.92}$$

which can be viewed as a modified Tanaka-Mori Lemma. One then recovers the original result as $f \to 0$.

Moreover, consider the Dirichlet problem. We can exactly satisfy a key assumption, the average strain theorem

$$\langle \epsilon \rangle_\Omega = \langle \epsilon^0 + \epsilon^d \rangle_\Omega = \epsilon^0 + \langle \epsilon^d \rangle_\Omega = \epsilon^0 \,, \tag{6.93}$$

since the average disturbance strain field in V is zero:

$$\langle \epsilon^d \rangle_\Omega = f \langle \epsilon^d \rangle_{\Omega_I} + (1-f) \langle \epsilon^d \rangle_{\Omega_E} = \left(f \langle \mathbb{S}^{I,D} \rangle_{\Omega_I} + (1-f) \langle \mathbb{S}^{E,D} \rangle_{\Omega_E} \right) : \epsilon^* = 0 \,. \tag{6.94}$$

Likewise, for the Neumann problem, the average stress theorem

$$\langle \boldsymbol{\sigma} \rangle_\Omega = \boldsymbol{\sigma}^0 \ , \tag{6.95}$$

is exactly satisfied since $\langle \boldsymbol{\sigma}^d \rangle_\Omega = f \langle \boldsymbol{\sigma}^d \rangle_{\Omega_I} + (1-f)\langle \boldsymbol{\sigma}^d \rangle_{\Omega_E} = 0$ due to

$$f \langle \mathbb{T}^{I,N} \rangle_{\Omega_I} + (1-f) \langle \mathbb{T}^{E,N} \rangle_{\Omega_E} = \mathbb{O} \ , \tag{6.96}$$

where $\mathbb{T}^{I,N}$ and $\mathbb{T}^{E,N}$ are the conjugate Neumann Eshelby tensors related to the Neumann Eshelby tensors by the expressions

$$\langle \mathbb{S}^{I,N} \rangle_{\Omega_I} + \langle \mathbb{T}^{I,N} \rangle_{\Omega_I} = \mathbb{I}^s \ , \quad \text{and} \quad \langle \mathbb{S}^{E,N} \rangle_{\Omega_E} + \langle \mathbb{T}^{E,N} \rangle_{\Omega_E} = \mathbb{O} \ . \tag{6.97}$$

where $\mathbb{I}^s$ is the fourth order symmetric unit tensor and $\mathbb{O}$ is the fourth order null tensor.

6.6 Improvements of classical homogenization methods

We now use the finite Eshelby tensors in two classical homogenization procedures to estimate effective material properties, namely, the homogenization for composites with dilute suspension and the Mori-Tanaka model.

6.6.1 *Dilute suspension model*

The dilute suspension method predicts two different effective elastic tensors depending on the different boundary conditions, e.g. [Nemat-Nasser and Hori (1999)]. We first consider the prescribed macro-strain B.C., i.e. the Dirichlet BVP ($\boldsymbol{u}^d = 0$ on ∂V), as discussed in Part I. The average stress consistency condition for the considered homogenization scheme (for prescribed eigenstrain within Ω_I as motivated in Part I) is

$$\mathbb{C}^I : (\boldsymbol{\epsilon}^0 + \langle \boldsymbol{\epsilon}^d \rangle_{\Omega_I}) = \mathbb{C} : (\boldsymbol{\epsilon}^0 + \langle \boldsymbol{\epsilon}^d \rangle_{\Omega_I} - \boldsymbol{\epsilon}^*) \ , \quad \forall \mathbf{x} \in \Omega_I \ . \tag{6.98}$$

Note that $\mathbb{C}^I$, $\mathbb{C}$, $\boldsymbol{\epsilon}^0$ and $\boldsymbol{\epsilon}^*$ are considered constant. From Eq. (6.98) we obtain

$$\boldsymbol{\epsilon}^0 + \langle \boldsymbol{\epsilon}^d \rangle_{\Omega_I} = \mathbb{A} : \boldsymbol{\epsilon}^* \ , \tag{6.99}$$

where $\mathbb{A} := (\mathbb{C} - \mathbb{C}^I)^{-1} : \mathbb{C}$. Consider the interior average of the disturbance strain field,

$$\langle \boldsymbol{\epsilon}^d \rangle_{\Omega_I} = \langle \mathbb{S}^{I,D} \rangle_{\Omega_I} : \boldsymbol{\epsilon}^* \ , \tag{6.100}$$

and substitute (6.100) into (6.99). This yields

$$\boldsymbol{\epsilon}^* = \left[\mathbb{A} - \langle \mathbb{S}^{I,D} \rangle_{\Omega_I} \right]^{-1} : \boldsymbol{\epsilon}^0 \ , \tag{6.101}$$

and consequently,

$$\langle \boldsymbol{\epsilon} \rangle_{\Omega_I} = \boldsymbol{\epsilon}^0 + \langle \boldsymbol{\epsilon}^d \rangle_{\Omega_I} = \mathbb{A} : \left[\mathbb{A} - \langle \mathbb{S}^{I,D} \rangle_{\Omega_I} \right]^{-1} : \boldsymbol{\epsilon}^0 \ . \tag{6.102}$$

Following the standard procedure e.g. [Nemat-Nasser and Hori (1999)], we find the estimate of the effective elasticity tensor for the prescribed macrostrain BC,

$$\boxed{\overline{\mathbb{C}} = \mathbb{C} - f\mathbb{C} : \left[\mathbb{A} - \langle \mathbb{S}^{I,D} \rangle_{\Omega_I}\right]^{-1}} \, . \tag{6.103}$$

The only difference between Eq. 6.103 and the classical solution for dilute suspension is that a different Eshelby tensor is used. Considering isotropic materials, the effective bulk and shear moduli become

$$\bar{\kappa} = \kappa - f\kappa\Big[\frac{1}{1 - \kappa^I/\kappa} - s_1^{I,D}\Big]^{-1}, \qquad \bar{\mu} = \mu - f\mu\Big[\frac{1}{1 - \mu^I/\mu} - s_2^{I,D}\Big]^{-1} . \tag{6.104}$$

For the prescribed macrostress boundary condition, the new Dilute Suspension estimate is

$$\boxed{\overline{\mathbb{D}} = \mathbb{D} + f\mathbb{D} : \left[\mathbb{A} - \langle \mathbb{S}^{I,N} \rangle_{\Omega_I}\right]^{-1}} \, , \tag{6.105}$$

where $\langle \mathbb{S}^{I,N} \rangle_{\Omega_I}$ is the interior average Neumann-Eshelby tensor. For isotropic composites, the corresponding effective bulk and shear moduli are,

$$\bar{\kappa}^{-1} = \kappa^{-1} + f\kappa^{-1}\Big[\frac{1}{1 - \kappa^I/\kappa} - s_1^{I,N}\Big]^{-1}, \; \bar{\mu}^{-1} = \mu^{-1} + f\mu^{-1}\Big[\frac{1}{1 - \mu^I/\mu} - s_2^{I,N}\Big]^{-1} . \tag{6.106}$$

Figure 6.5 shows the curves of the normalized bulk modulus, $\bar{\kappa}/\kappa$, and shear modulus, $\bar{\mu}/\mu$, in dependance of the volume fraction f of the inclusion. The material properties of the inclusion are chosen as $\kappa^I/\kappa = 10$, $\mu^I/\mu = 4$, with $\nu = 0.1$. We have plotted result (6.103) using the Dirichlet-Eshelby tensor $\mathbb{S}^{I,D}$ (dark) and (6.105) using the Neumann-Eshelby tensor $\mathbb{S}^{I,N}$ (light). We compare the new results with the conventional dilute suspension results using the infinite Eshelby tensor $\mathbb{S}^{I,\infty}$ in (6.103) (dashed line 2) and in (6.105) (dashed line 1).

From this figure we can observe the well-known result that the classical solution is not self-consistent, i.e $\bar{\mathbb{D}} \neq \bar{\mathbb{C}}^{-1}$. When we use the new finite Eshelby tensors this situation is significantly improved. For the effective bulk modulus, the new scheme is self-consistent, i.e. $(6.104)_1$ and $(6.106)_1$ are equal. The estimated effective shear modulus is not self-consistent, but it is quite close as shown in Fig. 6.5.

6.6.2 *A refined Mori-Tanaka model*

The original Mori and Tanaka model [Mori and Tanaka (1973)] is derived for an infinite RVE. In the following, we re-derive the Mori-Tanaka estimate for a two-phase composite in a finite RVE.

Realistically, the boundary condition of an RVE is neither a prescribed displacement boundary condition nor is it a prescribed traction boundary condition. One can thus define a 'General Boundary Eshelby Tensor' as the linear combination of the Dirichlet-Eshelby tensor and the Neumann-Eshelby tensor,

$$\mathbb{S}^{\bullet,F} = \alpha\mathbb{S}^{\bullet,D} + (1 - \alpha)\mathbb{S}^{\bullet,N} \, , \quad \bullet = I, \text{or } E \, . \tag{6.107}$$

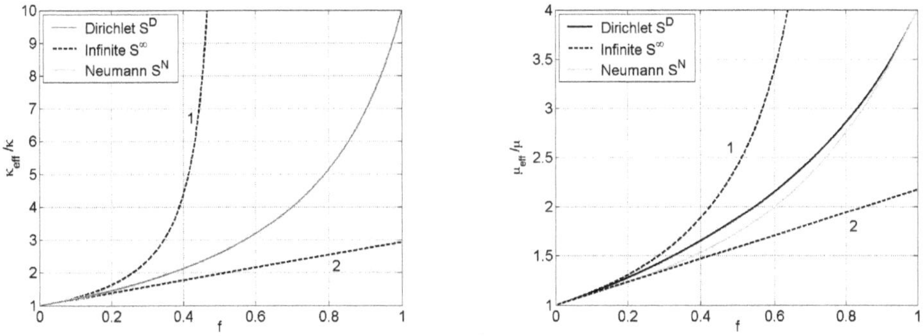

(a) (b)

Fig. 6.5 Effective moduli $\bar{\kappa}$, $\bar{\mu}$ (or κ_{eff} and μ_{eff}) obtained by using the dilute suspension method.

For detailed justification, derivation and discussion of this concept, readers are referred to Sauer et al [Sauer *et al.* (2007)] .

The essence of the Mori-Tanaka procedure is the following incremental homogenization procedure. Let us denote the current background strain of the RVE as $\langle\boldsymbol{\epsilon}^b\rangle_V$, which may or may not be the average strain of the RVE. Adding an inclusion (or a cluster of inclusions represented by a single inclusion) into the RVE, the new average strain $\langle\boldsymbol{\epsilon}\rangle$ in each phase will be the sum of the background strain and the disturbance strain,

$$\langle\boldsymbol{\epsilon}\rangle_{\Omega_I} = \langle\boldsymbol{\epsilon}^b\rangle_V + \langle\boldsymbol{\epsilon}^d\rangle_{\Omega_I} \ , \tag{6.108}$$

$$\langle\boldsymbol{\epsilon}\rangle_{\Omega_E} = \langle\boldsymbol{\epsilon}^b\rangle_V + \langle\boldsymbol{\epsilon}^d\rangle_{\Omega_E} \ . \tag{6.109}$$

The classical Tanaka-Mori Lemma states that $\langle\boldsymbol{\epsilon}^d\rangle_{\Omega_E} = 0$. This is only true when the RVE is infinite, since $\langle\mathbb{S}^{E,\infty}\rangle_{\Omega_I} = 0$. For a finite RVE, we have to take into account the change of the effective material properties in the matrix,

$$\langle\boldsymbol{\epsilon}\rangle_{\Omega_I} = \langle\boldsymbol{\epsilon}^b\rangle_V + \langle\mathbb{S}^{I,F}\rangle_{\Omega_I} : \boldsymbol{\epsilon}^* \ , \tag{6.110}$$

$$\langle\boldsymbol{\epsilon}\rangle_{\Omega_E} = \langle\boldsymbol{\epsilon}^b\rangle_V + \langle\mathbb{S}^{E,F}\rangle_{\Omega_E} : \boldsymbol{\epsilon}^* \ . \tag{6.111}$$

Consider the average stress consistency condition (for $\mathbf{x} \in \Omega_I$)

$$\mathbb{C}^I : \langle\boldsymbol{\epsilon}\rangle_{\Omega_I} = \mathbb{C} : \left(\langle\boldsymbol{\epsilon}\rangle_{\Omega_I} - \boldsymbol{\epsilon}^*\right) \ . \tag{6.112}$$

Solving Eqs. (6.112) for $\langle\boldsymbol{\epsilon}\rangle_{\Omega_I}$ yields

$$\langle\boldsymbol{\epsilon}\rangle_{\Omega_I} = \mathbb{A} : \boldsymbol{\epsilon}^* \ , \tag{6.113}$$

where $\mathbb{A} = (\mathbb{C} - \mathbb{C}^I)^{-1} : \mathbb{C}$. Considering Eq. (6.110), we can express the eigenstrain in terms of the background strain as

$$\boldsymbol{\epsilon}^* = [\mathbb{A} - \langle\mathbb{S}^{I,F}\rangle_{\Omega_I}]^{-1} : \langle\boldsymbol{\epsilon}^b\rangle_\Omega \ . \tag{6.114}$$

Considering the basic average equation of the strain

$$\langle \boldsymbol{\epsilon} \rangle_\Omega = f \langle \boldsymbol{\epsilon} \rangle_{\Omega_I} + (1-f) \langle \boldsymbol{\epsilon} \rangle_{\Omega_E} \,, \tag{6.115}$$

and substituting (6.110), (6.111) and (6.114) into (6.115), we can express the average strain $\langle \boldsymbol{\epsilon} \rangle_\Omega$ in terms of the background strain as

$$\langle \boldsymbol{\epsilon} \rangle_\Omega = \mathcal{A}^F : \langle \boldsymbol{\epsilon}^b \rangle_\Omega \,. \tag{6.116}$$

Here $\mathcal{A}^F$ is the concentration tensor defined as

$$\mathcal{A}^F = \left[\mathbb{A} - (1-f)\big(\langle \mathbb{S}^{I,F} \rangle_{\Omega_I} - \langle \mathbb{S}^{E,F} \rangle_{\Omega_E} \big) \right] : \left[\mathbb{A} - \langle \mathbb{S}^{I,F} \rangle_{\Omega_I} \right]^{-1} . \tag{6.117}$$

By virtue of Eqs. (6.113) and (6.114), the average stress field inside the inclusion can now be written as

$$\langle \boldsymbol{\sigma} \rangle_{\Omega_I} = \mathbb{C} : \left[\mathbb{A} - \mathbb{I}^{(4s)} \right] : \left[\mathbb{A} - \langle \mathbb{S}^{I,F} \rangle_{\Omega_I} \right]^{-1} : \langle \boldsymbol{\epsilon}^b \rangle_\Omega \,. \tag{6.118}$$

Applying the basic equation for mixture to the stress field,

$$\langle \boldsymbol{\sigma} \rangle_\Omega = f \langle \boldsymbol{\sigma} \rangle_{\Omega_I} + (1-f) \langle \boldsymbol{\sigma} \rangle_{\Omega_E} \,, \tag{6.119}$$

and substituting (6.110), (6.111) and (6.114) into (6.119), we can express the average stress $\langle \boldsymbol{\sigma} \rangle_\Omega$ in terms of the background strain as

$$\langle \boldsymbol{\sigma} \rangle_\Omega = \mathcal{B}^F : \langle \boldsymbol{\epsilon}^b \rangle_\Omega \,, \tag{6.120}$$

with

$$\mathcal{B}^F = \mathbb{C} : \left[\mathbb{A} - f\mathbb{I}^{(4s)} - (1-f)\big(\langle \mathbb{S}^{I,F} \rangle_{\Omega_I} - \langle \mathbb{S}^{E,F} \rangle_{\Omega_E} \big) \right] : \left[\mathbb{A} - \langle \mathbb{S}^{I,F} \rangle_{\Omega_I} \right]^{-1} . \tag{6.121}$$

Finally from $\langle \boldsymbol{\sigma} \rangle_\Omega = \overline{\mathbb{C}} : \langle \boldsymbol{\epsilon} \rangle_\Omega$, we obtain the effective elastic tensor,

$$\boxed{\overline{\mathbb{C}} = \mathbb{C} - f\mathbb{C} : \Big[\mathbb{A} - (1-f)\Big(\langle \mathbb{S}^{I,F} \rangle_{\Omega_I} - \langle \mathbb{S}^{E,F} \rangle_{\Omega_E} \Big) \Big]^{-1} \,, \quad \overline{\mathbb{D}} = \overline{\mathbb{C}}^{-1} .} \tag{6.122}$$

We note in passing that this model is self-consistent. The homogenization procedure with finite Eshelby tensors, $\mathbb{S}^{I,F}$ and $\mathbb{S}^{E,F}$, furnishes a refined Mori-Tanaka model. For isotropic two-phase composites, the corresponding formulas are

$$\bar{\kappa} = \kappa - f\kappa \Big[\frac{1}{1-\kappa^I/\kappa} - (1-f)\Delta s_1^F \Big]^{-1} , \quad \bar{\mu} = \mu - f\mu \Big[\frac{1}{1-\mu^I/\mu} - (1-f)\Delta s_2^F \Big]^{-1} . \tag{6.123}$$

Note that the differences $\Delta s_1^F = s_1^{I,F} - s_1^{E,F}$ and $\Delta s_2^F = s_2^{I,F} - s_2^{E,F}$ are given by Eqs. (6.86)-(6.88).

Figure 6.6 displays the profiles of the normalized effective moduli $\bar{\kappa}/\kappa$ and $\bar{\mu}/\mu$ over the volume fraction of the second phase. The same material data is used for the results shown in Fig. 6.5. In the case of the bulk modulus the dark, dashed and the light curves match exactly. Indeed, expression $(6.123)_1$ is mathematically identical when applying $\mathbb{S}^{\bullet,\infty}$, $\mathbb{S}^{\bullet,D}$ or $\mathbb{S}^{\bullet,N}$, since $\Delta s_1^\infty = \Delta s_1^D = \Delta s_1^N$ as noted in Eq. (6.86). For the shear modulus Eq. $(6.123)_2$ gives three distinct lines when applying $\mathbb{S}^{\bullet,\infty}$, $\mathbb{S}^{\bullet,D}$ or $\mathbb{S}^{\bullet,N}$.

Remarkably, when comparing Figs. 6.5 and 6.6, we find that the dark and light lines match exactly. In other words, it can be shown that by using the finite Eshelby tensors, the dilute suspension method (6.103) and (6.105) is equivalent to the Mori-Tanaka method (6.122), when using the corresponding $\mathbb{S}^{\bullet,D}$ and $\mathbb{S}^{\bullet,N}$. The finite Eshelby tensors $\mathbb{S}^{\bullet,D}$ and $\mathbb{S}^{\bullet,N}$ unify the previously distinct homogenization methods.

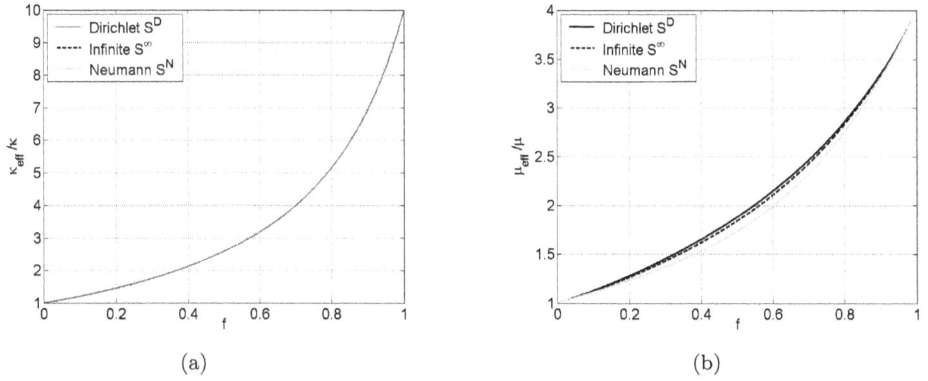

Fig. 6.6 Effective moduli $\bar{\kappa}$, $\bar{\mu}$ (or κ_{eff} and μ_{eff}) obtained by using the Mori-Tanaka method.

6.6.3 *Multiphase variational bounds*

One of the useful homogenization methods for composite materials are the Hashin-Shtrikman variational principles, which have been extensively used in deriving bounds for effective material properties. In the procedure of deriving the variational bounds, the Eshelby tensor is needed in order to estimate the disturbance strain field due to stress polarization or to estimate the disturbance stress field due to the eigenstrain.

Since the classical Eshelby tensor is obtained for an inclusion solution in an unbounded region, in principle, it can not be directly used in the derivation of the variational bounds of a composite with finite volume. In the past, additional probability arguments and approximations based on assumptions of the statistical nature of the inclusion distribution, have been employed to justify the use of the classical Eshelby tensor, e.g. [Willis (1981)].

In this section, we show that the finite Eshelby tensors are a perfect fit for the Hashin-Shtrikman variational principle. They can be directly used in combination with the Hashin-Shtrikman principles to derive variational bounds without resorting to additional statistical arguments.

6.6.3.1 *Eshelby tensors for multiphase composites*

To utilize the finite Eshelby tensors to represent different microstructure, a so-called spherical shell model is developed, that is a n-phase composite RVE modeled by n concentric spherical shells. To illustrate the model, we present the detailed study of a three-layer shell model (see Fig. 6.7).

For the three-layer shell model, the RVE consists of three concentric spherical shells, which are labeled as

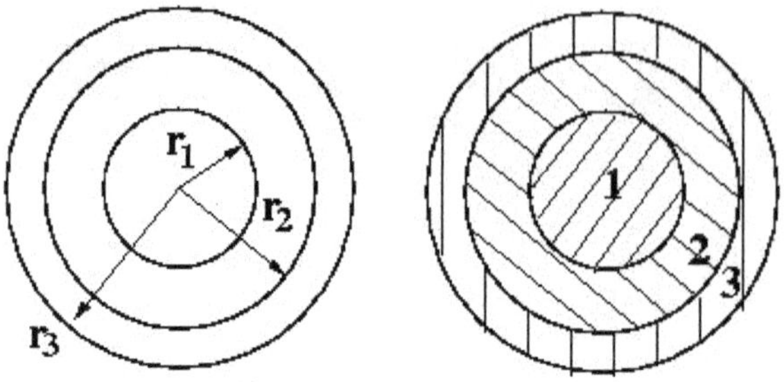

Fig. 6.7 A three-layer shell model.

$$\Omega_1(\mathbf{x}) = \{\mathbf{x} \mid |\mathbf{x}| < r_1\}, \ \Omega_2(\mathbf{x}) = \{\mathbf{x} \mid r_1 < |\mathbf{x}| < r_2\}, \ \Omega_3(\mathbf{x}) = \{\mathbf{x} \mid r_2 < |\mathbf{x}| < r_3\}.$$

Here the radius of the RVE is r_3, and the volume fraction of the three shells are,

$$f_1 = \left(\frac{r_1}{r_3}\right)^3, \ f_2 = \frac{r_2^3 - r_1^3}{r_3^3}, \ f_3 = \frac{r_3^3 - r_2^3}{r_3^3}, \ \text{and} \ f_1 + f_2 + f_3 = 1 . \tag{6.124}$$

To derive the Eshelby tensors for each shell, we consider three partially overlapped concentric spheres,

$$\Omega_I(\mathbf{x}) = \{\mathbf{x} \mid |\mathbf{x}| < r_1\}, \ \Omega_{II}(\mathbf{x}) = \{\mathbf{x} \mid |\mathbf{x}| < r_2\}, \ \Omega_{III}(\mathbf{x}) = \{\mathbf{x} \mid |\mathbf{x}| < r_3\}.$$

The interior and exterior Eshelby tensors for each sphere Ω_J are denoted as

$$\mathbb{S}^{J,F}(\mathbf{x}) := \begin{cases} \mathbb{S}^{I,F}(\mathbf{x}), & \forall \mathbf{x} \in \Omega_J, \quad J = I, II, III, \\ \mathbb{S}^{E,F}(\mathbf{x}), & \forall \mathbf{x} \in V/\Omega_J, \quad J = I, II, III . \end{cases} \tag{6.125}$$

Subsequently the average of the Eshelby tensor is required for each shell. We first denote the average of the Eshelby tensor of the overlapping spheres,

$$\mathbb{S}^{Jj,F} := \langle \mathbb{S}^{J,F} \rangle_{\Omega_j}, \quad J = I, II, III \ \text{and} \ j = 1, 2, 3, \tag{6.126}$$

where the first superscript J (Roman numbers) denotes the sphere, Ω_J, in which the eigenstrain is prescribed, and where the second superscript j (Arabic numbers) denotes the shell, Ω_j, over which the average is taken. Similarly we denote the average Eshelby tensor of the shell domains as

$$\mathbb{S}^{ij,F} := \langle \mathbb{S}^{i,F} \rangle_{\Omega_j}, \quad i = 1, 2, 3 \ \text{and} \ j = 1, 2, 3, \tag{6.127}$$

Again, the first subscript index, i, refers to the shell region, Ω_i, in which the eigenstrains are prescribed, and the second index, j, denotes the shell region, Ω_j, over which the average is taken. As shown in section 6.5, the average Eshelby tensors can be written as

$$\mathbb{S}^{ij,F} = s_1^{ij,F} \mathbb{E}^{(1)} + s^{ij,F} \mathbb{E}^{(2)}, \quad i, j = 1, 2, 3 . \tag{6.128}$$

The idea is to use the Eshelby tensors of three overlapping spheres to represent the Eshelby tensors of the shells via superposition. For the first spherical shell (the inner most shell) we write

$$\mathbb{S}^{11,F} = \mathbb{S}^{I1,F} = s_1^{11,F}\mathbb{E}^{(1)} + s_2^{11,F}\mathbb{E}^{(2)} ,$$
$$\mathbb{S}^{1j,F} = \mathbb{S}^{Ij,F} = s_1^{1j,F}\mathbb{E}^{(1)} + s_2^{1j,F}\mathbb{E}^{(2)} . \tag{6.129}$$

Here, $s_\alpha^{11,F}$ are the coefficients of the interior Eshelby tensor, whereas $s_\alpha^{1j,F}, j = 2,3$ are the coefficients of the exterior Eshelby tensors. Using superposition, the Eshelby tensors for the second and third spherical shells can be obtained by using the combination of the average Eshelby tensors of the three overlapping spheres,

$$\mathbb{S}^{2i,F} = \mathbb{S}^{II\,i,F} - \mathbb{S}^{I\,i,F}, \quad i = 1,2,3 , \tag{6.130}$$
$$\mathbb{S}^{3i,F} = \mathbb{S}^{III\,i,F} - \mathbb{S}^{II\,i,F}, \quad i = 1,2,3 . \tag{6.131}$$

Therefore, for $\alpha = 1,2$,

$$s_\alpha^{2i,F} = s_\alpha^{IIi,F} - s_\alpha^{I\,i,F}, \quad \text{and} \quad s_\alpha^{3i,F} = s_\alpha^{III\,i,F} - s_\alpha^{II\,i,F}, \quad i = 1,2,3 . \tag{6.132}$$

To this end, all the coefficients of the Eshelby tensors for each shell layer are expressed in terms the of the Eshelby coefficients for solid spheres, Ω_I, Ω_{II} and Ω_{III}, which are documented in the Appendix for a three-sphere RVE.

6.6.3.2 *Variational bounds for multiphase composites*

By using the multi-phase Eshelby tensor proposed in the previous section, a systematic, multi-variable optimization procedure is developed for multi-phase composites.

We consider the case that the finite spherical RVE is subjected to a displacement boundary condition, i.e.

$$\mathbf{u}(\mathbf{x}) = \boldsymbol{\epsilon}^0\mathbf{x} , \quad \forall \mathbf{x} \in \partial V . \tag{6.133}$$

The standard statement of the Hashin-Shtrikman principles may be expressed in the following form

$$\underline{I}_p(\mathbf{p}, \boldsymbol{\epsilon}^d) \leq \inf_{\boldsymbol{\epsilon}^d \in E} W(\boldsymbol{\epsilon}^d) \leq \bar{I}_p(\mathbf{p}, \boldsymbol{\epsilon}^d) , \tag{6.134}$$

where

$$W(\boldsymbol{\epsilon}) = \frac{1}{2|V|}\int_V \boldsymbol{\sigma} : \boldsymbol{\epsilon}\, dV , \tag{6.135}$$

is the strain energy density, and

$$I_p(\mathbf{p}, \boldsymbol{\epsilon}^d) = W_0(\boldsymbol{\epsilon}^0) - \frac{1}{2|V|}\int_V \left\{\mathbf{p} : (\mathbb{C} - \mathbb{C}_0)^{-1} : \mathbf{p} - \mathbf{p} : \boldsymbol{\epsilon}^d - 2\mathbf{p} : \boldsymbol{\epsilon}^0\right\} dV . \tag{6.136}$$

Here $\mathbb{C}$ is the elastic tensor of the composite and $\mathbb{C}_0$ is the elastic tensor of a comparison solid such that

$$I_p = \begin{cases} \bar{I}_p , & \text{if } \Delta\mathbb{C} = \mathbb{C} - \mathbb{C}_0 < 0 , \\ \underline{I}_p , & \text{if } \Delta\mathbb{C} = \mathbb{C} - \mathbb{C}_0 > 0 . \end{cases} \tag{6.137}$$

Here $W_0(\boldsymbol{\epsilon}^0)$ is the strain energy density of the comparison solid, $\mathbf{p}$ is the stress polarization and $\boldsymbol{\epsilon}^d$ is the disturbance strain field due to the stress polarization. They are related by the following subsidiary boundary value problem

$$\nabla \cdot (\mathbb{C}_0 : \nabla \mathbf{u}^d(\mathbf{x}) + \mathbf{p}(\mathbf{x})) = 0 \ , \quad \forall \mathbf{x} \in V \ , \quad \mathbf{u}^d(\mathbf{x}) = 0 \ , \quad \forall \mathbf{x} \in \partial V \ . \qquad (6.138)$$

We consider the composite to be made of n distinct phases and assume that each phase may be represented by a hollow spherical shell inside the RVE. The homogenization or statistical model of the composite is that any macro point of the composite is modeled as an RVE consisting of n distinct concentric spherical shells with domain Ω_i so that $\bigcup_i^n \Omega_i = V$ and $\bigcap_{i=1}^n \Omega_i = \emptyset$. The stress polarization is chosen as a piecewise constant tensorial field,

$$\mathbf{p}(\mathbf{x}) = \sum_{i=1}^{n} \mathbf{p}_i \chi(\Omega_i) \ , \quad \text{with} \ \ \chi(\Omega_i) = \begin{cases} 1 \ , & \forall \mathbf{x} \in \Omega_i \ , \\ 0 \ , & \forall \mathbf{x} \notin \Omega_i \ . \end{cases} \qquad (6.139)$$

Let us consider

$$\mathbf{p}_i = p_i \mathbb{I}^{(2)} + \tau_i \mathbb{J}^{(2)} \ , \qquad (6.140)$$

where $\mathbb{I}^{(2)}$ is the second order unit tensor and $\mathbb{J}^{(2)}$ is its counterpart, the so-called deviatoric unit tensor, both defined as

$$\mathbb{I}^{(2)} = \delta_{ij} \mathbf{e}_i \otimes \mathbf{e}_j \ , \quad \delta_{ij} = \begin{cases} 1 \ , i = j \\ 0 \ , i \neq j \end{cases} , \quad \mathbb{J}^{(2)} = \beta_{ij} \mathbf{e}_i \otimes \mathbf{e}_j \ , \quad \beta_{ij} = \begin{cases} 0 \ , i = j \\ 1 \ , i \neq j \end{cases} . \qquad (6.141)$$

Based on the finite spherical inclusion model, the average disturbance strain will be the summation of the average disturbance strain in each phase,

$$\langle \boldsymbol{\epsilon}^d \rangle_\Omega = - \sum_{i=1}^{n} \sum_{j=1}^{n} \mathbb{C}_0^{-1} : \langle \mathbb{S}^{ij,D} \rangle : \mathbf{p}_i \ . \qquad (6.142)$$

As shown in section 2, the average Eshelby tensor can be written as

$$\langle \mathbb{S}^{ij,D} \rangle = s_1^{ij,D} \mathbb{E}^{(1)} + s_2^{ij,D} \mathbb{E}^{(2)} \ . \qquad (6.143)$$

We choose the prescribed boundary field as

$$\boldsymbol{\epsilon}^0 = \bar{\epsilon} \ \mathbb{I}^{(2)} + \bar{\gamma} \ \mathbb{J}^{(2)} \ , \qquad (6.144)$$

so that we obtain

$$\begin{aligned} I_p &= W_0(\bar{\boldsymbol{\epsilon}}) - \frac{1}{2|V|} \int_V \{ \mathbf{p} : (\mathbb{C} - \mathbb{C}_0)^{-1} : \mathbf{p} - \mathbf{p} : \boldsymbol{\epsilon}^d - 2\mathbf{p} : \bar{\boldsymbol{\epsilon}} \} \, dV \\ &= \frac{9}{2} \kappa_0 \bar{\epsilon}^2 + 6\mu_0 \bar{\gamma}^2 - \sum_{i=1}^{n} \left(\frac{f_i p_i^2}{2(\kappa_i - \kappa_0)} + \frac{3 f_i \tau_i^2}{2(\mu_i - \mu_0)} \right) \\ &\quad - \sum_{i=1}^{n} \sum_{j=1}^{n} \left(\frac{f_i s_1^{ji,D} p_i p_j}{2\kappa_0} + \frac{3 f_i s_2^{ji,D} \tau_i \tau_j}{2\mu_0} \right) + \sum_{i=1}^{n} (3 f_i p_i \bar{\epsilon} + 6 f_i \tau_i \bar{\gamma}) \ . \end{aligned} \qquad (6.145)$$

We first let $\dfrac{\partial I_p}{\partial p_i} = 0$. One can thus find

$$-\frac{p_i}{\kappa_i - \kappa_0} - \frac{s_1^{ii,D} p_i}{\kappa_0} - \sum_{j \neq i} \frac{s_1^{ji,D} p_j}{2\kappa_0} + 3\bar{\epsilon} = 0, \quad \forall\, i = 1, 2, \cdots n \,. \tag{6.146}$$

Hence the stationary value of each p_i can be obtained through the following system of equations:

$$\begin{bmatrix} \ddots & & & & \\ \cdots\ \cdots & \left(\dfrac{s_1^{ii,D}}{\kappa_0} + \dfrac{1}{\kappa_i - \kappa_0}\right) & \cdots & \dfrac{s_1^{ji,D}}{2\kappa_0} & \cdots \\ & & \ddots & & \\ & & & \ddots & \\ & & & & \ddots \end{bmatrix} \begin{bmatrix} p_1 \\ \vdots \\ p_i \\ \vdots \\ p_j \\ \vdots \\ p_n \end{bmatrix} = 3\bar{\epsilon} \begin{bmatrix} 1 \\ \vdots \\ \vdots \\ 1 \\ \vdots \\ \vdots \\ 1 \end{bmatrix} . \tag{6.147}$$

We further let $\dfrac{\partial I_p}{\partial \tau_i} = 0$, which leads to

$$-\frac{3 f_i \tau_i}{\mu_i - \mu_0} - \frac{3 f_i s_2^{ii,D} \tau_i}{\mu_0} - \sum_{j \neq i} \frac{3 f_i s_2^{ji,D} \tau_j}{2\mu_0} + 6 f_i \bar{\gamma} = 0 \,, \tag{6.148}$$

or in matrix form,

$$\begin{bmatrix} \ddots & & & & \\ \cdots\ \cdots & \left(\dfrac{s_2^{ii,D}}{\mu_0} + \dfrac{1}{\mu_i - \mu_0}\right) & \cdots & \dfrac{s_2^{ji,D}}{2\mu_0} & \cdots \\ & & \ddots & & \\ & & & \ddots & \\ & & & & \ddots \end{bmatrix} \begin{bmatrix} \tau_1 \\ \vdots \\ \tau_i \\ \vdots \\ \tau_j \\ \vdots \\ \tau_n \end{bmatrix} = 2\bar{\gamma} \begin{bmatrix} 1 \\ \vdots \\ \vdots \\ 1 \\ \vdots \\ \vdots \\ 1 \end{bmatrix} . \tag{6.149}$$

Remark 6.6. In the past, when deriving variational bounds for multi-phase composites, the same infinite Eshelby tensor was used for all the phases (except the comparison phase) without discrimination. This procedure excludes the interactions among different phases at the outset. By applying the shell model, proposed in the last section, with the finite Eshelby tensor this interaction can now be taken into account.

6.6.3.3 *Two-phase composites*

We now consider an isotropic two-phase composite, with $\kappa_2 > \kappa_1$ and $\mu_2 > \mu_1$. For the effective bulk modulus, we find the following bound under the prescribed displacement boundary condition,

$$\kappa_1 + \frac{f_2}{\dfrac{1}{\kappa_2 - \kappa_1} + \dfrac{s_1^{22,D}}{\kappa_1}} \le \bar{\kappa} \le \kappa_2 + \frac{f_1}{\dfrac{1}{\kappa_1 - \kappa_2} + \dfrac{s_1^{11,D}}{\kappa_2}}\,, \tag{6.150}$$

where

$$s_1^{22,D} = \frac{(1+\nu_1)f_1}{3(1-\nu_1)}\,, \quad \text{and} \quad s_1^{11,D} = \frac{(1+\nu_2)f_2}{3(1-\nu_2)}\,. \tag{6.151}$$

A similar result can be derived for the Neumann boundary condition,

$$\kappa_1^{-1} + \frac{f_2}{\dfrac{1}{\kappa_2^{-1} - \kappa_1^{-1}} + \dfrac{1 - s_1^{22,N}}{\kappa_1^{-1}}} \ge \bar{\kappa}^{-1} \ge \kappa_2^{-1} + \frac{f_1}{\dfrac{1}{\kappa_1^{-1} - \kappa_2^{-1}} + \dfrac{1 - s_1^{11,N}}{\kappa_2^{-1}}}\,, \tag{6.152}$$

where

$$s_1^{22,N} = \frac{1 + \nu_1 + 2(1-2\nu_1)f_2}{3(1-\nu_1)}\,, \quad \text{and} \quad s_1^{11,N} = \frac{1 + \nu_2 + 2(1-2\nu_2)f_1}{3(1-\nu_2)}\,. \tag{6.153}$$

It can be shown, by algebraic manipulation, that the bounds (6.150) and (6.152) are identical. Furthermore, they are equal to the original Hashin-Shtrikman bounds, because the coefficients (6.151) and (6.153) are equal to those of the original infinite Eshelby Tensor.

Similarly, the bounds for the shear modulus can be obtained as

$$\mu_1 + \frac{f_2}{\dfrac{1}{\mu_2 - \mu_1} + \dfrac{s_2^{22,D}}{\mu_1}} \le \bar{\mu} \le \mu_2 + \frac{f_1}{\dfrac{1}{\mu_1 - \mu_2} + \dfrac{s_2^{11,D}}{\mu_2}}\,, \tag{6.154}$$

where

$$s_2^{22,D} = \frac{2(4-5\nu_1)f_1}{15(1-\nu_1)} - \frac{21 f_2 (1 - f_2^{2/3})^2}{10(1-\nu_1)(7-10\nu_1)}\,, \tag{6.155}$$

$$s_2^{11,D} = \frac{2(4-5\nu_2)f_2}{15(1-\nu_2)} - \frac{21 f_1 (1 - f_1^{2/3})^2}{10(1-\nu_2)(7-10\nu_2)}\,, \tag{6.156}$$

and

$$\mu_1^{-1} + \frac{f_2}{\dfrac{1}{\mu_2^{-1} - \mu_1^{-1}} + \dfrac{1 - s_2^{22,N}}{\mu_1^{-1}}} \ge \bar{\mu}^{-1} \ge \mu_2^{-1} + \frac{f_1}{\dfrac{1}{\mu_1^{-1} - \mu_2^{-1}} + \dfrac{1 - s_2^{11,N}}{\mu_2^{-1}}}\,, \tag{6.157}$$

where

$$s_2^{22,N} = \frac{2(4-5\nu_1) + (7-5\nu_1)f_2}{15(1-\nu_1)} + \frac{84 f_2^2 (1 - f_2^{2/3})^2}{10(1-\nu_1)(7+5\nu_1)}\,, \tag{6.158}$$

$$s_2^{11,N} = \frac{2(4-5\nu_2) + (7-5\nu_2)f_1}{15(1-\nu_2)} + \frac{84 f_1^2 (1 - f_1^{2/3})^2}{10(1-\nu_2)(7+5\nu_2)}\,. \tag{6.159}$$

Now the shear modulus bounds (6.154) and (6.157) are distinct, and they are different from the original Hashin-Shtrikman bounds based on the classical Eshelby

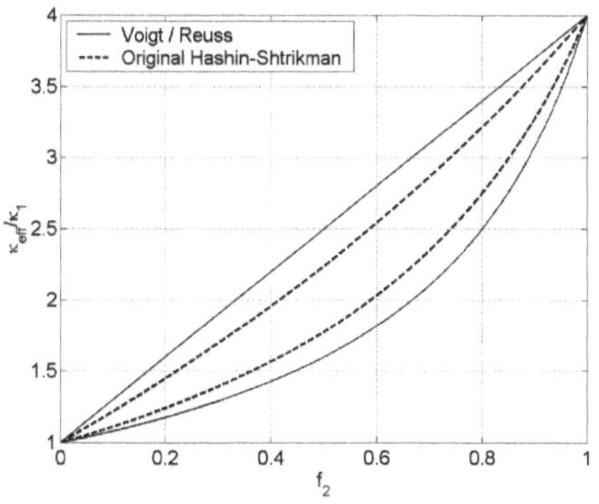

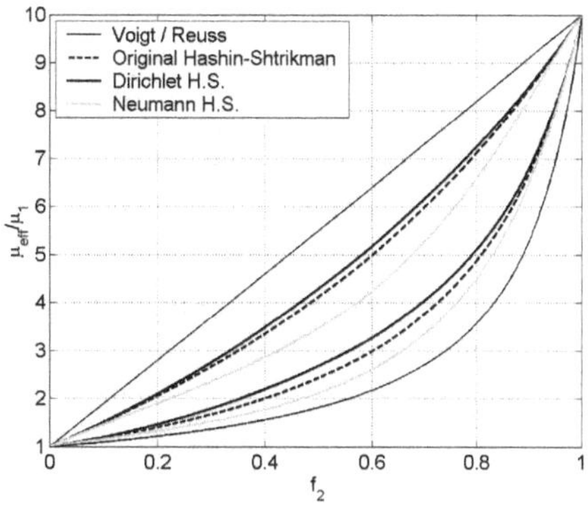

Fig. 6.8 Improved Hashin-Shtrikman Bounds for the effective bulk and shear moduli.

tensor in an unbounded RVE. The new variational bounds for both bulk and shear modulus are displayed in figure 6.8 with respect to f_2. The material data is chosen as $\kappa_2 = 4\kappa_1$, $\mu_2 = 10\mu_1$ and $\nu_1 = 0.3$ (implying $\nu_2 = 0.083$).

Figure 6.8 (a) shows that the boundary conditions have no effect on the bulk modulus, whose bounds coincide with the original HS bounds. On the other hand the boundary conditions do affect the variational bounds of the shear modulus. In Fig. 6.8 (b), the three sets of the variational bounds (Dirichlet, Neumann and the Original) for the shear modulus are juxtaposed in comparison. We note that the difference between these three pairs is solely caused by the second term in coefficients $s_2^{ii,D}$ and $s_2^{ii,N}$. Without the second term in Eqs. (6.155), (6.156) and (6.158), (6.159), the three sets of bounds will coincide.

Remark 6.7. There is a difference between material ordering, i.e. $\kappa_1 \leq \kappa_2 \leq \cdots \leq \kappa_n$ and geometric ordering, i.e. concentric spherical shells $r_1 \leq r_2 \leq \cdots \leq r_n$. Since one does not necessarily place the phase with the smallest material constants in the inner most region of the RVE, the combination of mappings between material ordering and geometric ordering is multiple. There are differences in the homogenization results due to these different combinations.

For a two-phase composite, there are two ways to place the phase which is not the comparison phase in an RVE: either in the interior of the RVE or in the exterior of the RVE. By alternating the material phase from the interior region of the RVE to the exterior region of the RVE, the interior homogenization becomes the exterior homogenization, and they correspond to different finite Eshelby tensors as seen in section 4. Therefore, in principle, we can obtain for each boundary condition two distinct pairs of the variational bounds, namely one corresponding to the interior eigenstrain and one corresponding to the exterior eigenstrain method. For isotropic composites, alternating the phase position has no effect on the variational bounds for the bulk modulus, because the bulk part of the interior eigenstrain Eshelby tensor equals the bulk part of the exterior eigenstrain Eshelby tensor and thus the two pairs of bounds coincide.

On the other hand, for the shear modulus, alternating the phase position yields new variational bounds. These are not shown in Fig. 6.8 since they will only deviate slightly from the bounds shown in the figure. Altogether we have two pairs of distinct variational bounds for shear modulus under each boundary condition.

For multi-phase composites ($n \geq 3$), the dependence on phase position may become more pronounced.

6.6.3.4 *Three-phase composites*

Consider a three-phase isotropic composite with $\kappa_3 > \kappa_2 > \kappa_1$ and $\mu_3 > \mu_2 > \mu_1$. To obtain the lower bound, we choose $\kappa_0 = \kappa_1$ and $p_1 = 0$. One can then solve the stationarity condition (6.147) for p_2 and p_3,

$$p_2 = 3\bar{\epsilon}\underline{p}_2, \quad \underline{p}_2 = \frac{1}{\Delta_{\ell 1}} \left(\frac{s_1^{33,D} - 0.5\, s_1^{32,D}}{\kappa_1} + \frac{1}{\kappa_3 - \kappa_1} \right), \tag{6.160}$$

$$p_3 = 3\bar{\epsilon}\underline{p}_3, \quad \underline{p}_3 = \frac{1}{\Delta_{\ell 1}} \left(\frac{s_1^{22,D} - 0.5\, s_1^{23,D}}{\kappa_1} + \frac{1}{\kappa_2 - \kappa_1} \right), \tag{6.161}$$

where

$$\Delta_{\ell 1} = \left(\frac{s_1^{22,D}}{\kappa_1} + \frac{1}{\kappa_2 - \kappa_1} \right) \left(\frac{s_1^{33,D}}{\kappa_1} + \frac{1}{\kappa_3 - \kappa_1} \right) - \frac{s_1^{32,D} s_1^{23,D}}{4\, \kappa_1^2}. \tag{6.162}$$

Similarly, one can solve (6.147) for the stationary values of p_1 and p_2 for the upper bound by setting $\kappa_0 = \kappa_3$ and $p_3 = 0$, i.e.

$$p_1 = 3\bar{\epsilon}\bar{p}_1, \quad \bar{p}_1 = \frac{1}{\Delta_{u1}} \left(\frac{s_1^{22,D} - 0.5\, s_1^{21,D}}{\kappa_3} + \frac{1}{\kappa_2 - \kappa_3} \right), \tag{6.163}$$

$$p_2 = 3\bar{\epsilon}\bar{p}_2, \quad \bar{p}_2 = \frac{1}{\Delta_{u1}} \left(\frac{s_1^{11,D} - 0.5\, s_1^{12,D}}{\kappa_3} + \frac{1}{\kappa_1 - \kappa_3} \right), \tag{6.164}$$

where

$$\Delta_{u1} = \left(\frac{s_1^{11,D}}{\kappa_3} + \frac{1}{\kappa_1 - \kappa_3} \right) \left(\frac{s_1^{22,D}}{\kappa_3} + \frac{1}{\kappa_2 - \kappa_3} \right) - \frac{s_1^{12,D} s_1^{21,D}}{4\, \kappa_3^2}. \tag{6.165}$$

Substituting the stationary values (6.160), (6.161), (6.163) and (6.164) into the Hashin-Shtrikman variational principle (6.134), we find the explicit variational

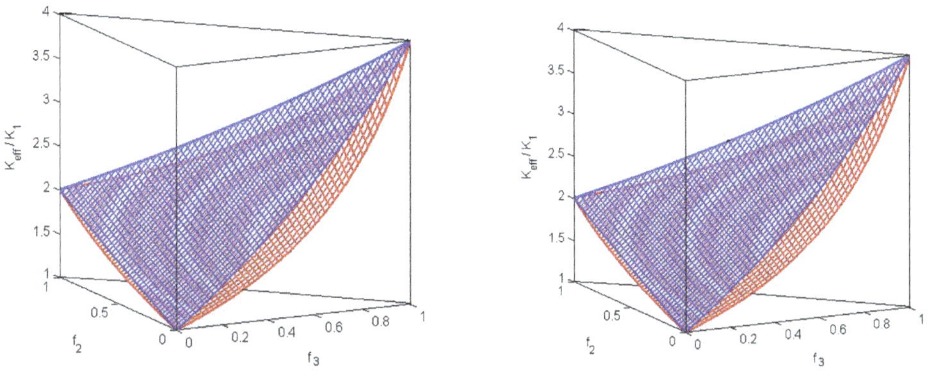

Fig. 6.9 Variational bounds for a 3-phase composite material: (a) Bounds for bulk modulus, and (b) Bounds for shear modulus.

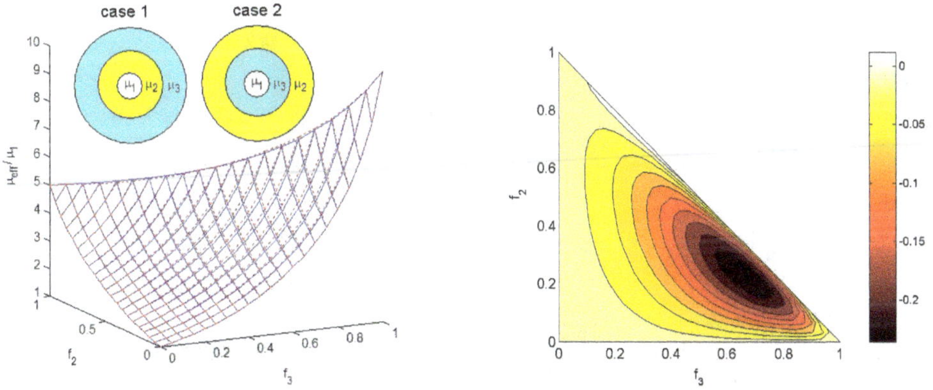

Fig. 6.10 Influence of phase position on 3-phase variational bounds.

bounds of the bulk modulus for three-phase composites,

$$
\begin{aligned}
&\kappa_1 - \Big(\frac{f_2\underline{p}_2^2}{(\kappa_2-\kappa_1)} + \frac{f_3\underline{p}_3^2}{(\kappa_3-\kappa_1)}\Big) - \frac{1}{\kappa_1}\Big(f_2\underline{p}_2^2 s_1^{22,D} + f_2\underline{p}_2\underline{p}_3 s_1^{32,D} + f_3\underline{p}_2\underline{p}_3 s_1^{23,D} \\
&+ f_3\underline{p}_3^2 s_1^{33,D}\Big) + 2\Big(f_2\underline{p}_2 + f_3\underline{p}_3\Big) \quad \le \bar{\kappa} \le \quad \kappa_3 - \Big(\frac{f_1\bar{p}_1^2}{(\kappa_1-\kappa_3)} + \frac{f_2\bar{p}_2^2}{(\kappa_2-\kappa_3)}\Big) \\
&- \frac{1}{\kappa_3}\Big(f_1\bar{p}_1^2 s_1^{11,D} + f_1\bar{p}_1\bar{p}_2 s_1^{21,D} + f_2\bar{p}_1\bar{p}_2 s_1^{12,D} + f_2\bar{p}_2^2 s_1^{22,D}\Big) \\
&+ 2\Big(f_1\bar{p}_1 + f_2\bar{p}_2\Big)\,, \qquad (6.166)
\end{aligned}
$$

Similarly, for the bounds of the shear modulus we have

$$\mu_1 - \Big(\frac{f_2\underline{\tau}_2^2}{(\mu_2-\mu_1)} + \frac{f_3\underline{\tau}_3^2}{(\mu_3-\mu_1)}\Big) - \frac{1}{\mu_1}\Big(f_2\underline{\tau}_2^2 s_2^{22,D} + f_2\underline{\tau}_2\underline{\tau}_3 s_2^{32,D} + f_3\underline{\tau}_2\underline{\tau}_3 s_2^{23,D}$$
$$+ f_3\underline{\tau}_3^2 s_2^{33,D}\Big) + 2\Big(f_2\underline{\tau}_2 + f_3\underline{\tau}_3\Big) \;\le \bar{\mu} \le\; \mu_3 - \Big(\frac{f_1\bar{\tau}_1^2}{(\mu_1-\mu_3)} + \frac{f_2\bar{\tau}_2^2}{(\mu_2-\mu_3)}\Big)$$
$$-\frac{1}{\mu_3}\Big(f_1\bar{\tau}_1^2 s_2^{11,D} + f_1\bar{\tau}_1\bar{\tau}_2 s_2^{21,D} + f_2\bar{\tau}_1\bar{\tau}_2 s_2^{12,D} + f_2\bar{\tau}_2^2 s_2^{22,D}\Big) + 2\Big(f_1\bar{\tau}_1 + f_2\bar{\tau}_2\Big) \;, \tag{6.167}$$

where

$$\underline{\tau}_2 = \frac{1}{\Delta_{\ell 2}}\Big(\frac{s_2^{33,D} - 0.5 s_2^{32,D}}{\mu_1} + \frac{1}{\mu_3-\mu_1}\Big)\,, \tag{6.168}$$

$$\underline{\tau}_3 = \frac{1}{\Delta_{\ell 2}}\Big(\frac{s_2^{22,D} - 0.5 s_2^{23,D}}{\mu_1} + \frac{1}{\mu_2-\mu_1}\Big)\,, \tag{6.169}$$

$$\bar{\tau}_1 = \frac{1}{\Delta_{u2}}\Big(\frac{s_2^{22,D} - 0.5 s_2^{21,D}}{\mu_3} + \frac{1}{\mu_2-\mu_3}\Big)\,, \tag{6.170}$$

$$\bar{\tau}_2 = \frac{1}{\Delta_{u2}}\Big(\frac{s_2^{11,D} - 0.5 s_2^{12,D}}{\mu_3} + \frac{1}{\mu_1-\mu_3}\Big)\,, \tag{6.171}$$

and

$$\Delta_{\ell 2} = \Big(\frac{s_2^{22,D}}{\mu_1} + \frac{1}{\mu_2-\mu_1}\Big)\Big(\frac{s_2^{33,D}}{\mu_1} + \frac{1}{\mu_3-\mu_1}\Big) - \frac{s_2^{23,D}s_2^{32,D}}{4\mu_1^2}\,, \tag{6.172}$$

$$\Delta_{u2} = \Big(\frac{s_2^{11,D}}{\mu_3} + \frac{1}{\mu_1-\mu_3}\Big)\Big(\frac{s_2^{22,D}}{\mu_3} + \frac{1}{\mu_2-\mu_3}\Big) - \frac{s_2^{12,D}s_2^{21,D}}{4\mu_3^2}\,. \tag{6.173}$$

Figure 6.9 shows the variational bounds for the effective bulk and shear modulus of a three-phase composite using the modulus ratios $\kappa_3 : \kappa_2 : \kappa_1 = 4 : 2 : 1$, $\mu_3 : \mu_2 : \mu_1 = 10 : 5 : 1$ and Poisson's ratio $\nu_1 = 0.3$. The unique features of variational bounds (6.166) and (6.167) are: (1) The boundary conditions are accurately taken into account without resorting to any approximation and ad hoc arguments; (2) Interaction among different phases, or in other words, the correlation among different phases are precisely taken into account by the cross term Eshelby tensor $\mathbb{S}^{ij,D}$, $i \neq j$, and this feature is absent in the classical HS bounds; (3) Microstructures of the composite are distinguished by mapping different combinations of the geometric ordering to the material ordering. For the bounds shown in Fig. 6.9, the geometric ordering coincides with the material ordering in ascending order, i.e. $(\kappa_1, \mu_1) \Rightarrow \Omega_1$, $(\kappa_2, \mu_2) \Rightarrow \Omega_2$, and $(\kappa_3, \mu_3) \Rightarrow \Omega_3$.

To examine the effect of the microstructure on the variational bounds, we exchange the material ordering within the domains Ω_2 and Ω_3. Figure 6.10 (a) shows a plot of the two lower bound surfaces of the shear modulus. The contour of the difference is shown in Fig. 6.10 (b). One can see that the maximum difference is about 0.2 , demonstrating the material ordering has little impact for this case.

In this section, the finite Eshelby tensors obtained are applied to develop various homogenization methods. It is shown that the special features of the finite Eshelby tensors can improve the accuracy of conventional homogenization methods and lead to more accurate predictions on effective material properties of composites.

For instance, we have found that for two-phase composites, there are at least two sets of Hashin-Shtrikman variational bounds corresponding to two different boundary conditions. This discovery may be instrumental and might even become the benchmark standard in numerical homogenization procedures. These new homogenization schemes will enrich the engineering homogenization repertoire and provide sharper estimates on effective material properties of multi-phase composites.

6.7 Application to multiscale finite element methods

6.7.1 *Variational multiscale eigenstrain formulation*

The finite Eshelby tensor has been found useful in formulating a multiscale computational method for linear elasticity. In the past, the multiscale methods are mainly based on element Green's function, which can only be obtained numerically in most cases. The recently proposed variational multiscale eigenstrain method (VMEM) ([Li *et al.* (2004, 2005b); Wang *et al.* (2005b)]) is to utilize micromechanics technique to construct the element Eshelby tensor that links the coarse scale residual to the fine scale strain. The new VMEM method provides analytical connection of multiscale solution to the Neumann-Eshelby problem, which we have discussed in previous sections.

6.7.1.1 *Coarse scale and fine scale formulation*

We now reformulate the variational eigenstrain multiscale formulation in the context of linear elasticity theory. Consider a simply connected domain, $V \in \mathbb{R}^d$ (d is the dimension of the physical space). Denote the body force b_i, prescribed displacement $u_i^0, \forall \mathbf{x} \in \Gamma_u$, and prescribed traction $t_i^0, \forall \mathbf{x} \in \Gamma_t$, where $\Gamma_u \bigcup \Gamma_t = \partial V$ and $\Gamma_u \cap \Gamma_t = \emptyset$. The variational statement of the boundary value problem of elastostatics is,

Find $\mathbf{u} \in \mathcal{S}$ *such that*

$$\int_V w_{(i,j)} \mathbb{C}_{ijk\ell}\, u_{(k,\ell)} dV = \int_V w_i b_i dV + \int_{\Gamma_t} w_i t_i^0 dS, \quad \forall \mathbf{w} = w_i \mathbf{e}_i \in \mathcal{V}\,. \tag{6.174}$$

where $\mathbb{C}_{ijk\ell}$ is the elastic tensor. The generalized Hooke's law $\sigma_{ij} = \mathbb{C}_{ijk\ell}\epsilon_{k\ell}$ is implied for Cauchy stress σ_{ij} and infinitesimal strain $\epsilon_{ij} = u_{(i,j)}$. Without elaboration, the trial function space $\mathcal{S}$ and the test function space $\mathcal{V}$ are defined here using standard notations in functional analysis (e.g. [Adams (1975)] or [Brenner and Scott (1994)]),

$$\mathcal{S} = \Big\{ \mathbf{u}(\mathbf{x}) \;\Big|\; \mathbf{u}(\mathbf{x}) \in [H^1(V)]^d, \mathbf{u} = \mathbf{u}^0, \ \forall \mathbf{x} \in \Gamma_u \Big\} \tag{6.175}$$

$$\mathcal{V} = \Big\{ \mathbf{w}(\mathbf{x}) \;\Big|\; \mathbf{w}(\mathbf{x}) \in [H^1(V)]^d, \mathbf{w} = \mathbf{0}, \ \forall \mathbf{x} \in \Gamma_u \Big\} \tag{6.176}$$

Define an abstract notation,

$$a(\mathbf{w},\mathbf{u}) : \mathcal{V}\times\mathcal{S}\rightarrow \mathrm{I\!R},\quad a(\mathbf{w},\mathbf{u}) := \int_V (\nabla\otimes\mathbf{w}) : \mathbf{C} : (\nabla\otimes\mathbf{u})dV, \tag{6.177}$$

and two linear forms

$$(\mathbf{w},\mathbf{b})_V : \mathcal{V}\times[H^{-1}(V)]^d \rightarrow \mathrm{I\!R},\quad (\mathbf{w},\mathbf{b})_V := \int_V \mathbf{w}\cdot\mathbf{b}dV \tag{6.178}$$

$$(\mathbf{w},\mathbf{t})_{\Gamma_t} : \mathcal{V}\times[H^{1/2}(V)]^d \rightarrow \mathrm{I\!R},\quad (\mathbf{w},\mathbf{t})_{\Gamma_t} := \int_{\Gamma_t} \mathbf{w}\cdot\mathbf{t}dS \tag{6.179}$$

The weak form (6.174) can be written in the compact form

$$a(\mathbf{w},\mathbf{u}) = (\mathbf{w},\mathbf{b})_V + (\mathbf{w},\mathbf{t}^0)_{\Gamma_t},\quad \forall\mathbf{w}\in\mathcal{V} \tag{6.180}$$

Following [Hughes *et al.* (1998)] , we assume that the exact solution of the weak form (6.180) can be decomposed into two solutions with different spatial resolutions, i.e.

$$\mathbf{u} = \bar{\mathbf{u}} + \mathbf{u}' \tag{6.181}$$

$$\mathbf{w} = \bar{\mathbf{w}} + \mathbf{w}' \tag{6.182}$$

where $\bar{\mathbf{u}}$ and $\mathbf{u}'$ represent the coarse scale and fine scale solutions. $\bar{\mathbf{w}}$ and $\mathbf{w}'$ represent coarse scale and fine scale trial functions respectively. Accordingly, the trial function space and test function space can be additively decomposed into a coarse scale space and a fine scale space, i.e. $\mathcal{S} = \bar{\mathcal{S}}\oplus\mathcal{S}'$. $\mathcal{V} = \bar{\mathcal{V}}\oplus\mathcal{V}'$. Since coarse scale spaces are to be solved numerically, $\bar{\mathcal{S}}$ and $\bar{\mathcal{V}}$ are finite dimensional spaces. On the other hand, the fine scale is to be sought analytically so $\mathcal{S}'$ and $\mathcal{V}'$ belong to infinite-dimensional function spaces.

Under two-scale decomposition and consider $\boldsymbol{w}$ and $\mathbf{u}$ are independent, the weak form (6.180) then becomes

$$a(\bar{\boldsymbol{w}},\bar{\mathbf{u}}) + a(\bar{\boldsymbol{w}},\mathbf{u}') = (\bar{\boldsymbol{w}},\mathbf{b})_V + (\bar{\boldsymbol{w}},\mathbf{t}^0)_{\Gamma_t},\quad \forall\bar{\boldsymbol{w}}\in\bar{\mathcal{V}} \tag{6.183}$$

$$a(\boldsymbol{w}',\bar{\mathbf{u}}) + a(\boldsymbol{w}',\mathbf{u}') = (\boldsymbol{w}',\mathbf{b})_V + (\boldsymbol{w}',\mathbf{t}^0)_{\Gamma_t},\quad \forall\boldsymbol{w}'\in\mathcal{V}' \tag{6.184}$$

6.7.1.2 *Fine scale solution*

To obtain fine scale solution analytically, we examine the two scale relation at a typical local domain, $\Omega_e^c \subset V \in \mathrm{I\!R}^2$. The local domain is chosen as a circular shape (superscript 'c'), and it is the smallest circle that encompasses the element e (subscript 'e'). Within Ω_e^c, we have

$$\mathbb{C}_{ijk\ell}u'_{k,\ell j}(\mathbf{x}) + \mathbb{C}_{ijk\ell}\bar{u}_{k,\ell j}(\mathbf{x}) + b_i(\mathbf{x}) = 0\ .\quad \forall\mathbf{x}\in\Omega_e^c \tag{6.185}$$

Note that this is not a boundary value problem, because around the boundary of the local domain, $\partial\Omega_e^c$, both displacement and traction are not prescribed.

To obtain the fine scale solution in the local domain, the above equilibrium equation is written in weak form,

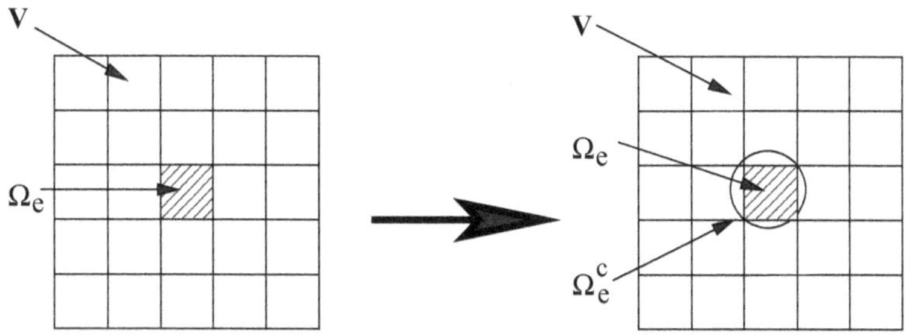

Fig. 6.11 Illustration of the concept of equivalent element domain Ω_e^c.

$$\int_{\Omega_e^c}\Big(\mathbb{C}_{mnk\ell}u'_{k,\ell n}(\mathbf{x})+\mathbb{C}_{mnk\ell}\bar{u}_{k,\ell n}(\mathbf{x})+b_m(\mathbf{x})\Big)G^{\infty}_{im}(\mathbf{y}-\mathbf{x})\;dV_x=0\;.\quad\forall\mathbf{x}\in\Omega_e^c \tag{6.186}$$

which reduces to the Somigliana identity [Somigliana (1886)] to express the fine scale solution as the following integral form,

$$u'_i(\mathbf{y})=\int_{\Omega_e^c}\Big(\mathbb{C}_{mnk\ell}\bar{u}_{k,\ell n}(\mathbf{x})+b_m(\mathbf{x})\Big)G^{\infty}_{im}(\mathbf{y}-\mathbf{x})dV_x$$
$$+\int_{\partial\Omega_e^c}\sigma'_{mn}(\mathbf{x})n_n(\mathbf{x})G^{\infty}_{im}(\mathbf{y}-\mathbf{x})dS_x+\int_{\partial\Omega_e^c}\sigma^{G_i^{\infty}}_{k\ell}(\mathbf{y}-\mathbf{x})n_\ell(\mathbf{x})u'_k(\mathbf{x})dS_x \tag{6.187}$$

where $u'_k(\mathbf{x})$ is the fine scale displacement, $\sigma'_{mn}(\mathbf{x})=\mathbb{C}_{mnkl}u'_{k,\ell}(\mathbf{x})$ is the corresponding fine scale stress component, $G^{\infty}_{im}(\mathbf{y}-\mathbf{x})$ is Green's function of the Navier equation of linear elasticity in an infinite domain, $\sigma^{G_i^{\infty}}_{k\ell}(\mathbf{y}-\mathbf{x})=\mathbb{C}_{mnk\ell}G^{\infty}_{im,n}(\mathbf{y}-\mathbf{x})$ is the stress of Green's function, and $n_\ell(\mathbf{x})$ is the out-normal of the surface $\partial\Omega_e^c$ at position $\mathbf{x}$. The subscript x in the term dV_x and dS_x of the above equation denotes that the integral is evaluated with respect to the variable $\mathbf{x}$. The derivative $G^{\infty}_{im,n}(\mathbf{y}-\mathbf{x})=\dfrac{\partial}{\partial y_n}G^{\infty}_{im}(\mathbf{y}-\mathbf{x})=-\dfrac{\partial}{\partial x_n}G^{\infty}_{im}(\mathbf{y}-\mathbf{x})$

It can be assumed that all the boundary contributions from the fine scale solution are small, and the following approximations are adopted:

$$\int_{\partial\Omega_e^c}\sigma'_{ij}(\mathbf{x})n_j(\mathbf{x})G^{\infty}_{im}(\mathbf{y}-\mathbf{x})dS_x\approx 0 \tag{6.188}$$

$$\int_{\partial\Omega_e^c}\sigma^{G_m^{\infty}}_{k\ell}(\mathbf{y}-\mathbf{x})n_\ell(\mathbf{x})u'_k(\mathbf{x})dS_x\approx 0. \tag{6.189}$$

In general, neglecting the boundary contribution can lead to considerable error in numerical computation. We will proceed without such an approximation. Integrating Eq. (6.187) by parts yields

$$u'_i(\mathbf{y})=\int_{\Omega_e^c}\mathbb{C}_{mnk\ell}\bar{u}_{k,\ell}(\mathbf{x})G^{\infty}_{im,n}(\mathbf{y}-\mathbf{x})dV_x+\int_{\Omega_e^c}b_m(\mathbf{x})G^{\infty}_{im}(\mathbf{y}-\mathbf{x})dV_x$$
$$+\int_{\partial\Omega_e^c}\sigma_{mn}(\mathbf{x})n_n(\mathbf{x})G^{\infty}_{im}(\mathbf{y}-\mathbf{x})dS_x+\int_{\partial\Omega_e^c}\sigma^{G_i^{\infty}}_{k\ell}(\mathbf{y}-\mathbf{x})n_\ell(\mathbf{x})u'_k(\mathbf{x})dS_x\;. \tag{6.190}$$

Note that by definition, $\sigma_{mn}(\mathbf{x}) = \bar{\sigma}_{mn}(\mathbf{x}) + \sigma'_{mn}(\mathbf{x})$.

Consider the weak form of the equilibrium equation, $\sigma_{mn,n}(\mathbf{x})+b_m(\mathbf{x}) = 0$, $\forall \mathbf{x} \in \Omega_e^c$,

$$\int_{\Omega_e^c} \big(\sigma_{mn,n}(\mathbf{x}) + b_m(\mathbf{x})\big) G_{im}^{\infty}(\mathbf{y}-\mathbf{x}) dV_x = 0 \ . \tag{6.191}$$

Integrating Eq. (6.191) by parts yields,

$$\int_{\partial\Omega_e^c} \sigma_{mn}(\mathbf{x}) n_n(\mathbf{x}) G_{im}^{\infty}(\mathbf{y}-\mathbf{x}) dS_x + \int_{\Omega_e^c} b_m(\mathbf{x}) G_{im}^{\infty}(\mathbf{y}-\mathbf{x}) dV_x$$
$$= -\int_{\Omega_e^c} \sigma_{mn}(\mathbf{x}) G_{im,n}^{\infty}(\mathbf{y}-\mathbf{x}) dV_x \tag{6.192}$$

Substituting Eq. (6.192) into Eq. (6.190), we have

$$u'_i(\mathbf{y}) = \int_{\Omega_e^c} \mathbb{C}_{mnk\ell}\Big(\bar{u}_{k,\ell}(\mathbf{x}) - u_{k,\ell}(\mathbf{x})\Big) G_{im,n}^{\infty}(\mathbf{y}-\mathbf{x}) dV_x$$
$$+ \int_{\partial\Omega_e^c} \sigma_{k\ell}^{G_i^{\infty}}(\mathbf{y}-\mathbf{x}) n_\ell(\mathbf{x}) u'_k(\mathbf{x}) dS_x \tag{6.193}$$

Involving various approximations, similar expression is also derived with the assumption that terms of boundary integration are neglected i.e., using approximation Eqs.(6.188,6.189). However, we emphasize that Eq.(6.193) is derived in an exact way. While the real displacement gradient $u_{k,\ell}(\mathbf{x})$ is unknown, we may approximate it by a posteriori estimate of the coarse scale computation. Here we choose the popular Zienkiewicz-Zhu *a posteriori* estimator, i.e.

$$\bar{u}_{(k,\ell)}(\mathbf{x}) - u_{(k,\ell)}(\mathbf{x}) \approx \bar{u}_{(k,\ell)}(\mathbf{x}) - u^Z_{(k,\ell)}(\mathbf{x}) \ . \tag{6.194}$$

In the above equation, $u^Z_{(k,\ell)}$ is the Zienkiewicz-Zhu recovery displacement gradient field, which will be discussed later. The discussion of the difference between the Z-Z recovery field and the coarse scale field may be viewed as the residual due to discretization. Borrowing the terminology from *Micromechanics*, we refer such residual strain as *the eigenstrain*, i.e.

$$\epsilon^*_{k\ell}(\mathbf{x}) := u^Z_{(k,\ell)}(\mathbf{x}) - \bar{u}_{(k,\ell)}(\mathbf{x}) \ . \tag{6.195}$$

Hence,

$$u'_i(\mathbf{y}) = -\int_{\Omega_e^c} \mathbb{C}_{mnk\ell} G_{im,n}^{\infty}(\mathbf{y}-\mathbf{x}) \epsilon^*_{k\ell}(\mathbf{x}) dV_x + \int_{\partial\Omega_e^c} \sigma_{k\ell}^{G_i^{\infty}}(\mathbf{y}-\mathbf{x}) n_\ell(\mathbf{x}) u'_k(\mathbf{x}) dS_x \tag{6.196}$$

Eq. (6.196) is *identical* to the statement of an inclusion embedded in a finite domain under prescribed Neumann boundary condition, as Eq. (6.7) in Section 4. Heuristically, for two-dimensional problem we take the finite Neumann-Eshelby tensors $\mathbb{S}^{I,N}_{ijk\ell}$ and $\mathbb{S}^{E,N}_{ijk\ell}$ for a circular inclusion in a circular RVE, and rewrite the fine scale solution in a compact form as

$$\epsilon'_{ij}(\mathbf{y}) = \mathbb{S}^{I,F}_{ijk\ell}(\mathbf{y})\ \epsilon^*_{k\ell}(\mathbf{y}) + \sum_{\mathbf{x}\notin\Omega_e} \mathbb{S}^{E,F}_{ijk\ell}(\mathbf{y}-\mathbf{x}) \epsilon^*_{k\ell}(\mathbf{x}), \quad \text{for } \mathbf{y} \in \Omega_e \tag{6.197}$$

The above formulation separates the contribution to the fine scale strain into two parts: the interior part and the exterior parts. Examining the expressions of the finite Neumann-Eshelby tensors, we find $S^{E,N}_{ijk\ell}$ is of order $\mathcal{O}(\rho_0^2)$ and $\rho_0 < 1$ for $\mathbf{y} \notin \Omega^c_e$. This clearly shows that the exterior contribution is of second order which is much smaller than the interior contribution and decays fast when $\mathbf{y}$ is away from Ω^c_e. Based on this fact, we make a further approximation by neglecting the exterior parts of the fine scale strain, which represents the interaction of discretization error among different elements

That leads to the expression:

$$\epsilon'_{ij}(\mathbf{y}) = \mathbb{S}^{I,N}_{ijk\ell}(\mathbf{y})\ \epsilon^*_{k\ell}(\mathbf{y}), \quad \text{for } \mathbf{y} \in \Omega_e \tag{6.198}$$

Since no exterior Eshelby tensor is present in the fine scale solution, we replace $\mathbb{S}^{I,N}_{ijk\ell}$ with $\mathbb{S}_{ijk\ell}$ without causing any confusion. A simplified fine scale solution may be expressed as follows

$$\epsilon'_{ij}(\mathbf{y}) = \sum_{e=1}^{n_{el}} \left\{ \mathbb{S}_{ijk\ell}(\mathbf{y}) \left(u^Z_{(k,\ell)}(\mathbf{y}) - \bar{u}^e_{(k,\ell)}(\mathbf{y}) \right) \chi(\Omega_e) \right\} \tag{6.199}$$

where $\chi(\Omega_e)$ is the characteristic function of the element Ω_e,

$$\chi(\Omega_e) = \begin{cases} 1 \ \mathbf{y} \in \Omega_e \\ 0 \ \mathbf{y} \in \Omega^c_e/\Omega_e \end{cases} \tag{6.200}$$

It worths pointing out that the above formulation has exactly the same format if approximations Eqs.(6.188) and (6.189) are invoked. In that case, $\mathbb{S}_{ijk\ell}$ is just replaced by the original infinite Eshelby tensor. In the next section, we will formulate multiscale finite element formulation using Eq. (6.199). We call it *(original) smart element method* if $\mathbb{S}_{ijk\ell}$ uses original infinite Eshelby tensor, and *modified smart element method* if $\mathbb{S}_{ijk\ell}$ uses Neumann-Eshelby tensor.

6.7.1.3 *Zienkiewicz-Zhu estimate*

C^0 continuity of the trial function space results in a discontinuous approximation of the coarse scale displacement gradient field across elements. More accurate displacement gradient field can be recovered from the coarse scale solution by various techniques. Following the procedure outlined in [Zienkiewicz and Zhu (1992a,b)], the recovered displacement gradient field can be interpolated as:

$$u^Z_{(k,\ell)}(\mathbf{x}) = \sum_{n=1}^{n_{ed}} N^n(\mathbf{x}) \tilde{u}^{Z,n}_{(k,\ell)} \tag{6.201}$$

where n_{ed} is the number of nodes in an element; $N^n(\mathbf{x})$ is the same basis function used for the interpolation of displacements; $\tilde{u}^{Z,n}_{(k,\ell)}$ is the nodal value of the displacement gradient. Superscript n is used here to denote a particular node.

It is further assumed that the nodal displacement gradient $\tilde{u}^{Z,n}_{(k,\ell)}$ belongs to a polynomial expansion of the same order as that present in the basis function over the element cluster ${\Omega_E}^{(n)}$ surrounding the node n under consideration,

$$\tilde{u}^p_{(k,\ell)}(\mathbf{x}) = \mathbf{P}(\mathbf{x})\mathbf{a} \tag{6.202}$$

where $\mathbf{a}$ is an unknown vector for the (k,ℓ) component of the strain. For a two dimensional triangle element mesh, we may choose the polynomial basis $\mathbf{P}(\mathbf{x}) = [1, x, y]$, and for two-dimensional quadrilateral element mesh, we choose $\mathbf{P}(\mathbf{x}) = [1, x, y, xy]$.

The unknown vector $\mathbf{a}$ is determined by minimizing the functional $E(\mathbf{a})$ defined on the element cluster $\Omega_E^{(n)}$ surrounding the node n under consideration

$$E(\mathbf{a}) = \int_{\Omega_E^{(n)}} \left(\tilde{u}^p_{(k,l)}(\mathbf{x}) - \bar{u}_{(k,l)}(\mathbf{x})\right)^2 dV = \int_{\Omega_E^{(n)}} \left(\mathbf{P}(\mathbf{x})\mathbf{a} - \bar{u}_{(k,l)}(\mathbf{x})\right)^2 dV \quad (6.203)$$

The minimization can be solved in matrix form as

$$\mathbf{a} = \mathbf{A}^{-1}\mathbf{b} \quad (6.204)$$

where

$$\mathbf{A} = \int_{\Omega_E{}^{(n)}} \mathbf{P}^T\mathbf{P}dV, \quad \text{and} \quad \mathbf{b} = \int_{\Omega_E^{(n)}} \mathbf{P}^T \bar{u}_{(k,l)} dV \quad (6.205)$$

The nodal displacement gradients are then obtained by substituting appropriate coordinates $\mathbf{x}^n$ of the node n into the polynomial expansion

$$\tilde{u}^{Z,n}_{(k,\ell)} = \tilde{u}^p_{(k,\ell)}(\mathbf{x}^n) = \mathbf{P}(\mathbf{x}^n)\mathbf{a} = \mathbf{P}^n\mathbf{a} \quad (6.206)$$

where $\mathbf{P}^n = \mathbf{P}(\mathbf{x}^n)$.

Therefore the recovered nodal displacement gradient is

$$\tilde{u}^{Z,n}_{(k,l)} = \mathbf{P}^n\mathbf{A}^{-1} \int_{\Omega_E^{(n)}} \mathbf{P}^T \bar{u}_{(k,l)} dV = \int_{\Omega_E^{(n)}} \mathbf{P}^n\mathbf{A}^{-1}\mathbf{P}^T \bar{u}_{(k,l)} dV \quad (6.207)$$

Note that $\mathbf{P}^n\mathbf{A}^{-1}\mathbf{P}^T$ is a scalar quantity, so the above equation states that the recovered nodal displacement gradient at a node n is a weighted average of the coarse scale displacement gradient over the element cluster $\Omega_E^{(n)}$ surrounding node n. One may also observe the nonlocal nature of the recovery. [Zienkiewicz and Zhu (1992a,b)] reported superconvergence of the recovered displacement gradient field for linear elasticity problems, i.e. an $O(h^2)$ convergence for both linear triangle and bilinear quadrilateral elements. So we can approximate the exact displacement gradient in Eq.(6.193) by the recovered field, since the latter is indeed accurate.

Substituting the above result into Eq. (6.201) and Eq. (6.199), the fine scale solution over element Ω_e can be expressed analytically in terms of the coarse scale solution as

$$\epsilon'_{ij}(\mathbf{x}) = \mathbb{S}_{ijk\ell}(\mathbf{x}) \left\{ \sum_{n=1}^{n_{ed}} N^n(\mathbf{x}) \left[\int_{\Omega_E^{(n)}} \mathbf{P}^n\mathbf{A}^{-1}\mathbf{P}^T \bar{u}_{(k,l)} dV \right] - \bar{u}_{(k,\ell)}(\mathbf{x}) \right\}, \quad \forall\, \mathbf{x} \in \Omega_e \quad (6.208)$$

6.7.1.4 *Modified smart element solution*

With the fine scale solution obtained, the coarse scale weak formulation,

$$a(\bar{\mathbf{w}}, \bar{\mathbf{u}}) + a(\bar{\mathbf{w}}, \mathbf{u}') = (\bar{\mathbf{w}}, \mathbf{f})_V + (\bar{\mathbf{w}}, \mathbf{t}^0)_{\Gamma_t} \tag{6.209}$$

can be solved. Substituting Eq. (6.208) into Eq. (6.209), we obtained

$$\mathop{\boldsymbol{A}}_{e=1}^{n_{el}} \int_{\Omega_e} \bar{w}^e_{(i,j)} \mathbb{C}_{ijkl} \left\{ \bar{u}^e_{(k,\ell)} + \mathbb{S}_{k\ell mn} \left[\sum_{n=1}^{n_{ed}} N^n(\mathbf{x}) \Big(\int_{\Omega_E^{(n)}} \mathbf{P}^n \mathbf{A}^{-1} \mathbf{P}^T \bar{u}^{e_j}_{(m,n)} dV_x \Big) - \bar{u}^e_{(m,n)}(\mathbf{x}) \right] \right\} dV_x = \mathop{\boldsymbol{A}}_{e=1}^{n_{el}} \left\{ \int_{\Omega_e} \bar{w}^e_i b_i dV_x + \int_{\Gamma_t \cap \partial\Omega_e} \bar{w}^e_i t^0_i dS_x \right\} \tag{6.210}$$

Using finite element spatial discretization, we can write (6.210) in a matrix form:

$$[\mathbf{K}][\mathbf{d}] = [\mathbf{R}]\ . \tag{6.211}$$

The global force vector $[\mathbf{R}]$ is the same as that in conventional finite element methods,

$$[\mathbf{R}] = \mathop{\boldsymbol{A}}_{e=1}^{n_{el}} \left\{ \int_{\Omega_e} [\mathbf{N}]_e^T [\mathbf{b}]_e dV_x + \int_{\partial\Omega_e \cap \Gamma_t} [\mathbf{N}]_e^T [\mathbf{t}^0]_e dS_x \right\} \tag{6.212}$$

where the symbol $\boldsymbol{A}$ is the so-called *element assembly operator* (see [Hughes (1987)]) . $[\boldsymbol{N}]_e$ and $[\boldsymbol{B}]_e$ are the element shape function matrix and shape function gradient matrix, respectively. $[\mathbf{b}]_e$ is the element body force vector and $[\mathbf{t}^0]_e$ is the element traction vector.

Because of nonlocal nature of the Zeinkiewicz-Zhu recovery procedure, elemental DOFs in forming the stiffness matrix $[\mathbf{K}]$ are coupled through element clusters. With slightly abuse of notations, the self-adaptive stiffness matrix $[\mathbf{K}]$ can be constructed by the following nested submatrix assemblage procedure,

$$[\mathbf{K}] = \sum_{r,s\in\Omega_e} \int_{\Omega_e} [\mathbf{B}_r]^T [\mathbf{C}] \left\{ [\mathbf{B}_s] + [\mathbf{S}] \Big(\sum_{t\in\Omega_E^{(s)}} [\mathbf{N}_s][\tilde{\mathbf{B}}_{st}] - [\mathbf{B}_s] \Big) \right\} dV_x \tag{6.213}$$

where r, s and t are nodal numbers in an element or in a cluster, and $[\boldsymbol{C}]$ & $[\mathbf{S}]$ are the matrix form of the elasticity tensor and the finite Eshelby tensor. $[\mathbf{B}_r]$ is the shape function gradient submatrix for node r. The $\sum$ in Eq. (6.213) should be understood as assembling the nodal submatrix (2×2 for 2D) to the corresponding global DOFs. Eq. (6.213) represents a nested assemblage procedure: $[\tilde{\mathbf{B}}_{st}]$, the weighted nodal gradient submatrix at node s, is assembled from each node t in the element cluster $\Omega_E^{(s)}$ surrounding s. $[\tilde{\mathbf{B}}_{st}]$ is then interpolated using shape function matrix $[\mathbf{N}_s]$ of s to form the recovery solution inside Ω_e. The difference of recovery strain and coarse strain is then filtered by the finite Eshelby matrix $[\mathbf{S}]$ to form the fine scale solution. Finally, nodal d.o.f for nodes r, s are assembled. All the matrices in Eq. (6.213) are formed explicitly for 2D plane strain formulation, with 2D Nuemann-Eshelby tensor formulated in [Wang *et al.* (2005a)]. Extension of these expressions to 3D formulation is obvious.

6.7.2 *Modal analysis of the modified smart element*

In this section we perform a modal analysis of the modified smart element to examine its intrinsic property to avoid volumetric locking at the incompressible limit. For the sake of simplicity, we choose a single square element, and the result will be compared to that of the standard 4-node quadrilateral (Q1) element.

Following the procedures presented in [Armero (2002)] and [Huerta and Fernandez-Mendez (2001)], the eigenvalues and eigenvectors of the element stiffness matrix are computed for the standard Q1 element and the modified smart element respectively. Please note that in the single element case, the element cluster at each node only includes the featured element itself. There will still be homogenization effect. This can be seen from Eq. (6.213): in this case, the element cluster, $\Omega_E^{(s)}$, only contains one element, nonetheless, the overall stiffness matrix is modified. Both of the stiffness matrices of the Q1 and the modified smart element are 8×8 for a quadrilateral setting. The eight eigenvectors of the stiffness matrices correspond to eight modes — three rigid body modes (Fig.6.12(a)(b)(c)) for translations and a rotation, three constant strain modes for the volumetric, stretch and shear modes (Fig.6.12(d)(e)(f)), and two hourglass modes (Fig.6.12(g)(h)) representing element bending. To study the locking effect, the eigenvalues are plotted against different

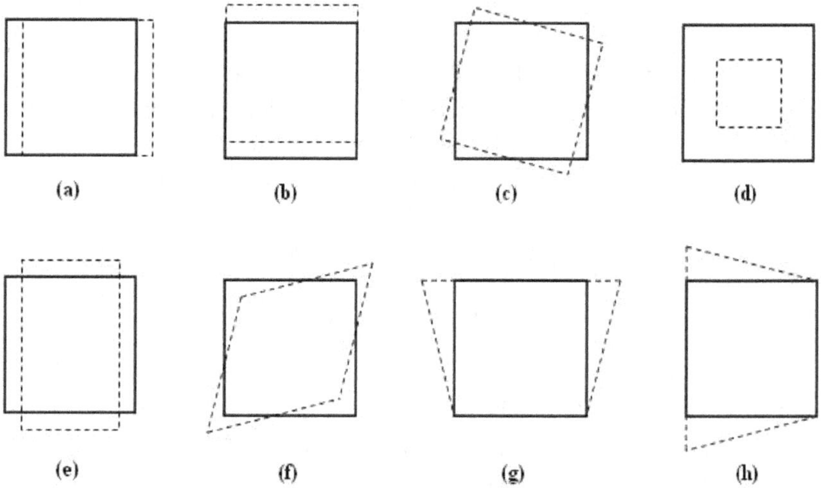

Fig. 6.12 Eight Modes of a Four-node Quadrilateral Element – Rigid body modes (a) x-direction translation (b) y-direction translation (c) rotation; Constant strain modes (d) volumetric (e) stretch (f) shear; Hourglass modes (g) bending I (h) bending II.

Poisson's ratio ν. As shown in Fig.6.13, the eigenvalues of $Q1$ element and modified smart element matrices appear to be very similar for the constant strain modes. Specifically, the eigenvalues associated with the stretch mode and the shear mode stay finite as $\nu \to 0.5$, so no volumetric locking will occur for this two modes. On the other hand, the eigenvalues associated with the volumetric mode both go to infinity as $\nu \to 0.5$, which means at the incompressible limit, an infinitely large force is needed to induce the displacements of this mode. This phenomenon is *expected* for the volumetric mode to preserve volume at the incompressible limit and is refereed as *physical locking*. So no volumetric locking occurs in the constant strain modes for both elements.

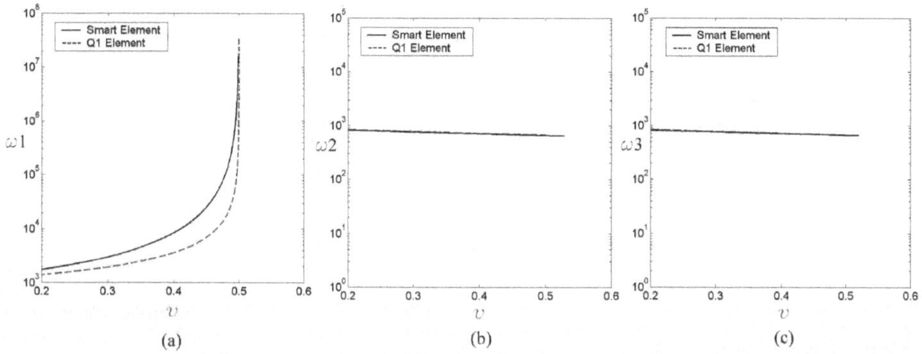

Fig. 6.13 Comparison of constant strain eigenvalues (a) volumetric mode (b) stretch mode (c) shear mode.

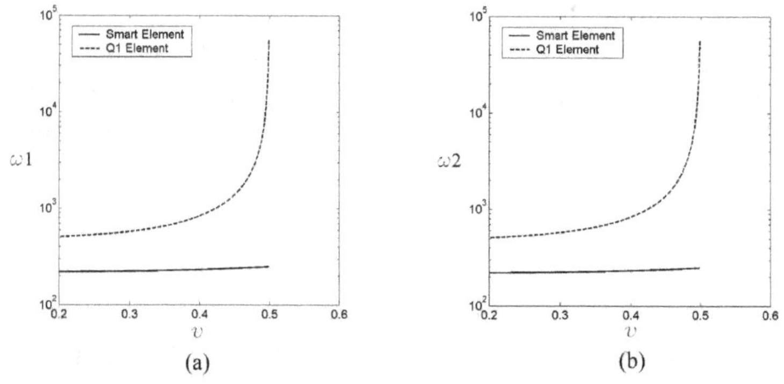

Fig. 6.14 Comparison of the hourglass eigenvalues (a) bending I (b) bending II.

Shown in Fig.6.14 are the eigenvalues of the two hourglass modes against ν. When $\nu \to 0.5$, the hourglass eigenvalues of $Q1$ element tend to infinity, which is referred as *non-physical locking* and is the reason for its volumetric locking phenomena at the incompressible limit. On the other hand, the hourglass eigenvalues of smart element stay finite as $\nu \to 0.5$. So with the smart element approximation,

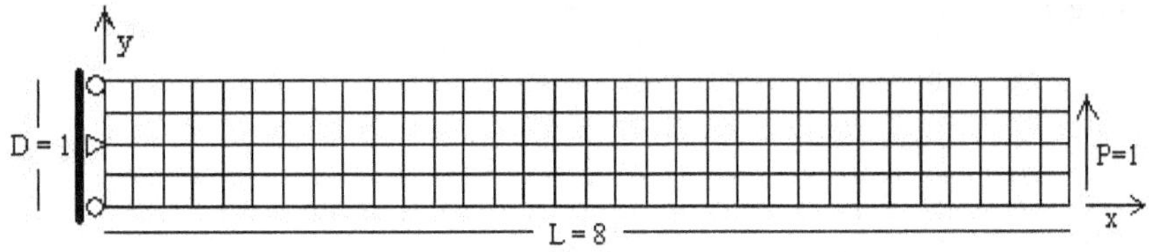

(a)

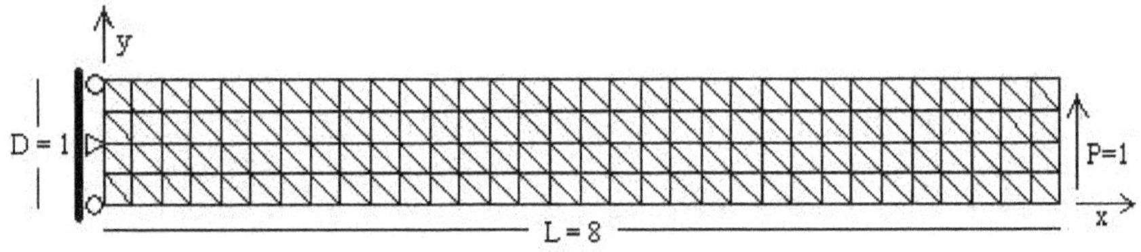

(b)

Fig. 6.15 A cantilever beam with (a) quadrilateral mesh, and (b) triangle mesh.

bending can occur at the incompressible limit, i.e. the modified smart element is free of volumetric locking.

6.7.3 *Numerical examples*

To demonstrate the improved performance of the modified smart element, three numerical examples have been carried out by using the modified formulation: the cantilever beam problem, the problem of a plate with a hole, and the L-shaped plate problem. Numerical solutions from conventional FEM method, original smart element and the modified smart element method are compared in each example.

To quantify the accuracy of the numerical solutions and the rate of convergence, we used error L_2 norm as defined in [Zienkiewicz and Taylor (2000)],

$$\eta_{L_2} = \frac{\|\mathbf{e}\|_{L_2}}{\|\mathbf{u}\|_{L_2}} \tag{6.214}$$

where

$$\|\mathbf{e}\|_{L_2} = \left[\int_V (\mathbf{u}-\mathbf{u}^h)^T(\mathbf{u}-\mathbf{u}^h)dV\right]^{\frac{1}{2}} \quad \text{and} \quad \|\mathbf{u}\|_{L_2} = \left[\int_V \mathbf{u}^T\mathbf{u}dV\right]^{\frac{1}{2}} \tag{6.215}$$

with $\mathbf{u}$ the exact solution and $\mathbf{u}^h$ the numerical solution.

6.7.3.1 *Cantilever beam example*

The exact solution for the bending problem of a cantilever beam subjected to end loading (Fig. 6.15) is given by Timoshenko and Goodier [Timoshenko and Goodier (1951)].

$$u_x = -\frac{Py}{6\bar{E}I}\left(y - \frac{D}{2}\right)\left[3x(2L-x) + (2+\bar{\nu})y(y-D)\right] \tag{6.216}$$

$$u_y = \frac{P}{6\bar{E}I}\left[x^2(3L-x) + 3\bar{\nu}(L-x)\left(y-\frac{D}{2}\right)^2 + \frac{4+5\bar{\nu}}{4}D^2x\right] \tag{6.217}$$

where

$$I = \frac{D^3}{12} \tag{6.218}$$

$$\bar{E} = \begin{cases} E & \text{for plane stress} \\ E/(1-\nu^2) & \text{for plane strain} \end{cases} \tag{6.219}$$

$$\bar{\nu} = \begin{cases} \nu & \text{for plane stress} \\ \nu/(1-\nu) & \text{for plane strain} \end{cases} \tag{6.220}$$

The corresponding stress field is

$$\sigma_{xx}(x,y) = -\frac{P}{I}(L-x)\left(y-\frac{D}{2}\right) \tag{6.221}$$

$$\sigma_{yy}(x,y) = 0 \tag{6.222}$$

$$\sigma_{xy}(x,y) = \frac{Py}{2I}\left(y-D\right) \tag{6.223}$$

The problem has been solved at the pseudo-incompressible limit for plane strain case, with Young's modulus $E = 1000$, Poisson's ratio $\nu = 0.499$, and zero body force, i.e. $b_m = 0.0$. The dimensions of the beam are: $L = 8.0$ and $D = 1.0$. Fig. 6.15 shows examples of structured quadrilateral and triangular meshes used in the analysis. In the computation, exact displacement solution is prescribed on $x = 0$ beam edge, while exact traction solution is applied along the edge $x = L$. The rest of the boundary is traction free.

The numerical results obtained via modified smart element method are compared to the exact solution, the conventional finite element solution and original smart element solution in Fig. 6.16. At the incompressible limit, conventional FEM solution locks as expected, while both smart elements show point-wise improvement, and the modified formulation has improved accuracy over the original scheme. In term of the overall error L_2 norm, Figs. 6.16(c) and (d) show that the convergence rate of the modified smart element is faster than the original formulation, and the rate of convergence is comparable with the Selective Reduced Integration scheme.

For this problem, we have compared the computational cost of the proposed method with the enhanced strain method [Simo and Rifai (1990)], which is proven to handle incompressibility well. Both codes are written in MATLAB and run in a PC with a Pentium-III processor. The results are summarized in the table 6.7.3.1. We can see the proposed method is computationally more expensive than the enhanced strain method.

6.7.3.2 *L-shaped plate*

The stress field in an L-shaped plane-elastic body of thickness t loaded by mode I symmetric loading is singular in the corner, e.g. [Williams (1952)]. Convergence

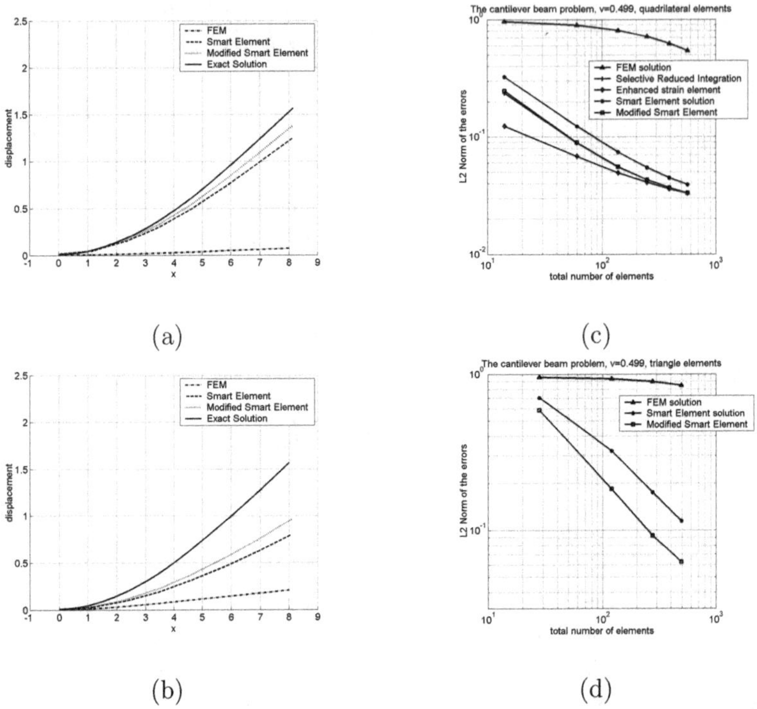

Fig. 6.16 Comparison of modified smart element results to conventional FEM solution, smart element solution and the exact solution: (a) Vertical displacement along center line $y = 0$, quadrilateral element mesh. (b) Vertical displacement along center line $y = 0$, triangle element mesh. (c) Log-Log convergence plot in term of the L_2 norm of the error, quadrilateral element. (d) Log-Log convergence plot in term of the L_2 norm of the error, triangle element.

Method	CPU Time (14 elements)	CPU Time (60 elements)
Smart Element	1.3180 sec.	7.9070 sec.
Enhanced Strain	0.3810 sec.	2.4030 sec.

study of this example via p-refinement and h-refinement of conventional finite element has been presented in [Szabo (1986)]. Here we will compare the performance of bilinear quadrilateral element and smart elements.

As shown in Fig.6.17(a), traction is prescribed along the exterior edges A-B-C-D-E of the plate according to the following exact stress solution,

$$\sigma_x = A_1\lambda_1 r^{\lambda_1-1}[(2 - Q_1(\lambda_1+1))\cos(\lambda_1-1)\theta - (\lambda_1-1)\cos(\lambda_1-3)\theta] \quad (6.224)$$

$$\sigma_y = A_1\lambda_1 r^{\lambda_1-1}[(2 + Q_1(\lambda_1+1))\cos(\lambda_1-1)\theta + (\lambda_1-1)\cos(\lambda_1-3)\theta] \quad (6.225)$$

$$\tau_{xy} = A_1\lambda_1 r^{\lambda_1-1}[(\lambda_1-1)\sin(\lambda_1-3)\theta + Q_1(\lambda_1+1)\sin(\lambda_1-1)\theta] \quad (6.226)$$

where A_1 is a generalized stress-intensity factor (an arbitrary number), $\lambda_1 =$

0.544483737, $Q_1 = 0.543075579$. r and θ serve as polar coordinates. It is obvious that stresses are singular as $r \to 0$.

Exact solution of corresponding displacement field is known as

$$u_x = \frac{A_1}{2\mu} r^{\lambda_1}[(\kappa - Q_1(\lambda_1 + 1))\cos\lambda\theta - \lambda_1\cos(\lambda_1 - 2)\theta] \tag{6.227}$$

$$u_y = \frac{A_1}{2\mu} r^{\lambda_1}[(\kappa + Q_1(\lambda_1 + 1))\sin\lambda\theta + \lambda_1\sin(\lambda_1 - 2)\theta] \tag{6.228}$$

where $\mu = E/(1 - 2\nu)$ is the shear modulus, $\kappa = 3 - 4\nu$, ν is Poisson's ratio. We choose $\nu = 0.3$, $A_1 = 1$, $a = 1$, $E = 10$ is the example.

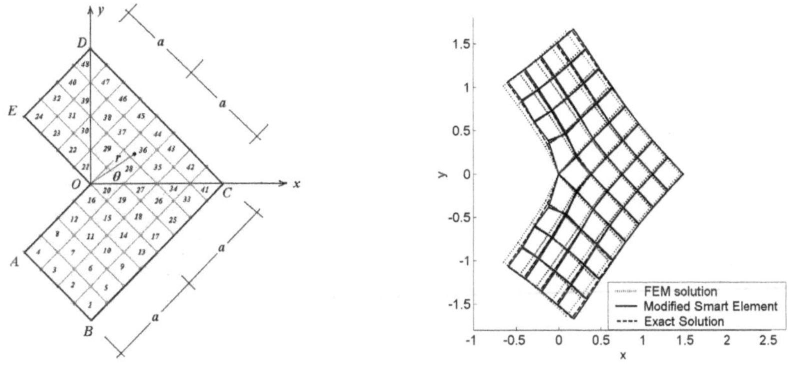

(a) (b)

Fig. 6.17 An L-shaped plate (a) quadrilateral mesh (b) deformed meshes.

Fig. 6.17 (b) shows the deformed meshes computed by finite element and modified smart element against the exact displacement field. Element-wise L_2 error distributions are compared for conventional finite element, original smart element, and modified smart element in Fig. 6.18 (a). One may find that there is obvious improvement in element-wise errors, if the smart element is used. In particular, the modified solution tends to smooth the error distribution further with fine scale boundary condition properly integrated. Uniformly refining the mesh, convergence rates for all these methods are shown in Fig. 6.18 (b). Again, modified smart element shows improved rate of convergence.

It is worth noticing the deformed mesh of modified smart element is slightly zigzagged along edge O-A in Fig. 6.17 (b). This might arise from the fact that the fine scale is under-integrated using 2×2 Gaussian integration. While not directly addressed here, it appears that special treatment may be needed for elements at the solid boundary.

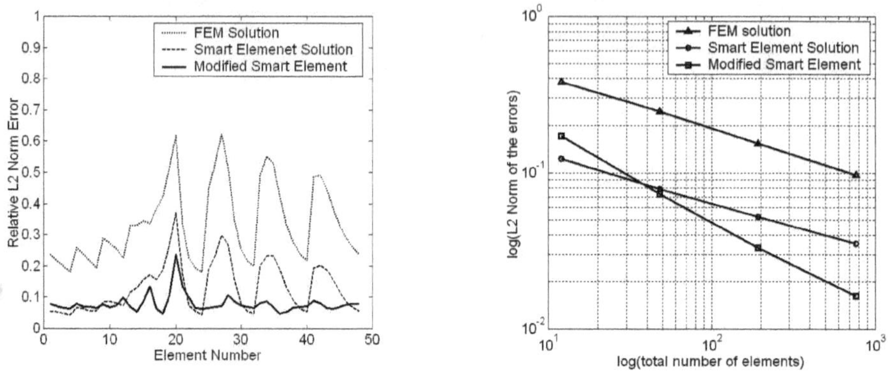

(a) (b)

Fig. 6.18 L-shaped Plate (a) element-wise comparison of L_2 norm of the error (b) Log-Log convergence plot in terms of the L_2 norm of the error.

6.8 Exercises

Problem 6.1. *Compute* $\mathbb{S}^{I,D}_{iijj}, \mathbb{S}^{I,N}_{iijj}, \mathbb{S}^{E,D}$, *and* $\mathbb{S}^{E,N}$. *Show that:*

(1.)

$$\mathbb{S}^{I,N}_{iijj} \geq \mathbb{S}^{I,D}_{iijj}, \quad \text{and ‘}\mathbb{S}^{E,N}_{iijj} \geq \mathbb{S}^{E,D}_{iijj},$$

and (2.)

$$\mathbb{S}^{I,D}_{iijj} - \mathbb{S}^{E,D}_{iijj} = \mathbb{S}^{I,N}_{iijj} - \mathbb{S}^{E,N}_{iijj} = \frac{1+\nu}{1-\nu} = \mathbb{S}^{I,\infty}_{iijj}$$

Chapter 7

MICROMECHANICAL DAMAGE THEORY

Initiation, growth, and coalescence of internal voids under imposed stress and strain have been frequently related to the fracture of ductile materials based on numerous experimental observations. In the late 1960's and early 1970's, micromechanics theories were used by several pioneer researchers to establish the relationship between growth of a void and imposed external stress and strain, which laid down the foundation to reveal the underlying mechanism for the failure of ductile materials.

In this chapter, we present the theories of several damage models that were considered by F. A. McClintock [McClintock (1968)], J.R. Rice and D. M. Tracey [Rice and Tracey (1969)], A. L. Gurson [Gurson (1977)], and B. Budiansky *et al.* [Budiansky *et al.* (1982)].

7.1 Spherical void growth in linear viscous solids

Consider a spherical void inside an unbounded linear viscous RVE under uniform remote stress, as shown in Fig. 7.1. The constitutive behavior of the viscous matrix at microscale can be described as the following rate-dependent expression,

$$\sigma_{ij} = L_{ijk\ell}\dot{\epsilon}_{k\ell} \tag{7.1}$$

where $L_{ijk\ell}$ resembles the linear isotropic elasticity tensor, and can be expressed using a viscosity-like parameter η, and the viscous Poisson's ratio ν as:

$$L_{ijk\ell} = \frac{2\eta\nu}{1-2\nu}\delta_{ij}\delta_{k\ell} + \eta(\delta_{ik}\delta_{j\ell} + \delta_{i\ell}\delta_{jk}) \tag{7.2}$$

Therefore the stress-strain rate relationship becomes

$$\sigma_{ij} = 2\eta\left(\dot{\epsilon}_{ij} + \frac{\nu}{1-2\nu}\dot{\epsilon}_{kk}\delta_{ij}\right) \tag{7.3}$$

In the case of incompressible viscous solid, the deviatoric stress and strain rate can be simplified as $s_{ij} = 2\eta\dot{e}_{ij}$.

Assume that the spherical void, Ω, inside an RVE has a radius of a. A uniform triaxial stress state is imposed at the remote boundary of the RVE, i.e.

$$t_i = \sigma_{ij}^{\infty} n_j, \quad \forall x \in \partial\Omega$$

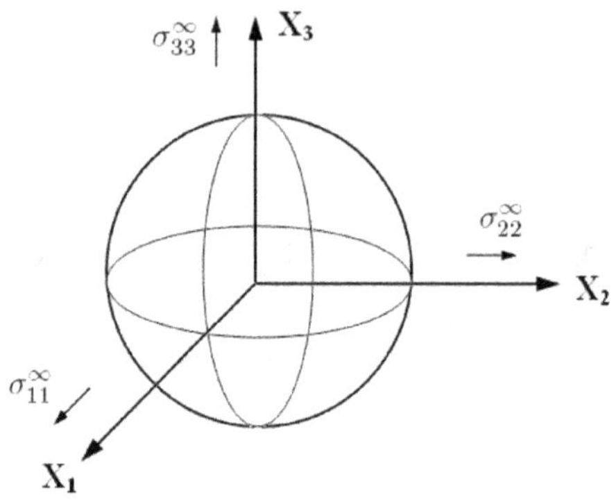

Fig. 7.1 A spherical void inside a linear viscous RVE.

Applying Eshelby's equivalent eigenstrain principle, the stress inside the void can be written as

$$\sigma_{ij} = L_{ijk\ell}\Big(\dot{\epsilon}_{k\ell}^{\infty} + \dot{\epsilon}_{k\ell}^{d} - \dot{\epsilon}_{k\ell}^{*}\Big), \quad \forall x \in \Omega$$

Note that the remote strain rate and remote stress are related as $\dot{\epsilon}_{ij}^{\infty} = D_{ijk\ell}\sigma_{k\ell}^{\infty}$, and $D_{ijk\ell} = L_{ijk\ell}^{-1}$. Since there is no stress inside the void, i.e., $\sigma_{ij} = 0$, we have the following relation for the strain-rate of the void $\dot{\epsilon}_{ij}^{\Omega}$,

$$\dot{\epsilon}_{ij}^{\Omega} = \dot{\epsilon}_{ij}^{\infty} + \dot{\epsilon}_{ij}^{d} = \dot{\epsilon}_{ij}^{*}, \quad \forall x \in \Omega$$

The above equation gives the physical meaning for the eigenstrain rate, i.e., the prescribed eigenstrain-rate should be the same as the stain-rate of the void. Moreover, one can find that

$$\sigma_{ij}^{\infty} = L_{ijk\ell}(\dot{\epsilon}_{k\ell}^{*} - \dot{\epsilon}_{k\ell}^{d})$$

By Eshelby's single inclusion solution, one can write the disturbance strain rate as

$$\dot{\epsilon}_{ij}^{d} = S_{ijk\ell}\dot{\epsilon}_{k\ell}^{*}$$

Although Eshelby's result was for linear elasticity and small strain deformation, the extension to the linear viscosity case is obvious as long as the strain rate is considered to be the current velocity strain. Therefore, strain-rate of the void can be related to the remote stresses σ_{ij}^{∞} according to

$$\sigma_{ij}^{\infty} = Q_{ijk\ell}\dot{\epsilon}_{k\ell}^{\Omega} \tag{7.4}$$

where the tensor $\mathbb{Q}$ is defined as

$$\mathbb{Q} := \mathbb{L} : (\mathbb{I}^{(4s)} - \mathbb{S}) \ .$$

It is straightforward to evaluate $\mathbb{Q}$ considering that

$$\mathbb{L} = 2\eta\frac{1+\nu}{1-2\nu}\mathbb{E}^{(1)} + 2\eta\mathbb{E}^{(2)}$$

$$\mathbb{S} = s_1\mathbb{E}^{(1)} + s_2\mathbb{E}^{(2)}$$

$$s_1 = \frac{1+\nu}{3(1-\nu)} \text{ and } s_2 = \frac{2(4-5\nu)}{3(1-\nu)}$$

We list several useful components of $\mathbb{Q}$ in the following,

$$Q_{1111} = Q_{2222} = Q_{3333} = -\frac{16\eta(1-2\nu)}{9(1-\nu)}$$

$$Q_{1122} = Q_{1133} = Q_{2233} = \frac{2\eta(7-5\nu)}{9(1-\nu)}$$

$$Q_{ii11} = Q_{ii22} = Q_{ii33} = \frac{4\eta(1+\nu)}{3(1-\nu)} .$$

Let the remote stress be biaxial loading

$$\sigma_{11}^{\infty} = \sigma_{22}^{\infty} = S, \qquad \sigma_{33}^{\infty} = T.$$

From (7.4), the remote stress can be related to the strain-rate of the void as

$$\begin{aligned} \sigma_{ii}^{\infty} &= Q_{ii11}\dot{\epsilon}_{11}^{\Omega} + Q_{ii22}\dot{\epsilon}_{22}^{\Omega} + Q_{ii33}\dot{\epsilon}_{33}^{\Omega} \\ &= \frac{4}{3}\frac{\eta(1+\nu)}{(1-\nu)}(\dot{\epsilon}_{11}^{\Omega} + \dot{\epsilon}_{22}^{\Omega} + \dot{\epsilon}_{33}^{\Omega}) = \frac{4\eta(1+\nu)}{3(1-\nu)}\frac{\dot{\Omega}}{\Omega} \end{aligned} \tag{7.5}$$

In the last line, the volumetric rate of the inclusion $\dot{\Omega}/\Omega = \dot{\epsilon}_{11}^{\Omega} + \dot{\epsilon}_{22}^{\Omega} + \dot{\epsilon}_{33}^{\Omega}$ is used. Finally, we conclude that the void growth rate depends only on the remote mean stress

$$2S + T = \frac{4\eta(1+\nu)}{3(1-\nu)}\frac{\dot{\Omega}}{\Omega}$$

The special case of an incompressible matrix can be obtained by assigning $\nu = 0.5$ to the above solution. Unless the remote stress is hydrostatic, i.e., $T = S$, the shape of the initially spherical void can not be retained over time. The void changes in the size and shape under various prescribed stress ratios S/T. B. Budiansky *et al.* derived the asymptotic solution of initially spherical void for all representative combinations of S and T. Depending on the remote stress ratio, the void can evolve into a cylinder, a spheroid, a needle, or even collapse into a crack. The asymptotic growth-rate can provide a rough estimate of the condition for void coalescence, and hence the propagation of ductile fracture.

7.2 McClintock solution to cylindrical void growth problem

The first landmark study on the void growth in nonlinear solids is pioneer work of F. A. McClintock [McClintock (1968)] . In his study, an analytical solution was provided for the void expansion rate of a long cylindrical void, pulled in the axial direction while subjected to transverse tension, see Fig.7.2. McClintock's solution assumes the material surrounding the void behaves as an incompressible, rigid-plastic material at the micro-level. The assumptions are reasonable considering void growth in ductile materials demands significant amount of plastic deformation, such that the elastic deformation can be ignored in the analysis. Long, roughly cylindrical voids are often observed in the neck of a tension bar under large deformation.

Although the solution is for an idealized symmetric geometry and an idealized material, the McClintock solution has been served as the benchmark example in many homogenization schemes for inelastic solids.

Consider the yield surface of the plastic material that can be described by the J_2 criterion (the von Mises criterion),

$$\mathcal{F} = J_2 - \frac{{\sigma_y}^2}{3} = \frac{1}{2} s_{ij} s_{ij} - \frac{{\sigma_y}^2}{3} = 0 \tag{7.6}$$

where s_{ij} is the deviatoric stress tensor, and the flow rule is defined as,

$$\dot{\epsilon}^p_{ij} = \dot{\lambda} \frac{\partial \mathcal{F}}{\partial s_{ij}} = \dot{\lambda} \frac{\partial f}{\partial s_{ij}} = \dot{\lambda} s_{ij} \tag{7.7}$$

The proportionality $\dot{\lambda}$ can be determined by submitting Eq. (7.7) into Eq. (7.6)

$$\dot{\lambda} = \sqrt{\frac{3}{2}} \frac{\sqrt{\dot{\epsilon}^p_{ij} \dot{\epsilon}^p_{ij}}}{\sigma_y} = \frac{3}{2} \frac{\dot{\bar{\epsilon}}^p}{\sigma_y} \tag{7.8}$$

where the effective plastic strain rate $\dot{\bar{\epsilon}}^p$ is defined as

$$\dot{\bar{\epsilon}}^p = \sqrt{\frac{2}{3} \dot{\epsilon}^p_{ij} \dot{\epsilon}^p_{ij}} \tag{7.9}$$

In cylindrical coordinate,

$$\dot{\bar{\epsilon}}^p = \sqrt{\frac{2}{3}\Big((\dot{\epsilon}^p_{rr})^2 + (\dot{\epsilon}^p_{\theta\theta})^2 + (\dot{\epsilon}^p_{zz})^2 \Big)} \tag{7.10}$$

Therefore, the constitutive relation at micro-level is,

$$\dot{\epsilon}^p_{ij} = \frac{3}{2} \frac{\dot{\bar{\epsilon}}^p}{\sigma_y} s_{ij} \tag{7.11}$$

Consider that the problem is antisymmetric and independent on the z coordinate in the transverse plane, so the velocity field can be assumed as

$$\dot{u}_r = \dot{u}(r), \;\; \dot{u}_\theta = 0, \dot{u}_z = \dot{\epsilon}_z z \;\; and \;\; \dot{\epsilon}_z = const. \, . \tag{7.12}$$

Hence, the incompressible condition yields the following differential equation for the displacement field $u(r)$,

$$\dot{\epsilon}_{rr} + \dot{\epsilon}_{\theta\theta} + \dot{\epsilon}_{zz} = \frac{d\dot{u}}{dr} + \frac{\dot{u}}{r} + \dot{\epsilon}_z = 0 \, . \tag{7.13}$$

Rewrite the above expression as

$$r \frac{d\dot{u}}{dr} + \dot{u} + r\dot{\epsilon}_z = 0 \; \Rightarrow \; d\Big(r\dot{u} \Big) = -r\dot{\epsilon}_z dr \tag{7.14}$$

Integrate over the radial direction from the surface of the void to the interior of the RVE,

$$\int_a^r d\Big(\rho \dot{u}(\rho) \Big) = - \int_a^r \rho \dot{\epsilon}_z d\rho \tag{7.15}$$

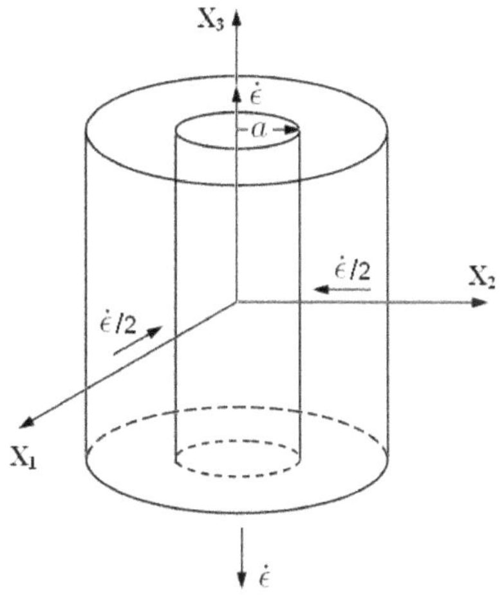

Fig. 7.2 A cylindrical void in an inelastic RVE.

Note that the variable ρ is the dummy variable. Considering $\dot{\epsilon}_z = \dot{\epsilon} = const.$, we have

$$r\dot{u}(r) - a\dot{a} = -\left(r^2 - a^2\right)\frac{\dot{\epsilon}_z}{2}$$

Finally we solve for the velocity field

$$\dot{u}_r = \dot{u}(r) = \frac{a^2}{r}\left(\frac{\dot{a}}{a} + \frac{\dot{\epsilon}_z}{2}\right) - \frac{\dot{\epsilon}_z r}{2}, \tag{7.16}$$

Direct calculation gives the following strain field,

$$\dot{\epsilon}_{rr} = \frac{d\dot{u}_r}{dr} = -\frac{a^2}{r^2}\left(\frac{\dot{a}}{a} + \frac{\dot{\epsilon}_z}{2}\right) - \frac{\dot{\epsilon}_z}{2} \tag{7.17}$$

$$\dot{\epsilon}_{\theta\theta} = \frac{\dot{u}_r}{r} = \frac{a^2}{r^2}\left(\frac{\dot{a}}{a} + \frac{\dot{\epsilon}_z}{2}\right) - \frac{\dot{\epsilon}_z}{2} \tag{7.18}$$

$$\dot{\epsilon}_{zz} = \frac{d\dot{u}_z}{dz} = \dot{\epsilon}_z \tag{7.19}$$

One can verify that the strain field satisfies the incompressibility condition $\dot{\epsilon}_{rr} + \dot{\epsilon}_{\theta\theta} + \dot{\epsilon}_{zz} = 0$. In cylindrical coordinate, the effective strain rate is

$$\dot{\bar{\epsilon}}^p = \sqrt{\frac{2}{3}\left((\dot{\epsilon}^p_{rr})^2 + (\dot{\epsilon}^p_{\theta\theta})^2 + (\dot{\epsilon}^p_{zz})^2\right)} = |\dot{\epsilon}_z|\sqrt{\frac{4}{3}\left(\frac{\dot{a}}{a\dot{\epsilon}_z} + \frac{1}{2}\right)^2\left(\frac{a^2}{r^2}\right)^2 + 1}$$

We define the following variables to further simplify the above equations,

$$x := \frac{a^2}{r^2}\frac{2}{\sqrt{3}}\left(\frac{\dot{a}}{a\dot{\epsilon}_z} + \frac{1}{2}\right) - \omega\frac{a^2}{r^2} \tag{7.20}$$

where

$$\omega := \frac{2}{\sqrt{3}}\left(\frac{\dot{a}}{a\dot{\epsilon}_z} + \frac{1}{2}\right) \tag{7.21}$$

Subsequently,

$$\dot{\epsilon}_{rr} = -\frac{\dot{\epsilon}_z}{2}(1+\sqrt{3}x) \tag{7.22}$$

$$\dot{\epsilon}_{\theta\theta} = -\frac{\dot{\epsilon}_z}{2}(1-\sqrt{3}x) \tag{7.23}$$

$$\dot{\bar{\epsilon}}^p = |\dot{\epsilon}_z|\sqrt{1+x^2} \tag{7.24}$$

We proceed to solve the stress field. Let

$$\sigma = \frac{1}{3}(\sigma_{rr} + \sigma_{\theta\theta} + \sigma_{zz}) \ . \tag{7.25}$$

We have deviatoric stresses

$$s_{rr} = \sigma_{rr} - \sigma \tag{7.26}$$

$$s_{\theta\theta} = \sigma_{\theta\theta} - \sigma \tag{7.27}$$

The constitutive equation (7.11) written in cylindrical coordinate becomes

$$\dot{\epsilon}_{rr} = \dot{\epsilon}^p_{rr} = \frac{3}{2}\frac{\dot{\bar{\epsilon}}^p}{\sigma_y}s_{rr} = \frac{3}{2}\frac{\dot{\bar{\epsilon}}^p}{\sigma_y}(\sigma_{rr} - \sigma) \tag{7.28}$$

$$\dot{\epsilon}_{\theta\theta} = \dot{\epsilon}^p_{\theta\theta} = \frac{3}{2}\frac{\dot{\bar{\epsilon}}^p}{\sigma_y}s_{\theta\theta} = \frac{3}{2}\frac{\dot{\bar{\epsilon}}^p}{\sigma_y}(\sigma_{\theta\theta} - \sigma) \tag{7.29}$$

Eqs.(7.28) - (7.29) leads to

$$\dot{\epsilon}_{\theta\theta} - \dot{\epsilon}_{rr} = \frac{3}{2}(\sigma_{\theta\theta} - \sigma_{rr})\frac{\dot{\bar{\epsilon}}^p}{\sigma_y} \tag{7.30}$$

Utilizing (7.30), it can be found that

$$\frac{\sigma_{\theta\theta} - \sigma_{rr}}{r} = \frac{2\sigma_y}{3r}\frac{\dot{\epsilon}_{\theta\theta} - \dot{\epsilon}_{rr}}{\dot{\bar{\epsilon}}^p} \tag{7.31}$$

Therefore the equilibrium equation becomes

$$\frac{d\sigma_{rr}}{dr} + \frac{\sigma_{rr} - \sigma_{\theta\theta}}{r} = \frac{d\sigma_{rr}}{dr} + \frac{2\sigma_y}{3r}\frac{(\dot{\epsilon}_{rr} - \dot{\epsilon}_{\theta\theta})}{\dot{\bar{\epsilon}}^p} = 0 \ . \tag{7.32}$$

Integrating above equation over the radius direction from the void surface to infinity,

$$\frac{1}{\sigma_y}\int_a^\infty d\sigma_{rr} = \frac{2}{3}\int_a^\infty \frac{(\dot{\epsilon}_{\theta\theta} - \dot{\epsilon}_{rr})}{\dot{\bar{\epsilon}}^p}\frac{dr}{r}$$

$$\Rightarrow \frac{\sigma_{rr}(\infty) - \sigma_{rr}(a)}{\sigma_y} = \frac{2}{3}\int_a^\infty \frac{(\dot{\epsilon}_{\theta\theta} - \dot{\epsilon}_{rr})}{\dot{\bar{\epsilon}}^p}\frac{dr}{r}$$

Consider the traction boundary condition,

$$\sigma_{rr}(a) = 0, \quad \text{and} \quad \sigma_{rr}(\infty) = \sigma_\infty \tag{7.33}$$

and strain field solution (7.22) and (7.23), we have

$$\frac{\sigma_{rr}(\infty)}{\sigma_y} = \frac{2}{3}\int_a^\infty \frac{(\dot{\epsilon}_\theta - \dot{\epsilon}_r)}{\dot{\bar{\epsilon}}^p}\frac{dr}{r} = \frac{2}{3}\int_a^\infty \frac{\sqrt{3}\dot{\epsilon}_z x}{|\dot{\epsilon}_z|\sqrt{1+x^2}}\frac{dr}{r} \tag{7.34}$$

Since $x = \omega a^2/r^2$, so $dx = -\frac{2}{r}xdr$, hence

$$\frac{dr}{r} = -\frac{1}{2}\frac{dx}{x} \ . \tag{7.35}$$

Making change of variable, we integrate (7.34), and notice that when $r = a,\ x = \omega$, and $r \to \infty,\ x \to 0$,

$$\begin{aligned}\frac{\sigma_\infty}{\sigma_y} &= \frac{2}{3}\int_a^\infty \frac{\sqrt{3}\dot{\epsilon}_z x}{|\dot{\epsilon}_z|\sqrt{1+x^2}}\frac{dr}{r} = \frac{1}{\sqrt{3}}\mathrm{sign}(\dot{\epsilon}_z)\int_0^\omega \frac{dx}{\sqrt{1+x^2}}\\ &= \frac{1}{\sqrt{3}}\mathrm{sign}(\dot{\epsilon}_z)\mathrm{arcsinh}(\omega)\end{aligned}$$

The inverse expression of the above result is

$$\frac{2}{\sqrt{3}}\Big(\frac{\dot{a}}{a\dot{\epsilon}_z} + \frac{1}{2}\Big) = \mathrm{sign}(\dot{\epsilon}_z)\sinh\Big[\frac{\sqrt{3}\sigma_\infty}{\sigma_y}\Big] \tag{7.36}$$

We obtain the relationship between void growth rate and remote stress value,

$$\frac{\dot{a}}{a} = \frac{\sqrt{3}}{2}|\dot{\epsilon}_z|\sinh\Big[\frac{\sqrt{3}\sigma_\infty}{\sigma_y}\Big] - \frac{1}{2}\dot{\epsilon}_z \tag{7.37}$$

A few comments about the McClintock solution are as follows:

(1) The McClintock solution is the only exact solution available for void growth in nonlinear solids; so far the non-linearity of field equations has excluded almost all exact solutions but the one solved by McClintock. Approximate solutions can be sought for spherical void growth in rigid-plastic materials, as presented by Rice and Tracey [Rice and Tracey (1969)] as well as other types of void growth [Golaganu *et al.* (1993, 1994)] .

(2) The McClintock solution reveals that the relative void growth rate per unit applied strain rate increases exponentially with the transverse remote stress. To illustrate the fact, we consider a finite cylindrical void with a height, H, and radius b. The volume of the cylinder is

$$\Omega = \pi a^2 H \ \Rightarrow\ \dot{\Omega} = 2\pi a\dot{a}H + \pi a^2\dot{H} \tag{7.38}$$

Thereby,

$$\frac{\dot{\Omega}}{\Omega} = 2\frac{\dot{a}}{a} + \dot{\epsilon}_z \tag{7.39}$$

and hence

$$\frac{\dot{\Omega}}{\Omega} = \sqrt{3}|\dot{\epsilon}_z|\sinh\Big[\frac{\sqrt{3}\sigma_\infty}{\sigma_y}\Big] \tag{7.40}$$

One may compare (7.40) with Budiansky *et al*'s linear viscous void solution,

$$\frac{\dot{\Omega}}{\Omega} = \frac{9}{4}\frac{1-\nu}{\eta(1+\nu)}\sigma_\infty \tag{7.41}$$

which states that the relative void growth rate is linearly proportional to the transverse remote stress.

(3) At the remote boundary, $r \to \infty$,

$$\dot{\epsilon}_{zz}^{\infty} = \dot{\epsilon}_z, \ \dot{\epsilon}_{rr}^{\infty} = \dot{\epsilon}_{\theta\theta}^{\infty} = -\frac{1}{2}\dot{\epsilon}_z \tag{7.42}$$

Hence the macro equivalent strain rate is

$$\dot{\bar{\epsilon}}^{\infty} = \sqrt{\frac{2}{3}\dot{\epsilon}_{ij}^{\infty}\dot{\epsilon}_{ij}^{\infty}} = \dot{\epsilon}_z \tag{7.43}$$

Bi-axial stress state is applied at the remote boundary, ∂V, i.e.

$$\Sigma_{11} = \Sigma_{22} = \sigma_{\infty}, \ \ \Sigma_{33} = T, \text{and } \ \Sigma_m = \frac{1}{3}\Big(2\Sigma_{11} + \Sigma_{33}\Big) \tag{7.44}$$

The von Mises criterion becomes

$$\begin{aligned}\Sigma_{eq} &= \sqrt{\frac{3}{2}\Big((\Sigma_{11} - \Sigma_m)^2 + (\Sigma_{22} - \Sigma_m)^2 + (\Sigma_{33} - \Sigma_m)^2\Big)} \\ &= |\Sigma_{33} - \Sigma_{11}| \le \sigma_y\end{aligned}$$

The yield surface is $|\Sigma_{33} - \Sigma_{11}| = \sigma_y$.
Under such condition, we can rewrite the void growth rate equation as

$$\frac{\dot{\Omega}}{\Omega|\dot{\epsilon}_z|} = \sqrt{3}\sinh\Big(\frac{\sqrt{3}\sigma_{\infty}}{\sigma_y}\Big) = \sqrt{3}\sinh\Big(\frac{\sqrt{3}\Sigma_{11}}{|\Sigma_{33} - \Sigma_{11}|}\Big) \ . \tag{7.45}$$

(4) Let the total volume of the RVE be

$$V = \Omega + V_{matrix} \tag{7.46}$$

and

$$\frac{dV}{dt} = \dot{V} = \frac{d\Omega}{dt} + \frac{dV_{matrix}}{dt} = \frac{d\Omega}{dt} \tag{7.47}$$

because the matrix is incompressible, $\dfrac{dV_{matrix}}{dt} = 0$. Define the volume fraction of the void as

$$f = \frac{\Omega}{V} \ . \tag{7.48}$$

Then

$$\dot{f} = \frac{\dot{\Omega}}{V} - \frac{\Omega}{V^2}\dot{V} = \frac{\dot{\Omega}}{V}\Big(\frac{V - \Omega}{V^2}\Big) = \frac{\dot{\Omega}}{V}(1 - f) = \frac{\dot{\Omega}}{\Omega}f(1 - f)$$

Finally, we can express the rate of volume fraction as

$$\dot{f} = \sqrt{3}f(1 - f)|\dot{\epsilon}_z|\sinh\Big(\frac{\sqrt{3}\Sigma_{11}}{|\Sigma_{33} - \Sigma_{11}|}\Big) \tag{7.49}$$

7.3 Gurson model

The significance of the McClintock solution is that it links the remote stress, or macro stress, with the void growth rate, and reveals that in a perfectly plastic RVE, the void growth rate is expedientially related to the macro-stress. Although, it can be argued that the notion of representative volume element is employed in the McClintock solution, it does provide new constitutive representation at macro-level.

Not long after the publication of the McClintock solution [McClintock (1968)], a young scientist at that time, A. L. Gurson , realized that there is more to explore in the cylindrical void model analyzed by McClintock. In fact, one can derive the plastic potential surface at macro-level by homogenized (meaning averaging in space) micro-stress distribution. Gurson achieved this goal in his Ph.D. dissertation [Gurson (1975)], which has become one of most cited works in inelastic constitutive modeling and micromechanics.

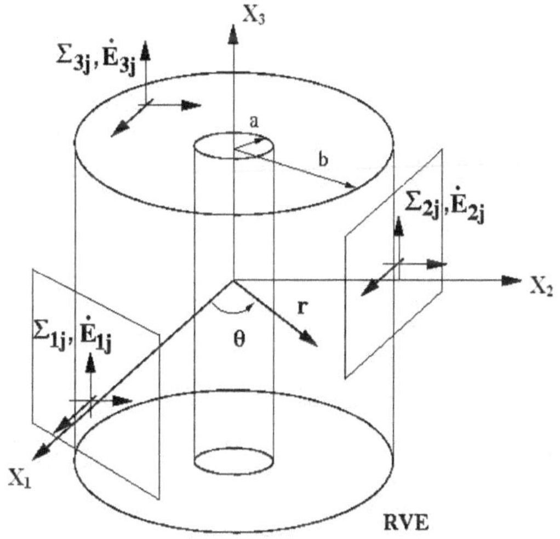

Fig. 7.3 A cylindrical void in a rigid-perfectly plastic von Mises RVE.

The main objective of the Gurson model is to find a macroscopic yield function in terms of macro-stress measures and volume fraction of voids in an RVE, i.e., we are looking for a function in the following form,

$$\mathcal{F}(\Sigma_{eq}, \Sigma_m, f) = 0 \tag{7.50}$$

where

$$\Sigma_{eq} = \sqrt{\frac{3}{2}\Sigma'_{ij}\Sigma'_{ij}}, \quad \Sigma'_{ij} = \Sigma_{ij} - \Sigma_m, \quad \text{and} \quad \Sigma_m = \frac{1}{3}\Sigma_{ii} \tag{7.51}$$

are macro-stresses which are defined as the spatial average of micro-stresses over the RVE. The Gurson model specifies the RVE to be a concentric cylinder of radius b encompassing a cylindrical void of radius a, as shown in Fig. 7.3.

The same as the McClintock solution, the governing equations for the matrix materials are,

(1) Equilibrium equation:

$$\frac{d\sigma_{rr}}{dr} + \frac{\sigma_{rr} - \sigma_{\theta\theta}}{r} = 0 \ . \tag{7.52}$$

(2) Perfectly-plastic constitutive relation follows von Mises flow rule:

$$s_{ij} = \frac{2}{3}\frac{\sigma_y}{\dot{\bar{\epsilon}}^p}\dot{\epsilon}_{ij} \tag{7.53}$$

(3) Incompressible condition of the matrix:

$$\dot{\epsilon}_{rr} + \dot{\epsilon}_{\theta\theta} + \dot{\epsilon}_{zz} = 0 \ . \tag{7.54}$$

Assume that the remote macro-stress loading is axisymmetric, i.e.

$$\sigma_{11}\Big|_{\partial V} = \Sigma_{11}, \quad \sigma_{22}\Big|_{\partial V} = \Sigma_{22}, \text{ and } \Sigma_{11} = \Sigma_{22} \tag{7.55}$$

$$\sigma_{33}\Big|_{\partial V} = \Sigma_{33} \tag{7.56}$$

Under the axisymmetric loading condition,

$$\begin{aligned} \Sigma_{eq} &= \sqrt{\frac{1}{2}\Big[(\Sigma_{11} - \Sigma_{22})^2 + (\Sigma_{33} - \Sigma_{11})^2 + (\Sigma_{33} - \Sigma_{22})^2\Big]} \\ &= |\Sigma_{33} - \Sigma_{11}|, \end{aligned} \tag{7.57}$$

$$\begin{aligned} \Sigma_m &= \frac{1}{3}(\Sigma_{11} + \Sigma_{22} + \Sigma_{33}) = \frac{1}{3}\Big(\Sigma_{\alpha\alpha} + \Sigma_{33}\Big) \\ &= \Sigma_{11} + \frac{1}{3}\Big(\Sigma_{33} - \Sigma_{11}\Big) = \frac{1}{2}\Sigma_{\alpha\alpha} + \frac{1}{3}\Sigma_{eq} \end{aligned} \tag{7.58}$$

where $\Sigma_{\alpha\alpha} = \Sigma_{11} + \Sigma_{22} = 2\Sigma_{11}$, or $\Sigma_{11} = \Sigma_{22} = \frac{1}{2}\Sigma_{\alpha\alpha}$. Therefore, we are essentially looking for the yielding effects due to Σ_{11} and $\Sigma_{33} - \Sigma_{11}$.

Consider the following axisymmetric kinematic pattern,

$$\dot{u}_r = \dot{u}(r), \quad \dot{u}_z(z) = \dot{E}_{33} z \ . \tag{7.59}$$

We have the following strain field from the McClintock solution in the previous section,

$$\dot{\bar{\epsilon}}^p = |\dot{\epsilon}_z|\sqrt{1 + x^2} \tag{7.60}$$

$$\dot{\epsilon}_{rr} = -\frac{\dot{\epsilon}_z}{2}(1 + \sqrt{3}x) \tag{7.61}$$

$$\dot{\epsilon}_{\theta\theta} = -\frac{\dot{\epsilon}_z}{2}(1 - \sqrt{3}x) \tag{7.62}$$

where

$$x := \frac{a^2}{r^2}\frac{2}{\sqrt{3}}\Big(\frac{\dot{a}}{a\dot{\epsilon}_z} + \frac{1}{2}\Big) = \omega\frac{a^2}{r^2} \tag{7.63}$$

$$\omega := \frac{2}{\sqrt{3}}\Big(\frac{\dot{b}}{b\dot{\epsilon}_z} + \frac{1}{2}\Big) \tag{7.64}$$

Since the matrix is a rigid-perfectly plastic von-Mises material, it obeys the following flow rule,

$$s_{ij} = \frac{2}{3}\frac{\sigma_y}{\dot{\bar{\epsilon}}^p}\dot{\epsilon}_{ij} \tag{7.65}$$

Therefore, we can write,

$$\begin{aligned} s_{rr} &= \frac{2}{3}\frac{\sigma_y}{\dot{E}_{33}\sqrt{1+x^2}}\dot{\epsilon}_{rr} = \frac{1}{3}\frac{\sigma_y}{\sqrt{1+x^2}}(-1-\sqrt{3}x) \\ s_{\theta\theta} &= \frac{2}{3}\frac{\sigma_y}{\dot{E}_{33}\sqrt{1+x^2}}\dot{\epsilon}_{\theta\theta} = \frac{1}{3}\frac{\sigma_y}{\sqrt{1+x^2}}(-1+\sqrt{3}x) \\ s_{zz} &= \frac{2}{3}\frac{\sigma_y}{\dot{E}_{33}\sqrt{1+x^2}}\dot{E}_{33} = \frac{2}{3}\frac{\sigma_y}{\sqrt{1+x^2}} \end{aligned}$$

We can then find that

$$\begin{aligned} s_{\theta\theta} - s_{rr} &= \frac{1}{3}\frac{\sigma_y}{\sqrt{1+x^2}}(2\sqrt{3}x) = \sigma_{\theta\theta} - \sigma_{rr} \\ s_{zz} - \frac{1}{2}(s_{rr} + s_{\theta\theta}) &= \frac{\sigma_y}{\sqrt{1+x^2}} = \sigma_{zz} - \frac{1}{2}(\sigma_{rr} + \sigma_{\theta\theta}) \end{aligned}$$

To this end, we are in a position to link the macro-stresses, Σ_{11}, $\Sigma_{33} - \Sigma_{11}$, and void volume fraction, f, together in a macro yield potential. We first link Σ_{11} and $|\Sigma_{33} - \Sigma_{11}|$ with remote strain rate, $\dot{E}_{ij}$. Consider the traction boundary conditions on the surface of the void and the surface of the RVE,

$$\sigma_{rr}(a) = 0, \quad \text{and} \quad \sigma_{rr}(b) = \frac{1}{2}\Sigma_{\alpha\alpha} = \Sigma_{11} \tag{7.66}$$

note that $\Sigma_{rr}(b) = \Sigma_{\theta\theta}(b) = \frac{1}{2}\Sigma_{\alpha\alpha}$.

1. Integrating equilibrium equation along the radius direction yields,

$$\Sigma_{11} = \sigma_{rr}(b) - \sigma_{rr}(a) = \int_a^b \frac{d\sigma_{rr}}{dr}dr = \int_a^b \frac{\sigma_{\theta\theta} - \sigma_{rr}}{r}dr \tag{7.67}$$

Since,

$$\frac{\sigma_{\theta\theta} - \sigma_{rr}}{r} = \frac{1}{3r}\frac{\sigma_y}{\sqrt{1+x^2}}(2\sqrt{3}x) \tag{7.68}$$

Use the following changes of variables,

$$x = \omega\Big(\frac{a}{r}\Big)^2 : \quad x \to [\omega, f\omega], \quad \text{when} \quad r \to [a, b]\ . \tag{7.69}$$

where $f = \dfrac{a^2}{b^2} = \dfrac{\Omega}{V}$. Therefore $\dfrac{dr}{r} = -\dfrac{1}{2}\dfrac{dx}{x}$, we have

$$\Sigma_{11} = -\Big(\frac{\sqrt{3}}{3}\sigma_y\Big)\int_{\omega}^{f\omega} \frac{dx}{\sqrt{1+x^2}}$$

Thereby,

$$\frac{\Sigma_{\alpha\alpha}}{2} = \frac{\sigma_y}{\sqrt{3}} \int_{f\omega}^{\omega} \frac{dx}{\sqrt{1+x^2}} \tag{7.70}$$

We then find that the in-plane hydrostatic stress can be written as

$$\boxed{\frac{\sqrt{3}}{2}\frac{\Sigma_{\alpha\alpha}}{\sigma_y} = \ln\left(\frac{\omega+\sqrt{1+\omega^2}}{f\omega+\sqrt{1+(f\omega)^2}}\right)} \tag{7.71}$$

2. Consider the fact that $\sigma_{11}+\sigma_{22}=\sigma_{rr}+\sigma_{\theta\theta}$, and $\Sigma_{11}=\Sigma_{22}=\frac{1}{2}\Sigma_{\alpha\alpha}$, we evaluate the macro-stresses by spatially homogenizing the micro-stresses over the RVE,

$$\begin{aligned}\Sigma_{33}-\Sigma_{11} &= \Sigma_{33}-\frac{1}{2}(\Sigma_{11}+\Sigma_{22}) = \frac{1}{V}\int_V \Big(\sigma_{zz}-\frac{1}{2}(\sigma_{xx}+\sigma_{yy})\Big)dV \\ &= \frac{1}{V}\int_V \Big(s_{zz}-\frac{1}{2}(s_{rr}+s_{\theta\theta})\Big)dV \\ &= \frac{1}{V}\int_{V_M} \Big(s_{zz}-\frac{1}{2}(s_{rr}+s_{\theta\theta})\Big)dV\end{aligned}$$

Recall that

$$s_{zz}-\frac{1}{2}(s_{rr}+s_{\theta\theta}) = \frac{\sigma_y}{\sqrt{1+x^2}}$$

and $dV = rdrd\theta dz$. We have

$$\Sigma_{33}-\Sigma_{11} = \frac{2\pi H}{\pi b^2 H}\int_a^b \frac{\sigma_y}{\sqrt{1+x^2}} rdr$$

Again, we make changes of variables in the integration, $rdr = -\dfrac{\omega a^2}{2}\dfrac{dx}{x^2}$, hence

$$\Sigma_{33}-\Sigma_{11} = \frac{2\sigma_y}{b^2}\int_a^b \frac{rdr}{\sqrt{1+x^2}} = f\omega\sigma_y \int_{f\omega}^{\omega} \frac{dx}{x^2\sqrt{1+x^2}}$$

Carrying out the integration, we have

$$\Sigma_{33}-\Sigma_{11} = \sigma_y\left(\sqrt{1+\omega^2 f^2} - f\sqrt{1+\omega^2}\right) \tag{7.72}$$

The above equation links the deviatoric macro-stress with macro-strain rate and void volume fraction,

$$\boxed{\frac{\Sigma_{eq}}{\sigma_y} = \sqrt{1+f^2\omega^2} - f\sqrt{1+\omega^2}} \tag{7.73}$$

ω in (7.71) and (7.73) could be eliminated after some algebra. It is easy to show that

$$\cosh\left(\frac{\sqrt{3}}{2}\frac{\Sigma_{\alpha\alpha}}{\sigma_y}\right) = \sqrt{1+(f\omega)^2(1+\omega^2)} - f\omega^2 \tag{7.74}$$

$$\left(\frac{\Sigma_{eq}}{\sigma_y}\right)^2 = 1 + f^2 - 2f\sqrt{1 + (f\omega)^2(1+\omega^2)} + 2f^2\omega^2 \tag{7.75}$$

Finally, we derived the yield function in terms of macro-stress and volume fraction as

$$\boxed{\mathcal{F}(\Sigma_{eq}, \Sigma_{\alpha\alpha}, f) = \left(\frac{\Sigma_{eq}}{\sigma_y}\right)^2 + 2f\cosh\Bigl(\frac{\sqrt{3}}{2}\frac{\Sigma_{\alpha\alpha}}{\sigma_y}\Bigr) - (1+f^2) = 0\ .} \tag{7.76}$$

On the other hand, (7.71) can be evaluated as

$$\sinh\Bigl(\frac{\sqrt{3}}{2}\frac{\Sigma_{\alpha\alpha}}{\sigma_y}\Bigr) = \omega(\sqrt{1+f^2\omega^2} - f\sqrt{1+\omega^2}) \tag{7.77}$$

Consider

$$\omega = \frac{\dot{\Omega}}{\Omega}\frac{1}{\sqrt{3}\dot{E}_{33}}, \quad \text{and} \quad \frac{\Sigma_{eq}}{\sigma_y} = \sqrt{1+f^2\omega^2} - f\sqrt{1+\omega^2} \tag{7.78}$$

Eq. (7.77) can be rewritten as

$$\left|\frac{\dot{\Omega}}{\Omega}\right| = \sqrt{3}\dot{E}_{33}\Bigl(\frac{\sigma_y}{\Sigma_{eq}}\Bigr)\sinh\Bigl(\frac{\sqrt{3}}{2}\frac{\Sigma_{\alpha\alpha}}{\sigma_y}\Bigr) \tag{7.79}$$

recovering the McClintock solution.

The yield function (7.76) has some distinctive features worth to be discussed. Figure 7.4 shows the yield function plotted in normalized 3D principal macro-stress space and normalized 2D mean/deviatoric macro-stress space. The potential function consists of a family of convex surfaces parameterized by the volume fraction of the void. For virgin material ($f = 0$), the potential function reduces to pressure-independent von Mises model, while presence of voids turn it into a pressure-dependent yield function. Furthermore, the yield function progressively reduce in size with increasing void volume fraction, representing the material degradation process due to void growth. The homogenization result obtained by Gurson marks a significant milestone in the development of micromechanics, because the outcome of the homogenization is fundamentally different from that of micro-elasticity theory. In micro-elasticity theory, the homogenized constitutive relations are virtually the same as the constitutive relations in microscale, i.e., both of them are linear elastic materials obeying the generalized Hook's law. The only differences in constitutive laws at different scales are the magnitude and spatial orientation of elastic constants. Whereas, in the Gurson model, a completely new constitutive relation at macro-level emerges from the homogenization, which represents a new philosophy: *Seeking for new physical laws and new mechanics in a route of homogenization.* This notion is so attractive, and it has remained the very ideal and ultimate objective of contemporary micromechanics and multiscale simulations.

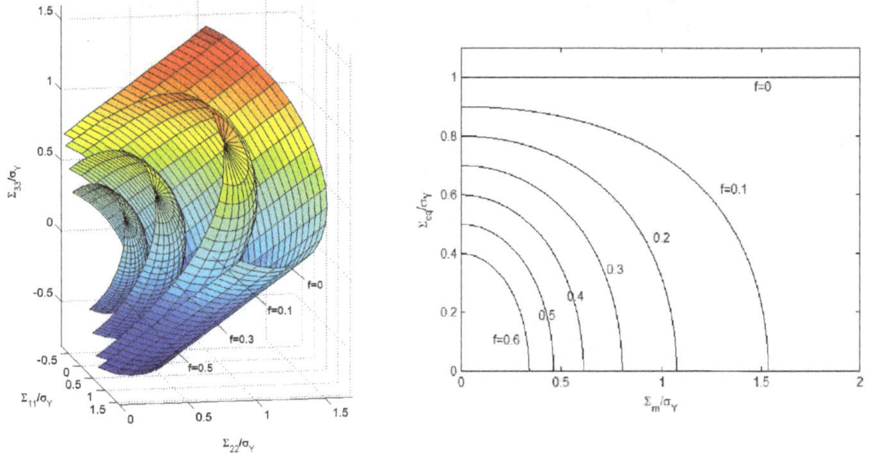

Fig. 7.4 Yield function of the Gurson model.

7.4 Gurson-Tvergaard-Needleman (GTN) model

The Gurson model has been used in computational mechanics to model ductile fracture, which is commonly referred to as the Gurson-Tvergaard-Needleman (GTN) model . Today the GTN model has become one of the primary damage models in computational mechanics and computational materials, and it has been applied to many engineering problems as a predictive modeling tool. In fact, the GTN model has been integrated in some major finite element commercial softwares, such as ABACUS, and its simulation results have been validated extensively by experiments.

At macrolevel, the flow potential of the standard GTN model reads as

$$\Phi = \Big(\frac{\Sigma_{eq}}{\bar{\sigma}_0}\Big)^2 + 2q_1 f^* \cosh\Big(\frac{3q_2\Sigma_m}{2\bar{\sigma}_0}\Big) - \Big(1 + q_1^2 f^{*2}\Big) = 0 \tag{7.80}$$

where

$$\Sigma_m = \frac{1}{3}\mathbf{\Sigma} : \mathbb{I}^{(2)} \tag{7.81}$$

$$\Sigma_{eq} = \Big(\frac{3}{2}\mathbf{\Sigma}_{dev} : \mathbf{\Sigma}_{dev}\Big)^{1/2}, \quad \text{with} \;\; \mathbf{\Sigma}_{dev} = \mathbf{\Sigma} - \Sigma_m \mathbb{I}^{(2)} \tag{7.82}$$

in which the macroscale Cauchy stress can be decomposed as

$$\mathbf{\Sigma} = \Sigma_m \mathbb{I}^{(2)} + \frac{2}{3}\Sigma_{eq}\mathbf{n}, \quad \text{with} \;\; \mathbf{n} := \frac{3}{2\Sigma_{eq}}\mathbf{\Sigma}_{dev} \,. \tag{7.83}$$

The first major difference between the Gurson model and the GTN model is: in GTN model the micro- constitutive relation is assumed to be a general, isotropic strain hardening plasticity; whereas in the original Gurson's model the material at micro-level is assumed to be perfectly plastic. In GTN model, the current yield stress, $\bar{\sigma}_0$ is described by a power law type of hardening relationship,

$$\bar{\sigma}_0 = f_h(\bar{\epsilon}^p) = \sigma_y \Big(1 - \frac{E}{\sigma_y}\bar{\epsilon}^p\Big)^N \tag{7.84}$$

where σ_y is the initial yield stress, E is Young's modulus, N is an exponent, and $\bar{\epsilon}^p$ is the microscope equivalent plastic strain that is determined by the condition: *The microscale plastic dissipation is equal to the macroscale plastic dissipation*, which can be expressed as

$$(1-f)\bar{\sigma}_0 \dot{\bar{\epsilon}}^p = \mathbf{\Sigma} : \mathbf{d}^p, \quad \text{and} \quad \dot{\bar{\epsilon}}^p = \frac{\mathbf{\Sigma} : \mathbf{d}^p}{(1-f)\bar{\sigma}_0} \tag{7.85}$$

where $\mathbf{d}^p$ is the macroscale plastic rate of deformation.

The second difference between the GTN model and the original Gurson model is the evolution of damage. In the original Gurson model, the void growth rate is governed by the McClintock solution; whereas in the GTN model, Tvergaard [Tvergaard (1981, 1982)] introduced the parameters, q_1 and q_2, to bring the predictions of the model into closer agreement with full numerical analysis of a periodic array of voids. Tvergaard and Needleman [Tvergaard and Needleman (1984)] proposed an equivalent damage parameter function, f^*, to replace the volume fraction of voids, f, in order to account for the effects of rapid void coalescence at failure. f^* can be evaluated as,

$$f^*(f) = \begin{cases} f & \text{if} \quad f \leq f_c \\ f_c + \dfrac{1/q_1 - f_c}{f_f - f_c}(f - f_c) & \text{if} \quad f_c < f \leq f_f \\ 1/q_1 & \text{if} \quad f > f_f \end{cases} \tag{7.86}$$

where f_c is the volume fraction of the void at the incidence of coalescence, and f_f is the void volume fraction at failure. In the GTN model, the void growth rate is partly due to the expansion of the existing voids and partly due to the nucleation of new voids,

$$\dot{f} = \dot{f}_{growth} + \dot{f}_{nucleation} \, . \tag{7.87}$$

Since the material of the matrix is incompressible, the growth rate of existing voids is

$$\dot{f}_{growth} = (1-f)\mathbf{d}^p : \mathbb{I}^{(2)} \tag{7.88}$$

There were several void nucleation mechanisms. For instance, the nucleation is controlled by microscale plastic strain [Chu and Needleman (1980)],

$$\dot{f}_{nucleation} = A_N \dot{\bar{\epsilon}}^p \tag{7.89}$$

and the coefficient A_N is a function of microscale plastic strain distribution

$$A_N(\bar{\epsilon}^p) = \frac{f_N}{s_N\sqrt{2\pi}} \exp\Big[-\frac{1}{2}\Big(\frac{\bar{\epsilon}^p - \epsilon_N}{s_N}\Big)\Big] \tag{7.90}$$

where f_N is the volume fraction of void nucleating particles, and ϵ_N is its mean value and s_N is the standard deviation.

Considering the macroscale flow rule

$$\mathbf{d}^p = \dot{\lambda}\frac{\partial \Phi}{\partial \boldsymbol{\Sigma}} = \dot{\lambda}\Big(\frac{1}{3}\frac{\partial \Phi}{\partial \Sigma_m}\mathbb{I}^{(2)} + \frac{\partial \Phi}{\partial \Sigma_{eq}}\mathbf{n}\Big) \tag{7.91}$$

we can define a decomposition of the plastic rate of deformation,

$$\mathbf{d}^p = \frac{1}{3}d_m\mathbb{I}^{(2)} + d_{eq}\mathbf{n} \tag{7.92}$$

where

$$d_m = \dot{\lambda}\Big(\frac{\partial \Phi}{\partial \Sigma_m}\Big), \quad \text{and} \quad d_{eq} = \dot{\lambda}\Big(\frac{\partial \Phi}{\partial \Sigma_{eq}}\Big) \tag{7.93}$$

One of the distinguished features of the GTN model is that we can link microscale state variable evolution equations with the macroscale variables. This can be shown by

$$\frac{\partial \bar{\epsilon}^p}{\partial t} = \frac{\Sigma_m d_m + \Sigma_{eq} d_{eq}}{(1-f)\bar{\sigma}_0} \tag{7.94}$$

$$\frac{\partial f}{\partial t} = (1-f)d_m + A_N\frac{\partial \bar{\epsilon}^p}{\partial t} \tag{7.95}$$

The GTN model has been used extensively to simulate material failures. The plots shown in Fig. 7.5 are from a meshfree simulation of ductile crack growth by using the GTN model [Simonsen and Li (2004); Li and Simonsen (2005)].

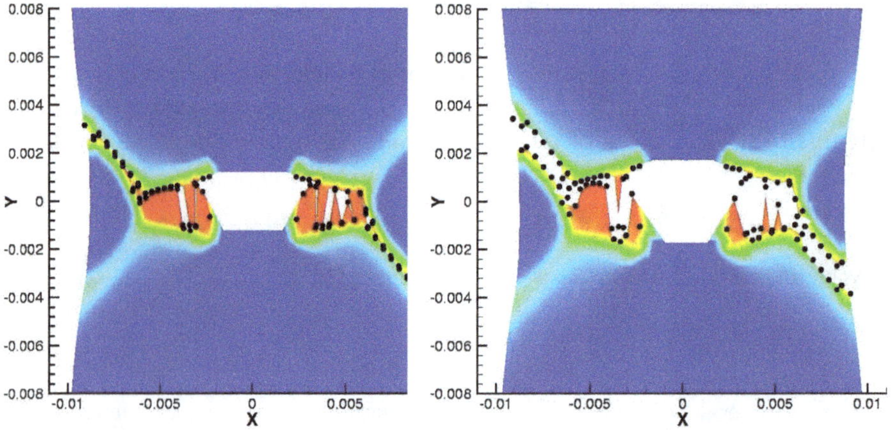

(a) (b)

Fig. 7.5 Meshfree simulations of ductile crack propagation via GTN model.

7.5 A cohesive micro-crack damage model

Failure mechanism due to void growth is supported by many experimental observations on failures of ductile materials (e.g. [McClintock (1968); Duva and Hutchinson (1984); Tvergaard (1990); Gologanu *et al.* (1995)], and [Pardoen and Hutchinson (2000)]). In previous sections, we have presented several classical models on this topic. On the other hand, in most brittle, quasi-brittle, and even some ductile materials, such as concrete, rocks, ceramics, and some metals, material's failure mechanism may be attributed to nucleation and coalescence of micro-cracks. In 2004, S. Li and G. Wang proposed a damage theory for quasi-brittle materials [Li and Wang (2004); Wang and Li (2004)]. The following presentation is the outline of that theory.

Although several micro-crack based damage models have been proposed to describe elastic damage processes (e.g. [Budiansky and O'Connell (1976); Hutchinson (1987a); Fleck (1991); Kachanov (1994); Krajcinovic (1996); Nemat-Nasser and Hori (1999)] and others) , only few micro-crack damage models are available for inelastic damage processes. W. Ju and his co-workers ([Ju and Tseng (1996); Ju and Sun (2001); Sun and Ju (2001)]) have applied micromechanics techniques to model effective elastoplastic behaviors of a composite with distributed inhomogeneities. In their study, qualitatively, the macro constitutive relation of the composite is virtually the same as the micro constitutive relation of the matrix and the inhomogeneity — the classical elasto-plastic (J_2) constitutive relations. Their objective is only to find quantitatively homogenized constitutive relation, which has the same form as the constitutive relation at microscale.

Since G. I. Barenblatt [Barenblatt (1959, 1962)] and D. S. Dugdale's pioneer contribution [Dugdale (1960)] cohesive crack models have been studied extensively. In applications, the assessment of overall damage effect due to cohesive defect distribution is important for studying material damage at macro-level. In other words, the density of the micro-crack distribution is a measure of damage state, whose evolution can characterize material degradation. In this section, micromechanics techniques are applied to study effective constitutive behaviors of a solid with randomly distributed cohesive cracks. New cohesive damage models are derived, which are based on homogenization of randomly distributed penny shaped cohesive cracks (Barenblatt-Dugdale type) in an elastic RVE.

A main hypothesis or approximation of the proposed cohesive damage model is : *the overall damage due to the permanent crack opening is only associated with average hydrostatic stress (spherical) state in an RVE, and the overall damage effect due to the average deviatoric stress can be neglected.*

Based on this hypothesis, the particular damage we are interested in is only susceptible to macro hydrostatic stress state and we neglect the damage effect due to shear deformation or macro deviatoric stress states. Based on this assumption and since mode II and mode III types of remote loading will not contribute to crack opening volume, they are absent in our analysis of material damage, though they definitely contribute surface separation and in general there are cohesive shear forces between two sliding crack surfaces.

7.5.1 *Average theorem for a cohesive RVE*

Since the cohesive crack is not a traction-free defect, we may need to re-examine traditional micro-mechanics averaging theory for traction-free defects in a solid. An averaging theorem for solids containing cohesive defects with constant cohesive traction would be useful for our purpose.

Define the macro stress tensor, Σ_{ij} as the volume average of micro stress tensor in an RVE,

$$\Sigma_{ij} := < \sigma_{ij} > = \frac{1}{V} \int_V \sigma_{ij} dV \tag{7.96}$$

We first consider the average stress in a three-dimensional (3D) elastic representative volume element with a single penny-shaped Barenblatt-Dugdale crack at the center of the RVE. We adopt the assumption that body force has no effect on material properties. The equilibrium equation inside an RVE takes the form,

$$\sigma_{ji,j} = 0, \quad \forall \, \mathbf{x} \in V \tag{7.97}$$

Assume that the prescribed traction on the remote boundary of the RVE (∂V_∞) are generated by a constant stress tensor σ_{ij}^∞. Let ∂V_{ec} denote the traction free part of a cohesive crack surface, and let ∂V_{pz} denote the cohesive part of the crack surface where constant traction force t_j is applied. Using divergence theorem, it is straightforward to show that

$$\begin{aligned}
< \sigma_{ij} > &= \frac{1}{V} \int_V \sigma_{ij} dV = \frac{1}{V} \int_V \Big(\sigma_{kj} x_i \Big)_{,k} dV \\
&= \frac{1}{V} \Big\{ \int_V \sigma_{kj}^\infty \delta_{ik} dV - \int_{\partial V_{ec}} 0 \cdot x_i n_k dS - \int_{\partial V_{pz}} \sigma_{kj} x_i n_k dS \Big\} \\
&= \sigma_{ij}^\infty - \frac{1}{V} \int_{\partial V_{pz}} \sigma_{kj} x_i n_k dS = \sigma_{ij}^\infty - \frac{1}{V} \int_{\partial V_{pz}} t_j x_i dS
\end{aligned} \tag{7.98}$$

where t_j is the constant cohesive traction.

Note that $\partial V_{pz} = \partial V_{pz+} \cup \partial V_{pz-}$ and $|\partial V_{pz+}| = |\partial V_{pz-}| = \frac{1}{2} |\partial V_{pz}|$, where subscript $'+'$ and $'-'$ are used to denote upper and lower part of the crack surfaces. So the last term in (7.98) becomes

$$\begin{aligned}
\frac{1}{V} \int_{\partial V_{pz}} t_j x_i dS &= \frac{1}{V} \Big(\int_{\partial V_{pz+}} t_j^+ x_i dS + \int_{\partial V_{pz-}} t_j^- x_i dS \Big) \\
&= \frac{1}{2V} (t_j^+ + t_j^-) x_i |\partial V_{pz}| = 0
\end{aligned} \tag{7.99}$$

where $t_j^+ = -t_j^-$ are the cohesive traction acting on ∂V_{pz+} and ∂V_{pz-} respectively.

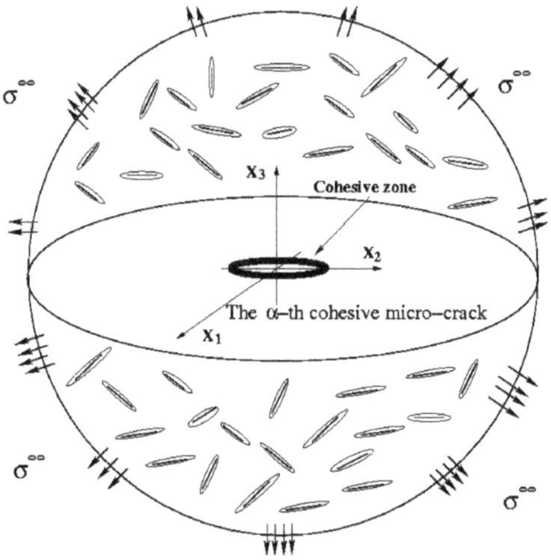

Fig. 7.6 Isotropic distribution of cracks with different orientations.

Therefore, the average stress inside the RVE is equal to the remote stress

$$\Sigma_{ij} =< \sigma_{ij} >= \sigma_{ij}^{\infty} \ . \tag{7.100}$$

By superposition, it is straightforward to generalize this result to an RVE with N cohesive cracks randomly distributed inside. As shown above,

$$< \sigma_{ij} >= \sigma_{ij}^{\infty} - \frac{1}{V} \sum_{\alpha=1}^{N} \int_{\partial V_{pz\alpha}} t_j^{(\alpha)} x_i dS = \sigma_{ij}^{\infty} \ . \tag{7.101}$$

Hence, the following averaging theorem follows:

Theorem 7.1. *Suppose*

(1) An elastic representative volume element contains N Barenblatt-Dugdale penny-shaped cracks with cohesive traction in the cohesive zones;
(2) The traction on the remote boundary of the RVE is generated by a constant stress tensor, i.e, $t_i^{\infty} = n_j \sigma_{ji}^{\infty}$, *and* $\sigma_{ij}^{\infty} = const.$.

Then, the macro stress tensor of an RVE equals the remote constant stress tensor, i.e.

$$\Sigma_{ij} =< \sigma_{ij} >= \sigma_{ij}^{\infty} \tag{7.102}$$

7.5.2 *Penny-shaped cohesive crack under uniform triaxial tension*

Before discussing the homogenization of three-dimensional (3D) cohesive cracks, we first outline the analytical solution of 3D cohesive penny-shaped crack in an RVE that is under uniform triaxial tension (see Fig. 7.7).

The penny-shaped Dugdale crack problem has been studied by several authors. The early contribution was made by L. Keer and T. Mura [Keer and Mura (1965)], who used the Tresca yield criterion linking the cohesive strength with the microscale yield stress. In their study, only uniaxial tension loading was considered. More recently, Chen and Keer [Chen and Keer (1991, 1993)] re-examined the problem, and they obtained the general solutions for a penny-shaped cohesive crack under mixed-mode loading. On the other hand, however, the problem has not been thoroughly

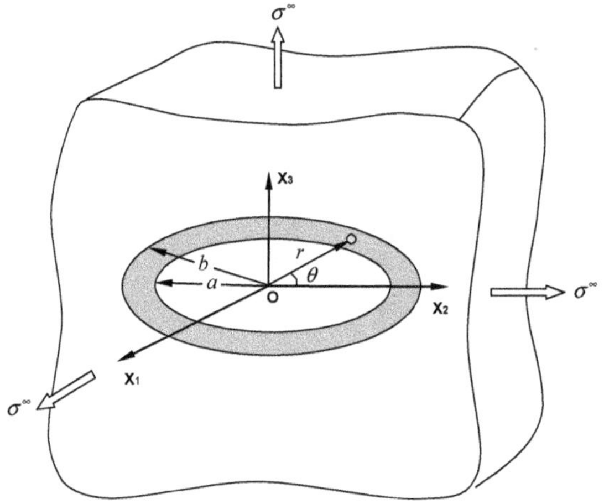

Fig. 7.7 A penny-shaped cohesive crack in representative volume element (the shaded region: cohesive zone–yielded ring).

examined from a micromechanics perspective. For example, the connection among the onset value of cohesive strength, micro yield stress in an RVE, and remote macro stress has not been made. By examining a cohesive penny-shaped crack model in an RVE, the study provides a link among cohesive strength, micro yield stress, and remote stresses on the boundary of an RVE, which provides a foundation for ensuing homogenization.

Consider a three-dimensional penny-shaped Dugdale crack of radius a with a ring-shaped cohesive zone with width $b - a$ in an RVE, which may be viewed as an infinite isotropic space by "*a micro-observer*" inside the RVE.

Let the outward normal to crack surface parallel to Z (X_3) axis (see Fig. 7.7) and a uniform triaxial tension stress is applied at the remote boundary of the RVE, $\sigma_{ij}^{\infty} = \sigma^{\infty}\delta_{ij}$ and $\sigma^{\infty} = \Sigma_m$ based on average theorem shown above. In cylindrical coordinate, the traction conditions on the remote boundary ∂V_{∞} and symmetric displacement boundary condition are expressed as

$$\sigma_{zz}\Big|_{\partial V_\infty} = \Sigma_m \tag{7.103}$$

$$\sigma_{rr}\Big|_{\partial V_\infty} = \Sigma_m \tag{7.104}$$

$$\sigma_{\theta\theta}\Big|_{\partial V_\infty} = \Sigma_m \tag{7.105}$$

$$u_z(r,\theta,0) = 0, \qquad b \le r,\ 0 \le \theta \le 2\pi \tag{7.106}$$

The stress distribution on the crack surface and cohesive zone is

$$\sigma_{zz}(r,\theta,0) = \sigma_0 H(r-a)\ , \quad 0 \le r \le b,\ 0 \le \theta \le 2\pi \tag{7.107}$$

where Σ_m is the remote stress, $H(r-a)$ is the Heaviside function, and σ_0 is the material's cohesive strength, the onset value for crack opening, and it is different from the micro yielding stress. The problem can be solved via superposition of two sub-problems: a trivial problem — an intact RVE in uniform triaxial tension state, i.e. $\forall\ \mathbf{x} \in V$,

$$\sigma_{zz}^{(0)} = \Sigma_m \tag{7.108}$$

$$\sigma_{rr}^{(0)} = \Sigma_m \tag{7.109}$$

$$\sigma_{\theta\theta}^{(0)} = \Sigma_m \tag{7.110}$$

$$\sigma_{rz}^{(0)} = \sigma_{r\theta}^{(0)} = \sigma_{z\theta}^{(0)} = 0 \tag{7.111}$$

and a crack problem — an RVE with a center crack that is subjected to the following boundary conditions (see Fig. 7.8).

$$\sigma_{zz}^{(c)}\Big|_{\partial V_\infty} = 0 \tag{7.112}$$

$$\sigma_{rr}^{(c)}\Big|_{\partial V_\infty} = 0 \tag{7.113}$$

$$\sigma_{\theta\theta}^{(c)}\Big|_{\partial V_\infty} = 0 \tag{7.114}$$

$$\sigma_{zz}^{(c)}(r,\theta,0) = -\Sigma_m + \sigma_0 H(r-a)\ , \quad 0 < r < b,\ 0 \le \theta \le 2\pi \tag{7.115}$$

$$u_z^{(c)}(r,\theta,0) = 0, \qquad b \le r,\ 0 \le \theta \le 2\pi \tag{7.116}$$

For the crack problem, the crack opening displacement is found as,

$$u_z(r) = \begin{cases} \dfrac{2}{\pi}\Big(\dfrac{1-\nu^*}{\mu^*}\Big)\Big(\Sigma_m\sqrt{b^2-r^2} - \sigma_0\displaystyle\int_a^b \frac{\sqrt{t^2-a^2}}{\sqrt{t^2-r^2}}dt\Big) & 0 < r < a \\[2ex] \dfrac{2}{\pi}\Big(\dfrac{1-\nu^*}{\mu^*}\Big)\Big(\Sigma_m\sqrt{b^2-r^2} - \sigma_0\displaystyle\int_r^b \frac{\sqrt{t^2-a^2}}{\sqrt{t^2-r^2}}dt\Big) & a < r < b \end{cases} \tag{7.117}$$

Within the yield ring ($z = 0$ and $a < r < b$) the stress distributions are found

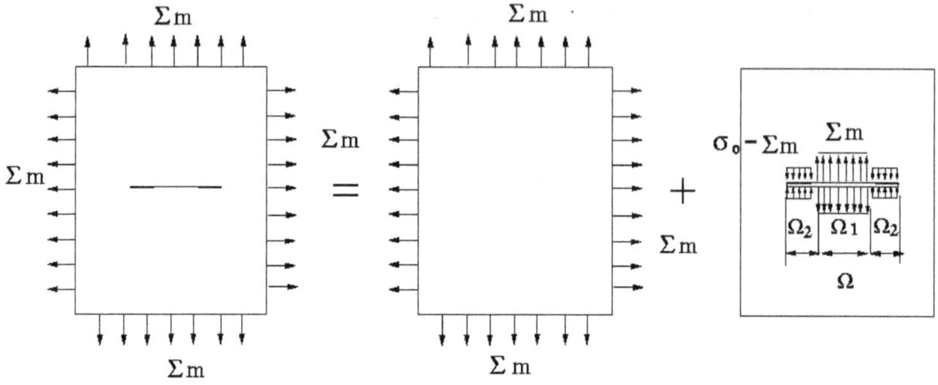

Fig. 7.8 Illustration of superposition of cohesive crack problem.

as,

$$\sigma_{zz}^{(c)} = \sigma_0 - \Sigma_m \tag{7.118}$$

$$\sigma_{rr}^{(c)} = -\frac{1+2\nu^*}{2}\Sigma_m + \left[\frac{1-2\nu^*}{2}\left(1+\frac{a^2}{r^2}\right) + 2\nu^*\right]\sigma_0 \tag{7.119}$$

$$\sigma_{\theta\theta}^{(c)} = -\frac{1+2\nu^*}{2}\Sigma_m + \left[\frac{1-2\nu^*}{2}(1-\frac{a^2}{r^2}) + 2\nu^*\right]\sigma_0 \tag{7.120}$$

$$\sigma_{rz}^{(c)} = \sigma_{r\theta}^{(c)} = \sigma_{z\theta}^{(c)} = 0 \tag{7.121}$$

where Poisson's ratio ν^*, to be viewed as either overall or matrix material property, is unspecified at time being, which may depend on subsequent homogenization procedures.

To ensure the stresses at crack tip to be finite, the size of the cohesive zone, a/b, remote stress Σ_m, and the cohesive stress, σ_0 are related through the following expression,

$$\frac{a}{b} = \sqrt{1 - \frac{(\Sigma_m)^2}{(\sigma_0)^2}} \quad \text{or} \quad \frac{\Sigma_m}{\sigma_0} = \sqrt{1 - \frac{a^2}{b^2}} \tag{7.122}$$

If we are mainly interested in inelastic deformation of quasi-brittle materials, we may assume that micro-scale yielding due to hydrostatic stress state is of small-scale yielding: $\frac{a^2}{b^2} \approx 1$. Therefore, $\left(\frac{a^2}{r^2} \approx 1\right)$, for $a \le r \le b$. The total stress distribution within the cohesive zone (7.118) - (7.121) may be approximated as

$$\sigma_{zz}^{(t)} = \sigma_{zz}^{(0)} + \sigma_{zz}^{(c)} = \sigma_0 \tag{7.123}$$

$$\sigma_{rr}^{(t)} = \sigma_{rr}^{(0)} + \sigma_{rr}^{(c)} = \frac{1-2\nu^*}{2}\Sigma_m + \sigma_0 \tag{7.124}$$

$$\sigma_{\theta\theta}^{(t)} = \sigma_{\theta\theta}^{(0)} + \sigma_{\theta\theta}^{(c)} = \frac{1-2\nu^*}{2}\Sigma_m + 2\nu^*\sigma_0 \tag{7.125}$$

$$\sigma_{rz}^{(t)} = \sigma_{r\theta}^{(t)} = \sigma_{z\theta}^{(t)} = 0 \tag{7.126}$$

It is assumed that inside the cohesive zone micro plastic yielding is controlled by the Huber-von Mises criterion . Therefore, we can link the cohesive strength, σ_0, with the yield stress of the virgin material, σ_y, by

$$\frac{1}{2}\Big[(\sigma^{(t)}_{rr} - \sigma^{(t)}_{zz})^2 + (\sigma^{(t)}_{\theta\theta} - \sigma^{(t)}_{zz})^2 + (\sigma^{(t)}_{rr} - \sigma^{(t)}_{\theta\theta})^2\Big] = \sigma_Y^2 \tag{7.127}$$

Substitute Eqs. (7.123)–(7.125) into (7.127) and solve for σ_0. The following quadratic equation may be obtained

$$4\Big(\frac{\sigma_0}{\Sigma_m}\Big)^2 - 2\Big(\frac{\sigma_0}{\Sigma_m}\Big) + 1 - \Big(\frac{2}{1-2\nu^*}\frac{\sigma_Y}{\Sigma_m}\Big)^2 = 0 \ , \tag{7.128}$$

which has two roots. The positive root is chosen to link the cohesive stress σ_0 with the yield stress in uniaxial tension σ_Y,

$$\frac{\sigma_0}{\Sigma_m} = \frac{1 + \sqrt{\Big(\frac{4}{1-2\nu^*}\frac{\sigma_Y}{\Sigma_m}\Big)^2 - 3}}{4} \tag{7.129}$$

7.5.3 *Effective elastic material properties of an RVE*

Define the *macro strain tensor*

$$\mathcal{E}_{ij} := \frac{\partial \bar{W}_c}{\partial \Sigma_{ij}} =: \bar{D}_{ijk\ell}\Sigma_{k\ell} = \bar{D}_{ijk\ell}\sigma^{\infty}_{k\ell} \tag{7.130}$$

where $\bar{W}_c$ is the overall complementary energy density of an RVE. $\Sigma_{ij} =< \sigma_{ij} >$ is the macro stress tensor previously defined, and $\bar{D}_{ijk\ell}$ is the effective compliance modulus.

Note that the macro strain in an RVE may not be the volume average strain in an RVE, that is $\mathcal{E}_{ij} \neq < \epsilon_{ij} >$. Furthermore Eq. (7.130) may not be a linear relationship, because $\bar{D}_{ijk\ell}$ depend on Σ_{ij} in general.

A common strategy for homogenization of randomly distributed defects is to find a so-called additional strain tensor, $\boldsymbol{\epsilon}^{(add)}$ (e.g. [Nemat-Nasser and Hori (1999); Lubarda and Krajcinovic (1993, 1994)]) such that

$$\mathcal{E}_{ij} = \epsilon^{(0)}_{ij} + \epsilon^{(add)}_{ij} \tag{7.131}$$

where $\epsilon^{(0)}_{ij} = D_{ijk\ell}\Sigma_{k\ell}$ and $D_{ijk\ell}$ is the elastic compliance of the corresponding virgin material. If the relationship between additional strain and macro stress can be found, $\epsilon^{(add)}_{ij} = H_{ijk\ell}\Sigma_{k\ell}$, where $H_{ijk\ell}$ is the added compliance due to microcracks, subsequently the effective elastic compliance modulus, $\bar{\mathbb{D}}$, can be deduced.

Energy method is applied to find an additional strain formula for cohesive cracks. The essence of energy methods is to find the energy release in a cohesive fracture process and hence to find the equivalent reduction of material properties. Nonetheless, the energy dissipation process in cohesive fracture is much more complicated than a purely elastic fracture process. It includes energy dissipation from both surface separation and plastic dissipation.

Although accurate determination of energy loss during a damage process requires an in-depth understanding of the physical process involved, an upper bound estimate may be made based on simplified assumptions. It is assumed that the total energy release of a cohesive crack is completely consumed in surface separation, which may not be true in cohesive fracture, because of plastic dissipation in the cohesive zone.

Subjected to unform triaxial loading, the total energy release of an RVE with a single penny-shaped cohesive micro-crack can be estimated as

$$\mathcal{R} = \int_{\Omega} \Sigma_m [u_z] dS - \int_{\Omega_2} \sigma_0 [u_z] dS \tag{7.132}$$

Carrying out the integration using crack displacement solution, the energy release estimate can be written as the following expression:

$$\mathcal{R} = \frac{16(1-\nu^*)}{3\mu^*} \sigma_0^2 a^3 \left(1 - \sqrt{1 - \left(\frac{\Sigma_m}{\sigma_0} \right)^2} \right) \tag{7.133}$$

Consider that there are N penny shaped cracks inside the RVE. The density of energy release of the RVE is estimated as sum of each crack contribution. Define the crack opening volume fraction as

$$f := \sum_{\alpha=1}^{N} \frac{4\pi a_\alpha^3}{3V} \beta, \tag{7.134}$$

where a_α is the radius of the α-th crack, and $4\pi a_\alpha^3/3$ is the volume of a sphere with radius a_α, and β is the ratio between the volume of permanent crack opening and the volume of total crack opening of a cohesive crack. For simplicity, we assume that this ratio is fixed for every crack inside an RVE. Obviously, $0 \le \beta \le 1$.

Utilizing (7.133) and (7.134), the density of energy release estimate can be written as

$$\begin{aligned} \frac{\mathcal{R}}{V} &= \frac{16(1-\nu^*)}{3\mu^* \beta} \sigma_0^2 \sum_{\alpha=1}^{N} \left(\frac{4\pi a_\alpha^3}{3V} \beta \right) \frac{3}{4\pi} \left(1 - \sqrt{1 - \left(\frac{\Sigma_m}{\sigma_0} \right)^2} \right) \\ &= \frac{4(1-\nu^*)}{\beta \pi \mu^*} \sigma_0^2 f \left(1 - \sqrt{1 - \left(\frac{\Sigma_m}{\sigma_0} \right)^2} \right), \quad \omega = 1, 2 \end{aligned} \tag{7.135}$$

The overall complementary energy density may then be expressed as the sum of complementary energy density of corresponding virgin material and the density of energy release due to microcrack distribution,

$$\begin{aligned} \bar{W}^c &= W^c + \frac{\mathcal{R}}{V} \\ &= \frac{1}{2} D_{ijk\ell} \sigma_{ij}^\infty \sigma_{k\ell}^\infty + \frac{4(1-\nu^*)}{\beta \pi \mu^*} \sigma_0^2 f \left(1 - \sqrt{1 - \left(\frac{\Sigma_m}{\sigma_0} \right)^2} \right) \end{aligned} \tag{7.136}$$

Based on definition (7.130) and the averaging theorem 7.1, for a given crack opening volume fraction, f, the macro strain tensor can be obtained as

$$\begin{aligned}\mathcal{E}_{ij} &= \frac{\partial \bar{W}^c}{\partial \Sigma_{ij}} = \frac{\partial \bar{W}^c}{\partial \sigma^{\infty}_{ij}} = D_{ijk\ell}\sigma^{\infty}_{k\ell} + \frac{\partial(\mathcal{R}/V)}{\partial \Sigma_m}\frac{\partial \Sigma_m}{\partial \sigma^{\infty}_{ij}} \\ &= D_{ijk\ell}\sigma^{\infty}_{k\ell} + \frac{4(1-\nu^*)}{3\beta\pi\mu^*} f \frac{\Sigma_m \delta_{ij}}{\sqrt{1-\left(\dfrac{\Sigma_m}{\sigma_0}\right)^2}}\end{aligned} \tag{7.137}$$

It may be noted that Eq. (7.137) is only valid when the RVE is under hydrostatic stress state, i.e., $\sigma^{\infty}_{ij} = \Sigma_m \delta_{ij}$. From (7.137), one can find an expression for additional strain

$$\epsilon^{(add)}_{ij} = \frac{4(1-\nu^*)}{3\beta\pi\mu^*} f \frac{\Sigma_m \delta_{ij}}{\sqrt{1-\left(\dfrac{\Sigma_m}{\sigma_0}\right)^2}} \tag{7.138}$$

A bona fide self-consistent scheme should take into account micro-crack interaction (see [Hill (1965c,b); Budiansky and O'Connell (1976)]). Since the micro-crack distribution is isotropic, the damaged RVE should also be considered as isotropic *at micro-level.* The micro-crack interaction effect could be captured by taking $\mu^* = \bar{\mu}$ and $\nu^* = \bar{\nu}$ in all above derivations, where $\bar{\mu}$ and $\bar{\nu}$ are effective shear modulus and effective Poisson's ratio in an RVE. Recast Eq. (7.138) into a more general form,

$$\boldsymbol{\epsilon}^{(add)} = \mathbb{H} : \boldsymbol{\Sigma}, \tag{7.139}$$

so

$$\mathcal{E} = \bar{\mathbb{D}} : \boldsymbol{\Sigma} = (\mathbb{D} + \mathbb{H}) : \boldsymbol{\Sigma}, \quad \text{where} \quad \bar{\mathbb{D}} = \mathbb{D} + \mathbb{H} \tag{7.140}$$

where $\mathbb{H}$ is an isotropic tensor, which may be written as

$$\mathbb{H} = \frac{h_1}{3}\mathbb{I}^{(2)} \otimes \mathbb{I}^{(2)} + h_2 \mathbb{I}^{(4s)} \tag{7.141}$$

where $\mathbb{I}^{(2)} = \delta_{ij}\mathbf{e}_i \otimes \mathbf{e}_j$, and $\mathbb{I}^{(4s)} = \frac{1}{2}(\delta_{ik}\delta_{j\ell} + \delta_{i\ell}\delta_{jk})\mathbf{e}_i \otimes \mathbf{e}_j \otimes \mathbf{e}_k \otimes \mathbf{e}_\ell$, and parameters h_1, h_2 are yet to be determined.

Decompose

$$\mathbb{D} = \frac{1}{3K}\mathbb{E}^{(1)} + \frac{1}{2\mu}\mathbb{E}^{(2)} \tag{7.142}$$

$$\bar{\mathbb{D}} = \frac{1}{3\bar{K}}\mathbb{E}^{(1)} + \frac{1}{2\bar{\mu}}\mathbb{E}^{(2)} \tag{7.143}$$

$$\mathbb{H} = (h_1 + h_2)\mathbb{E}^{(1)} + h_2\mathbb{E}^{(2)} \tag{7.144}$$

and consider

$$\frac{1}{3K} = \frac{(1-2\nu)}{E} \quad \text{and} \quad \frac{1}{3\bar{K}} = \frac{(1-2\bar{\nu})}{\bar{E}} \quad ; \tag{7.145}$$

$$\frac{1}{2\mu} = \frac{(1+\nu)}{E} \quad \text{and} \quad \frac{1}{2\bar{\mu}} = \frac{(1+\bar{\nu})}{\bar{E}} \quad . \tag{7.146}$$

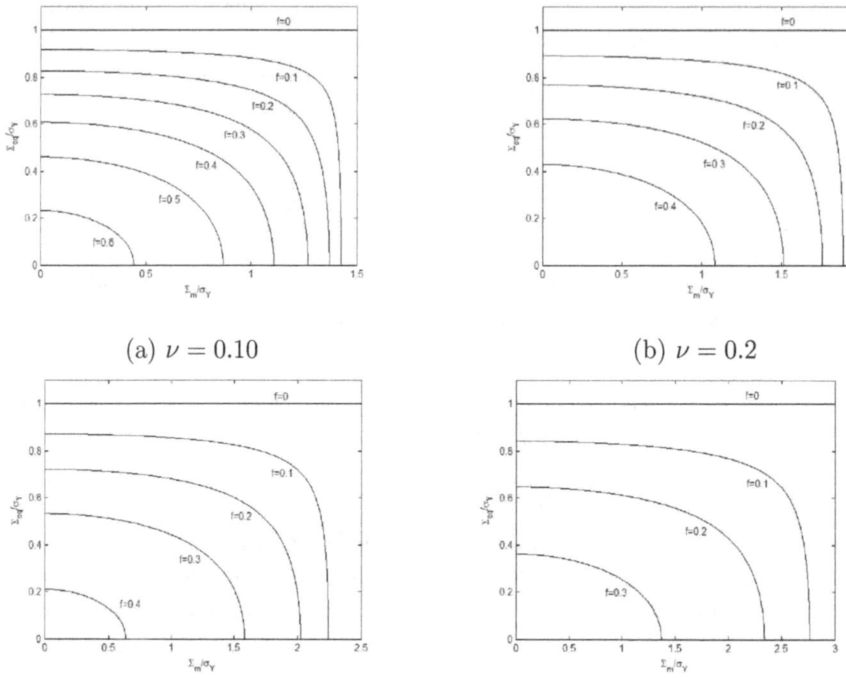

(a) $\nu = 0.10$ (b) $\nu = 0.2$

(c) $\nu = 0.25$ (d) $\nu = 0.3$

Fig. 7.9 Cohesive micro-crack damage model, $\Psi(\Sigma_{eq}, \Sigma_m, \mathbf{q})$, with different Poisson's ratios ($\beta = 1/3$).

The $\mathbb{H}$ tensor in Eq. (7.141) can not be uniquely determined, since on the remote boundary of the RVE, the traction stress state is hydrostatic, $\mathbf{\Sigma} = \Sigma_m \delta_{ij} \mathbf{e}_i \otimes \mathbf{e}_j$. Hence the information carried in (7.140) only admits one scalar equation,

$$\bar{\mathbb{D}} : \Sigma_m \mathbb{I}^{(2)} = \Big(\mathbb{D} + \mathbb{H}\Big) : \Sigma_m \mathbb{I}^{(2)} \tag{7.147}$$

Consider $\mathcal{E}_{ij} = \epsilon_{ij}^{(0)} + \epsilon_{ij}^{(add)}$ and identities, $\mathbb{E}^{(1)} : \mathbb{I}^{(2)} = \mathbb{I}^{(2)}$ and $\mathbb{E}^{(2)} : \mathbb{I}^{(2)} = \mathbf{0}$, and by virtue of (7.138) and (7.147), it can be shown that

$$\frac{1}{3\bar{K}} = \frac{1}{3K} + (h_1 + h_2) = \frac{1}{3K} + \frac{4(1-\bar{\nu})}{3\beta\pi\bar{\mu}} \frac{f}{\sqrt{1 - \left(\dfrac{\Sigma_m}{\sigma_0}\right)^2}} \tag{7.148}$$

There are two unknowns, $\bar{K}$ and $\bar{\mu}$, or equivalently h_1 and h_2 in Eq. (7.148). An additional condition is needed to uniquely determine $\bar{\mathbb{D}}$ or $\mathbb{H}$. Impose a restriction

$$\frac{\bar{K}}{K} = \frac{\bar{\mu}}{\mu} \tag{7.149}$$

This restriction guarantees the positive definiteness of the overall strain energy. It also implies that the relative reduction of the shear modulus is the same as that of the bulk modulus.

Consider (7.145) and (7.146). A direct consequence of (7.149) is $\bar{\nu} = \nu$, which leads to

$$\frac{1}{3\bar{K}} = \frac{1}{2\bar{\mu}}\left(\frac{1-2\bar{\nu}}{1+\bar{\nu}}\right) = \frac{1}{2\bar{\mu}}\left(\frac{1-2\nu}{1+\nu}\right), \quad \text{and} \quad \frac{1}{3\bar{K}} = \frac{1}{2\bar{\mu}}\left(\frac{1-2\nu}{1+\nu}\right)$$

Substitution of above expressions into (7.148) leads to estimates of effective elastic moduli

$$\frac{\bar{K}}{K} = \frac{\bar{\mu}}{\mu} = 1 - \frac{4\omega(1-\nu^2)}{3\beta\pi(1-2\nu)} \frac{f}{\sqrt{1-\left(\dfrac{\Sigma_m}{\sigma_0}\right)^2}}\,, \quad \omega = 1, 2 \tag{7.150}$$

7.5.4 *Micro-cohesive-crack damage models*

Homogenization of nonlinear problems is often difficult. Without proper statistical closure, averaging alone may not be sufficient to provide sensible results. In this model, it is postulated that there is a limit for the amount of distortional energy that a given material ensemble can store. This reflects in the following hypothesis on the condition of macro-yielding:

The macroscopic yielding of an RVE begins when the distortional strain energy density of an RVE reaches to a threshold. In other words, the maximum elastic distortional energy of an RVE is a material constant,

$$U_d \le U_d^{(cr)}\,. \tag{7.151}$$

It is noted that the above criterion is a reminiscence of the Hencky's maximum distortional energy principle in traditional infinitesimal plasticity. The criterion can be calibrated using an uniaxial tension test of the virgin material

$$U_d^{(cr)} = \frac{1}{6\mu}\sigma_Y^2 \tag{7.152}$$

Define the macro deviatoric stress tensor and its second invariant as

$$\Sigma'_{ij} = \Sigma_{ij} - \frac{1}{3}\Sigma_{kk}\delta_{ij}, \quad J_2 = \frac{1}{2}\Sigma'_{ij}\Sigma'_{ij} \tag{7.153}$$

and the equivalent macro stress $\Sigma_{eq} = \sqrt{3J_2}$. In a real damage evolution process, the above criteria take the following form

$$U_d = \frac{\Sigma_{eq}^2}{6\bar{\mu}} \le U_d^{(cr)}, \quad \text{where } \Sigma_{eq} := \sqrt{3J_2}. \tag{7.154}$$

Then the criterion of the maximum distortional energy density of an RVE becomes

$$\frac{\Sigma_{eq}^2}{\sigma_Y^2} = \frac{\bar{\mu}}{\mu} \tag{7.155}$$

Using (7.150), one may derive the following effective yielding potential,

$$\boxed{\Psi(\Sigma_{eq}, \Sigma_m, \mathbf{q}) = \frac{\Sigma_{eq}^2}{\sigma_Y^2} + \frac{2(1-\nu^2)}{3\beta\pi(1-2\nu)} \frac{f}{\sqrt{1-\left(\dfrac{\Sigma_m}{\sigma_0}\right)^2}} - 1 = 0\,.} \tag{7.156}$$

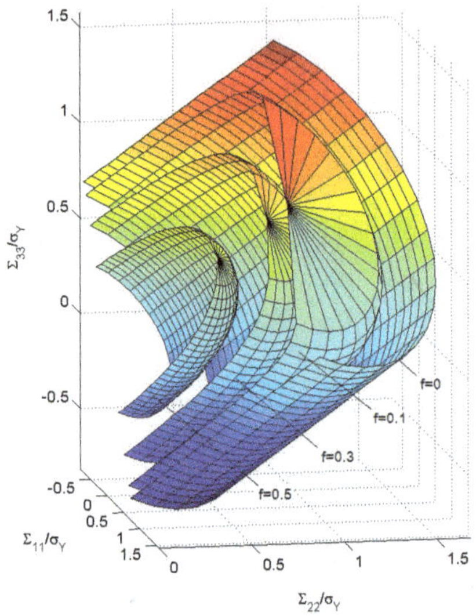

Fig. 7.10 The cohesive model Ψ ($\nu = 0.1$).

where Σ_{eq} and Σ_m are defined as the macro equivalent stress and mean stress, and $\mathbf{q}$ represents the other internal variables, which may be implicitly embedded in σ_Y. In terms of the ratio Σ_m/σ_Y, the effective yielding potential function of plastic flow Ψ can be finally recast as follows,

$$\boxed{\begin{aligned}\Psi(\Sigma_{eq}, \Sigma_m, \mathbf{q}) = \frac{\Sigma_{eq}^2}{\sigma_Y^2} + \frac{8(1-\nu^2)f}{3\beta\pi(1-2\nu)} \\ \cdot \frac{1 + \left[\left(\dfrac{4\sigma_Y}{(1-2\nu)\Sigma_m}\right)^2 - 3\right]^{1/2}}{\left[\left(1 + \left[\left(\dfrac{4\sigma_Y}{(1-2\nu)\Sigma_m}\right)^2 - 3\right]^{1/2}\right)^2 - 16\right]^{1/2}} - 1 = 0\end{aligned}} \tag{7.157}$$

The above pressure-sensitive yield function Ψ, is displayed in Figs. 7.9 and 7.10 with different Poisson's ratios. Compared to the Gurson model, both yield functions will reduce to J_2 plasticity when $f = 0$. For the limit case of infinitesimal amount of damage ($f \to 0$), the cohesive damage model predicts yielding of material when the stress ratio of hydrostatic stress and the cohesive strength approaches a finite value, i.e. $\Sigma_m \to \frac{4}{\sqrt{12}(1-2\nu)}$. For the Gurson model, , when the amount of damage is infinitesimal, the material will not yield unless the hydrostatic stress becomes infinite. So the cohesive model is a pressure-sensitive capped yield surface, which represent the feature of ductile fracture. It should also be mentioned that the self-consistent scheme based on the cohesive damage model will fail at $\nu = 0.5$, since for incompressible materials, uniform triaxial tension load will not be able to produce dilatational strain energy.

7.6 Frank A. McClintock

Fig. 7.11 Frank A. McClintock (photo provided by Dr. Xue Lian).

Dr. Frank A. McClintock is a professor emeritus at Massachusetts Institute of Technology. In 1991, he was elected to the National Academy of Engineering USA for his pioneering and sustained contributions to the understanding of the process of ductile fracture of engineering materials.

7.7 Exercises

Problem 7.1. *Consider a 2D circular RVE with sufficiently large number of cracks, sat N cracks, in an infinite solid. Consider the same material with the a circular inclusion (See Fig. 7.12).*

We say that the two systems are equivalent, if they have the same elastic energy density. Based on the Mori-Tanaka theory and Kachanov's additional strain formula, the two systems are equivalent if and only if

$$\begin{aligned}&\frac{1}{2}\boldsymbol{\sigma}^0 : [\mathbf{C}_0 : (\bar{\mathbf{C}} - \mathbf{C}_0)^{-1} : \mathbf{C}_0 + \mathbf{C}_0 : \mathbf{S}_0]^{-1} : \boldsymbol{\sigma}^0 \\ &\qquad = -\frac{1}{2V}\boldsymbol{\sigma}^0 : \sum_{i=1}^{N} \frac{1}{2} \int_{\partial\Omega_i} \Big(\mathbf{n}_i \otimes [\mathbf{u}_i] + (\mathbf{n}_i \otimes [\mathbf{u}_i])^T\Big) d\Gamma \qquad (7.158)\end{aligned}$$

First let $\sigma^0_{\alpha\beta} = \sigma^0_K \delta_{\alpha\beta}$, *and let* $\sigma^0_{G11} = -\sigma^0_{G22} = \sigma^0_G$ *and* $\sigma^0_{G\alpha\beta} = 0$, *if* $\alpha = \beta$.

Show

$$\frac{\bar{K}}{K_0} = \frac{1}{1 + \dfrac{1-\nu_0}{1-2\nu_0}\dfrac{\pi f}{1 - f\pi/2}} \tag{7.159}$$

$$\frac{\bar{G}}{G_0} = \frac{1}{1 + (1-\nu_0)\dfrac{\pi f}{1 - f\pi/4}} \tag{7.160}$$

where $f = \dfrac{1}{V}\sum_{i=1}^{N} a_i^2$.

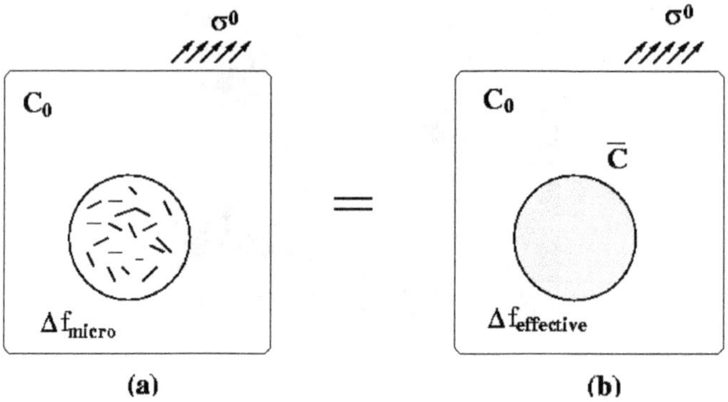

Fig. 7.12 Schematic diagram for energy balance equation.

Hint: **1.**

$$\mathbf{S}_0 = s_1\mathbf{E}^{(1)} + s_2\mathbf{E}^{(2)} \tag{7.161}$$

2. *The crack opening displacements (COD) for plane mode I and mode II problems are*

$$[u_i] = \sqrt{a^2 - x_1^2}\,\frac{4}{E'}\sigma^\infty_{i2}, \qquad |x_1| \le a \quad (i = 1, 2) \tag{7.162}$$

where

$$\frac{1}{E'} = \frac{\kappa + 1}{8\mu} = \begin{cases} \dfrac{1-\nu^2}{E} & \text{for plane strain} \\ \dfrac{1}{E} & \text{for plane stress} \end{cases} \tag{7.163}$$

Hints: see [Shen and Yi (2000a,b)].

Problem 7.2. *Consider a Dugdale-Barenblatt mode III crack. After stress reaches to the yield stress,* k_0, *material behavior is assumed to be perfectly elasto-plastic. The displacement fields are,*

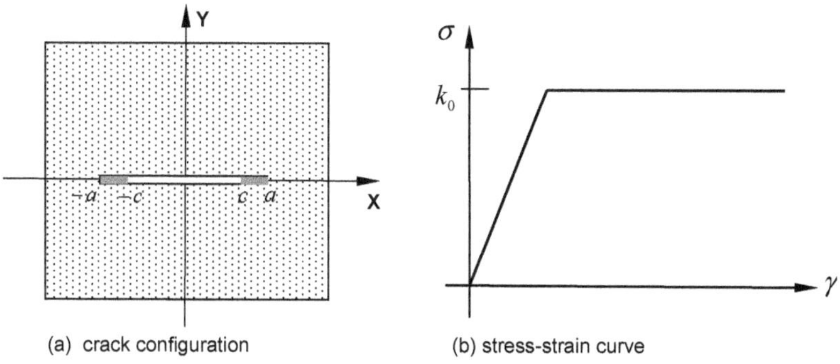

Fig. 7.13 The Dugdale-Barenblatt mode III crack.

$$u_x = 0, \quad u_y = 0, \quad u_z = w(x, y) \tag{7.164}$$

The corresponding stress components all vanish but,

$$\sigma_{xz} = \mu \frac{\partial w}{\partial x}, \quad \sigma_{yz} = \mu \frac{\partial w}{\partial y}. \tag{7.165}$$

The equilibrium equation,

$$\frac{\partial \sigma_{xz}}{\partial x} + \frac{\partial \sigma_{yz}}{\partial y} = 0, \tag{7.166}$$

becomes

$$\frac{\partial^2 w}{\partial x^2} + \frac{\partial^2 w}{\partial y^2} = 0 . \tag{7.167}$$

The mixed boundary value problem is described as

$$\sigma_{yz} = 0, \quad x_2 \to \infty \tag{7.168}$$

$$\sigma_{yz}(x, 0) = -s, \quad \forall \ |x| \leq c \tag{7.169}$$

$$\sigma_{yz}(x, 0) = k_0 - s, \quad \forall \ c < |x| \leq a \tag{7.170}$$

$$w(x, 0) = 0, \quad \forall \ |x| > a \tag{7.171}$$

Use Fourier cosine transform to find $w(x, 0)$.

Problem 7.3. *Consider prescribed displacement boundary condition for a RVE with traction-free defects, i.e.*

$$\mathbf{u}^0 = \mathbf{x} \cdot \boldsymbol{\epsilon}, \quad \forall \mathbf{x} \in \partial V \tag{7.172}$$

$$\mathbf{t}^0 = \mathbf{0} \ , \quad \forall \mathbf{x} \in \partial \Omega \tag{7.173}$$

Show

$$\bar{\sigma}^c = -\mathbf{C} : \left\{ \frac{1}{2V} \int_{\partial \Omega} (\mathbf{n} \otimes \mathbf{u} + \mathbf{u} \otimes \mathbf{n}) dS \right\} \tag{7.174}$$

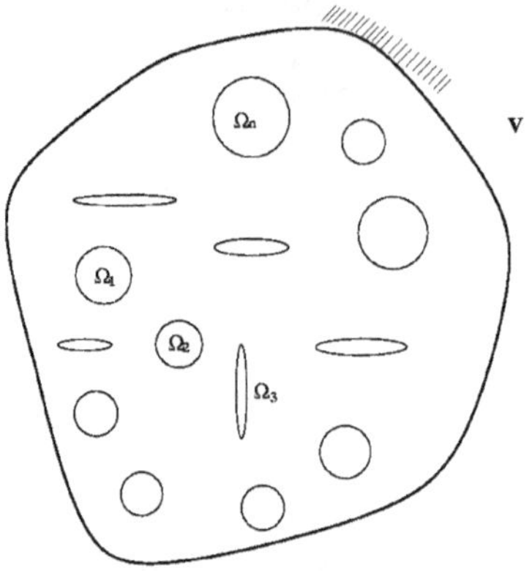

Fig. 7.14 A RVE with traction-free defects.

Problem 7.4. *Consider a penny-shaped micro-crack of radius a_α, lying in the x_1, x_2-plane with its center at the origin of the coordinate system. The unit normal of the positive crack surface, $\mathbf{n}^+ = \mathbf{e}_3$. Under the remote stress field, $\sigma_{13}^\infty = \sigma_{31}^\infty$, $\sigma_{23}^\infty = \sigma_{32}^\infty$, and $\sigma_{33}^\infty > 0$, the crack opening displacements (CODs) are*

$$\Big[u_i\Big] = \sqrt{a^2 - r^2}\frac{16(1-\nu^2)}{\pi E(2-\nu)}\sigma_{i3}^\infty, \quad r \le a_\alpha, (i = 1, 2) \tag{7.175}$$

$$\Big[u_3\Big] = \sqrt{a^2 - r^2}\frac{8(1-\nu^2)}{\pi E}\sigma_{33}^\infty, \quad r \le a_\alpha \tag{7.176}$$

where $r^2 = x_1^2 + x_2^2$.

Assume all the micro-cracks are lying on x_1-x_2 plane.

Suppose that the RVE is under the pure traction boundary condition. Show that the average strain inside RVE is,

$$< \boldsymbol{\epsilon} >_V = \boldsymbol{\epsilon}^0 + \boldsymbol{\epsilon}^{add} \tag{7.177}$$

where $\boldsymbol{\epsilon}^0 = \mathbf{D}^0 :< \boldsymbol{\sigma} >_V$ and the additional strain is

$$\boldsymbol{\epsilon}^{add} = \frac{1}{V}\sum_{i=1}^{N}\int_{\Omega_i}\frac{1}{2}\Big(\mathbf{n}\otimes\mathbf{u} + \mathbf{u}\otimes\mathbf{n}\Big)dS \tag{7.178}$$

Calculate additional strain for a penny shaped micro-crack;

Chapter 8

INTRODUCTION OF DISLOCATION THEORY

In material science, a dislocation may be defined as a disturbed region between two substantially perfect parts of a crystal. In elasticity theory, a dislocation is defined as the strong discontinuity of the displacement field. As a form of ubiquitous defect in solids, the dislocation-like defect has profound effects on materials microstructure as well as behaviors. The study of dislocation or the dislocation theory has been an important part of both micromechanics as well as nanomechanics.

In this Chapter, we shall first study dislocation theory within the framework of linear elasticity, and then we shall study the theory of the Peach-Koehler force. An introduce to discrete dislocation dynamics (DDD) is presented, and we present a thorough examination of Peierls-Nabarro model, i.e. a dislocation theory by considering lattice structure. Subsequently, we shall discuss one of the most important applications of dislocation theory: dislocations in thin films, which includes the Frenkel & Kontorova model and the Matthews & Blackeslee's equilibrium theory. At the end of this Chapter, we shall present a solution of surface dislocation.

8.1 Screw dislocation

A multiply-connected region is defined as a region which contains at least one irreducible circuit, i.e. a closed curve that can not be contracted to a single point without passing out of the region. Consider a multiply-connected region $\mathcal{V}$. A Volterra dislocation is defined as a discontinuity of the displacement field $\mathbf{u}$ or the rotation field $\boldsymbol{\omega}$ over a line segment S (2D) or surface S (3D), i.e.

$$\begin{aligned}\Big[\mathbf{u}\Big] &= \mathbf{u}(\mathbf{P}^+) - \mathbf{u}(\mathbf{P}^-) = \mathbf{b} + \mathbf{d} \times \mathbf{x}, \quad \mathbf{P} \in S \\ \Big[\boldsymbol{\omega}\Big] &= \boldsymbol{\omega}(\mathbf{P}^+) - \boldsymbol{\omega}(\mathbf{P}^-) = \mathbf{d} \end{aligned} \tag{8.1}$$

where $\mathbf{P}^+$ and $\mathbf{P}^-$ are material points on two sides of a line segment or a gliding surface S, and $\mathbf{b}$ is the Burgers vector that can be defined as

$$\mathbf{b} = \oint_C \Big(\boldsymbol{\epsilon}(\mathbf{y}) + (\mathbf{x} - \mathbf{y}) \times \Big[\nabla \times \boldsymbol{\epsilon}(\mathbf{y}) \Big]^T \Big) d\mathbf{y} \tag{8.2}$$

and the discontinuity in rigid-body rotation,

$$\mathbf{d} = -\oint_C \left(\nabla \times \boldsymbol{\epsilon}(\mathbf{y})\right)^T d\mathbf{y} \tag{8.3}$$

and $\boldsymbol{\epsilon}$ is the strain tensor.

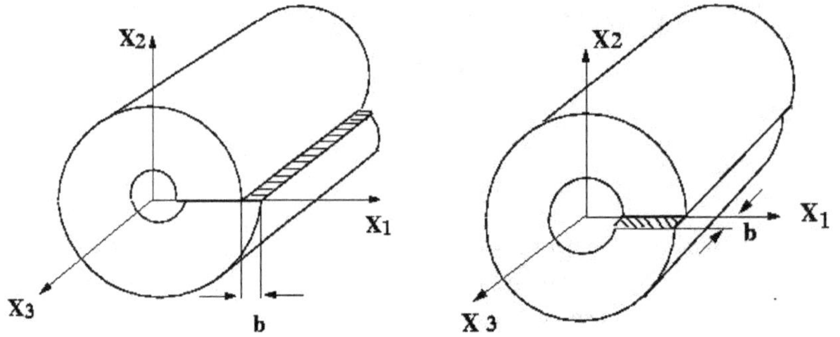

Fig. 8.1 Illustrations of dislocations: (a) an edge dislocation, and (b) a screw dislocation.

Historically, there is another type of dislocation: the Somigliana dislocations that are defined as discontinuities in the displacement field $\mathbf{u}$ as well as in traction field $\mathbf{t} = \mathbf{n} \cdot \boldsymbol{\sigma}$ over the gliding plane,

$$\left[\mathbf{u}\right] = \mathbf{u}^+ - \mathbf{u}^- = \mathbf{b}, \quad \forall \mathbf{x} \in S \tag{8.4}$$

$$\left[\mathbf{t}\right] = \mathbf{t}^+ - \mathbf{t}^- = 0, \quad \forall \mathbf{x} \in S \tag{8.5}$$

That is the traction is required to be continuous across the slip plane. However, the solution of such boundary-value problem is difficult, and people have not found any important applications of such dislocation model.

8.1.1 *The solution of a screw dislocation*

We first derive the solution for the screw dislocation. The kinematics of the screw dislocation belong to that of anti-plane problem:

$$u_1 = 0, \quad u_2 = 0, \quad \text{and} \quad u_3 = w(x, y)\ . \tag{8.6}$$

All the strain components are zero, except the out-plane shear strains

$$\epsilon_{xz} = \frac{1}{2}\frac{\partial w}{\partial x}, \quad \epsilon_{yz} = \frac{1}{2}\frac{\partial w}{\partial y}\ . \tag{8.7}$$

The corresponding non-zero shear stresses are

$$\sigma_{xz} = \mu \frac{\partial w}{\partial x} \tag{8.8}$$

$$\sigma_{xy} = \mu \frac{\partial w}{\partial y} \tag{8.9}$$

The non-trivial equilibrium equation

$$\frac{\partial \sigma_{xz}}{\partial x} + \frac{\partial \sigma_{yz}}{\partial y} + \frac{\partial \sigma_{zz}}{\partial z} = 0 \tag{8.10}$$

leads to the governing equation

$$\frac{\partial^2 w}{\partial x^2} + \frac{\partial^2 w}{\partial y^2} = \nabla^2 w = 0 \ . \tag{8.11}$$

We denote the displacement jump in w at $y = 0$ and $x > 0$ as b_z i.e. $\mathbf{b} = b_z \mathbf{e}_z$, and the jump condition may be expressed as

$$\lim_{\eta \to 0, x>0} \Big(w(x, -\eta) - w(x, \eta) \Big) = [w(x,0)] = b_z, \quad \eta > 0 \tag{8.12}$$

Use polar coordinates,

$$\nabla^2 w = \Big(\frac{\partial^2}{\partial^2 r} + \frac{1}{r}\frac{\partial}{\partial r} + \frac{1}{r^2}\frac{\partial^2}{\partial \theta^2} \Big) w = 0 \ . \tag{8.13}$$

Use the separation of variables and let

$$w(r,\theta) = f(r) g(\theta) \tag{8.14}$$

we have

$$\frac{r^2}{f(r)} \Big(\frac{d^2 f}{dr^2} + \frac{1}{r}\frac{df}{dr} \Big) + \frac{1}{g(\theta)} \frac{d^2 g}{d\theta^2} = 0 \ . \tag{8.15}$$

We then end with two ordinary differential equations,

$$\begin{cases} \dfrac{d^2 f}{dr^2} + \dfrac{1}{r}\dfrac{df}{dr} - \dfrac{n^2 f}{r^2} = 0 \\[2ex] \dfrac{d^2 g}{d\theta^2} + n^2 g(\theta) = 0 \end{cases} \tag{8.16}$$

If $n = 0$, one may find that

$$g(\theta) = A + B\theta \tag{8.17}$$

$$f(r) = C \ln r + D \tag{8.18}$$

For $n \neq 0$,

$$g(\theta) = C_n \cos n\theta + D_n \sin n\theta \tag{8.19}$$

$$f(r) = E_n r^n + F_n r^{-n} \tag{8.20}$$

where C_n, D_n, E_n, F_n are constants. The s is true because

$$\Big(\frac{d^2}{dr^2} + \frac{1}{r}\frac{d}{dr} - \frac{n^2}{r^2} \Big) r^n = \Big(n(n-1) + n - n^2 \Big) r^{n-2} \equiv 0 \ . \tag{8.21}$$

Because the displacement, w, has to be finite, we can only consider the case where $n = 0$. Again, because of the convergence requirement on the displacement field, $C = 0$; and because of the jump condition, $A = 0$.

By absorbing the constant D into the constant B, the displacement field is

$$w(r,\theta) = B\theta \tag{8.22}$$

Use the jump condition,

$$w(r,2\pi) - w(r,0) = b \tag{8.23}$$

one may find that $2\pi B = b$ and hence

$$B = \frac{b}{2\pi} \tag{8.24}$$

Finally,

$$w(r,\theta) = \frac{\theta b}{2\pi} = \frac{b}{2\pi}\arctan\Bigl(\frac{y}{x}\Bigr) \tag{8.25}$$

and

$$\frac{\partial w}{\partial x} = -\frac{b}{2\pi}\frac{y}{x^2+y^2} = -\frac{b\sin\theta}{2\pi r} \tag{8.26}$$

$$\frac{\partial w}{\partial y} = \frac{b}{2\pi}\frac{x}{x^2+y^2} = \frac{b\cos\theta}{2\pi r} \tag{8.27}$$

Consequently, the non-zero stress components are

$$\sigma_{xz} = -\Bigl(\frac{b\mu}{2\pi}\Bigr)\frac{y}{x^2+y^2} \tag{8.28}$$

$$\sigma_{yz} = \Bigl(\frac{b\mu}{2\pi}\Bigr)\frac{x}{x^2+y^2} \tag{8.29}$$

In the cylindrical coordinate,

$$\begin{bmatrix} \sigma_{rr} & \sigma_{r\theta} & \sigma_{rz} \\ \sigma_{\theta r} & \sigma_{\theta\theta} & \sigma_{\theta z} \\ \sigma_{zr} & \sigma_{z\theta} & \sigma_{zz} \end{bmatrix} = \begin{bmatrix} \cos\theta & \sin\theta & 0 \\ -\sin\theta & \cos\theta & 0 \\ 0 & 0 & 1 \end{bmatrix} \begin{bmatrix} 0 & 0 & \sigma_{xz} \\ 0 & 0 & \sigma_{yz} \\ \sigma_{zx} & \sigma_{zy} & 0 \end{bmatrix} \begin{bmatrix} \cos\theta & -\sin\theta & 0 \\ \sin\theta & \cos\theta & 0 \\ 0 & 0 & 1 \end{bmatrix}$$

The non-zero stress components are

$$\sigma_{rz} = \cos\theta\sigma_{xz} + \sin\theta\sigma_{yz} = 0 \tag{8.30}$$

$$\sigma_{\theta z} = -\sin\theta\sigma_{xz} + \cos\theta\sigma_{yz} = \frac{b\mu}{2\pi r}\ . \tag{8.31}$$

In the following, we calculate the self-energy of the screw dislocation in a hollow cylinder with the inner radius r_0 and the outer radius R. Note that the self-energy of a dislocation is defined as the strain energy contribution from the stress-strain field of the dislocation solution in an unbounded region.

Assume that the length of the hollow cylinder is L. The energy per unit length in z-direction is,

$$\begin{aligned} \frac{W}{L} &= \frac{1}{L}\int_V \frac{\sigma_{z\theta}^2}{2\mu}dV = \frac{1}{L}\int_0^L\int_0^{2\pi}\int_{r_0}^R \frac{\sigma_{z\theta}^2}{2\mu} r dr d\theta dz \\ &= \frac{b^2\mu}{4\pi}\int_{r_0}^R \frac{dr}{r} = \frac{b^2\mu}{4\pi}\ln\frac{R}{r_0}\ . \end{aligned} \tag{8.32}$$

First, as $R \to \infty$, $W/L \to \infty$. This shows that the self-energy of the dislocation depends on the size of the crystal. On the other hand, for a finite size crystal, the dislocation solution of unbounded domain does not hold true because of the image stress caused by the boundary.

Assume that the dislocation is far away from the boundary, the boundary effects are abated inside, one may choose the dimension of the crystal, say ℓ as R; in polycrystallines, one may choose the size of a grain as R, where the dislocation resides.

Second, as $r_0 \to 0$, $W/L \to -\infty$. This abnormality is due to the limitation of a linear elasticity model. Within five atomic spacing of a dislocation core, the linear elasticity model is no longer valid. In general, the length of the Burgers vector is close to the lattice spacing. Therefore, in practice, we usually choose $r_0 = 5b$ or $r_0 = b/\alpha$, $0 < \alpha < 1$ such that the elastic self-energy equals to

$$\frac{W}{L} = \frac{\mu b^2}{4\pi} \ln \frac{\ell}{5b}, \quad \text{or} \quad \frac{W}{L} = \frac{\mu b^2}{4\pi} \ln \frac{\alpha \ell}{b} \ . \tag{8.33}$$

By definition, the self-energy should include the core energy, i.e.

$$W^{self} = W^{elas} + W^{core} \tag{8.34}$$

The core energy is relatively small, but may not be negligible, because it is 10% to 20 % of the elastic self-energy. Overall, the linear elasticity theory gives a good estimate of self-energy. In latter section of this Chapter, we shall discuss the Peierls-Nabarro model, which provides a means to estimate the core energy.

8.1.2 *Image stress of a screw dislocation in a half space*

Consider a crystal occupying a half space $x \le 0$. Consider a screw dislocation located at the position $x = -\ell$ (see Fig. 8.2). The screw dislocation in an unbounded space gives the following stress distribution,

$$\sigma^{\infty}_{xz}(x, y) = -\frac{b\mu}{2\pi} \frac{y}{(x+\ell)^2 + y^2} \tag{8.35}$$

$$\sigma^{\infty}_{yz}(x, y) = \frac{b\mu}{2\pi} \frac{(x+\ell)}{(x+\ell)^2 + y^2} \ . \tag{8.36}$$

This solution does not satisfy the traction-free boundary condition at $x = 0$, because

$$\sigma^{\infty}_{xz}(0, y) = -\frac{b\mu}{2\pi} \frac{y}{\ell^2 + y^2} \neq 0 \ . \tag{8.37}$$

To enforce the traction-free boundary condition, we place a fictitious screw dislocation with the Burgers vector, $b' = -b$, at the position $x = \ell$, and it generates the following so-called image stress distribution:

$$\sigma^{I}_{xz}(x, y) = \frac{b\mu}{2\pi} \frac{y}{(x-\ell)^2 + y^2} \tag{8.38}$$

$$\sigma^{I}_{yz}(x, y) = -\frac{b\mu}{2\pi} \frac{(x-\ell)}{(x-\ell)^2 + y^2} \ . \tag{8.39}$$

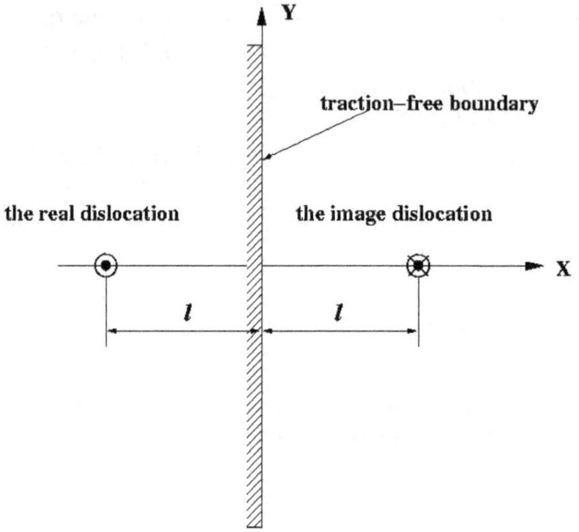

Fig. 8.2 An image screw dislocation.

The total stress distribution is then the superposition of the solution in the infinite space and the solution of image stress distribution, i.e. $\sigma^t_{ij} = \sigma^\infty_{ij} + \sigma^I_{ij}$, where the superscript, t, ∞, and I denote the total stress solution, the solution obtained in the infinite space, and the image stress solution, respectively.

By anti-symmetry, the traction-free boundary condition at $x = 0$ is then enforced,

$$\sigma^t_{xz}(0,y) = \sigma^\infty_{xz}(0,y) + \sigma^I_{xz}(0,y) = -\frac{by}{2\pi}\frac{y}{\ell^2+y^2} + \frac{by}{2\pi}\frac{y}{\ell^2+y^2} \equiv 0\,. \tag{8.40}$$

Remark 8.1. 1. Note that the image stresses at $x = -\ell$ and $y = 0$, i.e. the position of the real dislocation, are

$$\sigma^I_{xz}(-\ell,0) = 0, \quad \sigma^I_{yz}(-\ell,0) = \frac{b\mu}{4\pi\ell}\,. \tag{8.41}$$

2. When $|\mathbf{x}|, |\mathbf{y}| >> \ell$,

$$\sigma^t_{xz}(x,y) \approx 0, \quad \text{and} \quad \sigma^t_{yz}(x,y) \approx 0, \tag{8.42}$$

which means that outside the region of $\{(x,y) \,\Big|\, (x+\ell)^2 + y^2 \le \ell^2\}$, the total stress is almost negligible.

8.1.3 *Eshelby's twist: screw dislocation in a finite whisker*

Consider a screw dislocation in a finite cylinder (whisker) of radius R. One may find that the solution of a single screw dislocation in an infinite space actually satisfies the lateral boundary conditions of the problem:

$$\sigma_{\theta z} = \frac{\mu b}{2\pi r}, \quad \forall r \leq R \tag{8.43}$$

$$\sigma_{rr} = \sigma_{r\theta} = \sigma_{rz} = 0, \quad 0 \leq r \leq R \tag{8.44}$$

However, there is one problem: there are non-zero moments or torques at the two open ends of the cylinder, i.e.

$$M_z = \int_0^R \int_0^{2\pi} r\sigma_{\theta z} r dr d\theta = 2\pi \frac{\mu b}{2\pi} \int_0^R r dr = \frac{\mu b R^2}{2} \tag{8.45}$$

To negate the end moment, we superpose two ends moments with the opposite direction of $M_z^{'} = -M_z$ such that the total moments at the two ends of the cylinder become zero, and then based on Saint-Venant's principle we can declare the validity of the solution.

The superimposed moments at two ends will result in the following stress distribution that can be calculated by the elementary torsion formula,

$$\sigma_{\theta z}^{'} = \frac{M_z^{'} r}{J} = -\frac{\mu b r}{\pi R^2} \tag{8.46}$$

In the last equation, we used the fact that the polar moment of a circular region is $J = \pi R^4/2$.

Then the stress distribution in a whisker is

$$\sigma_{\theta z} = \frac{\mu b}{2\pi r} - \frac{\mu b r}{\pi R^2} \ . \tag{8.47}$$

where the extra term $-(\mu b r)/(\pi R^2)$ may be viewed as an equivalent image stress stemming from the superimposed boundary moment.

8.2 Edge dislocation

The edge dislocation problem can be solved as a plane strain problem.

Introduce the Airy stress function, such that

$$\sigma_{xx} = \frac{\partial^2 \psi}{\partial y^2}, \ \sigma_{yy} = \frac{\partial^2 \psi}{\partial x^2}, \ \text{and } \sigma_{xy} = -\frac{\partial^2 \psi}{\partial x \partial y} \ . \tag{8.48}$$

The in-plane equilibrium equations,

$$\frac{\partial \sigma_{xx}}{\partial x} + \frac{\partial \sigma_{yx}}{\partial y} = 0, \quad \text{and} \quad \frac{\partial \sigma_{xy}}{\partial x} + \frac{\partial \sigma_{yy}}{\partial y} = 0, \tag{8.49}$$

lead to the following bi-harmonic equation,

$$\nabla^2 \nabla^2 \psi - 0 \tag{8.50}$$

Let $\phi = \sigma_{xx} + \sigma_{yy} = \nabla^2 \psi$, then $\nabla^2 \nabla^2 \psi = \nabla^2 \phi = 0$, and in polar coordinates,

$$\Big(\frac{\partial^2}{\partial r^2} + \frac{1}{r}\frac{\partial}{\partial r} + \frac{1}{r^2}\frac{\partial^2}{\partial \theta^2} \Big) \phi = 0 \ . \tag{8.51}$$

Based on the general solution obtained in the previous subsection, ϕ has the following form,

$$\phi(r,\theta) = (\alpha_0 + \beta_0 \ln r) + \sum_{n=1}^{\infty}\Big(\alpha_n r^n + \beta_n r^{-n}\Big)\sin n\theta + \sum_{n=1}^{\infty}\Big(\gamma_n r^n + \delta_n r^{-n}\Big)\cos n\theta \tag{8.52}$$

Because of the defect configuration, for an edge dislocation, the region right above the dislocation core should be in compression, whereas the region right below the dislocation core should be in tension, i.e.

$$\phi(r_0, \pi/2) = \phi_{min}, \quad \text{and} \quad \phi(r_0, -\pi/2) = \phi_{max} \tag{8.53}$$

Considering the convergence requirement at the remote region, i.e. $(\phi \to 0, r \to \infty)$, the suitable choice of the solution can be written as follows,

$$\phi = \beta_1 r^{-1} \sin\theta \tag{8.54}$$

Note that the *cosine* terms drop out because of conditions in Eq. (8.53). Then $\nabla^2 \psi = \phi$ implies

$$\Big(\frac{\partial^2}{\partial r^2} + \frac{1}{r}\frac{\partial}{\partial r} + \frac{1}{r^2}\frac{\partial^2}{\partial \theta^2}\Big)\psi = \beta_1 r^{-1}\sin\theta \tag{8.55}$$

Let $\psi = h(r)\sin\theta$. One may find that

$$\Big(\frac{d^2}{dr^2} + \frac{1}{r}\frac{d}{dr} - \frac{1}{r^2}\Big)h = \frac{d}{dr}\Big(\frac{1}{r}\frac{d}{dr}(rh)\Big) = \beta_1 r^{-1}\,. \tag{8.56}$$

By integration, one can verify that a particular solution is

$$\psi_e = \frac{\beta_1}{2} r \sin\theta \ln r \;=\; \frac{\beta_1 y}{4}\ln(x^2+y^2) \tag{8.57}$$

Consider the jump condition,

$$\lim_{\eta\to 0} -\int_{-\infty}^{\infty}\Big[\epsilon_{xx}(x,\eta) - \epsilon_{xx}(x,-\eta)\Big]dx = b \tag{8.58}$$

One can determine the constant β_1,

$$\beta_1 = -\frac{\mu b}{\pi(1-\nu)} \quad \Rightarrow \quad \psi_e = -\frac{\nu b y}{4\pi(1-\nu)}\ln(x^2+y^2) \tag{8.59}$$

One can then find the stress components

$$\sigma_{xx} = -\frac{\mu b}{2\pi(1-\nu)}\frac{y(3x^2+y^2)}{(x^2+y^2)^2} \tag{8.60}$$

$$\sigma_{yy} = \frac{\mu b}{2\pi(1-\nu)}\frac{y(x^2-y^2)}{(x^2+y^2)^2} \tag{8.61}$$

$$\sigma_{xy} = \frac{\mu b}{2\pi(1-\nu)}\frac{x(x^2-y^2)}{(x^2+y^2)^2}, \quad \text{and} \tag{8.62}$$

$$\sigma_{zz} = \nu(\sigma_{xx} + \sigma_{yy}) \tag{8.63}$$

or in polar coordinates

$$\sigma_{rr} = \sigma_{\theta\theta} = -\frac{\mu b \sin\theta}{2\pi(1-\nu)r} \tag{8.64}$$

$$\sigma_{r\theta} = \frac{\mu b \cos\theta}{2\pi(1-\nu)r} \quad \sigma_{zz} = \nu(\sigma_{rr} + \sigma_{\theta\theta}) = -\frac{\mu b \nu \sin\theta}{\pi(1-\nu)r} \tag{8.65}$$

It is then easy to find the strain fields by simply applying Hooke's law of the plane strain condition,

$$\epsilon_{xx} = \frac{by}{2\pi} \frac{(\mu y^2 + (2\lambda + 3\mu)x^2}{(\lambda + 2\mu)(x^2 + y^2)^2} \tag{8.66}$$

$$\epsilon_{yy} = -\frac{by}{2\pi} \frac{((2\lambda + \mu)x^2 - \mu y^2)}{(\lambda + 2\mu)(x^2 + y^2)^2} \tag{8.67}$$

$$\epsilon_{xy} = -\frac{b}{2\pi(1-\nu)} \frac{x(x^2 - y^2)}{(x^2 + y^2)^2} \tag{8.68}$$

By neglecting all the integration constants, a straightforward integration of the above strain components gives

$$u(x,y) = -\frac{b}{2\pi}\left[\tan^{-1}\left(\frac{y}{x}\right) + \frac{\lambda+\mu}{\lambda+2\mu}\frac{xy}{x^2+y^2}\right] \tag{8.69}$$

$$v(x,y) = -\frac{b}{2\pi}\left[-\frac{\mu}{2(\lambda+2\mu)}\ln(x^2+y^2) + \frac{\lambda+\mu}{\lambda+2\mu}\frac{y^2}{x^2+y^2}\right] \tag{8.70}$$

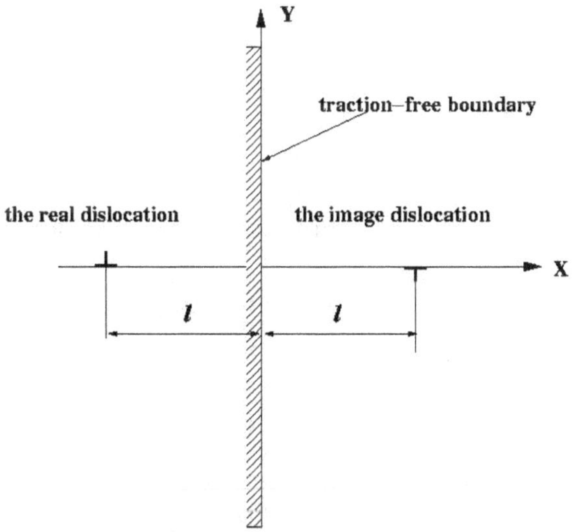

Fig. 8.3 An image edge dislocation.

8.2.1 *Image stress for an edge dislocation*

The solution of the image stress distribution for an edge dislocation is more complicated than that of a screw dislocation.

Consider an edge dislocation being placed at $x = -\ell$ inside a half space ($x < 0$). The solution obtained from the unbounded space,

$$\sigma_{xx}^{\infty} = -\frac{\mu b}{2\pi(1-\nu)} \frac{y(3(x+\ell)^2 + y^2)}{((x+\ell)^2 + y^2)^2} \tag{8.71}$$

$$\sigma_{yy}^{\infty} = \frac{\mu b}{2\pi(1-\nu)} \frac{y((x+\ell)^2 - y^2)}{((x+\ell)^2 + y^2)^2} \tag{8.72}$$

$$\sigma_{xy}^{\infty} = \frac{\mu b}{2\pi(1-\nu)} \frac{(x+\ell)((x+\ell)^2 - y^2)}{((x+\ell)^2 + y^2)^2} \tag{8.73}$$

will not satisfy the traction-free boundary condition at $x = 0$, i.e., $\sigma_{xx}(0, y) \neq 0$ and $\sigma_{xy}(0, y) \neq 0$.

If we place a fictitious dislocation at $x = \ell$ with an opposite Burgers vector $-b$, the following induced image stress fields,

$$\sigma_{xx}^{I} = \frac{\mu b}{2\pi(1-\nu)} \frac{y(3(x-\ell)^2 + y^2)}{((x-\ell)^2 + y^2)^2} \tag{8.74}$$

$$\sigma_{yy}^{I} = -\frac{\mu b}{2\pi(1-\nu)} \frac{y((x-\ell)^2 - y^2)}{((x-\ell)^2 + y^2)^2} \tag{8.75}$$

$$\sigma_{xy}^{I} = -\frac{\mu b}{2\pi(1-\nu)} \frac{(x-\ell)((x-\ell)^2 - y^2)}{((x-\ell)^2 + y^2)^2} \tag{8.76}$$

will cancel the normal stress on traction-free surface, i.e. $\sigma_{xx}^{\infty}(0, y) + \sigma_{xx}^{I}(0, y) = 0$, but they can not cancel the shear stress at $x = 0$. In fact,

$$\sigma_{xy}^{\infty}(0, y) + \sigma_{xy}^{I}(0, y) = \frac{\mu b}{\pi(1-\nu)} \frac{\ell(\ell^2 - y^2)}{(\ell^2 + y^2)^2} \neq 0\ . \tag{8.77}$$

To cancel the shear stress on traction-free surface, one has to superpose another stress field, such that the third stress fields satisfy the condition,

$$\sigma_{xx}^{'''}(0, y) = 0, \quad \text{and} \quad \sigma_{xy}^{'''}(0, y) = -\frac{\mu b}{\pi(1-\nu)} \frac{\ell(\ell^2 - y^2)}{(\ell^2 + y^2)^2}\ . \tag{8.78}$$

Consider the Airy stress function, $\Psi(x, y)$, which satisfies the bi-harmonic equation,

$$\nabla^2 \nabla^2 \Psi = 0 \tag{8.79}$$

Introduce the Fourier-sine and the Fourier-cosine transforms,

$$\bar{f}_s(\xi) = \frac{1}{\pi} \int_{-\infty}^{\infty} f(y) \sin(\xi y) dy, \quad f(y) = \int_0^{\infty} \bar{f}_s(\xi) \sin(\xi y) d\xi; \tag{8.80}$$

$$\bar{f}_c(\xi) = \frac{1}{\pi} \int_{-\infty}^{\infty} f(y) \cos(\xi y) dy, \quad f(y) = \int_0^{\infty} \bar{f}_c(\xi) \cos(\xi y) d\xi \tag{8.81}$$

Since σ_{xy} must be even in y, the Airy stress function, Ψ, is anti-symmetric in y. We apply the Fourier-sine transform to Eq. (8.79), and it yields an ordinary differential equation,

$$\frac{d^4\bar{\Psi}_s}{dx^4} - 2\xi^2\frac{d^2\bar{\Psi}_s}{dx^2} + \xi^4\bar{\Psi}_s = 0 \tag{8.82}$$

Solving (8.82) yields the following solution,

$$\bar{\Psi}_s(x,\xi) = (a_0(\xi) + a_1(\xi)x)\exp(\xi x) + (b_0(\xi) + b_1(\xi)x)\exp(-\xi x) \tag{8.83}$$

The boundary conditions solve the constant a_0, b_0, b_1 as follows,

$$\textbf{(1)}. \quad x \to -\infty, \quad \bar{\Psi}_s \to 0, \quad \Rightarrow b_0 = b_1 = 0; \tag{8.84}$$

$$\textbf{(2)}. \quad x = 0, \quad \sigma_{xx}(0,y) = 0, \quad \Rightarrow a_0 = 0\ . \tag{8.85}$$

Therefore, $\bar{\Psi}_s(x,\xi) = a_1(\xi)x\exp(\xi x)$, and

$$\Psi(x,y) = \frac{1}{\pi}\int_{\infty}^{\infty} a_1(\xi)x\exp(\xi x)\sin(\xi y)d\xi\ . \tag{8.86}$$

Using the boundary condition for the shear stress,

$$\begin{aligned} -\sigma_{xy}^{\prime\prime\prime}(0,y) = \Big(\frac{\partial^2\Psi}{\partial x\partial y}\Big)\Big|_{x=0} &= \int_0^\infty a_1(\xi)\xi\cos(\xi y)dy \\ &= \frac{\mu b}{\pi(1-\nu)}\frac{\ell(\ell^2-y^2)}{(\ell^2+y^2)^2} \end{aligned} \tag{8.87}$$

and the definition of the Fourier-cosine transform, one may find that

$$\begin{aligned} a_1(\xi)\xi &= \frac{1}{\pi}\int_{-\infty}^{\infty}\frac{\mu b}{\pi(1-\nu)}\frac{\ell(\ell^2-y^2)}{(\ell^2+y^2)^2}\cos(\xi y)dy \\ &= \frac{\mu b}{\pi^2(1-\nu)}\int_{-\infty}^{\infty}\frac{\ell(\ell^2-y^2)}{(\ell^2+y^2)^2}\exp(i\xi y)dy\ . \end{aligned} \tag{8.88}$$

The last line is because of $\int_{-\infty}^{\infty}\frac{\ell(\ell^2-y^2)}{(\ell^2+y^2)^2}\sin(\xi y)dy = 0$.
Using the residue theorem to evaluate the integral,

$$\begin{aligned} &\int_{-\infty}^{\infty}\frac{\ell(\ell^2-y^2)}{(\ell^2+y^2)^2}\exp(i\xi y)dy = 2\pi i\sum Res\ F(y_N)\Big|_{y_N=i\ell} \\ &= 2\pi i\Big(-\frac{i\xi\ell}{2}\exp(-\xi\ell)\Big) = \pi\xi\ell\exp(-\xi\ell)\ . \end{aligned} \tag{8.89}$$

we then find that

$$a_1(\xi) = \frac{\mu b\ell}{\pi(1-\nu)}\exp(-\xi\ell) \tag{8.90}$$

so that

$$\begin{aligned} \Psi(x,y) &= \frac{\mu b\ell}{\pi(1-\nu)}\int_0^\infty x\exp\left(\xi(x-\ell)\right)\sin(\xi y)d\xi \\ &= \frac{\mu b\ell xy}{\pi(1-\nu)[(x-\ell)^2+y^2]} \end{aligned} \tag{8.91}$$

note that above integration is integrable under the restriction that $x < \ell$. Therefore, we can find the stresses from the potential

$$\sigma_{xy}^{'''} = -\frac{\partial^2 \Psi}{\partial x \partial y} = -\frac{\mu b \ell}{\pi(1-\nu)}\left(\frac{\ell^2 + 3y^2 - x^2}{[(x-\ell)^2 + y^2]^2} - \frac{4y^2(y^2 + \ell^2 - x^2)}{[(x-\ell)^2 + y^2]^3}\right) \tag{8.92}$$

$$\sigma_{xx}^{'''} = -\frac{\partial^2 \Psi}{\partial y^2} = -\frac{2\mu b \ell x y}{\pi(1-\nu)}\left(\frac{3(\ell - x)^2 - y^2}{[(x-\ell)^2 + y^2]^6}\right) \tag{8.93}$$

Indeed, it can be found that on the traction free surface ($x = 0$),

$$\sigma_{xy}^{'''}(0, y) = -\frac{\mu b \ell}{\pi(1-\nu)} \frac{\ell^2 - y^2}{(\ell^2 + y^2)^2}, \quad \text{and} \quad \sigma_{xx}^{'''}(0, y) = 0 \ . \tag{8.94}$$

Moreover, since $\sigma_{xy}^{'''}(-\ell, 0) = 0$, the shear stress acting on the real dislocation due the traction-free boundary is only contributed by the stress applied by the image dislocation (the second dislocation), i.e.

$$\sigma_{xy}^{t}(-\ell, 0) = \sigma_{xy}^{I}(-\ell, 0) + \sigma_{xy}^{'''}(-\ell, 0) = \frac{\mu b}{4\pi(1-\nu)\ell} \ . \tag{8.95}$$

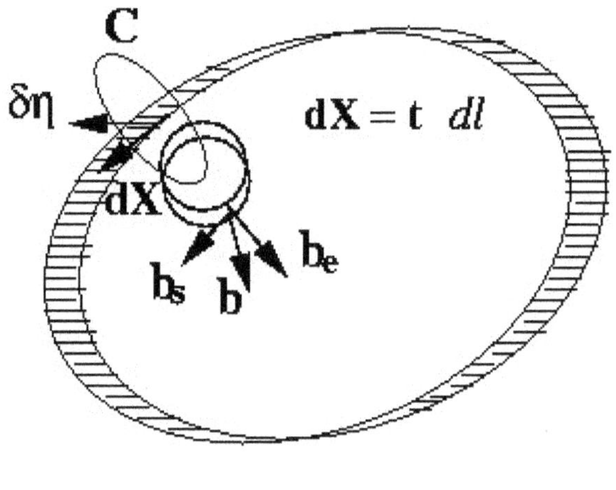

Fig. 8.4 A virtual displacement of a dislocation loop.

8.3 Peach-Koehler force

Consider a dislocation loop undergoing a virtual displacement $\delta\boldsymbol{\eta}$ (see Fig. 8.4). An infinitesimal dislocation line segment, $d\mathbf{X}$ will sweep through an area,

$$d\mathbf{A} = d\mathbf{X} \times \delta\boldsymbol{\eta} \ . \tag{8.96}$$

Note that the direction of $d\mathbf{A}$ is outward.

All the atoms on this area will be subjected to a discontinuous jump with the direction and the magnitude of the local , $\mathbf{b}$. The traction forces on the infinitesimal area can be expressed as $\boldsymbol{\sigma} \cdot d\mathbf{A}$. To be precise, it is

$$\boldsymbol{\sigma} \cdot d\mathbf{A} = \boldsymbol{\sigma} \cdot (d\mathbf{X} \times \delta\boldsymbol{\eta}) \tag{8.97}$$

If we assume that the work done by stresses yield decreases the potential energy of the dislocation,

$$d(\delta E) = -\mathbf{b} \cdot \boldsymbol{\sigma} \cdot (d\mathbf{X} \times \delta\boldsymbol{\eta}) \tag{8.98}$$

The change of the total energy due to the virtual displacement field is

$$\delta E = -\int_{\mathcal{L}} \mathbf{b} \cdot \boldsymbol{\sigma} \cdot (d\mathbf{X} \times \delta\boldsymbol{\eta}) = -\int_{\mathcal{L}} (\boldsymbol{\sigma} \cdot \mathbf{b}) \times \mathbf{t} d\ell \cdot \delta\boldsymbol{\eta} \tag{8.99}$$

where $d\mathbf{X} = \mathbf{t} d\ell$.

By definition, the decrease of the potential energy due to the virtual displacement field is the external virtual work done along the dislocation loop, i.e.

$$\delta E = -\mathbf{F} \cdot \delta\boldsymbol{\eta} = -\int_{\mathcal{L}} \mathbf{F}_\ell d\ell \cdot \delta\boldsymbol{\eta} \ , \tag{8.100}$$

where $\mathbf{F}_\ell$ is the force per unit length along the dislocation loop.

Hence, we obtained the celebrated Peach-Koehler equation [Peach and Koehler (1950)] ,

$$\mathbf{F} = \int_{\mathcal{L}} \Big(\boldsymbol{\sigma} \cdot \mathbf{b}\Big) \times \mathbf{t} d\ell, \quad \text{and} \quad \mathbf{F}_\ell = \Big(\boldsymbol{\sigma} \cdot \mathbf{b}\Big) \times \mathbf{t} \ . \tag{8.101}$$

where $\mathbf{F}_\ell$ is the force per unit length. In the case of a straight dislocation line, we often denote it as $\dfrac{\mathbf{F}}{L}$.

To simplify the computation, we denote

$$\mathbf{g} := \boldsymbol{\sigma} \cdot \mathbf{b}. \tag{8.102}$$

Then the Peach-Koehler force formula can be conveniently written into a matrix form,

$$\mathbf{F}_\ell = \mathbf{g} \times \mathbf{t} = \begin{vmatrix} e_1 & e_2 & e_3 \\ g_1 & g_2 & g_3 \\ t_1 & t_2 & t_3 \end{vmatrix} \ . \tag{8.103}$$

We now consider a few examples.

Example 8.1. This example is illustrated in Fig. 8.5 (a). We are examining the external forces exerted on a straight screw dislocation. In this case, the unit vector of the dislocation line is $\mathbf{t} = \mathbf{e}_z$, the Burgers vector is $\mathbf{b} = b\mathbf{e}_z$. The external stresses (other than self-stress) are specified as follows

$$\boldsymbol{\sigma} = \sigma_{xz} \mathbf{e}_x \otimes \mathbf{e}_z + \sigma_{zx} \mathbf{e}_z \otimes \mathbf{e}_x + \sigma_{yz} \mathbf{e}_y \otimes \mathbf{e}_z + \sigma_{zy} \mathbf{e}_z \otimes \mathbf{e}_y \ . \tag{8.104}$$

and

$$g_x = \sigma_{xz} b, \quad g_y = \sigma_{yz} b, \quad g_z = 0 \ . \tag{8.105}$$

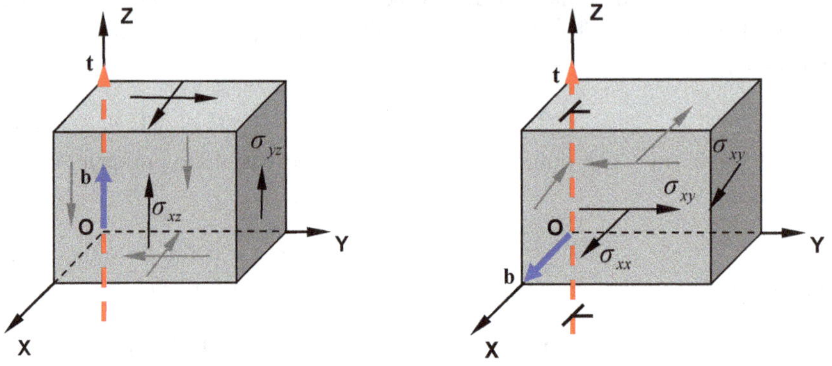

Fig. 8.5 External forces on dislocation. (a) A straight screw dislocation, (b) A straight edge dislocation.

Hence

$$\mathbf{F}_\ell = \mathbf{g} \times \mathbf{t} = \begin{vmatrix} \boldsymbol{e}_x & \boldsymbol{e}_y & \boldsymbol{e}_z \\ \sigma_{xz}b & \sigma_{yz}b & 0 \\ 0 & 0 & 1 \end{vmatrix} = \sigma_{yz}b\boldsymbol{e}_x - \sigma_{xz}b\boldsymbol{e}_y \ . \tag{8.106}$$

To interpret the meanings of this expression, we would say that the shear stress, σ_{yz}, moves the dislocation line to +X direction, whereas the shear stress σ_{xz} moves the dislocation line towards the negative direction of Y-axis, i.e. -Y direction.

Example 8.2. In the second example, we consider a straight edge dislocation. This example is illustrated in 8.5 (b). In this example, again $\mathbf{t} = \boldsymbol{e}_z$, but $\mathbf{b} = b\boldsymbol{e}_x$, and

$$\boldsymbol{\sigma} = \sigma_{xx}\boldsymbol{e}_x \otimes \boldsymbol{e}_x + \sigma_{xy}\boldsymbol{e}_x \otimes \boldsymbol{e}_y + \sigma_{yx}\boldsymbol{e}_y \otimes \boldsymbol{e}_x \ . \tag{8.107}$$

Thus,

$$g_x = \sigma_{xx}b, \quad g_y = \sigma_{yx}b, \quad \text{and} \quad g_z = 0 \ , \tag{8.108}$$

and

$$\mathbf{F}_\ell = \mathbf{g} \times \mathbf{t} = \begin{vmatrix} \boldsymbol{e}_x & \boldsymbol{e}_y & \boldsymbol{e}_z \\ \sigma_{xx}b & \sigma_{xy}b & 0 \\ 0 & 0 & 1 \end{vmatrix} = \sigma_{xy}b\boldsymbol{e}_x - \sigma_{xx}b\boldsymbol{e}_y \ . \tag{8.109}$$

This is to say that the shear stress, σ_{xy}, will move the dislocation line along the slip plane in the positive direction of X-axis. On the other hand, the normal stress, σ_{xx}, will make the dislocation line translating along its own direction. This is an nonconservative motion, because if the motion is admissible, one has to remove some material at one end of the dislocation line and add some material (atoms) at the other end of the dislocation line. In the literature, we refer such dislocation movement as "*climbing*".

From Eq. (8.109), one may find that if $\sigma_{xx} < 0$, which means the material is under compression, the Peach-Koehler force will squeeze the dislocation line up in Y-axis, and when $\sigma_{xx} > 0$ it will pull the material apart and let the dislocation line climbing down.

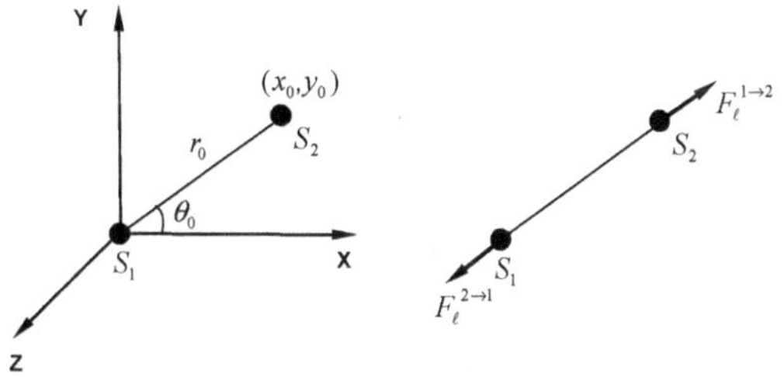

Fig. 8.6 Interactions of two parallel screw dislocations.

Example 8.3. In this example, we consider the interactions between two parallel screw dislocations along the Z-axis, $\mathbf{t} = \boldsymbol{e}_z$, S_1 and S_2. They have different Burgers vectors, i.e. $\mathbf{b}_1 = b_1 \boldsymbol{e}_z$ and $\mathbf{b}_2 = b_2 \boldsymbol{e}_z$. For the dislocation S_1 the stress field is

$$\sigma^I_{xz} = -\frac{\mu b_1}{2\pi}\frac{\sin\theta}{r}, \quad \sigma^I_{yz} = \frac{\mu b_1}{2\pi}\frac{\cos\theta}{r} \quad ; \tag{8.110}$$

and for the dislocation, S_2, the stress field is

$$\sigma^{II}_{xz} = -\frac{\mu b_2}{2\pi}\frac{(y-y_0)}{(x-x_0)^2+(y-y_0)^2} \quad , \tag{8.111}$$

$$\sigma^{II}_{yz} = \frac{\mu b_2}{2\pi}\frac{(x-x_0)}{(x-x_0)^2+(y-y_0)^2} \quad . \tag{8.112}$$

In this case, the Peach-Koehler force equation is

$$\mathbf{F}_\ell = \sigma_{yz}\boldsymbol{e}_x - \sigma_{xz}\boldsymbol{e}_y \ . \tag{8.113}$$

(1). Calculate the force, $\mathbf{F}^{1\to 2}_\ell$, which is the force exerted on the dislocation, S_2, by the dislocation, S_1. Let $r = r_0$ and $\theta = \theta_0$ in (8.110) and substitute them into (8.113). We have

$$\begin{aligned}
\mathbf{F}^{1\to 2}_\ell &= \sigma^I_{yz}\Big|_{(x_0,y_0)} b_2\boldsymbol{e}_x - \sigma^I_{xz}\Big|_{(x_0,y_0)} b_2\boldsymbol{e}_y \\
&= \frac{\mu b_1 b_2}{2\pi}\frac{\cos\theta_0}{r_0}\boldsymbol{e}_x + \frac{\mu b_1 b_2}{2\pi}\frac{\sin\theta_0}{r_0}\boldsymbol{e}_y \\
&= \frac{\mu b_1 b_2}{2\pi r_0}\Big(\cos\theta_0\boldsymbol{e}_x + \sin\theta_0\boldsymbol{e}_y\Big) = \frac{\mu b_1 b_2}{2\pi r_0}\bar{\mathbf{r}}_0 \ ,
\end{aligned} \tag{8.114}$$

where $\bar{\mathbf{r}}_0 = \mathbf{r}_0/|\mathbf{r}_0|$ is the unit vector in the $\mathbf{r}_0$ direction.

2. Calculate the force exerted on the dislocation S_1 by the dislocation S_2. In this case, we let $x = 0, y = 0$ in (8.111) and (8.112) and substitute them into (8.113),

$$
\begin{aligned}
\mathbf{F}_\ell^{2\to 1} &= \sigma_{yz}^{II}\Big|_{0,0} b_1 \boldsymbol{e}_x - \sigma_{xz}^{II}\Big|_{0,0} b_1 \boldsymbol{e}_y \\
&= -\frac{\mu b_1 b_2}{2\pi}\frac{\cos\theta_0}{r_0}\boldsymbol{e}_x - \frac{\mu b_1 b_2}{2\pi}\frac{\sin\theta_0}{r_0}\boldsymbol{e}_y \\
&= -\frac{\mu b_1 b_2}{2\pi r_0}\Big(\cos\theta_0 \boldsymbol{e}_x + \sin\theta_0 \boldsymbol{e}_y\Big) = -\frac{\mu b_1 b_2}{2\pi r_0}\bar{\mathbf{r}}_0 \,.
\end{aligned}
\tag{8.115}
$$

It is obvious that $\mathbf{F}_\ell^{1\to 2} = -\mathbf{F}_\ell^{2\to 1}$ (see Fig. 8.6).

We then conclude that when $\mathbf{b}_1$ and $\mathbf{b}_2$ are along the same direction, the two screw dislocations repel each other; if $b_1 b_2 < 0$, i.e. $\mathbf{b}_1$ and $\mathbf{b}_2$ are in opposite direction, then the two screw dislocations attract each other.

Remark 8.2. [Biot-Savart analogy]
In electro-magnetics, if there are two parallel wires having electric current passing through, the interaction force between the two wires can be calculated by the well-known Boit-Savart law,

$$
\mathbf{F}_\ell^i = \frac{I^i}{c}\Big(\mathbf{t} \times \mathbf{B}^j\Big), \quad i \neq j \ \text{ and } \ i, j = 1, 2 \tag{8.116}
$$

where $\mathbf{F}_\ell^i$ is the force exerted on the wire i by the magnetic field generated by the wire j; I^i is the electric current density in the wire i, while $\mathbf{B}^j$ is the magnetic induction flux density generated by the wire j, and c is the light speed in the medium.

In the Peach-Koehler equation, if we define the magnitude of the Burgers' vector of the i-th dislocation as $b^i = |\mathbf{b}^i|$. Let $\mathbf{G}^j := \boldsymbol{\sigma}^j \cdot \hat{\mathbf{b}}^i$, where $\hat{\mathbf{b}}^i = \dfrac{\mathbf{b}^i}{|\mathbf{b}^i|} = \dfrac{\mathbf{b}^i}{b^i}$, then

$$
\mathbf{g} = \boldsymbol{\sigma}^j \cdot \mathbf{b}^i = \boldsymbol{\sigma}^j \cdot b^i \hat{\mathbf{b}}^i = \mathbf{G}^j b^i \,. \tag{8.117}
$$

We can rewrite the Peach-Koehler force as

$$
\mathbf{F}_\ell^i = -b^i \left(\mathbf{t} \times \mathbf{G}^j\right) \,. \tag{8.118}
$$

It has a similar form as the Biot-Savart law. Since $-b^i$ is the analogy of I_i/c, we may call the strength of a Burgers vector as *the dislocation current density.* By the same token, we may call the stress projection due to the dislocation line $\mathbf{G}^j$, $j = 1, 2$ as the stress induction flux.

The only difference between (8.116) and (8.117) is the minus sign in (8.117). This is because in electro-magnetics. Two wires with the same (opposite) electric current direction attract (repel) each other, whereas two screw dislocation lines having the same (opposite) dislocation current direction repel (attract) each other.

8.4 Point defects

There are several different types of point defects in crystalline solids, including

- Vacancy, where an atom is missing;
- Interstitial, where an impurity atom of smaller size is in an irregular place of the lattice structure;
- Self-interstitial, where a matrix atom occupying a non-lattice site;
- Substitutional atom, an atom of a type other than the matrix atoms occupying at a lattice site.

Fig. 8.7 illustrates these point defects. From the viewpoint of continuum mechanics,

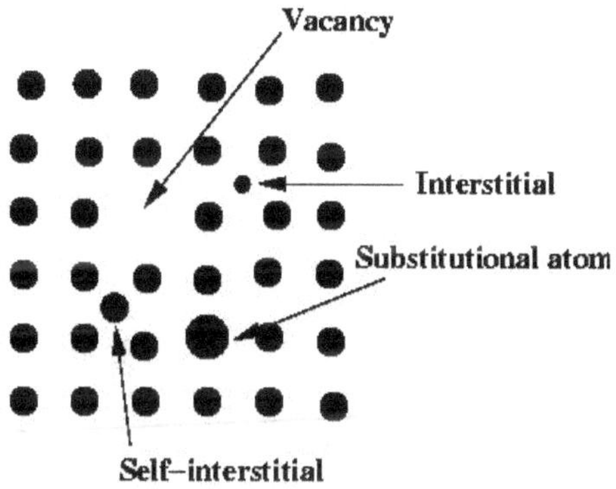

Fig. 8.7 Schematic illustration of some point defects in a lattice.

these point defects can be represented by either a volumetric dilation center or a volumetric contraction center, which produce a localized deformation fields. In continuum physics, these localized deformation fields may be characterized by a quantity called the *formation volume tensor* [Aziz (1997); Zhao *et al.* (1999b,a)]. When external forces are present, the point defect will alter the system's Gibbs free energy of formation,

$$\mathcal{G}^f = (\mathcal{U}^f - T\mathcal{S}^f) - W^{ex} \tag{8.119}$$

where $\mathcal{U}^f$ is the internal energy of formation, T is the temperature, $\mathcal{S}^f$ is the change in entropy. The second contribution is due to the work done by the external force over the deformation field of the point defect, which may be expressed as,

$$-W^{ex} = -\boldsymbol{\sigma}^{ex} : \mathbf{V}^f \tag{8.120}$$

In fact, the formation volumetric tensor is formally defined as

$$\mathbf{V}^f = \frac{\partial \mathcal{G}^f}{\partial \boldsymbol{\sigma}^{ex}} \tag{8.121}$$

A related quantity, the migration volume tensor,

$$\mathbf{V}^m = \frac{\partial \mathcal{G}^m}{\partial \boldsymbol{\sigma}^{ex}},$$

is associated with the free energy difference between the ground state and the excited state at which the defect moves from a lattice site into another in a crystal [Daw *et al.* (2001)]. These formation/migration volumetric tensors are of importance in studying defect diffusion and transport, because the defect diffusivity mainly depends on formation free-energies,

$$D = D_0 \exp\Big(-(\mathcal{G}^f + \mathcal{G}^m)/k_B T \Big) \tag{8.122}$$

where k_B is the Boltzman constant, and T is the absolute temperature.

8.4.1 *Displacement field induced by a point defect*

In continuum mechanics or in elasticity, a point defect may be modeled as the self-equilibrated point force dipole around a center of expansion or contraction. These dipoles are illustrated in Fig. 8.8. The force distribution corresponding to the arrangement of point forces can be written as

$$\mathbf{f}(\mathbf{x}) = \sum_{i=1}^{3} \Big(\mathbf{F}_i \delta(\mathbf{x} - \mathbf{x}' - \mathbf{d}_i \mathbf{e}_i) - \mathbf{F}_i \delta(\mathbf{x} - \mathbf{x}' + \mathbf{d}_i \mathbf{e}_i) \Big) \tag{8.123}$$

where $\delta(\cdot)$ is the Dirac's delta function, $\mathbf{F}_i = F_i \mathbf{e}_i, i = 1, 2, 3$, and $\mathbf{d}_i = d_i \mathbf{e}_i$; here $\mathbf{e}_i$ are the basis vectors for the Cartesian coordinate.

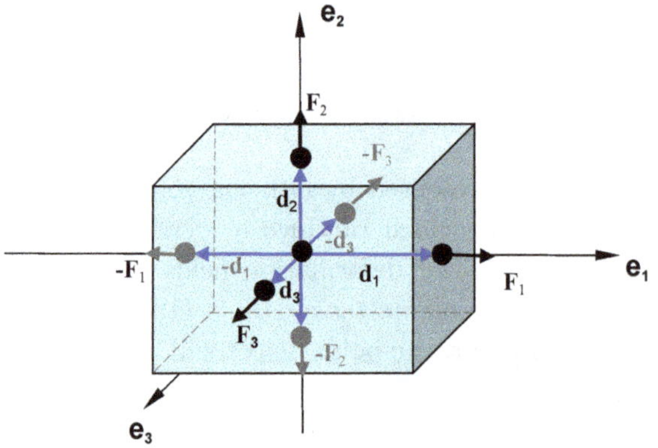

Fig. 8.8 Spatial force dipoles corresponding to the point defects.

Let us consider the case that the magnitudes of the forces tend to infinity and the magnitudes of the force spacings approach to zero while their products, i.e. the dipole strength, remain constants,

$$D_{ii} = \lim_{d_i \to 0} 2F_i d_i = const. \,, \quad i = 1, 2, 3$$

Define the dipole tensor,

$$\mathbf{D} := \lim_{\boldsymbol{d}_i \to 0} 2\mathbf{F}_i \otimes \mathbf{d}_i \tag{8.124}$$

We can then express the force distribution as

$$\mathbf{f}(\mathbf{x}) = \lim_{\boldsymbol{d}_i \to 0} \sum_{i=1}^{3} \Big(\mathbf{F}_i \delta(\mathbf{x} - \mathbf{x}' - d_i \mathbf{e}_i) - \mathbf{F}_i \delta(\mathbf{x} - \mathbf{x}' + d_i \mathbf{e}_i) \Big) \tag{8.125}$$

Based on the definition of the Gâteaux derivative or directional derivative, we have

$$\mathbf{f}(\mathbf{x}) = -\mathbf{F}_i \otimes \Big(\mathbf{d}_i \cdot \nabla_{\mathbf{x}} \delta(\mathbf{x} - \mathbf{x}') \Big) = -\mathbf{D} \cdot \nabla_{\mathbf{x}} \delta(\mathbf{x} - \mathbf{x}') \tag{8.126}$$

In indicial notation, the equilibrium equations of the crystal can be written

$$\mathbb{C}_{ijk\ell} \frac{\partial^2 u_{km}}{\partial x_j \partial x_\ell} - D_{ij} \frac{\partial \delta(\mathbf{x} - \mathbf{x}')}{\partial x_j} = 0 \tag{8.127}$$

Using Green's function (see Chapter 2) and integrating by parts, we find that the displacement field in an unbounded domain is,

$$\begin{aligned} u_i(\mathbf{x}) &= -\int_{\mathbb{R}^3} G_{ij}^{\infty}(\mathbf{x} - \mathbf{x}') D_{jk} \frac{\partial \delta(\mathbf{x}' - \mathbf{x}^{''})}{\partial x_k^{'}} dV_{\mathbf{x}'} \\ &= -\int_{\mathbb{R}^3} \Big(\frac{\partial}{\partial x_k} G_{ij}^{\infty}(\mathbf{x} - \mathbf{x}') \Big) D_{jk} \delta(\mathbf{x}' - \mathbf{x}^{"}) dV_{\mathbf{x}'} \\ &= -\Big(\frac{\partial}{\partial x_k} G_{ij}^{\infty}(\mathbf{x} - \mathbf{x}'') \Big) D_{jk} \end{aligned} \tag{8.128}$$

Note that here the identity $\partial G_{ij}^{\infty}(\mathbf{x} - \mathbf{x}^{'})/\partial x_k^{'} = -\partial G_{ij}^{\infty}(\mathbf{x} - \mathbf{x}^{'})/\partial x_k$ is used in the second line, and the definition of Dirac function is used in the last line. For an isotropic dipole, $D_{ij} = D\delta_{ij}$, the induced displacement field becomes,

$$u_i(\mathbf{x}) = -D \frac{\partial}{\partial x_j} G_{ij}^{\infty}(\mathbf{x} - \mathbf{x}') \tag{8.129}$$

where $\mathbf{x}'$ is the center of the dipole. One may then find the corresponding strain fields

$$\epsilon_{ij} = -\frac{D}{2} \Big(\frac{\partial^2 G_{ik}^{\infty}}{\partial x_j \partial x_k} + \frac{\partial^2 G_{jk}^{\infty}}{\partial x_i \partial x_k} \Big) \tag{8.130}$$

8.4.2 *Formation volume tensor*

Eshelby [Eshelby (1956)] probably was among the first to study formation volume tensor of a point defect. However, Eshelby only derived an expression for the trace of the formation volumetric tensor $tr[\mathbf{V}^f]$ in an unbounded domain, which he called the *total volume dilation*. The following derivation follows that of [Garikipati *et al.* (2006)]. As argued by Garikipati et al, the formation volume tensor has two parts: (1) the tensorial volume change due to strain relaxation of the crystal around the point defect, denoting as V^r_{ij} and (2) the addition of an atomic volume, for isotropic case, it is $\frac{1}{3}\Omega\delta_{ij}$. That is

$$V^f_{ij} = V^r_{ij} + \frac{1}{3}\Omega\delta_{ij} = \int_{B_{crys}} \epsilon_{ij} dV + \frac{1}{3}\Omega\delta_{ij} \tag{8.131}$$

where B_{crys} is a finite crystal domain. We are mainly interested in how to calculate V^r_{ij}.

Considering the relaxation of the crystal is in the linear elastic state, we can then write,

$$V^r_{ij} = \int_{B_{crys}} \epsilon_{ij} dV = \int_{B_{crys}} \mathbb{D}_{ijk\ell}\sigma_{k\ell} dV \tag{8.132}$$

where $\mathbb{D}_{ijk\ell}$ is the elastic compliance tensor. By virtue of divergence theorem,

$$\begin{aligned}\int_{B_{crys}} \sigma_{k\ell} dV &= \int_{B_{crys}} \Big[\frac{\partial(x_k\sigma_{m\ell})}{\partial x_m} - x_k\frac{\partial\sigma_{m\ell}}{\partial x_m}\Big] dV \\ &= \int_{B_{crys}} \Big[\frac{\partial(x_k\sigma_{m\ell})}{\partial x_m} - x_k D_{\ell m}\frac{\partial\delta(\mathbf{x}-\mathbf{x}')}{\partial x_m}\Big] dV\end{aligned} \tag{8.133}$$

where in the last line the equilibrium equation of an expansion dipole center is used, i.e.

$$\frac{\partial\sigma_{ij}}{\partial x_j} - D_{ij}\frac{\partial\delta(\mathbf{x}-\mathbf{x}')}{\partial x_j} = 0\,. \tag{8.134}$$

Since the dipole center is inside the finite crystal ball i.e. $\mathbf{x}' \notin \partial B_{crys}$, the second term in (8.134) becomes

$$\begin{aligned}-\int_{B_{crys}} x_k D_{\ell m}\frac{\partial\delta(\mathbf{x}-\mathbf{x}')}{\partial x_m} dV &= -\underbrace{\int_{\partial B_{crys}} x_k n_m D_{\ell m}\delta(\mathbf{x}-\mathbf{x}') dS}_{=0} \\ &+ \int_{B_{crys}} x_{k,m} D_{\ell m}\delta(\mathbf{x}-\mathbf{x}') dV = \delta_{km}D_{\ell m} = D_{\ell k}\end{aligned} \tag{8.135}$$

Substituting (8.133) and (8.135) into (8.132) or (8.131), we obtain an expression for the formation volume tensor,

$$V^f_{ij} = \mathbb{D}_{ijk\ell}D_{k\ell} + \mathbb{D}_{ijk\ell}\int_{\partial B_{crys}} x_k\sigma_{\ell m}n_m dS + \frac{1}{3}\Omega\delta_{ij} \tag{8.136}$$

For isotropic and surface traction-free crystal, the relaxation formation tensor becomes

$$V^r_{ij}\mathbb{D}_{ijk\ell}D\delta_{k\ell} = \mathbb{D}_{ijkk} \tag{8.137}$$

and its trace is

$$tr[\mathbf{V}^r] = D\mathbb{D}_{iijj}, \tag{8.138}$$

which is the well-known result derived by Eshelby [Eshelby (1956)].

8.5 Continuum theory of dislocation

One of the popular meso-scale simulations in solids is the discrete dislocation dynamics, which is often referred in the literature as DD or DDD. Since Kubin and Devincre's pioneer work, numerical simulations of dislocation dynamics has become an indispensable part of multiscale simulations. The current trend is to develop concurrent multiscale simulations to couple the atomistic molecular dynamics (MD) simulations with continuum based dislocation dynamics (DD) simulations. In this section, we shall briefly introduce the basic concepts and theories of dislocation dynamics.

8.5.1 *Volterra and Mura's formulas*

We begin the discussions with the displacement and the stress fields of the curved dislocations. The general theory of curved dislocations in anisotropic media was developed by Volterra [Volterra (1907)], de Wit [de Wit (1960)], and Mura [Mura (1963, 1968)]. The special case of curved dislocation in an isotropic medium was attributed to J. M. Burgers [Burgers (1939)] and Peach & Koehler [Peach and Koehler (1950)]. The presentation in this book is an adaptation of Mura's work with contemporary flavor.

Before we proceed to derive the Volterra and Mura's formulas, it is expedient to lay out some useful formulas. Consider a simply connected region, $\Omega \in \mathbb{R}^3$, with a smooth boundary. Define a characteristic function,

$$\chi(\mathbf{x}) = \begin{cases} 1, \mathbf{x} \in \Omega \\ 0, \mathbf{x} \notin \Omega \end{cases} \tag{8.139}$$

Consider a (slip) plane S that is characterized by its normal $\boldsymbol{n}$ and its distance to the origin of the coordinate, s. The Radon transform of $\chi(\mathbf{x})$ will be

$$\int_{\mathbb{R}^3} \chi(\mathbf{x}')\delta(s - \boldsymbol{n} \cdot \mathbf{x}')d\mathbf{x}' = \int_{S\cap\Omega} dS \tag{8.140}$$

if $\Omega = \mathbb{R}^3$, we have

$$\int_{\mathbb{R}^3} \chi(\mathbf{x}')\delta(s - \boldsymbol{n} \cdot \mathbf{x}')d\mathbf{x}' = \int_{\mathbb{R}^3} \delta(s - \boldsymbol{n} \cdot \mathbf{x}')d\mathbf{x}' = \int_S dS \tag{8.141}$$

Conceptually, we can generalize the Radon projection formula to a two-dimensional curved surface (2D manifold), S, i.e.

$$\int_\Omega f(\mathbf{x}')\delta(s - \boldsymbol{n} \cdot \mathbf{x}')d\mathbf{x}' = \int_{S\cap\Omega} f(\mathbf{x}')dS' \tag{8.142}$$

$$\int_{\mathbb{R}^3} f(\mathbf{x}')\delta(s - \boldsymbol{n} \cdot \mathbf{x}')d\mathbf{x}' = \int_S f(\mathbf{x}')dS' \tag{8.143}$$

or

$$\int_\Omega f(\mathbf{x}')\delta(S - \mathbf{x}')d\mathbf{x}' = \int_{S\cap\Omega} f(\mathbf{x}')dS' \tag{8.144}$$

$$\int_{\mathbb{R}^3} f(\mathbf{x}')\delta(S - \mathbf{x}')\mathbf{x}' = \int_S f(\mathbf{x}')dS' \tag{8.145}$$

where $\delta(S-\mathbf{x})$ is an abbreviation of $\delta(dist(S,\mathbf{x}))$ and $dist(S,\mathbf{x}) = \inf\{|\mathbf{x}-\mathbf{y}|, \forall \mathbf{y} \in S\}$.

Now we consider the following integral,

$$\int_S \delta(\mathbf{x}-\mathbf{x}')dS' \tag{8.146}$$

where $\delta(\mathbf{x}-\mathbf{x}')$ is Dirac's delta function in three-dimensional space. Based on Eq. (8.145), we have

$$\int_S \delta(\mathbf{x}-\mathbf{x}')dS' = \int_{\mathbb{R}^3} \delta(\mathbf{x}-\mathbf{x}')\delta(S-\mathbf{x}')d\mathbf{x}' = \delta(S-\mathbf{x}) \tag{8.147}$$

Assume that there is a dislocation loop embedded in an elastic continuum. To

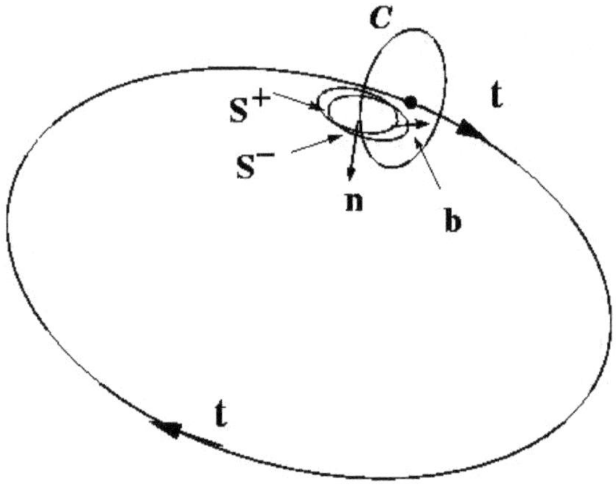

Fig. 8.9 Curved dislocation loop $\mathcal{L}$ and the Burgers circuit $\mathbb{C}$.

define a dislocation line, we take the tangent at a position $\mathbf{x}$ on the dislocation loop, $\boldsymbol{t}$, as the local direction of the dislocation. Obviously, $\boldsymbol{t}$ lies on the tangent plane at point $\mathbf{x}$. We denote the tangent plane at $\mathbf{x}$ as S, which is also the local slip plane. It is assumed that the upper plane of S (denoted by S^+) slips a distance $\mathbf{b}$ relative to its lower plane S^-. Choose a circuit around the vector $\boldsymbol{t}$ in a plane that is perpendicular to $\boldsymbol{t}$ (or $\boldsymbol{t}$ is the normal of the plane). Circle the circuit (the Burgers circuit) in a direction that makes $\boldsymbol{t}$ as a right-handed rotation vector.

In this definition, both the tangent vector $\boldsymbol{t}$ and the local Burgers vector, $\mathbf{b}$ could depend on the spatial location, though in the rest of the presentation, we assume that $\mathbf{b}$ is a constant vector. Note that the real slip plane may not be the tangent plane at $\mathbf{x}$, it could be a curved surface, but the tangent plane of the slip surface at the interception of Burgers circuit should coincide with the tangent plane of the dislocation loop at point $\mathbf{x}$.

To homogenize such dislocation field, one may assume that the total displacement gradient can be written as two parts,

$$u_{i,j} = \beta_{ij} + \beta^*_{ij} \tag{8.148}$$

where β_{ij} is elastic distortion, and β^* is equivalent eigen-distortion, or plastic distortion.

The total strain, ϵ_{ij}, elastic strain, e_{ij}, and eigenstrain, $\epsilon^*{}_{ij}$ can be expressed as

$$\epsilon_{ij} = \frac{1}{2}\Big(u_{i,j} + u_{j,i}\Big) \tag{8.149}$$

$$e_{ij} = \frac{1}{2}\Big(\beta_{ij} + \beta_{ji}\Big) \tag{8.150}$$

$$\epsilon^*_{ij} = \frac{1}{2}\Big(\beta^*_{ij} + \beta^*_{ji}\Big) \tag{8.151}$$

where the eigen-distortion is prescribed as

$$\beta^*_{ji} = -b_i n_j \delta(S - \mathbf{x}) \tag{8.152}$$

the normal vector $\mathbf{n}$ is pointing from S^+ to S^-.

The eigen-distortion caused by slip b_i of plane S^+ may be written as

$$\beta^*_{ji}(\mathbf{x}) = -b_i n_j \delta(S - \mathbf{x}) \tag{8.153}$$

Therefore,

$$\epsilon^*_{ij} = -\frac{1}{2}\Big(b_i n_j + b_j n_i\Big)\delta(S - \mathbf{x}) \tag{8.154}$$

Therefore,

$$\begin{aligned} u_i(\mathbf{x}) &= -\int_{\mathbb{R}^3} C_{j\ell mn}\epsilon^*_{mn}(\mathbf{y})G_{ij,\ell}(\mathbf{x}-\mathbf{y})d\mathbf{y} \\ &= \int_{\mathbb{R}^3} C_{j\ell mn}\epsilon^*_{mn}(\mathbf{y})\delta(S-\mathbf{y})G_{ij,\ell}(\mathbf{x}-\mathbf{y})d\mathbf{y} \\ &= \int_S C_{j\ell mn} b_m n_n G_{ij,\ell}(\mathbf{x}-\mathbf{y})dS_{\mathbf{y}} \end{aligned} \tag{8.155}$$

The above expression was derived by Volterra, and it is called Volterra formula [Volterra (1907)].

Differentiating (8.155) yields

$$u_{i,j}(\mathbf{x}) = \int_S C_{p\ell mn} b_m n_n G_{ip,\ell j}(\mathbf{x}-\mathbf{y})dS_{\mathbf{y}} \tag{8.156}$$

and the elastic distortion becomes

$$\beta_{ji}(\mathbf{x}) = \int_S C_{p\ell mn} b_m n_n G_{ip,\ell j}(\mathbf{x}-\mathbf{y})dS_{\mathbf{y}} + b_i n_j \delta(S - \mathbf{x}) \tag{8.157}$$

Mura showed [Mura (1963)] that the above surface integration can be written as a line integration,

$$\beta_{ji}(\mathbf{x}) = \oint_L e_{jnh} C_{pqmn} G_{ip,q}(\mathbf{x}-\mathbf{y}) b_m t_h d\ell_{\mathbf{y}} \tag{8.158}$$

which is termed as Mura's formula. More general and systematic approaches may be found in [Willis (1967b); Teodosiu (1982)].

To prove the equivalency between (8.158) and (8.157), we first consider Stokes' theorem of a third order tensor field, $\boldsymbol{A} = A_{jih}\boldsymbol{e}_j \otimes \boldsymbol{e}_i \otimes \boldsymbol{e}_h$.

$$\int_S \boldsymbol{n} \cdot (\nabla \times \boldsymbol{A}) dS = \oint \boldsymbol{t} \cdot \boldsymbol{A} d\ell \tag{8.159}$$

or in component form

$$\int_S e_{k\ell h} n_k A_{jih,\ell} dS = \oint t_h A_{jih} d\ell \tag{8.160}$$

Let $A_{jih} = e_{jnh} C_{pqmn} b_m G_{ip,q}$. We have

$$\begin{aligned} &\oint_L e_{jnh} C_{pqmn} b_m G_{ip,q}(\mathbf{x}-\mathbf{y}) t_h d\ell_{\mathbf{y}} \\ &= -\int_S e_{k\ell h} n_k \Big(e_{jnh} C_{pqmn} b_m G_{ip,q\ell}(\mathbf{x}-\mathbf{y}) \Big) dS_{\mathbf{y}} \end{aligned} \tag{8.161}$$

where $G_{ip,q\ell} = -\dfrac{\partial}{\partial x'_\ell} G_{ip,q}$. Utilizing the identity $e_{k\ell h} e_{jnh} = \delta_{kj}\delta_{\ell n} - \delta_{kn}\delta_{\ell j}$, one can obtain

$$\begin{aligned} &-\int_S (\delta_{kj}\delta_{\ell n} - \delta_{kn}\delta_{\ell j}) n_k b_m C_{pqmn} G_{ip,q\ell}(\mathbf{x}-\mathbf{x}') dS' \\ &= -\int_S \Big(n_j b_m C_{pqm\ell} G_{ip,q\ell}(\mathbf{x}-\mathbf{x}') - n_n b_m C_{pqmn} G_{ip,qj}(\mathbf{x}-\mathbf{x}') \Big) dS \\ &= \int_S \Big(n_j b_m \delta_{im} \delta(\mathbf{x}-\mathbf{x}') + n_n b_m C_{pqmn} G_{ip,qj}(\mathbf{x}-\mathbf{x}') \Big) dS' \\ &= \int_\Omega n_j b_i \delta(S-\mathbf{x}')\delta(\mathbf{x}-\mathbf{x}') d\mathbf{x}' + \int_S n_n b_m C_{pqmn} G_{ip,qj}(\mathbf{x}-\mathbf{x}') dS' \\ &= n_j b_i \delta(S-\mathbf{x}) + \int_S n_n b_m C_{pqmn} G_{ip,qj}(\mathbf{x}-\mathbf{x}') dS' \end{aligned} \tag{8.162}$$

Finally, we showed that (8.157) is equivalent to (8.158).

8.5.2 *The Burgers formula*

For isotropic materials, the Volterra formula can be simplified and explicitly expressed in terms of elementary line integrals, which are instrumental in contemporary discrete dislocation dynamics formulations.

To derive the Burgers formula, we start from the Volterra formula,

$$u_m(\mathbf{x}) = b_i \int_S C_{ijk\ell} G^\infty_{km,\ell}(\mathbf{x}-\mathbf{x}') dS'_j \tag{8.163}$$

where the surface S is the dislocation surface, which is a cap of dislocation line $C = \partial S$, and $dS'_j := n_j dS$.

For isotropic materials, both the elastic tensor and the Green's function are quite amiable

$$C_{ijk\ell} = \lambda\delta_{ij}\delta_{k\ell} + \mu(\delta_{ik}\delta_{j\ell} + \delta_{i\ell}\delta_{jk}) \tag{8.164}$$

$$G^\infty_{km}(\mathbf{x}) = \frac{1}{8\pi\mu}\left[\delta_{km} r_{,pp} - \frac{\lambda+\mu}{\lambda+2\mu} r_{,km}\right] . \tag{8.165}$$

Denote $\mathbf{R} = \mathbf{x} - \mathbf{x}'$ and $R = |\mathbf{x} - \mathbf{x}'| = \sqrt{(x_i - x_i')(x_i - x_i')}$.

Then,

$$C_{ijk\ell}G^{\infty}_{km,\ell}(\mathbf{R}) = (\lambda\delta_{ij}\delta_{k\ell} + \mu(\delta_{ik}\delta_{j\ell} + \delta_{i\ell}\delta_{jk}))\frac{1}{8\pi\mu}\Big[\delta_{km}R_{,pp\ell}$$
$$-\frac{\lambda+\mu}{\lambda+2\mu}R_{,km\ell}\Big] = \frac{1}{8\pi\mu}\Big\{\frac{\lambda\mu}{\lambda+\mu}\delta_{ij}R_{,ppm}$$
$$+\mu(\delta_{im}R_{,ppj} + \delta_{jm}R_{,ppi}) - 2\Big(\frac{\lambda+\mu}{\lambda+2\mu}\Big)\mu R_{,mij}\Big\} \tag{8.166}$$

Utilizing the identity,

$$\frac{\lambda}{\lambda+2\mu} = 2\frac{(\lambda+\mu)}{\lambda+2\mu} - 1\ ,$$

one may find that

$$b_iC_{ijk\ell}G^{\infty}_{km,\ell}(\mathbf{R}) = \frac{1}{8\pi\mu}\{\mu b_m R_{,ppj} + \mu(b_\ell R_{,pp\ell}\delta_{jm} - b_jR_{,ppm})$$
$$+ 2\Big(\frac{\lambda+\mu}{\lambda+2\mu}\Big)\mu\Big(b_jR_{,ppm} - b_iR_{,mij}\Big)\Big\} \tag{8.167}$$

Changing the dummy variable, we can then write

$$u_m(\mathbf{x}) = \frac{1}{8\pi}\int_S b_mR_{,ppj}dS_j' + \frac{1}{8\pi}\int_S\Big(b_\ell R_{,pp\ell}dS_m' - b_\ell R_{,ppm}dS_\ell'\Big)$$
$$+ \frac{1}{4\pi}\frac{\lambda+\mu}{\lambda+2\mu}b_j\int_S(R_{,pmp}dS_j - R_{,jmp}dS_p')\ . \tag{8.168}$$

Consider Stoke's theorem,

$$\int_S(\nabla\times\mathbf{A})\cdot d\mathbf{S} = \oint_{\partial S}\mathbf{A}\cdot d\boldsymbol{\ell}\ . \tag{8.169}$$

Let,

$$\nabla = \frac{\partial}{\partial x_m}\mathbf{e}_m,\quad \mathbf{A} = A_{,\ldots}\mathbf{e}_n, d\mathbf{S} = dS_k\mathbf{e}_k,\ \text{and}\ \ d\boldsymbol{\ell} = t_kd\ell\mathbf{e}_k = dx_k\mathbf{e}_k\ .$$

A special case of the Stoke's theorem is,

$$\int_S e_{mnk}\frac{\partial A_{,\ldots}}{\partial x_m}dS_k = \oint_{\partial S}A_{,\ldots}dx_n\ . \tag{8.170}$$

Change the free-index, $n \to k$,

$$-\int_S e_{mnk}\frac{\partial A_{,\ldots}}{\partial x_m}dS_n = \oint_{\partial S}A_{,\ldots}dx_k\ . \tag{8.171}$$

We then have

$$-e_{ijk}e_{mnk}\int_S A_{,\ldots m}dS_n = e_{ijk}\oint_{\partial S}A_{,\ldots}dx_k$$
$$-(\delta_{im}\delta_{jn} - \delta_{in}\delta_{jm})\int_S A_{,\ldots m}dS_n = e_{ijk}\oint_{\partial S}A_{,\ldots}dx_k \tag{8.172}$$

which eventually leads to the desired form,

$$\int_S \Big(A_{,\dots j} dS_i - A_{,\dots i} dS_j \Big) = e_{ijk} \oint_{\partial S} A_{,\dots} dx_k \ . \tag{8.173}$$

In (8.168), we may view $R_{,pp}$ as $A_{,pp}$ in the second integral and $R_{,mp}$ as $A_{,mp}$ in the third integral and then apply the Stoke's theorem (8.173) to (8.168),

$$b_\ell \int_S \Big(R_{,pp\ell} dS'_m - R_{,ppm} dS'_\ell \Big) = -b_\ell \int_S \Big(R_{,pp\ell'} dS'_m - R_{,ppm'} dS'_\ell \Big)$$

$$= -b_\ell \oint_C e_{m\ell k} R_{,pp} dx'_k$$

$$b_j \int_S \Big(R_{,pmp} dS'_j - R_{,pmj} dS'_p \Big) = -b_j \int_S \Big(R_{,pmp'} dS'_j - R_{,pmj'} dS'_p \Big)$$

$$= -b_j \oint_C e_{jpk} R_{,pm} dx'_k$$

We derive the Burgers formula,

$$u_m(\mathbf{x}) = \frac{1}{8\pi} \int_S b_m R_{,ppj} dS'_j - \frac{1}{8\pi} \int_C b_\ell e_{m\ell k} R_{,pp} dx'_k$$
$$- \frac{1}{8\pi(1-\nu)} \int_C b_j e_{jpk} R_{,mp} dx'_k \ . \tag{8.174}$$

In the last line, the identity $\dfrac{\lambda+\mu}{\lambda+2\mu} = \dfrac{1}{2(1-\nu)}$ is used. Consider the fact that

$$R_{,j} = \frac{x_j - x'_j}{R} = \frac{R_j}{R}, \quad \text{and} \quad R_{,mp} = \frac{\delta_{mp}}{R} - \frac{R_m R_p}{R^3}$$

hence

$$R_{,pp} = \frac{2}{R} \quad \text{and} \quad R_{,ppj} = \frac{-2R_j}{R^3} \ .$$

Therefore,

$$u_m(\mathbf{x}) = -\frac{1}{4\pi} \int_S \frac{b_m R_j}{R^3} dS'_j - \frac{1}{4\pi} \oint_C \frac{e_{m\ell k} b_\ell}{R} dx'_k$$
$$- \frac{1}{8\pi(1-\nu)} \oint_C e_{pjk} b_j \frac{\partial}{\partial x_m} \Big(\frac{R_p}{R} \Big) dx'_k \tag{8.175}$$

which can be put into an elementary vector form, i.e. the Burgers formula

$$\mathbf{u}(\mathbf{x}) = -\frac{\mathbf{b}}{4\pi}\Omega - \frac{1}{4\pi} \int_C \frac{\mathbf{b} \times d\boldsymbol{\ell}'}{R} - \frac{1}{8\pi(1-\nu)} \nabla \oint_C \frac{\mathbf{b} \times \mathbf{R} \cdot d\boldsymbol{\ell}'}{R} \ . \tag{8.176}$$

In (8.176), $d\boldsymbol{\ell}' = t_k d\ell \mathbf{e}_k = dx'_k \mathbf{e}_k$, and Ω is the so-called solid angle, which is defined as the surface area of a unit sphere covered by the surface's projection onto the sphere. In this case, the angle is subtended by the dislocation surface, S, i.e.

$$\Omega = \int_S \frac{R_j dS'_j}{R^3} = \int_S \frac{\mathbf{n} \cdot d\mathbf{S}'}{R^2} \tag{8.177}$$

where $\mathbf{n} := \mathbf{R}/R$ is a unit vector from the point $\mathbf{x}$ to the dislocation surface, S.

If the surface is a sphere, $d\mathbf{S} = R^2 d\boldsymbol{\omega}$ and

$$\Omega = \oint_{S_2} \frac{R^2 \mathbf{n} \cdot d\boldsymbol{\omega}}{R^2} = \oint_{S_2} \mathbf{n} \cdot d\boldsymbol{\omega} = \oint_{S_2} n_i n_i d\omega = 4\pi \ . \tag{8.178}$$

8.5.3 Peach-Koehler stress formula for dislocation loop

In this section, we discuss how to express stress field of a dislocation loop in terms of line integral. Taking derivatives of displacement fields expressed by the Burgers formula, we have

$$u_{m,\ell} = \frac{1}{8\pi}\int_S b_m R_{,ppj\ell} dS_j' - \frac{1}{8\pi}\oint_C e_{mnk} b_n R_{,pp\ell} dx_k'$$
$$= -\frac{1}{8\pi(1-\nu)}\oint_C e_{jpk} b_j R_{,mp\ell} dx_k' \tag{8.179}$$

In the above equation, only the first term is not a line integral. Nevertheless, we claim that

$$\int_S b_m R_{,pp\ell j} dS_j' = -8\pi\delta(S-\mathbf{x}) b_m n_\ell - b_m \oint_C e_{j\ell k} R_{,ppj} dx_k' \,.$$

Proof:

Apply Stokes' theorem,

$$\oint_C e_{ijk}\phi dx_k' = \int_S \Big[\phi_{,j} dS_i - \phi_{,i} dS_j'\Big] \tag{8.180}$$

to the above expression,

$$\oint_C e_{i\ell k} R_{,pp} dx_k' = \int_S \Big(R_{,pp\ell'} dS_j' - R_{,ppj'} dS_\ell' = \int_S \Big(R_{,ppj} dS_\ell' - R_{,pp\ell} dS_j'\Big) \tag{8.181}$$

Therefore,

$$\frac{\partial}{\partial x_j}\oint_C e_{j\ell k} R_{,pp} dx_k' = \int_S \Big[R_{,ppjj} dS_\ell' - R_{,pp\ell j} dS_j'\Big] \tag{8.182}$$

Consider

$$G^P(\mathbf{x}-\mathbf{x}') = \frac{1}{4\pi R}, \quad \text{and} \quad \nabla^2 G^P = -\delta(\mathbf{x}-\mathbf{x}'),$$

we then have

$$R_{,pp} = \frac{2}{R} = 8\pi G^P(\mathbf{x}-\mathbf{x}') \quad \text{and} \quad R_{,ppjj} = 8\pi\nabla^2 G^P(\mathbf{x}-\mathbf{x}') = -8\pi\delta(\mathbf{x}-\mathbf{x}').$$

Consequently,

$$b_m \oint_C e_{j\ell k} R_{,ppj} dx_k' = -8\pi b_m \int_S \delta(\mathbf{x}-\mathbf{x}') dS_\ell' - b_m \int_S R_{,pp\ell j} dS_j'$$

Use Radon transformation,

$$\int_S \delta(\mathbf{x}-\mathbf{x}') dS_\ell' = \int_S \delta(\mathbf{x}-\mathbf{x}') n_\ell dS$$
$$= \int_{\mathbb{R}^3} \delta(\mathbf{x}-\mathbf{x}') n_\ell \delta(S-\mathbf{x}') d\Omega' = \delta(S-\mathbf{x}) n_\ell \tag{8.183}$$

Hence, we verified the claim.

Note that $\beta^*_{m\ell} = -8\pi b_m n_\ell \delta(S - \mathbf{x})$, we again recover Mura's formula

$$\beta_{m\ell} = u_{m,\ell} - \beta^*_{m\ell} = -\frac{1}{8\pi}\oint_C e_{j\ell k} b_m R_{,ppj} dx'_k$$
$$-\frac{1}{8\pi}\oint_C e_{mnk} b_n R_{,pp\ell} dx'_k - \frac{1}{8\pi(1-\nu)}\oint_C e_{jpk} b_j R_{,mp\ell} dx'_k \tag{8.184}$$

Shifting the dummy indices, one may find that

$$e_{ij} = \frac{1}{2}(\beta_{ij} + \beta_{ji}) = \frac{1}{8\pi}\oint_C \Big\{ -\frac{1}{2}\Big(e_{jk\ell} b_i R_{,\ell} + e_{ik\ell} b_j R_{,\ell}$$
$$-e_{jk\ell} b_\ell R_{,i} - e_{ik\ell} b_\ell R_{,j}\Big) + \frac{1}{1-\nu} e_{mnk} b_n R_{,ijm} \Big\} dx'_k \tag{8.185}$$

Repeated using the e-δ identity $e_{pij} e_{pmn} = \delta_{im}\delta_{jn} - \delta_{in}\delta_{jm}$, one has

$$e_{jk\ell}(b_i R_{,\ell} - b_\ell R_{,i}) = e_{jk\ell}(\delta_{is}\delta_{\ell t} - \delta_{\ell s}\delta_{it}) b_s R_{,t} = e_{jk\ell} e_{i\ell p} e_{stp} b_s R_{,t}$$
$$= e_{pst} e_{jk\ell} e_{ip\ell} b_s R_{,t} = e_{pst}(\delta_{ji}\delta_{kp} - \delta_{jp}\delta_{ki}) b_s R_{,t}$$
$$= (e_{kst}\delta_{ji} - e_{jst}\delta_{ki}) b_s R_{,t} \tag{8.186}$$

Similarly, one may find,

$$e_{ik\ell}(b_j R_{,\ell} - b_\ell R_{,j}) = (e_{kst}\delta_{ij} - e_{ist}\delta_{kj}) b_s R_{,t} \tag{8.187}$$

which enable us to write

$$e_{ij} = \frac{1}{8\pi}\oint_C \Big\{ -b_s R_{,ppt}\Big[e_{kst}\delta_{ij} - \frac{1}{2} e_{ist}\delta_{kj} - \frac{1}{2} e_{jst}\delta_{ki}\Big]$$
$$+\frac{1}{1-\nu} e_{mnk} b_n R_{,ijm} \Big\} dx'_k \tag{8.188}$$

For linear isotropic elastic materials,

$$\sigma_{ij} = C_{ijk\ell} e_{k\ell}, \quad \text{and} \quad C_{ijk\ell} = \lambda\delta_{ij}\delta_{k\ell} + \mu(\delta_{ik}\delta_{j\ell} + \delta_{i\ell}\delta_{jk}) \tag{8.189}$$

Finally, one can obtain the Peach-Koehler formula for stress field of a dislocation loop,

$$\sigma_{ij} = \frac{\mu}{4\pi}\oint_C \Big(\frac{b_n}{2} R_{,mpp} + (e_{jmn} dx'_i + e_{imn} dx'_j)$$
$$+ \frac{b_n}{1-\nu} e_{kmn}(R_{,ijm} - \delta_{ij} R_{,ppm}) dx'_k \Big) \tag{8.190}$$

Considering,

$$R_{,ppm} = -\frac{2R_m}{R^3} = \frac{\partial}{\partial x_m}\Big(\frac{2}{R}\Big)$$
$$R_{,ijm} = \nabla'_m \cdot \Big(\nabla_i \otimes \nabla_j R)\Big) \tag{8.191}$$

One can re-write the Peach-Koehler formula in a vector form,

$$\boldsymbol{\sigma} = \frac{\mu}{4\pi}\oint_C (\mathbf{b}\times\nabla')\frac{1}{R}\otimes d\boldsymbol{\ell}' + \frac{\mu}{4\pi}\oint_C d\boldsymbol{\ell}'\otimes(\mathbf{b}\times\nabla')\frac{1}{R}$$
$$= -\frac{\mu}{4\pi(1-\nu)}\oint_C \nabla'\cdot(\mathbf{b}\times d\boldsymbol{\ell}')\cdot(\nabla\otimes\nabla - \mathbb{I}^{(2)}\nabla^2)R\,. \tag{8.192}$$

8.6 Discrete dislocation dynamics (DDD)

The first simulation was attempted in late 1980s and early 1990s by J. Lepinoux and L. Kubin [Lepinoux and Kubin (1987)], R. J. Amodeo and N. M. Ghoniem [Amodeo and Ghoniem (1990a,b)], and B. Devincre and M. Condat [Devincre and Condat (1992)] The simulations conducted then were the interactions among infinitely long straight dislocations. Since 1990s, more realistic DDD simulations have been proposed in situations that are involved with more complicated micro-structures [Ghoniem *et al.* (2000); Devincre and Kubin (1997a,b); Kubin *et al.* (1998); Rhee *et al.* (1998); Zbib *et al.* (1998, 2000)] among others.

8.6.1 *Galerkin weak form formulation*

There are several versions of discrete dislocation dynamics (DDD). In the following, we shall present one of the latest dislocation dynamics formulated by Ghoniem and his co-workers. The discrete dislocation dynamics formulated by Ghoniem, Sun, and their co-workers is based on the finite element Galerkin weak formulation. The following presentation is mainly based on a series papers by N. M. Ghoniem et al [Ghoniem (1999); Ghoniem *et al.* (2000); Ghoniem and Sun (2000); Ghoniem *et al.* (2003)].

In this approach, the formulation focus on simulating one dislocation loop among many different dislocation loops. To formulate the discrete dislocation dynamics, we employ the virtual work principle. For a given virtual displacement field, $\delta\mathbf{x}$, the virtual work will be balanced on the dislocation loop considered.

The internal virtual work defined here consists of the virtual work done by all the stresses acting on the dislocation loop. This includes the virtual work done by the stress fields of all other dislocation loops, the stress field due to external loads, and the self-stress field. The external virtual work defined here is mainly the virtual work done by the friction forces that resist the motion of the dislocation loop.

We first consider the virtual work due to all other internal stresses except the self-stress,

$$\begin{aligned}\delta W_{PK} &= \oint_C d\mathbf{F}_{PK}\cdot\delta\mathbf{x} = \oint_C \Big[(\mathbf{b}\cdot\boldsymbol{\Sigma})\times d\boldsymbol{\ell}\Big]\cdot\delta\mathbf{x} \\ &= \oint_C \Big(\mathbf{b}\cdot\boldsymbol{\Sigma}\times\mathbf{t}\Big)d\ell\cdot\delta\mathbf{x} = \oint_C (e_{ijk}\Sigma_{jm}b_m t_k\delta x_i)d\ell\ , \qquad (8.193)\end{aligned}$$

where $\mathbf{b}$ is the Burgers vector, $\mathbf{t}$ is the tangential vector along the dislocation loop, and

$$\Sigma_{ij} = \sigma^I_{ij} + \sigma^e_{ij} \qquad (8.194)$$

Here σ^I_{ij} are the stress fields of all other dislocation loops inside the solid, which can be expressed as

$$\sigma^I_{ij} = \frac{\mu}{4\pi}\oint_C b_n\Big[\frac{1}{2}R_{,mpp}(e_{jmn}dx'_i + e_{imn}dx'_j) + \frac{1}{1-\nu}e_{kmn}(R_{,ijm} - \delta_{ij}R_{,ppm})\Big]dx'_k \tag{8.195}$$

and σ^e_{ij} is the stress field due to externally applied loads.

Denote

$$f^{PK}_i = e_{ijk}\Sigma_{jm}b_m t_k. \tag{8.196}$$

One may write

$$\delta W_{PK} = \oint_C f^{PK}_i d\ell \delta x_i. \tag{8.197}$$

In principle, the virtual work done by the self stress field can be also expressed by Eq. (8.195). However, in that case, Eq. (8.195) would become a singular integral, which can be evaluated in the sense of Cauchy principal value.

Since the core of a dislocation loop has specific physical meanings, it would be appropriate to treat the virtual work of self-stress field separately. S. D. Gavazza and D. M. Barnett [Gavazza and Barnett (1976)] expressed the virtual work of the self-stress field of planar curved dislocation loop in terms of a single integral expression,

$$\delta W_{self} = \oint_C \Big\{\Big[E(\mathbf{t}) - \Big(E(\mathbf{t}) + E''(\mathbf{t})\Big)\ln\Big(\frac{8}{\epsilon\kappa}\Big)\Big]\kappa - J(L,p)\Big\}\mathbf{n}\cdot\delta\mathbf{x}d\ell + [dU]_{core} \tag{8.198}$$

where $E(\mathbf{t}) = \frac{1}{2}\sigma_{ij}(\mathbf{t})b_i n_j$, ϵ is related to the core size, κ is the curvature of the dislocation line, $J(L,p)$ is a non-local interaction term, and $[dU]_{core}$ is the virtual work contribution from the core of the dislocation loop. Since $[dU]_{core}$ is related to the dislocation mobility, this term may be absorbed into the friction force.

Let,

$$\mathcal{E}^{self} = \Big\{E(\mathbf{t}) - \Big(E(\mathbf{t}) + E''(\mathbf{t})\Big)\ln\Big(\frac{8}{\epsilon\kappa}\Big)\Big]\kappa - J(L,p)\Big\} \tag{8.199}$$

and

$$f^{self}_i = \mathcal{E}n_i \tag{8.200}$$

The total active forces acting on a dislocation loop are

$$f^T_i = f^{PK}_i + f^{self}_i \tag{8.201}$$

In many cases, it has to include the change of chemical potential induced *Osmotic force*. Originally, the Osmotic force is a diffusive force due to the imbalance of the solute concentration in a solution. In dislocation theory, the Osmotic force is referred to as a force that is created by diffusive motions due to the imbalance of chemical potentials of defects in the medium. When concentrations of defect species such as vacancy and interstitial are out of balance, they will create diffusive movements that will exert a force on dislocations, and this force is an Osmotic force. The *Osmotic force* is usually responsible for the dislocation loop climb, e.g. see [Wang *et al.* (1998)], because the change in chemical potential per vacancy or interstitial will cause the dislocation loop climbing, or causing the none-conservative dislocation loop movement.

When a dislocation loop starting to move, it has to overcome the friction forces that resist its motion. The friction forces consist of (1) extrinsic resistances due to alloying, impurity atoms, Peierls stress (this part of force coming from $[dU]_{core}$), etc., and (2) intrinsic friction forces that are due to the atomistic bond force in a surface separation (fracture) process. Empirically, one can always assume that the friction forces are proportional to the dislocation velocity, such that

$$\delta W^{friction} = \oint_C C_{ik} V_k d\ell \delta x_i = \oint_C (\mathbf{C} \cdot \mathbf{V}) d\ell \cdot \delta \boldsymbol{x} \tag{8.202}$$

where

$$\mathbf{V} = \frac{d\mathbf{x}}{dt} \tag{8.203}$$

and $\mathbf{C}$ is called the resistivity matrix, which has three independent components in an isotropic medium (two for glide motion and one for climb motion),

$$[C_{ik}] = \begin{bmatrix} C_1 & 0 & 0 \\ 0 & C_2 & 0 \\ C_1 & 0 & C_3 \end{bmatrix} \tag{8.204}$$

Then the principle of virtual reads

$$\delta W^{int} - \delta W^{fric} = 0, \quad \Rightarrow \quad \oint_C \Big(f_i^T - C_{ik} V_k \Big) d\ell \delta x_i = 0 \ . \tag{8.205}$$

8.6.2 *Finite element implementation*

Truncating the dislocation loop into N_s segments, and mapping each segment into a one-dimensional parametric space, i.e., $N_I : [\mathbf{x}_{I-1}, \mathbf{x}_I] \quad \rightarrow \quad u \in [0, 1]$. Thereby, for $\mathbf{x} \in N_I$,

$$d\ell = \sqrt{\Big(\frac{\partial x_i}{\partial u} \frac{\partial x_i}{\partial u} \Big)} du \tag{8.206}$$

Consider the finite element discretization,

$$x_i^h(u,t) = \sum_{m=1}^{N_{DF}} N_{im}(u) q_m(t) \tag{8.207}$$

where $N_{im}(u)$ is the finite element shape function. The discretized velocity field is

$$V_i^h = x_{i,t}^h = \sum_{m=1}^{N_{DF}} N_{im}(u) q_{m,t}(t) \ . \tag{8.208}$$

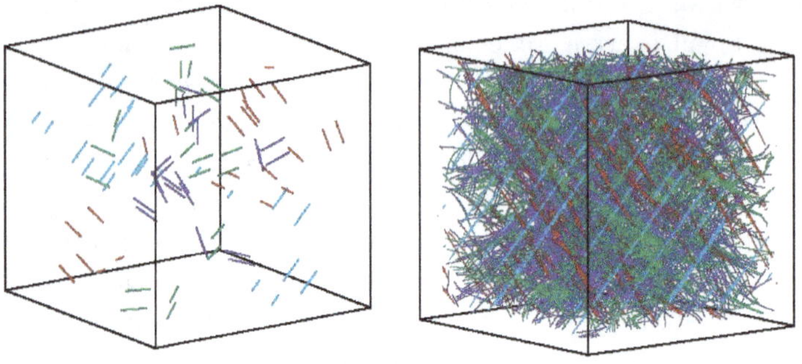

Simulation of plastic deformation of a 10 $\mu m \times 10\ \mu m$ volume. (a) Initial microstructure of dislocation. (b) Microstructure of dislocation at 0.3% strain. Different colors in the figures represent dislocations on different glide planes. From: Zhiqiang Wang, Nasr Ghoniema, Sriram Swaminarayanb and Richard LeSarb, A parallel algorithm for 3D dislocation dynamics, Journal of Computational Physics, V 219, Issue 2, pp. 608-621, 2006. Reproduced with permission from Elsevier Limited.

Fig. 8.10 Simulations of Discrete Dislocation Dynamics.

Denote the gradient of FEM shape function as $B_{im}(u) := N_{im,u}(u)$. The line integration element will be

$$d\ell = (x_\ell x_\ell)^{1/2} du = \Big(\sum_{p,s=1}^{N_{DF}} q_p q_s B_{\ell p}(u) B_{\ell s}(u) \Big)^{1/2} du \tag{8.209}$$

We can evaluate the internal stresses acting on the dislocation loop by numerical quadrature integration, i.e.

$$\sigma_{ij}^I = \frac{\mu}{4\pi} \sum_{\gamma=1}^{N_{loop}} \sum_{\beta=1}^{N_s} \sum_{\alpha=1}^{Q_{max}} b_n w_\alpha \Big[\frac{1}{2} R_{,mpp} (e_{jmn} x_{i,u} + e_{imn} x_{j,u}) + \frac{1}{1-\nu} e_{kmn} (R_{,ijm} - \delta_{ij} R_{,ppm}) x_{k,u} \Big] \tag{8.210}$$

where N_{loop} is the total number of dislocation loops, N_s is the total number of segments in each dislocation loop, and Q_{max} is the total number of quadrature points in a segment, and w_α is the quadrature weight.

Denote each segment of the dislocation loop as L_j. The discretized weak formulation is

$$\sum_{j=1}^{N_s} \sum_{\alpha=1}^{Q_{max}} \sum_{m=1}^{N_{DF}} N_{im}(u) \delta q_m \Big[f_i^T - C_{ik} \sum_{n=1}^{N_{DF}} N_{kn} \dot{q}_n \Big] \times \Big(\sum_{p,s=1}^{N_{DF}} q_p q_s B_{\ell p} B_{\ell s} \Big)^{1/2} w_\alpha = 0\,. \tag{8.211}$$

Define the generalized force vector,

$$f_m^h = \sum_{\alpha=1}^{Q_{max}} f_i^T N_{im}(u) \Big(\sum_{p,s=1}^{N_{DF}} q_p, q_s B_{\ell p} B_{\ell s} \Big)^{1/2} w_\alpha \tag{8.212}$$

and the resistivity matrix $\{\gamma_{mn}\}$, in which

$$\gamma_{mn} = \sum_{\alpha=1}^{Q_{max}} N_{im}(u) C_{ik} N_{kn}(u) \Big(\sum_{p,s=1}^{N_{DF}} q_p, q_s B_{\ell p} B_{\ell s} \Big)^{1/2} w_\alpha \tag{8.213}$$

Then, we can put the dislocation loop weak form into a matrix form,

$$\sum_{j=1}^{N_s} \Big[[\mathbf{f}]_j - [\gamma]_j \Big[\frac{d\mathbf{q}}{dt} \Big]_j \Big]^T \Big[\delta \mathbf{q} \Big]_j = 0 \, , \tag{8.214}$$

which leads to the global matrix formulation,

$$\Big[\Big[\mathbf{F} \Big] - \Big[\mathbf{\Gamma} \Big] \Big[\frac{d\mathbf{Q}}{dt} \Big] \Big]^T \Big[\delta \mathbf{Q} \Big] = \mathbf{0} \, , \tag{8.215}$$

where

$$\Big[\mathbf{F} \Big] = \mathbf{A}_{j=1}^{N_s} \Big[\mathbf{f} \Big]_j^{1 \times N_{DF}} \quad \text{and} \quad \Big[\mathbf{\Gamma} \Big] = \mathbf{A}_{j=1}^{N_s} \Big[\gamma \Big]_j^{N_{DF} \times N_{DF}} \tag{8.216}$$

Solving (8.215) yields,

$$\Big[\frac{d\mathbf{Q}}{dt} \Big] = \Big[\mathbf{\Gamma} \Big]^{-1} \Big[\mathbf{F} \Big] \tag{8.217}$$

Employing any desirable time stepping algorithm, one finds the updated dislocation loop configuration or position by

$$\Big[\mathbf{Q} \Big]_{n+1} = \Big[\mathbf{Q} \Big]_n + \Big[\mathbf{\Gamma} \Big]_{n+\alpha}^{-1} \Big[\mathbf{F} \Big]_{n+\alpha} \Delta t \tag{8.218}$$

where $0 \leq \alpha \leq 1$.

This is the state of the art discrete dislocation dynamics formulation.

8.7 Peierls-Nabarro model

Nanoscale mechanical activities are multiscale physical phenomena that are strongly influenced by materials atomistic or lattice structures in one hand, and on the other hand, they also exhibit many continuum features. There are two basic approaches to nanoscale mechanical problems. One is the so-call *bottom-up approach*, in which one builds a nanoscale model from first-principle, and the other approach is called *top-down approach*, in which one derives the nanomechanics model from continuum mechanics by incorporating microscale lattice structure into the continuum model. An archetype of the top-down approach is so-called Peierls-Nabarro model for dislocation mobilities.

In reality, there are many dislocations present in a crystal, however, those dislocations may not be mobile, and the crystal is still able to hold itself together. When a dislocation moves, it must overcome the lattice friction that resists the dislocation motion. Such lattice friction or resistance is a material property that relates to the strength of the material. The Peierls-Nabarro model explains the origin and the underline physics of such lattice friction, and moreover it provides a closed form expression to calculate the critical value of the lattice resistance to dislocation motions.

8.7.1 *Hilbert transform*

The Hilbert transform is a particular case of the Cauchy integral transforms. Let L be a closed smooth contour and $\phi(\zeta)$ be an arbitrary Holder continuous function specified on L and vanishing at infinity. Cauchy integral transforms are the following pair of mutually invertible integrals (e.g. Zhdanov [1984]),

$$\psi(\zeta_0) = P.V.\frac{1}{\pi i}\int_L \frac{\phi(\zeta)}{\zeta - \zeta_0} d\zeta \tag{8.219}$$

$$\phi(\zeta_0) = P.V.\frac{1}{\pi i}\int_L \frac{\psi(\zeta)}{\zeta - \zeta_0} d\zeta \tag{8.220}$$

where symbol $P.V.$ stands for the Cauchy principal value.

One special case of great value for applications is that L real axis, $Im(\psi(\zeta)) = g(x)$, $Re(\psi(\zeta)) = 0$, $Re(\phi(\zeta)) = f(x)$, and $Im(\phi(\zeta)) = 0$. That is $\phi(\zeta) = f(x) + i0$ and $\psi(\zeta) = 0 + ig(x)$. Here $f(x)$ and $g(x)$ are real functions of a real variable x satisfying the Holder condition for any finite x and vanishing at infinity. This special case of Cauchy integral transforms is the so-called *the Hilbert transforms*:

$$g(x) = \mathcal{H}(f(x)) = P.V.\frac{1}{\pi}\int_{\mathbb{R}^3} \frac{f(t)dt}{x - t} \tag{8.221}$$

$$f(x) = -\mathcal{H}(g(x)) = P.V. - \frac{1}{\pi}\int_{\mathbb{R}^3} \frac{g(t)dt}{x - t} \tag{8.222}$$

Note the position between x and t and position between ζ and ζ_0.

Hilbert transform table is available in many mathematics handbooks. In general, one can find Hilbert transform via Cauchy's residue theorem.

The following are a few examples:

$$\mathcal{H}\Big(\frac{1}{\pi(b-x)}\Big) = \frac{1}{\pi}\int_{\mathbb{R}^3}\Big(\frac{dt}{\pi(b-t)(x-t)}\Big) = \delta(x-b) \tag{8.223}$$

$$\mathcal{H}\Big(\frac{1}{(x^2+a^2)}\Big) = \frac{1}{\pi}\int_{\mathbb{R}^3}\Big(\frac{dt}{(t^2+a^2)(x-t)}\Big) = \frac{x}{a(x^2+a^2)} \tag{8.224}$$

$$\mathcal{H}\Big(\sin(bx)\Big) = \frac{1}{\pi}\int_{\mathbb{R}^3}\frac{\sin(bt)dt}{(x-t)} = -\cos(bx) \tag{8.225}$$

8.7.2 *Peierls-Nabarro dislocation model*

In the early development of dislocation theory, scientists were concerned with two important issues: (1) What is the size of a dislocation for a given Burgers vector? (2) How much force is needed to move a dislocation out of its stable position?

The second question is the so-called dislocation mobility, which is central to the understanding of the ductile material strength. The tries to answer this question. Before we discuss the Peierls-Nabarro model, we first examine the mechanical fields

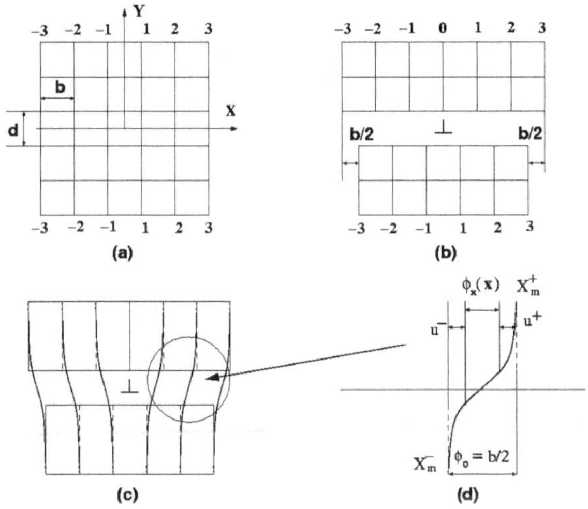

Fig. 8.11 The Peierls-Nabarro model.

of a straight edge dislocation (displacement fields are given up to a rigid body displacement),

$$u_x = \frac{b}{2\pi}\left[\tan^{-1}\frac{y}{x} + \frac{xy}{2(1-\nu)(x^2+y^2)}\right] \tag{8.226}$$

$$u_y = -\frac{b}{2\pi}\left[\frac{1-2\nu}{4(1-\nu)}ln(x^2+y^2) + \frac{x^2}{2(1-\nu)(x^2+y^2)}\right] \tag{8.227}$$

$$\sigma_{xx} = -\frac{\mu b}{2\pi(1-\nu)}\frac{y(3x^2+y^2)}{(x^2+y^2)^2} \tag{8.228}$$

$$\sigma_{yy} = \frac{\mu b}{2\pi(1-\nu)}\frac{y(x^2-y^2)}{(x^2+y^2)^2} \tag{8.229}$$

$$\sigma_{xy} = \frac{\mu b}{2\pi(1-\nu)}\frac{x(x^2-y^2)}{(x^2+y^2)^2} \tag{8.230}$$

As evident from the above equations, the stress fields are singular at the origin. Therefore the analytical solution presented above is no longer accurate near the core of the dislocation. To remove this singularity inside the dislocation core, Peierls

[Peierls (1940)] and [Nabarro (1947)] included the discrete atomic nature of the material and proposed the following lattice correction model.

The Peierls-Nabarro model(PN model) for a straight edge dislocation is described using two semi-infinite simple cubic crystals as shown in Fig. 8.11. The formal glide plane is $y = 0$. The two elastic half spaces are terminated on the planes $y \geq d/2$ and $y \leq -d/2$. At the middle of glide plane, a non-Hookean slab of width d (atomic spacing) joins the two half spaces. The symmetrical configuration indicated in Fig. 8.11 suggests that this is done by cutting the perfect crystal into two halves along the $y = 0$ plane, and inserting an additional layer of atoms in the upper half of the crystal space, which displaces the upper half crystal moving rigidly a distance $0.5b$ in both positive and negative x-direction, and we then re-weld the two half crystals.

Before the "re-welding", the initial dis-registry (misalignment) in x-direction of two vertical atom layers with respect to the upper and lower half crystal spaces is

$$\phi_x^0(x) := X_m^+ - X_m^- = \begin{cases} \dfrac{b}{2}, & x > 0 \\[2ex] -\dfrac{b}{2}, & x < 0 \end{cases} \qquad m = \pm 1, \pm 2, \cdots \pm \infty \tag{8.231}$$

After the re-welding, the misalignment, or the discontinuity, between the atom layer in the upper part of crystal and the same atom layer (m) of the lower part of the crystal becomes

$$\phi_x(x) = x_m^+ - x_m^- = X_m^+ + u^+(x) - (X_m^- + u^-(x))$$

$$\phi_x(x) = \begin{cases} \dfrac{b}{2} + u^+(x) - u^-(x), & x > 0 \\[2ex] -\dfrac{b}{2} + u^+(x) - u^-(x), & x < 0 \end{cases}$$

$$= \begin{cases} 2u_x(x) + \dfrac{b}{2}, & x > 0 \\[2ex] 2u_x(x) - \dfrac{b}{2}, & x < 0 \end{cases}$$

By antisymmetry, we assume that $u_x(x) = u^+(x) = -u^-(x)$.

At the remote boundary, dis-registry is enforced to be zero, i.e. there is no discontinuity at the remote boundary

$$\phi_x(x) \to 0, \text{ when } x \to \pm\infty \quad \Rightarrow \quad 2u_x(x) \pm \frac{b}{2} = 0, \ x \to \pm\infty \tag{8.232}$$

Therefore, $u_x(\pm\infty) = \mp\dfrac{b}{4}$. This implies that the total displacement along the interface should be

$$u_x(\infty) - u_x(-\infty) = \int_{-\infty}^{\infty} \left(\frac{du_x}{dx}\right)_{x=x'} dx' = -\frac{b}{2} \tag{8.233}$$

Based on Eshelby's interpretation [Eshelby (1949)], one may think that the Peierls-Nabarro model deploys a continuous edge dislocation distribution along the cohesive interface with its local Burgers vector density as $b'(x')$ to replace a single dislocation with a Burgers vector **b**. To make sure that these two dislocation systems are equivalent, we enforce the following condition on net Burgers vector equality,

$$-2\int_{infty}^{\infty}\left(\frac{du_x}{dx}\right)_{x=x'}dx' = \int_{-\infty}^{\infty} b'(x')dx' = b \tag{8.234}$$

From the above relation, one may derive that the distribution density of Burgers vector should be $b'(x') = -2\dfrac{du_x}{dx}(x')$.

The strains near the dislocation core are large, and therefore use of Hooke's law for the stresses is inappropriate. One the other hand, it is relevant to use the periodicity of the lattice, which implies σ_{xy} to be a periodic function of $\phi(x)$. We therefore assume that,

$$\sigma_{xy}(x,0) = C\sin\left(\frac{2\pi\phi_x}{b}\right) \tag{8.235}$$

When $\phi_x(x) << 1, \sigma_{xy}(x,0) \sim C\dfrac{2\pi\phi_x(x)}{b}$. Under small deformation limit, it is assumed that the cohesive law should comply to Hooke's law as well (is this a good assumption?), i.e.

$$\sigma_{xy}(x,0) = 2\mu\epsilon_{xy} = \frac{\mu\phi_x(x)}{d} = C\frac{2\pi\phi_x(x)}{b} \tag{8.236}$$

which determines the constant $C = \dfrac{\mu b}{2\pi d}$. Note that the shear strain inside the cohesive interface is (see Fig. 8.11)

$$\gamma_{xy} = \frac{\phi_x(x)}{d} \tag{8.237}$$

Thereby, one obtains that

$$\sigma_{xy}(x,0) = \frac{\mu b}{2\pi d}\sin\left(\pm\pi + \frac{4\pi u_x(x)}{b}\right) = -\frac{\mu b}{2\pi d}\sin\left(\frac{4\pi u_x(x)}{b}\right) \tag{8.238}$$

One can calculate the shear stress inside the cohesive strip due the continuously distributed dislocation via superposition. At $y = 0$,

$$\sigma_{xy}(x,0) = \frac{\mu}{2\pi(1-\nu)}\int_{\mathbb{R}}\frac{b'(t)dt}{x-t} = -\frac{\mu}{\pi(1-\nu)}\int_{\mathbb{R}}\frac{(du_x/dx)_{x=t}dt}{x-.t} \tag{8.239}$$

One may also derive the above integral equation based on the Boussinesq solution of linear elastic half space (e.g. Timoshenko and Goodier [Timoshenko and Goodier (1951)]).

Apparently, $\sigma_{xy}(x,0)$ is proportional to the Hilbert transform of du_x/dx. Thereby the inverse Hilbert transform gives

$$\frac{du_x}{dx} = \frac{(1-\nu)}{\mu}\int_{\mathbb{R}}\frac{\sigma_{xy}(t,0)dt}{x-t} \tag{8.240}$$

Integrating this yields,

$$u(x) = \frac{(1-\nu)}{\mu} \int_{\mathbb{R}} \sigma_{xy}(t,0) \ln|t-x| dt \tag{8.241}$$

Using (8.238) and (8.239), one can obtain the well-known Peierls-Nabarro integral equation for the unknown displacement field, $u_x(x)$,

$$\int_{\mathbb{R}} \frac{(du_x/dx)_{x=t} dt}{x-t} = \frac{b(1-\nu)}{2d} \sin \frac{4\pi u_x}{b} \tag{8.242}$$

which is a singular, nonlinear integral equation with the unknown function $u_x(x)$.

Luckily, the solution of the above integral equation can be found in closed form[1],

$$u_x(x) = -\frac{b}{2\pi} \tan^{-1} \frac{x}{r_c} \tag{8.243}$$

where $r_c = \dfrac{d}{2(1-\nu)}$, which is a parameter that characterizes the size of the dislocation core. When $|x| < r_c$, the dis-registry $\phi_x(x) > b/4$. At $x = r_c$, $u_x(r_c) = -b/8$ and $\phi_x(r_c) = b/4$.

Substituting (8.243) into (8.238) and utilizing the trigonometry identity

$$\tan^{-1}(y) = \sin^{-1}\Big(\frac{y}{\sqrt{1+y^2}}\Big)$$

one can find that

$$\sigma_{xy}(x,0) = \frac{\mu b}{2\pi(1-\nu)} \frac{x}{x^2+r_c^2} \tag{8.244}$$

On the other hand, by virtue of (8.243) the displacement gradient in x-direction is

$$\Big(\frac{du_x}{dx}\Big)_{x=t} = -\frac{b}{2\pi} \frac{r_c}{x^2+r_c^2} \tag{8.245}$$

and the Hilbert transform of the above expression is

$$\mathcal{H}\Big(\frac{du_x}{dx}\Big) = \mathcal{H}\Big(-\frac{br_c}{2\pi}\frac{1}{x^2+r_c^2}\Big) = -\frac{b}{2\pi}\frac{x}{x^2+r_c^2} \tag{8.246}$$

where the following Hilbert transform formula is used,

$$\mathcal{H}\Big(\frac{1}{x^2+r_c^2}\Big) = \frac{1}{r_c}\frac{x}{x^2+r_c^2}$$

Based on (8.239),

$$\sigma_{xy}(x,0) = -\frac{\mu}{(1-\nu)}\mathcal{H}\Big(\frac{du_x}{dx}\Big) = \frac{\mu b}{2\pi(1-\nu)}\frac{x}{x^2+r_c^2} \tag{8.247}$$

which is the same as the expression obtained above.

[1]This may be the reason why they took sine function as the cohesive law was to match the exact solution of this particular integral equation, which people had known before.

8.7.3 *Misfit energy and the Peierls force*

As we mentioned before, one of the motives to discuss the Peierls-Nabarro dislocation model [Peierls (1940); Nabarro (1947)] is to find the critical stress needed to move a dislocation from its stable position. This question cannot be answered by analyzing a Volterra dislocation.

To find the critical stress to move a dislocation, we first examine the stored elastic energy due to an edge dislocation. The total elastic energy stored induced by an edge dislocation may be divided into two parts: the energy stored inside the elastic crystal and the energy stored inside the cohesive layer. Since the two crystal half spaces maintain substantially perfect lattice structure, most of shear deformation is confined within the cohesive layer. For this reason, we call the energy stored inside the cohesive layer as the misfit energy.

The shear strain inside the cohesive zone is in fact an eigen shear strain, because it is the "shear strain" caused by the local jump. It is given as,

$$\gamma_{xy} = \frac{\phi_x(x)}{d} = \frac{2u_x(x) + (b/2)}{d}, \quad x > 0 \tag{8.248}$$

The misfit energy for a pair of atomic planes is,

$$\Delta W = -\frac{1}{2}\int_0^{\gamma_{xy}} \sigma'_{xy}(x,0)d\gamma'_{xy} b \cdot d = \int_{-b/4}^{u_x} \sigma_{xy} du_x b \cdot d \tag{8.249}$$

The factor of half is introduced in calculating the misfit energy because it is getting shared between two planes. Note that when $u(x) = -b/4 \rightarrow \gamma_{xy} = 0$. Therefore,

$$\begin{aligned}\Delta W(x) &= \frac{\mu b^2}{2\pi d}\int_{-b/4}^{u_x} \sin\Big(\frac{4\pi u_x}{b}\Big)du_x = \frac{\mu b^3}{8\pi^2 d}\cos\Big(\frac{4\pi u_x}{b}\Big)\Big|_{-b/4}^{u_x} \\ &= \frac{\mu b^3}{8\pi^2 d}\Big(1 + \cos\Big(\frac{4\pi u_x}{b}\Big)\Big)\end{aligned} \tag{8.250}$$

Substituting,

$$u_x = -\frac{b}{2\pi}\tan^{-1}\Big(\frac{x}{r_c}\Big) \tag{8.251}$$

into the above expression, we obtain the misfit energy for a pair of atomic planes as,

$$\Delta W = \frac{\mu b^3}{8\pi^2 d}\Big(1 + \cos\Big(2\tan^{-1}(\frac{x}{r_c})\Big)\Big) = \frac{\mu b^3}{4\pi^2 d}\frac{r_c^2}{x^2 + r_c^2} \tag{8.252}$$

Let the distance of the center of the dislocation from the nearest position of symmetry be $\xi = \alpha b$, where α is a variable. Then the position of all the atoms, on the two faces of the slip plane are defined by

$$x_m = \begin{cases} 2m\dfrac{b}{2} & \text{the upper half crystal} \\ (2m-1)\dfrac{b}{2} & \text{the lower half crystal} \end{cases} \tag{8.253}$$

and $m = 0, \pm 1, \pm 2, \pm 3, \cdots$ (see Fig. 8.12).

Then the total misfit energy is the summation,

$$
\begin{aligned}
W &= \sum_{m=-\infty}^{\infty} \Delta W(2m) + \Delta W(2m-1) \\
&= \sum_{n=0,\pm 2,\pm 4}^{\cdots} \frac{\mu b^3}{8\pi^2 d} \sum_{n=-\infty}^{+\infty} \Big(1 + \cos\Big(2\tan^{-1}(\alpha + 0.5n)(\frac{b}{r_c})\Big)\Big) \\
&+ \sum_{n=\pm 1,\pm 3}^{\cdots} \frac{\mu b^3}{8\pi^2 d} \sum_{n=-\infty}^{+\infty} \Big(1 + \cos\Big(2\tan^{-1}(\alpha + 0.5n)(\frac{b}{r_c})\Big)\Big)
\end{aligned}
\tag{8.254}
$$

which can be combined into a single expression, i.e. $x = (\alpha + 0.5n)b$ and $n = 0, \pm 1, \pm 2, \ldots.$ Therefore summing up over all the atomic planes we get the total misfit energy as

$$
W = \sum_{n=-\infty}^{+\infty} f(n) = \frac{\mu b^3}{8\pi^2 d} \sum_{n=-\infty}^{+\infty} \Big(1 + \cos\Big(2\tan^{-1}(\alpha + 0.5n)(\frac{b}{r_c})\Big)\Big) \tag{8.255}
$$

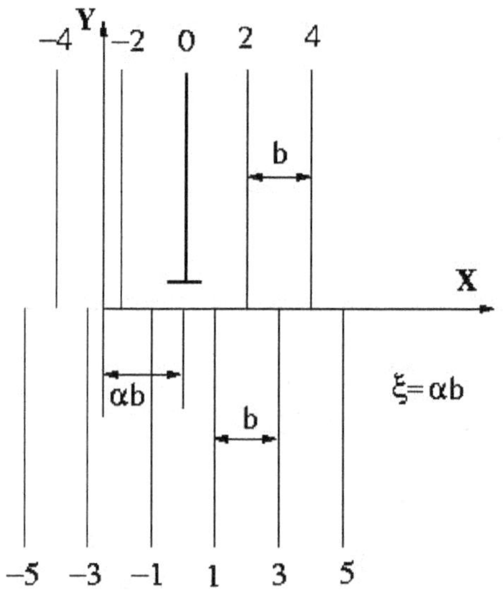

Fig. 8.12 The Nabarro counting scheme.

By using the Poisson's summation formula in harmonic analysis, we can write

$$
\sum_{n=-\infty}^{+\infty} f(n) = \sum_{n=-\infty}^{+\infty} \int_{-\infty}^{+\infty} f(x) \exp(-i2\pi x n) dx, \tag{8.256}
$$

If $f(x)$ is an even function, we can further write

$$
\sum_{n=-\infty}^{+\infty} f(n) = \sum_{n=-\infty}^{+\infty} \int_{-\infty}^{+\infty} f(x) cos(2\pi x n) dx, \tag{8.257}
$$

where we have used the fact that the function $f(n)$ is even in n. We can re-arrange the above expression as,

$$\sum_{n=-\infty}^{+\infty} f(n) = \int_{-\infty}^{+\infty} f(x)dx + 2\sum_{n=1}^{\infty}\int_{-\infty}^{+\infty} f(x)cos(2\pi xn)dx, \tag{8.258}$$

Therefore we can rewrite the total misfit energy from the equation ((8.255)) as,

$$\begin{aligned} W = &\frac{\mu b^3}{8\pi^2 d}\int_{-\infty}^{+\infty}(1+\cos(2\tan^{-1}z))dx \\ &+\frac{\mu b^3}{4\pi^2 d}\sum_{n=1}^{+\infty}\int_{-\infty}^{+\infty}(1+\cos(2\tan^{-1}z))cos\Big(2\pi n\Big(\frac{dz}{(1-\nu)b}-2\alpha\Big)\Big)dx \end{aligned} \tag{8.259}$$

where $z = (\alpha+\frac{x}{2})\frac{b}{r_c} = 2(1-\nu)(\alpha+\frac{x}{2})\frac{b}{d}$. Therefore $\frac{dz}{dx} = (1-\nu)\frac{b}{d}$ and $dx = \frac{d}{(1-\nu)b}dz$. Using these transformations and that $\cos(2\tan^{-1}z) = \frac{2}{1+z^2} - 1$, we get,

$$\begin{aligned} W = &\frac{\mu b^2}{4\pi^2(1-\nu)}\int_{-\infty}^{+\infty}\frac{1}{1+z^2}dz \\ &+\frac{\mu b^2}{2\pi^2(1-\nu)}\sum_{n=1}^{+\infty}\int_{-\infty}^{+\infty}\cos\Big(2\pi n\Big(\frac{dz}{(1-\nu)b}-2\alpha\Big)\Big)\frac{dz}{1+z^2} \end{aligned} \tag{8.260}$$

The first integral above can be calculated using the Cauchy residual theorem, that is we use the result:

$$\int_{-\infty}^{+\infty}\frac{1}{1+z^2}dz = 2\pi i Re(\frac{1}{1+z^2}) = \pi$$

where $Re(.)$ denotes the residual. Therefore the first term of the total misfit energy as $\frac{\mu b^2}{4\pi(1-\nu)}$. The second term in equation ((8.260)) can be further reduced to,

$$\frac{\mu b^2}{2\pi^2(1-\nu)}\sum_{n=1}^{+\infty}\cos(4\pi n\alpha)\int_{-\infty}^{+\infty}\cos\Big(\frac{2\pi nzd}{(1-\nu)b}\Big)\frac{dz}{1+z^2}$$

To evaluate this term we again use Cauchy residual theorem. Say $k = \frac{2\pi nd}{(1-\nu)b}$; then the integral in the above equation is equal to,

$$\int_{-\infty}^{+\infty}\frac{e^{ikz}}{1+z^2}dz$$

which is equal to πe^{-k}. Therefore we obtain the total misfit energy as,

$$W = \frac{\mu b^2}{4\pi(1-\nu)} + \frac{\mu b^2}{2\pi^2(1-\nu)}\sum_{n=1}^{+\infty}\pi e^{\frac{-4\pi r_c n}{b}}\cos(4\pi n\alpha) \tag{8.261}$$

The term in $n = 1$ dominates the sum, therefore we have,

$$W(\alpha) = \frac{\mu b^2}{4\pi(1-\nu)} + \frac{\mu b^2}{2\pi(1-\nu)}\exp\Big(-\frac{4\pi r_c}{b}\Big)\cos 4\pi\alpha \tag{8.262}$$

The corresponding force acting on dislocation is given by,

$$F = -\frac{1}{b}\frac{dW(\alpha)}{d\alpha} \tag{8.263}$$

Note that the dislocation moves a distance $-\alpha b$.

$dW(\alpha)/d\alpha$ reaches to maximum when $\sin 4\pi\alpha = 1$. From the relation that $\sigma_{xy} = F(b \times 1)$(unit thickness in z-direction), the critical shear stress to move the dislocation by one lattice site is

$$\sigma_p = \frac{2\mu}{(1-\nu)}\exp\Big(-\frac{4\pi r_c}{b}\Big) \tag{8.264}$$

where F is called the Peierls force and σ is called the Peierls stress, , which are required to move a dislocation over an energy barrier, which is also called as the Peierls barrier. The above expression of the Peierls stress was derived by Nabarro [Nabarro (1952, 1967)].

- **Joós-Dusebery formula**

A more physically realistic restoring stress is obtained if we use relative displacement (of the two half planes) instead of the lattice displacement in the above discussion. In the following, a more recent treatment of the PN model by B. Joós and M. Duesbery [Joós and Duesbery (1997)] is outlined, which considers the relative displacement instead of the independent lattice displacements in two half planes. We restrict our attention to the case of a straight edge dislocation. The new model predicts a Peierls stress which differs from the above mentioned expression by a factor of two in both the exponential and the coefficient of the exponential. This approach is also valid for the case of narrow dislocations. By $f(x)$ we define the displacement of the upper half of the crystal with respect to the lower half. If c is a constant, then $f(x-c)$ corresponds to a dislocation translated by c. For a discrete lattice this can be understood like this: if the dislocation is introduced at c, then the atomic planes at a position mb in the upper half of the crystal will experience a displacement of $f(mb-c)$ along the Burgers vector. The total misfit energy in this case can be written as:

$$W(c) = \frac{\mu b^3}{4\pi^2 d}\sum_{m=-\infty}^{+\infty}\Big(1+\cos\Big(2\tan^{-1}\big(\frac{mb-c}{r_c}\big)\Big)\Big) \tag{8.265}$$

Note the difference of a factor of $1/2$ in the expression of W from the earlier discussion. This is because we are no longer treating the two half planes independently, but we are using a relative displacement. Using further manipulations and substituting $\Gamma = r_c/b$ and $y = c/b$ we have,

$$W(y) = \frac{\mu b^2}{4\pi^2(1-\nu)}\sum_{m=-\infty}^{+\infty}\frac{\Gamma}{\Gamma^2+(m-y)^2} \tag{8.266}$$

$W(y)$ is an even periodic function of period 1. Using this information we can express the energy as the sum,

$$W(y) = \frac{a_0}{2} + \sum_{n=1}^{+\infty} a_n \cos 2\pi n y \tag{8.267}$$

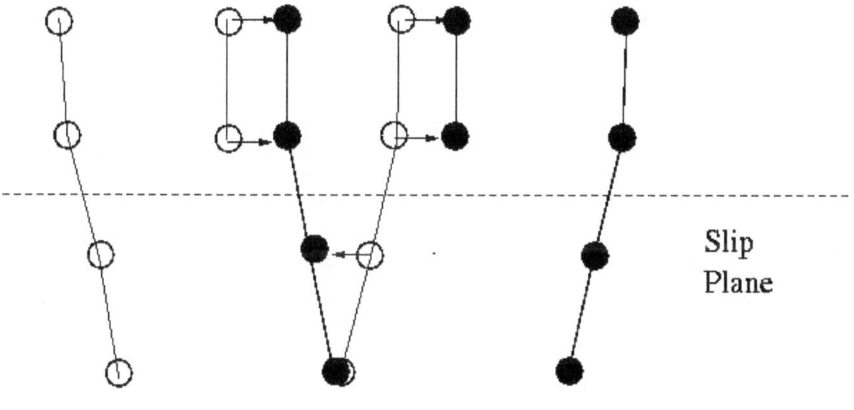

Fig. 8.13 Illustration of double counting scheme [Wang (1996)].

Where we can calculate the Fourier coefficients in the usual manner. After substituting the value of these Fourier coefficients, we get the expression for the total misfit energy as,

$$W(y) = \frac{\mu b^2}{4\pi(1-\nu)} + \frac{\mu b^2}{2\pi(1-\nu)} \sum_{n=1}^{+\infty} e^{-2\pi n \Gamma} \cos 2\pi n y \tag{8.268}$$

For the limit of wide dislocations ($\Gamma \gg 1$), only the first exponential term is kept. Then in the limit of wide dislocations we have,

$$W(c) = \frac{\mu b^2}{4\pi(1-\nu)} \Big(1 + 2\exp(\frac{-2\pi r_c}{b}) \cos \frac{2\pi c}{b}\Big) \tag{8.269}$$

From which we obtain, (using the relation $\sigma = \max\Big\{\frac{1}{b}\frac{dW}{dc}\Big\}$)

$$\sigma = \frac{\mu}{(1-\nu)} \exp\Big(-\frac{2\pi r_c}{b}\Big) \tag{8.270}$$

Note the difference between the above stress and the one obtained in Eq. (8.264).

- **Wang's double counting formula**

In the P-N model, when the center of the dislocation is translated by αb, all atoms both below and above the slip plane are assumed to be displaced by an amount αb at the same direction. However, Wang [Wang (1996)] found that during the actual translation of the dislocation line, the atoms above the slip plane move $\frac{\alpha b}{2}$ and the atoms below the slip plane move at the opposite direction by $\frac{\alpha b}{2}$. This is illustrated in Fig. 8.13, in which the previous dislocation position is marked by a set of "hollow atoms", and the translated dislocation position is marked by a set of "dark atoms". One can observe that as the dislocation line moves, the atoms above the slip plane move towards right, while the atoms below the slip plane move towards left. That is

$$x_n = \begin{cases} nb + \dfrac{\alpha b}{2} & y > 0 \\ & \qquad n = 0, \pm 1, \pm 2, \cdots\cdots \\ nb + \dfrac{b}{2} - \dfrac{\alpha b}{2} & y < 0 \end{cases} \tag{8.271}$$

Considering Eq. (8.252), we calculate the two summations that relates to the misfit energies in the top and the bottom half-crystals,

$$\Sigma_T = \sum_{-\infty}^{+\infty} \frac{1}{x_n^2 + r_c^2} = \sum_{\infty}^{\infty} \frac{1}{(nb + \dfrac{\alpha b}{2})^2} \tag{8.272}$$

$$\Sigma_B = \sum_{-\infty}^{+\infty} \frac{1}{x_n^2 + r_c^2} = \sum_{\infty}^{\infty} \frac{1}{(nb + \dfrac{b}{2} - \dfrac{\alpha b}{2})^2} \tag{8.273}$$

According to [Hansen (1975); Bromwich (1949)], the above summations have closed-form expressions,

$$\Sigma_T = \frac{\pi}{br_c} \frac{\sinh \dfrac{2\pi r_c}{b}}{\cosh \dfrac{2\pi r_c}{b} - \cos \pi\alpha} \tag{8.274}$$

$$\Sigma_B = \frac{\pi}{br_c} \frac{\sinh \dfrac{2\pi r_c}{b}}{\cosh \dfrac{2\pi r_c}{b} + \cos \pi\alpha} \tag{8.275}$$

Hence the total misfit energy will be

$$W(\alpha) = \sum_{-\infty}^{+\infty} \Delta W = \frac{Gb^3 r_c^2}{4\pi^2 d}(\Sigma_T + \Sigma_B) = \frac{\mu b^2}{4\pi} \frac{\sinh \dfrac{4\pi r_c}{b}}{\cosh^2 \dfrac{2\pi r_c}{b} - \cos^2 \pi\alpha} \tag{8.276}$$

The lattice friction is

$$\sigma_f = \frac{G}{8(1-\nu)} \frac{\sinh \dfrac{4\pi r_c}{b} \sin 2\pi\alpha}{(\cosh^2 \dfrac{2\pi r_c}{b} - \cos^2 \pi\alpha)^2} \tag{8.277}$$

At $\alpha = 1/4(x = b/4)$, the lattice friction reaches to its maximum, which is the Peierls stress,

$$\sigma_p = \frac{\mu}{1-\nu} \exp\left(-\frac{4\pi r_c}{b}\right) \tag{8.278}$$

The three different Peierls stress formulas are tabulated in Table 8.1.

[Nabarro (1967)]	[Joós and Duesbery (1997)]	[Wang (1996)]
$\dfrac{2\mu}{1-\nu}\exp\Big(-\dfrac{4\pi r_c}{b}\Big)$	$\dfrac{\mu}{1-\nu}\exp\Big(-\dfrac{2\pi r_c}{b}\Big)$	$\dfrac{\mu}{1-\nu}\exp\Big(-\dfrac{4\pi r_c}{b}\Big)$

Fig. 8.14 R. Peierls (left) and C. N. Yang (right) are in discussion at a Conference, 1969 (With the permission of the Princeton University Press).

8.7.4 *Variable core model*

In Eq. (8.247), the shear stress along the gliding plane is given as

$$\sigma_{xy}(x,0) = \frac{\mu b}{2\pi(1-\nu)}\frac{x}{x^2+r_c^2} \tag{8.279}$$

where r_c is the radius of the dislocation core. This suggests that the Peierls-Nabarro dislocation may be viewed as a Volterra dislocation with a variable Burgers' vector $b'(x)$ such that

$$\int_{-\infty}^{\infty} b'(x')dx' = b, \quad \text{and} \quad \mathcal{H}(b'(x)) = \frac{1}{\pi}\int_{-\infty}^{\infty}\frac{b'(\xi)}{x-\xi}d\xi = \frac{bx}{\pi(x^2+r_c^2)} \tag{8.280}$$

Based on (8.224), we find that

$$b'(x) = \frac{b}{\pi}\frac{r_c}{x^2+r_c^2} \tag{8.281}$$

The corresponding slip discontinuity along x-axis is

$$\phi(x) = \int_0^x b'(\xi)d\xi = \frac{b}{\pi}\frac{r_c}{x^2+r_c^2} \tag{8.282}$$

Recently, Lubarda and Markenscoff [Lubarda and Markenscoff (2006, 2007)] showed that the Peierls model may be viewed as a variable core model of the Volterra dislocation. For instance, one may superpose two semi-infinite dislocation walls, or two disclinations with distributed dislocation density $1/r_c$,

$$\sigma_{xy}(x,y) = \frac{\mu b}{2\pi(1-\nu)}\frac{x}{r_c}\Big[\frac{y}{x^2+y^2} - \frac{y-r_c}{x^2+(y-r_c)^2}\Big] \tag{8.283}$$

such that we can recover Eq. (8.279) on the glide plane ($y = 0$). Note that the core radius r_c here has different relation to lattice constants, unlike in the original Peierls model $r_c \neq \dfrac{d}{2(1-\nu)}$.

In fact, one can find the Airy stress function of the distributed Volterra dislocation by carrying out the following integration [Hirth and Lothe (1992)],

$$\begin{aligned}\Phi &= \frac{\mu}{2\pi(1-\nu)}\int_{-\infty}^{\infty} b'(x')y\ln\Big[(x-x')^2+y^2\Big]^{1/2}dx' \\ &= \frac{\mu b r_c y}{4\pi^2(1-\nu)}\int_{-\infty}^{\infty}\frac{ln\Big[(x-x')^2+y^2\Big]}{x'^2+r_c^2}dx' \end{aligned} \tag{8.284}$$

Its solution is

$$\Phi(x,y) = -\frac{\mu b}{4\pi(1-\nu)}y\ln\Big[x^2+\Big(y+sgn(y)r_c\Big)^2\Big] \tag{8.285}$$

Let $\zeta = sgn(y)r_c$. The complete stress distribution of the Peierls dislocation can be then found by differentiation of Airy stress function [Hirth and Lothe (1992); Lubarda and Markenscoff (2007)],

$$\sigma_{xy}(x,y) = \frac{\mu b}{2\pi(1-\nu)}\left\{\frac{x}{x^2+(y+\zeta)^2}-\frac{2xy(y+\zeta)}{[x^2+(y+\zeta)^2]^2}\right\} \tag{8.286}$$

$$\sigma_{xx}(x,y) = -\frac{\mu b}{2\pi(1-\nu)}\left\{\frac{(y+2\zeta)}{x^2+(y+\zeta)^2}+\frac{2x^2y}{[x^2+(y+\zeta)^2]^2}\right\} \tag{8.287}$$

$$\sigma_{yy}(x,y) = -\frac{\mu b}{2\pi(1-\nu)}\left\{\frac{y}{x^2+(y+\zeta)^2}-\frac{2x^2y}{[x^2+(y+\zeta)^2]^2}\right\} \tag{8.288}$$

and the displacement fields can be found as

$$u(x,y) = \frac{b}{2\pi}\Big(\tan^{-1}\frac{y+\zeta}{x}\mp\frac{\pi}{2}sgn(x)\Big)+\frac{b}{4\pi(1-\nu)}\frac{xy}{x^2+(y+\zeta)^2} \tag{8.289}$$

$$v(x,y) = -\frac{b(1-2\nu)}{8\pi(1-\nu)}\ln\frac{x^2+(y+\zeta)^2}{b^2}+\frac{b}{4\pi(1-\nu)}\frac{y(y+\zeta)}{x^2+(y+\zeta)^2} \tag{8.290}$$

The total strain energy of the variable core model can be calculated with the core energy evaluated in a large radius R along the glide plane,

$$E = \frac{1}{2}\int_{-R}^{R}\sigma_{xy}(x,0)\phi(x)dx + E_R,\quad E_R = \frac{\mu b^2}{8\pi(1-\nu)}, \tag{8.291}$$

which is

$$E = \frac{\mu b^2}{4\pi(1-\nu)}\ln\frac{e^{1/2}R}{2\rho} \tag{8.292}$$

Lubarda and Markenscoff [Lubarda and Markenscoff (2006, 2007)] assumed that the core radius is a periodic function of the dislocation glide distance Δ (see Fig. 8.15),

$$\rho(\Delta) = \frac{1}{2}(\rho_0+\rho_*)+\frac{1}{2}(\rho_0-\rho_*)\cos\frac{k\pi\Delta}{b},\quad k=2,4 \tag{8.293}$$

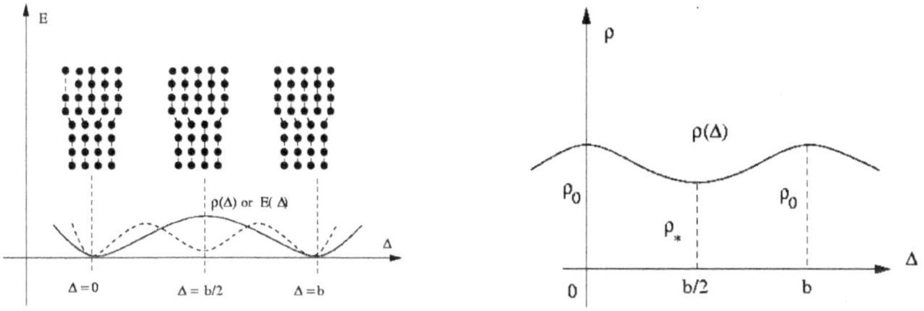

(a) (b)

Fig. 8.15 (a) An edge dislocation gliding within distance $0 \leq \Delta \leq b$. Three consecutive equilibrium configurations are shown with periodic variation of the core energy $E(\Delta)$; (b) Corresponding periodic variation of the core radius $\rho(\Delta)$ (After [Weertman and Weertman (1992); Lubarda and Markenscoff (2007)]).

with a period $2b/k, k = 2, 4$. Lubarda and Markenscoff assumed that the stable configuration has larger core size than that of the unstable equilibrium configuration, i.e. $\rho_0 > \rho_*$; this makes sense because $E_* > E_0$, see Eq. (8.292). This assumption is agreed with the analysis by the Weertmans see discussion in [Weertman and Weertman (1992)] pages 147-148.

One can then calculate the Peach-Koehler force following Eshelby's definition of material force,

$$F(\Delta) = \frac{dE}{d\Delta} = \frac{dE}{d\rho}\frac{d\rho}{d\Delta} = -\frac{\mu b^2}{4\pi(1-\nu)}\frac{1}{\rho}\frac{d\rho}{d\Delta} \tag{8.294}$$

From Eq. (8.293), we find that

$$\frac{d\rho}{d\Delta} = -\frac{k\pi}{2b}(\rho_0 - \rho_*)\sin\frac{k\pi\Delta}{b} \tag{8.295}$$

and hence

$$F(\Delta) = \frac{k\mu b}{8\pi(1-\nu)}\frac{\rho_0 - \rho_*}{\rho}\sin\frac{k\pi\Delta}{b} \tag{8.296}$$

To find the maximum value of $F(\Delta)$, we consider the stationary value,

$$\frac{dF}{d\Delta} = 0 \quad \Rightarrow \quad \sin\frac{k\pi\Delta}{b} = 2\sqrt{\frac{\rho_0\rho_*}{\rho_0 + \rho_*}} \tag{8.297}$$

The corresponding core radius is

$$\rho = 2\frac{\rho_0\rho_*}{\rho_0 + \rho_*}$$

and the maximum force acting on the Peierls-Nabarro dislocation is

$$F_{max} = \frac{k\mu b}{8(1-\nu)}\frac{\rho_0 - \rho_*}{\rho_0}\sqrt{\frac{\rho_0}{\rho_*}} \tag{8.298}$$

which is the critical Peach-Koehler force required to move a dislocation in a perfect crystal. If we choose $k = 2$ and

$$\frac{\rho_*}{\rho_0} = 1 - k_1 \exp\left(\frac{-k_2 \pi \rho_0}{b}\right) \tag{8.299}$$

One can find the Peierls stress as

$$\tau_{PS} = \frac{F_{max}}{b} \approx \frac{\mu k_1}{4(1-\nu)} \exp\left(\frac{-k_2 \pi \rho_0}{b}\right) , \tag{8.300}$$

which recovers the value of the classical Peierls stress, if we choose $k_1 = 8$ and $k_2 = 4$.

8.7.5 *Story of the Peierls-Nabarro model*

1. Account by E. Orowan

The following is an account on the discovery of Peierls-Nabarro model, which was given by the late Professor, Egon Orowan , of Massachusetts Institute of Technology, who was a well-known physicist and material scientist.

> *1937 I was invited to work at the University of Birmingham, in the Physics Department which had just taken over by M. L. E. Oliphant (now Sir Mark Oliphant). I felt that it would be urgent to know the width of the dislocation belt and the stress required to move it. The simplest assumption about this was the one made by Taylor, that the stress was zero; however, the extremely high yield stress of many hard materials such as diamond (which could be remarkably free from imperfections and thus could not contain too many dislocations) indicated that the most frequent cause of the hardness of crystalline materials was the high shear stress required to move a dislocation. I found that the width of the dislocation and the stress for moving it could be calculated, with a crude approximation, simply enough by assuming that the shearing force between the opposite shores of the slip plane in a dislocation was a sine function of the relative shear displacement (the initial tangent of the sine, of course, was given by the elastic modulus).*
>
> *One the other hand, displacement and shear traction at the surface of a half-space were connected by the equations of Boussinesq; equating the stresses and displacements of the sine approximation with those of Boussinesq led to an integral equation which was the solution of the problem It would have taken me days or weeks of study to solve it; fortunately I was a daily guest in the hospitable house of the brilliant theoretical physicist Rudolf Peierls. He solved the equation, if I remember well, within a few hours, and he also drove me to a conference at Bristol University in 1939 where I gave a paper and he gave another on the problem he had just solved.*
>
> *The calculation of the width of the dislocation and of the Peierls-Nabarro stress required for moving it was repeated and improved by Nabarro in 1947. The result was puzzling at first: the width calculated by Nabarro amounted to a few atomic spacings while Peierls obtains an order of magnitude of thousands of spacings. After some research in Birmingham and in Cambridge (where I was at the time) I discovered the sheet with Peierls's calculations in my desk; Peierls checked it and found that a factor of 2π was accidentally omitted in an exponent, which amounted to a factor of about 1000 in the result.*

Of course, the calculation with the sinusoidal approximation is useless in most interesting cases of directional bonds, in transition metals and the hard non-metallic crystals.

Excerpt from "The Peierls-Nabarro Force" from *The Sorby Centennial Symposium on the History of Metallurgy*, MSC, Vol. 27, 1963, pages 368-369. Reproduced with the permission from the Minerals, Metals & Material Society.

2. Account by R. Peierls:

I am glad to have this opportunity of making a statement that may help to set the record straight, though the correct facts have already been set out by Orowan.

I did not have in 1939, nor do I have today, any close knowledge or any deep understanding of the problems of dislocations. In 1939 Orowan asked me for help in the formulation and solution of an approximate model for a dislocation. He had a clear picture of the approximation to be used; namely, to assume that the interaction of the plane of atoms on either side of the slip plane with the adjacent half-space is given by the equations for an elastic continuum, and that the force across the slip plane is dominated by a sinusoidal term. The derivation of the integral equation resulting from this model is straightforward, and I was greatly surprised to find that the simplest function with the expected qualitative behavior turned out to be an exact solution of this integral equation. Orowan says that it would have taken him "days or weeks" to study this problem, and this may be a generous estimate; in any event there is no doubt he could have found the solution without difficulty.

When the result of the calculation which had been reported to a conference in Bristol, was published, I would have preferred to have this appear as a joint paper with Orowan, or perhaps as an appendix to a paper by him. However, he was not willing to agree to this, and at the time the matter did not seem of great importance. If I foreseen the attention this paper would receive, and the extent to which it would be quoted even today (which I did not fully realize until I heard some of the lectures at the Battelle Symposium), I would probably have pressed the point more strongly.

In 1947, Nabarro generalized, and for the purpose rederived the formula. It was then discovered that my calculation contained an error of a factor 2. Orowan mentions a factor 2π*, but in the interest of history accuracy, I must point out that this is an exaggeration. Actually, this error occurs in a large exponent, so that even factor 2 changes the magnitude of the critical stress by several orders of magnitude. It is a sobering thought that it could easily have been a factor* 2π*. Perhaps it is as well, in view of this, that the paper was published under my name, so that the responsibility for this slip can be correctly assigned; indeed this factor 2 would seem to be my only really original contribution to the subject.*

It would evidently be much more satisfactory if this force was known as the "Orowan-Nabarro Force." However, from one's general experience of the way in which the use of names becomes immutable one cannot feel very optimistic about the prospects of this happening.

Excerpt from *Commentary on "Peierls-Nabarro Force"* by R. Peierls in *Dislocation Dynamics*, Edited by A. R. Rosendfield, G. T. Hahn, A. L. Bement, Jr., and R. J. Jaffee, pages xiii-xiv, Battelle Institute Materials Science Colloquia, [1967], McGraw-Hill Book Co.

8.8 Dislocations in epitaxial thin films

The thin film is one of the most important nanoscale structures in modern technology. It is the primary configuration for integrated circuits, computer memories, e.g. RAM (the random access memory), micro-electrical mechanical sensors, and other nanoscale devices. Hence, the study of the mechanical, chemical, and electrical properties of the thin films has particular significance in nano-technologies.

The ancient Greek word $\epsilon\pi\iota$ (*epi*–placed or resting upon) and the word $\tau\alpha\xi\iota\zeta$ (*taxis* – arrangement) are the root of the modern word *epitaxy*, which describes an extremely important phenomenon exhibited by thin films. Epitaxy refers to a single-crystal film formation on top of a crystalline substrate and both have exactly the same crystal structure as the thin film. 90 % of thin films used in semi-conductor and computer industry, communication industry, and sensor and information industry are epitaxial thin films. Growing various defect-free epitaxial thin films has been the main challenge in semi-conductor industry in the past half century. In this section, we shall introduce the two basic dislocation models in thin-film mechanics.

8.8.1 *Frenkel & Kontorova and Frank & van der Merwe models*

One of the early atomistic model for dislocations is the Frenkel & Kontorova (FK) dislocation model [Frenkel and Kontorova (1938)], which was proposed in 1937. Since then, the FK model has become the prototype model to describe systems in which two competition between two incommensurate lengths determines the ground state energy. FK model has been applied to model a variety of physical phenomena including dislocation dynamics, nonlinear models for the DNA dynamics, Josephson Junctions, interfacial slip, surface and absorbed atomic layers, models for magnetic chains and hydrogen-bonded chains, among many others [Braun and Kivshar (2004)].

In the context of modeling dislocations, this model was studied in detailed by Frank and van der Merwe [Frank and van der Merwe (1949a,b)], and they applied it to study dislocation in or between an epitaxial thin film and a substrate. When studying misfit dislocations in a thin film, this model is also called FvdM theory.

In Frenkel & Kontorova model, the thin film is modeled as one dimensional monolayer spring with lattice spacing a_f, and the substrate is modeled as large slab with lattice spacing a_s, and $a_s \neq a_f$ and the lattice misfit is $\Delta = a_f - a_s$ (see Fig. 8.17).

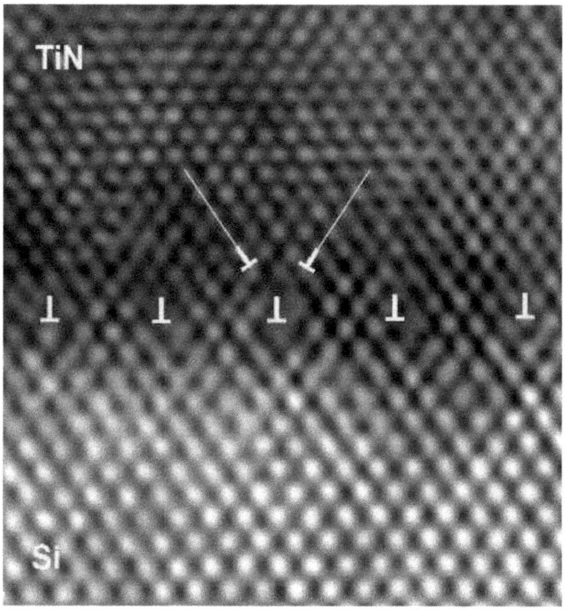

Fig. 8.16 Dislocation in an epitaxial TiN thin film on Si substrate (TEM image is kindly provided by Dr. Haiyan Wang of Texas A&M University).

The row of atoms in the thin film are under combined influence of harmonic forces between the nearest neighbors in the monolayer and non-linear interaction forces from substrate. Since the substrate is assumed much larger in dimension than the thin film, it is assumed to be rigid. The interaction between the thin film and substrate, or the force exerted on the thin film by the substrate is characterized by a sinusoidal potential with the amplitude $\frac{1}{2}W$ (see Fig. 8.17).

Mark the position (the open circle in Fig. 8.17) of the m-th atoms in the unstrained monolayer as

$$X_m = ma_f, \quad m = 0, \pm 1, \pm 2, \cdots \tag{8.301}$$

After attach the thin film onto the substrate, the thin film will be stretched to the position

$$x^i_m = ma_s = X_m + u^{mis}_m, \quad m = 0, \pm 1, \pm 2, \cdots . \tag{8.302}$$

where x^r_m is denoted as the spatial position of the m-th atom with respect to the substrate, and u^{mis}_m is the displacement of the atom due to the lattice misfit,

$$u^{mis}_m = m(a_s - a_f) .$$

During deformation, the actual spatial position of the m-th atom is

$$x_m = X_m + u^{mis}_m + u^s_m \tag{8.303}$$

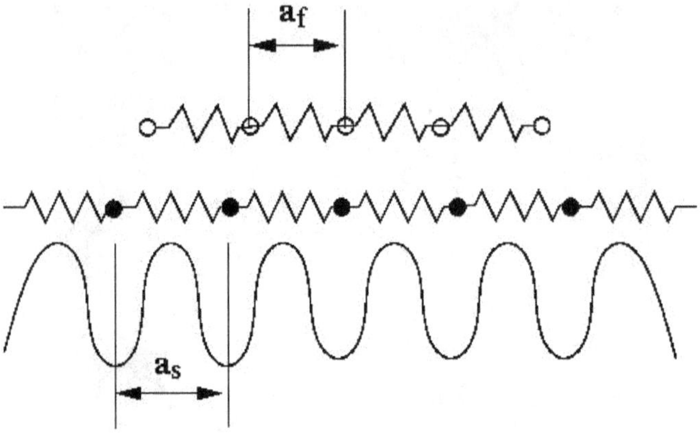

Fig. 8.17 Frank-van der Merwe dislocation thin film model.

or

$$u_m = x_m - x_m = u_m^{mis} + u_m^s \tag{8.304}$$

where u_m^s is the displacement of an atom measured with respect to substrate reference.

The total relative displacement between the two atoms is now

$$u_{m+1} - u_m = (u_{m+1}^s - u_m^s) - (a_f - a_s)\ . \tag{8.305}$$

The total potential energy of the system is

$$\Pi = \frac{1}{2}\sum_m \left\{ \mu(u_{m+1}^s - u_m^s - (a_f - a_s))^2 + W[1 - \cos\frac{2\pi u_m^s}{a}] \right\} \tag{8.306}$$

Let,

$$\xi_m = \frac{u_m^s}{a_s},\ \text{and}\ f = \frac{a_f - a_s}{a_s}\quad . \tag{8.307}$$

Hence

$$\Pi = \frac{1}{2}\sum_m \{\mu a^2(\xi_{m+1} - \xi_m - f)^2 + W[1 - \cos(2\pi\zeta_m)]\} \tag{8.308}$$

The equilibrium equation is derived from the stationary condition

$$\frac{dE}{d\xi_n} = 0,\quad n = 0, \pm 1, \pm 2, \cdots \Rightarrow$$

$$-\mu a^2(\xi_{n+1} - \xi_n + f) + \mu a^2(\xi_n - \xi_{n-1} + f) + W\pi \sin 2\pi\xi_n = 0\ , \tag{8.309}$$

i.e.

$$\Delta_n^2 \xi = (\xi_{n+1} - 2\xi_n + \xi_{n-1}) = \frac{\pi}{2\ell_0^2}\sin 2\pi\xi_n \tag{8.310}$$

where $\ell_0 = \sqrt{\mu a^2/2W}$.

The dynamics version of Eq. (8.310) is the finite-difference sine-Gordon equation,

$$\Delta_n^2 \xi - \frac{m_n}{\mu}\frac{d^2\xi_n}{dt^2} = \frac{\pi}{2\ell_0^2}\sin 2\pi\xi_n \tag{8.311}$$

If $\ell_0 >> 1$, one may use continuous approximation to replace the finite difference equation with a differential equation,

$$\Delta_n^2 \xi = \frac{d^2\xi_n}{dX_n^2}a_f^2 + \frac{2}{4!}\frac{d^4\xi}{dX_n^4}a_f^4 + \mathcal{O}(a_f^6) = \frac{d^2\xi}{dn^2} + \mathcal{O}(a_f^4) \tag{8.312}$$

Therefore, if we only consider static deformation, we have the following non-linear ordinary differential equation

$$\frac{d^2\xi}{dn^2} = \frac{\pi}{2\ell_0^2}\sin 2\pi\xi \ . \tag{8.313}$$

Consider the following boundary conditions,

$$\frac{d\xi}{dn}\Big|_{n=n_0} = \epsilon, \quad \text{and} \quad \xi\Big|_{n=n_0} = 0 \ . \tag{8.314}$$

One can integrate (8.310),

$$\left(\frac{d\xi}{dn}\right)^2 - \epsilon^2 = \frac{1}{2\ell_0^2}(1-\cos 2\pi\xi) \ , \tag{8.315}$$

which can be re-arranged as

$$\left(\frac{d\xi}{dn}\right)^2 = \frac{(1+\ell_0^2\epsilon^2)}{\ell_0^2}\left(1 - \frac{\cos^2\pi\xi}{1+\ell_0^2\epsilon^2}\right) \tag{8.316}$$

Changing of variable

$$\phi = \pi\xi - \frac{\pi}{2} \quad \text{and} \quad k = (1+\ell_0^2\epsilon^2)^{-1/2} \ . \tag{8.317}$$

one may transfer into the standard form of differential equations that can be solved by using the elliptic function and its integral,

$$\frac{d\phi}{dn} = \pm\left(\frac{\pi}{\ell_0 k}\right)(1-k^2\sin^2\phi)^{1/2} \tag{8.318}$$

Solutions of FK model:

1. Consider boundary condition

$$\epsilon = 0, \quad \text{and} \quad k = 1. \tag{8.319}$$

In this case, Eq. (8.316) is simplified to

$$\frac{d\xi}{dn} = \frac{1}{\ell_0}\sin\pi\xi \tag{8.320}$$

Assume at $n = 0$, $\xi(0) = 0.5$, and then

$$\frac{\pi}{\ell_0}\int_0^n dp = \pi\int_0^\xi \frac{d\zeta}{\sin\pi\zeta} \tag{8.321}$$

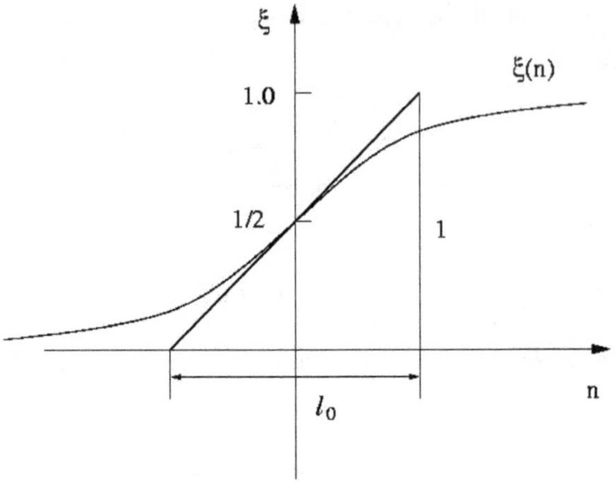

Fig. 8.18 A single dislocation solution of FKV model.

which yields the solution

$$\frac{\pi n}{\ell_0} = \ln\Big[\tan\Big(\frac{\pi\xi}{2}\Big)\Big] \tag{8.322}$$

Inversely,

$$\xi = \frac{2}{\pi}\tan^{-1}\Big[\exp\Big(\frac{\pi n}{\ell_0}\Big)\Big] \tag{8.323}$$

This solution represents a single dislocation far away from the remote boundary. We plot the positive solution in Fig. 8.18. One may find that at $\xi = 1/2$,

$$\frac{d\xi}{dn} = \frac{1}{\ell_0} \tag{8.324}$$

Since a unit change of ξ means a relative displacement in one lattice spacing a_s, Eq. (8.324) then implies that in a region of length ℓ_0 the number of troughs is one more than the number of atoms, i.e. there is extra plane of atoms in the substrate, which forms an edge dislocation. We call ℓ_0 as the effective length of the dislocation region.

2. General solution

The general static solution of sine-Gordon equation can be expressed by the elliptic function,

$$\Big(\frac{\pi}{\ell_0 k}\Big) = \int_0^{\phi}(1 - k^2\sin^2\psi)^{-1/2}d\psi = F(\phi, k) \tag{8.325}$$

where the upper limit ϕ is called the amplitude. The inverse relation of the above elliptic function is

$$\phi = am\Big(\frac{\pi n}{\ell_0 k}\Big) \tag{8.326}$$

or

$$\xi = \frac{1}{2} + \frac{1}{\pi} am\Big(\frac{\pi n}{\ell_0 k}\Big) \tag{8.327}$$

and

$$\frac{d\xi}{dn} = \frac{1}{\ell_0 k} dn\Big(\frac{\pi n}{\ell_0 k}\Big) = \frac{1}{\ell_0 k}(1 - k^2 \cos^2 \pi\xi)^{1/2} \tag{8.328}$$

At $\xi = \xi(0) = 1/2$,

$$\frac{d\xi}{dn} = \frac{1}{\ell_0 k} \tag{8.329}$$

i.e. $\ell_0 k$ is now the effective dislocation length.

Assume that $\xi(p) = 1.5$ and hence $\Delta\xi = \xi(p) - \xi(0) = 1$. The general solution of FK model is depicted on Fig. 8.19. Because $\Delta\xi/\Delta n = 1/p$, the number of atoms

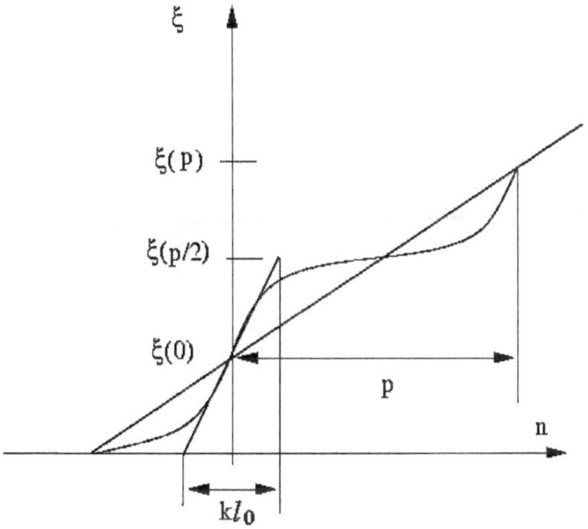

Fig. 8.19 The general solution of static sine-Gordon equation ($\frac{d\xi}{dn} \geq 0$).

per dislocation is

$$p = \frac{2\ell_0 k E(k)}{\pi} \tag{8.330}$$

where $E(k)$ is the following elliptic integral [Gradshteyn and Ryzhik (1994)],

$$E(k) = \int_0^{\pi/2} (1 - k^2 \sin^2 \psi)^{1/2} d\psi \tag{8.331}$$

The general solution indicates that there are many dislocations occurring simultaneously along the chain in a periodic fashion. In Fig. 8.20, we show the dislocation pattern created by the general solution. It would be interesting to examine the

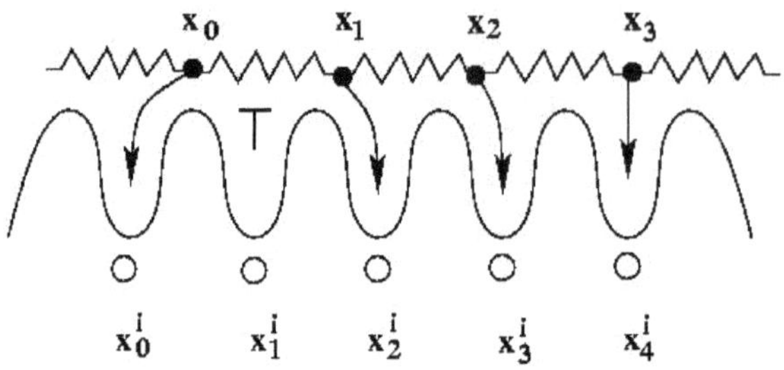

Fig. 8.20 Dislocation pattern for $p = 3$.

stability of Frenkel-Kontorova system. The potential energy of one dislocation is

$$\begin{aligned}\Pi &= W\ell_0^2\sum_{n=0}^{p-1}(\xi_{n+1}-\xi_n-f)^2+\frac{W}{2}\sum_{n=0}^{p-1}\Big(1-\cos 2\pi\xi_n\Big)\\ &= W\ell_0^2\int_0^p\Big(\frac{d\xi}{dn}-f\Big)^2dn+W\int_0^p\sin^2\pi\xi dn \end{aligned}\tag{8.332}$$

Considering

$$\frac{d\xi}{dn}=\frac{1}{\ell_0^2k^2}(1-k^2\cos^2\pi\xi)^{1/2}\tag{8.333}$$

one can write the potential energy per dislocation as

$$\Pi = W\ell_0^2\left\{\frac{4E(k)}{\pi k\ell_0}-\frac{2(1-k^2)K(k)}{\pi k\ell_0}-2f+pf^2\right\}\tag{8.334}$$

where

$$K(k)=\int_0^{\pi/2}(1-k^2\sin^2\psi)^{-1/2}d\psi$$

One may find that the potential energy consists of the contributions from both lattice misfit and dislocation misfit.

To examine the stability, let,

$$\frac{\partial\Pi}{\partial f}=W\ell_0^2(2-2pf)=0\ .\tag{8.335}$$

We find the critical lattice misfit,

$$f_{cr}=\frac{1}{p}=\frac{\pi}{2\ell_0 kK(k)}\tag{8.336}$$

When $k=1$,

$$f_{cr}=\frac{1}{p}=\frac{2}{\pi}\Big(\frac{W}{\mu a^2/2}\Big)^{1/2}\tag{8.337}$$

When lattice misfit $f > f_{cr}$, dislocations will spontaneously enter or depart from the monolayer chain.

8.8.2 *Matthews & Blackeslee's equilibrium theory*

Starting from 1974, Matthews and Blackeslee published a series of papers [Matthews and Blakeslee (1974, 1975, 1976)] , in which they proposed their equilibrium theory of dislocation relaxation mechanism for thin film growth. It was an immediate success, and it was soon received widespread attentions. Today, the Matthews-Blackeslee theory has become the fundamental theory for epitaxial thin film growth in semi-conductor industry. After Matthews & Blackeslee's contribution, many mechanicians have joined the research in the field e.g. [Freund (1987, 1990); Willis *et al.* (1990); Jain *et al.* (1992); Gosling and Willis (1993); Freund and Nix (1996); Freund and Suresh (2003)] . Today the Matthews-Blackeslee theory is often viewed as one of the corner stones of nano-mechanics.

In the following, we present a simple version of the Matthews-Blackeslee theory based on an award winning acceptance article by Professor W. D. Nix in 1989 [Nix (1989)].

Assume that the thin film is under homogeneous bi-axial plane stress load, i.e. in the film, $\epsilon_x = \epsilon_y = \epsilon$ and $\sigma_x = \sigma_y = \dfrac{E}{1-\nu}\epsilon$. The homogeneous misfit strain is due to the lattice misfit, i.e.

$$\epsilon = \frac{a_s - a_f}{a_f} \quad \text{or} \quad \epsilon = \frac{a_s - a_f}{a_s} \; . \tag{8.338}$$

The deformation of the substrate may be neglected. For a coherent thin film-substrate system, the strain energy per unit thin film area is (see Fig. 8.21)

$$E = \frac{2\mu(1+\nu)}{(1-\nu)}\epsilon^2 h = M\epsilon^2 h \; . \tag{8.339}$$

When the lattice misfit ϵ increases, it is energetically favorable to have dislocations

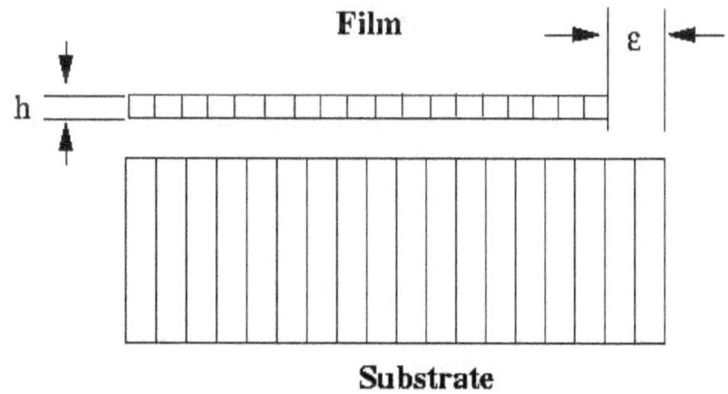

Fig. 8.21 Matthews & Blackeslee model (a).

present to relax the lattice misfit strain.

Consider a simplest scenario that there is periodically distributed edge dislocations distributed along the interface between the thin film and the substrate. The homogeneous distributed lattice misfit strain will be reduced to $f - b/S$ where

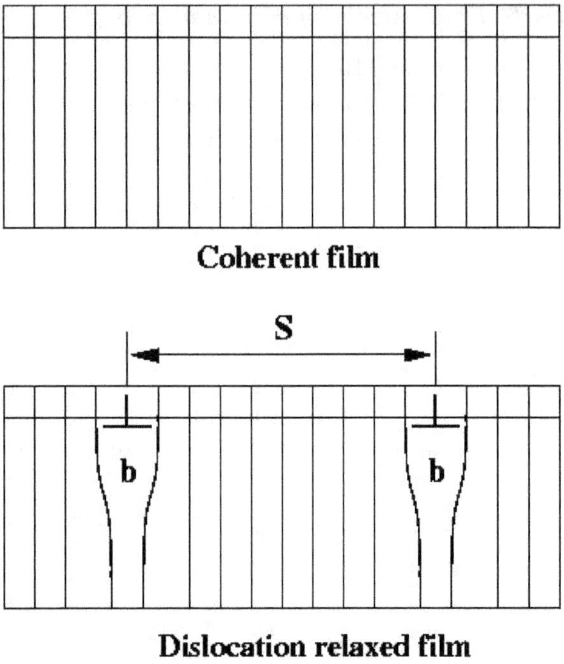

Fig. 8.22 Matthews & Blackeslee model (b).

S is the spacing between two edges dislocations. Then the elastic energy due to homogeneous deformation is

$$E_h = M\Big(\epsilon - \frac{b}{S}\Big)^2 h \tag{8.340}$$

Since there are two edge dislocations in an area $S \times 1$, the strain energy due to dislocation is

$$E_d = \frac{\mu b^2}{4\pi(1-\nu)} \ln\Big(\frac{\beta h}{b}\Big)\frac{2}{S} \tag{8.341}$$

The total energy is the summation of E_h and E_d,

$$E = M\Big(\epsilon - \frac{b}{S}\Big)^2 h + \frac{\mu b^2}{4\pi(1-\nu)} \ln\Big(\frac{\beta h}{b}\Big)\frac{2}{S} \tag{8.342}$$

The two competing effects will yield an equilibrium point at the bottom of energy well as shown in Fig. 8.23. We are seeking to find an equilibrium state that is defect-free, i.e. we are interested in an equilibrium state at which $b/S = 0$.

Consider the stationary condition,

$$\frac{\partial E}{\partial \frac{1}{S}} = -2Mh\Big(\epsilon - \frac{b}{S}\Big)b + \frac{\mu b^2}{2\pi(1-\nu)} \ln\Big(\frac{\beta h}{b}\Big)\Big|_{h=h_{cr}} = 0\ . \tag{8.343}$$

We can find a critical thickness h_{cr} of the thin film, below which the thin film will stay in a coherent state with the substrate. In other words, below such thickness, the thin film can be defect-free.

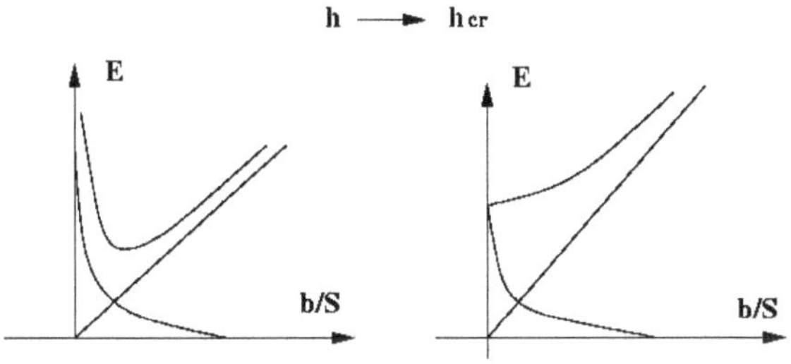

Fig. 8.23 Matthews & Blackeslee model (c).

From (8.343), one can find that the critical thickness can be determined from the following non-linear equation,

$$\frac{h_{cr}}{\ln\left(\dfrac{\beta h_{cr}}{b}\right)} = \frac{\mu b}{4\pi(1-\nu)M\epsilon} \tag{8.344}$$

8.8.3 *Mobility of screw dislocations in a thin film*

In a study of mechanical properties of thin films, Lee and Li [Lee and Li (2007, 2008)] proposed a half-space Peierls-Nabarro (HSPN) model to describe the mobility of dislocations along a thin film/substrate (half-space) interface. In thin film configuration, the dislocation is subjected to an image force due to the free surface, and there is interaction between the dislocation and the free surface, which will affect the mobility of dislocations.

Original P-N model is formulated for an interface between two half-spaces, it is only applicable for dislocation motions in bulk materials. The interaction between a free surface and a dislocation, which is important for the thin film configuration, is however incompatible with the original P-N model. It is believed that, the image force due to the free surface will interact with the dislocation in a thin film, which in turn alters the misfit energy landscape, and cause variance in core energy built-up. This important effect, to the best of authors' knowledge, has not been adequately taken into account in most dislocation mobility estimates in thin films, and most of the dislocation mobility estimates in thin films are still based on the original P-N model.

We consider a screw dislocation with a Burgers vector, $\mathbf{b} = b\mathbf{e}_z$, in the direction [001] of a crystal, and it is embedded in the interface between a layer of thin film and a substrate (see 8.24) with the same elastic shear modulus μ. Because of the free surface at $y = h$, the dislocation induced displacement field is different from that of an unbounded domain, such that the antisymmetric property of u_z ($u_z^+ = -u_z^-$), which is usually assumed for the case of the unbounded domain, is no longer held. To isolate the effect of the free surface, we split the displacements u_z into two parts, u_z^∞ and u_z^b. The former denotes the displacements caused by the distributed screw

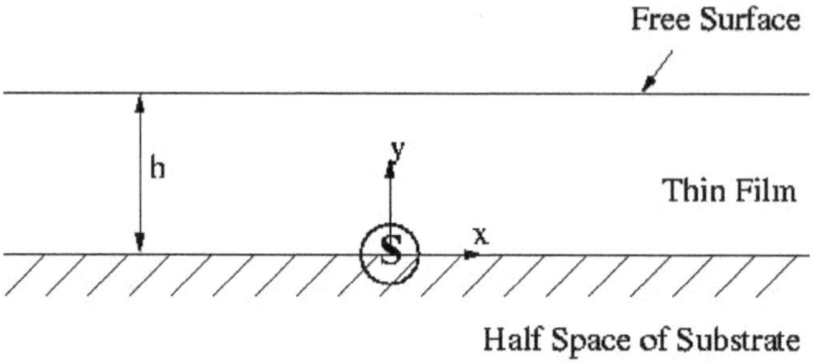

Fig. 8.24 A screw dislocation at the interface between a substrate and a thin film.

dislocations in the infinite domain, and the latter denotes the elastic displacements caused by the free surface at boundary $y = h$. We assume that the free surface can only cause a continuous displacement field across the interface at $y = 0$, and hence there is no discontinuity in $u_z^b(x)$, i.e. $\Delta u_z^b = 0$. The Burgers vector density can then be expressed as

$$b'(x) = \frac{d\Delta u_z^\infty}{dx} \ . \tag{8.345}$$

By applying the sinusoidal law [Peierls (1940)], the restoring force due to the displacement of a screw dislocation at the glide plane is

$$\sigma_{yz}^\infty(x, 0) = \tau_{\max} \sin \frac{2\pi \Delta u_z^\infty}{b} \ , \tag{8.346}$$

where Δu_z^∞ is the displacement jump caused by the distributed screw dislocation in the infinite domain, and

$$\tau_{\max} = \frac{\mu b}{2\pi d} = \frac{Db}{2w_h} \tag{8.347}$$

is chosen for the elastic limit to apply in small deformation theory. The symbol $D = \mu/2\pi$ is the normalized shear modulus with respect to 2π, $w_h = d/2$ is the half-width of the dislocation, and d is the thickness of the non-Hookean slab joining the thin film and the substrate. The superscript ∞ indicates that the stress does not include the stress due to the free surface. The coefficient of the above relation is obtained by virtue of infinitesimal deformation theory where Hooke's law applies. We assume that the core structure of the dislocation has not changed with the existence of the free surface, and the width of the dislocation, is independent of the film thickness h.

With above assumptions, we can use the original P-N solution in the analysis

$$\sigma_{yz}^\infty(x, 0) = \frac{Dbx}{x^2 + w_h^2} \ , \quad \Delta u_z^\infty(x, 0) = \frac{b}{\pi} \tan^{-1}\left(\frac{x}{w_h}\right) \ , \quad b'(x) = \frac{b}{\pi} \frac{w_h}{x^2 + w_h^2} \tag{8.348}$$

by enforcing $\Delta u_z^\infty(\pm\infty) = \pm b/2$ [Hirth and Lothe (1992)]. To find the shear stress corresponding to the misfit in the unbounded domain, we sum the contribution of the distributed dislocation along the gliding plane,

$$\sigma_{yz}^\infty(x,y) = D\int_{-\infty}^{\infty} b'(x')\frac{(x-x')dx'}{(x-x')^2+y^2} = \frac{Dbx}{x^2+(|y|+w_h)^2}\,. \tag{8.349}$$

To take into account the effect of the free surface, we generalize the concept of the image stress due to a single discrete dislocation to that of a distributed image dislocation layer due to a distributed dislocation layer (see Figure 8.25). It is

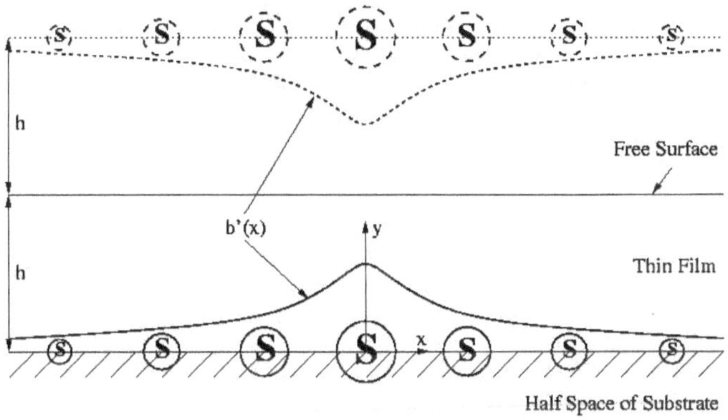

Fig. 8.25 A distributed screw image dislocations at $y = 2h$ (outside the physical domain).

assumed that the image screw dislocation with the opposite Burgers vector $(-\mathbf{b})$ is distributed along the horizontal line above the thin film at the height $y = 2h$ according to the standard P-N procedure. This image screw dislocation distribution causes an image stress in the lower half-space $(y \leq h)$,

$$\sigma_{yz}^I(x,y) = \frac{-Dbx}{x^2+(|y-2h|+w_h)^2}\,. \tag{8.350}$$

If we sum the shear stress due to these two screw dislocation distributions, we can obtain the total shear stress :

$$\begin{aligned}\sigma_{yz}(x,y) &= \sigma_{yz}^\infty(x,y)+\sigma_{yz}^I(x,y)\\ &= Dbx\left(\frac{1}{x^2+(|y|+w_h)^2}-\frac{1}{x^2+(|y-2h|+w_h)^2}\right)\,.\end{aligned} \tag{8.351}$$

One can verify that the total shear stress vanishes at $y = h$, i.e. the zero traction at the free surface is indeed satisfied. It is noted that since the proposed model requires a non-vanishing non-Hookean slab joining the thin film and the substrate, the thickness of the thin film for the model cannot vanish. The minimum thickness for the model to be valid should be $d/2$, i.e. $h_{\text{min}} = d/2$. The dislocation that corresponds to the minimum thickness h_{min} is understood as a surface dislocation.

To calculate the misfit energy, we adopt the discrete summation procedure by Joós and Duesbery [Joós and Duesbery (1997)]. The misfit energy density, which is the unit misfit energy stored in a volume element of height d, lattice spacing s, and unit depth in the z-direction, at the glide plane ($y = 0$), can then be written as

$$\begin{aligned}\Delta W(x) &= sd\int_0^{\gamma_{yz}} \sigma_{yz}(x,0)d\gamma_{yz} \\ &= -s\int_{b/2}^{\Delta u_z^\infty} \sigma_{yz}^\infty(x,0)d\Delta u_z^\infty - s\int_{b/2}^{\Delta u_z^\infty} \sigma_{yz}^I(x,0)d\Delta u_z^\infty \\ &= \Delta W^\infty(x) + \Delta W^I(x)\ . \end{aligned} \tag{8.352}$$

This misfit energy density is stored between a pair of atomic planes separated by a distance s. For the sake of comparison, we split the misfit energy density into two parts: $\Delta W^\infty(x)$ and $\Delta W^I(x)$. The former is the conventional misfit energy density, which can be calculated by standard procedure, i.e.

$$\Delta W^\infty(x) = -s\int_{b/2}^{\Delta u_z^\infty} \sigma_{yz}^\infty(x,0)d\Delta u_z^\infty = \frac{sDb^2}{4\pi w_h}\left(\cos\frac{2\pi\Delta u_z^\infty}{b} + 1\right)\ . \tag{8.353}$$

The latter is the misfit energy contribution due to the image stress $\sigma_{yz}^I(x,0)$. It is written as

$$\begin{aligned}\Delta W^I(x) &= -s\int_{b/2}^{\Delta u_z^\infty} \sigma_{yz}^I(x,0)d\Delta u_z^\infty \\ &= -\frac{sDb^2 w_h}{8\pi h(h+w_h)}\ln\frac{x^2+(2h+w_h)^2}{x^2+w_h^2}\ . \end{aligned} \tag{8.354}$$

Define two dimensionless parameters that depend on the film thickness h,

$$\rho(h) = \frac{4\pi h}{s}\ , \quad \kappa(h) = \frac{w_h}{h+w_h}\ . \tag{8.355}$$

We can then express the second misfit energy density as

$$\Delta W^I(x) = -\frac{Db^2\kappa(h)}{2\rho(h)}\ln\frac{x^2+(2h+w_h)^2}{x^2+w_h^2}\ . \tag{8.356}$$

Then the total misfit energy is the sum of the unit misfit energy along the glide plane over the lattice (over n), as shown below,

$$\begin{aligned}W(a) &= \sum_{n=-\infty}^{\infty} \Delta\tilde{W}(n,a) = \sum_{n=-\infty}^{\infty} (\Delta\tilde{W}^\infty(n,a) + \Delta\tilde{W}^I(n,a)) \\ &= W^\infty(a) + W^I(a)\ , \end{aligned} \tag{8.357}$$

where the symbol tilde ($\tilde{}$) indicates that the energy W takes the arguments n and a instead of x. After summation, we find that

$$\hat{W}^\infty(\alpha) = \frac{Db^2}{2}\frac{\sinh\xi}{\cosh\xi - \cos\alpha}\ . \tag{8.358}$$

$$\hat{W}^I(\alpha) = -\frac{Db^2\kappa(h)}{2\rho}\ln\left(\frac{\cosh(\xi+\rho(h)) - \cos\alpha}{\cosh\xi - \cos\alpha}\right) \tag{8.359}$$

where $\xi = 2\pi w_h/s$, $\alpha = 2\pi a/s$ are the two normalized dimensionless quantities. The symbol hat ($\hat{}$) indicates that the energy W takes the argument α instead of a. Combining both misfit energies yields the total misfit energy,

$$\hat{W}(\alpha) = \frac{Db^2}{2}\left(\frac{\sinh\xi}{\cosh\xi - \cos\alpha} - \frac{\kappa(h)}{\rho(h)}\ln\left(\frac{\cosh(\xi+\rho) - \cos\alpha}{\cosh\xi - \cos\alpha}\right)\right) . \quad (8.360)$$

The modified Peierls stress , which is the maximum lattice friction, for the screw dislocation at the interface of thin film/substrate system is then

$$\sigma_p = \frac{\mu b}{2s}\left(g_1(\alpha_m) - \frac{\kappa(h)}{\rho(h)}g_2(\alpha_m)\right)g_3(\alpha_m) , \quad (8.361)$$

where

$$g_1(\alpha_m) = \frac{\sinh\xi}{\cosh\xi - \cos\alpha_m} , \quad (8.362)$$

$$g_2(\alpha_m) = \frac{\cosh(\xi+\rho(h)) - \cosh\xi}{\cosh(\xi+\rho(h)) - \cos\alpha_m} , \quad (8.363)$$

$$g_3(\alpha_m) = \frac{\sin\alpha_m}{\cosh\xi - \cos\alpha_m} \quad (8.364)$$

and α_m maximizes the lattice friction. To compare the surface dislocation model

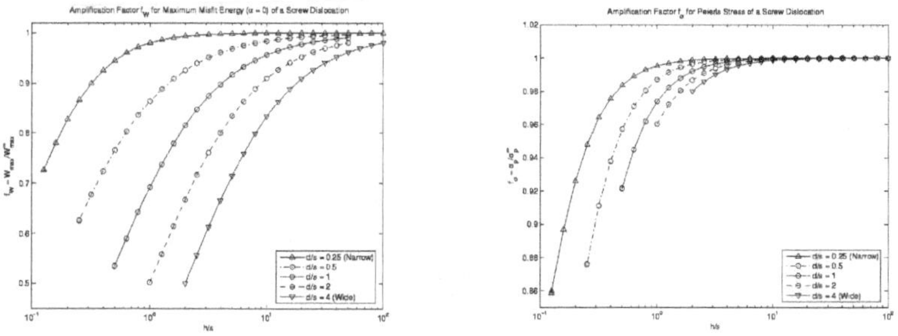

(a) (b)

Fig. 8.26 Amplification factors versus h/s for different d/b ratios: (a) Misfit Energy, and (b) Peierls Stress

with the bulk dislocation model, we define the amplification factor for misfit energy as the ratio of the maximum misfit energies, and it is given as

$$f_W(h) \equiv \frac{W_{\max}}{W_{\max}^{\infty}} = 1 - \left(\frac{\kappa(h)}{\rho(h)}\right)\frac{\cosh\xi - 1}{\sinh\xi}\ln\left(\frac{\cosh(\xi+\rho(h)) - 1}{\cosh\xi - 1}\right) \quad (8.365)$$

and we define the amplification factor for the Peierls stress σ_p as

$$f_\sigma(h) \equiv \frac{\sigma_p}{\sigma_p^{\infty}} = \frac{\left(g_3(\alpha_m)(g_1(\alpha_m) - \frac{\kappa(h)}{\rho(h)}g_2(\alpha_m)\right)}{g_3(\alpha_m^{\infty})g_1(\alpha_m^{\infty})} , \quad (8.366)$$

where the scalar α_m^∞ is the maximizer of the lattice friction σ^∞ in bulk material, which is given [Joós and Duesbery (1997)] as,

$$\alpha_m^\infty = \cos^{-1}\left(\frac{1}{2}\left(\sqrt{9+\sinh^2\xi} - \cosh\xi\right)\right) . \tag{8.367}$$

The maximizer α_m is different from α_m^∞ in general unless the thickness h is approaching infinity (unbounded domain).

To show how the misfit energy and the Peierls stress change as the thickness of the thin film increases, we plot the amplification factors f_W and f_σ versus the normalized thickness h/s in Fig. 8.26 for different d/s ratios. In these two plots, the middle three curves are plotted by using the exact amplification factors shown in (8.365) and (8.366). For the other two cases, we use the approximate amplification factors that are discussed in [Lee and Li (2007)].

Fig. 8.26(a) shows that the boundary effect on the misfit energy is quite significant. For the dislocation of core size $d/s = 1$, the reduction in the maximum misfit energy can be as high as about 50% for a surface dislocation. In general, for all sizes of dislocations, the reduction becomes less than 5% only when there are at least 100 layers of atoms between the dislocation and the free surface. In comparison with the misfit energy, the boundary effect on the Peierls stress of the dislocation is not that significant (see Fig. 8.26(b)). For the dislocation with core size $d/s = 1$, if there is at least one layer of atoms ($h/s = 1.5$) between the dislocation and the free surface, the reduction in the Peierls stress is less than 5%.

8.9 Exercises

Problem 8.1. *The stress fields of an edge dislocation (plane strain) are,*

$$\sigma_{rr} = \sigma_{\theta\theta} = \frac{\mu b}{2\pi}\frac{\sin\theta}{r} \tag{8.368}$$

$$\sigma_{zz} = \frac{\mu\nu b}{\pi(1-\nu)}\frac{\sin\theta}{r} \tag{8.369}$$

$$\sigma_{r\theta} = -\frac{\mu b}{2\pi(1-\nu)}\frac{\cos\theta}{r} \tag{8.370}$$

Find the self-strain energy density of the field in a hollow cylinder with outer radius R and inner radius r_0.

Problem 8.2. *In the most general case, two parallel dislocations shown in Fig. 8.27 have the Burgers vectors*

$$\mathbf{b}_1 = b_{1x}\mathbf{e}_x + b_{1y}\mathbf{e}_y + b_{1z}\mathbf{e}_z \tag{8.371}$$

$$\mathbf{b}_2 = b_{2x}\mathbf{e}_x + b_{2y}\mathbf{e}_y + b_{2z}\mathbf{e}_z \tag{8.372}$$

Find the interaction force between them.

Hint: (1) Find the interaction between two parallel edge dislocations; (2) Find the interaction between two climbing edge dislocations; (3) Find the interaction between

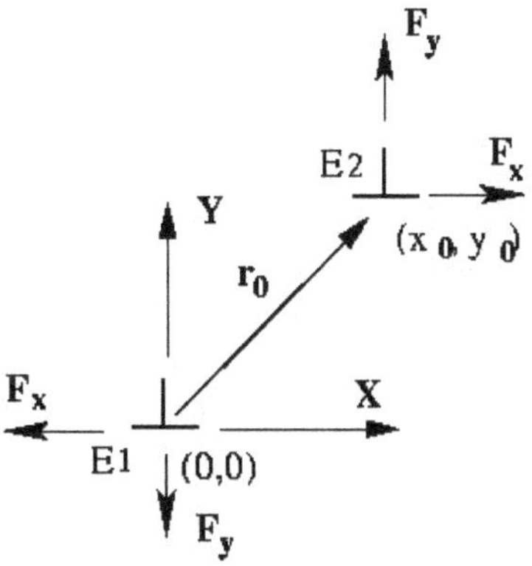

Fig. 8.27 Interactions between two parallel edge dislocations.

two screw dislocations; (4) superpose them all together, and (5) discuss the nature of the interactions, if you can make some senses out the solution.

Problem 8.3. *In the Peierls-Nabarro model, the distributed dislocation density is*

$$b'(x) = -2\frac{du_x}{dx}(x) \tag{8.373}$$

where

$$u_x(x) = -\frac{b}{2\pi}\tan^{-1}\frac{x}{r_c} \tag{8.374}$$

Since the Airy stress function for a classical edge dislocation is

$$\Psi = -\frac{\mu b y}{4\pi(1-\nu)}\ln(x^2+y^2) \tag{8.375}$$

Find the Airy stress function for the Peierls-Nabarro model by using superposition method, and find shear stress σ_{xy} *on the whole plane.*

Chapter 9

INTRODUCTION TO CONFIGURATIONAL MECHANICS

Since Eshelby's pioneer contribution on configurational mechanics in the 1950s [Eshelby (1951, 1956)], configurational mechanics has become a major research area in applied mechanics and material science. In this Chapter, we shall introduce some basic concepts of configurational mechanics and its relation to micro-mechanics.

9.1 Configurational force: Eshelby's energy-momentum tensor

The spatial environment in which we encounter and experience various physical phenomena in everyday life is called the physical space. If we neglect the effect of relativity, such a space is usually postulated to be a homogeneous and isotropic Euclidean space which implies that the properties of a closed system remain invariant under the rigid-body translation and rigid-body rotation. In fact, there are other and related invariant principles or properties, for example, invariant properties due to materials frame-indifference, or objectivity, e.g. [Truesdell and Noll (2004)], and invariance due to similarity of spatial scaling for materials that has no apparent micro-structures, etc. Moreover, If one considers a system in a homogeneous, reversible space-time ensemble, the properties of the system is also invariant in time.

The motions of matter in physical space can be described by mechanics physical laws, or differential equations. The general solutions of equation of motions may be expressed in the integral form. In their classical treatise on physics (Volume **1** *Mechanics* [Landau and Lifshitz (1965)]), Landau and Lifshitz wrote that:

> Not all integrals of the motion, however, are of equal importance in mechanics. There are some whose constancy is of profound significance, deriving from the fundamental homogeneity and isotropy of space and time.

In 1951 Jack Eshelby first found that indeed one of these invariant integrals in the context of continuum mechanics can be actually used as the measure of 'force' governing the equilibrium or the motion of topological defects in solids. Such topological defects include, point defects, e.g. the interstitial-vacancy; line defects, e.g. dislocations or declinations; and surface defects, e.g. cracks. The reason why the invariant integral is called as the 'material force' is because it is the measure of the change of the free-energy in a system due to the motion or evolution of defects. However the defect is not a matter, because they do not have masses. So the force acting on defects is not a force in physical space in the conventional sense of Newtonian mechanics, and the equation of motions for defects is not related to Newton's second law. To distinguish such differences, we call the force acting on defects as *the material force* or *configurational force*, which governs the evolution of topological defects in a continuum, such as motions of dislocations, growth of a crack, re-arrangement of lattice structures, and phase transformations in general. Consequently, the mechanics governs the defect evolution is called *the configurational mechanics*.

To explain what is configurational force, we first look at a real physical example: Consider that a solid contains an edge dislocation ($\mathbf{b} = b\mathbf{e}_x$) under external hydrostatic pressure, $\sigma_{11} = \sigma_{22} = \sigma_{33} = -p$, $p > 0$. This will cause the edge dislocation climbing, which in turn will change the system's total potential energy. While the edge dislocation climbs, it does not produce any volumetric strain, thus, the hydrostatic pressure never does any work in the process. Therefore, the conventional Newtonian force acting on the dislocation is not directly responsible for the potential energy change. In this case, the change or the decrease of the potential energy is due to the material force acting on the dislocation, which can be evaluated as

$$\mathbf{F}_\eta = -\frac{\partial W}{\partial \eta} = -pb\mathbf{e}_y, \tag{9.1}$$

where W is the potential energy, b is the Burgers vector, and $\mathbf{e}_y$ is the unit vector in y-direction. Note that $\mathbf{F}$ is not a conventional force that is acting in x-direction. Therefore, it is a generalized force related to the change of the lattice configuration due to the climbing of the dislocation.

Analogous to conventional Newtonian forces, configurational forces have their own force equilibrium from the viewpoint of energy balance or thermodynamics arguments [Gurtin (1995); Gurtin and Podio-Guidugli (1996); Gurtin (1999)]. Before we present Gurtin's elegant thermodynamics theory about the configurational force, we shall first follow the historic development to see how the subject has been evolved into a formal and rigorous theoretical framework.

9.1.1 *Eshelby's thought experiment*

We now compute the configurational force for a defect at equilibrium state. In order to evaluate the configurational force acting on a defect, we first calculate the change of potential energy due to the change of a defect's position or the change of material configuration.

The following is our adaptation of Eshelby's famous thought experiment on configurational force. The setting of Eshelby's thought experiment is a solid with a point defect, and it is subjected to external traction or displacement constraints

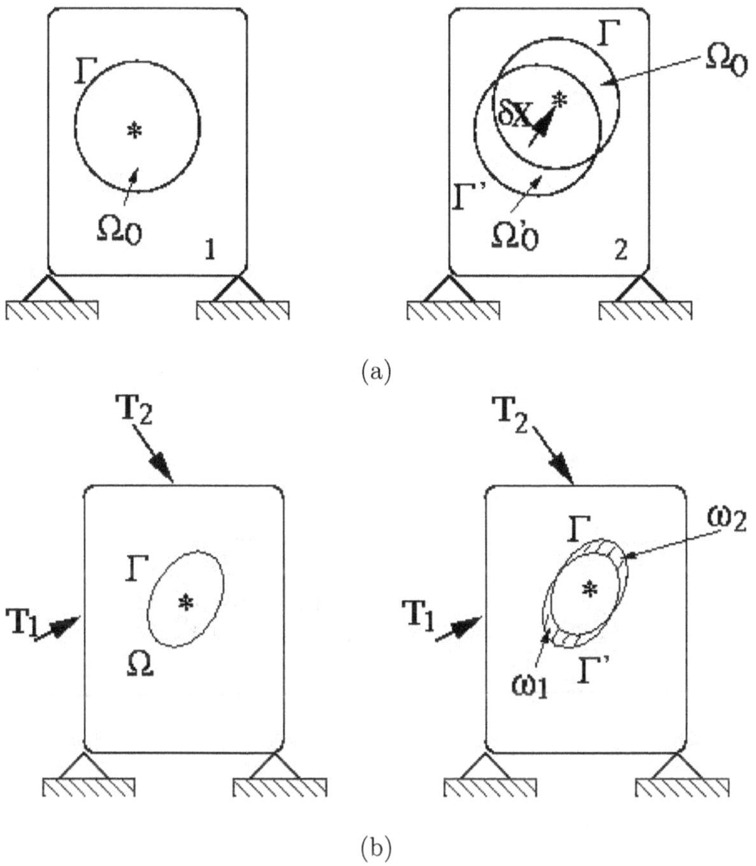

Fig. 9.1 A defect inside the local (refrential/spatial) configurations and the movement of the defect: (a) In the referential configuration, and (b) In the spatial configuration.

at the boundary. We denote the point defect as $\star$ inside the solid, and we link the defect $\star$ with its referential local configuration by embedding it into an arbitrarily chosen local volume Ω_0. We define the local configuration as the relative position of $\star$ inside Ω_0. We denote the boundary of the local volume as $\Gamma = \partial\Omega_0$ (see Fig. 9.1(a)).

The basic idea of Eshelby's thought experiment is to first alter the global configuration of the solid by changing the defect position, and then to compare the energy change in the control volume or the control mass.

Step 1. We first change the global configuration of the solid by changing the position of the defect an infinitesimal amount of displacement $\delta\mathbf{X}$ in referential configuration. Since the defect has been moved, the original local control volume containing $\star$ has an different local configuration, denoting as Ω'_0, which has the same material points as the original solid but now the relative defect position in it has changed, i.e. the local configuration has changed. When the defect, $\star$, moves to its new global referential position, we can choose the same local configuration (but with a different sets of material points), to identify it, i.e. we surround the defect $\star$ with the same local configuration Ω_0 again. Note that $\Omega_0 \in V_2$ is what M. E. Gurtin has identified as the so-called *migrating control volume*, and we are only interested in comparing energies between its original control volume and its migrating control volume in the same global configuration. Since, the material virtual displacement field represents a change of configurations, one may actually interpret this as the comparison of energies of the same control mass system in two different "*global configurations*", i.e. the global configurational V_1 and the global configurational V_2. There is a subtle point that one has to bear in mind: that is the local configurations or control volumes $\Omega_0 \in V_2$ and $\Omega_0 \in V_1$ are under the same mechanical state, i.e. stress and strain states. To Eshelby, this is obvious if both V_1 and V_2 are large enough, and $\delta \mathbf{X}$ is infinitesimal. Later Gurtin made an lucid and rigorous justification by the argument of observer objectivity in the referential configuration. Since the control volume system $\Omega_0 \in V_2$ is *equivalent* to the control volume system in $\Omega_0 \in V_1$, which is the same control mass system $\Omega_0' \in V_2$, the comparison of energies between the two control mass systems $\Omega_0 \in V_1$ and $\Omega_0' \in V_2$ is equivalent to the comparison of those between two control volume systems $\Omega_0 \in V_2$ and $\Omega_0' \in V_2$. This notion is illustrated in Fig. 9.1.

Step 2. Before calculating the difference of the energies stored in two different control volumes, we first load the body so that the solid deforms, and local material configurations are mapped to local spatial configurations. Denote a finite motion as ψ,

$$\phi : \Omega_0 \to \Omega \quad \text{and} \quad \phi : \Omega_0' \to \Omega' \tag{9.2}$$

Step 3. We then calculate the energy difference in two local control volumes Ω' and Ω of the same global configuration, or the energy difference in the same control mass of two different global configurations. Due to the change of global defect position, the stored energy difference in the bulk will be

$$\delta E_1 = \int_{\Omega'} W dV - \int_{\Omega} W dV \,. \tag{9.3}$$

From Fig. 9.2, one may observe that the area difference between Ω' and Ω is the difference of two crescent-shaped areas $\omega_1 - \omega_2$, i.e. adding the area ω_1 and removing the area ω_2. Hence the stored strain energy difference is

$$\delta E_1 = \int_{\omega_1} W dV - \int_{\omega_2} W dV \,. \tag{9.4}$$

Since $\delta \mathbf{X}$ is infinitesimal,

$$\omega_1 - \omega_2 = \int_{\omega_1 - \omega_2} dA = -\delta \mathbf{X} \cdot \int_{\Gamma} \mathbf{n} ds$$

as shown in Fig. 9.2. Therefore,

$$\delta E_1 = -\delta \mathbf{X} \cdot \int_{\Gamma} W d\ell \mathbf{n} = -\delta X_\ell \int_{\Gamma} W ds n_\ell \,. \tag{9.5}$$

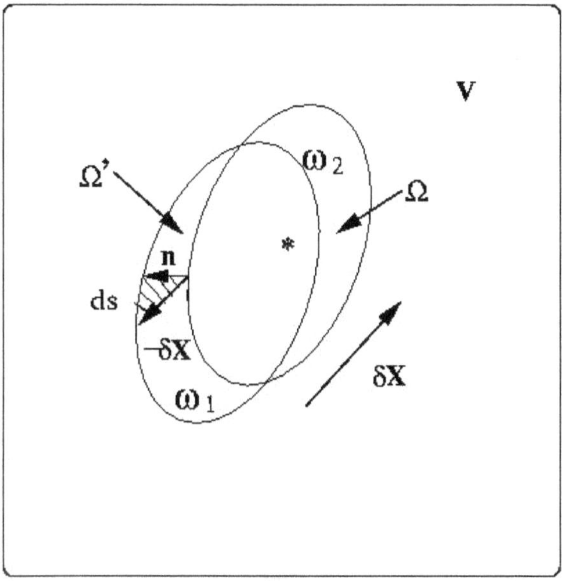

Fig. 9.2 Eshelby's imaginary operation.

Note that in this step, all the operations are performed in the referential configuration, and we are comparing the energy difference between two adjacent control volumes that differ by an infinitesimal translation.

Step 4. Calculating the energy difference between two adjacent control volumes can be figuratively interpreted as an operation of slice and restore [Eshelby (1951, 1975)], i.e. first cut the local control volume Ω from V_2 and then cut the local control volume Ω' from a replica of V_2 and then put the local control volume Ω' back into the hole in V_2 where the local configuration Ω is taken out (See Fig. 9.3). During the configuration change, the defect moves $+\delta\mathbf{X}$ from its original material position to the new material position, it may cause the relative material virtual displacement,

$$u_i(\Omega_0') - u_i(\Omega_0) = \delta u_i = -\frac{\partial u_i}{\partial X_j}\delta X_j \ . \tag{9.6}$$

Then the difference of the work done by the external forces at the boundaries of the control volume are:

$$\begin{aligned}\delta W^{ext} &= \int_{\Gamma'} u_i T_i ds - \int_{\Gamma} u_i T_i ds = \int_{\Gamma} \delta u_i \sigma_{ij} n_j ds \\ &= -\int_{\Gamma} u_{i,k}\sigma_{ij} n_j ds \delta X_k \ ,\end{aligned} \tag{9.7}$$

which will cause the decrease of the potential energy in the local configuration, i.e. $\delta E_2 = -\delta W^{ext}$.

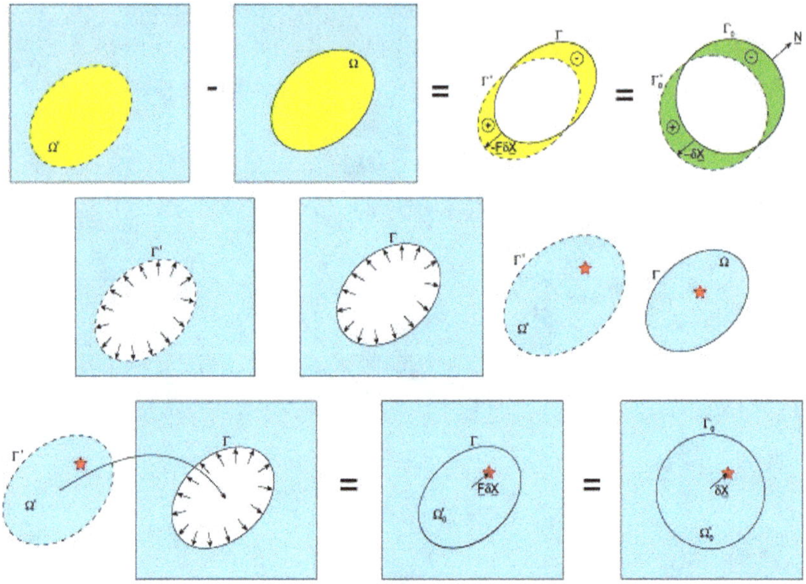

Fig. 9.3 The slice and restore operations (This figure was drawn by Dr. James Foulk III).

Then the total variation due to the change of configuration is,

$$\delta E = \delta E_1 + \delta E_2 = -\delta X_\ell \left\{ \oint_\Gamma (W n_\ell - u_{i,\ell}\sigma_{ij} n_j) ds \right\}$$
$$= = -\delta X_\ell \left\{ \oint_\Gamma \Big(W\delta_{\ell k} - u_{i,\ell}\sigma_{ik} \Big) n_k ds \right\}$$

To honor the tradition, the force on the defect is defined as the negative rate of increase of the total potential energy of the system, i.e.

$$\delta E = -\mathbf{F}^{inh} \cdot \delta \mathbf{X} = \frac{\partial E}{\partial \mathbf{X}} \cdot \delta \mathbf{X} \tag{9.8}$$

Therefore the configurational force acting on the defect is

$$F_\ell^{inh} = \oint_\Gamma \Big(W\delta_{\ell k} - u_{i,\ell}\sigma_{ik} \Big) n_k ds \ . \tag{9.9}$$

In two-dimensional case, the first component of F_ℓ^{inh} i.e.$\ell = 1$ is Rice's celebrated J-integral ,

$$F_1^{inh} = J = \oint_\Gamma \Big(W dx_2 - u_{i,1}\sigma_{ik} n_k ds \Big) \ , \tag{9.10}$$

which can be interpreted as the driving force of a crack that grows along x-axis.

The integrand of (9.9) is Eshelby's another celebrate tensor: the energy-momentum tensor. The name originates from the fact that the tensor is obtained by translating the local configuration. We denote it as

$$C_{\ell k} = W\delta_{\ell k} - u_{i,\ell}\sigma_{ik} \ . \tag{9.11}$$

Just as the Peach-Koehler force , Eshelby's energy momentum tensor was also inspired by an electromagnetic analogy. As Eshelby pointed out, "*the archetypal energy-momentum tensor is Maxwell's stress tensor in electromagnetic.*" We juxtapose the two for comparison,

$$\mathbf{C}^E = W\mathbb{I}^{(2)} - \mathbf{E}\otimes\mathbf{D} \tag{9.12}$$

$$\mathbf{C}^M = W\mathbb{I}^{(2)} - \nabla\mathbf{u}\otimes\boldsymbol{\sigma}\ . \tag{9.13}$$

where the superscripts, E and M, denote mechanical and electrical energy-momentum tensors respectively.

In the following, we shall show that the energy-momentum tensor is divergence-free in homogeneous solid without body force, which is the essence of path-independent invariant integrals.

The straightforward differentiation gives,

$$\begin{aligned}\frac{\partial C_{\ell k}}{\partial x_k} &= \frac{\partial W}{\partial \epsilon_{mn}}\frac{\partial \epsilon_{mn}}{\partial x_k}\delta_{\ell k} - u_{i,\ell k}\sigma_{ik} - u_{i,\ell}\sigma_{ik,k}\\ &= \sigma_{mn}u_{m,nk}\delta_{\ell k} - u_{i,\ell k}\sigma_{ik}\\ &= \sigma_{mn}u_{m,n\ell} - \sigma_{ik}u_{i,k\ell} = 0\ .\end{aligned} \tag{9.14}$$

Therefore, for homogenous solids,

$$F_\ell = \oint_\Gamma \mathbf{C}_{\ell k}n_k ds = \oint_\Gamma \Big(W\delta_{\ell k} - u_{i,\ell}\sigma_{ik})n_k ds = 0\ . \tag{9.15}$$

For inhomogeneous solids, the above statement is no longer true, this is because,

$$\frac{\partial \mathbb{C}_{ijmn}(\mathbf{x})}{\partial x_k} \neq 0,$$

and

$$\frac{\partial W}{\partial x_k} \neq \sigma_{mn}u_{m,nk}\ .$$

Suppose that there is a point defect at a material point ξ_i, and it may be represented by an equivalent inhomogeneous elastic stiffness tensor $\mathbb{C}_{ijk\ell}(\mathbf{x}-\boldsymbol{\xi})$, i.e.

$$\mathbb{C}_{ijk\ell}(\mathbf{x}-\boldsymbol{\xi}) = \begin{cases}\mathbb{C}^0_{ijk\ell}, & \forall\mathbf{x}\neq\boldsymbol{\xi}\\ \\ \mathbb{C}_{ijk\ell}(\boldsymbol{\xi}), & \forall\mathbf{x}=\boldsymbol{\xi}\end{cases} \tag{9.16}$$

Therefore, the total strain energy of the inhomogeneous body is

$$E = \frac{1}{2}\int_V \mathbb{C}_{ijk\ell}(\mathbf{x}-\boldsymbol{\xi})\epsilon_{ij}\epsilon_{k\ell}dV \tag{9.17}$$

By definition,

$$\begin{aligned}F_n^{inh} &= -\frac{\partial E}{\partial \xi_n} = -\frac{1}{2}\int_V \frac{\partial \mathbb{C}_{ijk\ell}}{\partial \zeta_n}\epsilon_{ij}\epsilon_{k\ell}dV\\ &= \frac{1}{2}\int_V \mathbb{C}_{ijk\ell,n}\epsilon_{ij}\epsilon_{k\ell}dV\\ &= \frac{1}{2}\int_V \Big[\Big(\mathbb{C}_{ijk\ell}\epsilon_{ij}\epsilon_{k\ell}\Big)_{,n} - 2\mathbb{C}_{ijk\ell}u_{i,j}u_{k,\ell n}\Big]dV\end{aligned}$$

Consider $\mathbb{C}_{ijk\ell}u_{i,j} = \sigma_{k\ell}$ and integration by parts for the second term of the integrand.

$$
\begin{aligned}
F_n^{inh} &= \int_V \Big[\Big(\frac{1}{2}\mathbb{C}_{ijk\ell}\epsilon_{ij}\epsilon_{k\ell}\Big)_{,n} - (\sigma_{k\ell}u_{k,n})_\ell + \sigma_{k,\ell\ell}u_{k,n}\Big]dV \\
&= \int_V \Big[\Big(\frac{1}{2}\mathbb{C}_{ijk\ell}\epsilon_{ij}\epsilon_{k\ell}\Big)_{,n} - (\sigma_{k\ell}u_{k,n})_{,\ell}\Big]dV \\
&= \oint_\Gamma \Big(W\delta_{n\ell} - u_{k,n}\sigma_{k\ell}\Big)n_\ell ds = \oint_\Gamma C_{n\ell}n_\ell ds \ .
\end{aligned}
\tag{9.18}
$$

In the following, we calculate the J-integral for a mode-III elastic crack, which is

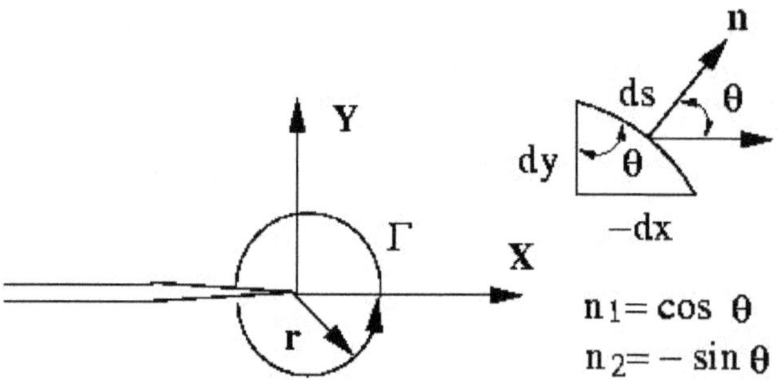

Fig. 9.4 Contour for J-integral around a crack tip.

the driving force for crack propagation.

Example 9.1. The asymptotic stress fields for a mode III crack is

$$
\sigma_{13} = -\frac{K_{III}}{\sqrt{2\pi r}}\sin\frac{\theta}{2}, \quad \sigma_{23} = \frac{K_{III}}{\sqrt{2\pi r}}\cos\frac{\theta}{2} \quad . \tag{9.19}
$$

We choose the integration contour $\Gamma : x_1 = r\cos\theta,\ x_2 = r\sin\theta, \quad -\pi \le \theta \le \pi$.

The J-integral reads as follows,

$$
\begin{aligned}
J &= \oint_\Gamma \Big(W dx_2 - \frac{\partial u_i}{\partial x_1}\sigma_{ik}n_k ds\Big) \\
&= \int_{-\pi}^{\pi} \Big(W r\cos\theta - \frac{\partial u_3}{\partial x_1}(\sigma_{31}n_1 + \sigma_{32}n_2) r d\theta\Big)
\end{aligned}
\tag{9.20}
$$

Consider $n_1 = \cos\theta$, $n_2 = -\sin\theta$, $\dfrac{\partial u_3}{\partial x_1} = 2\epsilon_{31} = \dfrac{\sigma_{31}}{\mu}$, and $W = K_{III}^2/(4\mu\pi r)$.

$$
\begin{aligned}
J &= \int_{-\pi}^{\pi} \frac{K_{III}^2}{4\mu\pi}\cos\theta d\theta - \frac{K_{III}^2}{2\pi\mu}\int_{-\pi}^{\pi}\Big(\sin^2\frac{\theta}{2}\cos\theta - \sin\frac{\theta}{2}\cos\frac{\theta}{2}\sin\theta\Big)d\theta \\
&= \frac{K_{III}^2}{2\pi\mu}\int_{-\pi}^{\pi}\Big(2\sin^2\frac{\theta}{2}\cos^2\theta - \sin^2\frac{\theta}{2}\Big(\cos^2\frac{\theta}{2} - \sin^2\frac{\theta}{2}\Big)\Big)d\theta \\
&= \frac{K_{III}^2}{2\pi\mu}\int_{-\pi}^{\pi}\sin^2\frac{\theta}{2}d\theta = \frac{K_{III}^2}{2\mu}
\end{aligned}
\tag{9.21}
$$

9.1.2 *Lessons from J.D. Eshelby*

The measure of your education is what you remember 15 years afterward, says one wiseacre. Well, it's been a little more than 15 years, and I don't think that I learned anything at the time, but the lectures I had from Professor J.D. (Jock) Eshelby still leave a mark.

Undergraduate students in materials science at Sheffield University were barely aware of the towering stature of this man, in the intellectual sense anyway. If you don't know who he was or what contributions he has made, then you probably have some serious holes in your own materials education, but you can still read on. A few Britishisms must be explained, though. First, the term "Jock" is used in the United Kingdom not for an athlete, as in the United States, but is a nickname commonly accorded to Scotsmen living in England; the U.S. sense could never apply to Jock Eshelby. Second the term "Faculty" in England is equivalent to a college in a U.S. University. Third, a professorship in the United Kingdom is a distinguished academic rank that has almost no equivalent in the United States. The closest would be a "leading professorship".

Way back then, Sheffield had a Faculty of Materials, with departments of Metallurgy, Ceramics, Glasses, Polymers, and the theory of materials. The department of the theory of materials was arguably a little top heavy. It had two professors, Eshelby and B.A.Bilby (whose name you should also know), one other lecturer, and a computer programmer. In a good year it had one undergraduate student.

Eshelby taught courses in elasticity and solid state bonding to the undergraduates in all of the departments, and his lecturing style was not particularly student-friendly. He did not work from notes. He would walk into the lecture hall, apparently already half-way through this lecture, pick up the chalk, and start writing on the board. Whether he was trying to show us how to solve Schrödinger's equation or develop the strain compatibility relations, the technique was always the same. He would clear a patch of board and start deriving a theorem. Running out of space, he would clear another patch, not necessarily connected with the first, and fill that up. Eventually, small pieces of the theorem would be scattered more-or-less at random across the chalkboard, stochastically mixed with the detritus of the previous lecture, and with random parts missing–erased to make space for more. It did not help that his writing was atrocious, and his speech sounded as though he had filled his cheeks with marbles before starting. On one occasion, one of my classmates managed to get the professor's attention (a challenge) and asked him if he could possibly write a little more clearly. For a few lines, the writing was four times as large, but still as illegible as before. Several lectures ended with Eshelby's discovery that he had mis-derived the theorem in question–a significant risk if you try to do it without notes, even if you are a bona fide genius. When this happened, he would stand back and survey the board. After a few moments, he would announce something like, "Well, there's a sign error there. You can correct it and work through to the result for yourselves." As if.

As time went by, our horror at his teaching style gave way to an understanding that the man was, in fact, a genius. Eccentric, yes but a genius. Apparently addicted to cheap cigars, he would smoke them down to the smallest butt, then draw a cherry pipe out of his pocket, and stuff the remains of the cigar into it, to be smoked until not a scrap of tobacco was left. He cared little for what people thought of him, I think, and did not pay much attention to the politics of academia and the scientific community. This resulted in an unconscionable delay in his being elevated to the rank of Fellow of the Royal Society, which does seem to have been a sore point. In one memorable lecture, he described all of the current theories on a particular topic, listing the names of their authors on an uncharacteristically cleared chalkboard. He then described what was wrong with each of their work, condemning the weak-mindedness of these "so-called scientists" in quite direct terms. Having disposed of their failed logic, he then wrote the magical letters "FRS" after each of the names. He was elected an FRS himself that year and did not repeat the performance as far as I can gather.

Eshelby's impact on material science is far, far out of proportion to the numbers of his publications. In total, he published less than 20 papers over his entire career[1]

A fine demonstration of the futility of today's obsession with publication-counting as a means of career assessment. Eshelby's work is characterized by real physical insight, complemented by elegant mathematical analysis (He was a professor of applied mathematics at Sheffield, in addition to being a professor of the theory of materials.) In contrast with his lectures, his written work is a model of clarity. Although he was a powerful mathematician, he felt that we should only engage in "mathematical weightlifting" if we could not reason our way to the desired result through simple physical logic. Goodness knows what he would have made of today's computer simulation techniques. I think he would probably have thought of them as the last desperate resort after both physical reasoning and mathematical analysis failed.

An insight into Eshelby's motivations was provided to us in an informal moment one day, sitting in the small but splendid museum of glassware belonging to the Faculty of Materials, in a traditional British tea break. The usually unapproachable Eshelby was unusually affable that day–perhaps he had just received word of his FRS election–but we fell into conversation and one undergraduate student asked him what had led to his being a "pure theoretician". He told us the story of a formative experience in his life. It seems that as a young teenager he had made a calculation of the thermal shock resistance of a piece of glass. This resulted from his mother's always using a thick cork pad beneath a coffee table. She explained the reason to him and he set to work calculating the effect of the anticipated thermal shock. A short

[1]This account may not be accurate. Eshelby published about 50 to 60 papers in his lifetime. A rather complete collection of his work was recently published in "Collected Works of J.D. Eshelby: The Mechanics of Defects and Inhomogeneities (Solid Mechanics & Its Applications)" edited by X. Markenscoff and A. Gupta [Markenscoff and Anurag (2006)]. but each of them is a classic.

while later, he came to his mother and announced that he had completed his analysis, and that table would withstand a sudden local rise to the boiling point of water. His mother, being a wise woman, advised him that the obvious experiment would not be forthcoming and that he was forbidden from performing it himself. Well, curiosity and the budding scientific mind got the better of his youthful judgement one day when he was alone in the house. He boiled a pan of water and place it at the center of the prized coffee table. In his own words, "Well, cracks flew in every direction, and I suddenly received a discouragement that from performing experiments that has lasted me the rest of my life."

*True to the creed of the theoretician, however, he refused to allow that the analysis was flawed, and instead blamed the experiment. " Of course, I knew immediately what was wrong. The d***d thing hadn't been annealed properly. It was FULL of residual stress!"*

By all accounts, this attack on the quality of the prized table did not endear him to his mother. Let all theorists beware of blaming the experiment lest they suffer similarly.

— By Alex King (MRS Bulletin, July 1999). Article reproduced by permission of the MRS Bulletin (www.mrs.org/bulletin).

9.1.3 *J-integral and energy release rate G*

In 1968, J.R. Rice [Rice (1968, 1986)] first explained the relationship between the J-integral and the energy release rate by crack, which provides insights and essential links between configurational force and fracture mechanics starting off a new era of configurational force mechanics.

In this section, we present an adaptation of Rice's elucidation on equivalence between J-integral and the energy release G (see [Kanninen and Popelar (1985); Gdoutos (1990)]). We consider a crack propagating in a nonlinear elastic solid. We further assume that the boundary of the elastic solid consists of prescribed traction boundary Γ_t and the prescribed displacement boundary Γ_u. That is $\Gamma_0 = \Gamma_t \bigcup \Gamma_u$. The external traction prescribed in Γ_t are independent from the evolving crack length a, and the variations of displacement field have to be kinematically admissible. For simplicity, we assume that the crack propagates along the x_1-axis direction, and it will remain along the same direction during the entire fracture process. In other words, we assume that *the crack extends in a self-similar manner.*

To describe how the crack moves, we attach a fixed Cartesian coordinate system (x_1, x_2, x_2) to the referential configuration of the solid, even though for a small deformation theory we may not be able to distinguish the difference between the referential configuration and current configuration, except the crack length. To capture the effect of an evolving crack length, we attach a moving coordinate system

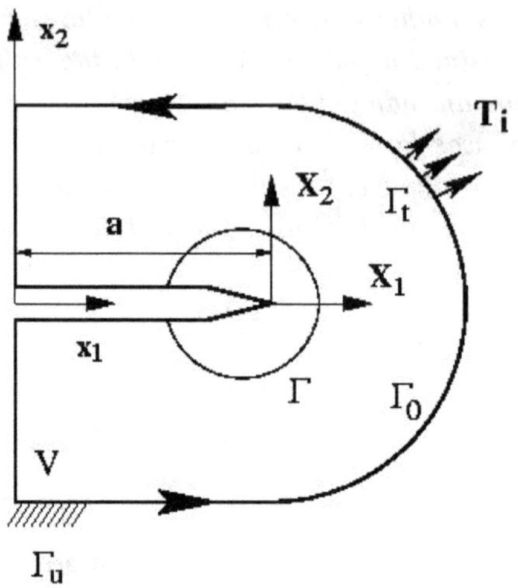

Fig. 9.5 A self-similar, planar crack extends in a finite nonlinear elastic solid.

at the tip of the crack (X_1, X_2) such that,

$$X_1 = x_1 - a, \quad X_2 = x_2 \tag{9.22}$$

Hence,

$$\frac{d}{da} = \frac{\partial}{\partial a} + \frac{\partial}{\partial X_1}\frac{\partial X_1}{\partial a} = \frac{\partial}{\partial a} - \frac{\partial}{\partial X_1} = \frac{\partial}{\partial a} - \frac{\partial}{\partial x_1} \tag{9.23}$$

The total potential energy in the elastic solid can be written as

$$\Pi(a) = \int_V w dV - \int_{\Gamma_t} T_k u_k dS \tag{9.24}$$

where w is the stored elastic strain energy density, and T_k is the prescribed traction on the boundary, which are related to the boundary condition,

$$\sigma_{ik} n_i = T_k, \quad \forall \mathbf{x} \in \Gamma_t$$

We now calculate the energy release due to crack extension by differentiation of the potential energy with respect to the crack length a,

$$\begin{aligned} \frac{d\Pi}{da} &= \int_V \frac{dw}{da} dV - \int_{\Gamma_t} T_k \frac{du_k}{da} dS \\ &= \int_V \Big(\frac{\partial w}{\partial a} - \frac{\partial w}{\partial x_1}\Big) dV - \int_{\Gamma_t} T_k \Big(\frac{\partial u_k}{\partial a} - \frac{\partial u_k}{\partial x_1}\Big) dS \end{aligned} \tag{9.25}$$

Since the solid is nonlinear elastic (which includes the linear elastic as a special case), we have

$$\frac{\partial w}{\partial a} = \frac{\partial w}{\partial \epsilon_{ij}}\frac{\partial \epsilon_{ij}}{\partial a} = \sigma_{ij}\frac{\partial \epsilon_{ij}}{\partial a} \tag{9.26}$$

By the principle of virtual work, an virtual extension of the crack will give

$$\delta W^{int} + \delta W^{ext} = \int_V \sigma_{ij}\delta\epsilon_{ij} dV - \int_{\Gamma_t} T_k \delta u_k dS$$
$$= \Big(\int_V \sigma_{ij}\frac{\partial \epsilon_{ij}}{\partial a} dV - \int_{\Gamma_t} T_k \frac{\partial u_k}{\partial a} dS\Big)\delta a = 0 \ . \tag{9.27}$$

Hence, the rate change of potential energy becomes,

$$\frac{d\Pi}{da} = -\Big(\int_V \frac{\partial w}{\partial x_1} dV - \int_{\Gamma_t} T_k \frac{\partial u_k}{\partial x_1} dS\Big) \tag{9.28}$$

By the divergence theorem, one finds that

$$\int_V \frac{\partial w}{\partial x_1} dV = \int_{\Gamma_0} w n_1 dS = \int_{\Gamma_0} w dx_2 \tag{9.29}$$

On the other hand, on the prescribed boundary Γ_u,

$$\frac{du_i}{da} = 0, \ \text{ and } \ \delta u_i = \frac{\partial u_i}{\partial a}\delta a = 0 \ \Rightarrow \ \frac{\partial u_i}{\partial x_1} = 0 \ , \forall \ \mathbf{x} \in \Gamma_u \tag{9.30}$$

Taking into consideration of (9.29) and (9.30), we can find that the J-integral along the boundary is exactly the negative rate change of potential energy due crack extension,

$$-\frac{d\Pi}{da} = J = \int_{\Gamma_0} \Big(w dx_2 - \sigma_{ik} n_k \frac{\partial u_i}{\partial x_1} dS\Big) \tag{9.31}$$

Because of path-independence of the J-integral, one can replace the contour Γ_0 by any contour Γ that encircles the crack tip. On the other hand, the definition of the energy release rate is the decrease of potential energy due to the crack extension, which is the same as the definition of configurational force in this context,

$$G = -\frac{d\Pi}{da}, \tag{9.32}$$

thus we have shown that the J-integral is exactly the same as the energy release rate during a self-similar crack extension, i.e.

$$J = G \ . \tag{9.33}$$

9.2 James R. Rice

Dr. J. R. Rice is Mallinckrodt Professor of Engineering Sciences and Geophysics at Harvard University. He has made fundamental contribution to configurational mechanics. He not only derived J-integral, but also was the first person who understands and explains in lucid terms how the J-integral is related to the energy release of a defect, *providing a sound and practical basis for the needed rapid development of inelastic fracture.*

Fig. 9.6 James R. Rice

9.3 Configurational compatibility

The original configurational force mechanics has two sources: the Peach-Koehler force for dislocations [Peach and Koehler (1950)] and Eshelby's energy-momentum tensor [Eshelby (1951)] for general defects in solids.

From a theoretical perspective, most conservation laws, or invariant integrals, in elasticity are the manifestations of symmetry properties of equilibrium equations or the balance of linear momentum, i.e. the equation of motion. There is hardly any conservation laws stemming from kinematics conditions.

In an effort to search compatibility conservation laws, [Li *et al.* (2007a)] has studied symmetry properties of kinematic conditions of continuum mechanics. Similar to the conservation laws arising from the symmetry properties of equilibrium condition, several classes of conservation laws that are purely based the symmetry properties of kinematics or compatibility conditions of a continuum, which are related to the continuum dislocation theory, have been found. In this section, we shall present an exposition of such theory.

9.3.1 *Continuum dislocation theory*

To fix the notation, we first introduce the convention used to describe dislocations. We identify a perfectly ordered state as a defect-free state. If defects are present, we are interested in a continuum description of defects. Considering dislocations as the main defect in this section, we obtain a continuum theory of dislocations. The incorporation of plasticity into the continuum was achieved by decomposing the total strain $\boldsymbol{\epsilon}$ of the plastic solid into an elastic part $\boldsymbol{\epsilon}^e$, which gives rise to stresses based on the general assumptions of elasticity theory, and a plastic or inelastic part $\boldsymbol{\epsilon}^p$, which changes the shape of the solid and leads to permanent deformation. In the infinitesimal case, this decomposition is given as

$$\boldsymbol{\epsilon} = \boldsymbol{\epsilon}^e + \boldsymbol{\epsilon}^p. \tag{9.34}$$

Contrary to $\boldsymbol{\epsilon}$, which is always a compatible field, $\boldsymbol{\epsilon}^e$ and $\boldsymbol{\epsilon}^p$ are, in general, not compatible fields. Since the plastic deformation is permanent, the elastic strain $\boldsymbol{\epsilon}^e$ no longer satisfies the compatibility equations, which will be affected by a defect distribution that constitutes inelastic deformations. The continuum defect theory is formulated with elastic or inelastic kinematic variables in order to represent the defect distribution. Even though the elastic strain $\boldsymbol{\epsilon}^e$ is a state quantity, which means that it can be uniquely measured at any time, it is not enough to describe the influence of dislocations in the body (Kröner, 1980) . The curvature, which serves as another state quantity, plays a dominant role in the continuum theory of defects, as it will be shown below. Following [Kröner (1980)], we introduce the corresponding anti-symmetric rotation tensors $\boldsymbol{\omega}$, $\boldsymbol{\omega}^e$, $\boldsymbol{\omega}^p$, and distortion tensors $\boldsymbol{\beta}$, $\boldsymbol{\beta}^e$, $\boldsymbol{\beta}^p$ as

$$\boldsymbol{\beta} = \boldsymbol{\epsilon} + \boldsymbol{\omega}, \quad \boldsymbol{\beta}^e = \boldsymbol{\epsilon}^e + \boldsymbol{\omega}^e \quad \text{and} \quad \boldsymbol{\beta}^p = \boldsymbol{\epsilon}^p + \boldsymbol{\omega}^p, \tag{9.35}$$

with the analogous decomposition as (9.34), namely

$$\boldsymbol{\omega} = \boldsymbol{\omega}^e + \boldsymbol{\omega}^p \quad \text{and} \quad \boldsymbol{\beta} = \boldsymbol{\beta}^e + \boldsymbol{\beta}^p. \tag{9.36}$$

and

$$\boldsymbol{\beta} := \nabla \otimes \mathbf{u}, \quad \boldsymbol{\epsilon} = \frac{1}{2}(\boldsymbol{\beta} + \boldsymbol{\beta}^T), \quad \text{and} \quad \boldsymbol{\omega} = \frac{1}{2}(\boldsymbol{\beta} - \boldsymbol{\beta}^T) \tag{9.37}$$

where $\mathbf{u}$ is the total displacement field. Note that the infinitesimal rotation tensor, $\boldsymbol{\omega}$, defined in this paper differs a minus sign with the conventional definition of the infinitesimal rotation in the literature.

In order to describe how a body is deformed by the total distortion $\boldsymbol{\beta}$, we may write the change of the total displacement vector

$$\mathbf{du} = \mathbf{dx} \cdot \boldsymbol{\beta} \quad \text{or} \quad du_j = \beta_{ij} dx_i \ . \tag{9.38}$$

The term distortion is used instead of displacement gradient, because the $\boldsymbol{\beta}$'s are gradients as in (9.38) only if the corresponding deformation is compatible. This is the case for the total distortion but in general does no longer hold for the elastic and plastic distortion.

To describe the defect state, we define the *geometrically necessary dislocation density* according to [Kröner (1980)] in terms of the plastic distortion tensor

$$\mathbf{d}^{GND} := \nabla \times \boldsymbol{\beta}^p, \quad \text{or in component form} \quad d_{ij}^{GND} = e_{ik\ell} \partial_k \beta^p_{\ell j} = e_{ik\ell} \beta^p_{\ell j,k} \tag{9.39}$$

where the permutation symbol $e_{ik\ell}$ is used. Note that (9.39) is Nye's definition for geometrically necessary dislocation density. Since the total distortion $\beta_{\ell j}$ has to remain compatible, namely $e_{ik\ell}\beta_{\ell j,k} = 0$ has to be satisfied which means that the body is not allowed to break, we can rewrite (9.39) as

$$\boldsymbol{\alpha} := -\nabla \times \boldsymbol{\beta}^e = \mathbf{d}^{GND} \tag{9.40}$$

As pointed out by Kröner, Eq. (9.39) is not a definition anymore, and it is a physical law. Here, we use $\boldsymbol{\alpha}$ representing the curl of negative elastic distortion, and the compatibility law is: $\boldsymbol{\alpha} = \mathbf{d}^{GND}$, which will be further discussed later.

The condition of the conservation of Burgers' vector follows directly from (9.40) as

$$\nabla \cdot \boldsymbol{\alpha} = 0, \tag{9.41}$$

which implies that dislocations do not end inside the body. The physical interpretation of (9.41) is the conservation or the balance of net Burgers' vector since

$$\oint_S \mathbf{n} \cdot \boldsymbol{\alpha} dS = \mathbf{b}. \tag{9.42}$$

where $\mathbf{b}$ is the Burgers' vector.

As usual we can express an anti-symmetric tensor by its axial (rotation) vector

$$\boldsymbol{\omega}^e = \boldsymbol{\mathcal{E}} : \boldsymbol{\theta} \left(\omega^e_{ij} = e_{ijk}\theta_k \right) \quad \text{or} \quad \boldsymbol{\theta} = \frac{1}{2}\boldsymbol{\mathcal{E}} : \boldsymbol{\omega}^e \left(\theta_k = \frac{1}{2} e_{ijk}\omega^e_{ij} \right), \tag{9.43}$$

where $\boldsymbol{\mathcal{E}} := e_{ijk}\mathbf{e}_i \otimes \mathbf{e}_j \otimes \mathbf{e}_k$ is the alternating tensor, e_{ijk} is the permutation symbol, and θ_k is the axial vector of the elastic rotation ω^e_{ij}. By virtue of (9.43), we can further write

$$e_{ik\ell}\omega^e_{\ell j,k} = e_{ik\ell}e_{\ell jm}\theta_{m,k} = \theta_{k,k}\delta_{ij} - \theta_{i,j}, \tag{9.44}$$

where the Kronecker symbol δ_{ij} has been used. Now we can use (9.35), (9.40), and (9.44) to obtain the following expression for the geometrically necessary dislocation density

$$\alpha_{ij} = -e_{ik\ell}\epsilon^e_{\ell j,k} + \theta_{i,j} - \theta_{k,k}\delta_{ij}. \tag{9.45}$$

If we introduce the curvature $\boldsymbol{\kappa}$ and the curl of the elastic strain $\boldsymbol{\zeta}$ as

$$\boldsymbol{\kappa} := \nabla \otimes \boldsymbol{\theta} \tag{9.46}$$

$$\boldsymbol{\zeta} := \nabla \times \boldsymbol{\epsilon}^e, \tag{9.47}$$

we can rewrite (9.45) as

$$\alpha_{ij} = -\zeta_{ij} + \kappa_{ji} - \kappa_{kk}\delta_{ij}. \tag{9.48}$$

Because $\epsilon^e_{\ell i} = \epsilon^e_{i\ell}$, $\zeta_{ii} = e_{ik\ell}\epsilon^e_{\ell i,k} \equiv 0$. We can link the trace of the curvature κ_{kk} with the trace of the geometric necessary dislocation density α_{kk} as $\kappa_{kk} = -\frac{1}{2}\alpha_{kk}$, which further allows us to write the inverse relation of (9.48) as

$$\kappa_{ij} = \zeta_{ji} + \alpha_{ji} - \frac{1}{2}\alpha_{kk}\delta_{ij}. \tag{9.49}$$

Summarizing (9.48) and (9.49), we have the following kinematic relations,

$$\boldsymbol{\alpha} = -\boldsymbol{\zeta} + \boldsymbol{\kappa}^T - \mathrm{tr}(\boldsymbol{\kappa})\mathbb{I}^{(2)} \quad \text{and} \quad \boldsymbol{\kappa} = \boldsymbol{\zeta}^T + \boldsymbol{\alpha}^T - \frac{1}{2}\mathrm{tr}(\boldsymbol{\alpha})\mathbb{I}^{(2)}. \tag{9.50}$$

where tr is the trace operator. [Nye (1953)] made an approximation to Eqs. (9.50) by neglecting the contribution from the curl of the elastic strain $\boldsymbol{\zeta}$, which is a valid approximation for small elastic strains. This approximation allows us to write Eqs. (9.50) as

$$\boldsymbol{\alpha} = \boldsymbol{\kappa}^T - \mathrm{tr}(\boldsymbol{\kappa})\mathbb{I}^{(2)} \quad \text{and} \quad \boldsymbol{\kappa} = \boldsymbol{\alpha}^T - \frac{1}{2}\mathrm{tr}(\boldsymbol{\alpha})\mathbb{I}^{(2)}. \tag{9.51}$$

which are known as Nye's theory.

9.3.2 *Continuum disclination theory*

For a self-contained exposition, we briefly outline the related disclination theory. In the literature, there are several different disclination theories. The one presented below is an adaption of the theory developed in Anthony [Anthony (1970)] and de Wit [deWit (1970)].

Consider (9.48) as the definition of dislocation density α_{ij}, meaning that

$$\boldsymbol{\alpha} \neq -\nabla \times \boldsymbol{\beta} \qquad \text{and} \qquad \boldsymbol{\kappa} \neq \nabla\boldsymbol{\theta} \ . \tag{9.52}$$

By superposition, the total dislocation density due to contortion and the total curvature due to dislocation as well as disclination are

$$\boldsymbol{\alpha} = \tilde{\boldsymbol{\alpha}} - \nabla \times \boldsymbol{\beta} \qquad \text{and} \qquad \boldsymbol{\kappa} = \tilde{\boldsymbol{\kappa}} + \nabla\boldsymbol{\theta} \ . \tag{9.53}$$

where the curvature $\tilde{\boldsymbol{\kappa}}$ is induced, or defined, by the disclination distribution $\boldsymbol{\phi}$, through a differential relation, i.e.

$$\phi_{ij} = -e_{ik\ell}\tilde{\kappa}_{\ell j,k} \qquad \text{or} \qquad \boldsymbol{\phi} = -\nabla \times \tilde{\boldsymbol{\kappa}} \tag{9.54}$$

Since $\nabla \times \nabla\boldsymbol{\theta} \equiv 0$, we obtain $\boldsymbol{\phi} = -\nabla \times \boldsymbol{\kappa}$. Due to the fact that the disclination density is anti-symmetric, $\phi_{ij} = -\phi_{ji}$, we can also use an axial vector to denote the disclination distribution, i.e.

$$\varphi_k = \frac{1}{2}e_{ijk}\phi_{ij} \qquad \text{or} \qquad \phi_{ij} = e_{ijk}\varphi_k \ . \tag{9.55}$$

It is readily to verify that

$$\varphi_k = \frac{1}{2}e_{ijk}\phi_{ij} = -\frac{1}{2}e_{ijk}e_{imn}\tilde{\kappa}_{nj,m} = -\frac{1}{2}\left(\tilde{\kappa}_{km,m} - \tilde{\kappa}_{nn,k}\right) . \tag{9.56}$$

If we take the divergence $\alpha_{ij,i}$, we obtain the following expression,

$$\alpha_{ij,i} = -e_{ik\ell}\epsilon_{\ell j,ki} + \kappa_{ji,i} - \kappa_{kk,j} + \tilde{\kappa}_{ji,i} - \tilde{\kappa}_{kk,j} = \tilde{\kappa}_{ji,i} - \tilde{\kappa}_{kk,j} = -2\varphi_j \tag{9.57}$$

We then obtain the dislocation continuity equation under the combined dislocation and disclination distributions,

$$\alpha_{ij,i} + 2\varphi_j = 0 \qquad \text{or} \qquad \nabla\cdot\boldsymbol{\alpha} + 2\boldsymbol{\varphi} = 0 \ . \tag{9.58}$$

Following similar arguments, we can also show that

$$\gamma_{ij,i} + 2\varphi_j = 0 \qquad \text{or} \qquad \nabla\cdot\boldsymbol{\gamma} + 2\boldsymbol{\varphi} = 0 \ , \tag{9.59}$$

because $\zeta_{ij,i} \equiv 0$. If the disclination density is zero, we recover the pure dislocation continuity equations, i.e. $\alpha_{ij,i} = 0$ or $\gamma_{ij,i} = 0$.

9.3.3 *Re-combination I: The generalized Nye theory*

A hidden beauty of Nye's theory is that it leads to a canonical 'dislocation potential' representation

$$W^d = \frac{1}{2}\boldsymbol{\alpha} : \boldsymbol{\kappa}$$

which allows us to express geometrically compatibility in terms of a defect potential function, because $\boldsymbol{\alpha}$ and $\boldsymbol{\kappa}$ are related by Eq. (9.50). Nevertheless, the Nye theory is only an approximation theory. Stimulated by Nye's dislocation potential for continuum kinematics, we are seeking for the exact defect potential representation of the kinematic relationships (9.50). When $\zeta_{ij} \neq 0$, there are three types of kinematic field variables involved in (9.50), which cannot lead to canonical defect potential representation. To obtain a meaningful result, we construct defect potentials by "recombining" — which is an operation opposite to decomposition — the kinematic fields.

To construct the defect potential, we first re-combine the geometric necessary dislocation density tensor α_{ij} with the curl of the elastic strain ζ_{ij} to define a new geometric object γ_{ij}

$$\gamma_{ij} := \zeta_{ij} + \alpha_{ij} \qquad \text{or} \qquad \boldsymbol{\gamma} := \boldsymbol{\zeta} + \boldsymbol{\alpha} \ . \tag{9.60}$$

which is the negative curl of the elastic rotation γ_{ij}.

Eq. (9.35) allows us to write the negative curl of the elastic rotation γ_{ij} as

$$\gamma_{ij} = -e_{ik\ell}\omega^e_{\ell j,k} \qquad \text{or} \qquad \boldsymbol{\gamma} = -\nabla \times \boldsymbol{\omega}^e \ . \tag{9.61}$$

Since the divergence of the geometrically necessary dislocation density α_{ij} vanishes by (9.41) and due to the commutability of partial derivatives, one can immediately find the following governing equation for the defined quantity,

$$\gamma_{ij,i} \equiv 0 \ . \tag{9.62}$$

Substituting (9.60) into (9.50) and using the fact that due to $\zeta_{kk} = 0$ we obtain $\gamma_{kk} = \alpha_{kk}$, we find the relation between the negative curl of the elastic rotation γ_{ij} and the curvature tensor κ_{ij},

$$\gamma_{ij} = \kappa_{ji} - \kappa_{kk}\delta_{ij}, \quad \kappa_{ij} = \gamma_{ji} - \frac{1}{2}\gamma_{kk}\delta_{ij} \tag{9.63}$$

which we denote as the generalized Nye relation. When $\zeta_{ij} = 0$, Eqs. (9.63) degenerate to the original Nye relation in terms of the geometrically necessary dislocation density α_{ij} and the curvature tensor κ_{ij}

$$\alpha_{ij} = \kappa_{ji} - \kappa_{kk}\delta_{ij}, \quad \kappa_{ij} = \alpha_{ji} - \frac{1}{2}\alpha_{kk}\delta_{ij} \tag{9.64}$$

Remark 9.1. In the generalized Nye theory, a defect measure γ_{ij} is defined in (9.60) by re-combination of the kinematic quantities in (9.50). γ_{ij} can be interpreted as a measure of defect density. The recombination of kinematic variables allows us to form the following defect potential function,

$$W^{(1)} = \frac{1}{2}\boldsymbol{\gamma} : \boldsymbol{\kappa}$$

This defect potential has very rich physical implications. It may be used to characterize both dislocation and disclination distributions.

9.3.4 *Re-combination II: The Kröner-deWit theory*

In Eqs. (9.50), we can define the elastic contortion K_{ij} by recombining the curvature tensor κ_{ij} with the curl of the elastic strain tensor ζ_{ij} such that

$$K_{ij} := \kappa_{ij} - \zeta_{ji} \quad \text{or} \quad \mathbf{K} := \boldsymbol{\kappa} - \boldsymbol{\zeta}^T \,. \tag{9.65}$$

where $(\,\cdot\,)^T$ denotes the transpose of $(\,\cdot\,)$. Noticing that the trace $\zeta_{ii} \equiv 0$ and therefore $K_{ii} = \kappa_{ii}$, we obtain the Kröner-deWit relation in terms of the geometrically necessary dislocation density α_{ij} and the contortion K_{ij} as

$$\alpha_{ij} = K_{ji} - K_{kk}\delta_{ij}, \quad \text{and} \quad K_{ij} = \alpha_{ji} - \frac{1}{2}\alpha_{kk}\delta_{ij} \tag{9.66}$$

which is an exact relation in dislocation theory. According to [deWit (1970)] , this re-combination was first proposed by Kröner [Kröner (1967)]. Again when $\zeta_{ij} = 0$, it degenerates to the original Nye theory. Comparing the Kröner-deWit relations in terms of the contortion (9.66) with the generalized Nye relations (9.63), one may interpret γ_{ij} as either the curl of elastic rotation, or a quantity generated purely by curvature κ_{ij} (geometrically exact), whereas α_{ij} is the dislocation density generated by the contortion K_{ij}.

With the second kinematic variable re-combination, we propose the second type of defect potentials that has two variants:

$$W^{(2a)} = \frac{1}{2}\boldsymbol{\alpha} : \mathbf{K}, \quad \text{and} \quad W^{(2b)} = \frac{1}{2}\boldsymbol{\alpha} : \mathbf{K}^T \tag{9.67}$$

When the GND density vanishes i.e. $\boldsymbol{\alpha} = 0$, this type of defect potential is a null field, and therefore this type of defect potentials may be a good measure for GND distributions.

If one assumes that $\kappa_{ij} \equiv 0$, we may write $W^{(2a)}$ and $W^{(2b)}$ in terms of $\boldsymbol{\zeta}$,

$$W^{(3a)} = \frac{1}{2}\boldsymbol{\zeta} : \boldsymbol{\zeta}^T, \quad \text{and} \quad W^{(3b)} = \frac{1}{2}\boldsymbol{\zeta} : \boldsymbol{\zeta} \tag{9.68}$$

In fact, one can construct other types of defect potential. For a detailed and complete theory of defect potentials, one may consult [Li (2008)].

9.3.5 *Compatibility conservation laws*

The recombination of kinematic field variables outlined in the previous section inspired us to construct various defect potentials. These defect potentials can be used to provide variational structures for the construction of compatibility conservation laws. This procedure is instrumental in leading to the discovery of *configurational compatibility*, a concept dual to the well established notion of the *configurational force*.

To illustrate the parallel structures between configurational forces and configurational compatibilities, we first outline some of the most important conservation laws of the Navier equations of linear elasticity, which are manifestations of the variational symmetries the elastic potential energy (e. g. see Knowles and Sternberg [Knowles and Sternberg (1972)]),

$$\Pi^{(0)}(\nabla \mathbf{u}) := \int_V W^{(0)}(\nabla \mathbf{u}) dV \quad \text{where} \quad W^{(0)} := \frac{1}{2}\sigma_{ij}\epsilon_{ij} \tag{9.69}$$

where ϵ_{ij} and σ_{ij} are the strains and stresses, respectively. Under the invariant transformation, it may result infinitely many conservation laws. We list the most important three conservation laws below [Knowles and Sternberg (1972); Budiansky and Rice (1973)]:

$$\begin{aligned} \mathbf{CL1}: \quad D^{(0)}_{k\alpha} &= W^{(0)}\delta_{k\alpha} - u_{\ell,\alpha}\sigma_{k\ell}, \\ \rightarrow \quad J_\alpha &= \oint_S C^{(0)}_{k\alpha} n_k dS \; ; \end{aligned} \tag{9.70}$$

$$\begin{aligned} \mathbf{CL2}: \quad G^{(0)}_{k\alpha} &= e_{\alpha\ell\beta}\Big(x_\ell E^{(0)}_{k\beta} + u_\ell \sigma_{k\beta}\Big) \\ \rightarrow \quad L_\alpha &= \oint_S G^{(0)}_{k\alpha} n_k dS \; ; \end{aligned} \tag{9.71}$$

$$\begin{aligned} \mathbf{CL3}: \quad H^{(0)}_k &= x_\alpha E^{(0)}_{k\alpha} - \frac{1}{2} u_\ell \sigma_{k\ell} \\ \rightarrow \quad M &= \oint_S H^{(0)}_k n_k dS \; ; \end{aligned} \tag{9.72}$$

We now construct conservational laws for continuum compatibility.

1. Conservation laws of the generalized Nye theory

Consider

$$\kappa_{ij} = \theta_{j,i} \tag{9.73}$$

$$\gamma_{ij} = \theta_{i,j} - \theta_{k,k}\delta_{ij} \tag{9.74}$$

The first defect potential may be expressed as

$$W^{(1)} = \frac{1}{2}\gamma_{ij}\kappa_{ij} = \frac{1}{2}\Big(\theta_{i,j}\theta_{j,i} - \theta_{k,k}\theta_{\ell,\ell}\Big) \tag{9.75}$$

Assume that the elastic rotation θ_i is prescribed on ∂V. Then the boundary value problem (9.76),

$$\gamma_{ij,i} = 0, \quad \forall \mathbf{x} \in V \quad \text{and} \quad \theta_i = \bar{\theta}_i, \quad \forall \mathbf{x} \in \partial V, \tag{9.76}$$

is equivalent to the stationary condition of the following functional

$$\Pi^{(1)}(\boldsymbol{\theta}_{,i}) := \int_V W^{(1)}(\boldsymbol{\theta}_{,i}) dV. \tag{9.77}$$

Taking the first variation of (9.77) and integration by parts results in

$$\delta\Pi^{(1)} = \int_V \frac{\partial W^{(1)}}{\partial \kappa_{ij}} \delta\kappa_{ij} dV = \int_V \gamma_{ij}\delta\theta_{j,i} dV = \int_{\partial V} \gamma_{ij} n_i \delta\theta_j dS - \int_V \gamma_{ij,i}\delta\theta_j dV$$

The stationary condition $\delta\Pi^{(1)} = 0$ then leads to

$$\gamma_{ij,i} = 0, \quad \forall \mathbf{x} \in V \quad \text{and} \quad \theta_i = \bar{\theta}_i, \quad \forall \mathbf{x} \in \partial V, \tag{9.78}$$

which is precisely (9.76). Therefore we may call (9.76) as the Euler-Lagrange equations of the fundamental integral (9.77).

Then based on the celebrated Noether's theorem (Appendix D), the following compatibility conservation laws hold or the following contour integrals are path independent:

$$S^{(1)}_{k\alpha} = W^{(1)}\delta_{k\alpha} - \theta_{\ell,\alpha}\gamma_{k\ell} \to L^{(1)}_{\alpha} = \oint_S S^{(1)}_{k\alpha} n_k dS \tag{9.79}$$

where $W^{(1)} = \frac{1}{2}\gamma_{ij}\kappa_{ij}$. The proof of this theorem can be found in [Li (2008)].

Remark 9.2. **1.** There is a direct analogue between the conservation law **1** with Eshelby's energy-momentum tensor given in Eq. (9.70) (**CL-1**), where the displacement u_ℓ corresponds to the lattice rotation θ_ℓ, displacement gradient $u_{\ell,\alpha}$ corresponds to the lattice curvature $\kappa_{\alpha\ell} = \theta_{\ell,\alpha}$, and the Cauchy stress $\sigma_{k\ell}$ corresponds to the negative curl of elastic rotation $\gamma_{k\ell}$. **2.** Since the energy momentum tensor in Eq. (9.70) can be derived as the invariance of coordinate translation of strain energy density, we denote the new conserved quantity $S^{(1)}_{k\alpha}$, which is obtained as the invariance of coordinate translation of a defect potential, as *compatibility-momentum tensor* or *defect-momentum tensor*.

In specific, the defect potential field will yield a true variational principle, if

$$\gamma_{ij,i} = 0, \quad \text{or} \quad \alpha_{ij,i} = 0 \ .$$

According to Eqs. (9.58) and (9.59), this implies that the constructed defect potential, $W^{(1)} = \frac{1}{2}\boldsymbol{\gamma} : \boldsymbol{\kappa}$, is a measure of the disclination density distribution. In other words, if the disclination distribution is a null field, i.e. $\boldsymbol{\varphi} = 0$, the contour integral in Eq. (9.79) is path-independent.

2. Conservation laws for the Kröner-deWit theory

Let

$$W^{(2)} = \frac{1}{2}\boldsymbol{\alpha} : \boldsymbol{K} = \frac{1}{2}\alpha_{ij}K_{ij} \tag{9.80}$$

Considering the definition that

$$\boldsymbol{\alpha} = -\nabla \times \boldsymbol{\beta}^e$$

one can show that the Euler-Lagrange equation of the following functional

$$\Pi^{(2)}(\nabla\boldsymbol{\beta}^e) := \int_V W^{(2)}(\nabla\boldsymbol{\beta}^e)dV. \tag{9.81}$$

is

$$\nabla \times \mathbf{K} = 0, \quad \Rightarrow \quad \nabla \times \boldsymbol{\epsilon}^e \times \nabla = 0 \tag{9.82}$$

whose physical validity renders $W^{(2)}$ variationally meaningful. By apply the Noether theorem (see Appendix D), one may find that the following conservation law of the variational principle,

$$S^{(2)}_{k\alpha} = W^{(2)}\delta_{k\alpha} - e_{kmi}K_{mj}\beta^e_{ij,\alpha} \to L^{(2)}_{\alpha} = \oint_S S^{(2)}_{k\alpha} n_k dS \tag{9.83}$$

3. Conservation laws for the Saint-Venant theory
The compatibility equations e.g. [Malvern (1969)],

$$e_{irk}e_{js\ell}\epsilon^e_{rs,\ell k} = 0, \quad \forall \mathbf{x} \in V \quad \text{and} \quad \epsilon^e_{ij} = \bar{\epsilon}_{ij}, \quad \forall \mathbf{x} \in \partial V \tag{9.84}$$

for infinitesimal deformation can be actually derived by the following variational principle [Tonti (1967)].

Let

$$W^{(3)} := \frac{1}{2}\boldsymbol{\zeta} : \boldsymbol{\zeta}^T = \frac{1}{2}\zeta_{ij}\zeta_{ji} = -\frac{1}{2}e_{irk}e_{js\ell}\epsilon^e_{rs,\ell}\epsilon^e_{ij,k} \tag{9.85}$$

and assume that ϵ_{ij} is prescribed on ∂V. The boundary value problem (9.84) is then equivalent to the stationary condition of the following functional,

$$\Pi^{(3)}(\nabla\boldsymbol{\epsilon}^e) := \int_V W^{(3)}(\nabla\boldsymbol{\epsilon}^e)dV \tag{9.86}$$

Then according to Noether's theorem for a tensorial field (Appendix D), the following strain compatibility conservation laws hold:

$$S^{(3)}_{k\alpha} = W^{(3)}\delta_{k\alpha} - e_{mki}\zeta_{jm}\epsilon^e_{ij,\alpha} \to L^{(3)}_{\alpha} = \oint_S S^{(4)}_{k\alpha}n_k dS \tag{9.87}$$

9.4 Multiscale energy-momentum tensor

We now apply the configurational compatibility tensor to study ductile fracture problem [Li (2007)]. We begin with construction of multiscale energy-momentum tensor. The idea and procedure of forming a multiscale energy momentum tensor are as follows. Assume that the total infinitesimal displacement field of a deformed solid can be decomposed into multiscale components,

$$\mathbf{u}(\mathbf{X}) = \bar{\mathbf{u}}(\mathbf{X}) + \mathbf{u}'(\mathbf{X}) \tag{9.88}$$

where $\bar{\mathbf{u}}$ is the coarse (macro-) scale displacement field, which may be viewed as a mean field; while $\mathbf{u}'$ is the fine (micro-) scale displacement field, which can be affected by the presence or fluctuation of defect distributions. Moreover, we assume that the fine scale defect distribution is localized so after homogenization the coarse-scale deformation field is simply a linear elastic solid. The total strain field can thus decompose to,

$$\boldsymbol{\epsilon} = \bar{\boldsymbol{\epsilon}} + \boldsymbol{\epsilon}' = \bar{\boldsymbol{\epsilon}} + \boldsymbol{\epsilon}^{e\prime} + \boldsymbol{\epsilon}^{p\prime} = \boldsymbol{\epsilon}^e + \boldsymbol{\epsilon}^p \tag{9.89}$$

where $\bar{\boldsymbol{\epsilon}} = \bar{\boldsymbol{\epsilon}}^e = Sym\Big(\nabla \otimes \bar{\mathbf{u}}\Big)$, and both $\boldsymbol{\epsilon}'_e$ and $\boldsymbol{\epsilon}'_p$ are incompatible elastic and plastic strains. Note that $\boldsymbol{\epsilon}^e = \bar{\boldsymbol{\epsilon}}^e + \boldsymbol{\epsilon}^{e\prime}$ and $\boldsymbol{\epsilon}^p = \boldsymbol{\epsilon}^{p\prime}$. We now propose the following Multiscale Free-Energy density,

$$W^m(\bar{\mathbf{u}}, \boldsymbol{\epsilon}^{e\prime}) = W^c(\bar{\mathbf{u}}) + W^f(\boldsymbol{\epsilon}^{e\prime}) \tag{9.90}$$

where the coarse scale strain energy density is

$$W^c = \frac{1}{2}\bar{\boldsymbol{\epsilon}}^e : \bar{\mathbf{C}} : \bar{\boldsymbol{\epsilon}}^e \tag{9.91}$$

and $\bar{\mathbf{C}}$ is the coarse scale elastic stiffness tensor. The fine scale free-energy density is

$$W^f(\epsilon^{e\prime}) := \frac{\mu\ell^2}{2}\boldsymbol{\zeta} : \boldsymbol{\zeta}^T = \frac{\mu\ell^2}{2}\zeta_{ij}\zeta_{ji}, \tag{9.92}$$

where $\zeta_{ij} := e_{ik\ell}\epsilon^e_{\ell j,k} = e_{ik\ell}\epsilon^{e\prime}_{\ell j,k}$ is the curl of elastic strain, $e_{ik\ell}$ is the permutation symbol, μ is the fine scale elastic shear modulus, and ℓ is a length scale or the Gauge length scale [Kleinert (1989)], and a coarse scale observer cannot distinguish any differences between two objects below such length scale. For instance, two integration pathes of the coarse scale make no difference whether it is on the circle or inside the circle in Fig. 9.7.

Note that the scaling factor, $\mu\ell^2$, makes the unit of $W^f(\epsilon^{e\prime})$ to energy density.

In the previous section, we have found that the defect potential (9.92) is variationally meaningful. That is: the first variation of the potential yields a stationary condition that is the Saint-Venant compatibility condition [Tonti (1967)],

$$\delta \int_V W^f(\epsilon^e)dV = 0 \quad \Rightarrow \quad e_{irk}e_{js\ell}\epsilon^e_{rs,\ell k} = 0 \ , \tag{9.93}$$

if the domain of interests is compatible or defect-free. Therefore the physical meaning of the potential (9.92) is a quantity that is proportional to the free-energy of the defect, in specific, the dislocation distribution.

We can then label $S^{(3)}_{k\alpha}$ as a fine scale compatibility-momentum tensor, i.e.

$$S^f_{k\alpha} = W^f\delta_{k\alpha} - \mu\ell^2 e_{mki}\zeta_{jm}\epsilon^e_{ij,\alpha} \tag{9.94}$$

We then form the multiscale energy-momentum tensor by an addictive composition of coarse scale energy-momentum tensor and the compatibility-momentum tensor,

$$S^m_{k\alpha} = S^c_{k\alpha} + S^f_{k\alpha} \tag{9.95}$$

in which

$$S^c_{k\alpha} = W^c\delta_{k\alpha} - \bar{u}_{\ell,\alpha}\bar{\sigma}_{k\ell} \tag{9.96}$$

is Eshelby's energy momentum tensor [Eshelby (1951, 1956)], $\bar{\sigma}_{ij}$ is the coarse scale Cauchy stress, $\bar{u}_\ell$ is the coarse scale displacement field, and $\delta_{k\alpha}$ is the Kronecker delta. Note that the subscript, $()_{,i}$, denotes the spatial derivative.

We argue that the stress measure is a macro quantity, and the strain measure can be micro quantity. Accordingly, the multiscale energy momentum tensor, $S^m_{k\alpha}$, has two-scale components: a coarse part to describe macroscale brittle cleavage surface separation and a fine scale part to describe dislocation motions.

By virtue of the equilibrium equation $\sigma_{ij,i} = 0$, one can show that Eshelby's energy momentum tensor obeying $S^c_{k\alpha,k} = 0$. Similarly via the compatibility condition (9.93), it is readily to show that $S^f_{k\alpha,k} = 0$, if there is no inelastic deformation in the solid. We may call $S^f_{k\alpha}$ as the compatibility-momentum tensor, because it is derived based on the symmetry condition of compatibility conditions [Li *et al.*

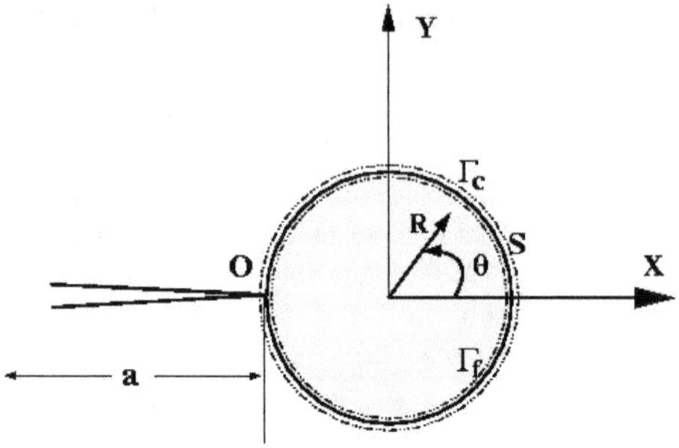

Fig. 9.7 Schematic illustration of a macroscopically brittle crack.

(2007a)] . Conceptually, configurational compatibility forms a duality pair with configurational force [Gurtin (1995)] .

We then calculate the multiscale configurational force,

$$\mathcal{L}_\alpha = \bar{J}_\alpha + L_\alpha \tag{9.97}$$

where the coarse scale configurational force is the J-integral [Rice (1968)],

$$\bar{J}_\alpha := \oint_{\Gamma_c} S^c_{k\alpha} n_k dS \tag{9.98}$$

where n_k is the surface normal of Γ_c, and the fine scale configurational force is the L-integral,

$$L_\alpha := \oint_{\Gamma_f} S^f_{k\alpha} n_k dS \tag{9.99}$$

which is the measure of configurational compatibility. Since both integrals are path-independent, and hence their linear combination is also path-independent. In the rest of this section, we denote the first component of $\mathcal{L}_\alpha$ as the multiscale L-integral, i.e. $L^m = \mathcal{L}_1$, which may represent the driving force for a macroscopically brittle crack moving in x_1 direction.

As an example, we now calculate the L^m-integral for a mode-III elasto-plastic crack growth problem by using the steady state solution of the Hula-McClintock (HM) [Hult and McClintock (1957)], which is under the assumption of small scale yielding. The integration contour, Γ_c, can be taken arbitrarily over the coarse scale field as long as it contains the crack tip; whereas the fine scale integration contour, Γ_f, is taken as the boundary of plastic, or process zone, region, i.e. S (see Fig. 9.7). In the calculation, we choose the contour for the coarse scale J-integral is slightly (infinitesimally) larger as S^+ (since it cannot see the fine scale), whereas the contour for the fine scale L-integral is slightly (infinitesimally) smaller as S^-.

Suppose the crack length is denoted as a, the remote stress is τ_∞. The multiscale driving force is

$$L^m = \left(\frac{\pi\tau_\infty^2}{2\mu}\right)a + \left(\frac{3\ell^2\pi\tau_0^4}{8\mu\tau_\infty^2}\right)\frac{1}{a} \,. \tag{9.100}$$

where we neglect the difference between the macro shear modulus and the micro shear modulus, i.e. $\bar{\mu} = \mu$.

To investigate the behaviors of the multiscale driving force, we perform a stability analysis of the equilibrium. For a macroscopically brittle fracture, we may assume a constant fracture resistance, i.e. $R = const.$ and $\dfrac{\partial R}{\partial a} = 0$. Thus $\dfrac{\partial L^m}{\partial a} < 0$ implies stable crack growth, or simply stability. The minimum may be found via the stationary condition,

$$\left.\frac{\partial L^m}{\partial a}\right|_{\tau_\infty} = \left(\frac{\pi\tau_\infty^2}{2\mu}\right) - \left(\frac{3\ell^2\pi\tau_0^4}{8\mu\tau_\infty^2}\right)\frac{1}{a^2} = 0 \tag{9.101}$$

Under load control, the minimum driving force to advance a crack and the stability point are given as

$$L^m_{min} = \frac{\pi\tau_\infty^2 a_{min}}{\mu}, \quad \text{and} \quad a_{min} = \frac{\sqrt{3}}{2}\left(\frac{\tau_0}{\tau_\infty}\right)^2\ell \tag{9.102}$$

The physical meaning of a_{min} is the crack length at instability.

We note that these findings indicate that an incompatible field will yield a minimum driving force. In addition, incompatibility enables stable crack growth during load control. To frame discussion with respect to the applied loading, we find the far-field stress at instability as

$$\tau_{\infty,min} = \left(\frac{3}{4}\right)^{1/4}\sqrt{\frac{\ell}{a_{min}}}\tau_0. \tag{9.103}$$

By assuming that $a_{min} \sim \mathcal{O}(\ell)$, the critical stress for brittle fracture at small scale may be estimated as $\tau_\infty^{cr} \sim 0.75\tau_0$ where τ_0 may be viewed as theoretical strength or the cohesive strength of the material. The driving force can be normalized by L^m_{min} as,

$$\frac{L^m}{L^m_{min}} = \frac{1}{2}\left(\frac{a}{a_{min}} + \frac{1}{a/a_{min}}\right) \tag{9.104}$$

In Fig. 9.8, we plot the multiscale driving force with different normalization against the normalized crack length. It can be seen from Fig. 9.8 that there is a well located minimum at a_{min}. This suggests that the driving force for crack growth at small scale cannot be zero, even if the crack length, a, approaches zero. The physical explanation for this is because the total energy release has two sources : (1) the strain energy release due to the surface separation at macroscale, and (2) misfit energy release, as a form of strain-gradient energy release, due to the change of the defect potential. The change of the defect potential can be interpreted as due to GND's absorption or release, deformation twinning, or other incompatible strain

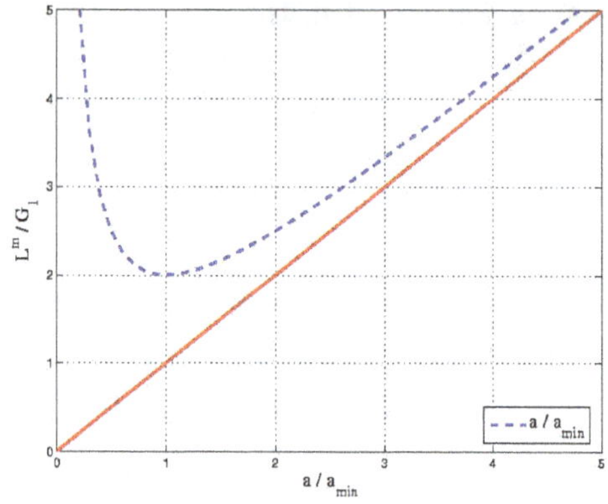

Fig. 9.8 The normalized driving force L^m/L^m_{min} vs. a/a_{min}.

field releases. At the small scale, even when there are not too many bond breaking, i.e. when $a \to 0$, dislocations may still be present. The competition of these two factors dictates overall behavior of the driving force.

A fundamental task of fracture mechanics is to determine the critical stress, τ_{cr}, under which the crack advances. Based on the Griffith criterion [Griffith (1921)], the critical stress can be obtained by an equilibrium condition — the balance of configurational force and resistance force.

Following Griffith's energetic argument, we equate the multiscale driving force to the resistance,

$$L^m = \left(\frac{\pi a}{2\mu}\right)(\tau_{cr})^2 + \left(\frac{3\ell^2 \pi \tau_0^4}{8\mu a}\right)\frac{1}{(\tau_{cr})^2} = 2\gamma_t \tag{9.105}$$

We observe that multiscale driving force has two parts: (1) the coarse scale part i.e. the release of elastic strain energy, or the value of the J-integral in an elastic medium, $\bar{J} = \dfrac{\pi a(\tau_{cr})^2}{2\mu}$ and (2) the fine scale part due to the release of the elastic free-energy stored inside the dislocation distribution zone, or the plastic zone, $L = \left(\dfrac{3\ell^2 \pi \tau_0^4}{8\mu a \tau_{cr}^2}\right)$. Note that in the multiscale Griffith equation the first part of the driving force may no longer be equal to the resistance due to surface separation, i.e. $2\gamma_s$. In other words, the strain energy release due to the reduction of the elastic potential in elastic region will not be solely consumed in surface separation. To expedite the analysis, we introduce a critical length scale, $\ell_{cr} := \dfrac{4}{\sqrt{3}}\left(\dfrac{\gamma_t \mu}{\pi \tau_0^2}\right)$, which is a function of the total resistance, elastic constant, and magnitude of yield stress. Hence its value depends on the resistance curve commonly referred to as the R-curve e.g. [Kanninen and Popelar (1985)]. If we scale the energy release with a reference resistance energy, $2\gamma_0 := \pi \ell \tau_0^2/2\mu$, which may be viewed as the fracture resistance that the theoretical strength of the material can offer for an ideally brittle crack, the ratio

$$I(\ell) := \frac{\bar{J}}{2\gamma_0} = \Big(\frac{a}{\ell}\Big)\Big(\frac{\tau_{cr}(a)}{\tau_0}\Big)^2 \tag{9.106}$$

is a function of the length scale ℓ. The symbol I is used in honor of G. R. Irwin . Subsequently, the multiscale Griffith equation (9.105) is normalized as

$$I(\ell) + \frac{3}{4}\frac{1}{I(\ell)} = \frac{4\gamma_t \mu}{\pi \ell \tau_0^2} = \sqrt{3}\Big(\frac{\ell_{cr}}{\ell}\Big) \tag{9.107}$$

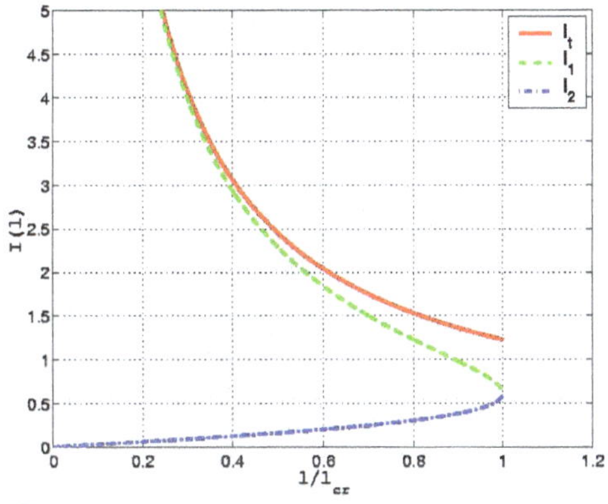

Fig. 9.9 Bifurcated solutions for energy releases $I_i(\ell), i = 1,2$ and $I_t(\ell)$ vs. ℓ/ℓ_{cr}.

Unlike the classical Griffith equation, the multiscale Griffith equation is a quadratic equation in terms of energy release $I(\ell)$. Consequently, the multiscale Griffith equation (9.107) yields two solutions:

$$I(\ell)_{1,2} = \frac{I_t(\ell)}{2}\Big[1 \pm \sqrt{1-(\frac{\ell}{\ell_{cr}})^2}\Big] = \Big(\frac{\gamma_t}{2\gamma_0}\Big)\Big[1 \pm \sqrt{1-(\frac{\ell}{\ell_{cr}})^2}\Big] \tag{9.108}$$

where $I_t(\ell)$ is the normalized total resistance at the equilibrium,

$$I_t(\ell) = \gamma_t/\gamma_0 = \sqrt{3}\Big(\frac{\ell_{cr}}{\ell}\Big). \tag{9.109}$$

To compare different energy releases, we plot the three normalized energy releases, $I_i(\ell), i = 1,2$ and $I_t = I_1 + I_2$ in Fig. 9.9. One may find that the first two solutions of I_i bifurcate at $\ell = \ell_{cr}$. Using the definition (9.106), we find the corresponding critical stresses as follows,

$$\tau_{1,2}^{cr} = \tau_0\sqrt{\frac{\ell I_{1,2}(\ell)}{a}} = \sqrt{\frac{4\mu\gamma_t}{a\pi}}\Big[\frac{1}{2}\Big(1 \pm \sqrt{1-\frac{\ell}{\ell_{cr}}}\Big)\Big]^{1/2} = \tau_I^{cr} f(\ell)_{1,2} \tag{9.110}$$

where $\tau_I^{cr} := \sqrt{\dfrac{4\mu\gamma_t}{a\pi}}$ denotes the critical Irwin stress, and the scaling factors are defined as

$$f_{1,2}(\ell) := \Big[\frac{1}{2}\Big(1 \pm \sqrt{1-\frac{\ell}{\ell_{cr}}}\Big)\Big]^{1/2} \leq 1.0 \tag{9.111}$$

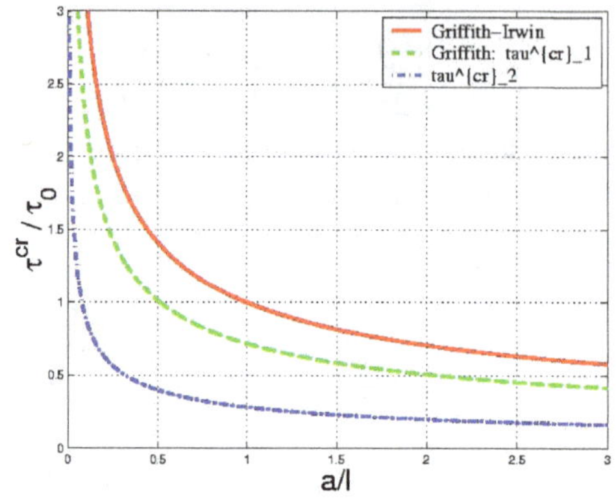

Fig. 9.10 The critical stresses vs. a/ℓ at $\ell/\ell_{cr} = 0.9$: (a) The Griffith-Irwin τ_I^{cr}/τ_0, (b) The 1st Multiscale (Griffith) solution τ_1^{cr}/τ_0, and (c) The 2nd Multiscale Solution τ_2/τ_0.

which are functions of the length scale parameter ℓ/ℓ_{cr}. In Fig. 9.10, the critical stresses corresponding to the multiscale Griffith criterion are compared with the Griffith-Irwin stress.

At equilibrium, the energy release solutions can be interpreted as either the driving forces or the resistances. To explore physical meanings of the two solutions, we examine the asymptotic expressions of the critical stresses related to $I_{1,2}(\ell)$:

$$\tau_1^{cr} = \tau_I^{cr} f_1(\ell) \approx \sqrt{\frac{4\gamma_t \mu}{\pi a}} + \mathcal{O}(\ell), \quad \text{and} \quad \tau_2^{cr} = \tau_I^{cr} f_2(\ell) \approx \sqrt{\frac{3\pi \ell^2 \tau_0^2}{16 a \gamma_t \mu}} + \mathcal{O}(\ell^2) \quad (9.112)$$

One can find that the stress corresponding to $I_1(\ell)$ is independent from the yield stress, τ_0. This indicates that the first solution, $I_1(\ell)$, may be related to the resistance to surface separation, i.e. $I_1(\ell) \sim \gamma_s/\gamma_0$, where γ_s is the resistance due to surface separation; Whereas one may find in (9.112) that τ_2 depends on the yield stress τ_0, and hence we identify that $I_2(\ell)$ corresponds to the resistance due to incompatible defect fields, or the dislocation field. That is $I_2(\ell) \sim \gamma_p/\gamma_0$, where γ_p denotes the energy dissipation due to the presence of dislocations. Fortuitously, the two roots of the multiscale Griffith equation (9.108) have an interesting property,

$$I_1(\ell) + I_2(\ell) = I_t(\ell) = \frac{\gamma_t}{\gamma_0} \quad (9.113)$$

Since under equilibrium condition both of I_1 and I_2 can be viewed as resistances as well, this suggests an addictive decomposition of fracture resistance,

$$\gamma_t = \gamma_s + \gamma_p \quad \Rightarrow \quad \tau_I^{cr} = \sqrt{\frac{4\mu(\gamma_s + \gamma_p)}{a\pi}} \quad (9.114)$$

which is the essential result of Irwin's multiscale theory of elastoplastic fracture under small scale yielding. To the best knowledge of the authors, this is the only rigorous justification of Irwin's theory [Irwin (1948)] by using multiscale analysis and continuum theory of dislocations [Li (2007)].

Since $\tau_I^{cr} \geq \tau_1^{cr}, \tau_2^{cr}$, it is natural to choose the Griffith-Irwin stress as the critical stress for its capture combined effects of τ_1^{cr} and τ_2^{cr} as exactly what G. R. Irwin did half century ago. One may view τ_1^{cr} as an approximation of the original Griffith stress for purely brittle fracture, and Fig. 9.10 shows how it compares with the Griffith-Irwin stress τ_I^{cr}.

9.5 Ekkehart Kröner (1919-2000)

Fig. 9.11 Ekkehart Kröner (The photo is kindly provided by Mrs. Kröner).

Dr. E. Kröner was a professor at the University of Stuttgart, and he was the head of the Institute für Theoretische und Angewandte Physik at the University of Stuttgart from 1969 to 1990s. His outstanding scientific contributions to statistical continuum mechanics of heterogeneous media, continuum defect theory in crystals and dislocation theory, and configurational mechanics, have laid foundations on contemporary micro-mechanics, defect mechanics, and configurational mechanics.

9.6 Exercises

Problem 9.1. *Define the* $J_k = \mathcal{F}_k^{inh}$ *and*

$$J = \mathcal{F}_1^{inh} = \oint_\Gamma \left(W dx_2 - \sigma_{ij} n_j \frac{\partial u_i}{\partial x_1} ds \right) . \tag{9.115}$$

Prove the relationship between J and the stress intensity factor for the mode-I crack propagation,

$$J = \frac{(1-\nu^2)K_I^2}{E} \tag{9.116}$$

Choose Γ as a circle of radius r and evaluate the integral as $r \to 0$. Note that the asymptotic stress and displacement fields for the mode-I crack are

$$\left\{ \begin{array}{c} \sigma_{11} \\ \sigma_{12} \\ \sigma_{22} \end{array} \right\} = \frac{K_I}{\sqrt{2\pi r}} \cos\frac{\theta}{2} \left\{ \begin{array}{c} 1 - \sin(\theta/2)\sin(3\theta/2) \\ \sin(\theta/2)\cos(3\theta/2) \\ 1 + \sin(\theta/2)\sin(3\theta/2) \end{array} \right\} \tag{9.117}$$

$$\left\{ \begin{array}{c} u_1 \\ u_2 \end{array} \right\} = \frac{K_I}{2\mu}\sqrt{\frac{r}{2\pi}} \left\{ \begin{array}{c} \cos(\theta/2)\Big(\kappa - 1 + 2\sin^2(\theta/2)\Big) \\ \sin(\theta/2)\Big(\kappa + 1 - 2\cos^2(\theta/2)\Big) \end{array} \right\} \tag{9.118}$$

where

$$\kappa = \left\{ \begin{array}{ll} 3 - 4\nu & \text{for plane strain} \\ (3-\nu)/(1+\nu) & \text{for plane stress} \end{array} \right. \tag{9.119}$$

Chapter 10

SMALL SCALE COARSE-GRAINED MODELS

There are two basic approaches to nanoscale mechanics problems: the top-down approach and the bottom-up approach. The bottom-up approach is an approach that builds its coarse grained model from atomistic models or first principle. The coarse graining procedure can be statistical averaging, simplified lattice dynamics, or matching and bridging different dynamics or mechanics at nanoscale. On the other hand, the top-down approach is through building a refined continuum model to describe the nanoscale mechanics problems. In this Chapter, we shall introduce several small scale mechanics models, which are representatives of both approaches.

10.1 Gurtin-Murdoch surface elasticity model

One of most important features of nanoscale mechanics is the effect of free surface, because as the scale becomes smaller and smaller the surface effect will start playing a key role in mechanical, chemical, thermal, as well as electrical interactions.

Currently, one of few continuum surface mechanics models that are available is the Gurtin-Murdoch surface mechanics model [Gurtin and Murdoch (1975a,b)]. The following presentation largely follows from Gurtin et al. [Gurtin *et al.* (1998)] and [Sharma and Ganti (2004); Mi and Kouris (2006)].

10.1.1 *Projection operator*

At small scale, the surface tension and surface curvature will have significant influences over mechanical behaviors of an interface or even a free surface. In 1975, Gurtin and Murdoch proposed a continuum interface model that is able to address some of those problems in certain degree. Define a projection operator,

$$\mathbf{P} = \mathbb{I}^{(2)} - \mathbf{n} \otimes \mathbf{n} \tag{10.1}$$

or in component form,

$$P_{ij} = \delta_{ij} - n_i n_j \tag{10.2}$$

Note that the projection tensor $\mathbf{P}$ is symmetric.

Example 10.1. Let $\mathbf{n} = \mathbf{e}_3$. In this case, the interface is X_1OX_2-plane. We have

$$\mathbf{P} = \delta_{ij}\mathbf{e}_i \otimes \mathbf{e}_j - \mathbf{e}_3 \otimes \mathbf{e}_3 = \mathbf{e}_1 \otimes \mathbf{e}_1 + \mathbf{e}_2 \otimes \mathbf{e}_2 \tag{10.3}$$

Or in matrix notation,

$$[\mathbf{P}] = \begin{bmatrix} 1\,0\,0 \\ 0\,1\,0 \\ 0\,0\,1 \end{bmatrix} - \begin{bmatrix} 0\,0\,0 \\ 0\,0\,0 \\ 0\,0\,1 \end{bmatrix} = \begin{bmatrix} 1\,0\,0 \\ 0\,1\,0 \\ 0\,0\,0 \end{bmatrix} \tag{10.4}$$

Example 10.2. Consider a spherical surface or interface,

$$f(\mathbf{x}) = \frac{x_i x_i}{R^2} - 1 = 0 \ .$$

The gradient of f is along the direction of the normal,

$$grad f = \frac{\partial f}{\partial x_i}\mathbf{e}_i = \frac{2x_i}{R^2}\mathbf{e}_i$$

and

$$\mathbf{n} = \frac{1}{\|grad f\|} grad f = \frac{x_i}{R}\mathbf{e}_i$$

Then

$$\mathbf{P} = [\delta_{ij} - \frac{x_i x_j}{R^2}]\mathbf{e}_i \otimes \mathbf{e}_j \tag{10.5}$$

Since $\mathbf{P}$ is symmetric, the projection of a vector $\mathbf{A}$ onto a surface is

$$\mathbf{A}^S = \mathbf{P} \cdot \mathbf{A} = \mathbf{A} \cdot \mathbf{P} = A_k P_{ki}\mathbf{e}_i = (A_i - n_i n_k A_k)\mathbf{e}_i \tag{10.6}$$

If $n_i = \delta_{i3}$ ($\mathbf{n} = \mathbf{e}_3$),

$$\mathbf{P} \cdot \mathbf{A} = A_1\mathbf{e}_1 + A_2\mathbf{e}_2 \ .$$

Consider $n_i = x_i/R$ i.e. $\mathbf{n} = \mathbf{e}_r$ (spherical surface) ,

$$\begin{aligned} \mathbf{A}^S &= \Big(A_i - \frac{x_i}{R}\frac{x_k A_k}{R}\Big)\mathbf{e}_i = \mathbf{A} - (\mathbf{A} \cdot \mathbf{n})\mathbf{n} \\ &= A_\theta \mathbf{e}_\theta + A_\phi \mathbf{e}_\phi \end{aligned} \tag{10.7}$$

Consider a second order tensor $\mathbf{B} = B_{ij}\mathbf{e}_i \otimes \mathbf{e}_j$. Let $\mathbf{n} = \mathbf{e}_3$. We have

$$\mathbf{B} \cdot \mathbf{P} = (B_{ij} - \delta_{i3}B_{3j})\mathbf{e}_i \otimes \mathbf{e}_j \quad \Rightarrow \quad \begin{bmatrix} B_{11} & B_{12} & B_{13} \\ B_{21} & B_{22} & B_{23} \\ 0 & 0 & 0 \end{bmatrix} \tag{10.8}$$

The projection of a second order tensor onto the X_1OX_2 plane is defined as

$$\mathbf{B}^S = \mathbf{P} \cdot \mathbf{B} \cdot \mathbf{P} = P_{ik}B_{k\ell}P_{\ell j}\mathbf{e}_i \otimes \mathbf{e}_j$$

For the case $\mathbf{n} = \mathbf{e}_3$,

$$[B^S_{ij}] = \begin{bmatrix} B_{11} & B_{12} & 0 \\ B_{21} & B_{22} & 0 \\ 0 & 0 & 0 \end{bmatrix} \tag{10.9}$$

Choose $\mathbf{B} = \mathbb{I}^{(2)}$. We define the projection of a unit second order tensor as a surface second order unit tensor denoted as

$$\mathbb{I}_S^{(2)} := \mathbf{P} \cdot \mathbb{I}^{(2)} \cdot \mathbf{P} \quad \Rightarrow \quad \begin{bmatrix} 1 & 0 & 0 \\ 0 & 1 & 0 \\ 0 & 0 & 0 \end{bmatrix} \tag{10.10}$$

Therefore $\mathbb{I}_S^{(2)} = \mathbf{P}$ is a second order surface unit tensor.

We now consider interface differentiation operator or gradient operator. It is defined as

$$\nabla_S = \mathbf{P} \cdot \nabla \quad \Rightarrow \quad (\delta_{ij} - n_i n_j) \frac{\partial}{\partial x_j} \quad \Rightarrow \quad \nabla - \mathbf{n}(\mathbf{n} \cdot \nabla) \tag{10.11}$$

Let $\mathbf{n} = \mathbf{e}_3$. One can show that

$$\nabla_S = \mathbf{P} \cdot \nabla = \frac{\partial}{\partial x_1} \mathbf{e}_1 + \frac{\partial}{\partial x_2} \mathbf{e}_2 \tag{10.12}$$

Let $\mathbf{v}$ being a vector field. By definition,

$$div_S \mathbf{v} = \nabla_S \cdot \mathbf{v} = tr(\nabla \otimes \mathbf{v}) \quad \Rightarrow \quad v_{i,k} P_{kj} \mathbf{e}_i \otimes \mathbf{e}_j \tag{10.13}$$

Note that by definition in this book the nabla operator is acting from behind. Then

$$div_S \mathbf{v} = tr(\nabla_S \mathbf{v}) = v_{i,k} P_{k,i}$$

Let $\mathbf{T}$ be a second order tensor. One has

$$div_S \mathbf{T} = tr(\nabla_S \mathbf{T}) \;\text{ or }\; \mathbf{T} \cdot \nabla_S \quad \Rightarrow \quad T_{ij,\ell} P_{\ell j} \mathbf{e}_i \tag{10.14}$$

Note that $\nabla_S \mathbf{T} = T_{ij,\ell} P_{\ell k} \mathbf{e}_i \otimes \mathbf{e}_j \otimes \mathbf{e}_k$, and the divergence operator is acting from behind. Therefore, it is the index j contracting with the index k.

In particular, if we let $\mathbf{T} = \mathbf{P}$,

$$div_S \mathbf{P} \;\Rightarrow\; (\delta_{ij} - n_i n_j)_{,\ell} P_{\ell j} = -(n_{i,\ell} P_{\ell j} n_j + n_i n_{j,\ell} P_{\ell j}) \tag{10.15}$$

One can show that

$$n_{i,\ell} P_{\ell j} n_j = n_{i,j} n_j - n_{i,\ell} n_\ell \equiv 0$$

and hence,

$$div_S \mathbf{P} = -\Big(div_S \mathbf{n}\Big)\mathbf{n} = -tr(\nabla_S \mathbf{n})\mathbf{n}$$

In differential geometry, it customarily defines the curvature tensor as

$$\mathbf{L} = \nabla_S \mathbf{n}, \quad \text{so} \quad div_S \mathbf{P} = -tr(\mathbf{L})\mathbf{n} \; . \tag{10.16}$$

For spherical interfacial surface, $\mathbf{n} = x_i / R \mathbf{e}_i$, and

$$\begin{aligned} \nabla_S \mathbf{n} \quad \Rightarrow \quad n_{i,k} P_{kj} &= \frac{1}{R} x_{i,k} (\delta_{kj} - n_i n_j) \\ &= \frac{1}{R} (\delta_{ik} \delta_{kj} - n_i n_j) = \frac{1}{R} P_{ij} \end{aligned} \tag{10.17}$$

Subsequently,

$$div_S \mathbf{n} = tr(\nabla_S \mathbf{n}) = \frac{1}{R} (\delta_{ii} - n_i n_i) = \frac{2}{R} = 2\kappa \tag{10.18}$$

It turns out that this is true for any smooth surface, i.e.

$$div_S \mathbf{n} = tr(\mathbf{L}) = 2\kappa \;\text{ and }\; div_S \mathbf{P} = -2\kappa \mathbf{n} = -\frac{2}{R} \mathbf{n} \; . \tag{10.19}$$

10.1.2 *Gurtin-Murdoch theory*

In small deformation version of the Gurtin-Murdoch theory, the surface and interface is modeled as a thin film or a thin membranae with diminishing thickness that adhere to the bulk material without slipping.

The surface strain is defined as the projection of bulk strain field onto the interface, i.e.

$$\boldsymbol{\epsilon}^S = \mathbf{P} \cdot \boldsymbol{\epsilon} \cdot \mathbf{P} \tag{10.20}$$

This is based on the assumption that displacements are continuous across the interface, and therefore kinematic variables on the interface should be equal to the projection of bulk kinematic variables onto the interface.

On the other hand, the static variables, i.e. the Cauchy stress may not be continuous across the interface. There will be a jump or discontinuity of traction across the interface. Therefore, one cannot link the surface stress with the bulk stress by projection, and they have to be related with surface constitutive relations, which are different from the bulk material properties, or they are not the projection of bulk material properties either.

In specific, the surface stresses are linked with surface tension and surface energy by the following expression,

$$\boldsymbol{\sigma}^S = \tau_0 \mathbb{I}_S^{(2)} + \frac{\partial \Gamma}{\partial \boldsymbol{\epsilon}^S} + \tau_0 \nabla_s \otimes \mathbf{u} \tag{10.21}$$

where $\boldsymbol{\sigma}^S$ is the surface stress tensor; $\boldsymbol{\epsilon}^S$ is the surface strain tensor; τ_0 is the surface tension, and Γ is the surface energy. The last term is interpreted in [Mi and Kouris (2006)] as

$$\nabla_S \otimes \mathbf{u} \to (u_i)^S_{,j} \mathbf{e}_i \otimes \mathbf{e}_j = u_{i,k} P_{kj} \mathbf{e}_i \otimes \mathbf{e}_j$$

However, it has been omitted or neglected by some authors, i.e. [Sharma and Ganti (2004)]. We note that in the Gurtin-Murdoch theory, the surface tension may be understood as the residual surface stress due to surface relaxation. This is because the atoms at and close to the surface lose some of their neighboring atoms, which will cause re-distribution of electric density distribution among atomic surface bonds, and it in turn results the change of bond length and subsequently the stress state as a macroscopic indicator. For infinitesimal deformation cases, Gurtin and Murdoch proposed the following quadratic form of the surface energy,

$$\Gamma = \frac{1}{2} \epsilon^S_{ij} C^S_{ijk\ell} \epsilon^S_{k\ell}, \quad i,j,k,\ell = 1,2 \tag{10.22}$$

in which is surface elastic tensor is related to surface tension as well,

$$C^S_{ijk\ell} = (\lambda^S + \tau_0)\delta_{ij}\delta_{k\ell} + \mu(\delta_{ik}\delta_{j\ell} + \delta_{i\ell}\delta_{jk}), \quad i,j,k,\ell = 1,2$$

Hence the surface constitutive relations are

$$\boldsymbol{\sigma}^S = \tau_0 \mathbb{I}_S^{(2)} + 2(\mu^S - \tau_0)\boldsymbol{\epsilon}^S + (\lambda^S + \tau_0) tr(\boldsymbol{\epsilon}^S) \mathbb{I}_S^{(2)}$$

Note that the unit of surface stresses and surface tensor is force per unit length.

Because the surface stress is obtained from the surface constitutive relation, its relation to bulk stress should come from surface balance condition. The surface equilibrium equations are as follows,

$$div_S \boldsymbol{\sigma}^S + [\sigma^B \cdot \mathbf{n}] = 0, \quad \forall \mathbf{x} \in S \tag{10.23}$$

where the jump operator is defined as

$$[\boldsymbol{\sigma}^B \cdot \mathbf{n}] = \Big(\boldsymbol{\sigma}^B(S_+) - \boldsymbol{\sigma}^B(S_-)\Big) \cdot \mathbf{n}$$

Denote the surface traction,

$$\mathbf{t}^+ = \boldsymbol{\sigma}^B(S_+) \cdot \mathbf{n}, \quad \text{and} \quad \mathbf{t}^- = \boldsymbol{\sigma}^B(S_-) \cdot \mathbf{n},$$

where $\mathbf{n} := \mathbf{n}^+$. We may write the surface equilibrium equations on a flat surface as

$$\frac{\partial \sigma_{11}^S}{\partial x_1} + \frac{\partial \sigma_{12}^S}{\partial x_2} + (t_1^+ - t_1^-) = 0 \tag{10.24}$$

$$\frac{\partial \sigma_{21}^S}{\partial x_1} + \frac{\partial \sigma_{22}^S}{\partial x_2} + (t_2^+ - t_2^-) = 0 \tag{10.25}$$

Remark 10.1. The underline physics principle of Eqs. (10.24)-(10.25) are the reminiscence of the well-known Laplace-Young law or the Laplace-Young equation for fluids.

10.1.3 *Spherical inclusion problem*

Sharma and Ganti [Sharma *et al.* (2003); Sharma and Ganti (2004)] applied the Gurtin-Murdoch theory to solve a spherical inclusion problem under the prescribed volumetric eigenstrain, i.e.

$$\epsilon_{11}^* = \epsilon_{22}^* = \epsilon_{33}^* = \epsilon^*$$

Because of spherical symmetry condition, the deformation is purely in the radial direction, and the three non-zero strain components are

$$\epsilon_{rr} = \frac{du}{dr}, \quad \epsilon_{\theta\theta} = \epsilon_{\phi\phi} = \frac{u}{r}$$

At the interface, $r = R_0$, we have

$$\epsilon_{\theta\theta}^S = \epsilon_{\phi\phi}^S = \frac{u}{R_0}$$

and the surface stress tensor becomes,

$$\begin{aligned}
\boldsymbol{\sigma}^S &= \begin{bmatrix} \tau_0 & 0 \\ 0 & \tau_0 \end{bmatrix} + 2(\mu^S - \tau_0) \begin{bmatrix} \epsilon_{\theta\theta}^S & 0 \\ 0 & \epsilon_{\phi\phi}^S \end{bmatrix} + (\lambda^S + \tau_0)(\epsilon_{\theta\theta}^S + \epsilon_{\phi\phi}^S) \begin{bmatrix} 1 & 0 \\ 0 & 1 \end{bmatrix} \\
&= (\tau_0 + (\lambda^S + \mu^S)(\epsilon_{\theta\theta}^S + \epsilon_{\phi\phi}^S)) \begin{bmatrix} 1 & 0 \\ 0 & 1 \end{bmatrix} = s\mathbf{P}
\end{aligned} \tag{10.26}$$

where $s := (\tau_0 + (\lambda^S + \mu^S)(\epsilon^S_{\theta\theta} + \epsilon^S_{\phi\phi}))$. Recall the fact that $\mathbb{I}^{(2)}_S = \mathbf{P}$.

In the bulk material, the equilibrium equations read as

$$\sigma^B_{ij,j} + f^S_i = C_{ijk\ell}u_{k,\ell j} - \Big(C_{ijk\ell}\epsilon^*_{k\ell}H(\Omega)\Big)_{,j} - [\sigma^B_{ik}n_k]\delta(S - \mathbf{n}\cdot\mathbf{x}) = 0\,. \tag{10.27}$$

Using the Somigliana identity,

$$u_i(\mathbf{x}) = \int_{\mathbb{R}^3} G^\infty_{ij}(\mathbf{x}-\mathbf{y})f_j(\mathbf{y})d\Omega_y$$

in which we identify the body force as

$$f_i = -\Big(C_{ijk\ell}\epsilon^*_{k\ell}\Big)_{,j} - [\sigma^B_{ik}n_k]\delta(S - \mathbf{n}\cdot\mathbf{x})$$

Integration by parts yields

$$u_i(\mathbf{x}) = -\int_\Omega C_{k\ell mn}\epsilon^*_{mn}G^\infty_{ik,\ell}(\mathbf{x}-\mathbf{y})d\Omega_y + \int_S G^\infty_{ij}(\mathbf{x}-\mathbf{y})[div_S\boldsymbol{\sigma}^B]_{,j}dS_y \tag{10.28}$$

in the last line Eq. (10.23) and the definition of the Radon transform are used.

For spherical symmetry, $\boldsymbol{\sigma}^S = s\mathbf{P}$; and we further assume that the surface stresses on the interface are uniform, i.e. $tr(\boldsymbol{\epsilon}^S) = const.$ or $s = const.$. Therefore,

$$div_S\boldsymbol{\sigma}^S = s\ div_S\mathbf{P} = -\frac{2s}{R}\mathbf{n}$$

Applying the divergence theorem to the second term of (10.28) yields,

$$u_i(\mathbf{x}) = -\int_\Omega C_{k\ell mn}\epsilon^*_{mn}G^\infty_{ik,\ell}(\mathbf{x}-\mathbf{y})d\Omega_y + \frac{2s}{R}\int_\Omega G^\infty_{ij,k}(\mathbf{x}-\mathbf{y})d\Omega_y$$

and hence,

$$\begin{aligned}
\epsilon_{ij} &= \frac{1}{2}(u_{i,j} + u_{j,i})\\
&= \left\{-\frac{1}{2}\int_\Omega C_{k\ell mn}\Big(G^\infty_{ik,\ell j}(\mathbf{x}-\mathbf{y}) + G^\infty_{jk,\ell i}(\mathbf{x}-\mathbf{y})\Big)d\Omega_y\right\}\epsilon^*_{mn}\\
&- \left(\frac{2s}{R_0}\right)\left\{-\frac{1}{2}\int_\Omega C_{k\ell mn}\Big(G^\infty_{ik,\ell j}(\mathbf{x}-\mathbf{y}) + G^\infty_{jk,\ell i}(\mathbf{x}-\mathbf{y})\Big)d\Omega_y\right\}D_{mnst}\delta_{st}
\end{aligned}$$

which can be cast into a succinct form,

$$\boldsymbol{\epsilon} = \mathbb{S} : \boldsymbol{\epsilon}^* - \frac{2s}{R_0}\mathbb{S} : \mathbb{D} : \mathbb{I}^{(2)}$$

by recognizing,

$$S_{ijmn} = \left\{-\frac{1}{2}\int_\Omega C_{k\ell mn}\Big(G^\infty_{ik,\ell j}(\mathbf{x}-\mathbf{y}) + G^\infty_{jk,\ell i}(\mathbf{x}-\mathbf{y})\Big)d\Omega_y\right\}\,.$$

Considering,

$$\mathbb{D} = \frac{1}{3K}\mathbb{E}^{(1)} + \frac{1}{2\mu}\mathbb{E}^{(2)}\ \ \text{and}\ \ \mathbb{E}^{(1)} : \mathbb{I}^{(2)} = \mathbb{I}^{(2)}$$

we have

$$\begin{aligned}\boldsymbol{\epsilon} &= \mathbb{S} : \boldsymbol{\epsilon}^* - \frac{2s}{3KR_0}\mathbb{S} : \mathbb{I}^{(2)} \\ &= \mathbb{S} : \boldsymbol{\epsilon}^* - \frac{K^S}{3KR_0}(\mathbb{S} : \mathbb{I}^{(2)})tr(\boldsymbol{\epsilon}^S) - \frac{2\tau_0}{3KR_0}(\mathbb{S} : \mathbb{I}^{(2)})\end{aligned} \tag{10.29}$$

where $K^S = 2(\lambda^S + \mu^S)$. Since the prescribed eigenstrain is dilatational, i.e. $\boldsymbol{\epsilon}^* = \epsilon^*\mathbb{I}^{(2)}$ and for the interior Eshelby tensor

$$\mathbb{S} = s_1\mathbb{E}^{(1)} + s_2\mathbb{E}^{(2)}$$

we have

$$\boldsymbol{\epsilon} = s_1\Big(\epsilon^* - \frac{K^S}{3KR_0}(\epsilon^S_{\theta\theta} + \epsilon^S_{\phi\phi}) - \frac{2\tau_0}{3KR_0}\Big)\mathbb{I}^{(2)} \tag{10.30}$$

where $s_1 = (1+\nu)/(3(1-\nu)) = 3K/(3K+4\mu)$.

One can then solve the strain components inside the inclusion,

$$\epsilon_{rr} = \epsilon_{\theta\theta} = \epsilon_{\phi\phi} = \frac{3K\epsilon^* - 2\tau_0/R_0}{3K + 4\mu + 2K^S/R_0}, \quad r < R_0 \tag{10.31}$$

Accordingly, one can also find the exterior solution,

$$\epsilon_{rr}(r) = -\Big[\frac{3K\epsilon^* - 2\tau_0/R}{4\mu + 3K + 2K^S/R_0}\Big]\Big(\frac{2R_0^3}{r^3}\Big), \quad r > R_0 \tag{10.32}$$

$$\epsilon_{\theta\theta}(r) = \epsilon_{\phi\phi}(r) = \Big[\frac{3K\epsilon^* - 2\tau_0/R}{4\mu + 3K + 2K^S/R_0}\Big]\Big(\frac{R_0^3}{r^3}\Big), \quad r > R_0 \tag{10.33}$$

by using the relationships,

$$\mathbf{S}^{E,\infty} : \mathbb{I}^{(2)} = \rho^3\frac{3K}{3K+4\mu}(\delta_{ij} - 3n_in_j)\mathbf{e}_i \otimes \mathbf{e}_j \tag{10.34}$$

and $\delta_{ij}\mathbf{e}_i \otimes \mathbf{e}_j = \mathbf{e}_r \otimes \mathbf{e}_r + \mathbf{e}_\theta \otimes \mathbf{e}_\theta + \mathbf{e}_\phi \otimes \mathbf{e}_\phi$ and $n_in_j\mathbf{e}_i \otimes \mathbf{e}_j = \mathbf{e}_r \otimes \mathbf{e}_r$.

10.2 Cohesive quasi-continuum finite element method

Traditionally, most macroscale constitutive relations are phenomenological relationships that are based on experimental observations and subsequent idealization. On the other hand, the objective or idealogy of Micromechanics is to obtain macroscale constitutive relations by homogenization, if adequate microscale information is available. In principle, if one knows all the microscale information, one may find the macroscale material response by statistical homogenization, even though such *statistical homogenization* can be formidable, and it may not be tractable at all.

In general, using statistical physics to obtain the material properties of continua from quantum mechanics is called as *coarse-graining*, i.e. one uses a lower-resolution coarse-grained model with right statistical behaviors to replace a first principle based description.

A popular coarse-grain models for crystalline solids is the so-called *Quasi-continuum model* proposed by Tadmor, Ortiz, and Phillips [Tadmor *et al.* (1996); Knap and Ortiz (2001); Miller and Tadmor (2002)] , which uses the embedded atom method and the semi-empirical atomistic potential as the building block to construct an atomically enriched hyperplastic potential energy density, and hence the constitutive relation for the quasi-continuum. The quasi-continuum method has been recently used a multiscale model in many applications. The quasi-continuum method has two computation algorithms: the local quasi-continuum method and the non-local quasi-continuum method. The local version of quasi-continuum method utilizes the Cauchy-Born rule in homogenization, and hence it applies to where the local deformation is uniform; while the nonlocal version was designed to simulate inhomogeneous local deformations. It has atomic resolution, and hence it is not really a coarse grain model. Because of this, its computational cost is more expensive than the local quasi-continuum method. In fact, it is comparable to that of MD simulations. In this book, we only discuss the local quasi-continuum method.

10.2.1 *Atomistic modeling*

Consider a finite Bravais lattice space, V_0, that consists of N_a number of atomic nuclei. In the undeformed configuration, i.e. the reference configuration, the lattice position of ℓ-th atom can be expressed as

$$\mathbf{X}_\ell = \mathbf{X}_0 + \ell_1 \mathbf{a}_1 + \ell_2 \mathbf{a}_2 + \ell_3 \mathbf{a}_3, \quad \ell_i = 0, \pm 1, \pm 2, \cdots \tag{10.35}$$

where $\mathbf{X}_0$ is the position of a reference atom, and $\mathbf{a}_j, j = 1, 2, 3$ are Bravais vectors.

After the deformation ϕ, the lattice space maps to a spatial configuration,

$$\varphi : V_0 \to V$$

Then the spatial position of the ℓ-th atom is determined as

$$\mathbf{x}_\ell = \mathbf{X}_\ell + \mathbf{u}_\ell$$

where $\mathbf{u}_\ell$ is the displacement of the ℓ-th atom. Here we borrow the expressions of continuum field theory to describe lattice motion by assuming that the motions of lattice is embedded in a high resolution continuous displacement field where not every point is a physical point unless it is an atom site,

$$\mathbf{u}(\mathbf{X}_\ell) \equiv \mathbf{u}_\ell \; .$$

The points surrounding atoms may be viewed as mathematical configurational points. In the rest of the text, we call this "continuous field" as the quasi-continuum or the ambient high resolution lattice continuum.

The total energy, E^{tot} for the crystalline solid is the summation of the atomistic potentials at all atom sites,

$$E^{tot} = \sum_{\ell=1}^{N_a} E_\ell(\mathbf{u}) \tag{10.36}$$

where E_ℓ is the site energy for atom ℓ, which depends on its relative position to the surrounding environment. For instance in the Embedded Atom Method (EAM) [Daw and Baskes (1984); Foiles *et al.* (1986)] , the site energy can be written as

$$E_\ell = E_\ell^{(0)}(\rho) + E_\ell^{(2)} + E_\ell^{(3)}$$

where

$$E_\ell^{(0)} = \Phi_\ell(\bar{\rho}_\ell)$$

is interpreted as an electron-density dependent embedding energy, and its argument, $\bar{\rho}_\ell$, is the spherically-averaged electron density at host atom ℓ, which is the superposition of electron densities from the adjacent atoms,

$$\bar{\rho}_\ell = \sum_{i\neq\ell} \rho_i(r^{i\ell}), \quad \text{where } r^{i\ell} = \sqrt{(\mathbf{x}_i - \mathbf{x}_\ell)\cdot(\mathbf{x}_i - \mathbf{x}_\ell)}$$

$E_\ell^{(2)}$ is the pair potential for two nuclei interaction,

$$E_\ell^{(2)} = \frac{1}{2}\sum_{i\neq\ell} \Phi_{i\ell}(r^{i\ell})$$

For simplicity, in the following we drop the subscript for the pair potential. $E_\ell^{(3)}$ is the three-body (nuclei) interaction potential, i.e.

$$E_\ell^{(3)} = \frac{1}{6}\sum_{i\neq\ell}\sum_{j\neq\ell,i} \Phi_{ij\ell}(r^{i\ell}, r^{j\ell})$$

By considering external loads acting on individual atoms, the total potential energy of the lattice system will be

$$U(\mathbf{u}) = E^{tot}(\mathbf{u}) - \sum_{\ell=1}^{N_a} f_\ell u_\ell \tag{10.37}$$

10.2.2 *Quasi-continuum method*

At small scales, atomistic effects will play a dominate role in material behaviors. To take into account the effects of atomistic microstructure on material responses, Tadmor, Ortiz, and Phillips [Tadmor *et al.* (1996)] proposed a coarse grained constitutive model, which is labeled as the *quasi-continuum model.* In quasi-continuum model, one first discretizes the lattice continuum by using finite element method (FEM), which is illustrated in Fig. 10.1. We can then approximate the displacement field by using FEM interpolation,

$$\mathbf{u}^h = \sum_{L=1}^{N_{node}} N_L(\mathbf{X})\mathbf{u}_L - \sum_{L=1}^{N_{rep}} N_L(\mathbf{X})\mathbf{u}_L \tag{10.38}$$

where $N_L(\mathbf{X})$ are FEM interpolation function. In the literature, each FEM node is called *a representative atom* or *rep-atom* in short. Thus the total number of nodes equal to the total number of representative atoms, $N_{node} = N_{rep}$.

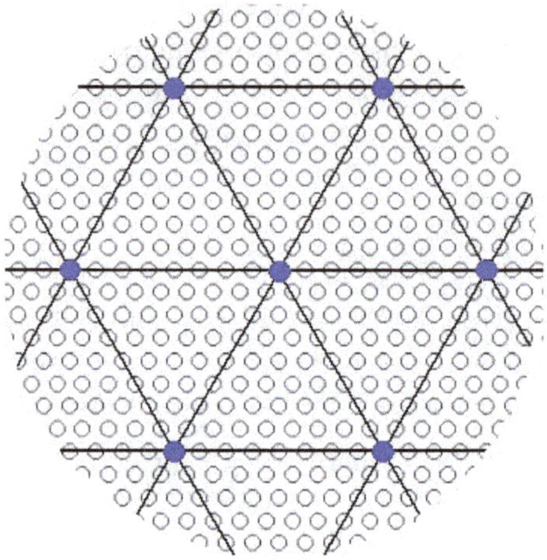

Fig. 10.1 Finite element mesh and Representative atoms.

In the lattice continuum, we can measure local deformation intensity by using deformation gradient

$$\mathbf{F} = \frac{\partial \mathbf{x}}{\partial \mathbf{X}}$$

Now we can approximate the total energy of the lattice system by the potential energy of the coarse grain model, or quasi-continuum model,

$$E^{tot} \approx E^{tot,h} = \sum_{e=1}^{N_{elem}} \Omega_e \mathcal{E}(\mathbf{F}_e) - \sum_{L=1}^{N_{node}} f_L u_L \tag{10.39}$$

where Ω_e is the volume of element e, N_{elem} is the total number of elements, $\mathbf{F}_e$ is the deformation gradient in the e-th element, and $\mathcal{E}(\mathbf{F}_e)$ is the strain energy density stored inside the element e.

To calculate the stored energy inside each element, we invoke another assumption of quasi-continuum method, that is: *The deformation gradient in each element will be a constant, i.e.* $\mathbf{F}_e = conts$. This assumption is usually called the Cauchy-Born rule (see [Born (1915); Born and Huang (1954)]), which requires the local deformation being uniform or homogeneous. Mathematically, the Cauchy-Born rule can be expressed in the following statement. *For any finite length vector in the reference configuration,* $\mathbf{R}^{ik} := \mathbf{X}_i - \mathbf{X}_k, \in \Omega_{e0}$, *it will be mapped into the spatial configuration as another finite length vector,* $\mathbf{r}^{ik} := \mathbf{x}_i - \mathbf{x}_k, \in \Omega_e$,

$$\mathbf{r}^{ik} = \mathbf{F}^e \cdot \mathbf{R}^{ik} \; . \tag{10.40}$$

It is often convenient to count bounds among different atoms by labeling them in a single sequence $\{k\}$ instead of counting bounds among atoms by the index of atoms. Denote the original line segment vector of the bound k as $\mathbf{R}_k \in \Omega_{e0}$ and its deformed counterpart as $\mathbf{r}_k \in \Omega_e$. Then the Cauchy-Born rule can be read as

$$\mathbf{r}_k = \mathbf{F}^e \cdot \mathbf{R}_k \ . \tag{10.41}$$

This is illustrated in Fig. 10.2. The key now is how to calculate the strain energy

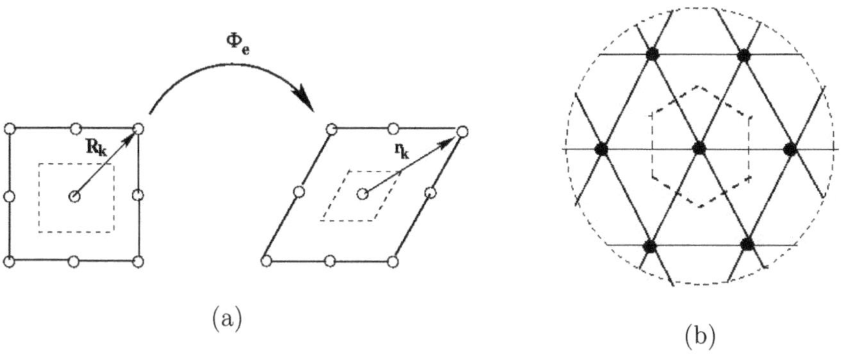

Fig. 10.2 (a) The Cauchy-Born Rule, and (b) Wigner-Seitz cell.

density in each element. Suppose in the e-th element, there are total N_e Neglecting the effects of electrons and three-body potential, we can write the strain energy density inside the e-th element as

$$\mathcal{E}(\mathbf{F}_e) = \frac{1}{2\Omega_0} \sum_{i \in S_0} \Phi(r_i(\mathbf{F}_e)) \tag{10.42}$$

where Ω_0 is the volume of representative unit cell, $\Phi(r_i)$ is the pair potential, and S_0 is the index set for all the atomic bonds inside the cell.

For Bravias lattices, a common choice of the unit cell is the Wigner-Seitz cell, see Fig. 10.2 (b). Since each cell only contains one atom, $\Omega_e = N_e \Omega_0$. And since the local deformation is homogeneous, all the atoms inside the element e experience the same deformation. Thus

$$\mathcal{E}(\mathbf{F}_e) := \frac{1}{\Omega_e}\Big(\frac{N_e}{2} \sum_{i \in S_0} \Phi(r_i(\mathbf{F}_e))\Big) = \frac{1}{2\Omega_0} \sum_{i \in S_0} \Phi(r_i(\mathbf{F}_e)) \tag{10.43}$$

where the factor $1/2$ comes from the fact each bond will count twice for the atoms at the two ends. For two-dimensional hexagonal lattices, there six bonds as shown in Fig. 10.2 (b), i.e. $S_e = \{1, 2, 3, 4, 5, 6\}$. Obviously for such choice, we can only count for the nearest neighbor interactions. In some quasi-continuum implementations e.g. [Park *et al.* (2006)], people may choose other types of representative cells that can take into account long range interactions. For simplicity, we choose the Wigner-Seitz primitive cell as the representative cell in this book.

Once the strain energy density is at hand, one can then find the stresses. For example, the second Piola-Kirchhoff stress,

$$\mathbf{S} = 2\frac{\partial \mathcal{E}}{\partial \boldsymbol{C}} = \frac{1}{\Omega_0} \sum_{k \in S_0} \frac{\partial \Phi}{\partial r_k} \frac{\partial r_k}{\partial \boldsymbol{C}} \tag{10.44}$$

where $r_k = \sqrt{\mathbf{r}_k \cdot \mathbf{r}_k}$. Considering

$$r_k = \sqrt{\mathbf{r}_k^T \cdot \mathbf{r}_k} = \sqrt{\mathbf{R}_k^T \cdot \mathbf{F}_e^T \cdot \mathbf{F}_e \cdot \mathbf{R}_k} = \sqrt{\mathbf{R}_k \cdot \mathbf{C}_e \cdot \mathbf{R}_k}$$

and the fact that $\mathbf{F}_e$ and $\boldsymbol{C}_e$ are constant tensors, we have

$$\frac{\partial r_k}{\partial \boldsymbol{C}_e} = \frac{1}{2r_k} \frac{\partial}{\partial \boldsymbol{C}_e} \Big(\mathbf{R}_k \cdot \boldsymbol{C}_e \cdot \mathbf{R}_k \Big) = \frac{1}{2r_k} \mathbf{R}_k \otimes \mathbf{R}_k \tag{10.45}$$

Finally, we have

$$\mathbf{S} = \rho_0 \sum_{k \in S_0} \frac{\partial \Phi}{\partial r_k} \frac{\mathbf{R}_k \otimes \mathbf{R}_k}{r_k} \tag{10.46}$$

where $\rho_0 = \Omega_0^{-1}$. Subsequently, we can find the first Piola-Kirchhoff stress, the Kirchhoff stress, and the Cauchy stress as,

$$\mathbf{P} = \rho_0 \sum_{k \in S_0} \frac{\partial \Phi}{\partial r_k} \frac{\mathbf{r}_k \otimes \mathbf{R}_k}{r_k} \tag{10.47}$$

$$\boldsymbol{\tau} = \rho_0 \sum_{k \in S_0} \frac{\partial \Phi}{\partial r_k} \frac{\mathbf{r}_k \otimes \mathbf{r}_k}{r_k} \tag{10.48}$$

$$\boldsymbol{\sigma} = \rho \sum_{k \in S_0} \frac{\partial \Phi}{\partial r_k} \frac{\mathbf{r}_k \otimes \mathbf{r}_k}{r_k} \tag{10.49}$$

where $\rho = \Omega(t)^{-1}$ and $\Omega(t)$ is the current volume of the Wigner-Seitz cell. It should be noted that the second elastic tangent modulus of the quasi-continuum,

$$\mathbb{C}^{SE} = 4\frac{\partial^2 \mathcal{E}}{\partial \boldsymbol{C} \partial \boldsymbol{C}} = \rho_0 \sum_{k \in S_0} \Big(\frac{\partial^2 \Phi}{\partial r_k^2} - \frac{1}{r_k} \frac{\partial \Phi}{\partial r_k} \Big) \frac{\mathbf{R}_k \otimes \mathbf{R}_k \otimes \mathbf{R}_k \otimes \mathbf{R}_k \otimes \mathbf{R}_k}{r_k^2} \tag{10.50}$$

is not positive definite, because $\mathcal{E}$ is not convex. Thus, the corresponding finite element formulation will be difficult to reach a global minimization.

10.2.3 *Interface Cauchy-Born rule*

Strictly speaking, the local quasi-continuum method cannot be even used to evaluate the interface stress at element boundary, because of the piecewise constant distribution of stresses due to the piecewise constant distribution of deformation gradient. This is in fact not a numerical problem; it actually hinges on the Cauchy-Born assumption or approximation.

If there is local inhomogeneous deformation, the local quasi-continuum method breaks down. On the other hand, the essence of Micromechanics or Nanomechanics is to study the defect motion, which inevitably causes inhomogeneous deformation at local level, for instance the motion of cracks and dislocations. In order to use such coarse-grain model to study defect motion, we have to make some amendments on the Cauchy-Born approximation for particular defect configurations. In this section, we introduce a theory of interface Cauchy-Born rule that may be able to simulate the motion of cracks and dislocations.

If we consider long range interaction, the strain energy density in an element may be written as

$$\mathcal{E} = \frac{N_e}{2\Omega_e} \sum_{j=1}^{M_b} \Phi(r^{ij}(\boldsymbol{F}_e)) \tag{10.51}$$

where $\Phi(r_{ij})$ is atomistic pair potential between atoms i and j, M_b is the number of atoms interacting with atom i in the bulk material. In general, $M_b << N_e$. Then the first Piola-Kirchhoff stress written as:

$$\boldsymbol{P} = \frac{\partial W^e}{\partial \boldsymbol{F}} = \frac{N_e}{2\Omega_e} \sum_{j=1}^{M_b} \frac{\partial \Phi(r^{ij})(\boldsymbol{F})}{\partial \boldsymbol{F}} = \frac{N_e}{2\Omega_e} \sum_{j=1}^{M_b} \Phi'(r^{ij}) \frac{\mathbf{r}^{ij} \otimes \mathbf{R}^{ij}}{r^{ij}} \tag{10.52}$$

We now consider local inhomogeneous deformation caused by strong discontinuity. We now consider a pair of separated surface element S_e^+ and S_e^-. They are the edges of two FE elements. Since we only need to formulate W_e^s for one of the edges, without losing generality, we shall do that for S^+. The surface energy density is defined as:

$$W^{es} = \frac{E^{es}}{A_{es}} \tag{10.53}$$

where E^{es} is the total surface energy within the surface element S_e^+, and A_{es} is the surface area. For E^{es}, we have:

$$E^{es} = \sum_{i=1}^{N_{es}} E_i^s \tag{10.54}$$

where N_{es} is the number of atoms within the surface element S_e^+ and

$$E_i^s = \frac{1}{2} \sum_{j=1}^{M_s} \Phi(r_s^{ij}) \tag{10.55}$$

where $\Phi(r_s^{ij})$ are atomistic potentials for surface atoms, r_s^{ij} are interatomic distances across the interface, i and j are indices for surface atoms only, and M_s is the number of atoms on the opposite surface element S_e^- that interact with atom i on the surface element S_e^+.

Remark 10.2. 1. It assume that the interface atomic distance, $r_s^{ij} = r_s^{ij}(\mathbf{\Delta}_d)$, is not a function of the local deformation gradient $\mathbf{F}$, but the nodal displacement separation (the jump) of associated surface elements. For simplicity, in the rest of paper, we drop the subscript, $_s$, unless it is important to indicate it. The nodal displacement separation is constant within each element. Therefore, this is essentially an **Interface Cauchy-Born Rule**, and we should further elaborated and substantiate this.

2. In calculating atomistic surface energy potential, as a preliminary study, we only consider the nearest neighbor interaction, and hence the effects of long-range interactions such as surface relaxation is neglected. Moreover, we neglect

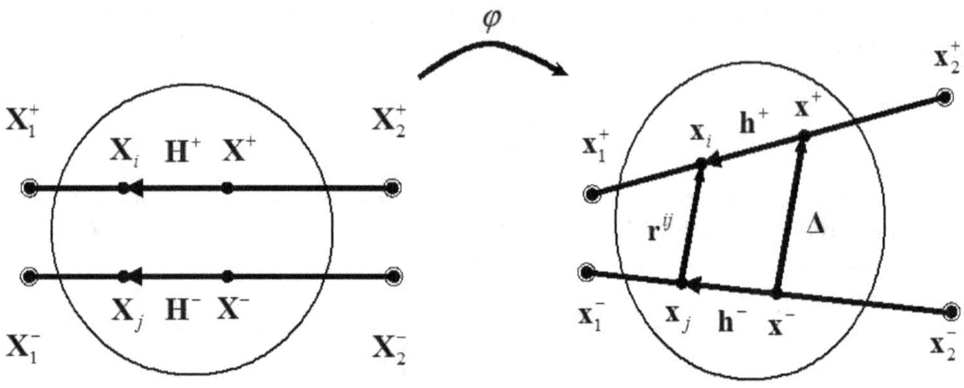

Fig. 10.3 Interface Kinematics: The bond separation between two cohesive surfaces.

surface effects due to atomistic interactions on the same surface. This lack of physical realism can be fixed, for instance, by using a so-called surface Cauchy-Born approach [Park *et al.* (2006)]. In fact, the procedure proposed by Park et al [Park *et al.* (2006)] is one way to calculate atomic surface potentials. It may be possible that we can still use the proposed approach, but choose an appropriate surface potential, $\Phi^{S}_{ij}(r^{ij}) \neq \Phi(r^{ij})$, to capture some of surface effects. In fact, there have been some specific surface potentials available in the literature, e.g. [Ibach (1997)].

Similar to what we did for element interior, we adopt the following quasi-continuum approximation:

$$E^{es} \approx N_{es} E^s_i \tag{10.56}$$

where N_{es} is the number of atoms on the surface element S^+_e. This approximation means that we assume that all atoms in S^+_e to have the same surface energy.

To derive the expression for the cohesive traction, we consider a FE discretization of the domain. The FE approximation of the displacement field is given by:

$$\boldsymbol{u}(\boldsymbol{X}) = \boldsymbol{N}(\boldsymbol{X})\boldsymbol{d} \tag{10.57}$$

where

$$\boldsymbol{N} = [N_{IJ}(\boldsymbol{X})]^{n_{dim} \times n_{dof}}$$

is the shape function matrix, n_{dim} is the number of dimension, n_{dof} is the number of degree of freedom, and $\boldsymbol{d}$ is the nodal displacement vector. We shall adopt the tensor notation during derivation. All matrices should be viewed as second-order tensors unless specified otherwise.

Consider a separation, or a jump, between a pair of cohesive surfaces as shown in Fig. 10.3. The jump can also be described by interpolation:

$$\boldsymbol{\Delta} = \boldsymbol{N}^s(\boldsymbol{X})\boldsymbol{\Delta}^d \tag{10.58}$$

where $\boldsymbol{N}^s(\mathbf{X})$ is the matrix of edge shape functions. The nodal jump vector $\boldsymbol{\Delta}^d$ is defined by:

$$\boldsymbol{\Delta}^d = \boldsymbol{d}^+ - \boldsymbol{d}^- \tag{10.59}$$

where $\boldsymbol{d}^+$ and $\boldsymbol{d}^-$ are nodal displacement vectors on S^+ and S^-, respectively. Moreover, the following relationship holds:

$$\boldsymbol{d}^+ = <\boldsymbol{d}> + \frac{1}{2}\boldsymbol{\Delta}^d \tag{10.60}$$

$$\boldsymbol{d}^- = <\boldsymbol{d}> - \frac{1}{2}\boldsymbol{\Delta}^d \tag{10.61}$$

where $<\boldsymbol{d}>= \frac{1}{2}(\boldsymbol{d}^+ + \boldsymbol{d}^-)$. So,

$$\frac{\partial \boldsymbol{d}^+}{\partial \boldsymbol{\Delta}^d} = \frac{1}{2} \tag{10.62}$$

$$\frac{\partial \boldsymbol{d}^-}{\partial \boldsymbol{\Delta}^d} = -\frac{1}{2} \tag{10.63}$$

In discrete case, the virtual work done by the cohesive force can be re-written as:

$$\int_{S_0} \boldsymbol{t}^{cohe} \cdot \delta \boldsymbol{\Delta} dS = \int_{S_0} \frac{\partial W^s}{\partial \boldsymbol{\Delta}^d} \cdot \delta \boldsymbol{\Delta}^d dS = \int_{S_0} \frac{\partial W^s}{\partial \boldsymbol{\Delta}^d} \cdot \delta \boldsymbol{d}^+ dS - \int_{S_0} \frac{\partial W^s}{\partial \boldsymbol{\Delta}^d} \cdot \delta \boldsymbol{d}^- dS \tag{10.64}$$

We can define the global *nodal* cohesive traction vector array from the above equation:

$$\boldsymbol{f}^{cohe} = \int_{S_0} \frac{\partial W^s}{\partial \boldsymbol{\Delta}^d} [\mathbf{N}^S]^T dS \tag{10.65}$$

where $[\mathbf{N}^S]$ is the global surface edge shape function matrix.

In computations, the quantity $\dfrac{\partial W^s}{\partial \boldsymbol{\Delta}^d}$ is calculated first at the element level. We first assume that in each element e there are $es = 1, \cdots, n_{selem}$ cohesive interfaces. Then in an element,

$$W^s = \sum_{es=1}^{n_{selem}} W^{es} H(S^{es})$$

where $H(S^{es})$ is the characteristic function or support function of the surface element S^{es}. Then for each cohesive interface, es, if we choose index i as the index of the representative atom, we have

$$\frac{\partial W^{es}}{\partial \boldsymbol{\Delta}^d} = \frac{N_{es}}{2A_{es}} \sum_{j=1}^{M_s} \Phi'(r^{ij}) \frac{\partial r^{ij}}{\partial \boldsymbol{\Delta}^d} = \frac{N_{se}}{2A_{es}} \sum_{j=1}^{M_s} \Phi'(r^{ij}) \frac{\mathbf{r}^{ij}}{r^{ij}} \frac{\partial \mathbf{r}^{ij}}{\partial \boldsymbol{\Delta}^d} \tag{10.66}$$

In order to calculate $\dfrac{\partial \mathbf{r}^{ij}}{\partial \boldsymbol{\Delta}^d}$, we need to establish *an interfacial kinematics* to represent the local inhomogeneous deformation due to the surface separation.

Refer to Fig. 10.3 for the following discussion. For a an arbitrary point $\boldsymbol{X}$ on the cohesive surface the jump is defined as $\boldsymbol{\Delta}(\boldsymbol{X})$. Since the jump is actually the stretch or separation of two adjacent atomic planes, the representative point $\mathbf{X}$ are in fact two points $\mathbf{X}^{\pm}$ in the reference configuration. Their images in deformed configuration are the points $\mathbf{x}^{+}$ and $\mathbf{x}^{-}$. However, for mesoscale quasi-continuum description, we view them as the *same* reference point.

A spatial point at atomic scale is not physically meaningful, if there is no atom residing at the point. So to define the jump at an arbitrary location needs a material point for reference. We use a pair of adjacent atoms $\mathbf{x}_i \in S^{+}$ and $\mathbf{x}_j \in S^{-}$ and the related the local position vectors $\mathbf{h}^{+}$ and $\mathbf{h}^{-}$ to locate $\mathbf{x}^{\pm}$, i.e.

$$\mathbf{x}^{+} = \mathbf{x}_i - \mathbf{h}^{+}, \quad \text{and} \quad \mathbf{x}^{-} = \mathbf{x}_j - \mathbf{h}^{-}.$$

which are the spatial difference between the atoms on $S^{\pm}$ and the spatial representative points $\mathbf{x}^{\pm}$. Note that physically the points $\mathbf{x}^{+}$ and $\mathbf{x}^{-}$ come from to reference points $\boldsymbol{X}^{+}$ and $\boldsymbol{X}^{-}$ which differ from an equilibrium atomic distance, $\mathbf{R}$.

With this local kinematic setup, we can express the bond separation in terms of FEM nodal displacement separation. We start from the following condition,

$$\mathbf{r}_s^{ij} = \boldsymbol{\Delta} + \boldsymbol{h}^{+} - \boldsymbol{h}^{-} \tag{10.67}$$

Here $\boldsymbol{\Delta}$ is the jump at the point $\boldsymbol{X}$, where we want to evaluate $\dfrac{\partial \mathbf{r}^{ij}}{\partial \boldsymbol{\Delta}^{d}}$. We can view $\boldsymbol{X}$ as positions of Gauss quadrature points to evaluate integrations in (10.64). Since elements in each half of the deformed body obey the Cauchy-Born rule, $\boldsymbol{h}^{+}$ and $\boldsymbol{h}^{-}$ can be written as:

$$\boldsymbol{h}^{+} = \boldsymbol{F}^{+}\boldsymbol{H}^{+} \tag{10.68}$$

$$\boldsymbol{h}^{-} = \boldsymbol{F}^{-}\boldsymbol{H}^{-} \tag{10.69}$$

Where $\boldsymbol{F}^{+}$ and $\boldsymbol{F}^{-}$ are local deformation gradients at different halves of the body and $\boldsymbol{H}^{+}$ and $\boldsymbol{H}^{-}$ are local interatomic position vectors in the undeformed configuration corresponding to the spatial position vectors, $\mathbf{h}^{\pm}$. The above equations can then be re-written as:

$$\boldsymbol{h}^{\pm} = \boldsymbol{F}^{\pm}\boldsymbol{H}^{\pm} = \left(\mathbf{1} + \boldsymbol{u}^{\pm}_{,\boldsymbol{X}}\right) \cdot \boldsymbol{H}^{\pm} = \mathbf{1} \cdot \boldsymbol{H}^{\pm} + \boldsymbol{B}^{\pm} \cdot \left(\boldsymbol{d}^{\pm} \otimes \boldsymbol{H}^{\pm}\right) \tag{10.70}$$

where $\boldsymbol{B}^{\pm}$ are two third-order tensors whose components are defined as

$$B^{\pm}_{ijA} = N^{\pm}_{ij,A} := \frac{\partial}{\partial X_A} N^{\pm}_{ij} \tag{10.71}$$

Here the capital letter A denotes material coordinates.

Considering the interfacial kinematic relation shown in Fig. 10.3, we have the following relations at an arbitrary point (a pair of points) of the interface,

$$\begin{aligned} \boldsymbol{\Delta} &= \mathbf{r}^{ij} + (\mathbf{h}^{-} - \mathbf{h}^{+}) = (\mathbf{x}_i - \mathbf{x}_j) + (\mathbf{h}^{-} - \mathbf{h}^{+}) \\ &= \mathbf{F}^{+} \cdot \mathbf{X}_i - \mathbf{F}^{-} \cdot \mathbf{X}_j + \mathbf{F}^{-} \cdot \mathbf{H}^{-} - \mathbf{F}^{+} \cdot \mathbf{H}^{+} \\ &= \mathbf{F}^{+} \cdot \mathbf{X}^{+} - \mathbf{F}^{-} \cdot \mathbf{X}^{-} \end{aligned} \tag{10.72}$$

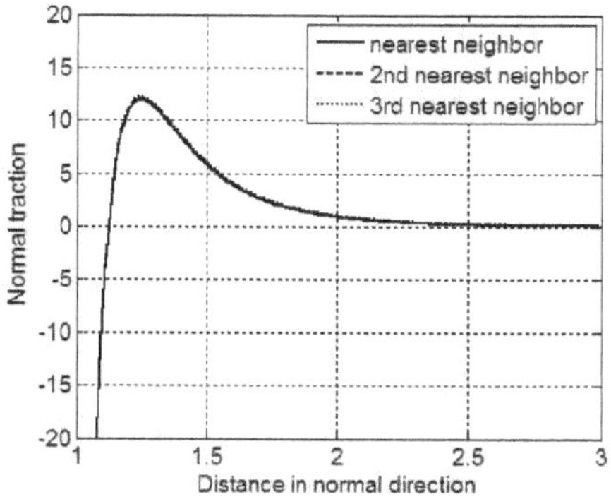

(a)

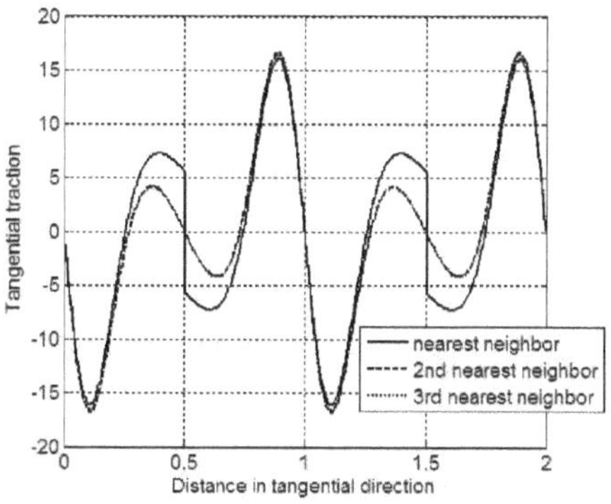

(b)

Fig. 10.4 Cohesive traction as a function of distances. (a) Normal traction, and (b) Tangential traction.

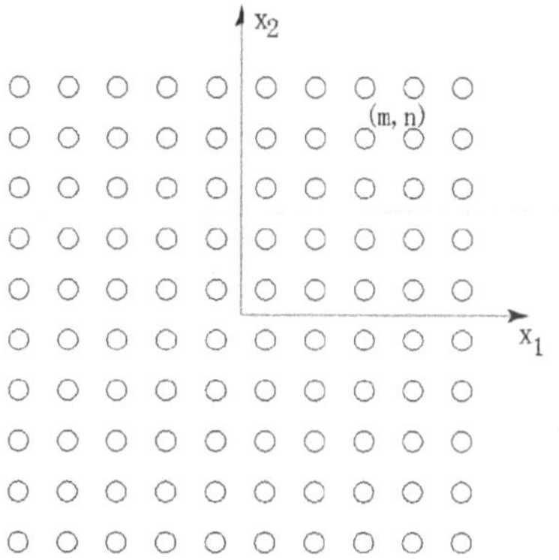

Fig. 10.5 A plane view of the cubic lattice.

Define

$$\bar{\mathbf{X}} := \frac{1}{2}\Big(\mathbf{X}^+ + \mathbf{X}^-\Big), \quad \text{and} \quad \mathbf{\Delta}_0 := \frac{1}{2}\Big(\mathbf{X}^+ - \mathbf{X}^-\Big). \tag{10.73}$$

We have an **Interface Cauchy-Born Rule** as follows:

$$\boxed{\mathbf{\Delta} = \Big(\mathbf{F}^+ + \mathbf{F}^-\Big)\cdot\mathbf{\Delta}_0 + \Big(\mathbf{F}^+ - \mathbf{F}^-\Big)\cdot\bar{\mathbf{X}}} \tag{10.74}$$

Now, we can calculate the derivative of $\mathbf{r}^{ij}$ with respect to element nodal jump vector, $\mathbf{\Delta}^d$,

$$\begin{aligned}\frac{\partial \mathbf{r}^{ij}}{\partial \mathbf{\Delta}^d} &= \frac{\partial \mathbf{\Delta}}{\partial \mathbf{\Delta}^d} + \frac{\partial \boldsymbol{h}^+}{\partial \mathbf{\Delta}^d} - \frac{\partial \boldsymbol{h}^-}{\partial \mathbf{\Delta}^d} = \boldsymbol{N}^s + \frac{\partial \boldsymbol{h}^+}{\partial \boldsymbol{d}^+}\frac{\partial \boldsymbol{d}^+}{\partial \mathbf{\Delta}^d} - \frac{\partial \boldsymbol{h}^-}{\partial \boldsymbol{d}^-}\frac{\partial \boldsymbol{d}^-}{\partial \mathbf{\Delta}^d} \\ &= \boldsymbol{N}^s + \frac{1}{2}\boldsymbol{B}^+\boldsymbol{H}^+ - \frac{1}{2}\boldsymbol{B}^-\boldsymbol{H}^-\end{aligned} \tag{10.75}$$

The final expression of $\frac{\partial W^{es}}{\partial \boldsymbol{\Delta}^d}$ reads:

$$\frac{\partial W^{es}}{\partial \boldsymbol{\Delta}^d} = \frac{N_{es}}{2A_{es}} \sum_{j=1}^{M_s} \Phi'(r^{ij}) \frac{\partial r^{ij}}{\partial \boldsymbol{\Delta}^d} = \frac{N_{es}}{2A_{es}} \sum_{j=1}^{M_s} \Phi'(r^{ij}) \frac{1}{r^{ij}} \cdot \left(\mathbf{r}^{ij} \cdot \boldsymbol{N}^s + \frac{1}{2} \mathbf{r}^{ij} \cdot \boldsymbol{B}^+ \boldsymbol{H}^+ - \frac{1}{2} \mathbf{r}^{ij} \cdot \boldsymbol{B}^- \boldsymbol{H}^- \right) \tag{10.76}$$

The matrix form of the above equation is:

$$\frac{\partial W^{es}}{\partial \boldsymbol{\Delta}^d} = \frac{N_{es}}{2A_{es}} \sum_{j=1}^{M_s} \frac{\Phi'(r^{ij})}{r^{ij}} \left([\boldsymbol{N}^{sT}]\mathbf{r}^{ij} + \frac{1}{2}[\boldsymbol{B}^+ \boldsymbol{H}^+]^T \mathbf{r}^{ij} - \frac{1}{2}[\boldsymbol{B}^- \boldsymbol{H}^-]^T \mathbf{r}^{ij} \right) \tag{10.77}$$

The components of matrices $\boldsymbol{B}^{\pm}\boldsymbol{H}^{\pm}$ are:

$$[\boldsymbol{B}^{\pm}\boldsymbol{H}^{\pm}]_{ij} = N^{\pm}_{ij,A} H^{\pm}_A \tag{10.78}$$

where $N^{\pm}_{ij}$ are components of $\boldsymbol{N}$. The force vector $\boldsymbol{f}^{cohe}$ can be obtained by integrating $\frac{\partial W^s}{\partial \boldsymbol{\Delta}^d}$.

Remark 10.3. From (10.76), we observe that when $\boldsymbol{B}^+\boldsymbol{H}^+ = \boldsymbol{B}^-\boldsymbol{H}^-$, the expression of nodal traction vector will reduce to:

$$\boldsymbol{f}^{cohe} = \mathop{\boldsymbol{A}}_{e=1}^{n_{elem}} \mathop{\boldsymbol{A}}_{es=1}^{n_{selem}} \int_{S_0^{es}} \frac{N_{es}}{2A_{es}} \sum_{j=1}^{M_s} \frac{\Phi'(r^{ij})}{r^{ij}} \boldsymbol{N}^{sT} \mathbf{r}^{ij} dS \tag{10.79}$$

This corresponds to the case when the deformed atomic position vectors are the same on S^+ and S^-. In the above equation $\boldsymbol{A}$ is the element assemble operator for both bulk elements and surface elements [Hughes (1987)]. Since in each element there may be several cohesive interfaces, i.e. $es = 1, \cdots, n_{selem}$, so we have used double assemble operators as a two-loop assembly for each bulk element and for each surface element within a bulk element.

In Fig. 10.4, we plot $\boldsymbol{f}^{cohe}$ against $\boldsymbol{\Delta}^d$. Of the two nodes, we shall plot $\boldsymbol{f}_1^{cohe}$. The two edges are chosen to be parallel to each other, so $\boldsymbol{\Delta}_1^d = \boldsymbol{\Delta}_2^d = \boldsymbol{\Delta}$. The interatomic distance is 1 and the edge length is 10. Lennard-Jones(LJ) potential is used. The expression for Φ is:

$$\Phi(r^{ij}) = 4\epsilon \left[\left(\frac{\sigma}{r^{ij}} \right)^{12} - \left(\frac{\sigma}{r^{ij}} \right)^{6} \right] \tag{10.80}$$

with $\sigma = \epsilon = 1$. When plotting Fig. 10.4(a), we fix Δ_t, the tangential component of $\boldsymbol{\Delta}$ to be zero (which means the node on S^+ is aligned longitudinally with a node on S^-) and vary Δ_n, the normal component. And Fig. 10.4(b) shows the result when we fix $\Delta_n = -1$ and vary Δ_t. We observe that as expected the normal traction shows a similar pattern as the interatomic force and the tangential traction has a period of the lattice constant. In both figures we show different cases when we

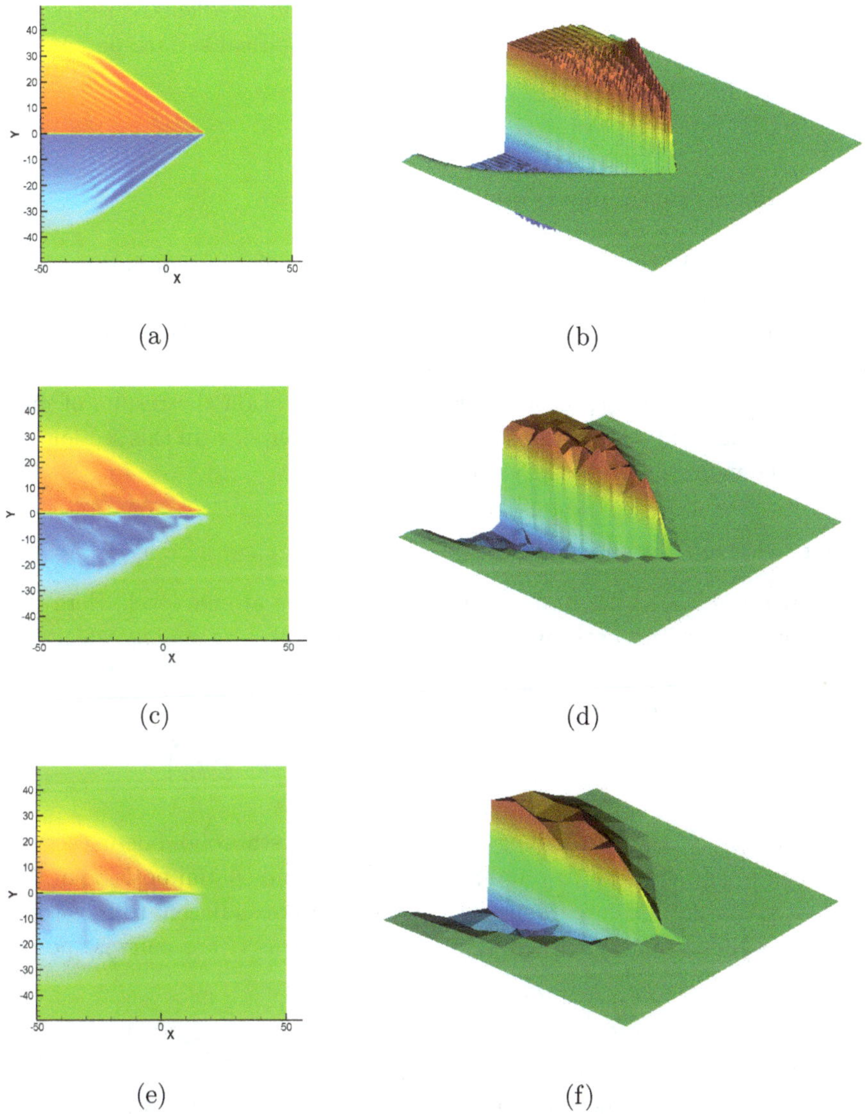

Fig. 10.6 Displacement profiles of a screw dislocation: (a) and (b) are MD simulation results; (c) and (d) are the cohesive CQC-FEM simulation results with 800 elements; (e) and (f) are the cohesive CQC-FEM simulation results with 200 elements (After [Liu *et al.* (2008)]).

include nearest neighbor only, up to 2nd nearest neighbor and up to 3rd nearest neighbor. We can see that while the nearest neighbor interaction is good enough for the normal traction, it yields a jump for the tangential traction when the node is in the middle of two atoms. The jump happens because with nearest neighbor interaction, the moving node "switches" its interacting partner in the middle and the force changes its sign. This artificial effect is eliminated when 2nd nearest neighbor interaction is considered. Xu and Needleman [Xu and Needleman (1994)] obtained a similar shape of cohesive relations by using an empirical surface potential.

10.3 Microscale or mesoscale stress

One of main objectives of nanomechanics is extrapolating statistical meaningful microscale information obtained in the atomistic simulation to correctly represent mesoscale or even macroscale quantities such as stress, thermal conductivity, specific heat, etc. A main link from microscale to macroscale is so-called *Virial Formula of Stress*, which has been extensively used in interpretation the results of molecular dynamics, quantum statistical mechanics, and astrophysics. In this section, we shall present the main theory of the virial stress.

10.3.1 *Virial stress*

Virial stress is a measure of mechanical stress at the atomic level. Here the word *virial* comes from the Latin word *vis* meaning for *force* or *energy*, which was first used by Clausius [Clausius (1870)] and Maxwell [Maxwell (1874)] to measure the local stress around an atom i,

$$\boldsymbol{\sigma}\Big|_{\mathbf{x}=\mathbf{x}_i} = \frac{1}{\Omega_i}\Big(-m_i \mathbf{v}_i \otimes \mathbf{v}_i + \frac{1}{2}\sum_{j \neq i} \mathbf{r}^{ij} \otimes \mathbf{f}_{ij}\Big) \tag{10.81}$$

where Ω_i is a volume surrounding the atom i, for crystal solid, one may view Ω_i as the Wigner-Seitz unit cell. m_i and $\mathbf{x}_i$ are the mass and spatial position of the atom i, $\mathbf{r}^{ij} := \mathbf{x}_i - \mathbf{x}_j$, and $\mathbf{x}_j$ are the positions of other atoms adjacent to the atom i, and $\mathbf{f}_{ji}$ denotes the interatomic force from atom j to atoms i. For pairwise potentials,

$$\mathbf{f}_{ij} = -\frac{\partial \Phi(r^{ij})}{\partial r^{ij}} \frac{\mathbf{r}^{ij}}{r^{ij}}, \quad \text{with} \quad r^{ij} := |\mathbf{r}^{ij}| \tag{10.82}$$

Here $\Phi(r^{ij})$ denotes the interatomic potential.

One may find that the construction of the viral stress is essentially the tensorial expansion of dot product of energy, which is the name virial comes from: The first term of (10.81) is the tensorial expansion of the kinetic energy of the atom i, i.e.

$$2T_i = m_i \mathbf{v}_i \cdot \mathbf{v}_i \quad \Rightarrow \quad m_i \mathbf{v}_i \otimes \mathbf{v}_i$$

and the second term of (10.81) is the tensorial expansion the potential energy of the atom i. The potential energy of the atom i is the product of the external force acting on it and the spatial position

$$\mathbf{f}_i \cdot \mathbf{x}_i, \quad \text{where } \mathbf{f}_i = \sum_{j=1}^{N} \mathbf{f}_{ij} \tag{10.83}$$

where $\mathbf{f}_{ij}$ is the force applied by atom j on the atom i, and hence the total potential energy stored in the system is

$$\sum_{i=1}^{N} \mathbf{f}_i \cdot \mathbf{x}_i = \sum_{i=1}^{N}\sum_{j=1}^{N} \mathbf{f}_{ij} \cdot \mathbf{x}_i = \frac{1}{2}\sum_{i=1}^{N}\sum_{j=1}^{N} \mathbf{f}_{ij} \cdot \mathbf{x}_i + \frac{1}{2}\sum_{j=1}^{N}\sum_{i=1}^{N} \mathbf{f}_{ij} \cdot \mathbf{x}_i$$
$$= \frac{1}{2}\sum_{i=1}^{N}\sum_{j=1}^{N} \mathbf{f}_{ij} \cdot \mathbf{x}_i + \frac{1}{2}\sum_{i=1}^{N}\sum_{j=1}^{N} \mathbf{f}_{ji} \cdot \mathbf{x}_j \tag{10.84}$$

Based on Newton's third law $\mathbf{f}_{ji} = -\mathbf{f}_{ij}$ and $\mathbf{f}_{ii} = 0$ i.e. no atom exerts force on its own, we have

$$\sum_{i=1}^{N} \mathbf{f}_i \cdot \mathbf{x}_i = \frac{1}{2}\sum_{i=1}^{N}\sum_{j\neq i} \mathbf{f}_{ij} \cdot (\mathbf{x}_i - \mathbf{x}_j) = \frac{1}{2}\sum_{i=1}^{N}\sum_{j\neq i} \mathbf{r}^{ij} \cdot \mathbf{f}_{ij} \tag{10.85}$$

Then one may express

$$\mathbf{f}_i \cdot \mathbf{x}_i = \frac{1}{2}\sum_{j\neq i} \mathbf{r}^{ij} \cdot \mathbf{f}_{ij}$$

Hence the tensorial expansion is

$$\frac{1}{2}\sum_{j\neq i} \mathbf{r}^{ij} \cdot \mathbf{f}_{ij} \quad \Rightarrow \quad \frac{1}{2}\sum_{j\neq i} \mathbf{r}^{ij} \otimes \mathbf{f}_{ij}\ . \tag{10.86}$$

Roughly speaking, the early interpretation of the physical meanings of the virial stress are: *the first term in (10.81) representing the pressure exerted on a fixed spatial surface by the motion of an atom flux, and the second term is due to the interatomic interaction across the same spatial space.*

Cheung and Yip [Cheung and Yip (1991)] have shown that on the free-surface of a crystal the normal component of the virial stress does not vanish for atoms on that surface. To fix the problem, the following average virial stress or the coarse-grained virial stress is proposed,

$$\boldsymbol{\sigma} = \frac{1}{\Omega}\sum_{\mathbf{x}_i\in\Omega}\Big(-m_i \mathbf{v}_i \otimes \mathbf{v}_i + \frac{1}{2}\sum_{j\neq i} \mathbf{r}^{ij} \otimes \mathbf{f}_{ij}\Big) \tag{10.87}$$

Nevertheless, the physical meaning or the physical interpretation of the virial stress has been a subject of debate for a long time. The focus of the debate is the physical meaning of the first term. Some think that the virial stress should be measured by *a material surface* rather than *a spatial surface*, so the first term should not be taken into account. Even based on the measurement of the material surface, there are still dispute on the first term. Irving and Kirkwood and Evans and Morrois argued that the fluctuation part of the atomistic velocity should be taken into account, which was later justified by Dommelen as the cross-over effect due to statistical fluctuation, because the definition of a material surface depends on time average. Today the standard definition of virial stress is

$$\boxed{\boldsymbol{\sigma} = \frac{1}{\Omega}\sum_{\mathbf{x}_i\in\Omega}\Big(-m_i (\mathbf{v}_i - \bar{\mathbf{v}}) \otimes (\mathbf{v}_i - \bar{\mathbf{v}}) + \frac{1}{2}\sum_{j\neq i} \mathbf{r}^{ij} \otimes \mathbf{f}_{ij}\Big)} \tag{10.88}$$

Although Zhou [Zhou (2003)] and some others believed that the first term due to kinetic energy should be dropped

$$\boldsymbol{\sigma} = \frac{1}{2\Omega} \sum_{\mathbf{x}_i \in \Omega} \sum_{j \neq i} \mathbf{r}^{ij} \otimes \mathbf{f}_{ij}, \tag{10.89}$$

this may not be a foregone conclusion, and we thought that the first term, which is due to fine scale thermal fluctuations, may represent the part of anharmonic thermal stress in the continuum.

Now we outline the proof of (10.88) under the condition that the mean drift velocity field is zero. The proof basically follows that of Lutsko [Lutsko (1988)] and Cormier et al [Cormier *et al.* (2001)].

Consider that there is a continuum field in the ambient space whose coordinate is labeled as **X**. On this field, we first define the linear momentum distribution

$$\bar{\mathbf{p}}(\mathbf{X}, t) = \sum_{i=1}^{N} \mathbf{p}_i \delta(\mathbf{X} - \mathbf{x}_i) \tag{10.90}$$

We can then define stress via local balance of linear momentum in spatial form,

$$\frac{\partial}{\partial t} \bar{\mathbf{p}}(\mathbf{X}, t) = -\nabla_{\mathbf{X}} \cdot \boldsymbol{\sigma} \tag{10.91}$$

Apply the Fourier transform to both (10.90) and (10.91). We have

$$\tilde{\bar{\mathbf{p}}} = \sum_i \mathbf{p}_i(t) \exp(-i\boldsymbol{\xi} \cdot \mathbf{x}_i(t)) \tag{10.92}$$

and

$$i\boldsymbol{\xi} \cdot \boldsymbol{\sigma} = \frac{d}{dt} \tilde{\bar{\mathbf{p}}} = \sum_i \Big(\dot{\mathbf{p}}_i - (i\boldsymbol{\xi} \cdot \dot{\mathbf{x}}_i) \mathbf{p}_i \Big) \exp(-i\boldsymbol{\xi} \cdot \mathbf{x}_i) \tag{10.93}$$

Consider the molecular dynamics of atom i satisfying Newton's Second Law,

$$\frac{d}{dt} \mathbf{p}_i = \mathbf{f}_i = \sum_{j \neq i} \mathbf{f}_{ij} = \sum_i -\frac{\partial \Phi(r^{ij})}{\partial r^{ij}} \frac{\mathbf{r}^{ij}}{r^{ij}} \tag{10.94}$$

and $\dot{\mathbf{x}} = \dfrac{\mathbf{p}_i}{m_i}$. We can rewrite (10.91) as

$$i\boldsymbol{\xi} \cdot \tilde{\boldsymbol{\sigma}} = \sum_i \Big(-i\boldsymbol{\xi} \cdot \frac{\mathbf{p}_i \otimes \mathbf{p}_i}{m_i} + \sum_{j \neq i} \mathbf{f}_{ij} \Big) \exp(-i\boldsymbol{\xi} \cdot \mathbf{x}_i) \tag{10.95}$$

Consider

$$\sum_i \sum_{j \neq i} \mathbf{f}_{ij} \exp(-i\boldsymbol{\xi} \cdot \mathbf{x}_i) = \frac{1}{2} \sum_i \sum_{j \neq i} \Big(\mathbf{f}_{ij} \exp(-i\boldsymbol{\xi} \cdot \mathbf{x}_i) + \mathbf{f}_{ji} \exp(-i\boldsymbol{\xi} \cdot \mathbf{x}_j) \Big)$$

$$= \frac{i\boldsymbol{\xi}}{2} \cdot \sum_i \sum_{j \neq i} \mathbf{r}^{ij} \otimes \mathbf{f}_{ij} \frac{1 - \exp(i\boldsymbol{\xi} \cdot \mathbf{r}^{ij})}{i\boldsymbol{\xi} \cdot \mathbf{r}^{ij}} \exp(-i\boldsymbol{\xi} \cdot \mathbf{x}_i) \tag{10.96}$$

We then have

$$\tilde{\boldsymbol{\sigma}} = \sum_i \Big(-\frac{\mathbf{p}_i \otimes \mathbf{p}_i}{m_i} + \sum_{j \neq i} \mathbf{r}^{ij} \otimes \mathbf{f}_{ij} \frac{1 - \exp(i\boldsymbol{\xi} \cdot \mathbf{r}^{ij})}{2i\boldsymbol{\xi} \cdot \mathbf{r}^{ij}} \Big) \exp(-i\boldsymbol{\xi} \cdot \mathbf{x}_i) \tag{10.97}$$

Consider the identity

$$\int_0^1 \exp(is\boldsymbol{\xi} \cdot \mathbf{r}^{ij}) ds = -\frac{1 - \exp(i\boldsymbol{\xi} \cdot \mathbf{r}^{ij})}{i\boldsymbol{\xi} \cdot \mathbf{r}^{ij}} \tag{10.98}$$

and the inverse Fourier transform,

$$\begin{aligned} &\frac{-1}{(2\pi)^3} \int_{\mathbb{R}^3} \Big[\int_0^1 \exp(is\boldsymbol{\xi} \cdot \mathbf{r}^{ij}) ds \Big] \exp\Big(i\boldsymbol{\xi} \cdot (\mathbf{X} - \mathbf{x}_i) \Big) d\boldsymbol{\xi} \\ &= -\int_0^1 \delta(\mathbf{X} - (\mathbf{x}_i + s\mathbf{r}^{ji})) ds =: B_\delta(\mathbf{x}_i, \mathbf{x}_j, \mathbf{X}) \end{aligned} \tag{10.99}$$

We have the expression of the local virial stress

$$\boldsymbol{\sigma}(\mathbf{X}, t) = \sum_i \Big(-\frac{\mathbf{p}_i \otimes \mathbf{p}_i}{m_i} \delta(\mathbf{X} - \mathbf{x}_i) + \sum_{j \neq i} \mathbf{r}_{ij} \otimes \mathbf{f}_{ij} B_\delta(\mathbf{x}_i, \mathbf{x}_j, \mathbf{X}) \Big) \tag{10.100}$$

If we average the average local stress in a finite sphere centered at $\mathbf{X}$, we obtain the expression given by Cormier et al [Cormier *et al.* (2001)],

$$\begin{aligned} < \boldsymbol{\sigma}(\mathbf{X}, t) > &= \frac{1}{\Omega} \int_\Omega \boldsymbol{\sigma}(\mathbf{X} + \mathbf{x}') d\Omega_{x'} \\ &= \sum_i \Big(-\frac{\mathbf{p}_i \otimes \mathbf{p}_i}{m_i} \chi_i(\mathbf{X}) - \frac{1}{2} \sum_{j \neq i} \frac{\partial \Phi(r^{ij})}{\partial r^{ij}} \frac{\mathbf{r}^{ij} \otimes \mathbf{r}_{ij}}{r^{ij}} \ell_{ij}(\mathbf{X}) \Big) \end{aligned} \tag{10.101}$$

where

$$\chi_i(\mathbf{X}) = \begin{cases} 1, & \forall \mathbf{x}_i \in \Omega \\ 0, & \text{otherwise} \end{cases} \tag{10.102}$$

and

$$\ell_{ij} = \frac{\overline{\mathbf{x}_i \mathbf{x}_j} \cap \Omega}{\overline{\mathbf{x}_i \mathbf{x}_j}} \,. \tag{10.103}$$

That is the fraction of line segment $\overline{\mathbf{x}_i \mathbf{x}_j}$ inside Ω, and hence $0 \leq \ell_{ij} \leq 1$.

10.3.2 *Hardy stress*

The above average virial stress approach implicitly imply that the following coarse graining approach:

$$\rho(\mathbf{X}, t) = \sum_{i=1}^N m_i \delta(\mathbf{x}_i - \mathbf{X}) \tag{10.104}$$

$$\bar{\mathbf{p}}(\mathbf{X}, t) = \sum_{i=1}^N m_i \mathbf{v}_i \delta(\mathbf{x}_i - \mathbf{X}) \tag{10.105}$$

$$\bar{\mathbf{v}}(\mathbf{X}, t) = \sum_{i=1}^N \mathbf{v}_i \delta(\mathbf{x}_i - \mathbf{X}) = \frac{\bar{\mathbf{p}}(\mathbf{X}, t)}{\rho(\mathbf{X}, t)} \tag{10.106}$$

Instead of using the Dirac delta function supported coarse graining approach, R. J. Hardy [Hardy (1982)] proposed or generalized the above coarse graining approach by replacing the Dirac delta function with a smoothed localization function, Ψ, such that

$$\rho(\mathbf{X},t) = \sum_{i=1}^{N} m_i \Psi_h(\mathbf{x}_i - \mathbf{X}) \tag{10.107}$$

$$\bar{\mathbf{p}}(\mathbf{X},t) = \sum_{i=1}^{N} m_i \mathbf{v}_i \Psi_h(\mathbf{x}_i - \mathbf{X}) \tag{10.108}$$

$$\bar{\mathbf{v}}(\mathbf{X},t) = \frac{\bar{\mathbf{p}}}{\rho(\mathbf{X},t)} \neq \sum_{i=1}^{N} \mathbf{v}_i \Psi_h(\mathbf{x}_i - \mathbf{X}) \tag{10.109}$$

$$E^0(\mathbf{X},t) = \sum_{i=1}^{N} \Big(\frac{1}{2} m_i \mathbf{v}_i \cdot \mathbf{v}_i + U_i \Big) \Psi_h(\mathbf{x}_i - \mathbf{X}) \tag{10.110}$$

where the capital letter $\mathbf{X}$ denotes the macroscopic position, $U_i = \frac{1}{2} \sum_{j=1}^{N} \Phi(r^{ij})$, and $\bar{\mathbf{p}}, \bar{\mathbf{v}}$ denote the coarse scale linear momentum and velocity; $\rho(\mathbf{X},t)$ denotes the density distribution that is intrinsically a coarse scale quantity, i.e. no fine scale counterpart. The smoothing function, $\Psi_h(\mathbf{x}) := \Psi(\mathbf{x}/h)$, and it satisfies the following conditions

$$(\mathbf{1.}) \quad \int_{\mathbb{R}^d} \Psi_h(\mathbf{x}) d\Omega_x = 1 \; ; \tag{10.111}$$

$$(\mathbf{2.}) \quad \Psi_h(\mathbf{x}) \to \delta(\mathbf{x}), \quad \text{as} \;\; h \to 0 \tag{10.112}$$

$$(\mathbf{3.}) \quad \Psi_h(\mathbf{x}) \in C_0^k(\mathbb{R}^d), d = 1,2,3, \;\; k \geq 1, \tag{10.113}$$

Consider Newton's second law

$$m_i \ddot{\mathbf{x}}_i = \mathbf{F}_i = \sum_{j \neq i} \mathbf{F}_{ij}$$

For simplicity, if we only consider the pair potential,

$$\mathbf{F}_{ij} = -\nabla_i \Phi(|\mathbf{r}^{ij}|) = -\Phi'(|\mathbf{r}^{ij}|) \frac{\mathbf{r}^{ij}}{r^{ij}} \tag{10.114}$$

where $\mathbf{r}^{ij} = \mathbf{x}_i - \mathbf{x}_j$ and $r_{ij} = |\mathbf{x}_i - \mathbf{x}_j|$. Recall the definition of the directional derivative,

$$\frac{\partial \Psi_h(\lambda \mathbf{r}^{ij} + \mathbf{x}_j - \mathbf{x})}{\partial \lambda} = -\mathbf{r}^{ij} \cdot \nabla_{\mathbf{x}} \Psi_h(\lambda \mathbf{r}^{ij} + \mathbf{x}_j - \mathbf{x})$$

We can also define the directional integration (compare with the Radon transform),

$$B(\mathbf{x},\mathbf{x}_j,\mathbf{X}) = \int_0^1 \Psi_h(\lambda \mathbf{r}^{ij} + \mathbf{x}_j - \mathbf{X}) d\lambda \tag{10.115}$$

It is readily to show the identity,

$$\Psi_h(\mathbf{x}_i - \mathbf{X}) - \Psi_h(\mathbf{x}_j - \mathbf{X}) = -\mathbf{r}^{ij} \cdot \nabla_{\mathbf{X}} B(\mathbf{x}_i, \mathbf{x}_j, \mathbf{X}) \tag{10.116}$$

Taking a time derivative of (10.110) with the aid of (10.116), one can find the energy equation,

$$\frac{\partial E^0}{\partial t} + \nabla_{\mathbf{X}} \cdot \Big[Q_V^0(\mathbf{X}, t) + Q_K^0(\mathbf{X}, t) \Big] = 0 \tag{10.117}$$

where the heat flux are defined as

$$\mathbf{Q}_V^0(\mathbf{X}, t) = \frac{1}{2} \sum_{i,j} (\mathbf{F}_{ij} \cdot \dot{\mathbf{x}}_i) \mathbf{r}^{ij} B(\mathbf{x}_i, \mathbf{x}_j, \mathbf{X}) \tag{10.118}$$

$$\mathbf{Q}_K^0(\mathbf{X}, t) = \frac{1}{2} \sum_{i} (m_i |\dot{\mathbf{x}}_i|^2 + U_i)) \dot{\mathbf{x}}_i \Psi_h(\mathbf{x}_i - \mathbf{X}) \tag{10.119}$$

By define the fine scale energy and heat flux as

$$E'(\mathbf{X}, t) = \frac{1}{2} \sum_{i} \Big(m_i |\dot{\mathbf{x}}_i - \bar{\mathbf{v}}|^2 + U_i \Big) \Psi_h(\mathbf{x}_i - \mathbf{X}) \tag{10.120}$$

$$\mathbf{Q}'_V(\mathbf{X}, t) = \frac{1}{2} \sum_{i,j} \Big(\mathbf{F}_{ij} \cdot (\dot{\mathbf{x}}_i - \bar{\mathbf{v}}(\mathbf{X}, t)) \Big) \mathbf{r}^{ij} B(\mathbf{x}_i, \mathbf{x}_j, \mathbf{X}) \tag{10.121}$$

$$\mathbf{Q}'_K(\mathbf{X}, t) = \frac{1}{2} \sum_{i} (m_i |\dot{\mathbf{x}}_i - \bar{\mathbf{v}}(\mathbf{X}, t)|^2 + U_i)) \dot{\mathbf{x}}_i \Psi_h(\mathbf{x}_i - \mathbf{X}) \tag{10.122}$$

It can be readily shown that

$$E^0 = E' + \frac{1}{2} \rho |\bar{\mathbf{v}}|^2, \tag{10.123}$$

$$\mathbf{Q}_V^0 = \mathbf{Q}'_V - \bar{\mathbf{v}} \cdot \boldsymbol{\sigma}_V \tag{10.124}$$

$$\mathbf{Q}_K^0 = \mathbf{Q}'_K - \bar{\mathbf{v}} \cdot \boldsymbol{\sigma}_K + E^0 \bar{\mathbf{v}} \tag{10.125}$$

From Eqs. (10.123)-(10.125), we can identify the macroscale Hardy stress,

$$\boxed{\begin{aligned} \boldsymbol{\sigma}(\mathbf{X}, t) &= \boldsymbol{\sigma}_V + \boldsymbol{\sigma}_K \\ &= - \sum_{i} \Big(m_i (\dot{\mathbf{x}}_i - \bar{\mathbf{v}}) \otimes (\dot{\mathbf{x}}_i - \bar{\mathbf{v}}) \Psi_h(\mathbf{x}_i - \mathbf{X}) \\ &\quad + \sum_{j \neq i} \frac{1}{2} \Phi'(r^{ij}) B(\mathbf{x}_i, \mathbf{x}_j, \mathbf{X}) \frac{\mathbf{r}^{ij} \otimes \mathbf{r}^{ij}}{r^{ij}} \Big) \end{aligned}} \tag{10.126}$$

which was first provided in [Hardy (1982)].

10.4 Exercises

Problem 10.1. *Consider a finite spherical RVE with the radius* $R_M >> R_0$ *having a concentric spherical inclusion with the radius* R_0. *A Gurtin-Murdoch interface is assumed between the inclusion and the matrix.*

Assume the material properties inside the inclusion are:

$$\lambda^\Omega, \ \mu^\Omega, \ \text{and} \ K^\Omega$$

the material properties in the matrix are

$$\lambda^M, \ \mu^M, \ \text{and} \ K^M$$

and the interface material properties are

$$\lambda^S, \ \mu^S, \ \text{and} \ K^S, \ \text{with the interface tension} \ \tau_0 \ .$$

Assume that loading of the RVE is spherical radial symmetric, i.e.

$$\sigma_{rr}\Big|_{r\to\infty} = \sigma^\infty$$

Find the displacement fields.

Can anyone find a solution satisfying the boundary condition ?

$$\sigma_{rr}\Big|_{r=R_M} = \sigma^\infty$$

Hints:
See [Sharma et al. *(2003)] .*

Problem 10.2. *Consider the Morse potential with the following parameters :*

$$\Phi = D\exp^{-2\alpha(r-ha)} - 2D\exp^{-\alpha(r-ha)} \tag{10.127}$$

where $D = 0.2703 \times 1.602 \times 10^{-19} Joule$ *and* $\alpha = 1.1646 \times 10^{10} m^{-1}$

Take: 1. $ha = 3.254856644271345 \times 10^{-10} m$

2. $ma = 26.981539 \times 1.66 \times 10^{-27} kg$

3. time step: $dt = 5 \times 10^{-13} sec$

4. Initial displacement: for $| x | \leq L_c$,

$$u_0(x, t = 0) = \frac{A}{A - u_c}(1 + b\cos\frac{2\pi x}{H})(A\exp^{-(\frac{x}{\sigma})^2} - u_c) \tag{10.128}$$

where $A = 0.6ha$, $\sigma = 20ha$, $H = \sigma/4$, $b = 0.01$, $L_c = 3\sigma$, *and* $u_c = A\exp^{-(Lc/\sigma)^2}$.

Use both Quasi-continuum method and classical molecular dynamics to simulate one-dimensional wave propagation, and compare their difference.

Chapter 11

PERIODIC MICROSTRUCTURE AND ASYMPTOTIC HOMOGENIZATION

In engineering applications, often times, we encounter situations where materials have periodic structure. Such examples are: solids with lattice structures, various composites with periodic micro-structure, for instance reticulated structures, DNAs, masonry structures, so forth. There are two types of methodologies in micromechanics analysis: (1) equivalent eigenstrain approach, and (2) asymptotic homogenization approach. We shall first introduce the equivalent eigenstrain approach. This approach was first used by T. Mura [Mura (1964)], and it was later systematically developed by S. Nemat-Nasser and his co-workers. For in-depth discussion, readers may consult the original papers in the following references [Nemate-Nasser *et al.* (1986, 1993); Nemat-Nasser and Hori (1999)].

11.1 Unit cell and Fourier series

Since we only consider solids with periodic micro-structures, we can label the minimum repeating unit of the solid as the Unit cell. By translating such unit cell throughout the space, one will be able to re-construct the entire solid object.

The unit cell is conceptually different from the Representative Volume Element (RVE) that is often used in micromechanical homogenization methods. The RVE contains a large number of inclusions such that it can statistically represent the overall composite properties. Accordingly, the inclusion distribution in the RVEs can only be accounted for in a statistical manner. Whereas for the unit cell, the microstructure is described exactly. Moreover, since the unit cell is the periodic unit in a periodic structure, it prompts us to use Fourier series to represent disturbance fields, such as displacements or strains, so that one may be able to solve the problem analytically, as will be demonstrated in the followings.

Consider a rectangular unit cell defined as the following local domain

$$Y := \left\{ \mathbf{x} \;\middle|\; -a_j \le x_j \le a_j, \quad j = 1, 2, 3 \right\} \tag{11.1}$$

where a_j is the half length of the unit cell in j-th direction.

For materials with periodic structures, material properties should be periodic functions as well i.e.

$$\mathbb{C}(\mathbf{x}+\boldsymbol{d}) = \mathbb{C}(\mathbf{x})$$

where $\boldsymbol{d} = \sum_{j=1}^{3} 2m_j a_j \boldsymbol{e}_j$, $j = 1, 2, 3$. Here m_j are arbitrary integers. The vector, $\boldsymbol{d}$, is not the minimum periodicity, unless $m_j = 1$. Under certain conditions, it is possible that displacement field may be periodic as well, i.e.

$$\boldsymbol{u}(\mathbf{x}+\boldsymbol{d}) = \boldsymbol{u}(\mathbf{x})$$

An immediate consequence is that strain field is also periodic,

$$\boldsymbol{\epsilon}(\mathbf{x}+\boldsymbol{d}) = \boldsymbol{\epsilon}(\mathbf{x})$$

Nevertheless, on the other hand, a periodic strain field does not necessarily produce a periodic displacement field. For instance, a constant strain field is periodic,

$$\boldsymbol{\epsilon}(\mathbf{x}+\boldsymbol{d}) = \boldsymbol{\epsilon}(\mathbf{x}) = \boldsymbol{\epsilon}^0, \quad \forall \boldsymbol{d} \in \mathbb{R}^3,$$

but it does not generate a periodic displacement field, instead $\boldsymbol{u}(\mathbf{x}) = \mathbf{x}\cdot\boldsymbol{\epsilon}^0$, and $\boldsymbol{u}(\mathbf{x}+\boldsymbol{d}) \neq \boldsymbol{u}(\mathbf{x})$.

A convenient mathematical tool to treat periodic functions is the Fourier series. Define a vector,

$$\boldsymbol{\xi} = \xi_j \boldsymbol{e}_j, \quad \text{and} \quad \xi_j = \frac{n_j \pi}{a_j}, \quad n_j = 0, \pm 1, \pm 2, \cdots, \cdots \tag{11.2}$$

and a countable set,

$$\Lambda = \left\{ \boldsymbol{\xi} = \xi_j \boldsymbol{e}_j \;\middle|\; \xi_j \frac{n_j \pi}{a_j}, n_j = 0, \pm 1, \pm 2, \cdots, \cdots, \right\} \tag{11.3}$$

For any real function, $f(\mathbf{x}) \in C^1(Y)$, $f(\mathbf{x})$ can be expanded into Fourier series,

$$f(\mathbf{x}) = \sum_{\boldsymbol{\xi}\in\Lambda} \mathcal{F}[f](\boldsymbol{\xi}) \exp(\mathrm{i}\boldsymbol{\xi}\cdot\mathbf{x}), \quad \mathrm{i} = \sqrt{-1}, \tag{11.4}$$

where the Fourier coefficient is

$$\mathcal{F}[f](\boldsymbol{\xi}) = \frac{1}{|Y|}\int_Y \boldsymbol{u}(\mathbf{x}) \exp(-\mathrm{i}\mathbf{x}\cdot\boldsymbol{\xi}) dV_{\mathbf{x}}$$

where $|Y|$ is the volume of the unit cell. For a rectangular unit cell, $|Y| = 8a_1a_2a_3$.

Recall the definition of Fourier series in an 1D interval, $[-a, a]$,

$$f(x) = \sum_{n=-\infty}^{\infty} \mathcal{F}[f](\xi) \exp\left(\mathrm{i}\frac{n\pi}{a}x\right), \quad n = 0, \pm 1, \pm 2, \cdots,$$

$$\mathcal{F}[f] = \frac{1}{2a}\int_{-a}^{a} f(x) \exp(-\mathrm{i}\frac{n\pi}{a}x) dx$$

and the orthogonal condition

$$\frac{1}{2a}\int_{-a}^{a} \exp(\mathrm{i}x\xi_m)\exp(-\mathrm{i}x\xi_n) dx = \delta_{mn}$$

where $\xi_n = \dfrac{n\pi}{a}$ and $\xi_m = \dfrac{m\pi}{a}$.

Accordingly, the three dimensional orthogonal conditions are

$$\frac{1}{|Y|}\int_Y \exp(\mathrm{i}\mathbf{x}\cdot\boldsymbol{\xi})\exp(-\mathrm{i}\mathbf{x}\cdot\boldsymbol{\zeta}) dV_{\mathbf{x}} = \begin{cases} 1, & \boldsymbol{\xi} = \boldsymbol{\zeta} \\ 0, & \boldsymbol{\xi} \neq \boldsymbol{\zeta} \end{cases}$$

where $\boldsymbol{\xi}, \boldsymbol{\zeta} \in \Lambda$, i.e.

$$\boldsymbol{\xi} = \xi_j \boldsymbol{e}_j = \frac{n_j\pi}{a_j}\boldsymbol{e}_j \quad \text{and} \quad \boldsymbol{\zeta} = \zeta_k \boldsymbol{e}_k = \frac{n_k\pi}{a_k}\boldsymbol{e}_k \;.$$

11.1.1 *Fourier transform of displacement field and strain field*

Suppose that displacement field is periodic. We may expand displacement field into Fourier series

$$\boldsymbol{u}(\mathbf{x}) = \sum_{\boldsymbol{\xi}\in\Lambda} \mathcal{F}[\boldsymbol{u}](\boldsymbol{\xi}) \exp(\mathrm{i}\mathbf{x}\cdot\boldsymbol{\xi}) \tag{11.5}$$

where

$$\mathcal{F}[\boldsymbol{u}](\boldsymbol{\xi}) = \frac{1}{|Y|}\int_Y \boldsymbol{u}(\mathbf{x}) \exp(-\mathrm{i}\mathbf{x}\cdot\boldsymbol{\xi}) dV_{\mathbf{x}}$$

or in component form

$$\mathcal{F}[u_i](\boldsymbol{\xi}) = \frac{1}{|Y|}\int_Y u_i(\mathbf{x}) \exp(-\mathrm{i}\mathbf{x}\cdot\boldsymbol{\xi}) dV_{\mathbf{x}}$$

Remark 11.1. In literature, the following expression is often used,

$$\boldsymbol{u}(\mathbf{x}) = \sum_{\boldsymbol{\xi}\in\Lambda'} \mathcal{F}[\boldsymbol{u}](\boldsymbol{\xi}) \exp(\mathrm{i}\mathbf{x}\cdot\boldsymbol{\xi})$$

where

$$\Lambda' = \left\{\boldsymbol{\xi} = \xi_j \boldsymbol{e}_j \;\middle|\; \xi_j = \frac{n_j\pi}{a_j},\;\; j = \pm 1, \pm 2, \cdots, \cdots\right\}$$

Note that the difference between index set Λ' and Λ is that $n_j \neq 0$, or $\boldsymbol{\xi} \neq 0$.

When $\boldsymbol{\xi} = 0$,

$$\mathcal{F}[\boldsymbol{u}](0) = \frac{1}{|Y|}\int_Y \boldsymbol{u}(x) dV_{\mathbf{x}}$$

which is the average displacement field.

On the other hand, if the composite undergoes a rigid body translation, $\boldsymbol{u}(\mathbf{x}) = \boldsymbol{u}^0$, which is not periodic, one may find that

$$\mathcal{F}[\boldsymbol{u}](0) = \boldsymbol{u}^0$$

Obviously, $u = u^0 \notin L^1(\mathbb{R})$ nor $u = u^0 \in L^2(\mathbb{R})$. Convergence issue may rise in mathematical manipulation. Anyway, rigid body translation is a trivial physical motion, we neglect its contribution in Fourier transform by restricting $\boldsymbol{\xi} \in \Lambda'$.

Now, we consider the Fourier transform of displacement gradient,

$$\nabla \otimes \boldsymbol{u}(\mathbf{x}) = \sum_{\boldsymbol{\xi}\in\Lambda} \mathcal{F}[\nabla \otimes \boldsymbol{u}](\boldsymbol{\xi}) \exp(\mathrm{i}\mathbf{x}\cdot\boldsymbol{\xi}) \tag{11.6}$$

and

$$\mathcal{F}[\nabla \otimes \mathbf{u}](\boldsymbol{\xi}) = \frac{1}{|Y|}\int_Y \nabla \otimes \boldsymbol{u}(\mathbf{x}) \exp(-\mathrm{i}\mathbf{x}\cdot\boldsymbol{\xi}) dV_{\mathbf{x}}$$

On the other hand, from (11.5), one may find that

$$\nabla \otimes \boldsymbol{u}(\mathbf{x}) = \sum_{\boldsymbol{\xi} \in \Lambda} \nabla \exp(\mathrm{i}\mathbf{x} \cdot \boldsymbol{\xi}) \otimes \mathcal{F}[\boldsymbol{u}](\boldsymbol{\xi})$$
$$= \mathrm{i} \sum_{\boldsymbol{\xi} \in \Lambda} \boldsymbol{\xi} \otimes \mathcal{F}[\boldsymbol{u}](\boldsymbol{\xi}) \exp(\mathrm{i}\mathbf{x} \cdot \boldsymbol{\xi}) \tag{11.7}$$

Comparing (11.6) with (11.7), we have

$$\mathcal{F}[\nabla \otimes \boldsymbol{u}](\boldsymbol{\xi}) = \mathrm{i}\boldsymbol{\xi} \otimes \mathcal{F}[\boldsymbol{u}](\boldsymbol{\xi}) \ .$$

Moreover, we may write Fourier series transform of strain field as

$$\boldsymbol{\epsilon}(\mathbf{x}) = \frac{\mathrm{i}}{2} \sum_{\boldsymbol{\xi} \in \Lambda} \Big(\boldsymbol{\xi} \otimes \mathcal{F}[\boldsymbol{u}](\boldsymbol{\xi}) + \mathcal{F}[\boldsymbol{u}](\boldsymbol{\xi}) \otimes \boldsymbol{\xi} \Big) \exp(\mathrm{i}\mathbf{x} \cdot \boldsymbol{\xi}) \tag{11.8}$$

From (11.8), we can deduce that

$$\mathcal{F}[\boldsymbol{\epsilon}](\boldsymbol{\xi}) = \frac{\mathrm{i}}{2} \Big(\boldsymbol{\xi} \otimes \mathcal{F}[\boldsymbol{u}](\boldsymbol{\xi}) + \mathcal{F}[\boldsymbol{u}](\boldsymbol{\xi}) \otimes \boldsymbol{\xi} \Big)$$

Hence

$$\mathcal{F}[\boldsymbol{\epsilon}](0) = \frac{1}{|Y|} \int_Y \boldsymbol{\epsilon}(\mathbf{x}) dV_{\mathbf{x}} = 0$$

which implies that the average of a periodic strain field is a null field.

11.1.2 *Fourier series transform of stress field*

Consider a periodic elastic stiffness tensor, $\mathbb{C}(\mathbf{x}+\boldsymbol{d}) = \mathbb{C}(\mathbf{x})$, which may be expanded into Fourier series,

$$\mathbb{C}(\mathbf{x}) = \sum_{\boldsymbol{\xi} \in \Lambda} \mathcal{F}[\mathbb{C}](\boldsymbol{\xi}) \exp(\mathrm{i}\mathbf{x} \cdot \boldsymbol{\xi}) \tag{11.9}$$

where

$$\mathcal{F}[\mathbb{C}] = \frac{1}{|Y|} \int_Y \mathbb{C}(\mathbf{x}) \exp(-\mathrm{i}\mathbf{x} \cdot \boldsymbol{\xi}) dV_{\mathbf{x}} \tag{11.10}$$

The corresponding stress field may then be written as

$$\boldsymbol{\sigma}(\mathbf{x}) = \mathbb{C}(\mathbf{x}) : \boldsymbol{\epsilon}(\mathbf{x}) = \left\{ \sum_{\boldsymbol{\xi} \in \Lambda} \mathcal{F}[\mathbb{C}](\boldsymbol{\xi}) \exp(\mathrm{i}\mathbf{x} \cdot \boldsymbol{\xi}) \right\} : \left\{ \sum_{\boldsymbol{\zeta} \in \Lambda} \mathcal{F}[\boldsymbol{\epsilon}](\boldsymbol{\zeta}) \exp(\mathrm{i}\mathbf{x} \cdot \boldsymbol{\zeta}) \right\}$$

Let $\boldsymbol{\eta} = \boldsymbol{\xi} + \boldsymbol{\zeta}$ or $\boldsymbol{\xi} = \boldsymbol{\eta} - \boldsymbol{\zeta}$. We have

$$\boldsymbol{\sigma}(\mathbf{x}) = \sum_{\boldsymbol{\eta} \in \Lambda} \left(\sum_{\boldsymbol{\zeta} \in \Lambda} \mathcal{F}[\mathbb{C}](\boldsymbol{\eta} - \boldsymbol{\zeta}) : \mathcal{F}[\boldsymbol{\epsilon}](\boldsymbol{\zeta}) \right) \exp(\mathrm{i}\mathbf{x} \cdot \boldsymbol{\eta})$$

and it is straightforward that

$$\mathcal{F}[\boldsymbol{\sigma}](\boldsymbol{\eta}) = \sum_{\boldsymbol{\zeta} \in \Lambda} \mathcal{F}[\mathbb{C}](\boldsymbol{\eta} - \boldsymbol{\zeta}) : \mathcal{F}[\boldsymbol{\epsilon}](\boldsymbol{\zeta})$$

If $\mathbb{C} = \mathbb{C}^0$ is a constant fourth order tensor,

$$\mathcal{F}[\mathbb{C}](\boldsymbol{\eta} - \boldsymbol{\zeta}) = \mathbb{C}^0, \quad \boldsymbol{\eta} = \boldsymbol{\zeta}, \qquad \text{and} \qquad \mathcal{F}[\mathbb{C}](\boldsymbol{\eta} - \boldsymbol{\zeta}) = 0, \quad \boldsymbol{\eta} \neq \boldsymbol{\zeta},$$

There is only one term left,

$$\mathcal{F}[\boldsymbol{\sigma}](\boldsymbol{\eta}) = \mathcal{F}[\mathbb{C}](0) : \mathcal{F}[\boldsymbol{\epsilon}](\boldsymbol{\zeta}) = \mathbb{C}^0 : \mathcal{F}[\epsilon](\boldsymbol{\eta}) \qquad \text{when} \quad \boldsymbol{\eta} = \boldsymbol{\zeta}.$$

Therefore,

$$\begin{aligned}\boldsymbol{\sigma}(\mathbf{x}) &= \sum_{\boldsymbol{\eta} \in \Lambda} \mathbb{C}^0 : \mathcal{F}[\epsilon](\boldsymbol{\eta}) \exp(\mathrm{i}\mathbf{x} \cdot \boldsymbol{\eta}) \\ &= \frac{\mathrm{i}}{2} \sum_{\boldsymbol{\eta} \in \Lambda} \mathbb{C}^0 : \Big(\boldsymbol{\eta} \otimes \mathcal{F}[\boldsymbol{u}](\boldsymbol{\eta}) + \mathcal{F}[\boldsymbol{u}](\boldsymbol{\eta}) \otimes \boldsymbol{\eta}\Big) \exp(\mathrm{i}\mathbf{x} \cdot \boldsymbol{\eta})\end{aligned}$$

Last, we evaluate the Fourier expansion,

$$\nabla \cdot \boldsymbol{\sigma} = \sum_{\boldsymbol{\xi} \in \Lambda} \mathcal{F}[\nabla \cdot \boldsymbol{\sigma}](\boldsymbol{\xi}) \exp(\mathrm{i}\mathbf{x} \cdot \boldsymbol{\xi})$$

Integration by parts yields

$$\begin{aligned}\mathcal{F}[\nabla \cdot \boldsymbol{\sigma}](\boldsymbol{\xi}) &= \frac{1}{|Y|} \int_Y \nabla \cdot \boldsymbol{\sigma}(\mathbf{x}) \exp(-\mathrm{i}\mathbf{x} \cdot \boldsymbol{\xi}) dV_{bx} \\ &= \frac{1}{|Y|} \int_Y \Big\{\nabla \cdot \Big(\boldsymbol{\sigma}(\mathbf{x}) \exp(-\mathrm{i}\mathbf{x} \cdot \boldsymbol{\xi})\Big) - \boldsymbol{\sigma} \cdot \Big(\nabla \exp(-\mathrm{i}\mathbf{x} \cdot \boldsymbol{\xi})\Big)\Big\} dV_{\mathbf{x}} \\ &= \frac{1}{Y} \Big\{\int_{\partial Y} \mathbf{n} \cdot \boldsymbol{\sigma}(\mathbf{x}) \exp(-\mathrm{i}\mathbf{x} \cdot \boldsymbol{\xi}) dS + i\boldsymbol{\xi} \int_Y \boldsymbol{\sigma}(\mathbf{x}) \exp(-\mathrm{i}\mathbf{x} \cdot \boldsymbol{\xi}) dV_{\mathbf{x}}\Big\} \\ &= \mathrm{i}\boldsymbol{\xi} \int_Y \boldsymbol{\sigma}(\mathbf{x}) \exp(-\mathrm{i}\mathbf{x} \cdot \boldsymbol{\xi}) dV_{\mathbf{x}}\end{aligned}$$

because

$$\int_{\partial Y} \mathbf{n} \cdot \boldsymbol{\sigma}(\mathbf{x}) \exp(-\mathrm{i}\mathbf{x} \cdot \boldsymbol{\xi}) dS = 0$$

by periodicity. In particular, when $\boldsymbol{\xi} = 0$,

$$\int_{\partial Y} \mathbf{n} \cdot \boldsymbol{\sigma}(\mathbf{x}) dS = 0$$

which stems from the fact that unit cell is in equilibrium.

11.2 Eigenstrain homogenization

Let $\mathbb{C}^M$ and $\mathbb{D}^M$ be elastic stiffness and compliance tensors in the matrix; and let $\mathbb{C}^\Omega$, $\mathbb{D}^\Omega$ be the effective stiffness and compliance tensors in the second phase, which is assumed to be distributed periodically in the composite. We are looking for effective stiffness and compliance tensors, $\bar{\mathbb{C}}$ and $\bar{\mathbb{D}}$.

Consider prescribed macro-strain boundary condition,

$$\boldsymbol{\epsilon} = \mathbf{x} \cdot \boldsymbol{\epsilon}^0, \quad \forall \mathbf{x} \in \partial V$$

The total strain may be written as

$$\epsilon_{ij} = \epsilon^0_{ij} + \epsilon^d_{ij}, \quad \forall \ \mathbf{x} \in V$$

The stress fields in the matrix and in the second phase are

$$\sigma^M_{ij} = C^M_{ijk\ell}(\epsilon^0_{k\ell} + \epsilon^d_{k\ell}), \quad \forall \mathbf{x} \in M = Y/\Omega$$
$$\sigma^\Omega_{ij} = C^\Omega_{ijk\ell}(\epsilon^0_{k\ell} + \epsilon^d_{k\ell}), \quad \forall \mathbf{x} \in \Omega$$

They satisfy the equilibrium equations,

$$\sigma^M_{ij,j} = 0, \quad \forall \ \mathbf{x} \in M \tag{11.11}$$
$$\sigma^\Omega_{ij,j} = 0, \quad \forall \ \mathbf{x} \in \Omega \tag{11.12}$$

and the displacement continuity condition at interface,

$$u^{d+}_i = u^{d-}_i, \quad \forall \ \mathbf{x} \in \partial\Omega \tag{11.13}$$

where $u^d_i(\mathbf{x})$ denotes the disturbance displacement field.

Consider an eigenstrain field,

$$\epsilon^*_{ij}(\mathbf{x}) = \epsilon^*_{ij}\chi(\Omega)$$

where $\chi(\Omega)$ is the characteristic function of the domain. Note that we assume that ϵ^*_{ij} are not uniform inside Ω.

Eshelby's equivalent inclusion principle reads as

$$\sigma^\Omega_{ij} = C^\Omega_{ijk\ell}(\epsilon^0_{k\ell} + \epsilon^d_{k\ell}) = C^M_{ijk\ell}(\epsilon^0_{k\ell} + \epsilon^d_{k\ell} - \epsilon^*_{k\ell}) \tag{11.14}$$

Substituting (11.14) into (11.12) yields

$$C^M_{ijk\ell}(\epsilon^0_{k\ell} + \epsilon^d_{k\ell} - \epsilon^*_{k\ell})_{,j} = 0, \quad \Rightarrow \quad C^M_{ijk\ell}u^d_{k,\ell j} = C^M_{ijk\ell}\epsilon^*_{k\ell,j} \tag{11.15}$$

Let,

$$\epsilon^*_{k\ell}(\mathbf{x}) = \sum_{\boldsymbol{\xi}\in\Lambda'} \mathcal{F}[\epsilon^*_{k\ell}](\boldsymbol{\xi})\exp(\mathrm{i}\boldsymbol{\xi}\cdot\mathbf{x}) = \sum_{\boldsymbol{\xi}\in\Lambda'} \hat{\epsilon}^*_{k\ell}\exp(\mathrm{i}\boldsymbol{\xi}\cdot\mathbf{x}) \tag{11.16}$$

where

$$\hat{\epsilon}^*_{k\ell} = \frac{1}{Y}\int_Y \epsilon^*_{k\ell}(\mathbf{x})\exp(-\mathrm{i}\boldsymbol{\xi}\cdot\mathbf{x})dV_{\mathbf{x}} = \frac{1}{Y}\int_\Omega \epsilon^*_{k\ell}\exp(-\mathrm{i}\boldsymbol{\xi}\cdot\mathbf{x})dV_{\mathbf{x}}$$

and

$$u^d_i(\mathbf{x}) = \sum_{\boldsymbol{\xi}\in\Lambda'} \mathcal{F}[u^d_i](\boldsymbol{\xi})\exp(i\boldsymbol{\xi}\cdot\mathbf{x}) = \sum_{\boldsymbol{\xi}\in\Lambda'} \hat{u}^d_i(\boldsymbol{\xi})\exp(\mathrm{i}\boldsymbol{\xi}\cdot\mathbf{x}) \tag{11.17}$$

where

$$\hat{u}^d_i(\boldsymbol{\xi}) = \frac{1}{|Y|}\int_Y u^d_i(\mathbf{x})\exp(-\mathrm{i}\boldsymbol{\xi}\cdot\mathbf{x})dV_{\mathbf{x}}$$

Note that uniform eigenstrain is excluded because it induces a divergent displacement field, i.e.

$$u^d_i(\mathbf{x}) = \epsilon^{*0}_{ij}x_j \quad \to \infty \quad \text{as} \ \mathbf{x} \to \infty$$

Substituting (11.16) and (11.17) into (11.15), we have

$$-C^M_{ijk\ell}\hat{u}^d_k\xi_\ell\xi_j = \mathrm{i}C^M_{ijk\ell}\hat{\epsilon}^*_{k\ell}\xi_j \tag{11.18}$$

Denote $K_{ik}(\boldsymbol{\xi}) = C^M_{ijk\ell}\xi_\ell\xi_j$ and $K^{-1}_{ik}(\boldsymbol{\xi}) = N_{ik}(\boldsymbol{\xi})/D(\boldsymbol{\xi})$.

$$\mathcal{F}[u^d_i](\boldsymbol{\xi}) := \hat{u}^d_i(\boldsymbol{\xi}) = -\mathrm{i}\frac{N_{ik}(\boldsymbol{\xi})}{D(\boldsymbol{\xi})}C^M_{k\ell mn}\epsilon^*_{mn}\xi_\ell \tag{11.19}$$

Recall,

$$\epsilon^d_{ij}(\mathbf{x}) = \frac{\mathrm{i}}{2}\sum_{\boldsymbol{\xi}\in\Lambda'}\Big(\xi_i\mathcal{F}[u^d_j](\boldsymbol{\xi}) + \mathcal{F}[u^d_i](\boldsymbol{\xi})\xi_j\Big)\exp(i\boldsymbol{\xi}\cdot\mathbf{x})$$

One can write

$$\begin{aligned}
\epsilon^d_{ij} &= \sum_{\boldsymbol{\xi}\in\Lambda'}\frac{1}{2}\Big(\xi_i\xi_\ell\frac{N_{jk}(\boldsymbol{\xi})}{D(\boldsymbol{\xi})}C^M_{k\ell mn} + \xi_j\xi_\ell\frac{N_{ik}(\boldsymbol{\xi})}{D(\boldsymbol{\xi})}C^M_{k\ell mn}\Big)\hat{\epsilon}^*_{mn}\exp(i\boldsymbol{\xi}\cdot\mathbf{x})\\
&= \sum_{\boldsymbol{\xi}\in\Lambda'} g_{ijmn}(\boldsymbol{\xi})\hat{\epsilon}^*_{mn}\exp(\mathrm{i}\boldsymbol{\xi}\cdot\mathbf{x})\\
&= \frac{1}{|Y|}\sum_{\boldsymbol{\xi}\in\Lambda'} g_{ijmn}(\boldsymbol{\xi})\int_Y \epsilon^*_{mn}(\mathbf{x}')\exp(-\mathrm{i}\boldsymbol{\xi}\cdot\mathbf{x}')dV_{\mathbf{x}'}\exp(\mathrm{i}\boldsymbol{\xi}\cdot\mathbf{x})
\end{aligned}$$

where a new fourth order tensor g_{ijmn} is defined as

$$g_{ijmn}(\boldsymbol{\xi}) = \frac{1}{2}\Big(\xi_i N_{jk}(\boldsymbol{\xi}) + \xi_j N_{ik}(\boldsymbol{\xi})\Big)\frac{C^M_{k\ell mn}\xi_\ell}{D(\boldsymbol{\xi})} \tag{11.20}$$

For isotropic materials,

$$\begin{aligned}
g_{ijk\ell}(\boldsymbol{\xi}) &= \frac{1}{2\xi^2}\Big[\xi_j(\delta_{i\ell}\xi_k + \delta_{ik}\xi_\ell) + \xi_i(\delta_{j\ell}\xi_k + \delta_{jk}\xi_\ell)\Big]\\
&\quad -\frac{1}{1-\nu}\frac{\xi_i\xi_j\xi_k\xi_\ell}{\xi^4} + \frac{\nu}{1-\nu}\frac{\xi_i\xi_j}{\xi^2}\delta_{k\ell}
\end{aligned} \tag{11.21}$$

Consider the dilute homogenization scheme,

$$\mathbb{C}^\Omega : (\boldsymbol{\epsilon}^0 + \boldsymbol{\epsilon}^d) = \mathbb{C}^M : (\boldsymbol{\epsilon}^0 + \boldsymbol{\epsilon}^d - \boldsymbol{\epsilon}^*)\ .$$

We have

$$\boldsymbol{\epsilon}^0 + \boldsymbol{\epsilon}^d = (\mathbb{C}^M - \mathbb{C}^\Omega)^{-1} : \mathbb{C}^M : \boldsymbol{\epsilon}^*$$

and subsequently,

$$\boldsymbol{\epsilon}^0 = \mathbb{A}^\Omega : \boldsymbol{\epsilon}^*(\mathbf{x}) - \boldsymbol{\epsilon}^d(\mathbf{x})$$

This leads to the following integral equation,

$$\begin{aligned}
&\epsilon^0_{ij} - A^\Omega_{ijmn}\epsilon^*_{mn}(\mathbf{x})\\
&+\sum_{\boldsymbol{\xi}\in\Lambda'} g_{ijmn}(\boldsymbol{\xi})\frac{1}{|Y|}\int_\Omega \epsilon^*_{mn}(\mathbf{x}')\exp(\mathrm{i}(\mathbf{x}-\mathbf{x}')\cdot\boldsymbol{\xi})dV_{\mathbf{x}'} = 0\ .
\end{aligned} \tag{11.22}$$

This equation is difficult to solve. Calculate the average $\dfrac{1}{|\Omega|}\displaystyle\int_\Omega (11.22) dV_{\mathbf{x}}$ in the inclusion. We define the average eigenstrain as

$$\bar{\boldsymbol{\epsilon}}^* = \frac{1}{|\Omega|}\int_\Omega \boldsymbol{\epsilon}^* d\Omega \ . \tag{11.23}$$

One has

$$\begin{aligned}\boldsymbol{\epsilon}^0 = \mathbb{A}^\Omega : \bar{\boldsymbol{\epsilon}}^* - \sum_{\boldsymbol{\epsilon}\in\Lambda'} \mathbf{g}(\boldsymbol{\xi}) : \Bigl(\frac{1}{|\Omega|}\int_\Omega \exp(\mathrm{i}\boldsymbol{\xi}\cdot\mathbf{x}) dV_{\mathbf{x}}\Bigr) \\ \cdot \Bigl(\frac{1}{|Y|}\int_\Omega \boldsymbol{\epsilon}^*(\mathbf{x}')\exp(-\mathrm{i}\boldsymbol{\xi}\cdot\mathbf{x}')dV_{\mathbf{x}'}\Bigr)\end{aligned}$$

where $\mathbf{g}(\boldsymbol{\xi})$ is the fourth order tensor: $\mathbf{g} = g_{ijk\ell}(\boldsymbol{\xi})\mathbf{e}_i\otimes\mathbf{e}_j\otimes\mathbf{e}_k\otimes\mathbf{e}_\ell$. Define a scalar function,

$$g_0(\boldsymbol{\xi}) = \frac{1}{|\Omega|}\int_\Omega \exp(\mathrm{i}\boldsymbol{\xi}\cdot\mathbf{x})dV_{\mathbf{x}} \tag{11.24}$$

The eigenstrain integral equation may be written as

$$\begin{aligned}\epsilon^0_{ij} - A^\Omega_{ijmn}\bar{\epsilon}^*_{mn} + \sum_{\boldsymbol{\xi}\in\Lambda'} g_{ijmn}(\boldsymbol{\xi})g_0(\boldsymbol{\xi}) \\ \cdot\Bigl(\frac{1}{|Y|}\int_\Omega \epsilon^*_{mn}(\mathbf{x}')\exp(-\mathrm{i}\boldsymbol{\xi}\cdot\mathbf{x}')dV_{\mathbf{x}'}\Bigr) = 0 \ .\end{aligned} \tag{11.25}$$

For prescribed macros stress boundary condition, one may be able to show that

$$\begin{aligned}\epsilon^0_{ij} - A^\Omega_{ijmn}\bar{\epsilon}^*_{mn} + \sum_{\boldsymbol{\xi}\in\Lambda'} g_{ijmn}(\boldsymbol{\xi})g_0(\boldsymbol{\xi}) \\ \cdot\Bigl(\frac{1}{|Y|}\int_\Omega \epsilon^*_{mn}(\mathbf{x}')\exp(-\mathrm{i}\boldsymbol{\xi}\cdot\mathbf{x}^{'})dV_{\mathbf{x}'}\Bigr) = 0 \ .\end{aligned} \tag{11.26}$$

where $\bar{\epsilon}_{ij} = D^M_{ijmn}\sigma^0_{mn}$.

The simplest approach to solve (11.25) is to replace $\boldsymbol{\epsilon}^*(\mathbf{x})$ inside the integral by its volume average, i.e., $\boldsymbol{\epsilon}^*(\mathbf{x}) \approx \bar{\boldsymbol{\epsilon}}^*$. Therefore,

$$\begin{aligned}\boldsymbol{\epsilon}^0 &= \mathbb{A}^\Omega : \bar{\boldsymbol{\epsilon}}^* - \sum_{\boldsymbol{\xi}\in\Lambda'} \mathbf{g}(\boldsymbol{\xi})g_0(\boldsymbol{\xi})\Bigl(\frac{1}{|Y|}\int_\Omega \exp(-\mathrm{i}\boldsymbol{\xi}\cdot\mathbf{x}')dV_{\mathbf{x}'}\Bigr) : \bar{\boldsymbol{\epsilon}}^* \\ &= \mathbb{A}^\Omega : \bar{\boldsymbol{\epsilon}}^* - \sum_{\boldsymbol{\xi}\in\Lambda'} f g_0(\boldsymbol{\xi})g_0(-\boldsymbol{\xi})\mathbf{g}(\boldsymbol{\xi}) : \bar{\boldsymbol{\epsilon}}^* \\ &= \mathbb{A}^\Omega : \bar{\boldsymbol{\epsilon}}^* - \sum_{\boldsymbol{\xi}\in\Lambda'} f G(\boldsymbol{\xi})\mathbf{g}(\boldsymbol{\xi}) : \bar{\boldsymbol{\epsilon}}^*\end{aligned}$$

where $G(\boldsymbol{\xi}) = g_0(\boldsymbol{\xi})g_0(-\boldsymbol{\xi})$.

Define the Eshelby tensor for periodic inhomogeneities,

$$S^\Omega_{ijmn} = \sum_{\boldsymbol{\xi}\in\Lambda'} f G(\boldsymbol{\xi}) g_{ijmn}(\boldsymbol{\xi}) \tag{11.27}$$

We recover the relationship between remote strain and eigenstrain (average eigenstrain be more precise),

$$\epsilon^0_{ij} = \left(A^{\Omega}_{ijmn} - S_{ijmn}\right)\bar{\epsilon}^*_{mn}$$

To this end, the homogenization of a composite with periodic microstructure can follow the same route as the homogenization of a composite with randomly distributed inhomogeneities, if one can find the corresponding Eshelby tensor. The key to evaluate Eshelby tensor is to find function, $G(\boldsymbol{\xi})$.

Example 11.1. Calculate $G(\boldsymbol{\xi})$ for a one-dimensional periodic unit cell as shown in Fig. 11.1.

One can show that

$$\begin{aligned} g_0(\xi) &= \frac{1}{2a}\int_{-a}^{a} \exp(i\xi x)dx = \frac{1}{2a}\frac{1}{i\xi}\exp(i\xi x)\Big|_{-a}^{a} \\ &= \frac{1}{2a\xi \mathrm{i}}\Big[\Big(\cos(\xi a) + \mathrm{i}\sin(\xi a)\Big) - \Big(\cos(\xi a) - \mathrm{i}\sin(\xi a)\Big)\Big] \\ &= \frac{1}{a\xi}\sin(\xi a) \end{aligned}$$

It is obvious that

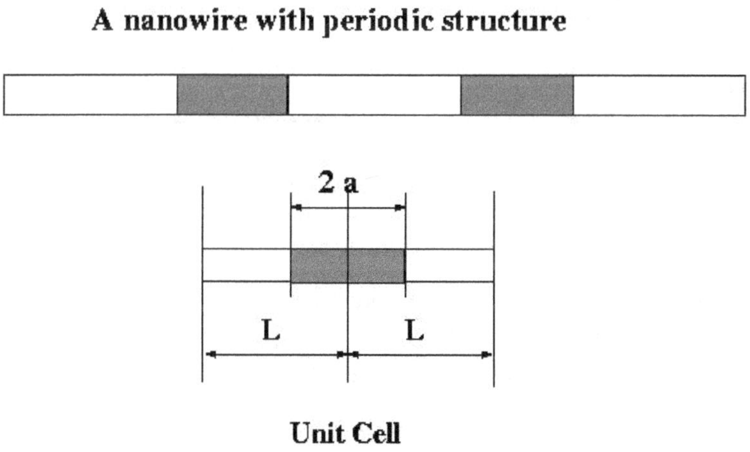

Fig. 11.1 An 1D model for a nanowire with periodic structure.

$$g_0(\ \xi) = g_0(\xi)$$

Hence

$$G(\xi) = \frac{1}{a^2\xi^2}\sin^2(\xi a)$$

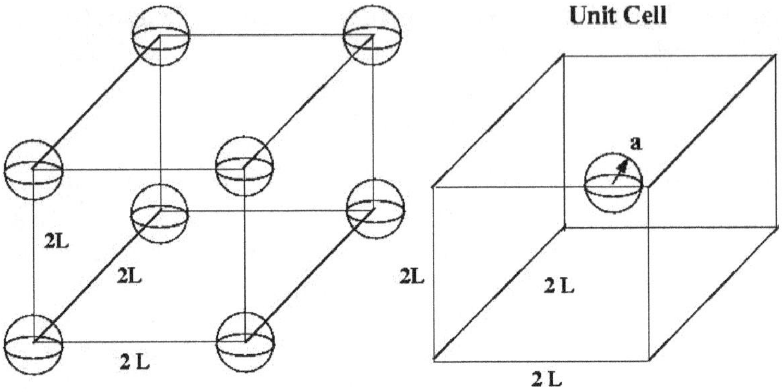

Fig. 11.2 Periodic distribution of spherical precipitates.

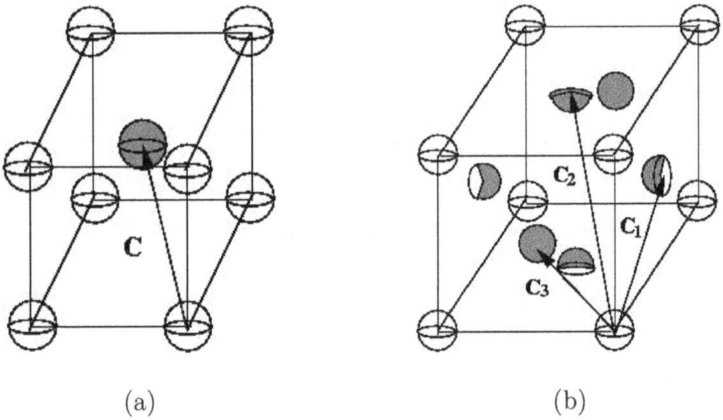

Fig. 11.3 Cluster of precipitates in various unit cells: (a) b.c.c. cluster, and (b) f.c.c cluster.

Example 11.2. In this example, we consider a spherical precipitate distribution in a cubic lattice as shown in Fig. (11.2).

The unit cell in this case is a $2L \times 2L \times 2L$ cubic region. There is a spherical ball with radius $r = a$ inside the unit cell.

Recall

$$\int_{\Omega} \exp(-\mathrm{i}\boldsymbol{\xi} \cdot \mathbf{x}) d\Omega = (2\pi)^{3/2} a^3 \frac{J_{3/2}(\eta)}{\eta^{3/2}}$$

where

$$\begin{aligned}
\eta = a|\xi| &= a\sqrt{\xi_1^2 + \xi_2^2 + \xi_3^2} \\
&= a\sqrt{\left(\frac{n_1\pi}{L}\right)^2 + \left(\frac{n_2\pi}{L}\right)^2 + \left(\frac{n_3\pi}{L}\right)^2} \\
&= \frac{\pi a}{L}\sqrt{n_1^2 + n_2^2 + n_3^2} = \frac{\pi a}{L}|\mathbf{n}|
\end{aligned}$$

Considering,

$$J_{3/2}(\eta) = \left(\frac{2}{\pi\eta}\right)^{1/2}(\eta^{-1}\sin\eta - \cos\eta) = \sqrt{\frac{2}{\pi}}\frac{1}{\eta^{3/2}}(\sin\eta - \eta\cos\eta)$$

one may write

$$\frac{1}{|\Omega|}\int_{\Omega}\exp(-\mathbf{i}\boldsymbol{\xi}\cdot\mathbf{x})d\Omega = \frac{3}{\eta^3}(\sin\eta - \eta\cos\eta)$$

and

$$G(\xi) = \frac{9}{a^6|\xi|^6}\Big[\sin(a|\xi|) - a|\xi|\cos(a|\xi|)\Big]^2 .$$

One may find that for BCC precipitate cluster,

$$g_0(-\boldsymbol{\xi}) = \frac{3}{\eta^3}(\sin\eta - \eta\cos\eta)\Big(1 + \exp(-i\boldsymbol{\xi}\cdot\mathbf{c})\Big)$$

and for FCC precipitate cluster,

$$g_0(-\boldsymbol{\xi}) = \frac{3}{\eta^3}(\sin\eta - \eta\cos\eta)\Big(1 + \exp(-i\boldsymbol{\xi}\cdot\mathbf{c}_1) + \exp(-i\boldsymbol{\xi}\cdot\mathbf{c}_2) + \exp(-i\boldsymbol{\xi}\cdot\mathbf{c}_3)\Big)$$

as shown in Fig. (11.3)

Example 11.3. Masonry is a two-phase material comprised of brick and mortar joints, normally arranged periodically. Studying the effective stiffness of the masonry is important for designing and retrofitting masonry structures.

For the periodic masonry structure, a unit area (called unit cell) is highlighted in Fig. 11.4, which represents the unit of periodicity in the horizontal and vertical directions. There are, of course, other possible choices for the unit cell.

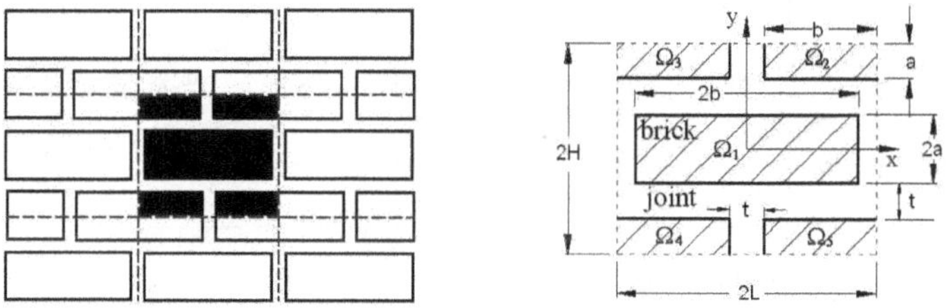

Fig. 11.4 Unit cells for masonry structure.

The selected unit cell (V) is a microstructured composite, it contains one complete brick unit, four one-quarter-sized brick units and mortar joints between bricks. In our analysis, the brick units are considered to be inclusions (Ω) and mortar is treated as matrix (M).

One can show that

$$g_0(\boldsymbol{\xi}) = \frac{1}{\Omega}\sum_{r=1}^{5}\int_{\Omega_r}\exp(i\boldsymbol{\xi}\cdot\mathbf{x})dx$$
$$= \frac{1}{2ab}\frac{1}{\xi_1\xi_2}\{\sin(\xi_1 b)\sin(\xi_2 a) + [\sin(\xi_1 L) - \sin(\xi_1(L-b))]$$
$$\cdot\ [\sin(\xi_2 H) - sin(\xi_2(H-a))]\}$$

A numerical example is considered below to illustrate the periodic eigenstrain method. The brick and mortar are assumed to be isotropic. The Young's modulus of the brick (E^b) and Poisson's ratio are 11,000 MPa and 0.20 respectively. The ratio of Young's moduli of the brick over the mortar (E^b/E^m) ranges from 1.1 to 11, and Poisson's ratio for the mortar is 0.25. The brick dimensions are 250 mm (length) × 55 mm (height). In Fig. 11.5, the ratios of effective stiffness over

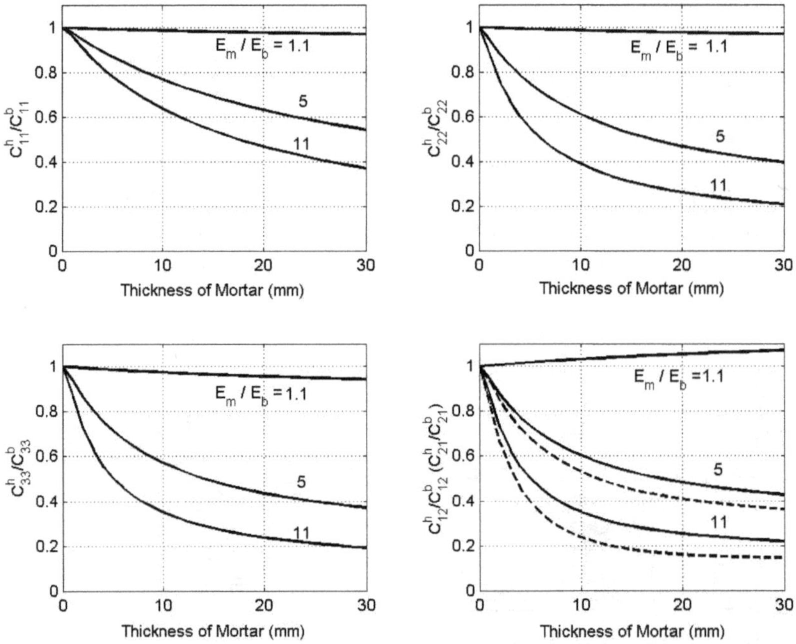

Fig. 11.5 Ratios between components of effective masonry stiffness (C^h) and brick stiffness (C^b) (After [Wang *et al.* (2007)]).

C^h the brick stiffness C^b are plotted against the mortar thickness t and the brick-mortar stiffness ratio E^b/E^m for each nonzero component. The mortar stiffness and thickness have significant effects on the overall properties of masonry. Moreover, all curves start from unity and asymptotically approach their theoretical limits as the thickness of mortar joints increases.

It is also interesting to point out that the resulting effective stiffness matrix of the masonry is not symmetric. In other words, the masonry composite is not exactly orthotropic. This property stems from the fact that the Eshelby tensor matrix is not symmetric in general, i.e. $S^{\Omega}_{1122} \neq S^{\Omega}_{2211}$. The ratio of C^h_{12}/C^b_{12} is plotted in the dashed lines against in solid lines C^h_{21}/C^b_{21} in Fig. 11.5, in which differences between these two components are evident, especially for the cases of high stiffness ratios. The finding is consistent with the finite element results presented by Ma [Ma *et al.* (2001)]. This feature also distinguishes the periodic eigenstrain model from most other models reported in literature, where orthotropy of masonry stiffness is usually assumed as a priori. In principle, the masonry composite should not be exactly orthotropic unless the RVE is infinitely large.

11.3 Introduction to asymptotic homogenization

The asymptotic method of homogenization is a systematic tool to find effective material properties or effective coefficients of a homogenized differential equation.

The main technique of asymptotic homogenization is the use of multiple-scale expansion. Often times, it involves with singular perturbation technique.

11.3.1 *One-dimensional model problem*

Consider an 1D model,

$$\frac{d}{dx}\Big(E\frac{du}{dx}\Big) = 0, \quad 0 < x < L \tag{11.28}$$

This equation can be viewed as either the deformation of 1D elastic bar, or 1D steady-state heat diffusion, etc.

Assume that the medium has periodic micro-structure that is varying at microscale, ℓ, which is the characteristic length of a unit cell. Therefore, the coefficient, E, is a periodic function of spatial variable. We also assume that at the interface of two different media in the unit cell the following continuity conditions hold,

$$[u] = 0, \quad \Big[E\frac{du}{dx}\Big] = 0 \ .$$

This 1D model problem has a very simple differential equation. An exact solution is possible. In general, for multiple dimension problems or nonlinear problems, analytical solutions may not be possible.

An important characteristics of this problem is the existence of two vastly different length scales: the microscale ℓ, which characterizes the dimension of the unit cell, and the macroscale L, which characterizes the global variations of external force or boundary data.

Suppose that one is more interested in the average variation over a region which is much greater than the typical period and less interested in the detailed variation over a local region. One may ask oneself that

Can one bypass the details to find macroscale governing equation ?

We define a small parameter $\epsilon = \frac{\ell}{L}$. Obviously, $\epsilon << 1$. To separate the effect of two scales, we introduce two coordinates: a fast coordinate and a slow coordinate, which are defined as

$$y \quad \text{and} \quad x = \epsilon y \tag{11.29}$$

You may suggest that the slow coordinate is slowed by small parameter, ϵ. Or vice versa,

$$x \quad \text{and} \quad y = \frac{x}{\epsilon} \tag{11.30}$$

You may suggest that the fast coordinate is speed up by a large parameter $\frac{1}{\epsilon}$.

Then, the field variable u may be expressed in a two-scale representation: $u = u(x, y)$ By using chain rule, we may write

$$\frac{d}{dy} = \frac{\partial}{\partial y} + \epsilon\frac{\partial}{\partial x} \tag{11.31}$$

or vice versa,

$$\frac{d}{dx} = \frac{\partial}{\partial x} + \frac{1}{\epsilon}\frac{\partial}{\partial y} \tag{11.32}$$

One can then rewrite Eq. (11.28) as

$$\frac{d}{dy}\left(E(y)\frac{du}{dy}\right) = 0, \quad 0 < y < L \tag{11.33}$$

It is clear that the coefficient has to be a periodic function of fast coordinate, i.e. $E = E(y)$.

Consider the following multi-scale expansion,

$$u(x, y) = u_0(x, y) + \epsilon u_1(x, y) + \epsilon^2 u_2(x, y) + \cdots \tag{11.34}$$

where $u_i(x, y)$ represents activity at i-th scale.

Applying (11.31) to (11.33) leads to the following partial differential equation,

$$\left(\frac{\partial}{\partial y} + \epsilon\frac{\partial}{\partial x}\right)\Big[E(y)\Big(\frac{\partial u_0}{\partial y} + \epsilon\Big[\frac{\partial u_0}{\partial x} + \frac{\partial u_1}{\partial y}\Big] + \epsilon^2\Big[\frac{\partial u_1}{\partial x} + \frac{\partial u_2}{\partial y}\Big]$$
$$\cdots\cdots\Big)\Big] = 0$$

A complete equilibrium implies that equilibrium holds in each scale,

$$\epsilon^0 : \ \frac{\partial}{\partial y}\Big[E(y)\frac{\partial u_0}{\partial y}\Big] = 0;$$

$$\epsilon^1 : \ \frac{\partial}{\partial y}\Big[E(y)\Big(\frac{\partial u_0}{\partial x} + \frac{\partial u_1}{\partial y}\Big)\Big] + E(y)\frac{\partial^2 u_0}{\partial x \partial y} = 0;$$

$$\epsilon^2 : \ \frac{\partial}{\partial y}\Big[E(y)\Big(\frac{\partial u_1}{\partial x} + \frac{\partial u_2}{\partial y}\Big)\Big] + E(y)\Big(\frac{\partial^2 u_0}{\partial x^2} + \frac{\partial^2 u_1}{\partial x \partial y}\Big) = 0;$$

$$\cdots\cdots$$

We first solve the zero-th order equation,

$$\frac{\partial}{\partial y}\Big(E(y)\frac{\partial u_0}{\partial y}\Big) = 0 \tag{11.35}$$

which only involves with the lowest scale field variable, $u_0(x, y)$.

Integrate (11.35) once,

$$E(y)\frac{\partial u_0}{\partial y} = A_1(x)$$

where $A_1(x)$ is a integration constant.

Integrating second time, we have

$$u_0(x, y) = A_1(x)\int_{y_0}^{y} \frac{d\tilde{y}}{E(\tilde{y})} + A_2(x)$$

Since $u_0(x, y)$ is periodic,

$$u_0(x, y_0) = u_0(x, y_0 + \ell) \ \Rightarrow \ A_2(x) = A_1(x)\int_{y_0}^{y_0+\ell} \frac{d\tilde{y}}{E(\tilde{y})} + A_2(x)$$

which implies that $A_1(x) = 0$.

This suggests that the leading-order displacement field only depends on the macro-scale variable,

$$u_0 = A_2(x) = u_0(x) \tag{11.36}$$

Now let's examine the first order differential equation,

$$\frac{\partial}{\partial y}\Big[E(y)\Big(\frac{\partial u_0}{\partial x} + \frac{\partial u_1}{\partial y}\Big)\Big] + E(y)\frac{\partial^2 u_0}{\partial x \partial y} = 0 \tag{11.37}$$

Based on (11.36), the last term in (11.37) vanishes.

To solve (11.37), we introduce the following partial separation of variable,

$$u_1(x, y) = Q(x, y)\frac{\partial u_0}{\partial x} + \bar{u}_1(x)$$

where $Q(x, y)$ is an unknown function.

Substitute the above expression into (11.37),

$$\frac{\partial}{\partial y}\Big\{E(y)\Big(\frac{\partial u_0}{\partial x} + \frac{\partial Q}{\partial y}\frac{\partial u_0}{\partial x}\Big)\Big\} =$$

$$\frac{\partial u_0}{\partial x}\frac{\partial}{\partial y}\Big\{E(y)\Big(1 + \frac{\partial Q}{\partial y}\Big)\Big\} = 0\ .$$

This leads to the so-called inhomogeneous canonical cell problem for unknown function, $Q(x,y)$,

$$\frac{\partial}{\partial y}\left\{E(y)\Big(1+\frac{\partial Q}{\partial y}\Big)\right\}=0,\quad \forall y\in(y_0,y_0+\ell) \tag{11.38}$$

$$[Q]=0,\quad \text{and}\quad \left[E(y)\Big(1+\frac{\partial Q}{\partial y}\Big)\right]=0\ ,\forall x \text{ at interface.} \tag{11.39}$$

Integrate (11.38) once,

$$E(y)\Big(1+\frac{\partial Q}{\partial y}\Big)=D_1(x)$$

$$\text{or}\quad \frac{\partial Q}{\partial y}=-1+\frac{D_1(x)}{E(y)}$$

where $D_1(x)$ is an integration constant.

Integrate second times,

$$Q(x,y)=-y+D_1(x)\int_{y_0}^{y}\frac{d\tilde{y}}{E(\tilde{y})}+D_2(x) \tag{11.40}$$

where $D_2(x)$ is another integration constant.

Since $Q(x,y)$ is y-periodic,

$$Q(x,y_0)=Q(x,y_0+\ell)$$

It leads to

$$-y_0+D_2(x)=-(y_0+\ell)+D_1(x)\int_{y_0}^{y_0+\ell}\frac{d\tilde{y}}{E(\tilde{y})}+D_2(x) \tag{11.41}$$

Eq. (11.41) is called the solvability condition for inhomogeneous problem for Q or u_1.

We then find that

$$D_1(x)=\frac{1}{\displaystyle\frac{1}{\ell}\int_{y_0}^{y_0+\ell}\frac{d\tilde{y}}{E(\tilde{y})}} \tag{11.42}$$

and hence

$$Q(x,y)=-y+\frac{\displaystyle\int_{y_0}^{y}\frac{d\tilde{y}}{E(\tilde{y})}}{\displaystyle\frac{1}{\ell}\int_{y_0}^{y_0+\ell}\frac{d\tilde{y}}{E(\tilde{y})}}+D_2(x) \tag{11.43}$$

Therefore,

$$u_1(x,y)=\left(-y+\frac{\displaystyle\int_{y_0}^{y}\frac{d\tilde{y}}{E(\tilde{y})}}{\displaystyle\frac{1}{\ell}\int_{y_0}^{y_0+\ell}\frac{d\tilde{y}}{E(\tilde{y})}}+D_2(x)\right)\frac{\partial u_0}{\partial x}+\bar{u}_1(x) \tag{11.44}$$

$$\frac{\partial u_1}{\partial y}=-\frac{\partial u_0}{\partial x}+\frac{1}{E(y)\Big(\displaystyle\frac{1}{\ell}\int_{y_0}^{y_0+\ell}\frac{d\tilde{y}}{E(\tilde{y})}\Big)}\frac{\partial u_0}{\partial x} \tag{11.45}$$

Next, we consider the differential equation at the second scale,

$$\epsilon^2: \quad \frac{\partial}{\partial y}\Big[E(y)\Big(\frac{\partial u_1}{\partial x}+\frac{\partial u_2}{\partial y}\Big)\Big]+E(y)\Big(\frac{\partial^2 u_0}{\partial x^2}+\frac{\partial u^2 u_1}{\partial x \partial y}\Big)=0\,. \tag{11.46}$$

Consider

$$\frac{\partial^2 u_1}{\partial x \partial y}=-\frac{\partial^2 u_0}{\partial x^2}+\frac{1}{E(y)\Big(\dfrac{1}{\ell}\displaystyle\int_{y_0}^{y_0+\ell}\frac{d\tilde{y}}{E(\tilde{y})}\Big)}\frac{\partial^2 u_0}{\partial x^2}$$

Eq. (11.46) becomes

$$\underbrace{\frac{\partial}{\partial y}\Big[E(y)\Big(\frac{\partial u_1}{\partial x}+\frac{\partial u_2}{\partial y}\Big)\Big]}_{function\ of\ y}+\underbrace{\frac{1}{\Big(\dfrac{1}{\ell}\displaystyle\int_{y_0}^{y_0+\ell}\frac{d\tilde{y}}{E(\tilde{y})}\Big)}\frac{\partial^2 u_0}{\partial x^2}}_{function\ of\ x}=0 \tag{11.47}$$

Hence, one can re-write

$$\frac{1}{\Big(\dfrac{1}{\ell}\displaystyle\int_{y_0}^{y_0+\ell}\frac{d\tilde{y}}{E(\tilde{y})}\Big)}\frac{\partial^2 u_0}{\partial x^2}=0 \;\Rightarrow\; \frac{\partial}{\partial x}\left\{\frac{1}{\Big(\dfrac{1}{\ell}\displaystyle\int_{y_0}^{y_0+\ell}\frac{d\tilde{y}}{E(\tilde{y})}\Big)}\frac{\partial u_0}{\partial x}\right\}=0\,. \tag{11.48}$$

This is the homogenized differential equation that governs the macroscale variation of the mean displacement field.

Compare the mean-field differential equation to the original differential equation,

$$\frac{d}{dy}\Big(E(y)\frac{du}{dy}\Big)=0$$

We conclude that the effective coefficient for the homogenized differential equation is

$$E_e=\frac{1}{\Big(\dfrac{1}{\ell}\displaystyle\int_{y_0}^{y_0+\ell}\frac{d\tilde{y}}{E(\tilde{y})}\Big)}=\Big\langle\frac{1}{E}\Big\rangle^{-1} \tag{11.49}$$

which is the harmonic mean of $E(y)$, or the estimate from Reuss bound.

Consider the unit cell shown in Fig. (11.6). One may find that

$$\begin{aligned}\frac{1}{\ell}\int_{y_0}^{y_0+\ell}\frac{dt}{E(t)}&=\frac{2}{\ell}\int_0^{\frac{1-f\ell}{2}}\frac{dt}{E_1}+\frac{1}{\ell}\int_0^{f\ell}\frac{dt}{E_2}\\&=\frac{\ell-f\ell}{\ell}\frac{1}{E_1}+\frac{f}{E_2}=\frac{(1-f)E_2+fE_1}{E_1E_2}\end{aligned}$$

and

$$E_e=\frac{1}{\dfrac{1}{\ell}\displaystyle\int_{y_0}^{y_0+\ell}\frac{dt}{E(t)}}=\frac{E_1E_2}{(1-f)E_2+fE_1} \tag{11.50}$$

The homogenized differential equation is,

$$\frac{d}{dx}\Big(E_e\frac{du_0}{dx}\Big)=0\,. \tag{11.51}$$

To sum up, asymptotic homogenization consists of the following steps:

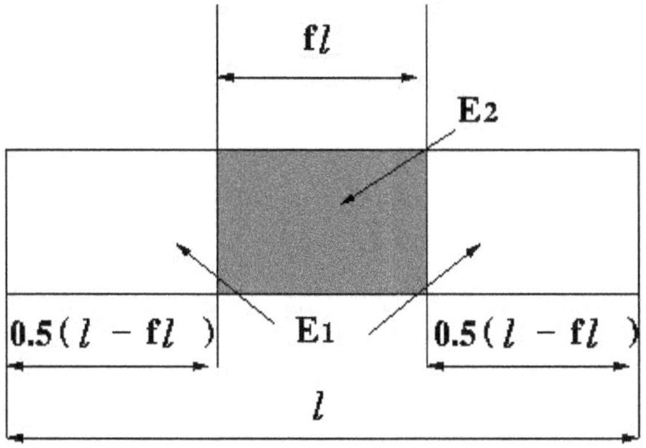

Fig. 11.6 One-dimensional unit cell.

Summary of Asymptotic Homogenization

(1) The objective of the homogenization is to find the average coefficients of the homogenized differential equation and find its solution;
(2) Identify the micro- and macroscales;
(3) Introduce multiple-scale variables and expansions, and deduce cell boundary-value problems (BVPs) at successive orders. The leading-order cell problem is homogeneous, i.e. $u_0 = u_0(\mathbf{x})$;
(4) Use linearity (or separation of variables) to express the next-order solution in terms of the leading-order solution and deduce an inhomogeneous canonical cell BVP;
(5) Require the solvability of the inhomogeneous cell problem;
(6) Find the differential equation that governs the macro-scale variation of the mean displacement or the evolution of the leading-order solution which includes the constitutive coefficients of the differential equation.

11.3.2 *A multiple dimension example*

Consider a 3D example,

$$A^\epsilon u = f, \quad \forall \mathbf{x} \in \Omega \tag{11.52}$$

$$u_\epsilon = 0, \quad \forall \mathbf{x} \in \partial\Omega \tag{11.53}$$

where

$$A^\epsilon = -\frac{\partial}{\partial x_i}\Big(a_{ij}(\frac{x}{\epsilon})\frac{\partial}{\partial x_j}\Big)$$

where $x = (x_1, x_2, x_3)$.

Define the fast coordinate,

$$y = \frac{x}{\epsilon}$$

as if y is speed-up by the large parameter $\frac{1}{\epsilon}$. We then can express the field variable as a function of two independent scales, $u_\epsilon(x) = u(x, y)$.

From chain rule, we have

$$\frac{d}{dx_i} = \frac{\partial}{\partial x_i} + \frac{\partial}{\partial y_i}\frac{\partial y_i}{\partial x_i} = \frac{\partial}{\partial x_i} + \frac{1}{\epsilon}\frac{\partial}{\partial y_i}$$

We can then expand the differential operator, A^ϵ, as

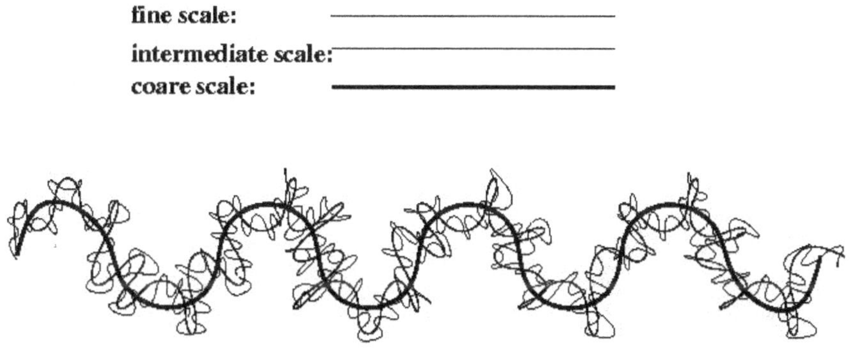

Fig. 11.7 Illustration of multiscale phenomena.

$$\begin{aligned}
A^\epsilon &= -\Big(\frac{\partial}{\partial x_i} + \frac{1}{\epsilon}\frac{\partial}{\partial y_i}\Big)\Big[a_{ij}(y)\Big(\frac{\partial}{\partial x_j} + \frac{1}{\epsilon}\frac{\partial}{\partial y_j}\Big)\Big] \\
&= -\epsilon^{-2}\Big[\frac{\partial}{\partial y_i}\Big(a_{ij}\frac{\partial}{\partial y_j}\Big)\Big] - \epsilon^{-1}\Big[\frac{\partial}{\partial x_i}a_{ij}(y)\frac{\partial}{\partial y_j} + \frac{\partial}{\partial y_i}a_{ij}(y)\frac{\partial}{\partial x_i}\Big] \\
&\quad -\epsilon^{0}\Big[\frac{\partial}{\partial x_i}a_{ij}(y)\frac{\partial}{\partial x_j}\Big] \\
&= \epsilon^{-2}A_1 + \epsilon^{-1}A_2 + \epsilon^{0}A_3
\end{aligned} \tag{11.54}$$

where

$$\begin{aligned}
A_1 &= -\Big[\frac{\partial}{\partial y_i}\Big(a_{ij}\frac{\partial}{\partial y_j}\Big)\Big] \\
A_2 &= -\Big[\frac{\partial}{\partial x_i}a_{ij}(y)\frac{\partial}{\partial y_j} + \frac{\partial}{\partial y_i}a_{ij}(y)\frac{\partial}{\partial x_j}\Big] \\
A_3 &= -\Big[\frac{\partial}{\partial x_i}a_{ij}(y)\frac{\partial}{\partial x_j}\Big]
\end{aligned}$$

Now we consider multiple scale expansion,

$$u_\epsilon(x) = u_0(x, y) + \epsilon u_1(x, y) + \epsilon^2 u_2(x, y) + \cdots \tag{11.55}$$

which decomposes or separates the activities at different scales.

Substituting both (11.55) and (11.54) into (11.52), we have

$$\begin{aligned}&\left(\epsilon^{-2}A_1+\epsilon^{-1}A_2+\epsilon^0A_3\right)\left(u_0+\epsilon u_1+\epsilon^2u_2+\cdots\right)=f\\&\epsilon^{-2}A_1u_0+\epsilon^{-1}\left(A_1u_1+A_2u_0\right)+\epsilon^0(A_1u_2+A_2u_1+A_3u_0)\\&+\cdots=f\end{aligned}\tag{11.56}$$

The total state equilibrium is equivalent to equilibrium states in each every scale. That is

$$\epsilon^{-2}:\ A_1u_0=0;\tag{11.57}$$

$$\epsilon^{-1}:\ A_1u_1+A_2u_0=0;\tag{11.58}$$

$$\epsilon^{0}:\ A_1u_2+A_2u_1+A_3u_0=f\tag{11.59}$$

$$\cdots\cdots$$

If one can solve differential equations at each scale, one can find out both local detailed information as well as global information.

As far as homogenization concern, we are looking for a homogenized differential equation that carries the overall information of fine scale.

Before we proceed further, we prove the following lemma.

Lemma 11.1. *If the differential equation,*

$$A_1u=F,\quad\forall\mathbf{y}\in Y$$

has a unique Y-periodic solution, the following equation holds

$$<F>=\frac{1}{|Y|}\int_Y F(y)dV_y=0\tag{11.60}$$

where $y=(y_1,y_2,y_3)$.

Proof:

By the assumption, one can assume that both u and F are Y-periodic, and

$$F(y)=\sum_{\xi\in\Lambda}\mathcal{F}[F](\xi)\exp(i\xi y)\tag{11.61}$$

$$u(x,y)=\sum_{\xi\in\Lambda}\mathcal{F}[u](\xi)\exp(i\xi y)\tag{11.62}$$

Hence

$$\begin{aligned}A_1u&=-\Big(\frac{\partial}{\partial y_i}a_{ij}(y)\frac{\partial}{\partial y_j}\Big)u\\&=-\sum_{\xi\in\Lambda}\mathrm{i}\xi_j\Big(\frac{\partial a_{ij}}{\partial y_i}+a_{ij}\mathrm{i}\xi_i\Big)\mathcal{F}[u](\xi)\exp(\mathrm{i}\xi y)\end{aligned}$$

Based on $A_1u=F$, one has

$$-\sum_{\xi\in\Lambda}\mathrm{i}\xi_j\Big(\frac{\partial a_{ij}}{\partial y_i}+i\xi_ia_{ij}\Big)\mathcal{F}[u]\exp(i\xi y)=\sum_{\xi\in\Lambda}\mathcal{F}[F](\xi)\exp(i\xi y)$$

$$\Rightarrow\mathcal{F}[F](\xi)=-\mathrm{i}\xi_j\Big(\frac{\partial a_{ij}}{\partial y_i}+i\xi_ia_{ij}\Big)$$

Therefore,

$$\mathcal{F}[F] = 0 \;\Rightarrow\; \mathcal{F}[F](0) = \frac{1}{|Y|}\int_Y F dV_y = 0 \; .$$

♣

To this end, we start to solve differential equations at each scale. At scale ϵ^{-2}, we have

$$A_1 u_0 = -\frac{\partial}{\partial y_i}\Big(a_{ij}(y)\frac{\partial}{\partial y_j}\Big)u_0 = 0$$

We claim that

$$u_0 = u_0(x) \; .$$

That is the leading-order expansion is only the function of slow scale variable.

Since u_0 is Y-periodic, we have

$$u_0 = \sum_{\xi\in\Lambda} \mathcal{F}[u_0](\xi)\exp(\mathrm{i}\xi y) \; .$$

Consequently,

$$A_1 u_0 = 0 \;\Rightarrow\; -\sum_{\xi\in\Lambda} \mathrm{i}\xi_j\Big(\frac{\partial a_{ij}}{\partial y_i} + \mathrm{i}\xi_i a_{ij}\Big)\mathcal{F}[u](\xi)\exp(\mathrm{i}\xi y) = 0 \; .$$

Then for $\xi \neq 0$, it is necessary

$$\mathcal{F}[u](\xi) = 0 \; . \tag{11.63}$$

Assume that

$$u_0 = c(x)Q(y) + \bar{u}_0(x)$$

Eq. (11.63) becomes

$$\begin{aligned}\mathcal{F}[u](\xi) &= \frac{1}{|Y|}\int_Y \Big(c(x)Q(y) + \bar{u}_0(x)\Big)\exp(-\mathrm{i}\xi y)dV_y \\ &= \frac{1}{|Y|}\int_Y \Big(c(x)Q(y)\Big)\exp(-\mathrm{i}\xi y)dV_y = 0 \end{aligned} \tag{11.64}$$

because $\displaystyle\int_Y \bar{u}_0(x)\exp(-\mathrm{i}\xi y)dV_y = 0$ when $\xi \neq 0$.

The only possibility that (11.64) holds is that $Q(y) = 1$ or $Q(y) = 0$. In either case, $u_0 = u_0(x)$. We proved our claim.

Next, we consider the differential equation at scale ϵ^{-1}:

$$A_1 u_1 + A_2 u_0 = 0 \; .$$

One can show that

$$A_2 u_0 = -\Big[\frac{\partial}{\partial x_i}\Big(a_{ij}(y)\Big)\frac{\partial}{\partial y_j} + \frac{\partial}{\partial y_i}\Big(a_{ij}(y)\Big)\frac{\partial}{\partial x_j}\Big]u_0(x) = -\frac{\partial a_{ij}}{\partial y_i}\frac{\partial u_0}{\partial x_j}$$

Hence

$$A_1 u_1 = \frac{\partial a_{ij}}{\partial y_i}\frac{\partial u_0}{\partial x_j} \tag{11.65}$$

This suggest the following separation of variable,

$$u_1(x,y) = U_k(y)\frac{\partial u_0}{\partial x_k} + \bar{u}_1(x) \tag{11.66}$$

and subsequently,

$$\begin{aligned} A_1 u_1 &= \Big(A_1 U_k(y)\Big)\frac{\partial u_0}{\partial x_k} \\ &= -\frac{\partial}{\partial y_i}\Big(a_{ij}(y)\frac{\partial U_k}{\partial y_j}\Big)\frac{\partial u_0}{\partial x_k} \end{aligned} \tag{11.67}$$

Combining (11.65) and (11.67), we find the canonical equation for a unit cell problem,

$$\frac{\partial a_{ik}}{\partial x_i} + \frac{\partial}{\partial y_i}\Big(a_{ij}(y)\frac{\partial U_k}{\partial y_j}\Big) = 0\,. \tag{11.68}$$

with the possible boundary conditions at interface of different phases,

$$\Big[U_k\Big] = 0, \quad \text{and} \quad \Big[\Big(a_{ik} + a_{ij}\frac{\partial U_k}{\partial x_j}\Big)n_i\Big] = 0 \tag{11.69}$$

We now consider the differential equation at ϵ^0 scale,

$$A_1 u_2 + A_2 u_1 + A_3 u_0 = f$$

which can be rewritten as

$$A_1 u_2 = f - \Big(A_2 u_1 + A_3 u_0\Big) \tag{11.70}$$

The condition that equation (11.70) has a unique periodic solution is that

$$< f - (A_2 u_1 + A_3 u_0) >= 0$$

That is

$$\frac{1}{|Y|}\int_Y \Big(A_2 u_1 + A_3 u_0\Big) dy = f \tag{11.71}$$

Consider

$$\begin{aligned} u_0 &= u_0(x) \\ u_1 &= U_j \frac{\partial u_0}{y_j} + \bar{u}_1(x) \end{aligned}$$

One can show that

$$A_3 u_0 = -a_{ij}\frac{\partial^2 u_0}{\partial x_i \partial x_j} \tag{11.72}$$

$$\begin{aligned} A_2 u_1 &= -\Big[\frac{\partial}{\partial x_i}\Big(a_{ij}(y)\frac{\partial}{\partial y_j} + \frac{\partial}{\partial y_i}\Big(a_{ij}(y)\frac{\partial}{\partial x_j}\Big]\Big(U_k(y)\frac{\partial u_0}{\partial x_k} + \bar{u}_1\Big) \\ &= -a_{ij}\frac{\partial U_k}{\partial y_j}\frac{\partial^2 u_0}{\partial x_i \partial x_k} - \frac{\partial}{\partial y_i}\Big(a_{ij}(y)U_k(y)\Big)\frac{\partial^2 u_0}{\partial x_j \partial x_k} \\ &\quad -\frac{\partial}{\partial y_i}\Big(a_{ij}(y)\Big)\frac{\partial \bar{u}_1}{\partial x_j} \end{aligned} \tag{11.73}$$

Change the dummy indices $j \leftrightarrow k$ in the first term of (11.73). We can write that

$$A_2 u_1 + A_3 u_0 = -\Big(a_{ij} + a_{ik}\frac{\partial U_j}{\partial x_k}\Big)\frac{\partial^2 u_0}{\partial x_i \partial x_j} - \frac{\partial}{\partial y_i}\Big(a_{ij}(y)U_k(y)\frac{\partial^2 u_0}{\partial x_j \partial x_k}\Big)$$
$$-\frac{\partial}{\partial y_i}\Big(a_{ij}(y)\frac{\partial \bar{u}_1}{\partial x_j}\Big)$$

Via divergence theorem,

$$\frac{1}{|Y|}\int_Y (A_2 u_1 + A_3 u_0) dy = -\frac{1}{|Y|}\int_Y \Big(a_{ij} + a_{ik}\frac{\partial U_j}{\partial x_k}\Big) dV_y \frac{\partial^2 u_0}{\partial x_i \partial x_j}$$
$$-\Big[(a_{ij}(y)U_k(y)u_{0,jk}(x)\Big] n_i - \Big[a_{ij}(y)\bar{u}_{1,j}\Big] n_i$$

By periodicity, the boundary terms will vanish. We then have

$$-\frac{1}{|Y|}\int_Y \Big(a_{ij} + a_{ik}\frac{\partial U_j}{\partial x_k}\Big) dV_y \frac{\partial^2 u_0}{\partial x_i \partial x_j} = f$$

Denote the effective coefficients as

$$\bar{a}_{ij} = \frac{1}{|Y|}\int_Y \Big(a_{ij} + a_{ik}\frac{\partial U_j}{\partial x_k}\Big) dV_y \tag{11.74}$$

and homogenized differential operator

$$A^H = -\frac{\partial}{\partial x_i}\Big(\bar{a}_{ij}\frac{\partial}{\partial x_j}\Big) \tag{11.75}$$

we finally derived the homogenized boundary-value problem,

$$A^H u_0 = 0, \quad \forall x \in \Omega \tag{11.76}$$
$$u_0 = 0, \quad \forall x \in \partial\Omega \tag{11.77}$$

Example 11.4. Consider a 2D steady-state heat transfer problem (see Fig. (11.8)),

$$\frac{\partial}{\partial x_\alpha}\Big(\lambda_{\alpha\beta}\Big(\frac{x_1}{\epsilon}\Big)\frac{\partial T^\epsilon}{\partial x_\beta}\Big) = 0, \quad \forall x \in D \tag{11.78}$$

where $T^\epsilon(x)$ is temperature field and $\lambda_{\alpha\beta}$ are heat conduction coefficients. We assume that the region $D = \Big\{(x_1, x_2) \,\Big|\, 0 \le x_1 \le \ell_1, \text{ and } 0 \le x_2 \le \ell_2\Big\}$ is thermally insulated in horizontal boundaries, i.e.

$$q_2 = \lambda_{2\beta}\frac{\partial T^\epsilon}{\partial x_\beta} = 0, \quad \forall x_2 = 0, \text{ and } x_2 = \ell_2 \tag{11.79}$$

Along the vertical boundaries of the region D, the heat flows are prescribed,

$$q_1 = \lambda_{1\beta}\frac{\partial T^\epsilon}{\partial x_\beta} = \mp q_0, \quad \forall x_1 = 0, \text{ and } x_1 = \ell_1 \tag{11.80}$$

Consider multiple expansion,

$$T^\epsilon(x) = T_0(x) + \epsilon T_1(x, y_1) + \cdots$$

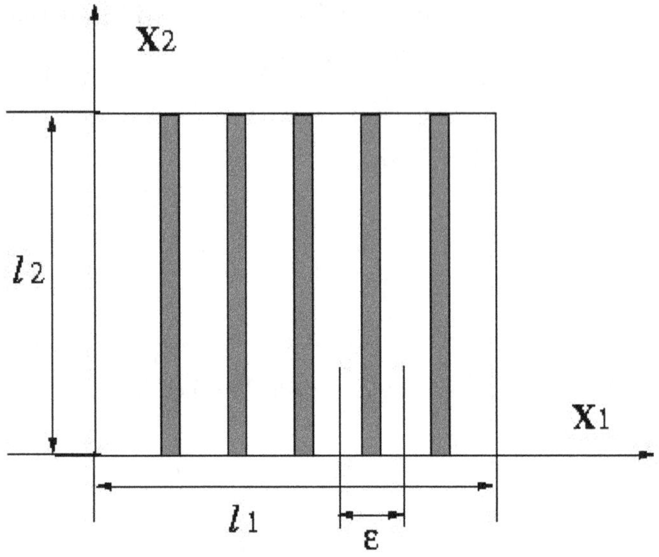

Fig. 11.8 An unilateral composite with periodic structure.

and the following separation of variable,

$$T_1(x, y_1) = U_\alpha(y_1)\frac{\partial T_0(x)}{\partial x_\alpha}, \quad \alpha = 1, 2$$

Note that first we assume that the mean temperature at this scale is zero, i.e. $\bar{T}_1(x) = 0$; and $U_\alpha(y_1)$ are Y-periodic functions that are the following 1D canonical cell problem,

$$-\frac{d}{dy_1}\Big(\lambda_{11}(y_1)\frac{dU_\alpha(y_1)}{dy_1}\Big) = \frac{d\lambda_{1\alpha}}{dy_1}, \quad \forall\, y_1 \in Y \tag{11.81}$$

$$\Big[U_\alpha\Big] = 0, \quad \text{and} \quad \Big[\lambda_{11}\frac{dU_\alpha}{dy_1}\Big] = 0, \quad \forall y_1 \text{ at interface.} \tag{11.82}$$

Integrate (11.81),

$$-\lambda_{11}(y_1)\frac{dU_\alpha(y_1)}{dy_1} = \lambda_{1\alpha}(y_1) - C_\alpha$$

$$\Rightarrow \quad \frac{dU_\alpha(y_1)}{dy_1} = -\frac{\lambda_{1\alpha}(y_1)}{\lambda_{11}(y_1)} + \frac{C_\alpha}{\lambda_{11}(y_1)}$$

where C_α are constants (note that they are not functions of x !).

Integrate second time,

$$U_\alpha(y_1) = -\int_0^{y_1} \lambda_{1\alpha}(\xi)\lambda_{11}^{-1}(\xi)d\xi + C_\alpha \int_0^{y_1} \lambda_{11}^{-1}(\xi)d\xi + D_\alpha$$

Note that we choose $D_\alpha = 0$, because the average temperature at scale ϵ^{-1} is assumed to be zero.

The solvability condition of the canonical cell problem requires $U_\alpha(y_1)$ as a Y-periodic function, i.e.

$$U_\alpha(0) = U_\alpha(\ell)$$

This condition allows us to determine the constants C_α,

$$C_\alpha = \frac{\displaystyle\int_0^\ell \lambda_{1\alpha}(\xi)\lambda_{11}^{-1}(\xi)d\xi}{\int_0^\ell \lambda_{11}^{-1}(\xi)d\xi} \tag{11.83}$$

In specific,

$$C_1 = \ell\left(\int_0^\ell \lambda_{11}^{-1}(\xi)d\xi\right)^{-1} \quad \text{and} \quad C_2 = \frac{\displaystyle\int_0^\ell \lambda_{12}(\xi)\lambda_{11}^{-1}(\xi)d\xi}{\displaystyle\int_0^\ell \lambda_{11}^{-1}(\xi)d\xi}$$

Consequently, we find the closed form solution for canonical cell problem,

$$U_1(y_1) = -y_1 + \frac{\displaystyle\int_0^{y_1} \lambda_{11}^{-1}(\xi)d\xi}{\displaystyle\int_0^\ell \lambda_{11}^{-1}(\xi)d\xi}\ell \tag{11.84}$$

$$U_2(y_1) = -\int_0^{y_1} \lambda_{12}(\xi)\lambda_{11}^{-1}(\xi)d\xi + \frac{\displaystyle\int_0^\ell \lambda_{12}(\xi)\lambda_{11}^{-1}(\xi)d\xi}{\displaystyle\int_0^\ell \lambda_{11}^{-1}(\xi)d\xi}\left(\int_0^{y_1} \lambda_{11}^{-1}(\xi)d\xi\right) \tag{11.85}$$

Define the effective heat conduction coefficients,

$$\bar{\lambda}_{ij} := \frac{1}{|Y|}\int_Y \left(a_{ij} + a_{ik}\frac{\partial U_j}{\partial x_k}\right)dy \ .$$

It is easy to find that

$$\begin{aligned}\bar{\lambda}_{11} &= \frac{1}{\ell}\int_0^\ell \left(\lambda_{11}(\xi) + \lambda_{11}(\xi)\frac{\partial U_1}{\partial y_1}(\xi)\right)d\xi \\ &= \frac{1}{\ell}\int_0^\ell \left(\lambda_{11} - \lambda_{11} + C_1\right)dy = \frac{1}{\ell}C_1 = \left(\int_0^\ell \lambda_{11}^{-1}(\xi)d\xi\right)^{-1}\end{aligned}$$

$$\begin{aligned}\bar{\lambda}_{12} &= \frac{1}{\ell}\int_0^\ell \left(\lambda_{12}(\xi) + \lambda_{11}(\xi)\frac{\partial U_2}{\partial y_1}(\xi)\right)d\xi \\ &= \frac{1}{\ell}\int_0^\ell \left(\lambda_{12}(\xi) - \lambda_{12}(\xi) + C_2\right)d\xi = \frac{\displaystyle\int_0^\ell \lambda_{12}(\xi)\lambda_{11}^{-1}(\xi)d\xi}{\displaystyle\int_0^\ell \lambda_{11}^{-1}(\xi)d\xi} = \bar{\lambda}_{21}\end{aligned}$$

and

$$\bar{\lambda}_{22} = \frac{1}{\ell}\int_0^\ell \Big(\lambda_{22}(\xi) + \lambda_{21}(\xi)\frac{\partial U_2}{\partial y_1}(\xi)\Big)d\xi$$
$$= \frac{1}{\ell}\int_0^\ell \Big(\lambda_{22}(\xi) - \lambda_{21}^2(\xi)\lambda_{11}^{-1}(\xi) + C_2\lambda_{12}(\xi)\lambda_{11}^{-1}(\xi)\Big)d\xi$$
$$= \frac{1}{\ell}\int_0^\ell \lambda_{22}(\xi)d\xi - \frac{1}{\ell}\int_0^\ell \lambda_{12}^2(\xi)\lambda_{11}^{-1}(\xi)d\xi + \frac{1}{\ell}\frac{\Big(\int_0^\ell \lambda_{12}(\xi)\lambda_{11}^{-1}(\xi)d\xi\Big)^2}{\int_0^\ell \lambda_{11}^{-1}(\xi)d\xi} \tag{11.86}$$

and the homogenized partial differential equation becomes

$$\bar{\lambda}_{11}\frac{\partial^2 T_0}{\partial x_1^2} + 2\bar{\lambda}_{12}\frac{\partial^2 T_0}{\partial x_1 \partial x_2} + \bar{\lambda}_{22}\frac{\partial^2 T_0}{\partial x_2^2} = 0\ .$$

11.4 Variational characterization

Recall the homogenization of conduction problem,

$$A^\epsilon u_\epsilon = f, \quad \forall x \in \Omega$$
$$u_\epsilon = 0, \quad \forall x \in \partial\Omega$$

Assume that

$$u^{(1)}(x,y) = U_k(y)\frac{\partial u^{(0)}(x)}{\partial x_k} \tag{11.87}$$

One can derive the following governing equations for the canonical cell problem,

$$\frac{\partial}{\partial y_k}\Big(a_{kj} + a_{k\ell}\frac{\partial U_j}{\partial y_\ell}\Big) = 0, \quad \forall y \in Y \tag{11.88}$$

with the proper interface and periodic conditions.

Subsequently, one can derive the effective coefficients for homogenized differential equation,

$$\bar{a}_{ij} = \frac{1}{|Y|}\int_Y \Big(a_{ij} + a_{i\ell}\frac{\partial U_j}{\partial y_\ell}\Big)dy = \frac{1}{|Y|}\int_Y a_{i\ell}\Big(\delta_{\ell j} + \frac{\partial U_j}{\partial y_\ell}\Big)dy \tag{11.89}$$

Based on (11.88), one may find that

$$-\frac{1}{|Y|}\int_Y \Big(a_{kj}(y) + a_{k\ell}(y)\frac{\partial U_j}{\partial y_\ell}\Big)U_i(y)dy = 0$$

Integration by parts yields

$$-\frac{1}{|Y|}\int_{\partial Y}\Big(a_{kj} + a_{k\ell}\frac{\partial U_j}{\partial y_\ell}\Big)U_i(y)n_k dS + \frac{1}{|Y|}\int_Y \Big(a_{kj} + a_{k\ell}\frac{\partial U_j}{\partial y_\ell}\Big)\frac{\partial U_i}{\partial y_k}dy$$
$$= \frac{1}{|Y|}\int_Y a_{k\ell}\Big(\delta_{\ell j} + \frac{\partial U_j}{\partial y_\ell}\Big)\frac{\partial U_i}{\partial y_k}dy = 0 \tag{11.90}$$

Adding (11.90) to (11.89), one may find that

$$\bar{a}_{ij} = \frac{1}{|Y|}\int_Y \Big(\delta_{\ell j} + \frac{\partial U_j}{\partial y_\ell}\Big)\Big(a_{i\ell}(y) + a_{k\ell}(y)\frac{\partial U_i}{\partial y_k}\Big)dy$$
$$= \frac{1}{|Y|}\int_Y a_{k\ell}(y)\Big(\delta_{ik} + \frac{\partial U_i}{\partial y_k}\Big)\Big(\delta_{\ell j} + \frac{\partial U_j}{\partial y_\ell}\Big)dy \qquad (11.91)$$

Eq. (11.91) links the effective coefficients of the homogenized equation with the variational characters of unit cell problem, which plays a significant role in Tartar's variational principle.

Consider constant vector, $\boldsymbol{\xi} = \xi_i \boldsymbol{e}_i$, or a flux vector of macro-scale variable.

We can form the following quadratic form,

$$\bar{a}_{ij}\xi_i\xi_j = \frac{1}{|Y|}\int_Y \xi_i\xi_j a_{k\ell}(y)\Big(\delta_{ik} + \frac{\partial U_i}{\partial y_k}\Big)\Big(\delta_{\ell j} + \frac{\partial U_j}{\partial y_\ell}\Big)dy$$
$$= \frac{1}{|Y|}\int_Y a_{k\ell}(y)\Big(\xi_k + \frac{\partial U_i\xi_i}{\partial y_k}\Big)\Big(\xi_\ell + \frac{\partial U_j\xi_j}{\partial y_\ell}\Big)dy \qquad (11.92)$$

Eq. (11.92) suggests that there exists a functional,

$$J(\mathbf{U}) = \frac{1}{Y}\int_Y a_{ij}(y)\Big(\xi_i + \frac{\partial U_k\xi_k}{\partial y_i}\Big)\Big(\xi_j + \frac{\partial U_\ell\xi_\ell}{\partial y_j}\Big)dy \qquad (11.93)$$

such that

$$\bar{a}_{ij}\xi_i\xi_j = \min_{\mathbf{U}\in H^1_\#(Y)} J\Big(\mathbf{U}\Big) \qquad (11.94)$$

where the function space $H^1_\#(Y)$[1] is defined as $H^1(Y)$ space of Y-periodic functions, i.e.

$$H^1_\#(Y) := \Big\{u \;\Big|\; u \text{ is Y} - \text{periodic}, \;\text{ and } u \in H^1(Y)\Big\}$$

that is

$$\int_Y (u^2 + |\nabla u|^2)dy < +\infty$$

To show this, we first show that the Euler-Lagrange equation of $J(\mathbf{U})$ is the governing equation of canonical cell problem.

Assume that a_{ij} is symmetric and real. It subsequently implies that a_{ij} is positive definite. Therefore,

$$\delta J = \frac{1}{|Y|}\int_Y a_{ij}(y)\Big(\frac{\partial\delta U_k\xi_k}{\partial y_i}\Big(\xi_j + \frac{\partial U_\ell\xi_\ell}{\partial y_j}\Big) + \Big(\xi_i + \frac{\partial U_k\xi_k}{\partial y_i}\Big)\frac{\partial\delta U_\ell\xi_\ell}{\partial y_j}\Big)dy$$
$$= \frac{2}{|Y|}\int_{\partial Y} a_{ij}(y)\Big(\xi_i + \frac{\partial U_k\xi_k}{\partial y_i}\Big)\delta U_\ell\xi_\ell n_j dS$$
$$-\frac{2}{|Y|}\int_Y \frac{\partial}{\partial y_j}\Big(a_{ij}(y)\Big(\xi_i + \frac{\partial U_k\xi_k}{\partial y_i}\Big)\Big)\delta U_\ell\xi_\ell dy = 0$$

[1] In music, the sign # is used to indicate that a note is to be raised by a half tone. Similar meaning implies here as well, i.e. a "half level higher" H^1 space.

By periodic conditions

$$\frac{2}{|Y|}\int_{\partial Y} a_{ij}(y)\Big(\xi_i + \frac{\partial U_k \xi_k}{\partial y_i}\Big)\delta U_\ell \xi_\ell n_j dS = 0,$$

it then leads to

$$\delta J = -\frac{2}{|Y|}\int_Y \frac{\partial}{\partial y_j}\Big(a_{ij}(y)\Big(\delta_{ik} + \frac{\partial U_k}{\partial y_i}\Big)\Big)\delta U_\ell \xi_k \xi_\ell dy = 0$$

and hence

$$-\frac{\partial}{\partial y_j}\Big(a_{ij}(y)\Big(\delta_{ik} + \frac{\partial U_k}{\partial y_i}\Big)\Big)\delta U_\ell = 0\ .$$

Consider $U_k = 0 \in H^1_\#(Y)$. One can find an upper bound for effective coefficient, $\bar{a}_{ij}$, i.e.

$$0 < \bar{a}_{ij}\xi_i\xi_j \le \Big(\frac{1}{|Y|}\int_Y a_{ij}(y)dy\Big)\xi_i\xi_j \tag{11.95}$$

or

$$\bar{a}_{ij} \le \frac{1}{|Y|}\int_Y a_{ij}(y)dy \tag{11.96}$$

This is the arithmetic mean or the so-called Voigt bound.

To find the lower bound, we have to enlarge the space $H^1_\#(Y)$. Consider function $\zeta_i \in L^2_\#(Y)$ and the mean value of ζ_i is zero, i.e.

$$\int_Y \zeta_i(y)dy = 0\ .$$

It is obvious that

$$\bar{a}_{ij}\xi_i\xi_j \ge \min_{\substack{\boldsymbol{\zeta}\in L^2_\#(Y)\ \text{and}\\ \int_Y \boldsymbol{\zeta}(y)dy=0}} J_c(\boldsymbol{\zeta}) \tag{11.97}$$

where

$$J_c(\boldsymbol{\zeta}) := \frac{1}{|Y|}\int_Y a_{ij}(\xi_i + \zeta_i(y))(\xi_j + \zeta_j(y))dy - 2C_k\Big(\int_Y \zeta_k(y)dy - 0\Big) \tag{11.98}$$

where C_k are Lagrange multipliers.

To find the minimizer in $L^2_\#(Y)$, we calculate the first variation of the functional, $J_c(\boldsymbol{\zeta})$,

$$\begin{aligned}\delta J_c &= \frac{2}{|Y|}\int_Y a_{ij}(y)(\xi_i + \zeta_i)\delta\zeta_j dy - 2\delta C_j\frac{1}{|Y|}\int_Y \zeta_j(y)dy\\ &\quad -2C_j\frac{1}{|Y|}\int_Y \delta\zeta_j(y)dy\\ &= \frac{2}{|Y|}\int_Y \Big(a_{ij}(y)(\xi_i + \zeta_i) - C_j\Big)\delta\zeta_j dy - 2\delta C_j\frac{1}{|Y|}\int_Y \zeta_j(y)dy = 0\end{aligned}$$

which yields Euler-Lagrangian equation and constrain conditions,

$$a_{ij}(\xi_j + \zeta_j) = C_i \tag{11.99}$$

$$\int_Y \zeta_j(y) dy = 0 \ . \tag{11.100}$$

Solving (11.99), we have

$$\xi_i + \zeta_i = a_{ij}^{-1} C_j \tag{11.101}$$

Average the above expression over the unit cell and considering the constraint condition (11.100),

$$\xi_i = < a_{ij}^{-1}(y) > C_j \tag{11.102}$$

which solves C_j in terms of ξ_i, i.e.

$$C_j = < a_{ji}^{-1}(y) >^{-1} \xi_i \tag{11.103}$$

The minimizer in $L^2_\#(Y)$ under the constraint is then

$$\begin{aligned}
\min_{\substack{\boldsymbol{\zeta} \in L^2_\#(Y) \text{ and} \\ \int_Y \boldsymbol{\zeta}(y)dy=0}} \quad J_c(\boldsymbol{\zeta}) &= \frac{1}{|Y|} \int_Y a_{ij}(\xi_i + \zeta_i)(\xi_j + \zeta_j) dy \\
&= \frac{1}{|Y|} \int_Y C_i(\xi_i + \zeta_i) dy = C_j \xi_j \\
&= < a_{ji}^{-1} >_Y^{-1} \xi_i \xi_j = \Big(\frac{1}{|Y|} \int_Y a_{ij}^{-1}(y) dy \Big)^{-1} \xi_i \xi_j
\end{aligned}$$

From the above estimate, we find a lower bound for effective coefficient, $\bar{a}_{ij}$, i.e.

$$\bar{a}_{ij} \geq \Big(\frac{1}{|Y|} \int_Y a_{ij}^{-1}(y) dy \Big)^{-1} \ . \tag{11.104}$$

which is the so-called Reuss bound.

11.5 Multiscale finite element method

In this section, we briefly introduce the theory of multiscale finite element method, which has become a very useful tool in numerical homogenizations. For more advanced and comprehensive treatments, readers are referred to the original work by T. Hou and X.-H. Wu [Hou and Wu (1997)], and a more recent work by G. Allaire and R. Brizzi [Allaire and Brizzi (2005)].

11.5.1 *Asymptotic homogenization of linear elasticity*

Consider a composite material with periodic structure and its elastic stiffness tensor satisfies the relation,

$$C_{ijk\ell}\Big(\frac{x}{\epsilon}\Big)\xi_{ij}\xi_{k\ell} = C_{ijk\ell}(y)\xi_{ij}\xi_{k\ell} \geq \alpha\xi_{ij}\xi_{ij}$$

where $\alpha > 0$.

Consider the following boundary value problem,

$$\frac{\partial \sigma^{\epsilon}_{ij}}{\partial x_j} + f_i = 0, \quad \forall x \in \Omega \tag{11.105}$$

$$\sigma^{\epsilon}_{ij} = C^{\epsilon}_{ijk\ell}u^{\epsilon}_{k,\ell} = C^{\epsilon}_{ijk\ell}e^{\epsilon}_{k\ell} \tag{11.106}$$

$$e^{\epsilon}_{k\ell} = \frac{1}{2}\Big(\frac{\partial u^{\epsilon}_k}{\partial x_\ell} + \frac{\partial u^{\epsilon}_\ell}{\partial x_k}\Big) \tag{11.107}$$

$$\sigma^{\epsilon}_{ij}n_j = t^0_i, \quad \forall x \in \Gamma_t \tag{11.108}$$

$$u^{\epsilon}_i = \bar{u}_i, \quad \forall x \in \Gamma_u \tag{11.109}$$

Consider multiple scale expansion,

$$u^{\epsilon}_i(x) = u^{(0)}_i(x,y) + \epsilon u^{(1)}_i(x,y) + \epsilon^2 u^{(2)}_i(x,y) + \cdots, \qquad y := \frac{x}{\epsilon}$$

Hence

$$\begin{aligned} u^{\epsilon}_{k,\ell} &= \frac{d}{dx_\ell}u^{\epsilon}_k = \Big(\frac{\partial}{\partial x_\ell} + \frac{1}{\epsilon}\frac{\partial}{\partial y_\ell}\Big)\Big(u^0_k + \epsilon u^1_k + \epsilon^2 u^2_k + \cdots\Big) \\ &= \epsilon^{-1}u^{(0)}_{Yk,\ell} + \epsilon^0(u^{(0)}_{Xk,\ell} + u^{(1)}_{Yk,\ell})) + \\ &\quad + \epsilon^1(u^{(1)}_{Xk,\ell} + u^2_{Yk,\ell}) + \cdots \end{aligned} \tag{11.110}$$

where the subscripts, X and Y, indicate whether the partial derivatives on the slow variable $x = (x_1, x_2, x_3)$ or on the fast variable $y = (y_1, y_2, y_3)$; and

$$\begin{aligned} \sigma^{\epsilon}_{ij}(x,y) &= C_{ijk\ell}(y)u^{\epsilon}_{k,\ell} \\ &= C_{ijk\ell}(y)\Big[\epsilon^{-1}u^{(0)}_{Yk,\ell} + \epsilon^0(u^{(0)}_{Xk,\ell} + u^{(1)}_{Yk,\ell}) + \epsilon(u^{(1)}_{Xk,\ell} + u^{(2)}_{Yk,\ell}) \\ &\quad + \cdots\Big] = \epsilon^{-1}\sigma^{(0)}_{ij} + \epsilon^0\sigma^{(1)}_{ij} + \epsilon^1\sigma^{(2)}_{ij} + \cdots \end{aligned} \tag{11.111}$$

In each scale, the constitutive relations are

$$\begin{aligned} \epsilon^{-1} &: \sigma^{(0)}_{ij} = C_{ijk\ell}(y)u^{(0)}_{Yk,\ell}; \\ \epsilon^{0} &: \sigma^{(1)}_{ij} = C_{ijk\ell}(y)(u^{(0)}_{Xk,\ell} + u^{(1)}_{Yk,\ell}); \\ \epsilon^{1} &: \sigma^{(1)}_{ij} = C_{ijk\ell}(y)(u^{(1)}_{Xk,\ell} + u^{(2)}_{Yk,\ell}); \\ &\cdots \end{aligned}$$

To derive equilibrium equation at different scales, one may write

$$\begin{aligned} \frac{\partial \sigma^{\epsilon}_{ij}}{\partial x_j} &= \Big(\frac{\partial}{\partial x_j} + \frac{1}{\epsilon}\frac{\partial}{\partial y_j}\Big)\sigma^{\epsilon}_{ij} + f_i = 0 \\ &= \Big(\frac{\partial}{\partial x_j} + \frac{1}{\epsilon}\frac{\partial}{\partial y_j}\Big)\Big(\epsilon^{-1}\sigma^{(0)}_{ij} + \epsilon^0\sigma^{(1)}_{ij} + \epsilon^1\sigma^{(2)}_{ij} + \cdots\Big) + f_i = 0 \end{aligned}$$

Consequently,

$$\epsilon^{-2}: \frac{\partial \sigma_{ij}^{(0)}}{\partial y_j} = 0; \tag{11.112}$$

$$\epsilon^{-1}: \frac{\partial \sigma_{ij}^{(0)}}{\partial x_j} + \frac{\partial \sigma_{ij}^{(1)}}{\partial y_j} = 0; \tag{11.113}$$

$$\epsilon^{0}: \frac{\partial \sigma_{ij}^{(1)}}{\partial x_j} + \frac{\partial \sigma_{ij}^{(2)}}{\partial y_j} + f_i = 0; \tag{11.114}$$

$$\epsilon^{s-1}: \frac{\partial \sigma_{ij}^{(s)}}{\partial x_j} + \frac{\partial \sigma_{ij}^{(s+1)}}{\partial y_j} = 0; \quad s = 2, 3, \cdots \tag{11.115}$$

and the boundary conditions are

$$\left(\epsilon^{-1}\sigma_{ij}^{(0)} + \epsilon^{0}\sigma_{ij}^{(1)} + \epsilon^{1}\sigma_{ij}^{(2)} + \cdots\right)n_j = t_i^0, \quad \forall x \in \Gamma_t \tag{11.116}$$

$$\left(u_i^{(0)} + \epsilon^1 u_i^{(1)} + \epsilon^2 u_i^{(2)} + \cdots\right) = 0, \ \forall x \in \Gamma_u \tag{11.117}$$

The boundary conditions in different scale are

$$\begin{array}{ll} \epsilon^{-1}: & \sigma_{ij}^{(0)} n_j = 0; \\ \epsilon^{0}: & \sigma_{ij}^{(1)} n_j = t_i^0; \\ \epsilon^{1}: & \sigma_{ij}^{(2)} n_j = 0; \\ & \cdots\cdots \end{array} \qquad \forall x \in \Gamma_t \tag{11.118}$$

and

$$\begin{array}{ll} \epsilon^{0}: & u_i^{(0)} = \bar{u}_i; \\ \epsilon^{1}: & u_i^{(1)} = 0; \\ \epsilon^{2}: & u_i^{(2)} = 0; \\ & \cdots\cdots \end{array} \qquad \forall x \in \Gamma_u \tag{11.119}$$

We first examine the leading order equilibrium equation and boundary condition,

$$\frac{\partial \sigma_{ij}^{(0)}}{\partial y_j} = 0$$

This yields

$$\sigma_{ij}^{(0)} = \sigma_{ij}^{(0)}(x)$$

On the other hand

$$\sigma_{ij}^{(0)} = C_{ijk\ell}(y)\frac{\partial u_k^{(0)}}{\partial y_\ell}$$

To commodate both conditions, we have to set

$$\sigma_{ij}^{(0)} = 0 \ . \tag{11.120}$$

and

$$u_i^{(0)} = u_i^{(0)}(x) \tag{11.121}$$

To solve the second order boundary-value problem, the following separation of variable is adopted

$$u_i^{(1)}(x,y) = \chi_i^{k\ell}(y)\frac{\partial u_k^{(0)}}{\partial x_\ell}(x) + \bar{u}_i^{(1)}(x) \tag{11.122}$$

where the unknown vector function, $\chi_i^{k\ell}(y)\mathbf{e}_i$, is often referred to as the *characteristic displacement field.* We further assume that

$$\sigma_{ij}^{(1)}(x,y) = \tilde{\sigma}_{ij}^{k\ell}(y)\frac{\partial u_k^0}{\partial x_\ell}(x) \tag{11.123}$$

where $\tilde{\sigma}_{ij}^{k\ell}$ is the fine scale stress field.

Consider

$$\frac{\partial u_i^{(1)}}{\partial y_j} = \frac{\partial \chi_i^{k\ell}}{\partial y_j}\frac{\partial u_k^{(0)}}{\partial x_\ell} \tag{11.124}$$

and

$$\sigma_{ij}^{(1)} = C_{ijk\ell}\Bigl(u_{Xk,\ell}^{(0)} + u_{Yk,\ell}^{(1)}\Bigr) = C_{ijk\ell}\Bigl(e_{Xk\ell}^{(0)} + u_{Yk,\ell}^{(1)}\Bigr)\,. \tag{11.125}$$

where $e_{Xk\ell}^{(0)} := \frac{1}{2}(u_{Xk,\ell}^{(0)} + u_{X\ell,k}^{(0)})$. We find that

$$\sigma_{ij}^{(1)} = C_{ijk\ell}\Bigl(T_{k\ell}^{mn} + \frac{\partial \chi_k^{mn}}{\partial y_\ell}\Bigr)u_{Xm,n}^{(0)} \tag{11.126}$$

where $T_{k\ell}^{mn} = \dfrac{1}{2}\Bigl(\delta_{km}\delta_{\ell n} + \delta_{kn}\delta_{\ell m}\Bigr)$, because $T_{k\ell}^{mn}u_{Xm,n}^{(0)} = e_{Xk\ell}^{(0)}$.

Accordingly,

$$\tilde{\sigma}_{ij}^{mn} = C_{ijk\ell}\Bigl(T_{k\ell}^{mn} + \frac{\partial \chi_k^{mn}}{\partial y_\ell}\Bigr)$$

Then the equilibrium equation on second scale (ϵ^{-1}) provides the governing equation for the canonical cell problem,

$$\frac{\partial \sigma_{ij}^{(1)}}{\partial y_j} = 0, \quad \Rightarrow \quad \frac{\partial \tilde{\sigma}_{ij}^{mn}}{\partial y_j}\frac{\partial u_m^{(0)}}{\partial x_n} = 0, \quad \Rightarrow \quad \frac{\partial \tilde{\sigma}_{ij}^{mn}}{\partial y_j} = 0\,. \tag{11.127}$$

More explicitly, the governing equation for canonical cell problem is

$$\boxed{\frac{\partial}{\partial y_j}\Bigl(C_{ijk\ell}\Bigl[T_{k\ell}^{mn} + \frac{\partial \chi_k^{mn}}{\partial y_\ell}\Bigr]\Bigr) = 0, \quad \forall y \in Y} \tag{11.128}$$

The related interface continuity conditions and periodic conditions are omitted here.

Consider the equilibrium equation at third scale (ϵ^0). We have

$$\frac{\partial \sigma_{ij}^{(2)}}{\partial y_j} = -\Bigl(f_i + \frac{\partial \sigma_{ij}^{(1)}}{\partial x_j}\Bigr) = F_i, \quad \forall\; y \in Y$$

The Fredholm alternative condition requires that

$$\frac{1}{|Y|}\int_Y F_i(y)dy = 0 \ .$$

This can be shown from the fact that

$$\frac{1}{|Y|}\int_Y \frac{\partial \sigma_{ij}^{(2)}}{\partial y_j}dy = \frac{1}{|Y|}\int_{\partial Y}\sigma_{ij}^{(2)} n_j dS = 0 \ .$$

Thereby,

$$\frac{1}{|Y|}\int_Y \Big(f_i + \frac{\partial \sigma_{ij}^{(1)}}{\partial x_j}\Big)dy = 0, \quad \Rightarrow \quad f_i + \frac{\partial}{\partial x_j} < \sigma_{ij}^{(1)} >_Y = 0 \ .$$

where

$$< \sigma_{ij}^{(1)} >_Y = < \tilde{\sigma}_{ij}^{k\ell}(y) >_Y \frac{\partial u_k^{(0)}}{\partial x_\ell} = C_{ijk\ell}^h \frac{\partial u_k^{(0)}}{\partial x_\ell} \tag{11.129}$$

and the homogenized elastic stiffness tensor is determined by the solution of the canonical cell problem,

$$C_{ijk\ell}^h = \frac{1}{|Y|}\int_Y C_{ijmn}(y)\Big[T_{mn}^{k\ell} + \frac{\partial \chi_m^{k\ell}}{\partial y_n}\Big]dy = \Big\langle \tilde{\sigma}_{ij}^{k\ell}\Big\rangle_Y \ . \tag{11.130}$$

The homogenized BVP is,

$$< \sigma_{ij} >_{,j} + f_i = 0, \quad \forall x \in \Omega \tag{11.131}$$

$$< \sigma_{ij} > n_j = t_i, \quad \forall x \in \Gamma_t \tag{11.132}$$

$$u_i^{(0)} = \bar{u}_i, \quad \forall x \in \Gamma_u \tag{11.133}$$

11.5.2 *Finite element formulation*

Choose $v_i \in H_\#^1(Y)$. Multiplying v_i with the leading order equilibrium equation (11.112) and integrating it over Y, we have

$$\int_Y \frac{\partial \sigma_{ij}^{(0)}}{\partial y_j} v_i d\Omega_y = 0, \quad \forall v_i \in H_\#^1(Y)$$

Integration by parts yields,

$$\int_Y \sigma_{ij}^{(0)} n_j v_i dS - \int_Y \sigma_{ij}^{(0)} \frac{\partial v_i}{\partial y_j} dV_y$$

$$= -\int_Y \sigma_{ij}^{(0)} \frac{\partial v_i}{\partial y_j} dV_y = -\int_Y C_{ijk\ell} \frac{\partial u_k^{(0)}}{\partial y_\ell}\frac{\partial v_i}{\partial y_j} d\Omega_y = 0 \ .$$

Let $v_i(x,y) = u_i^{(0)}(x,y)$. We have

$$\int_Y C_{ijk\ell} \frac{\partial u_k^{(0)}}{\partial y_\ell}\frac{\partial u_i^{(0)}}{\partial y_j} d\Omega_y = 0 \ . \tag{11.134}$$

Since $C_{ijk\ell}(y)$ is positive definite,

$$\frac{\partial u_i^{(0)}}{\partial y_j} = 0 , \quad \Rightarrow \quad u_i^{(0)} = u_i^{(0)}(x)$$

and consequently $\sigma_{ij}^{(0)} = 0$, as we have derived before.

Multiply Eq. (11.128) with a test function, $v_i \in H^1_\#(Y)$, and integrate them over Y. Integration by parts yields,

$$\begin{aligned}&\int_Y \frac{\partial}{\partial y_j}\Big[C_{ijk\ell}\Big(T_{k\ell}^{mn} + \frac{\partial \chi_k^{mn}}{\partial x_\ell}\Big)\Big] v_i dV_y \\ &= \int_{\partial Y}\Big[C_{ijk\ell}\Big(T_{k\ell}^{mn} + \frac{\partial \chi_k^{mn}}{\partial x_\ell}\Big)\Big] n_j v_i dS_y - \int_Y C_{ijk\ell}\Big(T_{k\ell}^{mn} + \frac{\partial \chi_k^{mn}}{\partial x_\ell}\Big)\frac{\partial v_i}{\partial y_j} dV_y \\ &= -\int_Y C_{ijk\ell}\Big(T_{k\ell}^{mn} + \frac{\partial \chi_k^{mn}}{\partial x_\ell}\Big)\frac{\partial v_i}{\partial y_j} dV_y = 0 .\end{aligned}$$

Consider the following parametric vector,

$$\mathbf{P}^{mn} = y_m \delta_{nk} \mathbf{e}_k = P_k^{mn} \mathbf{e}_k \tag{11.135}$$

One can show that

$$T_{k\ell}^{mn} = \frac{1}{2}\Big(\frac{\partial P_\ell^{mn}}{\partial y_k} + \frac{\partial P_k^{mn}}{\partial y_\ell}\Big) = P_{(k,\ell)}^{mn}$$

Therefore, the weak formulation for the canonical cell problem can be written as

$$\frac{1}{|Y|}\int_Y C_{ijk\ell}(y)\Big(P_{(k,\ell)}^{mn} + \chi_{(k,\ell)}^{mn}\Big) v_{(i,j)} dV_y = 0 . \tag{11.136}$$

Define the bilinear form

$$a_Y(\mathbf{u}, \mathbf{v}) = \frac{1}{|Y|}\int_Y C_{ijk\ell}(y) u_{(i,j)} v_{(k,\ell)} dV_y \tag{11.137}$$

The finite element formulation of canonical cell problem is:

Find $\boldsymbol{\chi}^{mn} \in H^1_\#(Y)$, such that

$$a_Y(\mathbf{P}^{mn} + \boldsymbol{\chi}^{mn}, \mathbf{v}) = 0, \quad \forall \;\; \mathbf{v} \in H^1_\#(Y) \tag{11.138}$$

Once $\chi_{k,\ell}^{mn}$ being determined, the effective elastic stiffness tensor can then be calculated based on definition

$$C_{stk\ell}^H = \frac{1}{|Y|}\int_Y C_{stmn}(y)(P_{m,n}^{k\ell} + \chi_{m,n}^{k\ell}(y)) dV_y \tag{11.139}$$

Consider the fact that

$$T_{st}^{ij} = P_{(s,t)}^{ij} = \frac{1}{2}\Big(\delta_{si}\delta_{tj} + \delta_{sj}\delta_{ti}\Big)$$

It is readily to show that

$$C_{stk\ell}^H T_{st}^{ij} = C_{stk\ell}^H \frac{1}{2}\Big(\delta_{si}\delta_{tj} + \delta_{sj}\delta_{ti}\Big) = C_{ijk\ell}^H \tag{11.140}$$

and

$$\begin{aligned} C^H_{ijk\ell} &= \frac{1}{|Y|}\int_Y C_{stmn}\Big(P^{k\ell}_{m,n} + \chi^{k\ell}_{m,n})T^{ij}_{st}dV_y \\ &= \frac{1}{|Y|}\int_Y C_{stmn}\Big(P^{k\ell}_{m,n} + \chi^{k\ell}_{m,n})P^{ij}_{(s,t)}dV_y \\ &= a_Y(\mathbf{P}^{k\ell} + \boldsymbol{\chi}^{k\ell}, \mathbf{P}^{ij}\Big)dV_y \end{aligned} \tag{11.141}$$

Finally, we define another function space,

$$\mathcal{V}_\Omega = \Big\{\mathbf{v}(x), x \in \Omega \;\Big|\; \mathbf{v}(x) \in [H^1(\Omega)]^d, d = dim\{\Omega\}, \text{ and }, \mathbf{v}(x)\Big|_{\Gamma_u} = 0\Big\}$$

The weak formulation for the following macro-level BVP,

$$\frac{\partial < \sigma^{(1)}_{ij} >_Y}{\partial x_j} + f_i = 0, \tag{11.142}$$

$$\text{where} \quad < \sigma^{(1)}_{ij} > = C^H_{ijk\ell}u^{(0)}_{(k,\ell)} \tag{11.143}$$

$$\text{and} \quad \frac{\partial}{\partial x_j}\Big[C^H_{ijk\ell}u^{(0)}_{(k,\ell)}\Big] + f_i = 0, \quad \forall x \in \Omega \tag{11.144}$$

$$< \sigma^{(1)}_{ij} > n_j = t^0_i, \quad \forall\, x \in \Gamma_t \tag{11.145}$$

$$u^{(0)}_i = \bar{u}_i, \quad \forall\, x \in \Gamma_u \tag{11.146}$$

is:

Find $\boldsymbol{u}^{(0)}(x) \in \mathcal{V}_\Omega$ such that

$$\int_\Omega C^H_{ijk\ell}u^{(0)}_{(k,\ell)}v_{(i,j)}dV_x = \int_\Omega f_i v_i dV_x + \int_{\Gamma_t} t^0_i v_i dS, \quad \forall\, \mathbf{v} \in \mathcal{V}_\Omega\,. \tag{11.147}$$

where $\mathbf{v} = v_i\mathbf{e}_i$.

Summary of Multiscale Finite Element Method

(1) Solve the canonical cell problem on Y first, i.e. find $\boldsymbol{\chi}^{k\ell}(y) \in H^1_\#(Y)$ by solving

$$a_Y\Big(\mathbf{P}^{k\ell} + \boldsymbol{\chi}^{k\ell}, \mathbf{v}\Big) = 0, \quad \forall \mathbf{v} \in H^1_\#(Y)$$

(2) Calculate macro-scale elastic stiffness tensor

$$C^H_{ijk\ell} = a_Y\Big(\mathbf{P}^{ij} + \boldsymbol{\chi}^{ij}, \mathbf{P}^{k\ell}\Big) \quad \text{and} \quad a_Y(\mathbf{u}, \mathbf{v}) := \frac{1}{Y}\int_Y C_{ijk\ell}(y)u_{i,j}v_{k,\ell}dV_y$$

(3) Solve the macro displacement field, $\mathbf{u}^{(0)}(\mathbf{x}) \in \mathcal{V}_\Omega$,

$$a^H_\Omega(\mathbf{u}^{(0)}, \mathbf{v}) = \Big\langle \mathbf{f}, \mathbf{v}\Big\rangle_\Omega + \Big\langle \mathbf{t}^0, \mathbf{v}\Big\rangle_{\Gamma_t} \quad \text{where} \quad a^H_\Omega(\mathbf{u}, \mathbf{v}) := \int_\Omega C^H_{ijk\ell}u_{i,j}v_{k,\ell}dV_y$$

where $\mathbf{v}$ is any function in $\mathcal{V}_\Omega$;

(4) Calculate the fine (local) scale stress distribution,

$$\sigma^{(1)}_{ij}(x, y) = C_{ijk\ell}(y)\Big(T^{mn}_{k\ell} + \chi^{mn}_{(k,\ell)}(y)\Big)\frac{\partial u^{(0)}_m}{\partial x_n}$$

11.6 G-, H-, and Γ- Convergence

Various notions of convergence are introduced in relation to asymptotic homogenization theory, such as Γ-convergence of E. De Giorgi [De Giorgi (1975, 1983)], the G-convergence of S. Spagnolo [Spagnolo (1968, 1975)], and the H-convergence of F. Murat, L. Tartar, and G. Francfort [Murat and Tartar (1997); Francfort and Murat (1986)]. These abstract mathematical notions provide powerful tools to analysis various numerical simulations of homogenization.

The question that we are interested in is: what is the limit in a homogenization process when the micro-scale approaches to zero (see Fig. 11.9) ? Does upscale homogenization will eventually converge to that limit ?

To answer this questions, we have to first define what do we mean by convergence, or convergence in what sense. In the following, we shall discuss main concepts and primary ideas of homogenization convergence. For more elaborated and rigorous presentations, readers may consult monographs by A. Cherkaev [Cherkaev (2000)] and G. Allaire [Allaire (2002)] on structural optimization and related references e.g. [Tartar (1990); Bensoussan *et al.* (1992)].

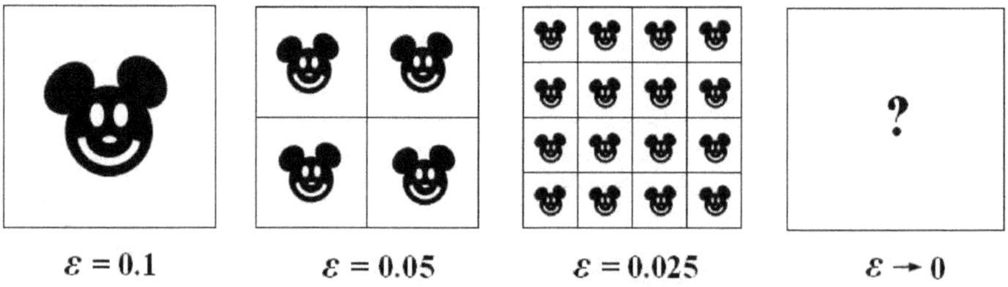

Fig. 11.9 Notion of convergence in homogenization.

11.6.1 *Strong convergence and weak convergence*

We first discuss the notion of strong convergence and weak convergence of functions in the Banach space .

Let Ω be an open set in $\mathbb{R}^d$. For $1 \leq p \leq +\infty$, the Lebesgue space $L^p(\Omega)$ of all measurable functions u in Ω is a Banach space endowed with the following norm,

$$\|u\|_{L^p(\Omega)} = \left(\int_\Omega |u|^p dx \right)^{1/p}, \quad \forall 1 \leq p < +\infty$$

When $p = \infty$, we define the so-called essential supremum

$$\|u\|_{L^\infty(\Omega)} = ess \sup_{x \in \Omega} |u(x)| := \inf_{\substack{Z \in \Omega \\ \mu(Z)=0}} \left\{ \sup_{x \in \Omega - Z} |u(x)| \right\}$$

Note that the physical meaning of $L^\infty(\Omega)$ space is that its occupant functions satisfying the condition $|u(x)| < \infty$ almost everywhere in Ω.

We use the short-handed notation, $\epsilon \to 0$ to denote a limit process of a sequence $\epsilon = \{\epsilon_1, \epsilon_2, \cdots \epsilon_n, \cdots\cdots\}$, and $\epsilon_n \to 0$ as $n \to \infty$.

The strong convergence of a function sequence, $u_\epsilon := \{u_{\epsilon_1}, u_{\epsilon_2}, \cdots, u_{\epsilon_n}, \cdots\}$, is measured by the distance in the particular normed space, i.e. a sequence, u_ϵ, is said to converge strongly in $L^p(\Omega)$ to a limit u_0, if

$$\lim_{\epsilon \to 0} \|u_\epsilon - u_0\|_{L^p(\Omega)} = 0 \ .$$

The strong convergence is denoted by an arrow, namely,

$$u_\epsilon \to u_0, \quad \text{in } L^p(\Omega) \ \text{ strongly}$$

On the other hand, the weak convergence is measured by a so-called weighted residual distance, which is associated with a weighting function, or test function in the dual space of the original norm space.

For the weak convergence in Lebesgue space $L^p(\Omega)$, the test function is in its dual space $L^{p'}(\Omega)$ with

$$\frac{1}{p} + \frac{1}{p'} = 1 \ .$$

Therefore, the formal statement of weak convergence in $L^p(\Omega)$, $1 \le p < +\infty$ is as follows: a sequence u_ϵ is said to converge weakly in $L^p(\Omega)$ to a limit u_0, if for any test function $\phi \in L^{p'}(\Omega)$, it satisfies

$$\lim_{\epsilon \to 0} \int_\Omega u_\epsilon(x)\phi(x)dx = \int_\Omega u(x)\phi(x)dx$$

The weak convergence is denoted by a harpoon, namely

$$u_\epsilon \rightharpoonup u_0 \ \text{ in } L^p(\Omega) \text{ weakly} \ .$$

An interesting property of weak convergence is its sequential relative compactness on bounded set. This means that for all bounded sequences, $\|u_\epsilon\|_{L^p(\Omega)} \le C$, there exists a subsequence $\Big(u_\epsilon\Big)_{\epsilon' > 0}$ and a limit u_0 such that $\Big(u_\epsilon\Big)_{\epsilon' > 0}$ converges weakly to u_0 in $L^p(\Omega)$, $1 < p < \infty$, which is not true for strong convergence.

Intuitively speaking, the strong convergence is more or less the usual pointwise convergence, while the weak convergence is a notion of convergence "in average" (up to a fluctuation of zero-mean).

If Ω is finite, we may choose test function

$$\phi(x) = \frac{1}{\Omega} \in L^{p'}(\Omega)$$

then $u_\epsilon(x) \rightharpoonup u_0(x)$ requires that

$$\lim_{\epsilon \to 0} \int_\Omega u_\epsilon = \frac{1}{\Omega}\int_\Omega u_\epsilon(x)dx = \frac{1}{\Omega}\int_\Omega u_0(x)dx$$

That is $\lim_{\epsilon \to 0} < u_\epsilon >_\Omega = < u_0 >_\Omega$. We state (without proof) the connection between strong convergence and pointwise convergence. This statement is false for weakly convergence.

Theorem 11.1.

(1) Let Ω be a bounded open set in $\mathbb{R}^d$. Let u_ϵ be a sequence converging strongly to a limit u_0 in $L^p(\Omega)$, $1 \leq p \leq +\infty$, i.e.

$$u_\epsilon(x) \rightarrow u_0(x)$$

Then there exists a subsequence, $u_{\epsilon'} \subset u_\epsilon$, and a function $h(x) \in L^p(\Omega)$ such that,

$$\lim_{\epsilon' \to 0} u_{\epsilon'}(x) = u_0(x), \quad \text{almost everywhere in } \Omega$$
$$|u_\epsilon(x)| \leq h(x), \quad \text{almost everywhere in } \Omega$$

(2) Assume that the sequence $u_\epsilon(x)$ is bounded in $L^p(\Omega)$ $(1 < p \leq \infty)$, and

$$\lim_{\epsilon \to 0} u_\epsilon(x) = u_0(x), \quad \text{almost everywhere in } \Omega$$

Then

$$u_\epsilon(x) \rightarrow u_0(x) \text{ in } L^q(\Omega) \ (1 \leq q < p) \ strongly\ .$$

To understand differences between strong convergence and weak convergence, we consider the following example.

Example 11.5. Let $u_\epsilon(x) = \sin\Bigl(\dfrac{x}{\epsilon}\Bigr)$, $p = 2$, and $\Omega = (1, 0)$. Choose test function $\phi(x) = 1$. We have

$$\int_0^1 u_\epsilon(x)\phi(x)dx = \int_0^1 \sin\Bigl(\frac{x}{\epsilon}\Bigr)dx = -\epsilon\cos\Bigl(\frac{x}{\epsilon}\Bigr)\Big|_0^1 = \epsilon\Bigl(1 - \cos\Bigl(\frac{1}{\epsilon}\Bigr)\Bigr)$$

As $\epsilon \to 0$, $u_\epsilon \rightharpoonup 0$, weakly in $L^2(\Omega)$, i.e. the weak limit of the sequence $u_\epsilon(x)$ is zero.

On the other hand, it seems that $u_\epsilon(x)$ has no strong limit in $L^2(\Omega)$. This is because

$$\|u_\epsilon\|_{L^2(\Omega)} = \sqrt{\int_0^1 u_\epsilon^2(x)dx} = \sqrt{\int_0^1 \sin^2\Bigl(\frac{x}{\epsilon}\Bigr)dx}$$
$$= \sqrt{\frac{1}{2}\int_0^1 \Bigl(1 - \cos\Bigl(\frac{2x}{\epsilon}\Bigr)\Bigr)dx} = \sqrt{\frac{1}{2}\Bigl(1 - \frac{\epsilon}{2}\sin^2(\frac{2}{\epsilon}\Bigr)}\ .$$

Suppose $u_\epsilon \rightarrow f(x)$ and $f(x) \in L^2(\Omega)$. Therefore,

$$\lim_{\epsilon \to 0}\int_0^1 \Bigl(\sin\frac{x}{\epsilon} - f(x)\Bigr)^2 dx = \int_0^1 \sin^2\Bigl(\frac{x}{\epsilon}\Bigr)dx - 2\int_0^1 \sin\Bigl(\frac{x}{\epsilon}\Bigr)f(x)dx$$
$$+ \int_0^1 f^2(x)dx = \frac{1}{2} + \int_0^1 f^2(x)dx \neq 0\ .$$

because $f(x) \in (L^2)'(\Omega)$.

Moreover, the fact that

$$\lim_{\epsilon\to 0}\int_0^1 \sin^2\Big(\frac{x}{\epsilon}\Big)dx = \frac{1}{2}$$

also indicates that the product of two weakly convergence sequences does not converge to the product of their weak limits. Otherwise,

$$\lim_{\epsilon\to 0}\int_0^1 \sin^2\Big(\frac{x}{\epsilon}\Big)dx = 0$$

because both $\sin\Big(\frac{x}{\epsilon}\Big) \rightharpoonup 0$ in $L^2([0,1])$.

It is worth noting that the product of two strong convergence sequences does converge to the product of the two limits strongly, but it maybe converge in a different Lebesgue space in general. For instance, if both $u_\epsilon \to u_0$ in $L^2(\Omega)$ strongly and $v_\epsilon \to v_0$ in $L^2(\Omega)$ strongly, then

$$\begin{aligned}\|u_\epsilon v_\epsilon - u_0 v_0\|_{L^2(\Omega)} &= \|(u_\epsilon - u_0)(v_\epsilon - v_0) + (u_\epsilon - u_0)v_0 + (v_\epsilon - v_0)u_0\|_{L^2(\Omega)}\\ &\le \Big(\|u_\epsilon - u_0\|_{L^2(\Omega)}\Big)^{1/2}\Big(\|v_\epsilon - v_0\|_{L^2(\Omega)}\Big)^{1/2}\\ &\quad + \|v_0\|^{1/2}_{L^2(\Omega)}\Big(\|u_\epsilon - u_0\|_{L^2(\Omega)}\Big)^{1/2} + \|u_0\|^{1/2}_{L^2(\Omega)}\Big(\|v_\epsilon - v_0\|_{L^2(\Omega)}\Big)^{1/2}\end{aligned}$$

Hence

$$u_\epsilon v_\epsilon \to u_0 v_0 \text{ in } L^2(\Omega) \text{ strongly .}$$

Unfortunately, the same is not true for the weakly convergent sequences. In our previous example,

$$u_\epsilon(x) = \sin\Big(\frac{x}{\epsilon}\Big) \rightharpoonup 0 \text{ in } L^2(\Omega) \text{ weakly}$$

but for $u_\epsilon(x) = v_\epsilon(x) = \sin\Big(\frac{x}{\epsilon}\Big)$

$$u_\epsilon(x)v_\epsilon(x) \rightharpoonup \frac{1}{2}\ ! \ \text{ in } L^p(\Omega)\ \ 1 \le p < +\infty\ .$$

Moreover, in practice, if $u_\epsilon \rightharpoonup u_0$ in $L^p(\Omega)$, and $J(u)$ is a nonlinear functional, say quadratic functional, $J : L^p(\Omega) \to \mathbb{R}$.

It is usually

$$J(u_\epsilon) \not\rightharpoonup J(u_0) \text{ in any sense !}$$

11.6.2 *G-convergence*

Consider our model homogenization BVP,

$$L^\epsilon u_\epsilon = f,\quad x \in \Omega,\quad \text{where } L^\epsilon = -\nabla\cdot \mathbf{A}(\frac{x}{\epsilon})\cdot\nabla$$

$$u_\epsilon\Big|_{\partial\Omega} = \bar{u},\quad \forall x \in \partial\Omega$$

where the heat conduction (or diffusion) coefficient $A_{ij}(y)$ are Y-periodic functions.

Suppose that solution of the above BVP can be found as

$$u_\epsilon(x) = \Big(L^\epsilon\Big)^{-1} f\ .$$

Obviously, $u_\epsilon \in H^1(\Omega)$ and $f \in H^{-1}(\Omega)$.

Recall the definition of Green's function . We have

$$u_\epsilon(x) = \left(L^\epsilon\right)^{-1} f = \int_\Omega G_\epsilon(x-y)f(y)dy$$

Suppose that there exists a weak limit $u_0(x)$ in $H^1(\Omega)$ such that

$$u_\epsilon(x) \rightharpoonup u_0(x) \text{ in } H^1(\Omega) \text{ weakly}$$

and the weak limit $u_0(x)$ has the representation,

$$u_0(x) = \int_\Omega G_0(x-y)f(y)dy =: \left(L_0\right)^{-1} f$$

Therefore, the weak convergence of $u_\epsilon(x)$, i.e. $u_\epsilon \rightharpoonup u_0(x)$, implies that

$$\int_{\Omega_x}\int_{\Omega_y}\Big(G_\epsilon(x-y) - G_0(x-y)\Big)f(y)dydx = 0, \quad \epsilon \to 0 \tag{11.148}$$

Change the order of integration, (11.148) yields

$$\int_{\Omega_y} f(y)\left(\int_{\Omega_x}\Big(G_\epsilon(x-y) - G_0(x-y)\Big)dx\right)dy = 0, \text{ as } \epsilon \to 0\ . \tag{11.149}$$

Equation (11.149) suggests that the weak convergence of Green's function, i.e. $G_\epsilon \rightharpoonup G_0$, which implies a special type of convergence of the differential operator sequence $L^\epsilon = -\nabla\cdot\mathbf{A}^\epsilon\cdot\nabla$. We call the convergence of differential operator sequence L_ϵ as the G-convergence,

$$L^\epsilon \overset{G}{\to} L_0 \tag{11.150}$$

in the sense of

$$G_\epsilon * f \rightharpoonup G_0 * f, \quad \text{in} \quad H^1(\Omega) \text{ weakly}\ .$$

Note that the symbol $*$ denotes the standard convolution.

In fact, the convergence of the differential operator sequence, $L^\epsilon = -\nabla\cdot A^\epsilon\cdot\nabla$, may be viewed as the convergence of matrix sequence, A^ϵ_{ij}, to its G-limit A^0_{ij}, or

$$A_{ij}\left(\frac{x}{\epsilon}\right) \overset{G}{\to} A^0_{ij}$$

Following [Allaire (2002)], we adopt the following definition of G-convergence .

Let $\mathcal{M}^s_d$ be the linear space of symmetric real matrices of order d. For any two positive constants $\alpha > 0$ and $\beta > 0$, we define a subspace of $\mathcal{M}^s_d$ made of coercive matrices with coercive inverse, namely,

$$\mathcal{M}^s_{\alpha,\beta} := \Big\{\{M_{ij}\} \in \mathcal{M}^s_d, \text{ such that } \alpha\xi^2 \le M_{ij}\xi_i\xi_j$$
$$\text{and } \beta\xi^2 \le M^{-1}_{ij}\xi_i\xi_j,\ \forall \xi \in \mathbb{R}^d\Big\}$$

Let Ω be a bounded open set in $\mathbb{R}^d$ and define the space $L^\infty(\Omega;\mathcal{M}^s_{\alpha,\beta})$ of admissible symmetric coefficient matrices.

We have the following definition of G-convergence.

Definition 11.1. A sequence of symmetric matrices, $A^{\epsilon} \in L^{\infty}(\Omega, \mathcal{M}^{s}_{\alpha,\beta})$ is said to be G-convergence to an homogenized, or G-limit, matrix $A^{0} \in L^{\infty}(\Omega, \mathcal{M}^{s}_{\alpha,\beta})$, if, for any $f \in H^{-1}(\Omega)$., the sequence solution $u_{\epsilon}(x)$ of the following model problem

$$-\nabla \cdot A^{\epsilon}\nabla u_{\epsilon} = f, \quad x \in \Omega$$
$$u_{\epsilon} = \bar{u}, \quad \forall x \in \partial\Omega$$

converges weakly in $H^{1}(\Omega)$ to the solution of the homogenized BVP,

$$-\nabla \cdot A^{0} \cdot \nabla u_{0} = f, \quad x \in \Omega$$
$$u_{0} = \bar{u}, \quad \forall x \in \partial\Omega$$

This definition makes sense because the following compactness theorem,

Theorem 11.2. *For any sequence $A^{\epsilon} \in L^{\infty}(\Omega; \mathcal{M}_{\alpha,\beta})$ of symmetric matrices, there exits a subsequence, $A^{\epsilon'} \subset A^{\epsilon}$, and a limit $A^{0} \in L^{\infty}(\Omega; \mathcal{M}_{\alpha,\beta})$ such that $A_{\epsilon'}$ G-converges to A^{0}.*

In the following examples, we want to show the differences between strong convergence, weak convergence, and G-convergence .

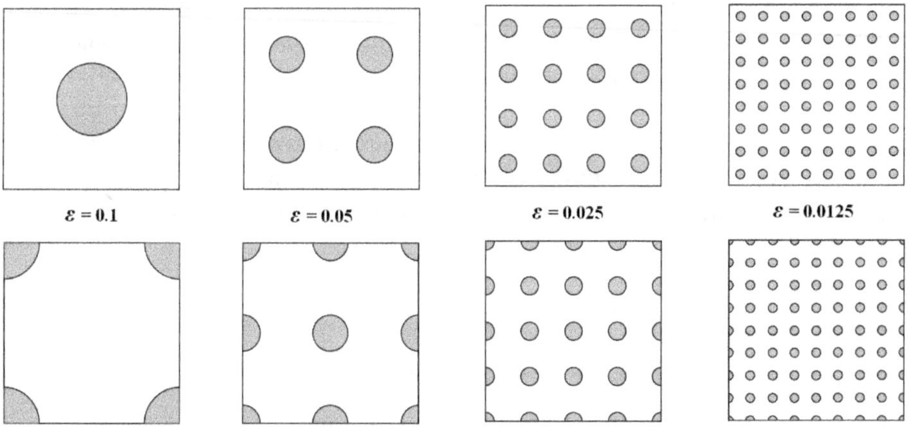

Fig. 11.10 The difference between strong convergence and G-convergence.

Example 11.6. In this example, suppose that we have two objects with the same macroscopic dimension but different microscopic structure, as shown in Fig. 11.10.

The diffusivity matrix coefficients are assumed to be isotropic

$$A_{ij} = a\delta_{ij}$$

We denote the diffusivity a in the white area as a_w and the diffusivity in the black area as a_b, and $a_b > a_w$.

We denote the first micro-structure (top row) as $\mathcal{S}_1^\epsilon$ and the second micro-structure (bottom row) as $\mathcal{S}_2^\epsilon$. Obviously, as $\epsilon \to 0$ the first sequence $a^\epsilon(\mathcal{S}_1^\epsilon)$ and the second sequence $a^\epsilon(\mathcal{S}_2^\epsilon)$ have the same G-limit, i.e.

$$a^0(\mathcal{S}_1^\epsilon) = a^0(\mathcal{S}_2^\epsilon) .$$

As one can verify that there is no pointwise convergence possibility: because at a given spatial point, for example, at the center of the configuration s $\mathbf{x}$, diffusivity converges to a_b in micro-structure $\mathcal{S}_1^\epsilon$, while diffusivity converges to a_w in micro-structure $\mathcal{S}_2^\epsilon$.

$$|a^0(\mathcal{S}_1^\epsilon) - a^0(\mathcal{S}_2^\epsilon)| = a_b - a_w > 0 .$$

Nevertheless, in this example, indeed, the weak convergence limit of the two layouts are the same

$$< a^\epsilon(\mathcal{S}_1^\epsilon) >_\Omega = < a^\epsilon(\mathcal{S}_2^\epsilon) >_\Omega$$

Example 11.7. In the second example, we would like to show a case that there are two micro-structure layouts with the same weak convergence limits, but different G-limits.

As shown in Fig. 11.11, we assume that in each unit cell, the areas of the black and white phases are the same, therefore the volume fraction of the two phases are the same in these two layouts. In the layout 1, all the "conductive" material are connected, therefore it is a better arrangement for heat conduction, whereas in the layout 2, all the " conductive" materials are isolated, disconnected, or insulated, it should be very hard for heat to diffuse from one point to another point.

Based on this argument, the two layouts should have different G-limit, and

$$a^0(\mathcal{S}_1^\epsilon) > a^0(\mathcal{S}_2^\epsilon) .$$

On the other hand, the two layouts have the same weak convergence limit,

$$< a^\epsilon(\mathcal{S}_1^\epsilon) >_\Omega = < a^\epsilon(\mathcal{S}_2^\epsilon) >_\Omega$$

as indicated above. This can be also illustrated for elastic composite materials. Assume that the strength of black phase is very strong, say it is almost rigid, while the white phase that has the same volume fraction as the black phase has much weaker strength. The composite corresponding to the unit cell layout 1 should be still very strong like a rigid solid, whereas the composite corresponding to the unit cell layout 2 has a weak strength matrix with a rigid inclusion, so its overall strength should be much lower than that of the unit cell layout 1. In mathematics term, they have different G-convergence limits while having the same weak convergence limit.

Example 11.8. In the third example, we would like to show a case in which two microstructure layouts have the same G-limit but different weak convergence limits.

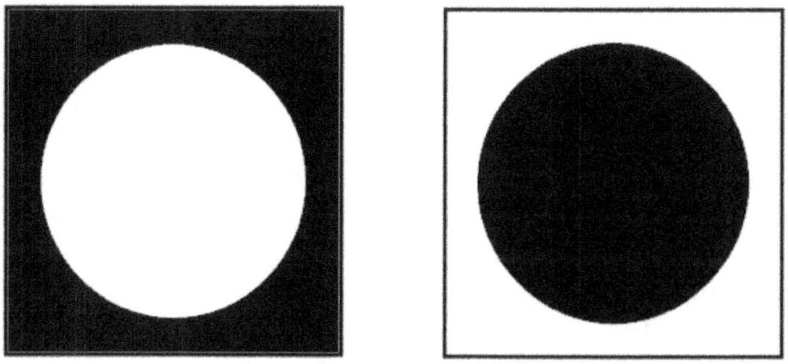

Fig. 11.11 The difference between weak convergence and G-convergence.

In this example, we fix the second layout of the previous example. Therefore, we know that the G-limit of the second layout will be bounded by the Voigt (upper) bound and the Reuss (lower) bound, i.e.

$$\text{Reuss bound} = \frac{2a_b a_w}{a_b + a_w} \leq a^0\left(\mathcal{S}_2^\epsilon\right) \leq \frac{1}{2}(a_b + a_w) = \text{Voigt bound}$$

We know change the first layout by increase the volume fraction of insolated white phase, f_w such that $f_w \in [0.5, 1)$ and $f_w \to 1$. Therefore, the G-limit of the first layout will be bounded by

$$\frac{1}{\dfrac{f_w}{a_w} + \dfrac{1 - f_w}{a_b}} \leq a^0(\mathcal{S}_1^\epsilon) \leq f_w a_w + (1 - f_w)a_b \tag{11.151}$$

Initially when $f_w = 0.5$ we have,

$$\frac{2a_w a_b}{a_w + a_b} < a^0\left(\mathcal{S}_2^\epsilon\right) < a^0\left(\mathcal{S}_1^\epsilon\right) < \frac{1}{2}(a_w + a_b)$$

If $a_w << a_b$, the Reuss bound for the second layout is approximately $2a_w$. From Eq. (11.151), one can see that as $f_w \to 1$, $a^0\left(\mathcal{S}_1^\epsilon\right) \to a_w$, therefore

$$a^0\left(\mathcal{S}_2^\epsilon\right) > a^0\left(\mathcal{S}_1^\epsilon\right), \quad \text{as} \quad f_w \to 1 \ .$$

This suggests that at certain volume fraction, $0.5 < f_w < 1.0$, the G-limits of the two layouts will be the same, i.e.

$$a^0(\mathcal{S}_1^\epsilon) = a^0(\mathcal{S}_2^\epsilon) \ .$$

In that case, since $f_w > 0.5$, the weak convergence limits of the two layouts will not be the same, i.e.

$$< a^0(\mathcal{S}_1^\epsilon) >_\Omega = f_w a_w + (1 - f_w)a_b \neq < a^0(\mathcal{S}_2^\epsilon) >_\Omega = 0.5(a_w + a_b) \ .$$

11.6.3 *H-convergence*

H-convergence is a generalization of G-convergence, in which, the differential operator A^ϵ, or its coefficient matrix, does not require to be symmetric anymore.

Definition 11.2 (Definition of H-Convergence). *A sequence of matrices* $\mathbf{A}^\epsilon$ *in* $L^\infty(\Omega, M_{\alpha,\beta})$ *is said to converge in the sense of homogenization, or simply H-convergence, to an homogenized limit, or H-limit, matrix* $\mathbf{A}^0 \in L^\infty(\Omega, M_{\alpha,\beta})$ *if, for any right hand side* $f \in H^{-1}(\Omega)$, *the sequence* u_ϵ *of solution of*

$$-\nabla \cdot \mathbf{A}^\epsilon \cdot \nabla u_\epsilon = f(x), \quad \forall x \in \Omega \tag{11.152}$$

$$u_\epsilon = \bar{u}, \quad \forall x \in \partial\Omega \tag{11.153}$$

satisfies

$$u_\epsilon(x) \rightarrow u_0(x) \;\; \text{weakly in } \mathrm{H}^1(\Omega) \tag{11.154}$$

$$\mathbf{A}^\epsilon \cdot \nabla u_\epsilon \rightarrow \mathbf{a}^* \cdot \nabla u_0 \;\; \text{weakly in } \left[L^2(\Omega)\right]^N \tag{11.155}$$

where u_0 *is the solution of the homogenized equation,*

$$-\nabla \cdot \mathbf{A}^0 \cdot \nabla u_0 = f(x), \quad \forall x \in \Omega \tag{11.156}$$

$$u_0 = \bar{u}, \quad \forall x \in \partial\Omega \tag{11.157}$$

11.6.4 *Γ-convergence*

For a large class of elliptical BVPs, each BVP under consideration has one-to-one correspondence to a variational principle. The well-known Lax-Milgram theorem guarantees the equivalence between the two.

Therefore, the convergence of differential operators may imply a possible convergence of the corresponding functional in the related function spaces.

Definition 11.3 (Definition of Γ-Convergence). *Let* X *be a functional space endowed with a norm* $\|\cdot\|_d$. *Let* ϵ *be a sequence of positive indexes which goes to zero. Let* F_ϵ *be a sequence of functional defined on* X *with values in* $\mathbb{R}$. *The sequence* F_ϵ *is said to be* Γ*-convergence to a limit functional* F_0 *if, for any function* $x \in X$,

(1) all sequences x_ϵ *converging to* x *satisfy*

$$F_0(x) \leq \lim_{\epsilon \rightarrow 0} \inf_{x \in X} F_\epsilon(x_\epsilon)$$

and

(2) there exists at least one sequence x_ϵ *converging to* x, *such that*

$$F_0(x) = \lim_{\epsilon \rightarrow 0} F_\epsilon(x_\epsilon)$$

Example 11.9 (An Example of Γ-Convergence). *Consider the following diffusion problem, with diffusion coefficient matrix,* $\mathbf{A}^{\epsilon}$ *is symmetric and Y-periodic,*

$$-\nabla \cdot \mathbf{A}\Big(\frac{x}{\epsilon}\Big)\nabla u_{\epsilon} = f, \quad \forall x \in \Omega \tag{11.158}$$

$$u_{\epsilon}(x) = 0, \quad \forall\, x \in \partial\Omega \tag{11.159}$$

The BVP (1) and (2) is equivalent to the following variational problem:

Find $u_{\epsilon} \in H_0^1(\Omega)$ *such that*

$$\inf_{u \in H_0^1} J(u) = \inf_{u \in H_0^1} \Big(\frac{1}{2}\int_{\Omega} \nabla u \cdot \mathbf{A}\Big(\frac{x}{\epsilon}\Big) \cdot \nabla u dx - \int_{\Omega} f u dx\Big)$$

Therefore. the Γ*-convergence of* $J_{\epsilon}(u)$ *(with respect to the strong topology of* $L^2(\Omega)$*) is equivalent to the homogenization of the PDE (1)-(2).*

11.7 Toshia Mura

Fig. 11.12 Toshio Mura

Dr. T. Mura is a Walter P. Murphy Professor Emeritus in the Department of Mechanical Engineering, Northwestern University, Evanston, Illinois, USA. He received his MS and PhD from the University of Tokyo, and he was elected as a member of the National Academy of Engineering (NAE) in 1986 for initiating and promoting micromechanics to bridge the gap between metal physics and engineering mechanics.

11.8 Exercises

Problem 11.1. *Show that for isotropic materials the fourth-order tensor,*

$$g_{ijk\ell}(\boldsymbol{\xi}) = \frac{1}{2\xi^2}\Big[\xi_j(\delta_{i\ell}\xi_k + \delta_{ik}\xi_\ell) + \xi_i(\delta_{j\ell}\xi_k + \delta_{jk}\xi_\ell)\Big] - \frac{1}{1-\nu}\frac{\xi_i\xi_j\xi_k\xi_\ell}{\xi^4} + \frac{\nu}{1-\nu}\frac{\xi_i\xi_j}{\xi^2}\delta_{k\ell}\Big] \,. \tag{11.160}$$

Problem 11.2. *Consider cuboidal region of inelastic strain (eigenstrain) due to solute segregation forming cuboidal precipitates. The precipitate subdomain (or inclusion) has the dimension* $2a \times 2a \times 2a$, *and the unit cell (U) has the dimension* $2L \times 2L \times 2L$. *The eigenstrain is assumed to have a constant value* ϵ *within each inclusion, and be zero outside the inclusion,*

$$\varepsilon^*_{ij} = \begin{cases} \delta_{ij}\varepsilon, & \forall \mathbf{x} \in \Omega; \\ 0\,, & \forall \mathbf{x} \in U/\Omega, \end{cases} \tag{11.161}$$

where

$$U = \Big\{\mathbf{x}\ \Big|\ -L \le x_i \le L, \quad i = 1,2,3\Big\} \tag{11.162}$$

$$\Omega = \Big\{\mathbf{x}\ \Big|\ -a \le x_i \le a, \quad i = 1,2,3\Big\}\ , \text{and} \qquad a < L \tag{11.163}$$

Find :
(a) the disturbed displacement field $u_1(\mathbf{x})$ *(Hint: [Mura (1987)] pages: 20-21).*
(b) $G(\boldsymbol{\xi}) = g_0(\boldsymbol{\xi})g_0(-\boldsymbol{\xi})$.

Problem 11.3. *Consider the following boundary-value problem in a medium with periodic structure,*

$$-\frac{\partial^2 u_\epsilon}{\partial x_i \partial x_i} = f, \quad \forall x \in \Omega \tag{11.164}$$

$$u_\epsilon = 0, \quad \forall x \in \partial\Omega \tag{11.165}$$

$$\frac{\partial u_\epsilon}{\partial n} = 0, \quad \forall x \in \Gamma \tag{11.166}$$

where Γ *is the interface between the matrix and inhomogeneous phase.*
Show that the homogenized differential equation is

$$-q_{ik}\frac{\partial^2 u_0}{\partial x_i \partial x_k} = f, \quad \forall x \in \Omega$$

with effective coefficients q_{ik} *defined as*

$$q_{ik} = \frac{1}{|Y|}\int_Y \Big(\delta_{ik} + 2\frac{\partial U_k}{\partial y_i}\Big) dy$$

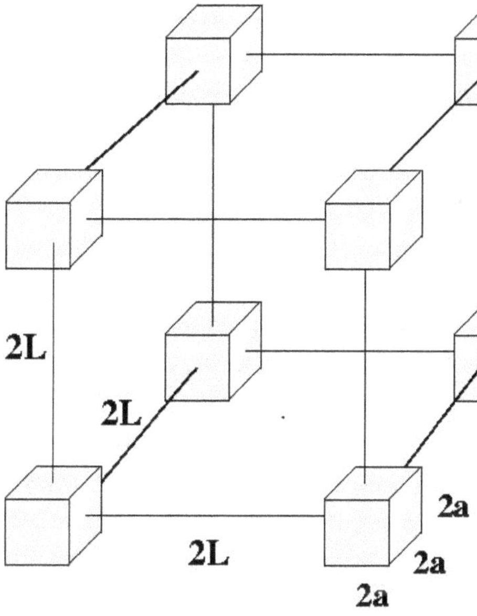

Fig. 11.13 Distribution of periodic precipitates.

and the associated canonical cell problem is

$$\frac{\partial^2 U_k}{\partial y_i \partial y_i} = 0, \quad \forall y \in Y \tag{11.167}$$

$$\frac{\partial U_k}{\partial y_i} n_i = n_k, \quad \forall y \in S \tag{11.168}$$

Appendix A

Appendix of Chapter 6

A.1 Integration formulas

In this appendix the solution of the fourteen integrals listed in Eqs. (A.15)-(A.21) and (A.22)-(A.28) is given. The procedure is similar to the two dimensional case reported in Li et al. [Li *et al.* (2005c)]. Considering $\mathbf{x} + r\bar{\mathbf{r}} = \mathbf{y}$, where $\mathbf{y} \in \partial\Omega$, we have

$$\bar{r}_i = \frac{A}{r}(n_i - t\,\bar{x}_i), \quad \text{or} \quad n_i = \frac{r}{A}\bar{r}_i + t\,\bar{x}_i\ . \tag{A.1}$$

Recall that $t = |\mathbf{x}|/A$. The relations defined in Eq. (A.1) are illustrated in both Fig. 6.1 and Fig. A.1(a). The surface integration over the RVE is performed w.r.t.

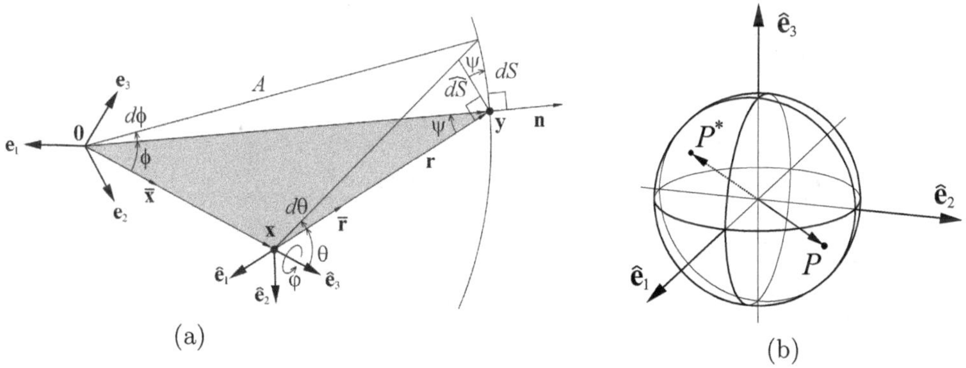

Fig. A.1 (a) Relation between dS, $d\phi$ and $d\theta$; (b) unit sphere.

the surface of a unit sphere, S_2, centered at point $\mathbf{x}$. According to Fig. A.1, we define a new basis $\hat{\mathbf{e}}_i$ at $\mathbf{x}$ such that $\hat{\mathbf{e}}_3 = \bar{\mathbf{x}}$. Vector $\bar{\mathbf{r}}$ is then described by the spherical coordinates φ and θ, i.e. $\bar{\mathbf{r}} = [\cos\varphi\sin\theta \;\; \sin\varphi\sin\theta \;\; \cos\theta]^T$.

Denote dS as the surface element of $\partial\Omega$ (the outer surface of the RVE). The projection of dS to the perpendicular direction of $\bar{\mathbf{r}}$ is denoted by $\widehat{dS}$, and is given by $\widehat{dS} = r^2 \sin\theta \, d\theta \, d\varphi$. It is related to dS by

$$dS = \frac{\widehat{dS}}{\cos\psi} = \frac{r^2}{\cos\psi}\sin\theta \, d\theta \, d\varphi = \frac{r^2}{\cos\psi} dS_2\ , \tag{A.2}$$

where $dS_2 = \sin\theta \, d\theta \, d\varphi$ is the surface element on the unit sphere S_2. Considering the shaded triangle (**Oxy**) in Fig. A.1, we can find that

$$\frac{A}{r} = \frac{1}{\sqrt{1 - 2t\cos\phi + t^2}} , \tag{A.3}$$

and

$$\cos\psi = \sqrt{1 - t^2 \sin^2\theta} . \tag{A.4}$$

Furthermore from $y_i y_i = A^2$, one can derive the relation

$$r = A\Big(- t\cos\theta + \sqrt{1 - t^2\sin^2\theta}\Big) . \tag{A.5}$$

Figure A.1(b) shows that for every point P on the surface of the unit sphere there exists a point P^* such that $\bar{\mathbf{r}}(P) = -\bar{\mathbf{r}}(P^*)$. Thus any function, $\mathcal{L}^o(\bar{\mathbf{r}}) = \bar{r}_i, \bar{r}_i\bar{r}_j, \bar{r}_i\bar{r}_j\bar{r}_m, ...$, which is odd in $\bar{\mathbf{r}}$, satisfies $\mathcal{L}^o(\bar{\mathbf{r}}(P)) = -\mathcal{L}^o(\bar{\mathbf{r}}(P^*))$, and therefore the integration of an odd function of $\bar{\mathbf{r}}$, i.e. $\mathcal{L}^o(\bar{\mathbf{r}})$, over the surface of the sphere will be zero. In particular, by applying eqs. (A.2) and (A.4), we find that

$$\int_{\partial\Omega} \frac{\mathcal{L}^o(\bar{\mathbf{r}})}{r^2} dS = \int_{S_2} A \frac{\mathcal{L}^o(\bar{\mathbf{r}})}{\sqrt{1 - t^2\sin^2\theta}} \, dS_2 = 0 . \tag{A.6}$$

Note that $\sin^2\theta$ is an even function in $\bar{\mathbf{r}}$, i.e. $\sin^2\theta(\bar{\mathbf{r}}) = \sin^2\theta(-\bar{\mathbf{r}})$. Further, we denote an even function of $\bar{\mathbf{r}}$ as $\mathcal{L}^e(\bar{\mathbf{r}})$, if $\mathcal{L}^e(\bar{\mathbf{r}}(P)) = \mathcal{L}^e(\bar{\mathbf{r}}(P^*))$. Then, by virtue of eqs. (A.2), (A.4) and (A.5), it follows that

$$\int_{\partial\Omega} \frac{\mathcal{L}^e(\bar{\mathbf{r}})}{r} dS = \int_{S_2} A\left(1 - \frac{t\cos\theta}{\sqrt{1 - t^2\sin^2\theta}}\right)\mathcal{L}^e(\bar{\mathbf{r}}) dS_2 = A \int_{S_2} \mathcal{L}^e(\bar{\mathbf{r}}) \, dS_2 , \tag{A.7}$$

because $\cos\theta$ is an odd function in $\bar{\mathbf{r}}$. Using eqs. (A.7) and (A.6) we obtain the following seven elemental integrals:

$$(1) \quad \int_{\partial\Omega} \frac{1}{r} dS_y = 4\pi A , \tag{A.8}$$

$$(2) \quad \int_{\partial\Omega} \frac{\bar{r}_i}{r^2} dS_y = 0 , \tag{A.9}$$

$$(3) \quad \int_{\partial\Omega} \frac{\bar{r}_i\bar{r}_j}{r} dS_y = A \int_{S_2} \bar{r}_i\bar{r}_j \, dS_u = \frac{4\pi}{3} A\delta_{ij} , \tag{A.10}$$

$$(4) \quad \int_{\partial\Omega} \frac{\bar{r}_i\bar{r}_j\bar{r}_m}{r^2} dS_y = 0 , \tag{A.11}$$

$$(5) \quad \int_{\partial\Omega} \frac{\bar{r}_i\bar{r}_j\bar{r}_m\bar{r}_n}{r} dS_y = A \int_{S_2} \bar{r}_i\bar{r}_j\bar{r}_m\bar{r}_n \, dS_u$$

$$= \frac{4\pi}{15} A(\delta_{ij}\delta_{mn} + \delta_{im}\delta_{jn} + \delta_{in}\delta_{jm}) , \tag{A.12}$$

$$(6) \quad \int_{\partial\Omega} \frac{\bar{r}_i\bar{r}_j\bar{r}_m\bar{r}_n\bar{r}_r}{r^2} dS_y = 0 , \tag{A.13}$$

$$(7) \quad \int_{\partial\Omega} \frac{\bar{r}_i\bar{r}_j\bar{r}_m\bar{r}_n\bar{r}_r\bar{r}_s}{r} dS_y = A \int_{S_2} \bar{r}_i\bar{r}_j\bar{r}_m\bar{r}_n\bar{r}_r\bar{r}_s \, dS_u$$

$$= \frac{4\pi}{105} A\Big(\delta_{ij}\delta_{mn}\delta_{rs} + \delta_{im}\delta_{jn}\delta_{rs} + \delta_{in}\delta_{jm}\delta_{rs} + \delta_{ir}\delta_{mn}\delta_{js} + \delta_{is}\delta_{mn}\delta_{jr}$$

$$\delta_{ij}\delta_{mr}\delta_{ns} + \delta_{im}\delta_{jr}\delta_{ns} + \delta_{in}\delta_{jr}\delta_{ms} + \delta_{ir}\delta_{mj}\delta_{ns} + \delta_{is}\delta_{mj}\delta_{nr}$$

$$\delta_{ij}\delta_{ms}\delta_{nr} + \delta_{im}\delta_{js}\delta_{nr} + \delta_{in}\delta_{js}\delta_{mr} + \delta_{ir}\delta_{ms}\delta_{nj} + \delta_{is}\delta_{mr}\delta_{nj}\Big) . \tag{A.14}$$

Using these seven elemental integrals and Eq. (A.1) we obtain all the integrals listed in Eqs. (A.15)-(A.21) and (A.22)-(A.28).

$$\text{(I)} \int_{\partial\Omega} \frac{1}{r^2}\bar{r}_k n_k \, dS_y = 4\pi \;, \tag{A.15}$$

$$\text{(I)} \int_{\partial\Omega} \frac{1}{r^2}\bar{r}_i n_j \, dS_y = \frac{4\pi}{3}\delta_{ij} \;, \tag{A.16}$$

$$\text{(III)} \int_{\partial\Omega} \frac{1}{r^2}\bar{r}_i\bar{r}_j\bar{r}_k n_k \, dS_y = \frac{4\pi}{3}\delta_{ij} \;, \tag{A.17}$$

$$\text{(IV)} \int_{\partial\Omega} \frac{1}{r^2}\bar{r}_k n_k n_m n_n \, dS_y = \frac{4\pi}{15}(5 - 3t^2)\delta_{mn} + \frac{12\pi}{5}t^2\bar{x}_m\bar{x}_n \;, \tag{A.18}$$

$$\text{(V)} \int_{\partial\Omega} \frac{1}{r^2}\bar{r}_i\bar{r}_j(\bar{r}_m n_n + \bar{r}_n n_m) dS_y = \frac{8\pi}{15}(\delta_{ij}\delta_{mn} + \delta_{im}\delta_{jn} + \delta_{jm}\delta_{in}) \;, \tag{A.19}$$

$$\begin{aligned}\text{(VI)} &\int_{\partial\Omega} \frac{1}{r^2}(\bar{r}_i n_j + \bar{r}_j n_i) n_m n_n \, dS_y \\ &= \frac{\pi}{105}\Big[(56 - 24t^2)(\delta_{ij}\delta_{mn} + \delta_{im}\delta_{jn} + \delta_{jm}\delta_{in}) + 120t^2\delta_{ij}\bar{x}_m\bar{x}_n \\ &- 48t^2\bar{x}_i\bar{x}_j\delta_{mn} + 36t^2(\delta_{im}\bar{x}_j\bar{x}_n + \delta_{in}\bar{x}_j\bar{x}_m + \delta_{jm}\bar{x}_i\bar{x}_n + \delta_{jn}\bar{x}_i\bar{x}_m)\Big] \;,\end{aligned} \tag{A.20}$$

$$\begin{aligned}\text{(VII)} &\int_{\partial\Omega} \frac{1}{r^2}\bar{r}_i\bar{r}_j\bar{r}_k n_k n_m n_n \, dS_y \\ &= \frac{\pi}{105}\Big[(28 - 20t^2)(\delta_{ij}\delta_{mn} + \delta_{im}\delta_{jn} + \delta_{jm}\delta_{in}) + 100t^2\delta_{ij}\bar{x}_m\bar{x}_n \\ &+ 16t^2\bar{x}_i\bar{x}_j\delta_{mn} - 12t^2(\delta_{im}\bar{x}_j\bar{x}_n + \delta_{in}\bar{x}_j\bar{x}_m + \delta_{jm}\bar{x}_i\bar{x}_n + \delta_{jn}\bar{x}_i\bar{x}_m)\Big] \;,\end{aligned} \tag{A.21}$$

$$\text{(VIII)} \int_{\partial\Omega} \frac{1}{r^2}\bar{r}_i dS_y = 0 \;, \tag{A.22}$$

$$\text{(IX)} \int_{\partial\Omega} \frac{1}{r^2}n_i n_k \bar{r}_k dS_y = \frac{8\pi}{3}\, t\, \bar{x}_i \;, \tag{A.23}$$

$$\text{(X)} \int_{\partial\Omega} \frac{1}{r^2}n_i n_j \bar{r}_k dS_y = \frac{4\pi}{15}t\,(3\bar{x}_i\delta_{jk} + 3\bar{x}_j\delta_{ik} - 2\bar{x}_k\delta_{ij}) \;, \tag{A.24}$$

$$\text{(XI)} \int_{\partial\Omega} \frac{1}{r^2}n_i\bar{r}_j\bar{r}_k n_p\bar{r}_p dS_y = \frac{4\pi}{15}t\,(4\bar{x}_i\delta_{jk} - \bar{x}_j\delta_{ik} - \bar{x}_k\delta_{ij}) \;, \tag{A.25}$$

$$\begin{aligned}\text{(XII)} &\int_{\partial\Omega} \frac{1}{r^2}n_i n_j n_k n_p \bar{r}_p dS_y \\ &= \frac{\pi}{105}\Big[t(56 - 48t^2)(\bar{x}_i\delta_{jk} + \bar{x}_j\delta_{ik} + \bar{x}_k\delta_{ij}) + 240\, t^3\, \bar{x}_i\bar{x}_j\bar{x}_k\Big] \;,\end{aligned} \tag{A.26}$$

$$\text{(XIII)} \int_{\partial\Omega} \frac{1}{r^2}\bar{r}_i n_p\bar{r}_p n_q\bar{r}_q dS_y = 0 \;, \tag{A.27}$$

$$\begin{aligned}\text{(XIV)} \int_{\partial\Omega} \frac{1}{r^2}n_i n_j\bar{r}_k n_p\bar{r}_p n_q\bar{r}_q dS_y &= \frac{\pi}{105}\Big[t\,(84 - 80t^2)\,(\bar{x}_i\delta_{jk} + \bar{x}_j\delta_{ik}) \\ &- t(56 - 32t^2)\bar{x}_k\delta_{ij} + 64t^3\, \bar{x}_i\bar{x}_j\bar{x}_k\Big] \;,\end{aligned} \tag{A.28}$$

A.2 Table of Eshelby tensor coefficients for three-layer shell model

In this section, a complete list of coefficients for the average Eshelby tensors of a three-sphere RVE is documented. The notation of these coefficents are explained and defined in Section 6.6.3.

$$
\begin{aligned}
s_1^{I1,D} &= \frac{(1+\nu)(1-f_1)}{3(1-\nu)} \\
s_1^{I1,N} &= \frac{(1+\nu)+2(1-2\nu)f_1}{3(1-\nu)} \\
s_2^{I1,D} &= \frac{2(4-5\nu)(1-f_1)}{15(1-\nu)} - 21\gamma_u[f_1](1-f_1^{2/3}) \\
s_2^{I1,N} &= \frac{2(4-5\nu)+(7-5\nu)f_1}{15(1-\nu)} + 21\gamma_t[f_1](1-f_1^{2/3}) \\
s_1^{I2,D} &= -\frac{(1+\nu)f_1}{3(1-\nu)} \\
s_1^{I2,N} &= \frac{2(1-2\nu)f_1}{3(1-\nu)} \\
s_2^{I2,D} &= -\frac{2(4-5\nu)f_1}{15(1-\nu)} - 21\gamma_u[f_1]\left(1-\frac{(f_1+f_2)^{5/3}-f_1^{5/3}}{f_2}\right) \\
s_2^{I2,N} &= \frac{(7-5\nu)f_1}{15(1-\nu)} + 21\gamma_t[f_1]\left(1-\frac{(f_1+f_2)^{5/3}-f_1^{5/3}}{f_2}\right) \\
s_1^{I3,D} &= -\frac{(1+\nu)f_1}{3(1-\nu)} \\
s_1^{I3,N} &= \frac{2(1-2\nu)f_1}{3(1-\nu)} \\
s_2^{I3,D} &= -\frac{2(4-5\nu)f_1}{15(1-\nu)} + 21\gamma_u[f_1]\frac{(f_1+f_2)(1-(f_1+f_2)^{2/3})}{f_3} \\
s_2^{I3,N} &= \frac{(7-5\nu)f_1}{15(1-\nu)} - 21\gamma_t[f_1]\frac{(f_1+f_2)(1-(f_1+f_2)^{2/3})}{f_3} \\
s_1^{II1,D} &= \frac{(1+\nu)f_3}{3(1-\nu)} \\
s_1^{II1,N} &= \frac{(1+\nu)+2(1-2\nu)(f_1+f_2)}{3(1-\nu)} \\
s_2^{II1,D} &= \frac{2(4-5\nu)f_3}{15(1-\nu)} - 21\gamma_u[f_1+f_2](1-f_1^{2/3})
\end{aligned}
$$

$$
\begin{aligned}
s_2^{II1,N} &= \frac{2(4-5\nu)+(7-5\nu)(f_1+f_2)}{15(1-\nu)} + 21\gamma_t[f_1+f_2](1-f_1^{2/3}) \\
s_1^{II2,D} &= \frac{(1+\nu)f_3}{3(1-\nu)} \\
s_1^{II2,N} &= \frac{(1+\nu)+2(1-2\nu)(f_1+f_2)}{3(1-\nu)} \\
s_2^{II2,D} &= \frac{2(4-5\nu)}{15(1-\nu)}f_3 - 21\gamma_u[f_1+f_2]\left(1-\frac{(f_1+f_2)^{5/3}-f_1^{5/3}}{f_2}\right) \\
s_2^{II2,N} &= \frac{2(4-5\nu)+(7-5\nu)(f_1+f_2)}{15(1-\nu)} + 21\gamma_t[f_1+f_2]\left(1-\frac{(f_1+f_2)^{5/3}-f_1^{5/3}}{f_2}\right) \\
s_1^{II3,D} &= -\frac{(1+\nu)(f_1+f_2)}{3(1-\nu)} \\
s_1^{II3,N} &= \frac{2(1-2\nu)(f_1+f_2)}{3(1-\nu)} \\
s_2^{II3,D} &= -\frac{2(4-5\nu)}{15(1-\nu)}(f_1+f_2) + 21\gamma_u[f_1+f_2]\frac{(f_1+f_2)(1-(f_1+f_2)^{2/3})}{f_3} \\
s_2^{II3,N} &= \frac{7-5\nu}{15(1-\nu)}(f_1+f_2) - 21\gamma_t[f_1+f_2]\frac{(f_1+f_2)(1-(f_1+f_2)^{2/3})}{f_3} \\
s_1^{III3,D} &= 0 \\
s_1^{III3,N} &= 1 \\
s_2^{III3,D} &= 0 \\
s_2^{III3,N} &= 1 \\
s_1^{III1,D} &= 0 \\
s_1^{III1,N} &= 1 \\
s_2^{III1,D} &= 0 \\
s_2^{III1,N} &= 1 \\
s_1^{III2,D} &= 0 \\
s_1^{III2,N} &= 1 \\
s_2^{III2,D} &= 0 \\
s_2^{III2,N} &= 1
\end{aligned}
$$

where

$$
\gamma_u[x] := \frac{x(1-x^{2/3})}{10(1-\nu)(7-10\nu)}, \quad \text{and} \quad \gamma_t[x] := \frac{4x(1-x^{2/3})}{10(1-\nu)(7+5\nu)}\,. \tag{A.29}
$$

A.3 Finite Eshelby tensors for circular inclusion

The detailed procedures on how to find the finite Eshelby tensor for a 2D circular inclusion in a circular RVE under prescribed Dirichlet and Neumann boundaries have been reported in Li et al. [Li *et al.* (2005c)] and Wang et al. [Wang *et al.* (2005a)] respectively.

For a circular inclusion embedded in a circular RVE under prescribed traction boundary condition, the closed-form expressions for the Neumann-Eshelby tensors for interior and exterior points are given as follows

$$\begin{aligned}
\mathbb{S}^{I,N}_{ijmn}(\mathbf{x}) = \frac{1}{8(1-\nu)} \Big\{ & \Big[(4\nu-1)(1-\rho_0^2) - 3\rho_0^2(1-\rho_0^2)(1-4\nu\, t^2)\Big]\delta_{ij}\delta_{mn} \\
&+ \Big[(3-4\nu) + \rho_0^2 + 3\rho_0^2(1-\rho_0^2)(1-2\,t^2)\Big](\delta_{im}\delta_{jn} + \delta_{in}\delta_{jm}) \\
&+ \Big[12(1-2\nu)\rho_0^2(1-\rho_0^2)\, t^2\Big]\delta_{ij}\bar{x}_m\bar{x}_n \Big\}, \quad \forall \mathbf{x} \in \Omega_I
\end{aligned} \tag{A.30}$$

$$\begin{aligned}
\mathbb{S}^{E,N}_{ijmn}(\mathbf{x}) = \frac{\rho_0^2}{8(1-\nu)} \Bigg\{ & \Bigg[-2(1+2\nu)\Big(\frac{1}{t^2}+1\Big) + 12\nu t^2 + \rho_0^2\left(\frac{9}{t^4} + 3 - 12\nu t^2\right)\Bigg]\delta_{ij}\delta_{mn} \\
&+ \Bigg[\frac{2}{t^2} + 4 - 6t^2 - 3\rho_0^2\Big(\frac{1}{t^4} + 1 - 2t^2\Big)\Bigg](\delta_{im}\delta_{jn} + \delta_{in}\delta_{jm}) \\
&+ \Bigg[4\Big(\frac{1+2\nu}{t^2} + 3(1-2\nu)\, t^2\Big) - 12\rho_0^2\Big(\frac{1}{t^4} + (1-2\nu)t^2\Big)\Bigg]\delta_{ij}\bar{x}_m\bar{x}_n \\
&+ \Bigg[\frac{4}{t^2}\Big(1 - \frac{3\rho_0^2}{t^2}\Big)\Bigg]\delta_{mn}\bar{x}_i\bar{x}_j + \Bigg[\frac{8}{t^2}\Big(\frac{3\rho_0^2}{t^2} - 2\Big)\Bigg]\bar{x}_i\bar{x}_j\bar{x}_m\bar{x}_n \Bigg\}, \\
& \forall \mathbf{x} \in \Omega/\Omega_I
\end{aligned} \tag{A.31}$$

where $\bar{x}_i = x_i/|\mathbf{x}|$ is the unit normal vector, $t = |\mathbf{x}|/H_o$ is the normalized radial position, $\rho_0 = a/H_0$ is the ratio of the inclusion and RVE radius. The resulting interior/exterior Neumann-Eshelby tensor is position dependent and "radially isotropic", in contrast to the well-known infinite space Eshelby tensor which is constant and isotropic.

The average interior Neumann-Eshelby tensor from Eq. (A.30) has the following form,

$$\langle \mathbb{S}^{I,N}_{ijmn} \rangle_{\Omega_e} = \frac{1 + (1-2\nu)f}{2(1-\nu)}\, \mathbb{E}^{(1)}_{ijmn} + \frac{(3-4\nu) + f(4 - f(6-3f))}{4(1-\nu)}\, \mathbb{E}^{(2)}_{ijmn} \tag{A.32}$$

in which $f = \rho_0^2 = a^2/H_0^2$, the tensorial bases $E^{(1)}_{ijmn}$ and $E^{(2)}_{ijmn}$ span an 4th order isotropic tensor space, and they are defined as

$$\mathbb{E}^{(1)}_{ijmn} = \frac{1}{2}\delta_{ij}\delta_{mn} \tag{A.33}$$

$$\mathbb{E}^{(2)}_{ijmn} = \frac{1}{2}\Big(\delta_{im}\delta_{jn} + \delta_{in}\delta_{jm} - \delta_{ij}\delta_{mn}\Big) \tag{A.34}$$

One may compare this result with the interior Eshelby tensor of an infinite space, which is given as

$$\mathbb{S}^{I,\infty}_{ijmn} = s^{\infty}_1 \, \mathbb{E}^{(1)}_{ijmn} + s^{\infty}_2 \, \mathbb{E}^{(2)}_{ijmn} = \frac{1}{2(1-\nu)} \mathbb{E}^{(1)}_{ijmn} + \frac{3-4\nu}{4(1-\nu)} \mathbb{E}^{(2)}_{ijmn} \, . \tag{A.35}$$

Appendix B

Noether's Theorems

B.1 Noether's theorem for a vector field

Consider the functional

$$\Pi(\boldsymbol{\theta}, \boldsymbol{\theta}_{,i}) = \int_{\Omega} W(\boldsymbol{\theta}, \boldsymbol{\theta}_{,i}) d\Omega_{\mathbf{x}} \tag{B.1}$$

Assume that we are given an r-parameter family of transformations on the coordinate variables x_i and θ_i,

$$\bar{x}_i = \bar{x}_i(\mathbf{x}, \boldsymbol{\theta}, \mathbf{s}) \quad \text{and} \quad \bar{\theta}_i = \bar{\theta}_i(\mathbf{x}, \boldsymbol{\theta}, \mathbf{s}) \tag{B.2}$$

where $\mathbf{s} = (s_1, s_2, ..., s_r)$ such that for a general vector field θ_i

$$\bar{x}_i \Big|_{\mathbf{s}=0} = \bar{x}_i(\mathbf{x}, \boldsymbol{\theta}, \mathbf{0}) = x_i \quad \text{and} \quad \bar{\theta}_i \Big|_{\mathbf{s}=0} = \bar{\theta}_i(\mathbf{x}, \boldsymbol{\theta}, \mathbf{0}) = \theta_i \ . \tag{B.3}$$

Definition B.4.

(1) The fundamental integral (B.1) is absolutely invariant under the r-parameter family of transformations (B.2) if and only if

$$\Pi(\bar{\boldsymbol{\theta}}, \bar{\boldsymbol{\theta}}_{,i}) = \Pi(\boldsymbol{\theta}, \boldsymbol{\theta}_{,i}). \tag{B.4}$$

(2) The fundamental integral (B.1) is infinitesimal invariant under the r-parameter family of transformations (B.2) if and only if

$$\int_{\bar{\Omega}} W(\bar{\theta}_{i,k}) d\Omega_{\bar{\mathbf{x}}} - \int_{\Omega} W(\theta_{i,k}) d\Omega_{\mathbf{x}} = o(\mathbf{s}) \ \forall \ \mathbf{s} \in \mathbb{R}^r. \tag{B.5}$$

(3) The strong form (local form) of the global condition (B.5) is given as

$$W\Big(\bar{x}_i, \bar{\theta}_i, \bar{\theta}_{i,k}\Big) det\Big(\frac{\partial \bar{\mathbf{x}}}{\partial \mathbf{x}}\Big) - W\Big(x_i, \theta_i, \theta_{i,k}\Big) = o(\mathbf{s}) \ . \tag{B.6}$$

Using Taylor's theorem, one can expand the r-parameter family of transformations, (B.2) in terms of a small vector variable $\{s\}_\alpha, \alpha = 1, 2, \cdots, r$,

$$\bar{x}_i = x_i + \varphi_{i\alpha}(\mathbf{x}\boldsymbol{\theta}) s_\alpha + o(\mathbf{s}) \quad \text{and} \quad \bar{\theta}_i = \theta_i + \xi_{i\alpha}(\mathbf{x}, \boldsymbol{\theta}) s_\alpha + o(\mathbf{s}), \tag{B.7}$$

where

$$\varphi_{i\alpha}(\mathbf{x}, \boldsymbol{\theta}) := \frac{\partial \bar{x}_i}{\partial s_\alpha}\Big|_{\mathbf{s}=\mathbf{0}} \quad \text{and} \quad \xi_{i\alpha}(\mathbf{x}, \boldsymbol{\theta}) := \frac{\partial \bar{\theta}_{ij}}{\partial s_\alpha}\Big|_{\mathbf{s}=\mathbf{0}} \tag{B.8}$$

Theorem B.3 ([Logan (1977)]). *The fundamental integral (B.1) is infinitesimally invariant, if the following condition holds*

$$\left\{ \frac{\partial W}{\partial x_i} \varphi_{i\alpha} + \frac{\partial W}{\partial \theta_i} \xi_{i\alpha} + \frac{\partial W}{\partial \theta_{i,k}} \left(\frac{d\xi_{i\alpha}}{dx_k} - \theta_{i,\ell} \frac{d\varphi_{\ell\alpha}}{dx_k} \right) \right\} + W \frac{d\varphi_{i\alpha}}{dx_i} = 0,$$
$$\forall \mathbf{x} \in \Omega, \quad \text{and} \quad \alpha = 1, 2, \cdots, r. \tag{B.9}$$

The above set of r identities should also be satisfied if the fundamental integral (B.1) is absolutely invariant.

Theorem B.4. *Assume that the fundamental integral $\Pi(\boldsymbol{\theta}, \boldsymbol{\theta}_{,i})$ is infinitesimally invariant under the r-parameter family of transformations (B.2). Then the following conservation laws hold*

$$\frac{d}{dx_k} \left[\left(W \delta_{k\ell} - \theta_{i,\ell} \frac{\partial W}{\partial \theta_{i,k}} \right) \varphi_{\ell\alpha} + \frac{\partial W}{\partial \theta_{i,k}} \xi_{i\alpha} \right] = 0, \quad \alpha = 1, \cdots, r. \tag{B.10}$$

This theorem can be found in many monographs such as [Logan (1977)], [Olver (1986)] or [Ibragimov (1994)].

B.2 Noether's theorem for a tensorial field

Consider the functional

$$\Pi(\boldsymbol{\lambda}, \boldsymbol{\lambda}_{,i}) = \int_\Omega W(\boldsymbol{\lambda}, \boldsymbol{\lambda}_{,i}) d\Omega_{\mathbf{x}} \tag{B.11}$$

Assume that we are given an r-parameter family of transformations on the coordinate variable x_i and the Cartesian tensor field λ_{ij},

$$\bar{x}_i = \bar{x}_i(\mathbf{x}, \boldsymbol{\lambda}, \mathbf{s}) \qquad \text{and} \qquad \bar{\lambda}_{ij} = \bar{\lambda}_{ij}(\mathbf{x}, \boldsymbol{\lambda}, \mathbf{s}) \tag{B.12}$$

where $\mathbf{s} = (s_1, ..., s_r)$ such that for a general second order Cartesian tensor λ_{ij}

$$\bar{x}_i \Big|_{\mathbf{s}=0} = \bar{x}_i(\mathbf{x}, \boldsymbol{\lambda}, \mathbf{0}) = x_i \quad \text{and} \quad \bar{\lambda}_{ij} \Big|_{\mathbf{s}=0} = \bar{\lambda}_{ij}(\mathbf{x}, \boldsymbol{\lambda}, \mathbf{0}) = \lambda_{ij}. \tag{B.13}$$

Definition B.5.

(1) The fundamental integral (B.11) is absolutely invariant under the r-parameter family of transformations (B.12) if and only if

$$\Pi(\bar{\boldsymbol{\lambda}}, \bar{\boldsymbol{\lambda}}_{,i}) = \Pi(\boldsymbol{\lambda}, \boldsymbol{\lambda}_{,i}). \tag{B.14}$$

(2) The fundamental integral (B.11) is infinitesimal invariant under the r-parameter family of transformations (B.12) if and only if

$$\int_{\bar{\Omega}} W(\bar{\lambda}_{ij,k}) d\Omega_{\bar{\mathbf{x}}} - \int_\Omega W(\lambda_{ij,k}) d\Omega_{\mathbf{x}} = o(\mathbf{s}) \ \forall \ \mathbf{s} \in \mathbb{R}^r. \tag{B.15}$$

(3) The strong form (local form) of the global condition (B.15) is given as

$$W\Big(\bar{x}_i, \bar{\lambda}_{ij}, \bar{\lambda}_{ij,k}\Big) det\Big(\frac{\partial \bar{\mathbf{x}}}{\partial \mathbf{x}}\Big) - W\Big(x_i, \lambda_{ij}, \lambda_{ij,k}\Big) = o(\mathbf{s}) \ . \tag{B.16}$$

Using Taylor's theorem, one can expand the r-parameter family of transformations, (B.12) in terms of a small vector variable $\{s\}_\alpha, \alpha = 1, 2, \cdots, r$,

$$\bar{x}_i = x_i + \varphi_{i\alpha}(\mathbf{x}, \boldsymbol{\lambda}) s_\alpha + o(\mathbf{s}) \quad \text{and} \quad \bar{\lambda}_{ij} = \lambda_{ij} + \xi_{ij\alpha}(\mathbf{x}, \boldsymbol{\lambda}) s_\alpha + o(\mathbf{s}), \tag{B.17}$$

$$\text{where} \quad \varphi_{i\alpha}(\mathbf{x}, \boldsymbol{\lambda}) := \frac{\partial \bar{x}_i}{\partial s_\alpha}\Big|_{\mathbf{s}=\mathbf{0}} \quad \text{and} \quad \xi_{ij\alpha}(\mathbf{x}, \boldsymbol{\lambda}) := \frac{\partial \bar{\lambda}_{ij}}{\partial s_\alpha}\Big|_{\mathbf{s}=\mathbf{0}} \tag{B.18}$$

Theorem B.5. *The fundamental integral (B.11) is infinitesimally invariant, if the following condition holds*

$$\left\{ \frac{\partial W}{\partial x_i} \varphi_{i\alpha} + \frac{\partial W}{\partial \lambda_{ij}} \xi_{ij\alpha} + \frac{\partial W}{\partial \lambda_{ij,k}} \left(\frac{d\xi_{ij\alpha}}{dx_k} - \lambda_{ij,\ell} \frac{d\varphi_{\ell\alpha}}{dx_k} \right) \right\} + W \frac{d\varphi_{i\alpha}}{dx_i} = 0,$$
$$\forall \mathbf{x} \in \Omega, \quad \text{and} \quad \alpha = 1, 2, \cdots, r. \tag{B.19}$$

The above sets of r identities should also be satisfied if the fundamental integral (B.11) is absolutely invariant.

Theorem B.6. *Assume that the fundamental integral $\Pi(\boldsymbol{\lambda}, \boldsymbol{\lambda}_{,i})$ is infinitesimally invariant under the r-parameter transformations (B.12). Then the following conservation laws hold*

$$\frac{d}{dx_k}\left[\left(W \delta_{k\ell} - \lambda_{ij,\ell} \frac{\partial W}{\partial \lambda_{ij,k}} \right) \varphi_{\ell\alpha} + \frac{\partial W}{\partial \lambda_{ij,k}} \xi_{ij\alpha} \right] = 0, \quad \alpha = 1, \cdots, r. \tag{B.20}$$

This theorem is proved in [Li *et al.* (2005a)] for a symmetric second order tensorial field. We verified that it holds for a general second order tensorial field as well.

According to Noether's theorem for a vector field under an r-parameter family of transformations (B.10), where we now choose $\Pi(\boldsymbol{\theta}, \boldsymbol{\theta}_{,i}) = \Pi^{(1)}(\boldsymbol{\theta}_{,i})$, the variational-symmetric conservation laws of the generalized Nye theory can be expressed in the general form,

$$\frac{dA_{k\alpha}}{dx_k} = 0, \quad \alpha = 1, 2, \cdots, r \tag{B.21}$$

where the conserved quantities are

$$A_{k\alpha} = \left(W^{(1)} \delta_{k\ell} - \theta_{i,\ell} \frac{\partial W^{(1)}}{\partial \theta_{i,k}} \right) \varphi_{\ell\alpha} + \frac{\partial W^{(1)}}{\partial \theta_{i,k}} \xi_{i\alpha}, \quad k = 1, 2, 3 \tag{B.22}$$

We outline the details of the proof as follows:

A1. Coordinate translation:
Let

$$\bar{x}_i = x_i + s_i \quad \text{and} \quad \bar{\theta}_i = \theta_i\ . \tag{B.23}$$

Then the corresponding infinitesimal generators for $\alpha = 1, 2, 3$ are

$$\varphi_{i\alpha} = \frac{\partial \bar{x}_i}{\partial s_\alpha}\Big|_{\mathbf{s}=0} = \delta_{i\alpha} \quad \text{and} \quad \xi_{i\alpha} = \frac{\partial \bar{\theta}_i}{\partial s_\alpha}\Big|_{\mathbf{s}=0} = 0. \tag{B.24}$$

One may verify that the r-invariant conditions (B.9) are satisfied under coordinate translation (B.23). We then obtain the conservation laws **A1** due to coordinate translation $S^{(1)}_{k\alpha}$ and the corresponding path-independent integrals $L^{(1)}_{\alpha}$ as

$$S^{(1)}_{k\alpha} = W^{(1)}\delta_{k\alpha} - \theta_{\ell,\alpha}\gamma_{k\ell} \quad \rightarrow \quad L^{(1)}_{\alpha} = \oint_S S^{(1)}_{k\alpha} n_k dS. \tag{B.25}$$

According to Noether's theorem for a tensorial field (B.20), where we now choose $\boldsymbol{\lambda} = \boldsymbol{\beta}$ and $\Pi(\boldsymbol{\lambda}, \boldsymbol{\lambda}_{,i}) = \Pi^{(2)}(\boldsymbol{\beta}_{,i})$, the variational-symmetric conservation laws of the Kröner-deWit theory can be expressed in the general form,

$$\frac{dB_{k\alpha}}{dx_k} = 0, \quad \alpha = 1, 2, \cdots, r \tag{B.26}$$

where the conserved quantities are

$$B_{k\alpha} = \left(W^{(2)}\delta_{k\ell} - \beta_{ij,\ell}\frac{\partial W^{(2a)}}{\partial \beta_{ij,k}}\right)\varphi_{\ell\alpha} + \frac{\partial W^{(2)}}{\partial \beta_{ij,k}}\xi_{ij\alpha}, \quad k = 1, 2, 3. \tag{B.27}$$

We outline the details of the proof as follows:

B1. Coordinate translation:
Let

$$\bar{x}_i = x_i + s_i \quad \text{and} \quad \bar{\beta}_{ij} = \beta_{ij}. \tag{B.28}$$

Then the corresponding infinitesimal generators for $\alpha = 1, 2, 3$ are

$$\varphi_{i\alpha} = \frac{\partial \bar{x}_i}{\partial s_\alpha}\bigg|_{\mathbf{s}=0} = \delta_{i\alpha} \quad \text{and} \quad \xi_{ij\alpha} = \frac{\partial \bar{\beta}_{ij}}{\partial s_\alpha}\bigg|_{\mathbf{s}=0} = 0. \tag{B.29}$$

One may verify that the r-invariant conditions (B.19) are satisfied under coordinate translation (B.28). We then obtain the conservation laws **B1** due to coordinate translation $S^{(2)}_{k\alpha}$ and the corresponding path-independent integrals $L^{(2)}_{\alpha}$ as

$$S^{(2)}_{k\alpha} = W^{(2)}\delta_{k\alpha} + e_{mki}K_{mj}\beta_{ij,\alpha} \quad \rightarrow \quad L^{(2)}_{\alpha} = \oint_S S^{(2)}_{k\alpha} n_k dS. \tag{B.30}$$

According to the result of Noether's theorem for a tensorial field (B.20), where we now choose $\boldsymbol{\lambda} = \boldsymbol{\epsilon}$ and $\Pi(\boldsymbol{\lambda}, \boldsymbol{\lambda}_{,i}) = \Pi^{(3)}(\boldsymbol{\epsilon}_{,i})$, the variational conservation laws based on the Saint-Venant principle can be expressed in the general form,

$$\frac{dC_{k\alpha}}{dx_k} = 0, \quad \alpha = 1, 2, \cdots, r \tag{B.31}$$

where the conserved quantities are

$$C_{k\alpha} = \left(W^{(3)}\delta_{k\ell} - \epsilon_{ij,\ell}\frac{\partial W^{(3)}}{\partial \epsilon_{ij,k}}\right)\varphi_{\ell\alpha} + \frac{\partial W^{(3)}}{\partial \epsilon_{ij,k}}\xi_{ij\alpha}, \quad k = 1, 2, 3. \tag{B.32}$$

We outline the details of the proof as follows:

C1. Coordinate translation:
Let

$$\bar{x}_i = x_i + s_i \quad \text{and} \quad \bar{\epsilon}_{ij} = \epsilon_{ij}. \tag{B.33}$$

Then the corresponding infinitesimal generators for $\alpha = 1, 2, 3$ are

$$\varphi_{i\alpha} = \frac{\partial \bar{x}_i}{\partial s_\alpha}\bigg|_{s=0} = \delta_{i\alpha} \quad \text{and} \quad \xi_{ij\alpha} = \frac{\partial \bar{\epsilon}_{ij}}{\partial s_\alpha}\bigg|_{s=0} = 0. \tag{B.34}$$

One may verify that the r-invariant conditions (B.19) are satisfied under coordinate translation (B.33). We then obtain the conservation laws due to coordinate translation $S^{(3)}_{k\alpha}$ and the corresponding path-independent integrals $L^{(3)}_\alpha$ as

$$S^{(3)}_{k\alpha} = W^{(3)}\delta_{k\alpha} - e_{mki}\zeta_{jm}\epsilon_{ij,\alpha} \quad \rightarrow \quad L^{(3)}_\alpha = \oint_S S^{(3)}_{k\alpha} n_k dS. \tag{B.35}$$

Bibliography

Adams, R. A. (1975). *Sobolev Spaces* (Academic Press, New York · San Francisco · London).

Allaire, G. (2002). *Shape Optimization by the Homogenization Method* (Springer-Verlag, New York).

Allaire, G. and Brizzi, R. (2005). A multiscale finite element method for numerical homogenization, *Multiscale Modeling & Simulation* **4**, pp. 790–812.

Amodeo, R. J. and Ghoniem, N. M. (1990a). Dislocation dynamics. i. a proposed methodology for deformation micromechanics, *Physical Review B* **41**, pp. 6958–6966.

Amodeo, R. J. and Ghoniem, N. M. (1990b). Dislocation dynamics. ii. applications to the formation of persistent slip bands, planar arrays, and dislocations cells, *Physical Review B* **41**, pp. 6968–6976.

Anthony, K. H. (1970). Die theorie der disklinationen, *Archive for Rational Mechanics and Analysis* **39**, pp. 43–88.

Armero, F. (2002). On the locking of standard finite elements, Ce232 classnotes, University of California at Berkeley, Berkeley CA.

Aziz, M. J. (1997). Thermodynamics of diffusion under pressure and stress: Relation to point defect mechanisms, *Applied Physics Letters* **70**, p. 2810.

Bakhvalov, N. S. and Panasenko, G. P. (1984). *Homogenization: Averaging Processes in Periodic Media* (Kluwer, Dordrecht).

Barenblatt, G. I. (1959). The formation of equilibrium cracks during brittle fracture, *Journal of Applied Mathematics & Mechanics, [English translation PMM]* **23 (434)**, p. 622.

Barenblatt, G. I. (1962). Mathematical theory of equilibrium cracks in brittle fracture, in *Advances in Applied Mechanics*, Vol. 7 (Academic Press, New York), pp. 55–129.

Barnett, D. M. (1972). The precise evaluation of derivatives of the anisotropic elastic Greeen's functions, *Phys. Stat. Sol. (b)* **49**, pp. 741–748.

Bensoussan, A., Boccardo, L. and Murat, F. (1992). H convergence for quasi-linear elliptic equations with quadratic growth, *Applied Mathematics and Optimization* **26**, pp. 253–272.

Bensoussan, A., Lions, J. L. and Papanicolaou, G. (1978). *Asymptotic Analysis for Periodic Structures* (North-Holland, Amsterdam).

Bishop, J. and Hill, R. (1951). A theory of the plastic distortion of a polycrystalline aggregate under combined stresses, *Philosophical Magazine* **42**, pp. 414–427.

Born, M. (1915). *Dynamik der Krystallgitter* (Teubner, Leipzid/Berlin).

Born, M. and Huang, K. (1954). *Dynamical Theory of Crystal Lattices* (Clarendon Press, Oxford).

Braun, O. M. and Kivshar, Y. S. (2004). *The Frenkel-Kontorova Model* (Springer, Berlin).

Brenner, S. C. and Scott, L. R. (1994). *The Mathematical Theory of Finite Element Methods* (Springer-Verlag, New York).

Bromwich, T. (1949). *An Introduction to the Theory of Inifinite Series*, 2nd edn. (MaCmillan, London).

Budiansky, B. (1965). On the elastic moduli of some heterogeneous materials, *Journal of the Mechanics and Physics of Solids* **13**, pp. 223–227.

Budiansky, B., Hutchinson, J. W. and Slutsky, S. (1982). Void growth and collapse in viscous solids, in H. G. Hopkins and M. J. Sewell (eds.), *Mechanics of Solids, The Rodney Hill 60th Anniversary Volume* (Pergamon Press), pp. 13–45.

Budiansky, B. and O'Connell, R. J. (1976). Elastic moduli of a cracked solid, *International Journal of Solids and Structures* **12**, pp. 81–97.

Budiansky, B. and Rice, J. R. (1973). Conservation laws and energy-release rates, *ASME Journal of Applied Mechanics* **40**, pp. 201–203.

Burgers, J. M. (1939). Some consideration on the field of stress connected with dislocation in a regular crystal lattice, *Proc. Kon. Ned. Akad. Wetenschjap.* **42**, pp. 293–324.

Buryachenko, V. A. (2007). *Micromechanics of Heterogeneous Materials* (Springer, New York).

Calvert, P. (1999). Nanotube composites: A recipe for strength, *Nature* **399**, pp. 210–211.

Chen, W. R. and Keer, L. M. (1991). Fatigue crack growth in mixed mode loading, *ASME Journal of Engineering Materials and Technology* **113**, pp. 222–227.

Chen, W. R. and Keer, L. M. (1993). Mixed-mode fatigue crack propagation of penny-shaped cracks, *ASME Journal of Engineering Materials and Technology* **115**, pp. 365–372.

Cherkaev, A. (2000). *Variational Methods for Structural Optimization* (Springer, New York).

Cheung, K. S. and Yip, S. (1991). Atomic-level stress in an inhomogeneous system, *Journal of Applied Physics* **70**, pp. 5688–5690.

Christensen, R. M. (1979). *Mechanics of Composite Materials* (Wiley-Interscience, New York, N.Y.).

Chu, C. C. and Needleman, A. (1980). Void nucleation effects in biaxially stretched sheets, *ASME Journal of Engineering Materials and Technology* **102**, pp. 249–256.

Clausius, R. (1870). On a mechanical theory applicable to heat, *Philosophical Magazine* **40**, pp. 122–127.

Cormier, J., Rickman, J. M. and Delph, T. (2001). Stress calculation in atomistic simulations of perfect and imperfact solids, *Journal of Applied Physics* **89**, pp. 99–104.

Cristescu, N. D., Craciun, E.-M. and E., S. (2004). *Mechanics of Elastic Composites* (Chapman & Hall, CRC).

Dahlquist, G. and Björck, A. (1974). *Numerical Methods* (Prentice-Hall, Englewood Cliffs, New Jersey).

Daw, M. and Baskes, M. I. (1984). Embedded-atom method: Derivation and application to impurities, surfaces, and other defects in metals, *Physical Review B* **29**, pp. 6443–6453.

Daw, M., Foiles, S. M. and Baskes, M. (1993). The embedded-atom-method: a review of theory and applications, *Material Science Report* **9**, pp. 251–310.

Daw, M. S., Windl, W., Carlson, N. N. and Laudon, M. (2001). Effect of stress on dopant and defect diffusion in S_i: A general treatment, *Physical Review B* **64**, p. 045205.

De Giorgi, E. (1975). Sulla convergenza di alcune successioni di integrali del tipo dell'areo, *Rendi Conti di Mat.* **8**, pp. 277–294.

De Giorgi, E. (1983). G-operators and γ-convergence, in *Proceedings of the interna-*

tional congress of mathematicians, PWN Polish Scientific Publishers (North Holland, Warsazwa), pp. 1175–1191.

de Wit, C. T. (1960). The continuum theory of stationary dislocations, *Solid State Physics* **10**, pp. 249–292.

Devincre, B. and Condat, M. (1992). Model validation of a 3d simulation of dislocation dynamics: discretization and line tension effects, *Acta Metall. Mater.* **40**, pp. 2629–2637.

Devincre, B. and Kubin, L. (1997a). Mesoscopic simulations of dislocations and plasticity, *Materials Science and Engineering* **A234-236**, pp. 8–14.

Devincre, B. and Kubin, L. (1997b). The modelling of dislocation dynamics: Elastic behavior versus core properties, *Philosophical Transactions: Mathematical, Physical and Engineering Science* **355**, pp. 2003–2012.

deWit, R. (1970). Linear theory of static disclinations, in J. Simmons, R. deWit and R. Bullough (eds.), *Fundamental Aspects of Dislocation Theory*, Vol. 1 (National Bureau of Standards Special Publication, Washington), pp. 651–673.

Dormieux, L., Kondo, D. and Ulm, F.-J. (2006). *Microporomechanics* (John Wiley & Sons Ltd., West Sussex, England).

Drugan, W. and Willis, J. (1996). A micromechanics-based nonlocal constitutive equations and estimates of representative volume element size for elastic composites, *Journal of Mechanics and Physics of Solids* **44**, pp. 497–524.

Dugdale, D. S. (1960). Yielding of steel sheets containing slits, *Journal of Mechanics and Physics of Solids* **8**, pp. 100–104.

Duva, J. D. and Hutchinson, J. W. (1984). Constitutive potentials for dilutely voided nonlinear materials, *Mechanics of Materials* **3**, pp. 41–54.

Ekeland, I. and Temam, R. (1976). *Convex analysis and variational problems* (North-Holland Pub. Co., New York, Amsterdam).

Eshelby, J. (1949). Edge dislocations in anisotropic materials, *Philosophical Magzine* **40**, pp. 903–912.

Eshelby, J. D. (1951). The force on an elastic singularity, *Phil. Trans. Roy. Soc.* **87**, pp. 12–111.

Eshelby, J. D. (1956). The continuum theory of lattice defects, in F. Seitz and D. Turnbull (eds.), *Solid State Physics*, Vol. 3 (Academic Press, New York), pp. 79–144.

Eshelby, J. D. (1957). The determination of the elastic field of an ellipsoidal inclusion and related problems, *Proceedings of Royal Society of London, A* **241**, pp. 376–396.

Eshelby, J. D. (1959). The elastic field outside an ellipsoidal inclusion, *Proceedings of Royal Society of London* **252**, pp. 561–569.

Eshelby, J. D. (1961). Elastic inclusions and inhomogeneities, in N. I. Snedden and R. Hill (eds.), *Progress in Solid Mechanics*, Vol. 2 (North-Holland), pp. 89–104.

Eshelby, J. D. (1975). The elastic energy-momentum tensor, *Journal of Elasticity* **5**, pp. 321–335.

Fisher, F. T., Bradshaw, R. D. and Brinson, L. (2002). Effects of nanotube waviness on the modulus of nanotube-reinforced polymers, *Applied Physics Letters* **80**, pp. 4647–4649.

Fleck, N. (1991). Brittle fracture due to an array of microcracks, *Proceedings of Royal Society of London, A* **A432**, pp. 55–76.

Foiles, S. M., Baskes, M. I. and Daw, M. S. (1986). Embedded-atom-method functions for the fcc metals c_u, a_g, a_u, n_i, p_d, p_t, and their alloys, *Physical Review B* **33**, pp. 7983–7991.

Francfort, G. and Murat, F. (1986). Homogenization and optimal bounds in linear elasticity, *Arch. Rat. Mech. Anal.* **94**, pp. 307–334.

Frank, F. C. and van der Merwe, J. H. (1949a). One-dimensional dislocations. i. static theory, *Proceedings of the Royal Society of London* **198**, pp. 205–216.

Frank, F. C. and van der Merwe, J. H. (1949b). One-dimensional dislocations. ii. misfitting monolayers and oriented overgrowth, *Proceedings of the Royal Society of London* **198**, pp. 216–225.

Frenkel, Y. I. and Kontorova, T. (1938). The model of dislocation in solid body, *Zh. Eksp. Teor. Fiz.* **13**, p. 1.

Freund, L. B. (1987). The stability of a dislocation threading dislocation in a strained layer on a substrate, *Journal of Applied Mechanics* **54**, pp. 553–557.

Freund, L. B. (1990). The driving force for glide of a threading dislocation in a strained epitaxial layer on a substrate, *Journal of the Mechanics and Physics of Solids* **38**, 5, pp. 657–679.

Freund, L. B. and Nix, W. D. (1996). Critical thickness condition for a strained compliant substrate/epitaxial film system, *Applied Physics Letters* **69**, 2, pp. 173–175.

Freund, L. B. and Suresh, S. (2003). *Thin film materials : stress, defect formation, and surface evolution* (Cambridge University Press, New York).

Gao, D. Y. (2000). *Duality principles in nonconvex systems* (Kluwer Academic Publishers, Dordrecht·Boston ·London).

Garikipati, K., Falk, M., Bouville, M., Puchala, B. and Narayanan, H. (2006). The continuum elastic and atomistic viewpoints on the formation volume and strain energy of a point defect, *Journal of the Mechanics and Physics of Solids* **54**, pp. 1929–1951.

Gavazza, S. D. and Barnett, D. M. (1976). The self-force on a planar dislocation loop in anisotropic linear-elastic medium, *Journal of the Mechanics and Physics of Solids* **24**, pp. 171–185.

Gdoutos, E. E. (1990). *Fracture Mechanics Criteria and Applications* (Kluwer Academic Publisher, Dordrecht/Boston/London).

Ghoniem, N. M. (1999). Curved parametric segments for the stress field of 3d dislocation loops, *ASME Journal of Engineering Materials and Technology* **121**, pp. 136–142.

Ghoniem, N. M., Busso, E., Kioussis, N. and Huang, H. (2003). Multiscale modeling of nanomechanics and micromechanics: an overview, *Philosophical Magazine* **83**, pp. 3475–3528.

Ghoniem, N. M. and Sun, L. Z. (2000). Fast-sum method for the elastic field of three-dimensional dislocation ensemble, *Physical Review B* **60**, pp. 128–140.

Ghoniem, N. M., Tong, S. and Sun, L. Z. (2000). Parametric dislocation dynamics: A thermodynamics-based approach to investigations of mesoscopic plastic deformation, *Physical Review B* **61**, pp. 913–927.

Golaganu, M., Leblond, J.-B. and Devaux, J. (1993). Approximate models for ductile metals containing non-spherical voids — case of axisymmetric prolate ellipsoidal cavities, *Journal of Mechanics and Physics of Solids* **41**, pp. 1723–1754.

Golaganu, M., Leblond, J.-B. and Devaux, J. (1994). Approximate models for ductile metals containing non-spherical voids — case of axisymmetric oblate ellipsoidal cavities, *ASME, Journal of Engrg. Mater. Tech.* **116**, pp. 290–297.

Gologanu, M., Leblond, J. B. and Devaux, J. (1995). Recent extensions of gurson's model for porous ductile metals, in P. P. Suquet (ed.), *Continuum Micromechanics* (Springer-Verlage, Berlin), pp. 61–130.

Gosling, T. J. and Willis, J. R. (1993). The energetics of dislocation array stability in strained epitaxial layers, *Journal of Applied Physics* **73**, pp. 8297–8303.

Gradshteyn, I. and Ryzhik, I. M. (1994). *Table of Integrals, Series, and Products*, 5th edn. (Academic Press, Boston).

Griffith, A. (1921). The phenomena of rupture and flow in solids, *Philosophical Transac-*

tions of the Royal Society, London **A221**, pp. 163–197.

Gurson, A. L. (1975). *Plastic flow and fracture behavior of ductile materials incorporating void nucleation, growth and interaction*, Ph.D. thesis, Brown University, Rhode Island.

Gurson, A. L. (1977). Continuum theory of ductile rupture by void nucleation and growth: Part i — yield criteria and flow rules for porous ductile media, *ASME Journal of Engineering Materials and Technology* **99**, pp. 3–15.

Gurtin, M. E. (1995). The nature of configurational forces, *Archive for Rational Mechanics and Analysis* **131**, pp. 67–100.

Gurtin, M. E. (1999). *Configurational Forces as Basic Concepts of Continuum Mechanics* (Springer-Verlag, New York).

Gurtin, M. E. and Murdoch, I. (1975a). A continuum theory of elastic material surface, *Archive of Rational Mechanics and Analysis* **57**, pp. 291–323.

Gurtin, M. E. and Murdoch, I. (1975b). A continuum theory of elastic material surface, *Archive of Rational Mechanics and Analysis* **59**, pp. 389–390.

Gurtin, M. E. and Podio-Guidugli, P. (1996). Configurational forces and the basic laws for crack propagation, *Journal of Mechanics and Physics of Solids* **44**, pp. 905–927.

Gurtin, M. E., Weissmuller, J. and Larche, F. (1998). A general theory pf curved deformable interfaces in solids at equilibrium, *Philosophical Magazine, A* **78**, pp. 1093–1109.

Hahn, H. T. and Tsai, S. W. (1980). *Introduction to Composite Materials* (Technomic Publishing Company, Inc., Lancaster, Pennsylvania).

Hansen, E. R. (1975). *A Table of Series and Products* (Prentice-Hall, Englewood Cliffs, N.J.).

Hardy, R. J. (1982). Formulas for determining local properties in molecular dynamics simulations: Shock waves, *Journal of Chemical Physics* **76**, pp. 622–628.

Hashin, Z. and Shtrikman, S. (1961). Note on a variational approach to the theory of composite elastic materials, *The Franklin Institute Laboratories* **271**, pp. 336–341.

Hashin, Z. and Shtrikman, S. (1962a). On some variational principlrd in anisotropic and nonhomogeneous elasticity, *Journal of Mechanics and Physics of Solids* **10**, pp. 335–342.

Hashin, Z. and Shtrikman, S. (1962b). A variational approach to the theory of the elastic behavior of polycrystals, *Journal of Mechanics and Physics of Solids* **10**, pp. 343–352.

Hill, R. (1963). New derivations of some elastic extremum principles, in *Progress in Applied Mechanics— The Prager Anniversary Volume* (Macmillan, New York), pp. 99–106.

Hill, R. (1965a). Continuum micro-mechanics of elastoplastic polycrystals, *Journal of Mechanics and Physics of Solids* **13**, pp. 89–101.

Hill, R. (1965b). A self-consistent mechanics of composite materials, *Journal of Mechanics and Physics of Solids* **13**, pp. 213–222.

Hill, R. (1965c). Theory of mechanical properties of fibre-strengthened materials–iii. self-consistent model, *Journal of Mechanics and Physics of Solids* **13**, pp. 189–198.

Hirth, J. P. and Lothe, J. (1992). *Theory of dislocations*, 2nd edn. (Krieger Pub. Co., Malabar, FL).

Holzapfel, G. A. (2000). *Nonlinear Solid Mechanics: A continuum approach for engineering* (John Wiley & Sons, LTD, Chichester · New York).

Hou, T. and Wu, X.-H. (1997). A multiscale finite element method for elliptic problems in composite materials and porous media, *Journal of Computational Physics* **134**, pp. 169–189.

Huerta, A. and Fernandez-Mendez, S. (2001). Locking at the incompressible limit for the element-free galerkin method, *Int. J. Numer. Meth. Engrg.* **51**, pp. 1361–1383.

Hughes, T. (1987). *The Finite Element Method: Linear Static and Dynamic Finite Element Analysis* (Prentice Hall, New York).

Hughes, T., Feijoo, G. R., Mazzei, L., and Quincy, J.-B. (1998). The variational multiscale method – a paradigm for computational mechanics, *Computer Methods in Applied Mechanics and Engineering* **166**, pp. 3–24.

Hult, J. A. H. and McClintock, F. A. (1957). Elastic-plastic stress and strain distributions around sharp notches under repeated shear, in *Proceedings of the 9th International Congress for Applied Mechanics*, Vol. 8 (University of Brussels), pp. 51–58.

Hutchinson, J. W. (1987a). Crack tip shielding by micro-cracking in brittle solids, *Acta Metallurgica* **35**, pp. 1605–1619.

Hutchinson, J. W. (1987b). Micro-mechanics of damage in deformation and fracture, Technical University of Denmark, Denmark.

Ibach, H. (1997). The role of surface stress in reconstruction, epitaxial growth and stabilization of mesoscopic structures, *Surface Science Reports* **29**, pp. 193–263.

Ibragimov, N. H. (1994). *Lie Group Analysis of Differential Equations*, Vol. 1-2-3 (CRC Press, Boca Raton).

Irwin, G. R. (1948). Fracture dynamics, in *Fracturing of Metals* (American Society for Metals, Cleveland, O.H.), pp. 147–166.

Jain, S. C., Gosling, T. J., Willis, J. R., Totterdell, D. H. J. and Bullough, R. (1992). A new study of critical layer thickness, stability and strain relazation in pseudomorphic $ge_x si_{1-x}$ strained eilayers, *Philosohical Magazine A* **65**, pp. 1151–1167.

Jones, D. (1966). *Generalized Functions* (McGraw Hill, New York).

Jones, J. (1924a). On the determination of molecular fields-i. from the variation of the viscosity of a gas with temperature, *Proceedings of the Royal Society (London)* **106A**, pp. 441–462.

Jones, J. (1924b). On the determination of molecular fields-ii. from the equation of state of a gas, *Proceedings of the Royal Society (London)* **106A**, pp. 463–.

Joós, B. and Duesbery, M. (1997). The peierls stress of dislocations: An analytical formula, *Physical Review Letters* **78**, pp. 266–269.

Ju, J. W. and Sun, L. Z. (1999). A novel formulation for the exterior point eshelby's tensor of an ellipsoidal inclusion, *ASME Journal of Applied Mechanics* **66**, pp. 570–574.

Ju, J. W. and Sun, L. Z. (2001). Effective elastoplastic behavior of metal matrix composites containing randomly located aligned spheroidal inhomogeneities. part i: micromechanics-based formulation, *International Journal of Solids and Structures* **38**, pp. 183–201.

Ju, J. W. and Tseng, K. H. (1996). Effective elastoplastic behavior of two-phase ductile matrix composites: a micromechanical framework, *International Journal of Solids and Structures* **33**, pp. 4267–4291.

Kachanov, M. (1994). Elastic solids with many cracks and related problems, in J. W. Hutchinson and T. Y. Wu (eds.), *Advances in Applied Mechanics*, Vol. 32 (Academic Press, New York), pp. 259–445.

Kang, H. and Milton, G. W. (2008). Solutions to the pólya-szegö conjecture and the weak eshelby conjecture, *Archival Rational Mechanics and Analysis* **188**, pp. 93–116.

Kanit, T., Forest, S., Galliet, I., Mounoury, V. and Jeulin, D. (2003). Determination of the size of the representative volume element for random composites: Statistical and numerical approach, *International Journal of Solids and Structures* **40**, pp. 3647–3679.

Kanninen, M. F. and Popelar, C. H. (1985). *Advanced Fracture Mechanics* (Oxford University Press, New York Clarendon Press, Oxford).

Keer, L. M. and Mura, T. (1965). Stationary crack and continuous distributions of dislocations, in T. Yokobori, T. Kawasaki and J. L. Swedlow (eds.), *Proceedings of The First International Conference on Fracture*, 1 (The Japanese Society for Strength and Fracture of Materials), pp. 99–115.

Kelvin, L. (1948). Note on the integration of the equations of equilibrium of an elastic solid, in Cambridge and D. M. Journal (eds.), *Mathematical and Physical Papers*, Vol. 1 (Cambridge University Press), pp. 97–98.

Kim, S. and Karrila, S. J. (1991). *Microhydrodynamics: Principles and selected applications* (Butterworth-Heinemann, Boston).

Kinoshita, N. and Mura, T. (1984). Eigenstrain problems in a finite elastic body, *SIAM Journal on Applied Mathematics* **44**, pp. 524–535.

Kleinert, H. (1989). *Gauge fields in condensed matter*, Vol. II (World Scientific, Singapore).

Knap, J. and Ortiz, M. (2001). An analysis of the quasicontinuum method, *Journal of Physics and Mechanics of Solids* **49**, pp. 1899–1923.

Knopp, K. (1990). *Theory and Application of Infinite Series* (Dover Publications, Inc., New York).

Knowles, J. K. and Sternberg, E. (1972). On a class of conservation laws in linear and finite elastostatics, *Archive for Rational Mechanics and Analysis* **44**, pp. 187–211.

Krajcinovic, D. (1996). *Damage mechanics* (Elsevier, Amsterdam, New York).

Kröner, E. (1967). Interrelations between various branches of continuum mechanics, in E. Kröner (ed.), *IUTAM Symposium Mechanics of Generalized Continua* (Springer, Stuttgart), pp. 330–340.

Kröner, E. (1980). Continuum theory of defects, in R. B. et al. (ed.), *Physics of Defects* (Springer-Verlag, Berlin), pp. pp. 215–312.

Kröner, E. (1986). Statistical modeling, in J. Gittus and J. Zarka (eds.), *Modeling Small Deformation of Polycrystals* (Elsevier Applied Science Pub., New York), pp. 229–291.

Kröner, E. (1990). Modified green's function in the theory of heterogeneous and/or anisotropic linearly elastic media, in G. Weng, M. Taya and H. Abe (eds.), *Micromechanics and Inhomogeneity, The Toshio Mura 65th Anniversary Volume* (Springer, New York), pp. 599–622.

Kubin, L., Devincre, B. and Tang, M. (1998). Mesoscopic modelling and simulation of plasticity in fcc and bcc crystals: Dislocation intersections and mobility, *Journal of Computer-Aided Materials Design* **5**, pp. 31–54.

Landau, L. D. and Lifshitz, E. M. (1965). *Mechanics* (Pergamon Press, Oxford, New York).

Lee, C.-L. and Li, S. (2007). A half-space peierls-nabarro model and the mobility of screw dislocations in a thin film, *Acta Materialia* **55**, pp. 2149–2157.

Lee, C.-L. and Li, S. (2008). The size effect of thin films on the peierls stress of edge dislocations, *Mathematics and Mechanics of Solids* **In Press**, pp. 1–15.

Lepinoux, J. and Kubin, L. (1987). The dynamic organization of dislocation structures: a simulation, *Scr. Metall.* **21**, pp. 833–838.

Li, S. (2000). The micromechanics of classical plates: A congruous estimate of overall elastic stiffness, *International Journal of Solids and Structures* **37**, pp. 5599–5628.

Li, S. (2007). A multiscale Griffith criterion, *Philosophical Magazine Letters* **87**, pp. 945–965.

Li, S. (2008). On variational symmetry of defect potentials and multiscale configurational force, *Philosophical Magazine* **In Press**, pp. 1–24.

Li, S., Gupta, A., Liu, X. and Mahyari, M. (2004). Variational eigenstrain multiscale finite element method, *Computer Methods in Applied Mechanics and Engineering* **193**, pp. 1803–1824.

Li, S., Gupta, A. and Markenscoff, X. (2005a). Conservation laws of linear elasticity in stress formulations, *Proceedings of the Royal Society of London, A* **461**, pp. 99–116.

Li, S., Linder, C. and Foulk III, J. (2007a). On configurational compatibility and multiscale energy-momentum tensor, *Journal of Mechanics and Physics and Solids* **55**, pp. 980–1000.

Li, S., Liu, X. and Gupta, A. (2005b). Smart element method I. zienkiewicz-zhu feedback, *International Journal for Numerical Methods in Engineering* **62**, pp. 1264–1294.

Li, S., Sauer, R. and Wang, G. (2005c). Circular inclusion in a finite elastic domain. i. the dirichelt-eshelby problem, *Acta Mechanica* **179**, pp. 67–90.

Li, S., Sauer, R. and Wang, R. (2007b). The eshelby tensors in a finite spherical domain : I. theoretical formulations, *ASME Journal of Applied Mechanics* **74**, pp. 770–783.

Li, S. and Simonsen, C. B. (2005). Meshfree simulations of ductile crack propagation, *International Journal of Computational Engineering Science* **6**, pp. 1–25.

Li, S. and Wang, G. (2004). On damage theory of a cohesive medium, *International Journal of Engineering Science* **42**, pp. 861–885.

Li, S., Wang, G. and Sauer, R. (2007c). The eshelby tensors in a finite spherical domain : II. applications to homogenization, *ASME Journal of Applied Mechanics* **74**, pp. 784–797.

Lighthill, M. J. (1964). *Introduction to Fourier Analysis and Generalized Functions* (Cambridge University Press, Cambridge, U.K.).

Liu, L. (2008). Solutions to the eshelby conjectures, *Proceedings of Royal Society of London* **In Press**, pp. 1–21.

Liu, L., James, R. D. and Leo, P. (2007). Periodic inclusion-matrix microstructures with constant field inclusions, *Metallurgical and Materials Transactions, A* **38**, pp. 781–787.

Liu, X., Li, S. and Sheng, N. (2008). A cohesive finite element for quasi-continua, *Computational Mechanics* **41**, p. Online.

Logan, J. D. (1977). *Invariant Variational Principles* (Academic Press, New York).

Lubarda, V. and Krajcinovic, D. (1993). Damage tensors and crack density distribution, *International Journal of Solids and Structures* **30**, pp. 2859–2877.

Lubarda, V. and Krajcinovic, D. (1994). Tensorial representation of the effective elastic properties of the damages material, *International Journal of Damage Mechanics* **3**, pp. 38–56.

Lubarda, V. A. and Markenscoff, X. (2006). Variable core model and the peierls stress for the mixed (screw-edge) dislocation, *Applied Physics Letters* **89**, p. 151923.

Lubarda, V. A. and Markenscoff, X. (2007). Configurational force on a lattice dislocation and the peierls stress, *Archive of Applied Mechanics* **77**, pp. 147–154.

Luo, H. A. and Weng, G. J. (1987). On eshelby's inclusion problem in a three-phase spherically concentric solid, and a modification of mori-tanaka's method, *Mechanics of Materials* **6**, pp. 347–361.

Lutsko, J. F. (1988). Stress and elastic constant in anisotropic solids: Molecular dynamics techniques, *Journal of Applied Physics* **64**, pp. 1152–1154.

Ma, G., Hao, H. and Lu, Y. (2001). Homogenization of masonry using numerical simulations, *ASCE Journal of Engineering Mechanics* **127**, pp. 421–431.

Malvern, L. E. (1969). *Introduction to the Mechanics of a Continuous Mdedium* (Prentice-Hall, New Jersey).

Maradudin, A. (1958). Screw dislocations and discrete elastic theory, *Journal of Physics and Chemistry of Solids* **9**, pp. 1–20.

Maradudin, A. A., Montroll, E. W., Weiss, G. H. and Ipatova, I. P. (1971). *Theory of Lattice Dynamics in the Harmonic Approximation*, 2nd edn. (Academic Press, New

York and London).

Maranganti, R. and Sharma, P. (2005). A review of strain field calculations in embedded quantum dots and wires, in M. Rieth and W. Schommers (eds.), *Handbook of Theoretical and Computational Nanotechnology*, Vol. 1 (American Scientific Publishers), pp. 1–44.

Markenscoff, X. (1998). Inclusions with constant eigenstress, *Journal of Mechanics and Physics of Solids* **46**, pp. 2297–2301.

Markenscoff, X. and Anurag, G. (eds.) (2006). *Collected Works of J. D. Eshelby, Solid Mechanics and Its Applications*, Vol. 133 (Springer, Berlin).

Marsden, J. and Hughes, T. (1983). *Mathematical Foundations of Elasticity* (Prentice-Hall, Englewood Cliffs, NJ.).

Martin, P. A., Richardson, J. D., Gary, L. J. and Berger, J. R. (2002). On green's function for a three-dimensional exponentially graded elastic solid, *Proceedings of Royal Society of London, A* **458**, pp. 1931–1947.

Martinsson, P. and Rodin, G. (2002). Asymptotic expansions of lattice green's functions, *Proceedings of Royal Society of London, A* **458**, pp. 2609–2622.

Matthews, J. W. and Blakeslee, A. E. (1974). Defects in epitaxial multilayers i. misfit dislocations, *Journal of Crystal Growth* **27**, pp. 118–125.

Matthews, J. W. and Blakeslee, A. E. (1975). Defects in epitaxial multilayers ii. dislocation pile-ups, threading dislocations, slip lines and cracks, *Journal of Crystal Growth* **27**, pp. 273–280.

Matthews, J. W. and Blakeslee, A. E. (1976). Defects in epitaxial multilayers iii. preparation of almost perfect multilayers, *Journal of Crystal Growth* **32**, pp. 265–273.

Maxwell, J. C. (1874). Van der waals on continuity of the gaseous and liquid states, *Nature* **6**, pp. 477–480.

Mazilu, P. (1972). On the theory of linear elasticity in statically homogeneous media, *Rev. Roum. Math. Pures et Appl.* **17**, pp. 261–273.

McClintock, F. A. (1968). A criterion for ductile fracture by the growth of holes, *ASME Journal of Applied Mechanics* **35**, pp. 363–371.

Mi, C. and Kouris, D. A. (2006). Nanoparticles under the influence of surface/interface elasticity, *Journal of Mechanics of Materials and Structures* **1**, pp. 763–791.

Miller, R. and Tadmor, E. B. (2002). The quasicontinuum method: Overview, applications and current directions, *Journal of Computer-Aided Materials Design* **9**, pp. 203–239.

Milstein, F. (1982). Crystal elasticity, in H. Hopkins and M. Sewell (eds.), *Mechanics of Solids: The Rodney Hill 60th Anniversary Volume* (Pergamon Press Ltd., Elmsford, New York), pp. 417–452.

Milstein, F. and Hill, R. (1977). Theoretical properties of cubic crystals at arbitrary presure: I. density and bulk modulus, *Journal of the Mechanics and Physics of Solids* **25**, pp. 457–477.

Milstein, F. and Hill, R. (1978). Theoretical properties of cubic crystals at arbitrary presure: Ii. shear modulus, *Journal of the Mechanics and Physics of Solids* **26**, pp. 213–239.

Milstein, F. and Hill, R. (1979a). Divergences among the born and classical stability criteria for cubic crystals under hydrostatic loading, *Physical Review Letters* **43**, pp. 1411–1413.

Milstein, F. and Hill, R. (1979b). Theoretical properties of cubic crystals at arbitrary presure: Iii. stability, *Journal of the Mechanics and Physics of Solids* **27**, pp. 255–279.

Milton, G. (2002). *The Theory of Composites* (Cambridge University Press, Cambridge, United Kingdom).

Mori, T. and Tanaka, K. (1973). Average stress in matrix and average elastic energy of materials with misfiting inclusion, *Acta Metallurgica* **21**, pp. 571–574.

Mura, T. (1963). Continuous distribution of moving dislocations, *Philosophical Magazine* **8**, pp. 843–857.

Mura, T. (1964). Periodic distributions of dislocations, *Proceedings of Royal Society of London, A* **A280**, pp. 528–544.

Mura, T. (1968). The continuum theory of dislocations, in H. Herman (ed.), *Advances in Materials Research*, Vol. 3 (Interscience Publishers, John Wiley and Sons, New York), pp. 1–108.

Mura, T. (1987). *Micromechanics of Defects in Solids*, second, revised edition edn. (Kluwer Academic Pub., Dordrecht/Boston/London).

Mura, T. (1997). The determination of the elastic field of a polygonal star shaped inclusion, *Mechanics Research Communications* **24**, pp. 473–482.

Mura, T. (2000). Some new problems in micromechanics, *Material Science and Engineering, A.* **A285**, pp. 224–228.

Murat, F. and Tartar, L. (1997). Calculus of variations and homogenization, in A. V. Cherkaev and R. V. Kohn (eds.), *Topics in the mathematical modeling of composite materials*, Vol. 31 (Birkhauser, Boston), pp. 139–173, progress in nonlinear differential equations and their applications.

Nabarro, F. (1947). Dislocations in a simple cubic lattice, *Proc. Phys. Soc. Lond.* **59**, pp. 256–272.

Nabarro, F. (1952). The mathematical theory of stationary dislocations, *Adv. Phys.* **1**, pp. 269–394.

Nabarro, F. R. N. (1967). *Theory of crystal dislocations* (Oxford University Press, London).

Nemat-Nasser, S. and Hori, M. (1990). Elastic solids with micro defects, in G. Weng, M. Taya and H. Abe (eds.), *Micromechanics and in homogeneity — The Toshio Mura anniversary volume* (Springer-Verlag, New York), pp. 297–320.

Nemat-Nasser, S. and Hori, M. (1999). *Micromechanics: overall properties of heterogeneous materials* (Elsevier, Amsterdam-Lausanne-New York).

Nemate-Nasser, S., Iwakuma, T. and Accorsi, M. (1986). Cavity growth and grain boundary sliding in polycrystalline solids, *Mechanics of Materials* **5**, pp. 317–332.

Nemate-Nasser, S., Yu, N. and Hori, M. (1993). Bounds and estimates of overall moduli of composites with periodic microstructures, *Mechanics of Materials* **15**, pp. 163–181.

Nix, W. D. (1989). Mechanical properties of thin films, *Metallurgical Transactions, A* **20A**, pp. 1989–2217.

Nye, J. F. (1953). Some geometrical relations in dislocated crystals, *Acta Metallurgica* **1**, pp. 153–162.

Odegard, G., Gates, T. S., Wise, K. E., Park, C. and Siochi, E. J. (2003). Constitutive modeling of nanotube-reinforced polymer composites, *Composite Science and technology* **63**, pp. 1671–1687.

Olver, P. J. (1986). *Applications of Lie Groups to Differential Equations* (Springer, Berlin).

Ostoja-Starzewski, M. (2002). Microstructural randomness verse representative volume element in thermomechanics, *ASME Journal of Applied Mechanics* **69**, pp. 25–30.

Ovid'ko, I. A. and Scheinerman, A. G. (2005). Elastic fields of inclusion in nanocomposite solids, *Reviews on Advanced Materials Science* **9**, pp. 17–33.

Pardoen, T. and Hutchinson, J. W. (2000). An extended model for void growth and coalescence, *Journal of the Mechanics and Physics of Solids* **48**, pp. 2467–2512.

Park, H., Klein, P. and Wagner, G. (2006). A surface cauchy-born model for nanoscale materials, *International Journal for Numerical Methods in Engineering* **68**, pp. 1072–1095.

Peach, M. and Koehler, J. (1950). The forces exerted on dislocations and the stress fields produced by them, *Physical Review* **80**, pp. 436–439.

Peierls, R. (1940). The size of a dislocation, *Proc. Phys. Soc. Lond.* **52**, pp. 34–37.

Phan-Thien, N. and Kim, S. (1994). *Microstructures in Elastic Media Principles and Computational Methods* (Oxford University Press, New York · Oxford).

Pólya, G. and Szegö, G. (1951). *Isoperimetric Inequalities in Mathematical Physics* (Princeton University Press, Princeton, NJ).

Ponte Castañeda, P. (1991). The effective mechanical properties of nonlinear isotropic composites, *Journal of Mechanics and Physics of Solids* **39**, pp. 45–71.

Ponte Castañeda, P. (1992a). A new variational principle and its application to nonlinear heterogeneous systems, *SIAM Journal on Applied Mathematics* **52**, pp. 1321–1341.

Ponte Castañeda, P. (1992b). New variational principle in plasticity and their its application to composite materials, *Journal of Mechanics and Physics of Solids* **40**, pp. 1757–1788.

Ponte Castañeda, P. and Suquet, P. (1998). Nonlinear composites, in *Advances in Applied Mechanics*, Vol. 34 (Academic Press, San Diego London Boston), pp. 171–302.

Qu, J. and Cherkaoui, M. (2006). *Fundamentals of Micromechanics of Solids* (John Wiley & Sons Inc. Hoboken, New Jersey).

Reddy, J. N. (2003). *Mechanics of Laminated Composite Plates and Shells* (CRC Publication Ltd, Boca Raton).

Rhee, M., Zbib, H. M., Hirth, J. P., Huang, H. and de la Rubia, T. (1998). Models for long-/short-range interactions and cross slip in 3d dislocation simulation of bcc single crystals, *Modelling Simul. Mater. Sci. Eng.* **6**, pp. 467–492.

Rice, J. R. (1968). A path independent integral and the approximate analysis of strain concentration by notches and cracks, *ASME Journal of Applied Mechanics* **35**, pp. 379–386.

Rice, J. R. (1986). Mathematical analysis in the mechanics of fracture, in H. Liebowitz (ed.), *Fracture: An Advanced Treatise*, Vol. 2 (Academic Press, New York), pp. 191–311.

Rice, J. R. and Tracey, D. M. (1969). On ductile enlargement of voids in triaxial stress fields, *Journal of the Mechanics and Physics of Solids* **17**, pp. 201–217.

Rodin, G. J. (1996). Eshelby's inclusion problem for polygons and polyhedra, *Journal of Mechanics and Physics of Solids* **44**, pp. 1977–1995.

Ru, Q.-C. and Schiavone, P. (1996). On the elliptic inclusion in anti-plane shear, *Mathematics and Mechanics of Solids* **1**, pp. 327–333.

Rudin, W. (1991). *Funcational analysis* (McGraw-Hill, New York).

Sanchez-Palencia, E. (1980). *Non-homogeneous Media and Vibration Theory, Lecture Notes in Physics, Vol. 127* (Springer-Verlag, Berlin).

Sauer, R., Wang, G. and Li, S. (2007). The composite eshelby tensors and their applications to homogenization, *Acta Mechanica* **Online**, pp. 1–30.

Seitz, F. (1987). *The modern theory of solids*, new ed edn. (Dover Pubs., New Yor).

Sendeckyj, G. (1970). Elastic inclusion problems in plane elastostatics, *International Journal of Solids and Structures* **6**, pp. 1535–1543.

Sharma, P. and Ganti, S. (2004). Size-dependent eshelby's tensor for embedded nano-inclusions incorporating surface/interface energies, *ASME Journal of Applied Mechanics* **71**, pp. 663–671.

Sharma, P., Ganti, S. and Bhate, N. (2003). Effect of surface on the size-dependent elastic state of nano-inhomogeneities, *Applied Physics Letters* **82**, pp. 535–537.

Shen, L. and Yi, S. (2000a). Approximate evaluation for effective elastic moduli of cracked solids, *International Journal of Fracture* **106**, pp. L15–L20.

Shen, L. and Yi, S. (2000b). New solutions for effective elastic moduli of cracked solids, *International Journal of Solids and Structures* **37**, pp. 3525–3534.

Shi, D.-L., Feng, X.-Q., Huang, Y.-Y., Hwang, K.-C. and Gao, H. (2004). The effect of nanotube waviness and agglomeration on the elastic property of carbon nanotube-reinforced composites, *ASME Journal of Engineering Materials and Technology* **126**, pp. 250–257.

Simo, J. and Rifai, M. (1990). A class of mixed assumed strain methods and the method of incompatible modes, *International Journal for Numerical Methods in Engineering* **29**, pp. 1595–1638.

Simonsen, B. C. and Li, S. (2004). Meshfree modeling of ductile fracture, *International Journal for Numerical Methods in Engineering* **60**, pp. 1425–1450.

Somigliana, C. (1886). Sopra l'equilibrio di un corpo elastico isotropo, *Il Nuovo Ciemento* **19**, pp. 84–90.

Spagnolo, S. (1968). Sulla convergenza di soluzioni di equazione paraboliche ed ellitiche, *Ann. Sc. Norm. Sup. Pisa.* **22**, pp. 577–597.

Spagnolo, S. (1975). Convergence in energy for elliptic operators, in B. Hubbard (ed.), *Numerical solutions of partial differential equations III Synspade*, Vol. III (Academic Press, New York), pp. 469–498.

Steinhaus, H. (1983). *Mathematical Snapshots* (Oxford, Oxford).

Stott, M. J. and Zarembra, E. (1980). Quasiatoms: An approach to atoms in nonuniform electronic system, *Physcal Review B* **22**, p. 1564.

Sun, L. Z. and Ju, J. W. (2001). Effective elastoplastic behavior of metal matrix composites containing randomly located aligned spheroidal inhomogeneities. part ii: applications, *International Journal of Solids and Structures* **38**, pp. 203–225.

Szabo, B. (1986). Mesh design for the p-version of the finite element method, *Computer Methods in Applied Mechanics and Engineering* **55**, pp. 181–197.

Tadmor, E., Ortiz, M. and Phillips, R. (1996). Quasicontinuum analysis of defects in solids, *Philosophical Magazine, A* **73**, pp. 1529–1563.

Talbot, D. and Willis, J. (1985). Variational principles for inhomogeneous non-linear media, *IMA Journal of Applied Mathematics* **35**, pp. 39–54.

Talbot, D. and Willis, J. (1987). Bounds and self-consistent estimates for the overall properties of nonlinear composites, *IMA Journal of Applied Mathematics* **39**, pp. 215–240.

Talbot, D. and Willis, J. (1998). Upper and lower bounds for the overall response of an elastoplastic composite, *Mechanics of Materials* **28**, pp. 1–8.

Tanaka, K. and Mori, T. (1972). Note on volume integrals of the elastic field around an ellipsoidal inclusion, *Journal of Elasticity* **2**, pp. 199–200.

Tang, Z. and Postle, R. (2000). Mechanics of three-dimensional braded structures for composite materials – part i: fabric structure, *Composite Structures* **49**, pp. 451–459.

Tang, Z. and Postle, R. (2001). Mechanics of three-dimensional braded structures for composite materials – part ii: prediction of the elastic moduli, *Composite Structures* **51**, pp. 451–457.

Tartar, L. (1990). On mathematical tools for studying homogenisation, osillations and concentration effects in partial differential equations, *Proceedings of the Royal Society of Edinburgh. A. Mathematical and Physical Sciences* **115**, pp. 193–230.

Teodosiu, C. (1982). *Elastic models of crystal defects* (Springer-Verlag, Berlin).

Thostenson, E. T., Ren, Z. and Chou, T. (2001). Advances in science and technology of carbon nanotubes and their composites: A review, *Composite Science and Technology* **61**, pp. 1899–1912.

Timoshenko, S. and Goodier, J. N. (1951). *Theory of Elasticity*, 2nd edn. (McGraw-Hill, New York).

Tonti, E. (1967). Variational principles in elastostatics, *Meccanica* **2**, pp. 201–208.

Torquato, S. (1997). Effective stiffness tensor of composite media — i. exact series expansions, *Journal of Mechanics and Physics of Solids* **45**, pp. 1421–1448.

Truesdell, C. and Noll, W. (2004). *The nonlinear field theorie of mechanics*, *Encyclopedia of Physics*, Vol. 3/3 (Springer-Verlag, Berlin), ed. Flügge, S.

Tvergaard, V. (1981). Influence of voids on shear band instabilities under plane strain conditions, *International Journal of Fracture* **17**, pp. 389–407.

Tvergaard, V. (1982). On localization in ductile materials containing voids, *International Journal of Fracture* **18**, pp. 237–251.

Tvergaard, V. (1990). Material failure by void growth to coalescence, in J. W. Hutchinson and T. Y. Wu (eds.), *Advances in Applied Mechanics*, Vol. 27 (Academic Press, New York), pp. 83–151.

Tvergaard, V. and Needleman, A. (1984). Analysis of the cup-cone fracture in a round tensile bar, *Acta Metallurgica* **32**, pp. 157–169.

Volterra, V. (1907). Sur l'équilibre del corps élastiques multiplement connexes, *Ann. Ec. Norm* **24**, pp. 401–517.

Walpole, L. J. (1981). Elastic behavior of composite materials: Theoretical foundations, in C.-S. Yih (ed.), *Advances in Applied Mechanics*, Vol. 21 (Academic Press, New York), pp. 169–242.

Wang, C. Y. and Achenbach, J. D. (1995). Three-dimensional time-harmonic elastodynamic green's functions for anisotropic solids, *Proceedings of Royal Society of London, A* **449**, pp. 441–458.

Wang, C. Y. and Achenbach, J. D. (1996). Lamb's problem for solids of general anisotropy, *Wave Motion* **24**, pp. 227–242.

Wang, G. and Li, S. (2004). On micromechanics theory of cohesive crack distribution, *Theoretical and Applied Fracture Mechanics* **42**, pp. 303–316.

Wang, G., Li, S., Nguyen, H.-N. and Sitar, N. (2007). Effective elastic stiffness for periodic masonry structures via eigenstrain homogenization, *ASCE Journal of Materials in Civil Engineering* **19**, pp. 269–277.

Wang, G., Li, S. and Sauer, R. (2005a). Circular inclusion in a finite elastic domain. II. the neumann-eshelby problem, *Acta Mechanica* **179**, pp. 91–110.

Wang, G., Liu, X., Li, S. and Sitar, N. (2005b). Smart element method II. finite eshelby formulation, *International Journal for Numerical Methods in Engineering* **64**, pp. 1303–1333.

Wang, H., Ng, G. I. and Hopgood, A. A. (1998). Kinetics of non-radiative-defect-related degradation in $g_a a_s$/al $g_a a_s$ heterojunction bipolar transistors, *J. Phys. D: Appl. Phys.* **31**, pp. 3168–3171.

Wang, J. (1996). A new modification of the formulation of pererls stress, *Acta Materialia* **44**, pp. 1541–1546.

Wang, Y., Tomanek, D. and Bertsch, G. F. (1991). Stiffness of a solid composed of c60 clusters, *Physical Review B* **44**, pp. 6562–6565.

Weertman, J. and Weertman, J. R. (1992). *Elementary Dislocation Theory* (Oxford University Press, New York Oxford).

Weng, G. J. (1984). Some elastic properties of reinforced solids with special reference to isotropic ones containing spherical inclusions, *International Journal of Engineering Science* **22**, pp. 845–856.

Weng, G. J. (1990). The theoretical connection between mori-tanaka's theory and the hashin-shtrikman-walpole bounds, *International Journal of Engineering Science* **28**, pp. 1111–1120.

Williams, M. (1952). Stress singularities resulting from various boundary conditions in angular corners of plates in extension, *Journal of Applied Mechanics, ASME* **19**, pp. 526–528.

Willis, J. (1967a). Boussinesq problems for an anisotropic half-space, *Journal of Mechanics and Physics of Solids* **15**, pp. 331–339.

Willis, J. R. (1967b). Secod order effects of dislocations in anisotropic media, *International Journal of Engineering Science* **5**, pp. 171–190.

Willis, J. R. (1968). The stress field around an elliptical crack in an anisotropic elastic medium, *International Journal of Engineering Science* **6**, pp. 253–263.

Willis, J. R. (1981). Variational and related methods for the overall properties of composites, in C.-S. Yih (ed.), *Advances in Applied Mechanics*, Vol. 21 (Academic Press, New York), pp. 1–78.

Willis, J. R., Jain, S. C. and Bullough, R. (1990). The energy of an array of dislocations: implications for strain relaxation in semiconductor heterostructures, *Philosophical Magazine, A* **62**, pp. 115–129.

Xu, X.-P. and Needleman, A. (1994). Numerical simulations of fast crack growth in brittle solids, *Journal of the Mechanics and Physics of Solids* **42**, pp. 1397–1434.

Zbib, H., de la Rubia, T. D., Rhee, M. and Hirth, J. P. (2000). 3d dislocation dynamics: stress-strain behavior and hardening mechanisms in fcc and bcc metals, *Journal of Nuclear Materials* **276**, pp. 154–165.

Zbib, H., Rhee, M. and Hirth, J. (1998). On plastic deformation and the dynamics of 3d dislocations, *Int. J. Mech. Sci.* **40**, pp. 113–127.

Zhao, Y., Aziz, M. J., Gossman, H. J., Mitha, S. and Schiferl, D. (1999a). Activation volume for antimony diffusion in silicon and implications for strained films, *Applied Physics Letters* **75**, p. 941.

Zhao, Y., Aziz, M. J., Gossman, H. J., Mitha, S. and Schiferl, D. (1999b). Activation volume for boron diffusion in silicon and implications for strained films, *Applied Physics Letters* **74**, p. 31.

Zhou, M. (2003). A new look at the atomic level virial stress — on continuum-molecular system equivalence, *Proceedings of Royal Society of London, A* **459**, pp. 2347–2392.

Zienkiewicz, O. C. and Taylor, R. (2000). *The Finite Element Method I. The Basis*, Vol. 1, fifth edition edn. (Butterworth-Heinemann).

Zienkiewicz, O. C. and Zhu, J. (1992a). The superconvergent patch recovery and a posteriori error estimates: Part i. the recovery technique, *International Journal for Numerical Methods in Engineering* **33**, pp. 1331–1364.

Zienkiewicz, O. C. and Zhu, J. (1992b). The superconvergent patch recovery and a posteriori error estimates: Part ii. error estimates and adaptivity, *International Journal for Numerical Methods and Engineering* **33**, pp. 1365–1382.

SUBJECT INDEX

AUTHORS INDEX

www.ingramcontent.com/pod-product-compliance
Lightning Source LLC
Chambersburg PA
CBHW080117220326
41598CB00032B/4875